1st North American Regional Conference on Healthy Buildings (HB 2015)

Innovation in a Time of Energy Uncertainty and Climate Adaptation

Boulder, Colorado, USA
19-22 July 2015

ISBN: 978-1-5108-2128-6

Printed from e-media with permission by:

Curran Associates, Inc.
57 Morehouse Lane
Red Hook, NY 12571

Some format issues inherent in the e-media version may also appear in this print version.

Printed by Curran Associates, Inc. (2016)

For permission requests, please contact International Society of Indoor Air Quality and Climate (ISIAQ) at the address below.

International Society of Indoor Air Quality and Climate (ISIAQ)
2548 Empire Grande
Santa Cruz CA 95060
USA

Phone: (831) 426-0148
Fax: (831) 426-6522

info@isiaq.org

Additional copies of this publication are available from:

Curran Associates, Inc.
57 Morehouse Lane
Red Hook, NY 12571 USA
Phone: 845-758-0400
Fax: 845-758-2633
Email: curran@proceedings.com
Web: www.proceedings.com

Contents

M.1. Indoor Air Quality in Homes **1**

Measuring The Ozone Penetration Factor In A Residence Under Infiltration Conditions
- Haoran Zhao, Brent Stephens 1

Operation Of Residential HVAC Systems: Implications For Healthy Buildings
- Marianne Touchie, Jeffrey A Siegel 5

The Impact Of Heating Process On PM2.5 Concentration In Cold Rural Areas Of Northern China
- Xueyan Zhang, Bin Chen 10

Measuring Flame Retardant Emissions From Spray Polyurethane Foam In A Home
- Dustin Poppendieck, Angelica Connor 17

Phthalate Levels In Green-Renovated Subsidized Housing
- Robin Dodson, Julia Udesky, and Ruthann Rudel 21

Health Benefits Of Living In Green, Low-Income Multi-Family Housing
- Meryl D. Colton, Jose Guillermo Cedeno Laurent, Piers MacNaughton, John Kane, Mae Bennett-Fripp, John Spengler, Gary Adamkiewicz 24

Ventilation Provision And Outcomes In Mainstream Contemporary New-Building Flats In London, UK
- Tim Sharpe, Antoinette Charles 29

Comparison Of Indoor And Outdoor Air Quality In Latino Communities In Peru, Chile, And USA
- Wyatt M. Champion, Francisco Barraza, Hctor Jorquera, and Lupita D. Montoya . . . 33

M.2. Building Materials and Furnishings **37**

Carpet Dust Culturable Fungi And Bacteria Results Used To Evaluate Carpet Maintenance
- Kent A Candee, Krista J Scott, Craig S Black, Dave W Havick 37

Emission of Phthalates and Phthalate Alternatives from Vinyl Flooring and Crib Mattress Covers: The Influence of Temperature
- Yirui Liang and Ying Xu 41

Laboratory and field performance of interior finish Materials using formaldehyde scavenging technology
- Stanley D. Gatland II, Douglas Gehring, and Davy Vennekens 45

Mass-Based Size Distribution Of House Dust Loading On Residential Floorings
- Yan Ma, Yilin Tian, and Andrea R. Ferro 49

Safeguarding Our Resources And Indoor Air Gypsum Wallboard Closed-Loop Recycling In An Era Of Increased Chemical Transparency
- Danielle Wagner, Zhongqi (Joshua) Cheng, Amanda Kaminsky, Sydney Mainster, Geoffrey Brock, Ronald Spalter, and Catherine Bobenhausen 54

Investigation Of Particle-Mediated Gas-Phase Transport Of Phthalates From Vinyl Flooring
- Yaoxing Wu, Eric Vejerano, Cong Liu, Steven S. Cox, Mingjie Xie, Linsey C. Marr, and John C. Little 59

Resuspension Of Deposited Particles Induced By Jets At Different Surface Dust Loads
- Xiong Mei, Tengfei (Tim) Zhang and Shugang Wang 63

Experimental Study Of Active Flows And Passive Materials For Indoor Air Quality Control
- Denise Mauney, Christella Suwongso, Miles Abarr, Wil W. Srubar, Lupita D. Montoya . 67

M.3. Indoor Air Exposures 71

Assessing Wildfire Exposure In Homes Near Wildfires
- Ernest R. Crutcher III and Heidie K. Crutcher-Bettes 71

Air Pollution Exposure Model For Individuals (EMI) In Health Studies: Evaluation For Ambient PM2.5
- Michael S Breen, Thomas Long, Bradley Schultz, Ronald Williams, Jennifer Richmond-Bryant 75

Improving The Indoor Air Quality By Using A Surface Emissions Trap For Exposure Reduction
- Lennart Larsson, Pawel Markowicz, and Johan Mattsson 79

Statistical Evaluation Of Human Sensory Response To Acetone
- Tunga Salthammer, Nicole Schulz, Ramona Stolte and Erik Uhde 83

M.4. Filtration 86

Challenges And Opportunities For Filter Forensics
- Daniel Haaland and Jeffrey A Siegel 86

Bioaccumulation on HVAC Filters in University Buildings in Singapore
- Irvan Luhung, Yan Wu, Victor WC Chang, and William W Nazaroff 90

Modeling The Impact Of Residential HVAC Filtration On Indoor Particles Of Outdoor Origin
- Parham Azimi, Dan Zhao, and Brent Stephens 94

Pollutant Exposures And Asthma: What Can HVAC Filter Dust Tell Us?
- Juan P. Maestre, Sandra Dedesko, Kelly Royse, Ying, Xu, Sharon Horner, Jeffrey Siegel, Kerry A. Kinney 98

M.5. Energy and Environmental Quality 102

Recommissioning A Zero-Carbon Multifamily Residential Buildings Ventilation System: A Necessity
- Jelle Laverge, Eline Himpe, Marc Delghust, Stijn VAN DE Puttel and Julio Vaillant Rebollar 102

College Students Perceptions Of Household Energy Efficiency
- Suchismita Bhattacharjee, Sandeep Langar, and Maggie Hogan 106

Indoor Environmental Resilience: A Review And Discussion
- Andrew Persily and Steven Emmerich 110

Air quality in data centers: people vs. The machines
- Chris Muller, Chuck Arent 114

M.6. Ventilation 122

A General Method To Predict Ventilation Requirements While Incorporating Building Material Emission Testing
- Wei Ye, Doyun Won, Xu Zhang, and Qi Zhou 122

Assessment Of The Implications Of Natural And Mechanical Ventilation On Human Health In The Residential Sector
- Sandra M Bernal 126

Time-resolved Aerosol and Fluorescent Bioaerosol Concentrations in an Air-Conditioned and Mechanically Ventilated Office in Singapore
- Jin Zhou, Victor W.-C. Chang, and William W Nazaroff 130

Testing Of A Downflow Ventilation System For High Risk Infectious Disease Isolation Rooms
- Blanca Beato Arribas, Catherine J Noakes, P. Andrew Sleigh 134

M.7. Sensing **138**
Open Source Building Science Sensors (OSBSS): An Open Source Sensor Network For Indoor Environmental Data Collection
- Akram S Ali, Torkan Fazli, Joseph CP Huan, Deion Debose, Boyang Dong, and Brent Stephens 138
Development Of An Experimental System For Assessing Indoor Bioaerosol Transport And Control
- Stephanie Kunkel, Parham Azimi, Brent Stephens 142
A Simple Method For Measuring Gas-Phase SVOC Concentration In Equilibrium With The Material Phase
- Yaoxing Wu, Steven S. Cox, Mingjie Xie, Linsey C. Marr, and John C. Little 146
Wireless Monitoring Of The Indoor Air Environment In A Winery During Harvest
- Nicolas M Madrid, David J Killeen, Andr Knoesen, David Mills and Roger B Boulton . 150

T.1. Transport and Chemistry **154**
Sources And Mitigation Of Dioxin And Dioxin-Like Compounds In Indoor Dust
- Patricia V Cline 154
Comparison Of Advective And Diffusive Transport Of SVOCS Through Cloth For Indoor Conditions
- Alexander Korff, Melissa Buechlein, Madison Gibler, Glenn Morrison 158
Pollution Transport By The Human Convection Flow Impact Of The Room Air Temperature And Seated Body Inclination
- Dusan Licina, Arsen Melikov, Chandra Sekhar, Kwok Wai Tham 162
The Influence Of Temperature On The Fate And Transport Of Indoor Phthalates: A Case Study In A Test House
- Chenyang Bi and Ying Xu 166
Investigating Indoor Chemistry Of Terpinolene Through The Use Of A New Derivatization Agent
- J.R. Wells, Jason E. Ham, Stephen R. Jackson, and Joel C. Harrison 170
Validation Of Pm Concentration Estimates Using Contam Multizone Indoor Air Quality Model With Resuspension Deposition Module
- Lisa D. Bramwell, Yilin Tian and Andrea R. Ferro 174
Secondary Organic Aerosol From -Terpineol Ozonolysis
- Yanan Yang, Michael S Waring 178
Changes In Outdoor Aerosol Chemical Composition Upon Transport Into The Indoor Environment
- Anita M Johnson, Michael S Waring, and Peter F Decarlo 182

T.2. Mold and Moisture **186**
A Practical Guide For Acceptable Indoor Airborne Mold Levels (Sporetrap) With Emphasis On Aspergillus-Penicillium
- Michael R. Meyer 186

Control Of Mold Contamination In The Bathroom
- Satomi Yanaso, Hiroshi Yamagishi, Takayuki Hasegawa, and Kosuke Tanaka 190
Investigating And Avoiding Moisture-Related Problems In Existing Buildings
- Lewis G. Harriman, Terry R. Brennan 194
Airborne Bacteria And Fungi Concentrations In Airtight Contemporary Dwellings
- Grinne M McGill, John E Moore, Tim Sharpe, Damian G Downey and Lukumon Oyedele 198
Moisture Measurement And Implications For Fungal Growth In Drywall
- Sandra Dedesko, Jeffrey Siegel 202
The IVAIRE Study: Correlations Between Ventilation And Mycological Parameters In Canadian Single Family Homes With Ashtmatic Children
- Daniel G Aubin, Doyun Won, Hans Schleibinger, Wenping, Yang, Pierre Lajoie, Denis Gauvin, Vronique Gingras, Jean-Marc Leclerc and Don Fugler 206
Synthetic Origin Of Microbial Volatile Organic Compounds
- Hyunok Choi, Norbert Schmidbauer, Carl-Gustaf Bornehag 210
Summary Of Sloan Symposium HEALTHY BUILDINGS 2015-EUROPE
- Hal Levin, Martin Tubel, and Mark Hernandez 217

T.3. IAQ in Hospitals **221**
Experimental And CFD Investigation Of Hospital Operating Room Air Distribution
- Zhiqiang (John) Zhai, James McNeill and Jean Hertzberg 221
The Role Of Surfaces In The Transmission Of Bioaerosols From Source To Patient In Hospital Rooms
- Marco-Felipe. King, Catherine J. Noakes, and P. Andrew Sleigh 225
Hospital Surge Capacity: Practical Aspects Of Temporary Isolation Ward Design
- Nicholas Clements, Shelly L Miller, Shobha S Subhash, Aaron E Eagan 233
Exploring Spatial Planning And Functional Program Impact On Microbial Diversity And Distribution In Two South African Hospital Microbiomes
- Jako A Nice, Piet Vosloo 237

T.4. Urban and Indoor Environments **241**
Model And Measurement Derived Estimates Of Residential PM2.5 Infiltration In 10 US Cities
- Jennifer M. Logue, Neil E. Klepeis, and Brett C. Singer 241
Urban Microbiome Pilot Study: Parks And Parking Lots
- Gwynne . Mhuireach, Josh Ladau, James F. Meadow, G.Z. Brown, Bart R. Johnson, Jessica L. Green 245
Environment Parameters Relevant To Microbes Spanning Human Urbanization In Sub-Tropical Climate
- Humberto Cavallin, Jean F. Ruiz, Atila Novoselac, Henrique Pereira, Luciana C. Paulino, Raphael Rios, Jean H. Perez., Martin J. Blaser, Maria Gloria Dominguez-Bello 249
Controlling Pollutants During New Construction To Reduce Exposure Upon Occupancy By Sensitive Individuals
- Carl Grimes 253

T.5. IAQ Challenges and Strategies 257
Understanding The Challenges Of Assessing "Good" Air Quality In The Indoor Environment
- Jennifer M. Logue, Max H. Sherman, and Brett C. Singer . . . 257
PM2.5: An Emerging Issue For Standard Of Care In Indoor Spaces
- Terry Brennan . . . 261
High Performance IAQ Specification For Net Zero Energy Homes
- Anthony Bernheim, Alfred Hodgson, and Andrew Persily . . . 265
Weatherization Ventilation Strategy As A Factor In Indoor Humidity
- William Rose, Stacy Gloss, Paul Francisco, Zachary Merrin, Salvatore Cali, and David Jacobs . . . 269

T.6. VOCs 273
A Study On TVOC Concentration Emitted By Plywood: Finish, Age And Environment
- E.V.H. Dia and L.M.T. Bo-ot . . . 273
A New Approach To Estimate Emission Factors Of Semi-Volatile Organic Compounds Based On Their Physical Properties
- Jiping Zhu, Peter Mochungong, and Julie S. Obrien . . . 277
New VOC Compounds In The Indoor Environment From Thermoplastic Filaments For 3D Printing
- Olaf Wilke, Stefan Seeger . . . 281
Characterizing Uncertainty Of A Formaldehyde Reference Standard
- Dustin Poppendieck . . . 283

W.1. Microbiome 287
Indoor Microbiomes Change When We Try To Control Them
- Alexander Mahnert, Christine Moissl-Eichinger, Henry Mller, Alexander J. Probst, Roscel Amor Ortega, Parag Vaishampayan, Kasthuri Venkateswaran, Gabriele Berg . . . 287
Building Characteristics As Determinants Of Moisture In Water-Damaged Homes A Study Design
- Michal P Spilak, Iman Sylvain, Rachel Adams and Michael S Waring . . . 291
Microbial Communities In House Dust And Asthma: A Birth Cohort Study
- Rachel I Adams, Martin Tubel, Pirkka Kirjavainen, Balamuralikrishna Jayaprakash, Anne Karvonen, Anne Hyvrinen, and Juha Pekkanen . . . 295
Effect Of Ultraviolet Germicidal Irradiation On Microbial Loading Of HVAC Heat Exchangers In Humid Versus Dry Entering Conditions
- Julia C Luongo, Shelly L Miller . . . 298
Survey Of Green Building Water Systems Reveals Elevated Water Age And Microbial Concerns
- William J. Rhoads, Amy Pruden, Marc A. Edwards . . . 302
Temporal Air Exchange Rates And Relative Humidity May Affect Airborne Microbiomes More Than Weatherization Efforts
- Mytien T Nguyen, Denina Hospodsky, Catherine Spirito, and Largus T Angenent . . . 307

W.2. Architecture and Energy 311
An Experimental Study On Heat Gain In Window With An External Shutter
- Esam M Alawadhi . . . 311

The Energy Consequences Of Excess Static Pressure In Central Residential Heating And Air-Conditioning Systems
- Torkan Fazli, Rou Yi Yeap, Brent Stephens 316
A Survey On Evaluation Of Workers And Their Thermal Environment In Offices
- Koichi Ikeda, Hikaru Enomoto, and Hitoshi Takaono 320
Development Of An Automatic Thermal Control System Using Human Facial Skin Temperature
- Bo Yi, Joon-Ho Choi, Douglas Noble, Kyle Konis 324
Energy Efficiency Improvements And Indoor Environmental Quality Through Climate-Responsive Architecture
-Shahryar Habibi 328
Human Factors In HVAC: Thermal Comfort And Indoor Air Quality For Lifetime Housing
- Robert Bean 333

W.3. IAQ in Schools **337**
Changes In IAQ Observed When Operating A Corona Discharge Air Cleaner In A Classroom Ventilation Unit
- Todd R. Crawford 337
USAs Missing Mission: Strategies To Protect Children From Poor IAQ In Schools
- Claire L Barnett 342
Crossover Study Of Fluorescent Bioaerosol Measurements In Elementary Classrooms With And Without Upper-Room Ultraviolet Germicidal Irradiance
- Chunxiao Su, Josephine Lau, and Fang Yu 347
Performance Of Filters With Different Efficiencies In Classrooms With Fan-Coil Air-Conditioing System
- Ailu Chen, Qingliang Cao, Jin Zhou and Victor W.-C. Chang 351
Low EMF Best Practices For Indoor Environmental Quality In School Classrooms
- Antoinette W. Stein, Devra Davis, Alex Stadtner, Karl Riley 355
Indoor Particulate Matter Assessment In Urban Kindergartens In Al Ain, United Arab Emirates
- George O. Odhiambo 370

Monday Poster Session **374**
The Cost Optimal Methodology Of Dwelling Buildings In Slovakia
- Lucia Borisov 374
Improvement Of Perceived Temperature Under 28°C Of Air Conditioning Control Temperature
- Yoshihito Kurazumi, Kenta Fukagawa, Emi Kondo, Tomonori Sakoi, Kunihito Tobita, Yoshiaki Yamato 378
Indoor Microclimate And Natural Ventilation In Double Transparent Facade
- Kristna Valoukov 382
A Simplified Model For Temperature Prediction In A Ventilated Wall Module
- T.M.J. Rakotomahefa, Tengfei (Tim) Zhang and Shugang Wang 386
How Reliable Are Standard (undisturbed) Airborne Mold (sporetrap) Tests? The Usefulness Of Disturbed Tests For More Accurate Assessments
- Michael R. Meyer 390

Thermal Performance Of Evaluation On Energy Efficient Paraffinic PCM Based Shape Stabilized Composites For Energy Saving
- Okyoung Chung, Su-Gwang Jeong, and Sumin Kim 394
Impact Of Water Damage On Microbial Community In Residential Buildings
- Iman Sylvain, Michal P Spilak, Michael S Waring and Rachel I Adams 398
Field Survey Of Indoor Environment And Its Effect On Human Health During Heating Period In Northern China
- Bin Chen, Min Zhou, Yu Chen, Xueyan Zhang, Yuhang Shu, Yuanbin Zhu, Yizhou Sang, and Joe R. Zhao 402
Modeling The Dispersion Of Evaporating Spittle Droplet In A Typical Waiting Room Of A Hospital
- Qiuyue Wang, Shinsuke Kato 407
Assessment Of Thermal Environment During Kitchen Work
- Kunihito Tobita, Yoshihito Kurazumi, Yoshiaki Yamato, Kenta Fukagawa 411
Odor Thresholds For 2,4,6 -Trichloroanisole Reviewed In The Context Of Indoor Air Quality
- Stephanie A Juran, Steven Nordin, Gunnar Johanson, Johnny C Lorentzen 415
The Effect Of Human Density On The Risk Of Infection In A Hospital Ward
- Marcelo Pereira, Lidia Morawska, Luke Knibbs, Congrong He, Graham Johnson 419
The Influence Of Street Canyon Design On Hospital Air Quality
- Marco-Felipe. King and Catherine J. Noakes 423
Priority Chemicals Emitted From Composite Wood Materials
- Doyun Won, Wenping Yang, Joan Wong, Cheryl Khoury and Hans Schleibinger 427
A Study Of Shading Device Orientation On The Natural Ventilation And Daylight Harvesting Potential In A Double Skin Facade
- Jeehwan Lee, Jae D. Chang and Hongyi Cai 432
The Relation Of Psychosocial And Demographic Characteristics With Energy Consumption
- Suchismita Bhattacharjee 436
Modeling Of Thermal Comfort Using Designbuilder
- Lucie Dobiov1, Daniel Adamovsk 440
Formaldehyde Sorption To Porous Media For Air Quality Applications
- Anne Wrobetz, Megan Matteazzi and Lupita D Montoya 444
Indoor Environment Quality And Nabers Ie Ratings: A 2014 Case Study For A Commercial Office Building Portfolio Of 25 Australian Buildings
- Tim M Callinan, Jack J Noonan, Vyt P Garnys 448
Web-Based Wireless Sensor System For Remote Air Environment Monitoring In A Commercial Winery
- Nicolas M Madrid, David J Killeen, Roger B Boulton and Andr Knoesen 452
Exposure Assessment To E-Cigarettes Part 1: Literature Review On Carbonyl Compounds Generation From E-Cigarettes And Affecting Factors
- Wenhao Chen, Ping Wang, Jeff Fowles, Toshiki Matsuo1,, Dennis Shusterman, Alfred T. Hodgson, Kazuhide Ito, Kazukiyo Kumagai 456
Exposure Assessment To E-Cigarettes Part 2: A Pilot Laboratory Study On Formation Of Volatile Carbonyls From Propylene Glycol A Major E-Cigarette Carrier Solvent
- Wenhao Chen, Ping Wang, Jeff Fowles, Toshiki Matsuo, Dennis Shusterman, Alfred T. Hodgson, Kazuhide Ito, Kazukiyo Kumagai 460

Exposure Assessment To E-Cigarettes Part 3: Preliminary Numerical Prediction Of Contaminant Distributions In Human Respiratory Tract Models
- Kazuhide Ito, Masato Yamashita, Sung-Jun Yoo, Toshiki Matsuo, Wenhao Chen, Ping Wang, Jeff Fowles, Kazukiyo Kumagai 464

Tuesday Poster Session **468**

Thermal Comfort During Summer In A High-Speed Railway Station In Cold Region Of China
- Gang Liu, Chenyi Lin, Yanbin Zhuo, Deheng Guo, Rui Dang 468
Controlling Indoor CO2 With A Solid Sorbent: Kinetics And Capacity
- Elliott T. Gall, Victor W.-C. Chang, William W Nazaroff 474
Water Droplets As Agents To Enhance The Removal Efficiency Of An Acoustic Aerosol Removal System
- Wai T Yuen, Sau C Fu, and Christopher Y H Chao 478
Estimates Of HVAC Filtration Efficiency For Fine And Ultrafine Particles Of Outdoor Origin
- Dan Zhao, Parham Azimi, and Brent Stephens 481
Study Design: Impact Of Ventilation And Cleaning On The Microbiology Of The Built Environment In Tribal Homes
- Richard Shaugnessy, Jordan Peccia, David Reisdorph, and Ulla Haverinen-Shaughnessy 485
Charactering Airborne Allergens And Human Allergic Responses In A University History Museum
- Yan Wu, Maosheng Yao and Victor W.-C. Chang 489
Study On Ventilation Strategy For Removing Dust Particles
- Yang Lu, Hiroshi Yoshino 493
From A Lab Scale To The Building Scale: Formaldehyde Removal By A Plant Module
- Ahu Aydogan Akseli, Anna Dyson, and Lupita D. Montoya 497
A Statistical Study Of Indoor Acoustical Perceptions Based On Urban Traffic Noise Transmitted Via Operable Windows
- Jeehwan Lee and Jae D. Chang 501
Determination Of Diffusion Parameters Using A Semi-Dynamic Dual Chamber Method
- Guoqing He, Barbora Krejcirikova, Carsten Rode, and Jianshun Zhang 505
A High Efficiency Instrument For Collecting Airborne Particles Down To 10 Nanometers On Solid Surfaces Or Into Liquid
- Arantzazu Eiguren-Fernandez, Nick Buglack, Ning Li, Pat Keady and Susanne Hering . 509
Resuspension Of Bacterial Spore Particles From Duct Surfaces
- Paricheher Salimifard, Paul Kremer, Donghyun Rim, James D. Freihaut 513
Black Carbon And PM2.5 Filtration In Classroom Ventilation Systems In Salt Lake City, Utah And Las Vegas, Nevada
- Paul T. Roberts, David L. Vaughn, Steven G. Brown 518
The Experimental Study Of The Regenerate Characteristic Effect On The Liquid Dehumidification System Performance
- Yulieh Wu, Kunhua Lyu, and Pohan Lin 522
Airtightening At Foundation-House Interface As A Means Of Radon Reduction In Living Space
- William Rose, Zachary Merrin, Paul Francisco and Stacy Gloss 526

Removing Acetaldehyde In Indoor Air Using Mn/TiO2 Nanoparticle OZCO Catalyst
- Minsu Kim, Eunseuk Park, Hyounduk Jung, and Jongsoo Jurng 530
Enhanced Catalytic Decomposition Of VOCs Using Synthesized MnOx/TiO2 With Adsorption-desorption
- Hyounduk Jung, Minsu Kim, Eunseuk Park, and Jongsoo Jurng 532
Exploring The Effectiveness Of Simple Fan Filter Systems In Ventilating An Enclosed Non-Airconditioned Space
- Ron Tan, Joanna Shen, Gayatri Sankaran, Theodore Prawiradiraja, Hooi Ming Yap, Ramona Gutierrez, Kwok Wai Tham, Chandra Sekhar, Kok Wai Cheong, Wei-Chung Chang, Keng Been Ang, Clement Yew Yi Tan and Lee Ching Ng 534
CFD Modeling And Parameterization Of The Photocatalytic Oxidation Process In Indoor Environments
- Eisaku Sumiyoshi, Takahiro Yamaguchi, Kiyoshi Yamamoto, Eunsu Lim and Kazuhide Ito 539
A Multi-Scale Exposure Concentration Analysis In A Large Factory Space Using Computational Fluid Dynamics Technique
- Alicia Murga, Sung-Jun Yoo, Kazuhide Ito 543
Towards Assessing The Viability Of Indoor Dust Microbiome Using Flow Cytometry
- Clarisse M. Betancourt Romn, Erica M. Hartmann, Jessica L. Green 547

Author Index **549**

1st North American Regional Conference

PROCEEDINGS
HEALTHY BUILDINGS 2015 AMERICA

Innovation in a Time of Energy Uncertainty and Climate Adaptation

ISIAQ
INTERNATIONAL SOCIETY OF INDOOR AIR QUALITY AND CLIMATE

University of Colorado **Boulder**

http://hb2015-america.org/

Welcome to *Healthy Buildings 2015 America*

Dear Friends and Colleagues:

Welcome to *Healthy Buildings 2015 America*, the 1st North American Regional Conference in the *Healthy Building* series. We greatly appreciate that the International Society of Indoor Air Quality and Climate chose the University of Colorado Boulder to organize this unique conference. We hope that our efforts yield a wonderful experience for you.

In planning HBA15 we reminded ourselves often of several important ideas.

Healthy Buildings is a unique forum for built environment researchers and professionals to engage with innovative projects, products, and services and to meet and collaborate with colleagues working on the pressing global challenge of making buildings healthy, energy efficient, and sustainable.

Healthy Buildings attracts researchers and other professionals from the fields of indoor air quality, built environments, HVAC, health sciences, public health policy, urban planning, mechanical engineering, architecture, building design and management, and more.

Healthy Buildings sustains ISIAQ's mission to support the creation of healthy, comfortable, and productive indoor environments. ISIAQ believes this is achievable by advancing the science and technology of indoor air quality and climate as it relates to indoor environment design, construction, operation, maintenance, air quality measurement, and health sciences.

ISIAQ's major role is to facilitate international and interdisciplinary communication and information exchange, as well as develop, adapt, and maintain codes, standards, and guidelines for the improvements of indoor air quality and climate.

During the conference there will be over 140 podium and poster presentations with a heavy emphasis on the indoor microbiome, climate change, energy efficiency, building materials and emissions, sensing, mission critical environments and more. There will be 14 workshops with which to engage in discussion and learn about a new field. There are seven fantastic plenary speakers with diverse backgrounds ranging from mechanical engineer to biologist to architect.

We encourage you to take advantage of all that the city of Boulder and the surroundings have to offer. It is a one of the top 10 livable cities in the US. Located in the Front Range at the edge of the Rockies, it is place to indulge in outdoor activities including rock climbing, hiking, and running, as well as top rated restaurants. The local community is dedicated to the responsible and effective use of renewable energy, green building practices, preserving open space, alternative transportation, and the environment.

There have been many who have worked hard these past two years to make HBA15 a possibility. CU Conference Services has been exceptional. We have had a great Scientific Committee with over 100 members who reviewed papers. Our Advisory Committee provided invaluable guidance, direction and new ideas. The students were excited to support the conference and plan a few special outings.

Your presence is the real key to the success of HBA15. We are grateful for your attendance and for your contributions, ranging from sharing scientific discoveries, practical guidance, networking, and friendship. Enjoy!

Cheers!
Shelly L Miller
Conference President

Zhiqiang (John) Zhai
Conference Vice-President

Welcome message from the President of ISIAQ

Greetings to participants of *Healthy Buildings 2015 America*! As President of the International Society of Indoor Air Quality and Climate (ISIAQ), I am looking forward to meeting many of you here in beautiful Boulder, Colorado. I am especially grateful to the organizers: Shelly Miller and John Zhai. They sacrificed much time to create a program that is rich with thought-provoking presentations and workshops.

Healthy Buildings 2015 America is the first regional *Healthy Buildings* conference to take place in the United States, and is organized with a focus on the challenges and solutions unique to North America. Each *Healthy Buildings* conference helps me re-connect to the practice of indoor air quality while engaging my scientific curiosity and energizing my creativity. I wish the same for you and for all participants. ISIAQ is a highly multidisciplinary organization that promotes the free exchange of ideas, technical expertise and scientific discovery. I hope to continue our long-standing tradition of providing members opportunities for learning from one another, developing collaborations and enlightening the world about the challenges and advances in improving indoor air quality and climate. How, you may ask, can I get more involved? Consider joining one of the ISIAQ Scientific and Technical Committees, mentor a younger ISIAQ member, or start early and plan your technical submission for next-year's *Indoor Air* conference. Your contributions make a difference!

With best regards,

Glenn Morrison
President of ISIAQ
Missouri University of Science and Technology, USA

Thank you for the generous support of the

Alfred P. Sloan Foundation

for *Healthy Buildings 2015 America.*

With their support we were able to provide two Workshops on research and methods from their Microbiology of the Built Environment Program, as well as the networking breakfast.

www.sloan.org

ALFRED P. SLOAN FOUNDATION
1934

Alfred P. Sloan
FOUNDATION

Map Engineering Center

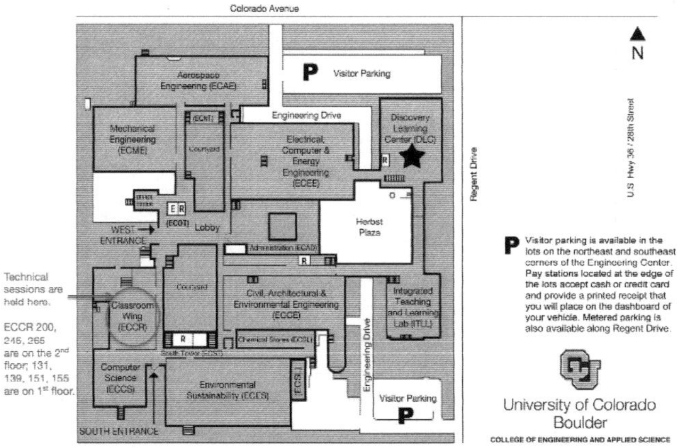

Map CU Boulder Campus

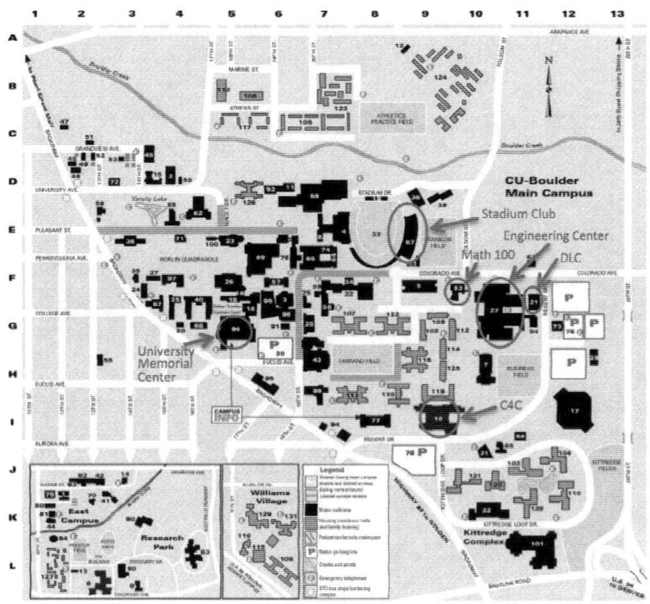

http://www.colorado.edu/campusmap/

Map City of Boulder

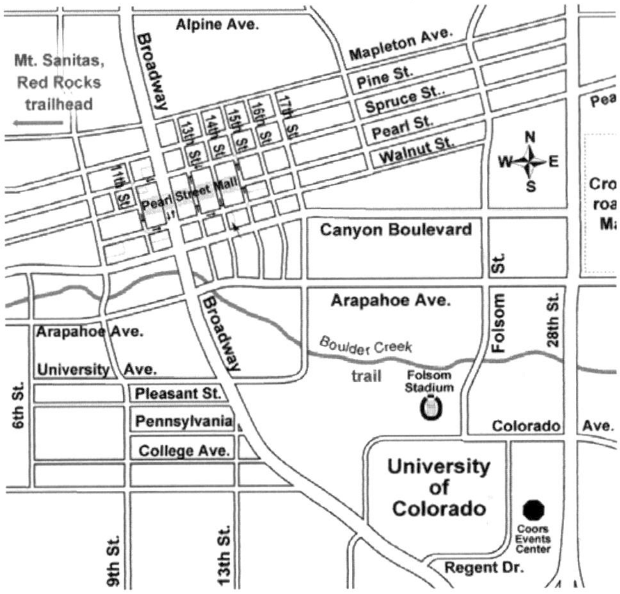

Map Boulder and Surroundings

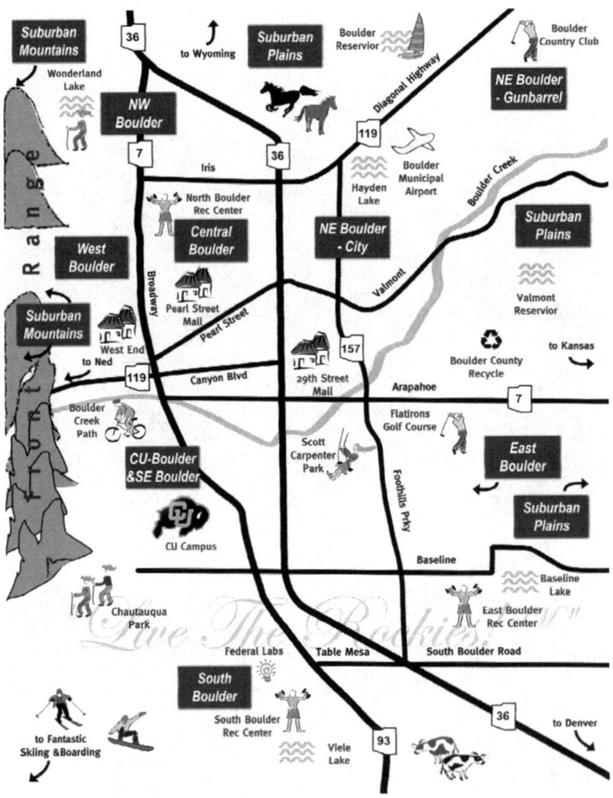

Locations and Other Information

Registration

- Sunday 1 PM – 8 PM University Memorial Center Glenn Miller Lounge
- Mon – Wed 8 AM – 9 AM Math 100 Courtyard
- Mon – Tues 9 AM – 4 PM Conference Headquarters ECCR 131

Technical Sessions

- Engineering Center Classroom Wing ECCR 151, 155, 200, 245, 265 (rooms starting with 1 are on 1st floor; rooms starting with 2 are on 2nd floor)

Plenaries

- Sunday Opening Plenary University Memorial Center West Ballroom
- Mon – Wed Plenaries Math 100

Lunch

- Center for Community (C4C)

Opening Reception

- University Memorial Center Room 235

Happy Hour / Posters

- Stadium Club

Speaker Lounge

- ECCR 139

Conference Headquarters

- ECCR 131

First Aid / Medical

- Conference Headquarters ECCR 131
- Wardenburg Medical Center on CU Campus

Parking Questions

- Conference Headquarters ECCR 131

Website

- http://www.hb2015-america.org

Twitter Hashtag

- #HBA15

Guidelines for Presenters

Podium and Workshops:

- Please sign in at the Speaker's Lounge ECCR 139 at least 2 hours before your session to upload your presentation file and to have the presentation checked for commercialism[1]
- A computer will be available in the podium rooms for presentations. Please plan on using the room's computer and not your own computer
- Please bring a backup copy of your presentation (in USB) to the session room (you don't need to copy it to the room's computer again unless it is requested by the session chair)
- Please arrive at the session room at least 10 min early to meet the session chair and other speakers
- Please prepare and provide a 2-sentence bio intro of the presenter to the session chair and you don't have to introduce yourself again before your presentation
- Please conclude your presentation in 12 min and you will be given a hint at 10 min by the chair
- Brief questions and answers (3 min) will be arranged after each talk when time allows, up to the chair's jurisdiction

Poster:

- Your allocated poster size is 46" X 46" (this is the maximum area you can use)
- There is no firm format requirement on the poster, but rather creativity and visibility are strongly suggested (and thus please use more figures and tables and use larger font if possible; 18 pt. should be the minimum font size)
- You will find the poster room and poster location (paper number) when you check in at the conference
- Pins and tape will be provided at the poster session site (Stadium Club)
- Please hang your poster before the Monday poster session (i.e. before 5:15 pm Monday)
- Please leave your poster hanging for both Monday and Tuesday's poster session
- Please be present at your poster on your designated presentation day
- Please remove your poster after the end of the poster session on Tuesday; any remaining poster will be destroyed

---

[1]The commercialism policy is based on ASHRAE's policy: Use of commercial names and logos shall not be done in ways that imply ISIAQ endorsement, approval or certification. The inclusion of commercial information shall be done in a fair and unbiased way so as to avoid explicit promotion of a product or commercial entity.

Schedule at a Glance

Sunday Jul 19

- 1 PM - 8 PM: Registration – University Memorial Center
- 3 PM - 4 PM: STC meetings – Engineering Center
- 5 PM - 8 PM: Opening session – University Memorial Center

Monday Jul 20

- 8 AM - 9 AM: Registration – Math 100 Courtyard
- 9 AM - 4 PM: Registration – Conference Headquarters ECCR 131
- 9 AM - 10 AM: Plenaries – Math 100
- 10 AM - 10:30 AM: Coffee Break – Engineering Center Lobby
- 10:30 AM - 12:30 PM: Technical Sessions – Engineering Center
- 12:30 PM - 1:15 PM: Lunch – Center for Community (C4C)
- 1:30 PM - 2:15 PM: ISIAQ Annual General Meeting – Math 100
- 2:30 PM - 3:30 PM: Technical Sessions – Engineering Center
- 3:30 PM - 4 PM: Coffee Break – Engineering Center Lobby
- 4 PM - 5 PM: Technical Sessions – Engineering Center
- 5:15 PM - 6:30 PM: Happy Hour/Posters – Stadium Club

Tuesday Jul 21

- 7:45 AM - 8:45 AM: Sloan Breakfast – Discovery Learning Center Lobby
- 8 AM - 9 AM: Registration – Math 100 Courtyard
- 9 AM - 10 AM: Plenaries – Math 100
- 9 AM - 4 PM: Registration – Conference Headquarters ECCR 131
- 10 AM - 10:30 AM: Coffee Break – Engineering Center Lobby
- 10:30 AM - 12:30 PM: Technical Sessions – Engineering Center
- 12:30 PM - 1:30 PM: Lunch – C4C
- 1:45 PM - 2:45 PM: Technical Sessions
- 2:45 PM - 3:15 PM: Coffee Break – Engineering Center Lobby
- 3:15 PM - 4:15 PM: Technical Sessions
- 4:30 PM - 6 PM: Happy Hour/Posters – Stadium Club

Wednesday Jul 22

- 8 AM - 9 AM: Registration – Math 100 Courtyard
- 9 AM - 10 AM: Plenaries – Math 100
- 10 AM - 10:30 AM: Coffee Break – Engineering Center Lobby
- 10:30 AM - 12 PM: Technical Sessions – Engineering Center
- 12 PM - 12:45 PM: Closing Session
- 2 PM - 3 PM: Campus Tours – Engineering Center Lobby

Podiums at a Glance

Monday Jul 20

- 10:30 AM-12:30 PM: **M.1. Indoor Air Quality in Homes** – ECCR 155
  Chair: Christopher Y. H. Chao
- 10:30 AM-12:30 PM: **M.2 Building Materials and Furnishings** – ECCR 151
  Chair: Suchismita Bhattacharjee
- 2:30 PM-3:30 PM: **M.3 Indoor Air Exposures** – ECCR 200
  Chair: Michal P. Spilak
- 2:30 PM-3:30 PM: **M.4 Filtration** – ECCR 155
  Chair: Marco-Felipe King
- 2:30 PM-3:30 PM: **M.5 Energy and Environmental Quality** – ECCR 151
  Chair: Julia C. Luongo
- 4:00 PM-5:00 PM: **M.6 Ventilation** – ECCR 155
  Chair: Catherine J. Noakes
- 4:00 PM-5:00 PM: **M.7 Sensing** – ECCR 151
  Chair: Seema Bhangar

Tuesday Jul 21

- 10:30 AM-12:30 PM: **T.1 Transport and Chemistry** – ECCR 155
  Chair: Jelle Laverge
- 10:30 AM-12:30 PM **: T.2 Mold and Moisture** – ECCR 151
  Chair: Josephine Lau
- 1:45 PM-2:45 PM : **T.3 IAQ in Hospitals** – ECCR 155
  Chair: Brent Stephens
- 1:45 PM-2:45 PM: **T.4 Urban and Indoor Environments** – ECCR 151
  Chair: Lupita D. Montoya
- 3:15 PM-4:15 PM: **T.5 IAQ Challenges and Strategies** – ECCR 155
  Chair: Andrew Hoisington
- 3:15 PM-4:15 PM: **T.6 VOCs** – ECCR 151
  Chair: Robert Bean

Wednesday Jul 22

- 10:30 AM-12:00 PM: **W.1 Microbiome** – ECCR 155
  Chair: Kerry Kinney
- 10:30 AM-12:00 PM: **W.2 Architecture and Energy** – ECCR 151
  Chair: Nicholas Clements
- 10:30 AM-12:00 PM: **W.3 IAQ in Schools** – ECCR 200
  Chair: Andrea R. Ferro

Workshops at a Glance

Monday Jul 20

- 10:30 AM-12:30 PM: **M.W.1 Alfred P. Sloan Foundation Dissemination and Integration of Microbiology of the Built Environment Research** – ECCR 265
  Sponsored by the Alfred P. Sloan Foundation, researchers from the Sloan Program on the Microbiology of the Built Environment will present and synthesize key results that are most relevant to practitioners attending the conference.
- 10:30 AM-12:30 PM: **M.W.2 Healthy Buildings: Reducing the Use of "Six Classes" of Harmful Chemicals** – ECCR 245
  Join Arlene Blum for a discussion of "Six Classes" that contain many of the chemicals of concern commonly found in building materials and consumer products, the science and policy of flame retardants and other hazardous chemicals, and the national impacts of California's changing flammability standards.
- 10:30 AM-12:30 PM: **M.W.3 Real-World Building Moisture Investigations: Tips, Tricks and Traps** – ECCR 200
  Three experienced investigators will share some of their favorite tips, traps, tools and techniques for different types of building moisture investigations. The exchange of experiences and ideas about better ways to investigate building moisture problems is encouraged.
- 2:30 PM-3:30 PM: **M.W.4 NIOSH Moisture Assessment Tools and Studies in Schools** – ECCR 245
  Researchers will lay out assessment tools and approaches available for investigating moisture and dampness in schools. The workshop will include discussions of the NIOSH dampness and Mold Assessment Tool (DMAT) and the latest developments in the 50 Elementary Schools Study.
- 4:00 PM-5:00 PM: **M.W.5 Indoor Environment Quality and Productivity: Agreeing on a Productivity Index for IEQ** – ECCR 200
  Attendees will discuss how to enhance the importance of indoor environmental quality for all building stakeholders through the measurement of occupant performance and productivity, with the goal of proposing an international method for productivity measurement in indoor workplaces.
- 4:00 PM-5:00 PM: **M.W.6 IESO-ASHRAE Standard for Assessment of Mold in Schools** – ECCR 245
  Panelists will discuss existing tools/protocols/guidance already to assess moisture/mold in schools, and refer to the new proposed IESO-ASHRAE Standard for moisture assessment. Discussion will focus on what is needed for schools, given the existing guidance available.

Tuesday Jul 21

- 10:30 AM-12:30 PM: **T.W.1 Alfred P. Sloan Foundation Microbiology of the Built Environment Study Methods** – ECCR 265
  Sponsored by the Alfred P. Sloan Foundation, researchers will demonstrate the various sampling and analysis methods used in investigations of the microbiology of the built environment.
- 10:30 AM-12:30 PM: **T.W.2 School Building Condition Impacts on Occupant Health, Academic Achievement, Fiscal and Environmental Sustainability & Social Justice Using a Data-Science Approach to the Collection, Analysis & Use of Information** – ECCR 200
  Attendees will discuss: Why do our schools look like they do? What do deteriorated conditions mean for occupant health, educational outcomes, fiscal and environmental sustainability and social justice? How can we better understand and fix these issues using new approaches in Big Data, Open Data APIs and the Internet of Things to collect data about building conditions and impacts on health, achievement etc?
- 10:30 AM-12:30 PM: **T.W.3 New Changes to the LEED Certification and Their Impact on the Indoor Air Workshop** – ECCR 245
  The newest version of the benchmark for high-performance buildings, LEED v4, will be discussed, which includes changes in material emission/selection and in indoor air quality testing requirements. We will examine the owner's perspective, the new IAQ testing, and a practical point of view.
- 1:45 PM-2:45 PM: **T.W.4 What Does the Indoor Environmental Community Have To Do to Prepare for Climate Change** – ECCR 245
  The changing global climate brings with it new challenges for IEQ researchers and professionals. We will discuss the nature of those challenges and their specific implications for researchers and professionals in the indoor environmental quality fields with a focus on health.
- 3:15 PM-4:15 PM: **T.W.5 Complexity of Moisture and Microbial Investigations and Remediation within Hotels** - ECCR 200
  The many complex issues investigators confront when completing moisture and microbial investigations with hotels will be discussed. Multiple building materials, sampling methodology, and not disrupting hotel operations are a few of the challenges presented during a typical investigation.
- 3:15 PM-4:15 PM: **T.W.6 US EPA Indoor Air and Climate Change Star Grantees Workshop** – ECCR 245
  The US EPA Indoor Air and Climate Change Star Grantees will discuss the shared challenges of their projects and opportunities for collaborations.

Wednesday Jul 22

- 10:30 AM-12:00 PM: **W.W.1 What Can Indoor CO_2 Concentration Measurements Tell Us?** – ECCR 245
Some of the most valuable uses of carbon dioxide (CO_2) measurements indoors and some of the challenges that users of those measurements face will be presented. The workshop is intended to focus on how to best use CO_2 in research and indoor building environmental assessments.
- 10:30 AM-12:00 PM: **W.W.2 Measurements of Semi-Volatile Organic Compound (SVOC) Emission Parameters** – ECCR 265
Building materials, consumer products and furnishings contain a vast array of chemicals that are emitted to indoor environments where people spend most of their time. Studies to improve the understanding of SVOC emissions from The University of Texas at Austin, Health Canada, US EPA, and Virginia Tech will be discussed.

Sunday, July 19, 2015

1:00 PM - 8:00 PM

Registration

University Memorial Center Glenn Miller Lounge

3:00 PM - 4:00 PM

Meeting: **STC 13, Microorganisms**

ECCR 155, Engineering Center

Open to Everyone

3:00 PM - 4:00 PM

Meeting: **STC 21, Ventilation**

ECCR 151, Engineering Center

Open to Everyone

5:00 PM - 5:15 PM

Event: **Welcome – Sam Weaver, Boulder City Council, "Advancing Efficient, Safe Buildings"**

University Memorial Center West Ballroom

5:15 PM - 6:00 PM

S.PL.1 *Opening Plenary* – **Chuck Kutscher, "Climate Change: The Latest Science, Why It's Serious, and What We Can Do About It"**

University Memorial Center West Ballroom

Dr. Charles (Chuck) Kutscher is Director of the Buildings and Thermal Systems Center at the National Renewable Energy Laboratory in Golden, Colorado. He has led research in solar heating and cooling, energy efficiency, solar industrial process heat, power plant cooling systems, and concentrating solar power. He is a Fellow of the American Solar Energy Society (ASES) and served as ASES Chair in 2000-2001. He was the Chair of two major conferences: the SOLAR 2006 national solar energy conference and the 2012 World Renewable Energy Forum. Dr. Kutscher is editor of the 200-page ASES report, Tackling Climate Change in the U.S., which details how energy efficiency and six renewable energy technologies can greatly reduce U.S carbon emissions by 2030. He is an adjunct professor at the University of Colorado Boulder where he has taught courses in engineering and "Climate Change Solutions." He has a B.S. in physics from the State University of New York at Albany, an M.S. in nuclear engineering from the University of Illinois, and Ph.D. in mechanical engineering from the University of Colorado Boulder.

6:00 PM - 8:00 PM

Event: **Welcome Reception**

University Memorial Center Room 235

Monday, July 20, 2015

8:00 AM - 9:00 AM

Registration

Math 100 Courtyard

9:00 AM - 4:00 PM

Registration

Conference Headquarters ECCR 131

9:00 AM - 9:30 AM

M.PL.1 *Plenary* – **Arlene Blum, "How Science Can Impact Policy and Purchasing: Flame Retardants and other Classes of Harmful Chemicals"**

Math 100

Arlene Blum PhD, biophysical chemist, author, and mountaineer is a Visiting Scholar in Chemistry at the University of California, Berkeley and executive director of the Green Science Policy Institute. The Institute brings government, industry, scientists and citizens groups together worldwide to support chemical policies to protect human health and the global environment. Blum's research and policy work with the Institute has contributed to preventing the use toxic flame retardants and other harmful chemicals in children's sleepwear, furniture, electronics, and other products globally. Blum's awards include selection as an American Association for the Advancement of Science Fellow, selection by the UK Guardian as one of the world's 100 most inspiring women and National Women's History Project selection as one of 100 Women Taking the Lead to Save Our Planet, and election to the Hall of Mountaineering Excellence. More information can be found at www.greensciencepolicy.org and www.arleneblum.com.

9:30 AM - 10:00 AM

M.PL.2 *Plenary* – **Jessica Greene, "Next-generation Indoor Environmental Quality'**

Math 100

Jessica is the Director of the Biology and the Built Environment Center (BioBE), and Associate Professor, Department of Biology, University of Oregon. The vision of the BioBE Center, funded by the Alfred P. Sloan Foundation, is to develop a hypothesis-driven, evidence-based approach to understand the built environment microbiome. Our goal is to optimize the design and operation of buildings to promote both human health and environmental sustainability, with an emphasis on green healthcare design. Dr. Green's research lab applies theoretical, computational and empirical approaches to study complex biological systems in a wide range of habitats: our bodies, cities, forests, and oceans. Her research uses interdisciplinary approaches at the interface of environmental genomics, ecology, phylogenetics, mathematics, and informatics.

10:00 AM - 10:30 AM

Event: **Coffee Break**
Engineering Center Lobby

10:30 AM - 12:30 PM

M.W.1 *Workshop:* **Alfred P. Sloan Foundation Dissemination And Integration Of Microbiology Of The Built Environment Research**

Chairs: *Shelly Miller and Brent Stephens*

Speakers: *Noah Fierer, Jack Gilbert, Jordan Peccia, Bill Nazaroff, Jessica Green and Kerry Kinney*

ECCR 265, Engineering Center

10:30 AM - 12:30 PM

M.W.2 *Workshop:* **Healthy Buildings: Reducing The Use Of "Six Classes" Of Harmful Chemicals**

Chair: *Arlene Blum*

ECCR 245, Engineering Center

10:30 AM - 12:30 PM

M.W.3 *Workshop:* **Real-World Building Moisture Investigations: Tips, Tricks And Traps**

Chair: *Laura Kolb*

Speakers: *Terry Brennan, Lew Harriman and Lan Chi Nguyen Weekes*

ECCR 200, Engineering Center

10:30 AM - 12:30 PM

M.1 *Podium:* **Indoor Air Quality in Homes**
ECCR 155, Engineering Center

Chair: *Christopher Y. H. Chao*

208 Ventilation Provision And Outcomes In Mainstream Contemporary New-building Flats In London, UK
Tim Sharpe, *Antoinette Charles*

175 The Impact Of Heating Process On PM2.5 Concentration In Cold Rural Areas Of Northern China
Xueyan Zhang, ***Bin Chen***

187 Measuring Flame Retardant Emissions From Spray Polyurethane Foam In A Home
Dustin Poppendieck, *Angelica Connor*

189 Phthalate Levels In Green-renovated Subsidized Housing
Robin E. Dodson, *Julia Udesky, Ruthann Rudel*

205 Health Benefits Of Living In Green, Low-income, Multi-family Housing
***Meryl D. Colton**, Jose GC Laurent, Piers MacNaughton, John Kane, Mae Bennett-Fripp, John Spengler, Gary Adamkiewicz*

166 Operation Of Residential HVAC Systems: Implications For Healthy Buildings
*Marianne Touchie, **Jeffrey Siegel***

154 Measuring The Ozone Penetration Factor In A Residence Under Infiltration Conditions
***Haoran Zhao**, Brent Stephens*

226 Comparison Of Indoor And Outdoor Air Quality In Latino Communities In Peru, Chile, And USA
***Wyatt M. Champion**, Francisco Barraza, Héctor Jorquera, Lupita D. Montoya*

10:30 AM - 12:30 PM

M.2 *Podium:* **Building Materials and Furnishings**
ECCR 151, Engineering Center

Chair: *Suchismita Hattacharjee*

10 Carpet Dust Culturable Fungi And Bacteria Results Used To Evaluate Carpet Maintenance
***Kent A. Candee**, Krista J Scott, Craig S Black, Dave W. Havick*

20 Emission Of Phthalates And Phthalate Alternatives From Vinyl Flooring And Crib Mattress Covers: The Influence Of Temperature
*Yirui Liang, **Ying Xu***

215 Safeguarding Our Resources And Indoor Air – Gypsum Wallboard Closed-loop Recycling In An Era Of Increased Chemical Transparency
***Danielle Wagner**, Zhongqi (Joshua) Cheng, Amanda Kaminsky, Sydney Mainster, Geoffrey Brock, Ronald Spalter, Catherine Bobenhausen*

196 Laboratory And Field Performance Of Interior Finish Materials Using Formaldehyde Scavenging Technology
*Stanley D. Gatland, Douglas Gehring, **Davy Vennekens***

213 Mass-based Size Distribution Of House Dust Loading On Residential Floorings
*Yan Ma, Yilin Tian, **Andrea R. Ferro***

222 Investigation Of Particle-mediated Gas-phase Transport Of Phthalates From Vinyl Flooring
*Yaoxing Wu, Eric Vejerano, Cong Liu, Steven Cox, Mingjie Xie, Linsey Marr, **John Little***

240 Resuspension Of Deposited Particles Induced By Jets At Different Surface Dust Loads
*Xiong Mei, **Tengfei Zhang**, Shugang Wang*

294 Experimental Study of Active Flows and Passive Materials for Indoor Air Quality Control
Denise Mauney, Christella Suwongso, Miles Abarr, Wil W. Srubar, ***Lupita D. Montoya***

12:30 PM - 1:15 PM

Lunch session

Center for Community (C4C)

1:30 PM - 2:15 PM

Meeting: **ISIAQ Annual General Meeting**

Math 100

Open to Everyone

2:30 PM - 3:30 PM

M.W.4 *Workshop:* **NIOSH Moisture Assessment Tools And Studies In Schools**

Chair: *Richard Shaughnessy*

Speakers: *Jean Cox-Ganser and Ju-Hyeong Park*

ECCR 245, Engineering Center

2:30 PM - 3:30 PM

M.3 *Podium:* **Indoor Air Exposures**

ECCR 200, Engineering Center

Chair: *Michal P. Spilak*

24 Assessing Wildfire Exposure In Homes Near Wildfires
Ernest R. Crutcher*, Heidie K. Crutcher-Bettes*

45 Air Pollution Exposure Model For Individuals (EMI) In Health Studies: Evaluation For Ambient PM2.5
Michael Breen*, Thomas Long, Bradley Schultz, Ronald Williams, Jennifer Richmond-Bryant*

143 Improving The Indoor Air Quality By Using A Surface Emissions Trap For Exposure Reduction
Lennart Larsson*, Pawel Markowicz, Johan Mattsson*

147 Statistical Evaluation Of Human Sensory Response To Acetone
Tunga Salthammer*, Nicole Schulz, Ramona Stolte, Erik Uhde*

2:30 PM - 3:30 PM

M.4 *Podium:* **Filtration**

ECCR 155, Engineering Center

Chair: *Marco-Felipe King*

47 Challenges And Opportunities For Filter Forensics
Daniel Haaland*, Jeffrey Siegel*

157 Bioaccumulation On HVAC Filters In University Buildings In Singapore
Irvan Luhung, *Yan Wu, Victor Wei Chung Chang, William W Nazaroff*

164 Modeling The Impact Of Residential HVAC Filtration On Indoor Particles Of Outdoor Origin
Parham Azimi, *Dan Zhao, Brent Stephens*

214 Pollutant Exposures And Asthma: What Can HVAC Filter Dust Tell Us?
Juan P. Maestre, *Sandra Dedesko, Kelly Royse, Ying Xu, Sharon Horner, Jeffrey Siegel, Kerry A. Kinney*

2:30 PM - 3:30 PM

M.5 *Podium:* **Energy and Environmental Quality**
ECCR 151, Engineering Center

Chair: *Julia C. Luongo*

163 Recommissioning A Zero-carbon Multifamily Residential Building's Ventilation System: A Necessity
Jelle Laverge, *Eline Himpe, Marc Delghust, Stijn Van de putte, Julio Vaillant Rebolar*

191 College Students' Perceptions Of Household Energy Efficiency
Suchismita Bhattacharjee, *Sandeep Langar, Maggie Hogan*

245 Indoor Environmental Resilience: A Review And Discussion
Andrew Persily, *Steven Emmerich*

247 Air Quality In Data Centers: People Vs. The Machines
Christopher Muller, *Chuck Arent*

3:30 PM - 4:00 PM

Event: **Coffee Break**
Engineering Center Lobby

4:00 PM - 5:00 PM

M.W.5 *Workshop:* **Indoor Environment Quality And Productivity: Agreeing On A Productivity Index For IEQ**

Chairs: *Vyt Garnys and Tim Callinan*

ECCR 200, Engineering Center

4:00 PM - 5:00 PM

M.W.6 *Workshop:* **IESO-ASHRAE Standard For Assessment Of Mold In Schools**

Chair: *Richard Shaughnessy*

Speakers: *Mike Mcguinness and Lan Chi Nguyen Weekes*

ECCR 245, Engineering Center

4:00 PM - 5:00 PM

M.6 *Podium:* **Ventilation**

ECCR 155, Engineering Center

Chair: *Catherine J. Noakes*

29 A General Method To Predict Ventilation Requirements While Incorporating Building Material Emission Testing
Wei Ye, *Doyun Won, Xu Zhang, Qi Zhou*

41 Time-resolved Aerosol And Fluorescent Bioaerosol Concentrations In An Air-conditioned And Mechanically Ventilated Office In Singapore
Jin Zhou, *Victor W.-C. Chang, William W Nazaroff*

33 Assessment Of The Implications Of Natural And Mechanical Ventilation On Human Health In The Residential Sector
Sandra M. Bernal

145 Testing Of A Downflow Ventilation System For High Risk Infectious Disease Isolation Rooms
Blanca Beato Arribas, ***Catherine J. Noakes****, P. Andrew Sleigh*

4:00 PM - 5:00 PM

M.7 *Podium:* **Sensing**

ECCR 151, Engineering Center

Chair: *Seema Bhangar*

210 Open Source Building Science Sensors (OSBSS): An Open Source Sensor Network For Indoor Environmental Data Collection
Akram Ali, Torkan Fazli, Joseph Huan, Deion Debose, Boyang Dong, ***Brent Stephens***

211 Development Of An Experimental System For Assessing Indoor Bioaerosol Transport And Control
Stephanie Kunkel, *Parham Azimi, Brent Stephens*

221 A Simple Method For Measuring Gas-phase SVOC Concentration In Equilibrium With The Material Phase
Yaoxing Wu, *Steven Cox, Mingjie Xie, Linsey Marr, John Little*

235 Wireless Monitoring Of The Indoor Air Environment In A Winery During Harvest
Nicolas M. Madrid, David J. Killeen, André Knoesen, David Mills, ***Roger Boulton***

5:15 PM - 6:30 PM

M.P *Poster:* **Happy Hour and Posters**

Stadium Club

13 The Cost Optimal Methodology Of Dwelling Buildings In Slovakia
Lucia Borisová

16 Improvement Of Perceived Temperature Under 28°C Of Air Conditioning Control Temperature
Yoshihito Kurazumi, *Kenta Fukagawa, Emi Kondo, Tomonori Sakoi, Kunihito Tobita, Yoshiaki Yamato*

18 Indoor Microclimate And Natural Ventilation In Double Transparent Facade
Kristyna Valouskova

32 A Simplified Model For Temperature Prediction In A Ventilated Wall Module
T.M.J. Rakotomahefa, ***Tengfei (Tim) Zhang***, *Shugang Wang*

34 How Reliable Are Standard (undisturbed) Airborne Mold (sporetrap) Tests? The Usefulness Of Disturbed Tests For More Accurate Assessments
Michael R. Meyer

36 Thermal Performance Of Evaluation On Energy Efficient Paraffinic PCM Based Shape Stabilized Composites For Energy Savings
Okyoung Chung, Su-Gwang Jeong, ***Sumin Kim***

39 Impact Of Water Damage On Microbial Community In Residential Buildings
Iman Sylvain, *Michal Spilak, Michael Waring, Rachel I. Adams*

51 Field Survey Of Indoor Environment And Its Effect On Human Health During Heating Period In Northern China
Bin Chen, Min Zhou, ***Yu Chen***, *Xueyan Zhang, Yuhang Shu, Yuanbing Zhu, Yizhou Sang, Joe R. Zhao*

52 Modeling The Dispersion Of Evaporating Spittle Droplet In A Typical Waiting Room Of A Hospital
Qiuyue Wang, *Shinsuke Kato*

151 Assessment Of Thermal Environment During Kitchen Work
Kunihito Tobita, *Yoshihito Kurazumi, Yoshiaki Yamato, Kenta Fukagawa*

162 Odor Thresholds For 2,4,6-trichloroanisole Reviewed In The Context Of Indoor Air Quality
Stephanie A. Juran, *Steven Nordin, Gunnar Johanson, Johnny C. Lorentzen*

173 The Effect Of Human Density On The Risk Of Infection In A Hospital Ward
Marcelo L. Pereira, *Lidia Morawska, Luke Knibbs, Congrong He, Graham Johnson*

182 The Influence Of Street Canyon Design On Hospital Air Quality
Marco-Felipe King, *Catherine J. Noakes*

184 Priority Chemicals Emitted From Composite Wood Materials
Doyun Won, *Wenping Yang, Joan Wong, Cheryl Khoury, Hans Schleibinger*

185 A Study Of Shading Device Orientation On The Natural Ventilation And Daylight Harvesting Potential In A Double Skin Facade
Jeehwan Lee, *Jae D. Chang, Hongyi Cai*

192 The Relation Of Psychosocial And Demographic Characteristics With Energy Consumption
Suchismita Bhattacharjee

194 Modeling Of Thermal Comfort Using Designbuilder
Lucie Dobiášová, *Daniel Adamovský*

223 Formaldehyde Sorption To Porous Media For Air Quality Applications
Anne Wrobetz, Megan Matteazzi, ***Lupita D. Montoya***

225 Indoor Environment Quality And NABERS IE Ratings: A 2014 Case Study For A Commercial Office Building Portfolio Of 25 Australian Buildings
Tim M. Callinan, *Jack J. Noonan, Vyt P. Garnys*

233 Web-Based Wireless Sensor System For Remote Air Environment Monitoring In A Commercial Winery
Nic Madrid, *David Killeen, Roger Boulton, Andre Knoesen*

273 Exposure Assessment To E-cigarettes Part 1: Literature Review On Carbonyl Compounds Generation From E-cigarettes And Affecting Factors
Wenhao Chen, *Ping Wang, Jeff Fowles, Toshiki Matsuo, Dennis Shusterman, Alfred T. Hodgson, Kazuhide Ito, Kazukiyo Kumagai*

283 Exposure Assessment To E-cigarettes Part 2: A Pilot Laboratory Study On Formation Of Volatile Carbonyls From Propylene Glycol – A Major E-cigarette Carrier Solvent
Wenhao Chen, *Ping Wang, Jeff Fowles, Toshiki Matsuo, Dennis Shusterman, Alfred T. Hodgson, Kazuhide Ito, Kazukiyo Kumagai*

284 Exposure Assessment To E-cigarettes Part 3: Preliminary Numerical Prediction Of Contaminant Distributions In Human Respiratory Tract Models
Kazuhide Ito, *Masato Yamashita, Sung-Jun Yoo, Toshiki Matsuo, Wenhao Chen, Ping Wang, Jeff Fowles, Kazukiyo Kumagai*

Tuesday, July 21, 2015

7:45 AM - 8:45 AM

Event: **Sloan Foundation Networking Breakfast**

Discovery Learning Center Lobby

All conference attendees involved in Sloan funded research or microbiome related work are invited to attend this breakfast to meet each other and network

---

8:00 AM - 9:00 AM

Registration

Math 100 Courtyard

9:00 AM - 4:00 PM

Registration

Conference Headquarters ECCR 131

9:00 AM - 9:30 AM

T.PL.1 *Plenary* – **Ellen Tohn, "The Role of Buildings in Health and Energy Policies"**

Math 100

Ellen Tohn is the Founder and Principal of Tohn Environmental Strategies, is a nationally recognized expert in environmental health, healthy housing, green building, indoor air quality and lead poisoning prevention. Ms. Tohn also has extensive sector experience working on clean air, solid and hazardous waste, toxic substances, and comparative risk. Ms. Tohn works with housing developers, property owners, managers and architects to create green and healthy housing. She has assisted national/regional and local health advocates catalyze effective and lasting policy solutions; contributed to numerous Federal and state guidance documents; developed Federal and local healthy housing and lead training courses; and designed and managed environmental health research studies resulting in peer-reviewed publications and influencing Federal regulatory actions. She recently directed a project to develop Guidance for EPA's Energy Star Indoor Air Quality Specifications and serves as an advisor on indoor air quality issues to the US Green Building Council's LEED for Homes standards setting process and numerous other green building programs. Ms. Tohn received her BA from Cornell University and holds a Masters of City Planning from the Massachusetts Institute of Technology. She is a member of the Wayland MA Energy Initiatives Advisory Committee.

9:30 AM - 10:00 AM

T.PL.2 *Plenary* – **Bronwyn Barry, "Looking at IAQ as a Team Sport: The Passive House Playbook"**

Math 100

Bronwyn Barry has worked in the green building industry for over 20 years. After earning her bachelors degree in South Africa, she moved to the US where she apprenticed at various architecture firms. Ms. Barry founded her own residential design practice in 2000, which she ran for 8 years. Her projects were selected for early Built It Green Home Tours and won multiple awards. During this time, she helped found the Green Remodeler's Guild, now called the Green Building Professionals Guild, run by Build It Green. Prior to this, Ms. Barry trained and led California's first All-Women Construction Crew at Mt Diablo Habitat for Humanity, where she integrated green building concepts into all chapter projects. More information can be found at http://oneskyhomes.com/team/leadership.

10:00 AM - 10:30 AM

Event: **Coffee Break**

Engineering Center Lobby

10:30 AM - 12:30 PM

T.W.1 *Workshop:* **Alfred P. Sloan Foundation Microbiology Of The Built Environment Study Methods**

Chair: *Shelly Miller*

Speakers: *Rachel Adams, Seema Bhangar, Jeff Siegel, Brent Stephens and Kyle Bibby*

ECCR 265, Engineering Center

10:30 AM - 12:30 PM

T.W.2 *Workshop:* **School Building Condition Impacts On Occupant Health, Academic Achievement, Fiscal And Environmental Sustainability, & Social Justice Using A Data-Science Approach To The Collection, Analysis & Use Of Information**

Chairs: *Jerry Roseman and Neil Roseman*

ECCR 200, Engineering Center

10:30 AM - 12:30 PM

T.W.3 *Workshop:* **New Changes To The LEED Certification And Their Impact On The Indoor Air Workshop**

Chairs: *Lan Chi Nguyen Weekes*

Speakers: *Dave Kahane, Steven Thweatt and Charlene Bayer*

ECCR 245, Engineering Center

10:30 AM - 12:30 PM

T.1 *Podium:* **Transport and Chemistry**

ECCR 155, Engineering Center

Chair: *Jelle Laverge*

152 Investigating Indoor Chemistry Of Terpinolene Through The Use Of A New Derivatization Agent
J. R. Wells, *Jason E. Ham, Stephen R. Jackson, Joel C. Harrison*

159 Secondary Organic Aerosol from α-Terpineol Ozonolysis
Yanan Yang, ***Michael Waring***

161 Changes In Outdoor Aerosol Chemical Composition Upon Transport Into The Indoor Environment
Anita Johnson, ***Michael Waring****, Peter DeCarlo*

55 The Influence Of Temperature On The Fate And Transport Of Indoor Phthalates: A Case Study In A Test House
Chenyang Bi*, Ying Xu*

153 Validation Of PM Concentration Estimates Using Contam Multizone Indoor Air Quality Model With Resuspension Deposition Module
Lisa Bramwell, Yilin Tian, ***Andrea R. Ferro***

53 Pollution Transport By The Human Convection Flow – Impact Of The Room Air Temperature And Seated Body Inclination
Dusan Licina*, Arsen Melikov, Chandra Sekhar, Kwok Wai Tham*

28 Sources And Mitigation Of Dioxin And Dioxin-like Compounds In Indoor Dust
Patricia V. Cline

49 Comparison Of Advective And Diffusive Transport Of SVOCs Through Cloth For Indoor Conditions
Alexander Korff, ***Melissa Buechlein****, Madison Gibler, Glenn Morrison*

10:30 AM - 12:30 PM

T.2 *Podium:* **Mold and Moisture**

ECCR 151, Engineering Center

Chair: *Josephine Lau*

30 A Practical Guide For Acceptable Indoor Airborne Mold Levels (sporetrap) With Emphasis On Aspergillus-Penicillium
Michael R. Meyer

37 Control Of Mold Contamination In The Bathroom
Satomi Yanaso*, Hiroshi Yamagishi, Takayuki Hasegawa, Kosuke Tanaka*

234 Synthetic Origin Of Microbial Volatile Organic Compounds
Hyunok Choi, *Norbert Schmidbauer, Carl-Gustaf Bornehag*

44 Investigating And Avoiding Moisture-related Problems In Existing Buildings
Lewis Harriman, *Terry Brennan*

207 Moisture Measurement And Implications For Fungal Growth In Drywall
Sandra Dedesko, *Jeffrey Siegel*

148 Airborne Bacteria And Fungi Concentrations In Airtight Contemporary Dwellings
Gráinne McGill, John E. Moore, ***Tim Sharpe****, Damian G. Downey, Lukumon Oyedele*

209 The IVAIRE Study: Correlations Between Ventilation And Mycological Parameters In Canadian Single Family Homes With Ashtmatic Children
Daniel Aubin, **Doyun Won**, *Hans Schleibinger, Wenping Yang, Pierre Lajoie, Denis Gauvin, Veronique Gingras, Jean-Marc Leclerc, Don Fugler*

220 Summary Of Sloan Symposium At Healthy Buildings 2015 – Europe
Hal Levin, *Martin Täubel, Mark T. Hernandez*

12:30 PM - 1:30 PM

Lunch session

C4C

1:45 PM - 2:45 PM

T.W.4 *Workshop:* **What Does The Indoor Environmental Community Have To Do To Prepare For Climate Change**

Chairs: *Hal Levin and Kazukiyo Kumagai*

Speakers: *William W Nazaroff, William J. Fisk*

ECCR 245, Engineering Center

1:45 PM - 2:45 PM

T.3 *Podium:* **IAQ in Hospitals**

ECCR 155, Engineering Center

Chair: *Brent Stephens*

2 Experimental And CFD Investigation Of Hospital Operating Room Air Distribution
Zhiqiang (John) Zhai, *James Mcneill, Jean Hertzberg*

193 The Role Of Surfaces In The Transmission Of Bioaerosols From Source To Patient In Hospital Rooms
Marco-Felipe King, *Catherine J. Noakes, P. Andrew Sleigh*

200 Hospital Surge Capacity: Practical Aspects Of Temporary Isolation Ward Design
Nicholas Clements, *Shelly L. Miller, Shobha S. Subhash, Aaron E. Eagan*

287 Exploring Spatial Planning And Functional Program Impact On Microbial Diversity And Distribution In Two South African Hospital Microbiomes
Jako A. Nice, *Piet Vosloo*

1:45 PM - 2:45 PM

T.4 *Podium:* **Urban and Indoor Environments**
ECCR 151, Engineering Center

Chair: *Lupita D. Montoya*

199 Model And Measurement Derived Estimates Of Residential PM2.5 Infiltration In 10 US Cities
Jennifer M. Logue, *Neil Klepeis, Brett C. Singer*

204 Urban Microbiome Pilot Study: Parks And Parking Lots
Gwynhwyfer Á. Mhuireach, *Joshua Ladau, James F. Meadow, G.Z. Brown, Bart Johnson, Jessica Green*

228 Environment Parameters Relevant To Microbes Spanning Human Urbanization In Sub-tropical Climate
Humberto Cavallin, Jean F. Ruiz, ***Atila Novoselac***, *Henrique Pereira, Luciana C. Paulino, Raphael Rios, Jean H. Perez, Martin J. Blaser, Maria Gloria Dominguez-Bello*

239 Controlling Pollutants During New Construction To Reduce Exposure Upon Occupancy By Sensitive Individuals
Carl Grimes

2:45 PM - 3:15 PM

Event: **Coffee Break**
Engineering Center Lobby

3:15 PM - 4:15 PM

T.W.5 *Workshop:* **Complexity of Moisture and Microbial Investigations And Remediation Within Hotels**

Chairs: *Richard Shaughnessy*

Speakers: *Brad Caddick and Peter Galoski*

ECCR 200, Engineering Center

3:15 PM - 4:15 PM

T.W.6 *Workshop:* **US EPA Indoor Air And Climate Change Star Grantees Workshop**

Chairs: *Vito Ilaqua and Laura Kolb*

Speakers: *Shelly Miller, Brent Stephens, Glenn Morrison, GZ Brown, Jessica Green, Howard Kipen, Petros Koutrakis, Brian Lamb, Christopher Ueijo, Olga Wilhelmi, David Sailor*

ECCR 245, Engineering Center

3:15 PM - 4:15 PM

T.5 *Podium:* **Indoor Air Quality Challenges and Strategies**

ECCR 155, Engineering Center

Chair: *Maj Andrew Hoisington*

198 Understanding The Challenges Of Assessing “Good” Air Quality In The Indoor Environment
***Jennifer M. Logue**, Max H. Sherman, Brett C. Singer*

250 Weatherization Ventilation Strategy As A Factor In Indoor Humidity
***William Rose**, Stacy Gloss, Paul Francisco, Zachary Merrin, Salvatore Cali, David Jacobs*

230 PM2.5: An Emerging Issue For Standard Of Care In Indoor Spaces
Terry Brennan

244 High Performance IAQ Specification For Net Zero Energy Homes
*Anthony Bernheim, Alfred Hodgson, **Andrew Persily***

3:15 PM - 4:15 PM

T.6 *Podium:* **VOCs**

ECCR 151, Engineering Center

Chair: *Robert Bean*

26 A New Approach To Estimate Emission Factors Of Semi-volatile Organic Compounds Based On Their Physical Properties
***Jiping Zhu**, Peter Mochungong, Julie S. O'Brien*

22 A Study On TVOC Concentration Emitted By Plywood: Finish, Age And Environment
*E. V. H. Dia, **L. M. T. Bo-ot***

181 New VOC Compounds In The Indoor Environment From Thermoplastic Filaments For 3D Printing
***Olaf Wilke**, Stefan Seeger*

190 Characterizing Uncertainty Of A Formaldehyde Reference Standard
Dustin Poppendieck

4:30 PM - 6:00 PM

T.P *Poster:* **Happy Hour and Posters**

Stadium Club

23 Thermal Comfort During Summer In A High-speed Railway Station In Cold Region Of China

*Gang Liu, **Chenyi Lin**, Yanbin Zhuo, Deheng Guo, Rui Dang*

42 Controlling Indoor CO2 With A Solid Sorbent: Kinetics And Capacity

***Elliott T. Gall**, Victor W.-C. Chang, William W. Nazaroff*

54 Water Droplets As Agents To Enhance The Removal Efficiency Of An Acoustic Aerosol Removal System

***Wai T. Yuen**, Sau Chung Fu, Christopher Y. H. Chao*

150 Estimates Of HVAC Filtration Efficiency For Fine And Ultrafine Particles Of Outdoor Origin

***Dan Zhao**, Parham Azimi, Brent Stephens*

170 Study Design: Impact Of Ventilation And Cleaning On The Microbiology Of The Built Environment In Tribal Homes

***Richard Shaughnessy**, Jordan Peccia, David Reisdorph, Ulla Haverinen-Shaughnessy*

171 Charactering Airborne Allergens And Human Allergic Responses In A University History Museum

***Yan Wu**, Maosheng Yao, Victor W.-C. Chang*

178 Study On Ventilation Strategy For Removing Dust Particles

***Yang Lu**, Hiroshi Yoshino*

180 From A Lab Scale To The Building Scale: Formaldehyde Removal By A Plant Module

***Ahu Aydogan Akseli**, Anna Dyson, Lupita D. Montoya*

186 A Statistical Study Of Indoor Acoustical Perceptions Based On Urban Traffic Noise Transmitted Via Operable Windows

***Jeehwan Lee**, Jae D. Chang*

212 Determination Of Diffusion Parameters Using A Semi-dynamic Dual Chamber Method

***Guoqing He**, Barbora Krejcirikova, Carsten Rode, Jianshun Zhang*

218 A High Efficiency Instrument For Collecting Airborne Particles Down To 10 Nanometers On Solid Surfaces Or Into Liquid

*Arantzazu Eiguren-Fernandez, Nick Buglack, Ning Li, **Patricia B. Keady**, Susanne Hering*

229 Resuspension Of Bacterial Spore Particles From Duct Surfaces
Parichehr Salimifard, *Paul Kremer, Donghyun Rim, James D. Freihaut*

232 Black Carbon And PM2.5 Filtration In Classroom Ventilation Systems In Salt Lake City, Utah And Las Vegas, Nevada
Paul T. Roberts, *David L. Vaughn, Steven G. Brown*

243 The Experimental Study Of The Regenerate Characteristic Effect On The Liquid Dehumidification System Performance
Yulieh Wu, *Kunhua Lyu, Pohan Lin*

251 Airtightening At Foundation-house Interface As A Means Of Radon Reduction In Living Space
William Rose, *Zachary Merrin, Paul Francisco, Stacy Gloss*

254 Removing Acetaldehyde In Indoor Air Using Mn/TiO2 Nanoparticle OZCO Catalyst
Minsu Kim, Eunseuk Park, Hyounduk Jung, ***Jongsoo Jurng***

255 Enhanced Catalytic Decomposition Of VOCs Using Synthesized MnOx/TiO2 With Adsorption-desorption
Hyounduk Jung, Minsu Kim, Eunseuk Park, ***Jongsoo Jurng***

256 Exploring The Effectiveness Of Simple Fan Filter Systems In Ventilating An Enclosed Non-air-conditioned Space
Ron Tan, *Joanna Shen, Gayatri Sankaran, Theodore Prawiradiraja, Hooi Ming Yap, Ramona Gutierrez, Kwok Wai Tham, Chandra Sekhar, Kok Wai Cheong, Wei-Chung Chang, Keng Been Ang, Clement Yew Yi Tan, Lee Ching Ng*

285 CFD Modeling And Parameterization Of The Photocatalytic Oxidation Process In Indoor Environments
Eisaku Sumiyoshi, *Takahiro Yamaguchi, Kiyoshi Yamamoto, Eunsu Lim, Kazuhide Ito*

286 A Multi-scale Exposure Concentration Analysis In A Large Factory Space Using Computational Fluid Dynamics Technique
Alicia Murga, *Sung-Jun Yoo, Kazuhide Ito*

293 Towards Assessing The Viability Of Indoor Dust Microbiome Using Flow Cytometry
Clarisse M. Betancourt Román, *Erica M. Hartmann, Jessica L. Green*

Wednesday, July 22, 2015

8:00 AM - 9:00 AM

Registration

Math 100 Courtyard

9:00 AM - 9:30 AM

W.PL.1 *Plenary* – **Lan Chi Nguyen Weekes, "Legionella in Building Water Systems: what do we know and what next?"**

Math 100

Lan Chi is the Senior Mechanical Engineer of the company, and one of the founders of InAIR Environmental Ltd. Her business philosophy and guiding principle is to address clients' concerns by providing them with expert and unbiased information and practical solutions, so the clients can make informed decisions about their indoor environment. Ms. Weekes holds a Bachelor's degree in Mechanical Engineering for L'Ecole Polytechnique of Montreal and a Masters of Applied Science (Building Environment) from Concordia University. For more information see http://www.inairenvironmental.ca/partners.html.

9:30 AM - 10:00 AM

W.PL.2 *Plenary* – **John Adgate, "Healthy Buildings and Healthy People: What We Know and What We'd Like to Learn"**

Math 100

John L. Adgate, Ph.D is Professor and Chair of the Department of Environmental and Occupational Health at the Colorado School of Public Health, University of Colorado. His exposure science research focuses on improving public health and epidemiological studies by documenting the magnitude and variability of human exposure to air pollutants, pesticides, metals, and allergens. Dr. Adgate's research projects have included evaluation of methods to reduce lead and allergen exposure exploration of longitudinal exposure to indoor and outdoor air pollutants, and, most recently, assessing the environmental and human health impacts of unconventional oil and gas development and the impact of climate change on indoor environments. Dr. Adgate has served on multiple U.S.EPA Science Advisory Panels. He was also a member of US Institute of Medicine's Committee on Research Ethics in Housing Related Health Hazard Research in Children and the National Research Council's 2011 Committee on Indoor Air and Climate Change. Dr. Adgate received a B.Sc. in biology from Calvin College, an M.S.P.H. from the University of North Carolina at Chapel Hill, and a Ph.D. in Environmental Health Sciences jointly awarded by Rutgers University and the University of Medicine and Dentistry of New Jersey.

10:00 AM - 10:30 AM

Event: **Coffee Break**

Engineering Center Lobby

10:30 AM - 12:00 PM

W.W.1 *Workshop:* **What Can Indoor CO2 Concentration Measurements Tell Us?**

Chair: *Hal Levin*

Speakers: *Bill Fisk, Jeff Siegel and Jelle Laverge*

ECCR 245, Engineering Center

10:30 AM - 12:00 PM

W.W.2 *Workshop:* **Measurements Of Semi-Volatile Organic Compound (SVOC) Emission Parameters**

Chairs: *Ying Xu and Jiping Zhu*

Speakers: *John Little and Xiaoyu Liu*

ECCR 265, Engineering Center

10:30 AM - 12:00 PM

W.1 *Podium:* **Microbiome**

ECCR 155, Engineering Center

Chair: *Kerry Kinney*

168 Effect Of Ultraviolet Germicidal Irradiation On Microbial Loading Of HVAC Heat Exchangers In Humid Versus Dry Entering Conditions
Julia C. Luongo, *Shelly L. Miller*

156 Building Characteristics As Determinants Of Moisture In Water-damaged Homes – A Study Design
Michal P. Spilak, *Iman Sylvain, Rachel Adams, Michael S. Waring*

165 Microbial Communities In House Dust And Asthma: A Birth Cohort Study
Rachel I. Adams, *Martin Täubel, Pirkka Kirjavainen, Balamuralikrishna Jayaprakash, Anne Karvonen, Anne Hyvarinen, Juha Pekkanen*

149 Indoor Microbiomes Change When We Try To Control Them
Alexander Mahnert, *Christine Moissl-Eichinger, Henry Müller, Alexander J. Probst, Roscel A. Ortega, Parag A. Vaishampayan, Kasthuri J. Venkateswaran, Gabriele Berg*

258 Temporal Air Exchange Rates And Relative Humidity May Affect Airborne Microbiomes More Than Weatherization Efforts
Mytien T. Nguyen, *Denina Hospodsky, Catherine Spirito, Largus T. Angenent*

253 Survey Of Green Building Water Systems Reveals Elevated Water Age And Microbial Concerns
William J. Rhoads, *Amy Pruden, Marc Edwards*

10:30 AM - 12:00 PM

W.2 *Podium:* **Architecture and Energy**

ECCR 151, Engineering Center

Chair: *Nicholas Clements*

15 An Experimental Study On Heat Gain In Window With An External Shutter
Esam Alawadhi

167 The Energy Consequences Of Excess Static Pressure In Central Residential Heating And Air-conditioning Systems
***Torkan Fazli**, Rou Yi Yeap, Brent Stephens*

248 Human Factors In HVAC: Thermal Comfort And Indoor Air Quality For Lifetime Housing
Robert Bean

174 A Survey On Evaluation Of Workers And Their Thermal Environment In Offices
***Koichi Ikeda**, Hikaru Enomoto, Hitoshi Takaono*

201 Development of an Automatic Thermal Control System Using Human Facial Skin Temperature
***Bo Yi**, Joon-Ho Choi, Douglas Noble, Kyle Konis*

10:30 AM - 12:00 PM

W.3 *Podium:* **IAQ in Schools**

ECCR 200, Engineering Center

Chair: *Andrea R. Ferro*

19 Changes In Classroom Indoor Air Quality Caused By A Corona Discharge Air Cleaner
Todd R. Crawford

249 Indoor Particulate Matter Assessment In Urban Kindergartens In Al Ain, United Arab Emirates
George O. Odhiambo

25 USA's Missing Mission: Strategies To Protect Children From Poor IAQ In Schools
Claire Barnett

176 Performance Of Filters With Different Efficiencies In Classrooms With Fan-coil Air-conditioning System
***Ailu Chen**, Qingliang Cao, Jin Zhou, Victor W.-C. Chang*

57 Crossover Study Of Fluorescent Bioaerosol Measurements In Elementary Classrooms With And Without Upper-room Ultraviolet Germicidal Irradiance
*Chunxiao Su, **Josephine Lau**, Fang Yu*

227 Low EMF Best Practices For Indoor Environmental Quality In School Classrooms; The Importance Of Implementing Existing NEC Electrical Building Codes And Other Feasible Best Practices To Reduce Non-ionizing Radiation Exposures From Electromagnetic Fields From Wiring And Other Common Electrical Devices To Protect Health And Safety Of Children And Teachers

Antoinette W. Stein, *Devra Davis, Alex Stadtner, Karl Riley*

12:00 PM - 12:45 PM

Event: **Closing Session**

Math 100

2:00 PM - 3:00 PM

Event: **Campus Tours**

Engineering Center Lobby

Notes

Notes

Measuring the ozone penetration factor in a residence under infiltration conditions

Haoran ZHAO,[1] Brent STEPHENS, Ph.D.,[1,*]

[1]Department of Civil, Architectural and Environmental Engineering, Illinois Institute of Technology, Chicago, IL USA

*Corresponding author email: brent@iit.edu

Keywords: Ozone, infiltration factor, indoor chemistry, building envelope

SUMMARY

Recent experiments have demonstrated that outdoor ozone reacts with materials inside residential building enclosures. However, test methods to measure ozone penetration factors remain limited. We developed an improved method for measuring the ozone penetration factor (*P*) in residences under natural infiltration conditions and applied it in an unoccupied apartment unit. Twenty-four repeated measurements were made and results were explored to (1) evaluate the accuracy and repeatability of the new procedure using multiple solution methods, (2) compare results from "interference-free" and conventional UV absorbance ozone monitors, and (3) compare results against those from a prior test method requiring artificial depressurization. The mean (±SD) estimate of *P* was 0.54±0.10 across a wide range of conditions using the new method with an interference-free monitor. This work represents the first known measurements of ozone penetration factors into a residential building operating under natural infiltration conditions and provides a new method for widespread use.

INTRODUCTION

Indoor exposures to ozone and byproducts from ozone reactions with surfaces and other compounds indoors are of concern for health (Weschler 2006). Most previous work estimating the indoor proportion of outdoor ozone has assumed that ozone penetrates through building enclosures 100% efficiently (Weschler 2006; Weschler 2000; Chen et al. 2011; Waring 2014; Gall et al. 2011; Georgopoulos et al. 2005). However, limited experimental and modeling investigations have demonstrated otherwise (Stephens et al. 2012; Walker and Sherman 2013; Liu and Nazaroff 2001). We previously developed a method for measuring ozone penetration factors through building enclosures and applied it in eight single-family homes (Stephens et al. 2012). However, the previous experimental method suffered from issues necessitated the use of an artificial depressurization process (i.e., blower door) to overcome high detection limits of the ozone monitor used at the time (2B Technologies Model 205). In this work, we developed and applied an improved method for measuring ozone penetration factors in residences under natural conditions using a new commercially available NO-scrubbed ozone monitor (2B Technologies Model 211) in an unoccupied and sparsely furnished test apartment in Chicago, IL. Twenty-four repeated measurements of ozone penetration factors were made, providing a dataset for the following: (1) an exploration of the accuracy and repeatability of the refined test method; (2) side-by-side comparisons between the 2B Technologies Model 205 and 211 ozone monitors during natural infiltration tests; (3) comparisons of results from natural infiltration experiments to those conducted with a blower door installed to test the validity of the original method; and (4) an exploration of various methods of solving for both ozone decay rate constants and penetration factors.

METHODOLOGIES

Measurements were conducted in *studioE* (the Suite for Testing Urban Dwellings and their Indoor and Outdoor Environments), an unoccupied and sparsely furnished apartment unit on the third floor of Carman Hall on the main campus of Illinois Institute of Technology in Chicago, IL. The improved test method involves alternately measuring indoor and outdoor ozone concentrations at high time-resolution (i.e., 2 minutes indoor and 1 minute outdoor) using an automated electronic switching valve (Swagelok SS-43GXS4-42DCX). Indoor ozone concentrations are first elevated by briefly operating ozone generators indoors. The air exchange rate during testing is simultaneously measured using injection and decay of CO_2 as a tracer gas. The air exchange rate (AER) was estimated by regressing the natural logarithm of the tracer gas concentrations versus time (ASTM E 741 2006). We used a dynamic mass balance approach to model the indoor ozone concentration in the well-mixed environment and in the absence of indoor sources (Equation 1). Three methods were used to solve for ozone decay rate constants and penetration factors and compared.

$$\frac{dC_{in}}{dt} = P\lambda C_{out} - (\lambda + k)C_{in} \tag{1}$$

In the steady-state solution, *P* was estimated using a steady-state condition (Equation 2). The first order ozone decay rate constant (*k*) was estimated using a log-linear regression to the initial portion of indoor decay that was not affected by outdoor ozone sources (Equation 3). This portion was identified by plotting the natural logarithm of indoor ozone concentration versus time, and used only the initial data that yielded $R^2 > 0.99$.

$$P = \frac{C_{in}}{C_{out}}\frac{\lambda + k}{\lambda} \tag{2}$$

$$-\ln\frac{C_{in,t}}{C_{in,t=0}} = (\lambda + k)t \tag{3}$$

The two-parameter analytical solution utilized a non-linear regression to estimate both *P* and *k* simultaneously (Equation 4). Outdoor ozone concentrations were assumed to remain constant during each test period, similar to Stephens et al. (2012).

$$C_{in,t} = C_{in,t=0}e^{-(\lambda+k)t} + \frac{P\lambda\overline{C_{out}}}{\lambda + k}\left(1 - e^{-(\lambda+k)t}\right) \tag{4}$$

Because outdoor ozone concentrations were not always constant during the test periods, a one-parameter discretized solution (Equation 5) was also used to solve for *P* with prior knowledge of *k* from Equation 3.

$$C_{in,t} = P\lambda C_{out,t}\Delta t + (1 - (\lambda + k)\Delta t)C_{in,t-1} \tag{5}$$

Uncertainty in each of these methods was estimated by error propagation with a combination of instrument uncertainty, relative standard deviations of average concentration measurements, and standard errors of regression coefficients, as most appropriate for each solution method.

RESULTS AND DISCUSSION

The Model 205 ozone monitor yielded higher steady-state I/O ozone concentration ratios than the Model 211: median of 0.14 compared to 0.09. Data from both ozone monitors were used to solve for *P* and *k* using the three different solution methods (Figure 1). While all three solution methods yielded similar estimates of *P* with the same ozone monitor, the Model 205 monitor resulted in a mean (± SD) best estimate of penetration factor of ~1.10±0.36 compared to only 0.59±0.11 estimated using data from the Model 211 monitor, suggesting that the 205 monitor is less reliable in this environment.

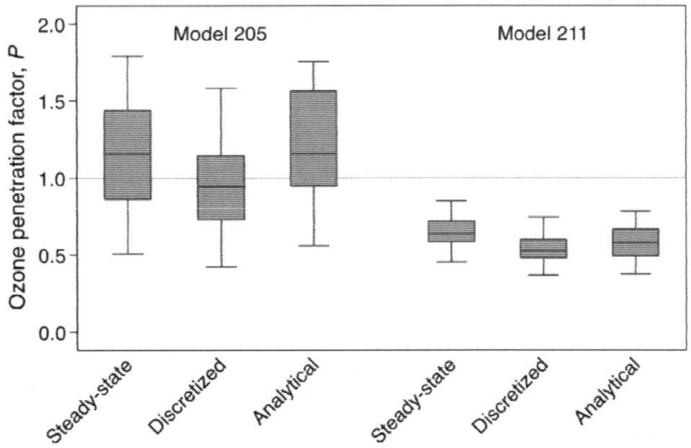

Figure 1. Estimates of ozone penetration factors (*P*) across 21 replicate experiments under natural infiltration conditions using three solution methods and two ozone monitors (2B Technologies Models 205 and 211).

The mean estimates (±SD) of ozone penetration factors (*P*) over all 21 tests in the test apartment were 0.65±0.11, 0.59±0.11, and 0.54±0.10 using the steady-state, two-parameter analytical, and one-parameter discretized solutions, respectively. The steady-state method had an average uncertainty (±SD) of 0.19±0.05, or 29±5% on a relative basis (Figure 2a), suggesting that either of the two dynamic solution methods are most appropriate to use for reducing uncertainty. As a measure of model fit between the two dynamic solution methods, the mean squared error (MSE) was calculated for all 21 natural infiltration experiments using data from both the analytical and discretized solution methods (Figure 2b). The mean (±SD) MSE of the two-parameter analytical and one-parameter discretized regressions was 0.15±0.11and 0.10±0.04, respectively, indicating that the discretized solution method provided a better model fit for the observed indoor concentration data by using the time-resolved outdoor ozone data.

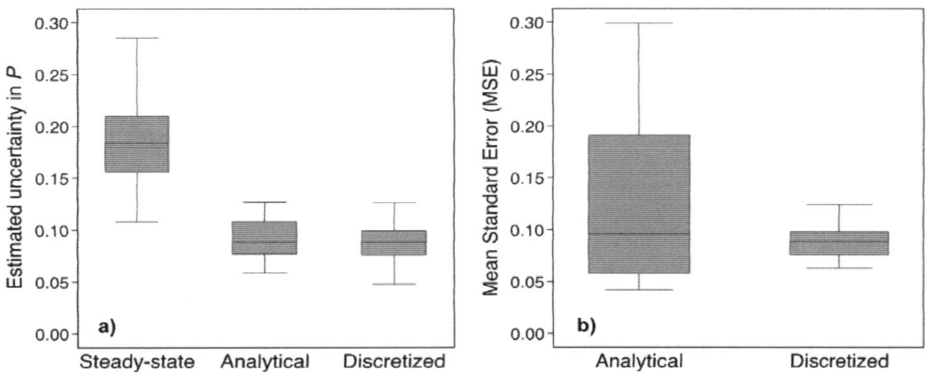

Figure 2. Range of uncertainty estimates using the Model 211 ozone monitor in 21 natural infiltration experiments: (a) estimated uncertainty in ozone penetration factor estimated by each solution method, and (b) mean standard error (MSE) for the analytical and discretized solution methods.

CONCLUSIONS

This work represents the first known measurements of ozone penetration factors into a residential building operating under natural infiltration conditions and provides a new method for widespread use.

ACKNOWLEDGEMENTS

We would like to acknowledge Jihad Zeid for his contributions to the measurements; Will Ollison and the American Petroleum Institute for funding a portion of this work.

REFERENCES

ASTM E 741. 2006. "Standard Test Method for Determining Air Change in a Single Zone by Means of a Tracer Gas Dilution."

Chen, Chun, Bin Zhao, and Charles J. Weschler. 2011. "Assessing the Influence of Indoor Exposure to 'Outdoor Ozone' on the Relationship between Ozone and Short-Term Mortality in U.S. Communities." *Environmental Health Perspectives*, November. doi:10.1289/ehp.1103970.

Gall, Elliott T., Richard L. Corsi, and Jeffrey A. Siegel. 2011. "Barriers and Opportunities for Passive Removal of Indoor Ozone." *Atmospheric Environment* 45 (19): 3338–41. doi:10.1016/j.atmosenv.2011.03.032.

Georgopoulos, Panos G, Sheng-Wei Wang, Vikram M Vyas, Qing Sun, Janet Burke, Ram Vedantham, Tom McCurdy, and Halük Özkaynak. 2005. "A Source-to-Dose Assessment of Population Exposures to Fine PM and Ozone in Philadelphia, PA, during a Summer 1999 Episode." *Journal of Exposure Analysis and Environmental Epidemiology* 15 (5): 439–57. doi:10.1038/sj.jea.7500422.

Liu, D, and W.W. Nazaroff. 2001. "Modeling Pollutant Penetration across Building Envelopes." *Atmospheric Environment* 35 (26): 4451–62. doi:10.1016/S1352-2310(01)00218-7.

Stephens, Brent, Elliott T Gall, and Jeffrey A Siegel. 2012. "Measuring the Penetration of Ambient Ozone into Residential Buildings." *Environmental Science & Technology* 46 (2): 929–36. doi:10.1021/es2028795.

Walker, Iain S., and Max H. Sherman. 2013. "Effect of Ventilation Strategies on Residential Ozone Levels." *Building and Environment* 59 (January): 456–65. doi:10.1016/j.buildenv.2012.09.013.

Waring, Michael S. 2014. "Secondary Organic Aerosol in Residences: Predicting Its Fraction of Fine Particle Mass and Determinants of Formation Strength." *Indoor Air* 24 (January): 376–89. doi:10.1111/ina.12092.

Weschler, Charles J. 2000. "Ozone in Indoor Environments: Concentration and Chemistry." *Indoor Air* 10 (4): 269–88. doi:10.1034/j.1600-0668.2000.010004269.x.

———. 2006. "Ozone's Impact on Public Health: Contributions from Indoor Exposures to Ozone and Products of Ozone-Initiated Chemistry." *Environmental Health Perspectives*, 1489–96.

OPERATION OF RESIDENTIAL HVAC SYSTEMS: IMPLICATIONS FOR HEALTHY BUILDINGS

Marianne Touchie[1,2], Jeffrey A Siegel[1,3*]
[1]Department of Civil Engineering, University of Toronto, Ontario, Canada
[2]Toronto Atmospheric Fund, Toronto, Ontario, Canada
[3]Dalla Lana School of Public Health, University of Toronto, Ontario, Canada

[*]Corresponding email: jeffrey.siegel@utoronto.ca

Keywords: Heating and cooling, runtime, flow rate, filtration

SUMMARY

Over two-thirds of North American homes rely on central forced air systems for heating and/or cooling. Many approaches to achieving healthy buildings, including filtration and ventilation/infiltration control rely on the operation of these systems. However, operational parameters such as system runtimes and flow rates are poorly understood and can critically impact the performance of any strategy to improve indoor environmental quality. This paper summarizes data from several studies that have measured field results and suggests that, even in extreme climates, system runtimes and flow rates are smaller than are typically assumed. These factors can limit any benefit from air cleaning devices and suggest that HVAC factors need to be accounted for when designing healthy buildings.

INTRODUCTION

Most North American homes use central forced air systems to provide heating and cooling. Despite the prevalence of these systems, there is a lack of data about runtimes and flow rates, which can be used to estimate the performance of air cleaning and ventilation strategies. Some previous modelling investigations which required runtime (e.g., Riley et al., 2002, Chen et al., 2012) have had to rely on estimates based on exterior temperatures or other factors. However, system controls such as thermostat set points and system capacity relative to building size both contribute to total runtime. As such, it is important to base runtime assumptions on actual monitored performance which captures all of these influencing factors. Similar to runtimes, relatively little is known about air flow rates through systems when they are operating. The purpose of this paper is to provide some measurement of these parameters and assess their impact on air cleaning strategies.

METHODOLOGIES

A number of field investigations have been conducted previously to characterize the performance of heating and air conditioning systems across the U.S and Canada. Data from eight field studies have been collated to generate a larger sample size from which to draw improved runtime and flow rate estimates.

The data were collected primarily from single-family homes in Texas, the Pacific Northwest, and in Toronto, Canada. The average date of construction in the sample was 1982, ranging from as early as 1915 to 2010, although the largest sample in the study was all new homes (built 2007-2010). Heating system types included natural gas furnaces, electric heaters and heat pumps. The runtimes for both heating and cooling systems were measured using different methods including logging of supply and/or return temperatures and fan power or current draw. **Table 1** summarizes the details about the studies and the homes included.

Table 1 Summary of Study Details

Study	Location	Number of homes	Median Volume [m3]	Median Heat/Cool Capacity [kW]	Seasons Sampled[1]	Sample Time [hrs]
PECAN	TX	621	415	19.3/10.6	All	1782
HUD	TX	11[2]	331	10.0/8.8	W/Sp/Su	680
HEATAG	ID, MT, OR, WA	25	415	14.8/-	W	-
Stephens[3]	TX	17	323	-/10.6	Su	184
MAPTAG	ID, WA	9	400	15.8/-	W	30
GASTAG	WA	8[4]	308	21.1/-	W	2011
ASHTAG	WA	5	282	10.8/-	W	22
Toronto	ON	5	-	-	W	888

[1]W=Winter, Sp=Spring, Su=Summer, [2]11 homes with flow measurement, 47 homes with run time measurement
[3]Stephens et al., (2011), [3]Homes were tested before/after duct retrofits.

In the studies where measurements were recorded in multiple seasons, multiple data points exist for each site. All data points have been included in this analysis. The median effective recirculation rate was calculated as the product of the median system air flow rate and the median system runtime divided by the median house volume. This recirculation rate was then used in a mass-balance to evaluate the effectiveness of different filtration approaches.

RESULTS AND DISCUSSION

Error! Reference source not found. shows the system runtime for each dataset including both heating and cooling. Outliers (outside of 1.5 times the interquartile range) are excluded from the plots but not from data analysis. Median system runtimes ranged from 7% to 46% between the studies. For studies where seasonal data were available, the maximum runtimes were observed in the main heating or cooling season, while the minimum runtimes were observed during the shoulder seasons, as expected. These data are consistent with other runtime data in the literature (e.g., Thornburg et al., 2004). The median winter heating runtime for the studies with heating-dominated loads (HEATAG, MAPTAG, Toronto) was 37.8%. The median summer cooling runtime for the Austin studies with cooling-dominated loads (PECAN, HUD and Stephens) was 27.3%. For comparison, using the approach in Chen et al. (2012), which is based on average outdoor temperatures, a typical summer cooling runtimes would be 93%.

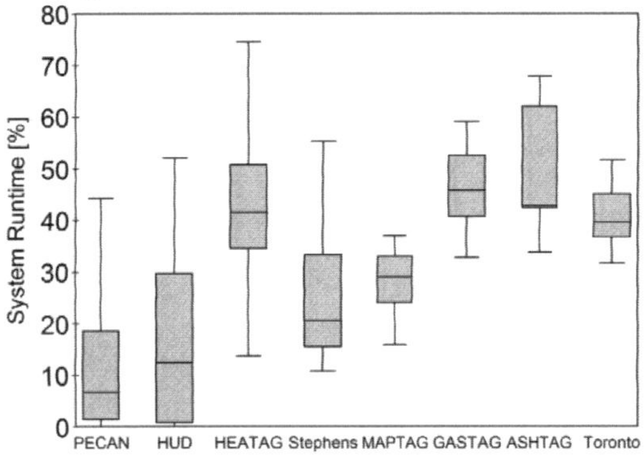

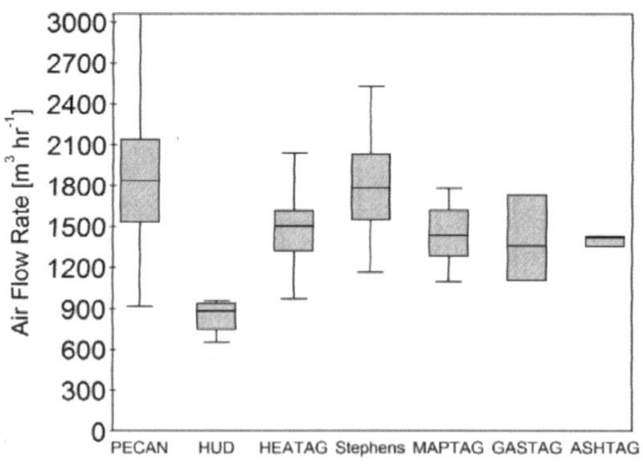

Figure 1: System Runtimes Figure 2: Air Flow Rates

Figure 2 shows the air flow rates for each dataset. Median flow rates ranged from 884 m^3hr^{-1} to 1835 m^3hr^{-1} in each study. The mean air flow rate from the PECAN data (611 m^3hr^{-1} per ton of cooling) was comparable to the findings of a study by Proctor (1997) (584 m^3hr^{-1} per ton) while the HUD data, from a similar climate, yielded a significantly lower mean flow rate (316 m^3hr^{-1} per ton). However all of these values are lower than the manufacturer recommendation of 680 m^3hr^{-1} per ton (Proctor 1997). Heating fan speeds are generally lower than cooling fan speeds and the numbers here are generally consistent with those in the literature (e.g., Francisco and Palmiter, 2003).

Figure 3 calculates filter effectiveness using a steady-state mass balance shown in Eqn 1.

$$H = 1 = \frac{\lambda + \beta + \eta\tau\frac{Q_r}{V}}{\lambda + \beta + \eta_{base}\tau\frac{Q_r}{V}} = \frac{\lambda + \beta + \eta\lambda_r}{\lambda + \beta + \eta_{base}\lambda_r} \quad (1)$$

Where λ is the air exchange rate, β is the deposition loss rate, η is filter efficiency (the subscript *base* refers to a base case reference filter efficiency, MERV 3, in the analysis below), τ is the runtime (i.e. Figure 1), Q_r is the HVAC flow rate (i.e., Figure 2), V is the house volume, and λ_r is the effective recirculation rate. The input parameters were adapted from the studies above as well as various sources including filter efficiency (MacDonald, 2013), air exchange rate (Persily et al., 2010), deposition loss (Meng et al, 2005 and Chen et al., 2012). Effectiveness was calculated relative to a typical low efficiency MERV 3 furnace filter for three higher efficiency filters and for both $PM_{2.5}$ and PM_{10}.

The median effective recirculation rate from several of the studies in Table 1 are shown on Figure 3. The largest effective recirculation rate was for the HEATAG homes (winter in the Pacific Northwest). The MAPTAG homes (same climatic conditions, but newer manufactured homes) had a median recirculation rate of just over 1 hr^{-1}. The PECAN (newer homes in Texas) showed very different conditions in the different seasons with very small values for all, except the summer cooling system. The other data sets fell within the range of these studies with similar patterns of heating and cooling homes.

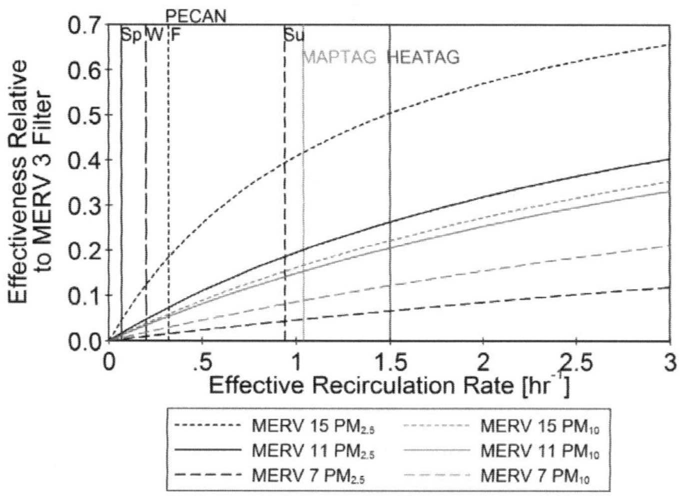

Figure 3: Impact of Effective Recirculation Rate on Filter Effectiveness

There are several important aspects of Figure 3. The first is that the measured values are often smaller than what has been used in simulations in the literature. Examples include 4 hr^{-1} (Riley et al., 2002), 3.1 hr^{-1} (Chen et al., 2012 for Austin, TX in summer) and approximately 0.5 hr^{-1} (Klepeis and Nazaroff, 2006). The second is that within a climate, runtimes vary considerably over the course of a year. The third is that the effectiveness of an air cleaner is strongly dependent on the recirculation rate, particularly for higher efficiency filters. The fourth is that, for small recirculation rates (typical for swing seasons), the change in effectiveness for going to a higher efficiency air cleaner can be very small. These findings would impact air cleaning for other contaminants as well suggesting considerable caution is needed when assessing effectiveness without appropriate HVAC data.

CONCLUSIONS

The data presented here indicate that runtimes and flows rates of residential heating and cooling systems, even in extreme climates, are lower than often assumed. In turn, the resulting recirculation rates suggest lower air cleaning effectiveness than have typically been calculated. Therefore, to achieve a specified indoor concentration of a contaminant or to assess health impacts, designers and researchers should use more realistic HVAC operation data.

ACKNOWLEDGEMENTS

The authors gratefully acknowledge the contribution of data from a number of researchers including Atila Novoselac (UT Austin), Paul Francisco (UIUC), and Bob Davis (Ecotope).

REFERENCES

Chen C, Zhao B, Weschler CJ. (2012) Indoor Exposure to "Outdoor PM10" Assessing It's Influence on the Relationship Between PM10 and Short-term Mortality in U.S. Cities. *Epidemiology*, **23**, 6, pp. 870-878.

Francisco, P. W., & Palmiter, L. (2003). Field evaluation of a new device to measure air handler flow. *ASHRAE Transactions*, **109**, 403-412.

Klepeis NE, Nazaroff WW (2006) Modeling Residential Exposure to Secondhand Tobacco Smoke. *Atmospheric Environment*, **40**, 4393-4407.

Macdonald, B (2011) PM Based Air Filter Efficiency and Rating. Presentation to ASHRAE SSPC 62.2, June 24, 2011. Received from author: brucesfilterteststandards@live.com.

Meng, Q.Y., et al. (2005) Influence of ambient (outdoor) sources on residential indoor and personal PM2.5 concentrations. *J. of Exposure Anal. and Environ. Epidem.* **15**, 17-28.

Persily, A., Musser, A. and Emmerich, S.J. (2010) Modeled infiltration rate distributions for U.S. housing, *Indoor Air*, **20**, 473-485.

Proctor JP. (1997) Field Measurements of New Residential Air Conditioners in Phoenix, Arizona. *ASHRAE Transactions.* **103**, 1-10.

Riley WJ, McKone TE, Lai AC, Nazaroff WW. (2002) Indoor particulate matter of outdoor origin: importance of size-dependent removal mechanisms. *Environmental Science and Technology.* **36**, 200-207.

Stephens, B., Siegel, J.A. and Novoselac, A. (2011) Operational characteristics of residential and light-commercial air-conditioning systems in a hot and humid climate zone. *Building and Environment*, **46**, 1972-1983.

Thornburg JW, Rodes, CE, Lawless PA, Stevens CD, Williams RW (2004) A pilot study of the influence of residential HAC duty cycle on indoor air quality. *Atmospheric Environment*, **38**, 1567-1577

THE IMPACT OF HEATING PROCESS ON PM2.5 CONCENTRATION IN COLD RURAL AREAS OF NORTHERN CHINA

Xueyan Zhang, Bin Chen*

Laboratory of Building Environment and New Energy Resources, School of Civil Engineering, Dalian University of Technology, Dalian, China.

*Corresponding email: chenbin@dlut.edu.cn

Keywords: Rural houses, PM2.5 emissions, Heating systems, Air quality

SUMMARY

In Northern China, most traditional heating systems, heated by mini-stove in the kitchen, usually take agricultural residues as fuels resources. Besides, burning cave under the ground-floor of a rural house is also widely used. Exhausted flue gases from these heating systems contain many harmful components and particulate matters discharge into atmosphere and usually seep into the room, which is crisis for human health. In this paper, the emission process of inhalable particles have been measured during the heating and cooking period in three rural households in Fuxin city, Liaoning province have been investigated. The measurements were correlated to factors such as living habits, indoor air temperature, indoor relative humidity, biomass fuels, and ventilation methods. In addition, the impact of outdoor wind speed and predominant wind direction on the PM2.5 emission has been studied.

INTRODUCTION

About 3 billion people in developing countries using solid fuels combustion for cooking food and heating rooms such as dung, wood, and agricultural residues (C.L. Li et al. 2012). Combustion of these fuels is generally inefficient, producing high levels of particulate matter less than 0.25 μm in size (PM 2.5) and CO (C. Armendariz-Arnez et al. 2010). Their toxic effects have been extensively documented (M. Zheng et al. 2005; Y.X. Zhang et al. 2007; Q. Zhang et al. 2009). Although many investigations in China have examined the impact of air pollution, these studies have largely been confined to major urban areas, Missing are data for rural areas where combustion of agricultural residue in caves is the major source of heating. Particularly concern of PM 2.5 levels discharged into outdoor atmosphere and flow into the house. This study could fill the pollution research gap.

METHODOLOGIES

In this study, pollution concentrations of the three typical rural houses in northern china heated by traditional heating systems have been measured. In addition, indoor and outdoor PM2.5, PM10 concentration and CO concentration have been measured and analyzed. These three houses were located in Shuangcheng village, Fuxin city, Liaoning Province, shown in Table 1. The study period was from Dec.6, 2013 to Dec.7, 2013.

According to the testing requirements in "Ambient Air Quality Standard"(DEP 2012), SHINYEI PM2.5 Sensor assessment monitors have been used for recording with 10 minutes interval, which have highly testing accuracy about 85%, and broader measurement range from 0 $\mu g/m^3$ to 200 $\mu g/m^3$. The CO concentration has been recorded by indoor air quality-

TSI7545. The indoor/outdoor temperature and relative humidity have been measured by thermo recorder TR-72U. Before recording, the doors and windows of the room have been closed for more than 24 hours. If the area of one heating room was less than 50m^2, the testing point could only be one, and the height of the testing point could be 1.2m.

Table1. Basic information of measured rural houses in Fuxin rural areas (Hapeng location)

Test period	Dec.6, 2013 ~ Dec.7, 2013		
Outdoor climate	Temp: -25~4℃; RH: 20~55%; Velocity: 0.35~1.5m/s; Wind direction: Southeast		
Measured house	1# house	2# house	3# house
Construction time	1996	1994	2006
Photo			
Measured room	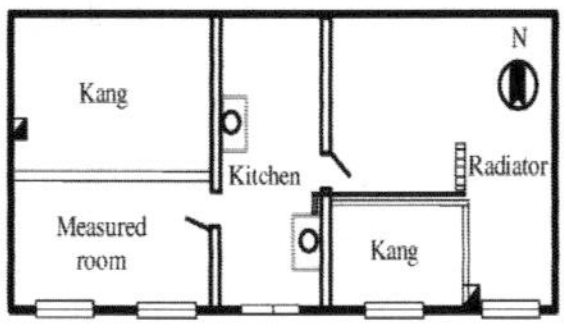	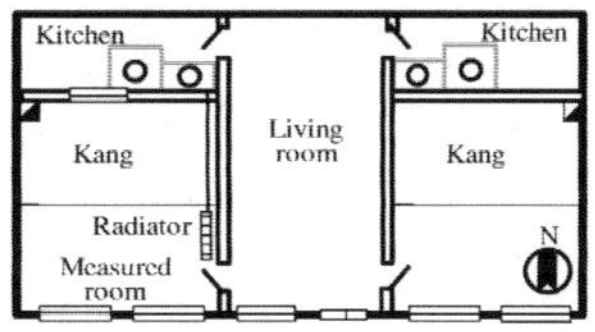	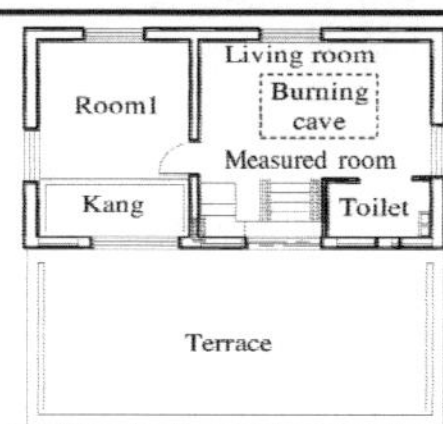
Areas of measured room	20m^2	18 m^2	24 m^2
Heating methods	 Kang	 Kang and radiators	 Burning cave

RESULTS AND DISCUSSION

Factors influencing on the PM2.5 concentration

From Figure 1, the concentration of PM2.5 inside and outside rural houses were 2 to 4 times exceeding the secondary limits in Chinese national standards (DEP 2012), at about 80% of the total time was exceeding obviously. Since the fuels smoldering, if the sealed property of the heating system is poor, CO and other harmful gases could infiltrate in the heating room. And the indoor and outdoor concentrations of harmful gases could be increased significantly. The higher levels generally occurred at the cooking time on the morning at 06:30 and in the afternoon at 18:00. However, the decline stage of PM2.5 concentration was from 1:00 to 6:00 during the sleeping stage. The significant differences among these results were mainly occurred due to biomass types and different burning approaches for heating and cooking both hourly and over longer periods during the 24-hour test.

The main influencing factors on the PM2.5 concentration are summarized as follows: (1) Indoor heating systems in different rural residents are intermittent mode combined with cooking and heating, resulting in a cyclical variation of PM2.5 concentrations; (2) The impact of indoor thermal environment, relative humidity on PM2.5 diffusion could be carried out; (3) Various fuel types and combustion process in different energy applications bring out different PM2.5 concentrations; (4) Residential space layout influencing on the air flow rate and emission process, resulting in a interrelated variation of PM2.5 concentration between adjacent rooms; (5) Under the condition of ventilation through windows, PM2.5concentration of the heating room was greatly influenced by cold air flow rate, air flow path and outdoor PM2.5 concentration.

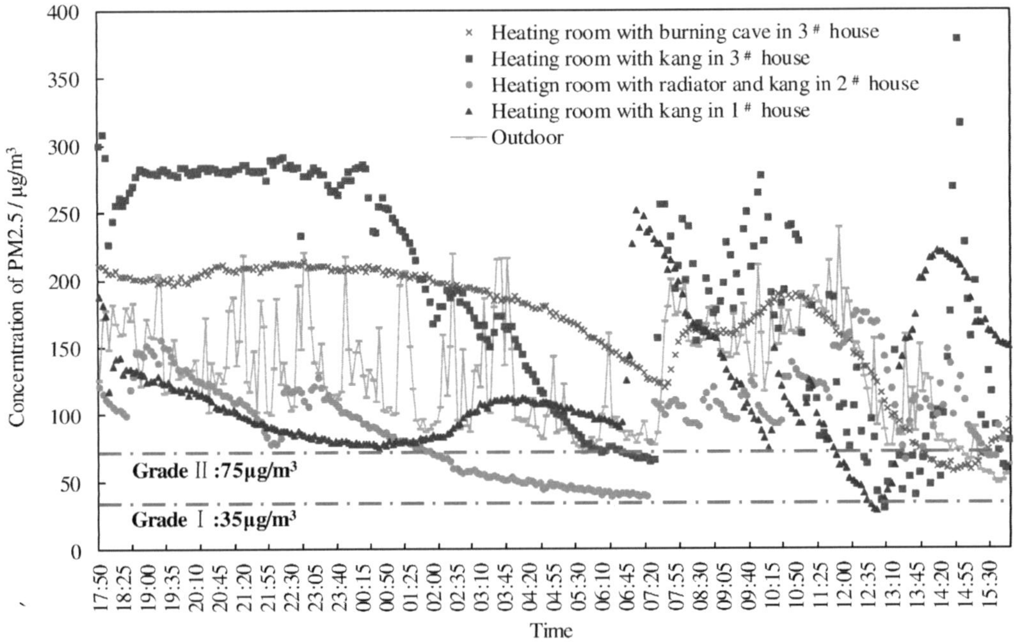

Figure 1. Variations of PM2.5 concentrations（DEC.6~ DEC.7. 2013）

Indoor temperature and relative humidity impact on PM2.5 concentration

As can be seen in Figure 2, the indoor PM2.5concentrations were reached to the largest value at 15mins~20mins after igniting the stove, and reach to the lowest value in the early morning at about 7:20 after sleeping all the night. As shown in Figure 2, the concentration of PM2.5 was gradually increased as indoor air temperature was increased, and decreased as indoor temperature was gradually reduced. The proportion between them was positive, and correlation coefficients were calculated according to the formulas in mathematical statistics (Q. Zhang 2009), the results are respectively 0.045 and 0.081. However, the concentration of PM2.5 was gradually decreased as indoor relative humidity increased. As a result, there is an agglomeration phenomenon. According to the results in Figure 2, the effect of indoor air temperature fluctuation on PM2.5 concentration was less than indoor relative humidity. While the indoor temperature difference was less than 1℃, indoor air relative humidity was increased from 60% to 80%, the decay rate of PM2.5 concentration was increased 2 times more than which under the relative humidity of 40%. As fitting analysis, there is almost no relationship between the indoor thermal environment and PM2.5 concentration.

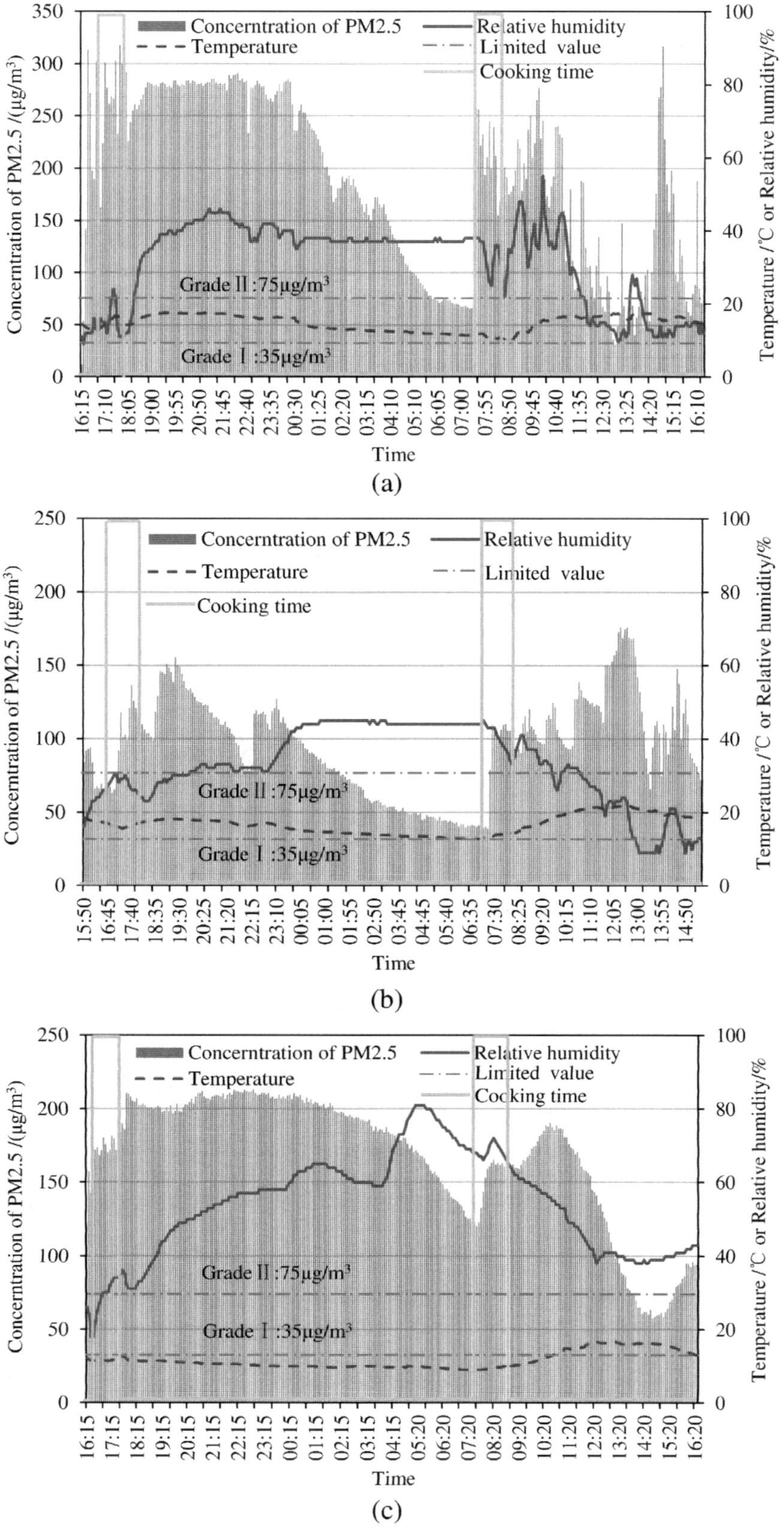

Figure 2. Indoor temperature and relative humidity influencing on PM2.5 concentration ((a) Measured room in 1# house; (b) Measured room in 2# house; (c) Measured room in 3# house)

Ventilation modes impact on PM2.5 concentration

During the cooking period about one hour, the mixture of straws and woods were used as fuels, whose combustion mode was full burning, and the combustion efficiency was 100%. From the Figure 3, the variation range of the positive correlation coefficient was from 0.75 to 1.63 between the PM2.5concentration of a kitchen and which in a bedroom. While ventilation was taken out during cooking period through the door, the variation range of the positive correlation coefficient was from 0.81 to 1.87 between the PM2.5concentration outside the house and which in a bedroom. It illustrates that traditional plane layout of rural house is so simple for pollutants directly emission into the bedroom from the kitchen. Besides, the PM2.5 concentration of the outdoor atmosphere could be quickly increased while cooking.

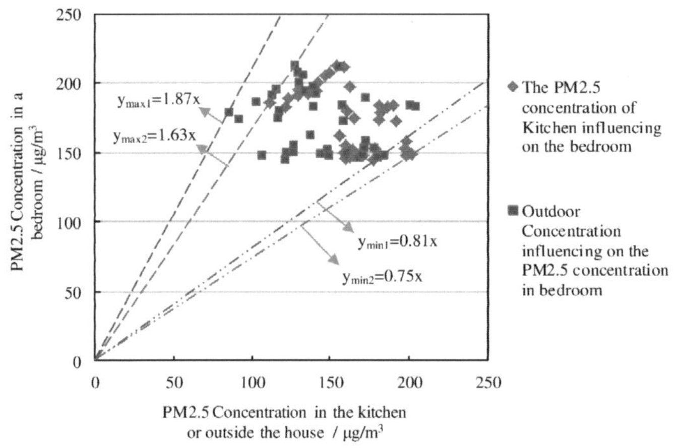

Figure 3. The concentration of PM2.5 in kitchen influencing on which in bedroom (measured house in Figure 1(b), during the cooking period)

Variation of outdoor PM2.5 concentration

As shown in Figure 4, the PM2.5 concentration is up to 2.72-3.15 mg/m^3, while the temperature of exhaust gas was 59.3℃, and the flux of the heating system was 0.45 L/min.

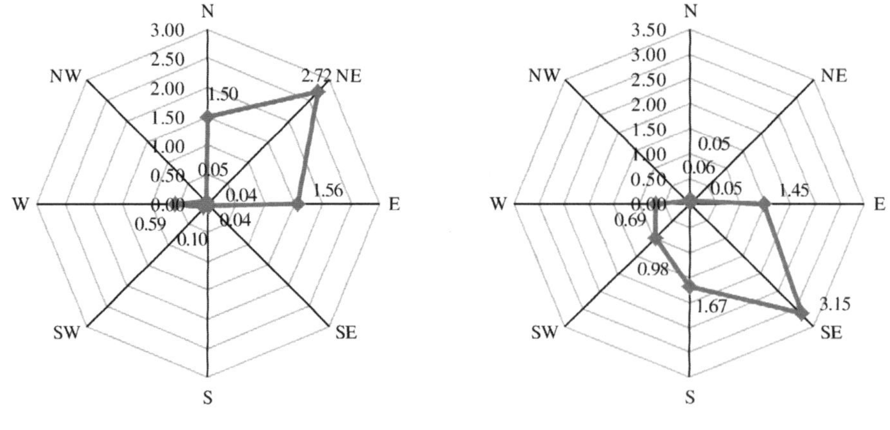

Figure 4. The impact of predominant wind direction on outdoor PM2.5 concentration (Unit: mg/m^3)

The PM2.5 diffusion from the chimney is significant influenced by the predominant wind direction in cold winter. The impact of PM2.5 emission in nearby urban areas could be estimated and calculated. The wind velocity is from 0.5 m/s to 1.5 m/s, and typical chimney diameter is 100mm, the amount of flue gas in heating and cooking is approximately from 14 to 42 $m^3/(h\cdot household)$. PM2.5 emission is from 1.536 million households of Fuxin. Assuming the PM2.5 concentration is 125 $\mu g/m^3$, with the emission from 13 to 41 kg/day and assuming the heating period in winter contains 176 days, the total emission of PM2.5 would be between 2.4 t and 7.2 t per house.

As shown in Figure 5, the PM2.5 concentration during the cooking period is highest to 3.15 mg/m^3. The PM2.5 diffusion from the smoke exhausting system of the heating house is also significant influenced by the wind speed in cold winter. The correlation coefficient between PM2.5 concentration and wind speed is about 0.23.

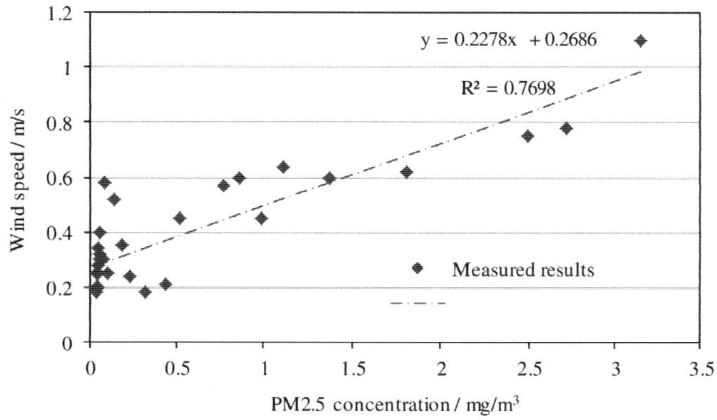

Figure 5. The impact of wine speed on outdoor PM2.5 concentration

Improving method

To reduce the pollutants and harmful exhausting gas, some optimizations has mainly taken as following methods. (1) Burning cave is designed as an independent, sealed and insulated one in order to prevent flue gas leakage and reduce large heat loss through the soil and air around the burning cave. (2) Taken the burning cave under the center of the entire house and raised the entire floor by 0.3m are benefit for heating internal air and heating the entire house via radiant floor heating system. (3) The heat transfer medium—flue gas in traditional burning cave is replaced by hot air in the optimized mode so that surface temperature tends to be more uniform. (4) The thin metal slice is taken as a middle layer inside the elevated floor to enhance heat transfer.

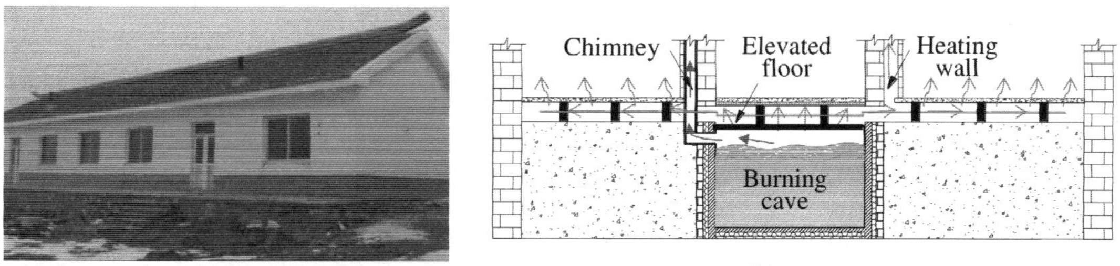

(a) (b)

Figure 6. The demo house and a new heat transport method under optimization design ((a) demo house; (b) a new heat transport method)

Figure 6 illustrates the new structure and heat transfer process of the optimized house integrated with new burning cave heating, which has been built in Dalian rural area, 2013. Its thermal performance has been improved. The main strategy is that using one burning cave to heat the entire house, and taking a clarification system on the chimney.

CONCLUSIONS

In this paper, through field measurements and calculation, some significant conclusions could be described as follows: (1) The indoor PM2.5 concentration of a rural house is 2 to 4 times more than the secondary limits of 75μg/m^3 in Chinese standard. (2) The upper and lower ratio of PM2.5 concentration between kitchen and bedroom respectively are 0.75 and 1.63, and which between the bedroom and the outside was from 0.81 to 1.87. (3) PM2.5 pollution from the rural area such as Fuxin city surrounding the urban areas could be between 2.4 t and 7.2 t per house, and would not be overlooked. (4) A low-cost way and better heat transfer ways for improve rural living environment and ecological environment to local conditions has be sought, and could reach the desired performance.

ACKNOWLEDGEMENT

The work was supported by a NSFC (No.51178073) and Key Projects in the National Science & Technology Pillar Program during the Twelfth Five-year Plan Period (No.2012BAJ02B05) of China.

REFERENCES

C.L. Li, S.C. Kang, P.F. Chen, et al. (2012). Personal PM2.5 and indoor CO in nomadic tents using open and chimney biomass stoves on the Tibetan Plateau. *Atmospheric Environment.*2012, vol. 59, pp. 207-213.

C. Armendariz-Arnez, R.D. Edwards, M. Johnson, et al. (2010). Indoor particle size distributions in homes with open fires and improved Patsari cook stoves. *Atmospheric Environment.* 2010, vol. 44, pp. 2881-2886.

M. Zheng, L.G. Salmon, J.J. Schauer, et al. (2005). Seasonal trends in PM2.5 source contributions in Beijing, China. *Atmospheric Environment.* 2005, vol. 39, pp. 3967-3976.

Y.X. Zhang, M. Shao, Y.H. Zhang, et al. (2007). Source profiles of particulate organic matters emitted from cereal straw burnings. *Journal of Environmental Sciences-China.* 2007, vol. 19, pp. 167-175.

Q. Zhang, D.G. Streets, G.R. Carmichael., et al. (2009). Asian emissions in 2006 for the NASA INTEX-B mission. *Atmospheric Chemistry and Physics.* 2009, vol. 9, pp. 5131-5153.

Department of environmental protection. *Ambient air quality standards* (GB3095-2012).

MEASURING FLAME RETARDANT EMISSIONS FROM SPRAY POLYURETHANE FOAM IN A HOME

Dustin POPPENDIECK[1,*] Angelica CONNOR[1,2]

[1]Indoor Air Quality and Ventilation Group, Engineering Laboratory, 100 Bureau Drive, Mail Stop 8633, Gaithersburg, MD 20899-8633

[2]Currently at Drexel University, 3141 Chestnut Street, Philadelphia, PA 19104

[*]Corresponding email: dustin.poppendieck@nist.gov

Keywords: Spray polyurethane foam, emissions, TCPP

SUMMARY

The use of spray polyurethane foam (SPF) insulation in the United States is increasing. The primary flame retardant used in SPF, Tris (1-chloro-2-propyl) phosphate (TCPP), has been detected in micro-chamber emission experiments investigating SPF. However, due to the use of TCPP in furniture, SPF has not previously been positively identified as a source of indoor TCPP concentrations. This research measured airborne TCPP concentrations in a furniture-free residential test facility that contained 15 m^2 of exposed, two-year-old, open cell SPF.

INTRODUCTION

Spray polyurethane foam (SPF) insulation is increasingly being used in both new construction and retrofits. SPF is a unique building product in two ways: 1) SPF reduces both convective and conductive heat loss through the building envelope, and 2) SPF is created on site through the reaction of two sets of chemicals. Chemicals used to make SPF include polyols, isocyanates, amines, surfactants, amine polyols, alkanolamines, blowing agents and flame retardants. SPF can contain more than 8 % flame retardant (Sebroski 2012). Recent research (Poppendieck et. al. 2015) has shown than when a sample of new open cell, high pressure SPF was tested in micro-chamber environments at 40 °C and 100 mL/min, the flame retardant Tris (1-chloro-2-propyl) phosphate (TCPP) was emitted at nearly a constant concentration (400 $\mu g/m^3$ to 500 $\mu g/m^3$). However, it is unclear how TCPP emissions emission factors from SPF micro-chamber experiments relate to TCPP sources and concentrations in buildings. TCPP is also used in products that contain polyurethane foam such as furniture, mattresses and sound insulation within consumer products. Hence, if TCPP is measured in air of an occupied house containing SPF, the TCPP source cannot be solely attributed to the insulation or the furniture.

The National Institute of Standards and Technology (NIST) built in 2012 a net-zero energy residential test facility (NZERTF) to support the development and adoption of cost-effective net-zero energy designs and technologies, construction methods, and building codes. The design and construction of the NZERTF are described in Pettit et al. (2014). The NZERTF is a two-story, detached home with an unfinished basement and attic within the building thermal envelope. The garage is not attached. The house is similar in size (242 m^2 for occupied floors, 485 m^2 inside the building envelope including the attic and basement) and aesthetics to homes in the surrounding communities. To achieve the net-zero energy goals, several technologies are employed, including a high efficiency heat pump, a solar hot water system, a heat recovery ventilator (HRV), and a 10.2 kW photovoltaic system. To comply with the outdoor air requirements in ASHRAE Standard 62.2-2010 (2010) the HRV was sized to deliver 137 m^3 h^{-1} of outdoor air. Special attention was paid to the design and construction of the highly insulated and airtight building envelope. Roughly 15 m^2 of high pressure, open cell SPF was used to insulate the basement rim joists. The basement is unfinished and the SPF is not covered by any finishing material. The house has no carpet and is not furnished other than permanently installed cabinetry. Hence, if TCPP is present in the indoor air of the house and not measured in the outdoor air it can likely be attributed to the SPF. This work sought to measure airborne TCPP concentrations in the NZERTF.

METHODOLOGIES

The first floor and basement of the NZERTF were sampled for TCPP over a period of two months. The NZERTF TCPP sampling involved two Tenax sorbent tubes in series. The first tube is used to quantify the TCPP concentration and the second to evaluate if there was breakthrough through the first. If TCPP breakthrough to the second tube was found, the data was not used. For each sampling event three sets of tubes were prepared.

Each tube set was sampled at 50 mL/min using a mass flow controller sampling system. Sampling times varied from 52 min to 216 min (average 155 min). The tubes were separated and spiked with internal standard (1.0 µL of 1.25 mg Toluene D-8/mL of methanol). Blank tubes (with internal standard) were run between the samples to quantify any carryover between samples. Samples were analyzed using a thermal desorption-gas chromatography/mass spectrometer system (TD-GC/MS).

When used in field applications TCPP typically consists of three isomers: tris(1-chloro-2-propyl) phosphate (≈66%), bis(1-chloro-2-propyl) (2-chloropropyl) phosphate (≈30%) and (1-chloro-2-propyl) bis(2-chloropropyl) phosphate (≈4%). The relative response ratios of the three isomers on the tubes with TCPP and the subsequent blanks were summed to determine the total response ratio. The combined relative response ratio was then integrated using a five point standard curve (20 ng, 30 ng, 50 ng, 70 ng and 90 ng). Typically, only the first two isomers were detected and only the first two isomers were quantified.

The 13 standard curve R-square values averaged 0.98 for the first and second isomer. On days when a standard curve was not run, check standards were run. The instrument detection limit was 8.65 ng and the method detection limit was 0.71 µg/m^3 to 2.86 µg/m^3 depending on the sample volume. Only values above the method detection limit for the corresponding sampling volume are shown below.

Samples were run over a period of two months in the summer of 2014. The initial thermostat set point (located on the first floor) was 23.9 °C, a setting that had been maintained for weeks prior to the analysis. The thermostat was raised to 32.2 °C and maintained at the temperature for a period of seven days. Temperatures in the basement were several degrees cooler than the thermostat set points. Temperature values shown in the following table and figure are 12-hour average readings from a thermocouple located in the center of the open basement.

To ensure that there we no sources of TCPP other than the SPF in the basement, small samples of a variety of materials with foam components were placed in a micro-chamber at 40 °C and sampled for TCPP using the same Tenax sorbent tubes and TD-GC/MS analysis. The sampled materials include rigid expanded polystyrene insulation, duct insulation, and two varieties of pipe insulation. No TCPP was detected from any of these materials (method detection limit 2.0 μg TCPP/g material m^3 air to 6.3 μg TCPP/g material m^3 air).

RESULTS AND DISCUSSION

Samples were taken in the basement and on the first floor with the HVAC (Heating, Ventilating, and Air Conditioning) operating under normal conditions (typical air change rates: 0.15 h^{-1} to 0.22 h^{-1}, Poppendieck et. al. 2015). Samples were also taken outdoors and no TCPP was detected. Table 1 shows that the average TCPP concentration for the basement samples was nearly twice that of the first floor samples. These data lends credence to the source of the TCPP being in the basement. As mentioned above, none of the other measured materials in the basement contained TCPP and there have never been other potential sources of TCPP, such as furniture or mattresses, in the NZERTF. This information indicates TCPP is being emitted from the SPF located in the basement, is transported to the living areas, and is measurable in the indoor air under normal operating conditions.

Table 1: Average TCPP concentrations measured in the NIST NZERTF.

Location	Average Temperature (°C)	Number of Samples (n)	Average Concentration (μg/m³)	Relative Standard Deviation
1st Floor	23.7	9	1.5*	7.0%
Basement	21.0	12	2.8	9.9%

*First floor samples ranged from 13 ng to 16 ng per sorbent tube. This is below the lowest standard, but above the instrument detection limit of 8.65 ng determined according to "Definition and procedure for the determination of the method detection limit – Revision 1.11" Pt. 136, App. B 40 CFR Ch. I (7–1–03 Edition).

Direct comparisons of measured concentrations (Table 1) to other residences is of limited use given the unique conditions of the NZERTF. First, SPF was not the primary insulation used in the house. SPF was only sprayed on 15 m^2 of the house exterior, a relatively small fraction of the greater than 600 m^2 of the building envelope. Other SPF application scenarios may involve a larger fraction of the building envelope. Second, the SPF was directly exposed to the basement air. In many SPF applications there are finishing products between the SPF and the occupied space. These products, such as drywall, can inhibit the transfer of TCPP to the occupied space. Third, there were no other sources of TCPP (furniture and mattresses) present in the house. Fourth, this work did not quantify the sorption and or re-emission of TCPP from dust or any other materials. Finally, the toxicological relevance of these concentrations is beyond the scope of this paper. There is limited data on chronic exposure to low TCPP concentrations (Farhat et. al. 2013). Current efforts to expand this knowledge are underway by other researchers.

Previous work (Poppendieck et. al. 2015) has shown a strong correlation between TCPP concentration and temperature when sampled in controlled micro-chamber experiments. To see if this relationship also held true in a full scale residence, the temperature of the NZERTF was raised for seven days (average basement temperate 28.5 °C). The results in Figure 1 illustrate the TCPP concentrations in the NZERTF are also strongly correlated to indoor temperature. There was minimal correlation with outdoor temperature (R square 0.03). The average outdoor temperature was 23 °C during both sets of indoor temperature experiments.

These data are consistent with the micro chamber data that demonstrate a relationship between temperature and emission rate of TCPP from SPF insulation. More research is needed to determine building envelope temperatures where SPF is applied and the fate and transport of TCPP through wall finishing materials.

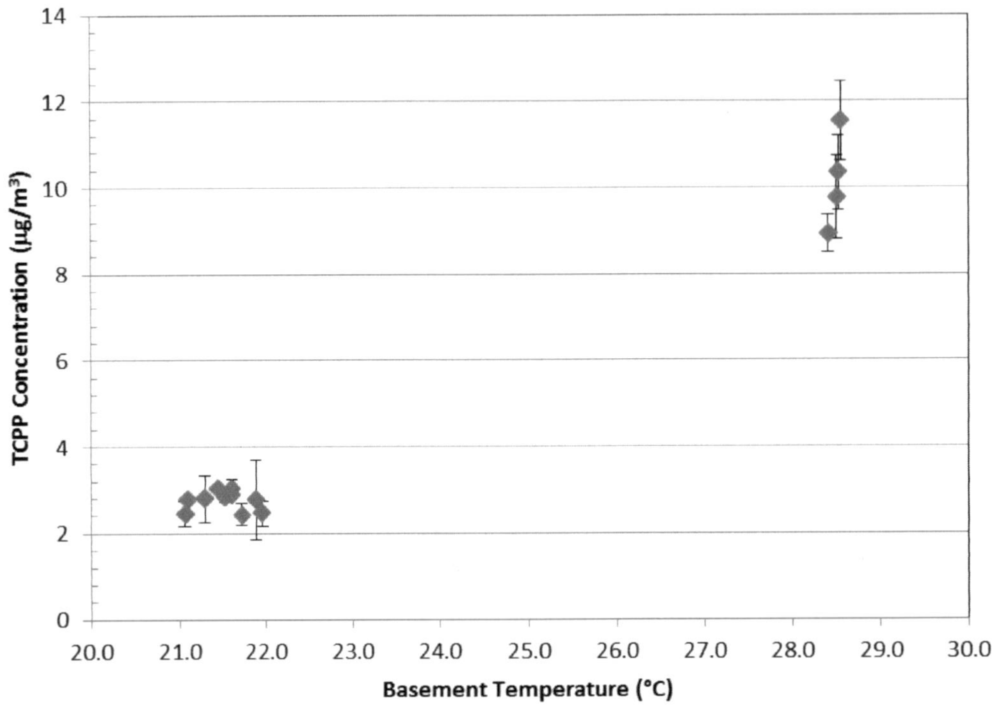

Figure 2: TCPP concentrations measured in the NIST NZERTF basement at various temperatures. Error bars show two standard deviations of triplicate sampling.

CONCLUSIONS

This research indicates that under the tested conditions, airborne TCPP concentrations can be measured in the NZERTF and the source of the TCPP is likely exposed, two-year-old, open-cell SPF.

REFERENCES

ASHRAE. Standard 62.2-2010: Ventilation and Acceptable Indoor Air Quality in Low-Rise Residential Buildings. Atlanta: American Society of Heating, Refrigerating and Air-Conditioning Engineers, Inc.; 2010.

Farhat, A.; Crump, D.; Chiu, S.; Williams, K. L.; Letcher, R. J.; Gauthier, L. T.; Kennedy, S. W. (2013). "In Ovo effects of two organophosphate flame retardants--TCPP and TDCPP--on pipping success, development, mRNA expression, and thyroid hormone levels in chicken embryos." Toxicol Sci 134(1): 92-102.

Pettit B, Gates C, Fanney AH, Healy W. Design Challenges of the NIST Net Zero Energy Residential Test Facility. TN-1847. Gaithersburg, MD: National Institute of Standards and Technology; 2014.

Poppendieck, D. Ng, L., Persily, A., Hodgson, A. Long Term Air Quality Monitoring in a Net-Zero Energy Residence Designed with Low Emitting Interior Products. Submitted Atmospheric and Environment. April 2015.

Poppendieck, D. Nabinger, S., Schlegel, M., Connor, A., Persily, A., Biggs, M., Thomas, T. Emissions from Open Cell Spray Polyurethane Foam Insulation. Submitted Building and Environment. May 2015.

Sebroski, J. Research Report for Measuring Emissions from Spray Polyurethane Foam (SPF) Insulation. Center for the Polyurethanes Industry (CPI) of the American Chemistry Council, September 4, 2012

PHTHALATE LEVELS IN GREEN-RENOVATED SUBSIDIZED HOUSING

Robin DODSON[1,*], Julia UDESKY[1], and Ruthann RUDEL[1]

[1]Silent Spring Institute, Newton, MA USA

[*]Corresponding email: dodson@silentpspring.org

Keywords: Phthalates, building materials, green housing, asthma, biomonitoring

SUMMARY

Phthalates are a group of chemicals used in plastics and personal care products. Exposure to phthalates has been associated with adverse reproductive development and allergy/asthma. We measured phthalates in indoor air, kitchen floor wipes, and urine samples to evaluate the impact of green renovations on residential exposures. Phthalates were found in all homes and children, and urinary metabolite levels were correlated with their parents in air and wipe samples for diethyl phthalate (DEP), di-n-butyl phthalate (DBP), and butyl benzyl phthalate (BBP). This suggests that the indoor environment is an important source of exposure for these phthalates. Green renovated units had lower BBP and bis(2-ethylhexyl) phthalate (DEHP) air concentrations than non-renovated units, presumably due to differences in building materials.

INTRODUCTION

Phthalates are ubiquitous in the indoor environment and used in many building materials and consumer products (Rudel et al., 2009). Phthalate exposures have been associated with altered reproductive development, and several epidemiological studies have suggested an association with allergic diseases such as asthma (Braun et al., 2013). To evaluate the impact of green renovations on residential phthalate levels, we measured phthalates in indoor air, kitchen floor wipes and urine from asthmatic children living in homes with and without green renovations.

METHODOLOGIES

We collected indoor air and kitchen floor wipes from homes, and urine samples from children enrolled in CDC's Green Housing Study (GHS), a longitudinal study of asthmatic children aged 7-12 living in recently "green" renovated and non-renovated subsidized housing. Samples were collected 6 months and 12 months after renovation in the first two GHS study sites: Boston, MA and Cincinnati, OH.

We actively collected air samples from the child's bedroom over 4-5 days (target flow rate 2.5 L/min). Samples were analyzed for 6 phthalates: diethyl phthalate (DEP); di-n-butyl phthalate (DBP); butyl benzyl phthalate (BBP); bis(2-ethylhexyl) phthalate (DEHP); di-cyclohexyl phthalate (DCHP); and di-iso-nonyl phthalate (DINP).

Isopropyl alcohol-wetted wipes were collected from kitchen floors. Wipes were analyzed for 7 phthalates: DEP, DBP, BBP, DEHP, DCHP, di-isoheptyl phthalate (DIHP), and DINP.

We collected morning void and spot urine samples from each child. We prioritized analysis of morning void samples, but analyzed spot samples when morning voids were not available. Urine samples were analyzed for 13 metabolites: monomethyl phthalate (MMP); monoethyl phthalate (MEP); mono-n-butyl phthalate (MBP); mono-iso-butyl phthalate (MiBP); monobenzyl phthalate (MBzP); mono-2-ethyl hexyl phthalate (MEHP); mono-2-ethyl-5-hydroxyhexyl phthalate (MEHHP); mono-2-ethyl-5-oxohexyl phthalate (MEOHP); mono-2-ethyl-5-carboxypentyl phthalate (MECPP); mono-3-carboxypropyl phthalate (MCPP); mono carboxyisooctyl phthalate (MCOP); mono carboxyisononyl phthalate (MCNP); and cyclohexane-1,2-dicarboxylic acid mono hydroxyl isononyl ester (MHiNCH).

RESULTS AND DISCUSSION

As in previous studies, we found phthalates in 100% of homes and participants. We found DBP and DEP, the more volatile phthalates, at the highest concentrations in air. In contrast, we found BBP, DEHP, and DINP, higher molecular weight phthalates, at the highest mass loadings in kitchen wipes. We found MEP, the monoester metabolite of DEP, at the highest mean concentration in urine. The urinary phthalates levels measured in our study were generally higher compared to NHANES data for 7-12 year olds in the 2009-2010 cycle.

The parent-metabolite pairs of DEP-MEP, DBP-MBP, and BBP-MBzP were significantly positively correlated in air and urine ($p<0.05$, Figure 1) and in wipes and urine ($p<0.05$), indicating the home represents an important contribution to total exposure for these phthalates. Only DEP and BBP were significantly positively correlated in air and wipes.

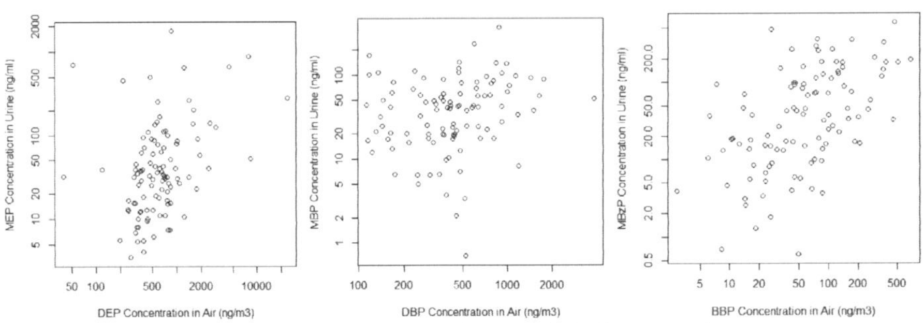

Figure 1. Measured indoor air and urinary phthalate levels.

We found significantly higher BBP and DEHP air concentrations in non-renovated versus green renovated homes based on mixed-effects models for repeated measures, with households as random effects ($p<0.05$). City-stratified mixed-effects models of wipe data revealed significantly lower BBP in Boston green units, consistent with findings in air, but higher BBP in Cincinnati green units. This may be due to flooring types used in the different types of units in each city. Metabolites of BBP and DEHP were higher in children living in control versus green homes; however, the differences were not significant.

CONCLUSIONS

These data are some of the first measurements of phthalates in green-renovated homes and provide an opportunity to evaluate the impact of renovations on indoor environmental quality, which is critical since substantial investments in green housing should not have unintended consequences on exposures and health.

ACKNOWLEDGEMENT

This work is funded by the U.S. Housing and Urban Development (MAHHU0005-12).

REFERENCES

Braun, J. M., S. Sathyanarayana, et al. (2013). "Phthalate exposure and children's health." Curr Opin Pediatr **25**(2): 247-254.

Rudel, R. A. and L. J. Perovich (2009). "Endocrine disrupting chemicals in indoor and outdoor air." Atmospheric Environment **43**(1): 170-181.

HEALTH BENEFITS OF LIVING IN GREEN, LOW-INCOME MULTI-FAMILY HOUSING

Meryl D. Colton[1], Jose Guillermo Cedeno Laurent[1], Piers MacNaughton[1], John Kane[2], Mae Bennett-Fripp, John Spengler[1], Gary Adamkiewicz[1]*

1. Department of Environmental Health, Harvard School of Public Health; Boston, MA
2. The Boston Housing Authority; Boston, MA
3. The Committee for Boston Public Housing; Boston, MA

[*]Corresponding email: mcolton@mail.harvard.edu

Keywords: Indoor air quality, Asthma, Sick building syndrome, Green buildings

SUMMARY

BACKGROUND: The indoor environment has been implicated in a range of poor health outcomes. These associations are particularly relevant in low-income settings where the prevalence of morbidities and environmental pollutants are elevated.
METHODS: We used questionnaires and a visual inspection to compare sick building syndrome (SBS) symptoms and asthma-related morbidity among residents in green and conventional low-income multi-family units.
RESULTS: We found that adults living in green units reported 1.4 (95%CI: 0.66, 2.05) fewer out of 14 possible SBS symptoms than those living in control homes. Furthermore, we found that asthmatic children living in green homes had lower odds of experiencing asthma symptoms (p=0.05), an asthma attack (p=0.03), a hospital visit (p=0.03), or missed school (p=0.02).
CONCLUSIONS: Participants living in green homes had improved health outcomes. Green housing may provide the greatest value in resource-poor settings where green construction has the potential to simultaneously reduce harmful exposures, promote resident health, and reduce operational costs.

INTRODUCTION

The indoor environment has been implicated in a range of poor health outcomes, most notably, asthma (IOM, 2000). These associations are particularly relevant in low-income settings where the prevalence of morbidities (including asthma) and the presence of environmental pollutants are elevated (Adamkiewicz et. al. , 2011). An approach to improving indoor environmental quality and health is to incorporate these goals into construction (Spengler and Chen, 2000) and renovation practices (Noris et. al. , 2013). Construction with the aims of increasing energy efficiency and using environmentally friendly materials—widely defined as 'green' construction—has the potential to both improve resident health and provide savings in housing costs. While the tightening of building envelopes to increase energy efficiency has been associated with an increase in a host of non-specific symptoms labeled 'sick building syndrome' (SBS) symptoms (Sundell et. al. , 2011, Clausen et. al. , 2011), current green building guidelines aim to avoid these health problems; however, these outcomes have not been rigorously tested. In this manuscript, we report differences in key health outcomes between residents living in green and conventional (control) public housing units in Boston, Massachusetts.

METHODOLOGIES

Study design. We asked residents from public housing developments operated by the Boston Housing Authority (BHA) to participate in a questionnaire and home visual inspection that was performed by trained research technicians. Participants were recruited from three public housing developments: Old Colony (green and conventional), Washington Beech (green), and Ruth Lillian Barkly (conventional). At the time of the study, over 600 units at Old Colony remained conventional (control) units within the same 20-acre site. The conventional buildings at Ruth Lillian Barkly were similar in building age and design to the conventional buildings at Old Colony. Both green sites received Leadership in Energy and Environmental Design (LEED) certification from US green building council.

A subset of 600 addresses was randomly chosen from the complete list of units at each development, distributed proportionally to the number of units in each housing group so as to obtain a representative sample. Participants were eligible if they were over 18 years of age and spoke English, Spanish, Mandarin Chinese or Cantonese. If more than one person in the apartment matched the selection criteria, we interviewed the first available, eligible participant. All participants were administered informed consent and compensated for their participation in accordance with the Harvard School of Public Health Institutional Review Board. Approximately one year after the initial visit, we administered the same questionnaire and visual inspection of the residential unit with each participant.

Statistical analysis. We tested differences in the occurrence of each SBS symptoms using multivariate, generalized marginal models. We calculated a Bonferonni adjusted p-value to account for multiple testing. We also calculated the total sum of SBS symptoms for each participant and tested differences in the mean sum between those living in each housing type using multivariate linear mixed effects models with random intercepts for each participant. For the adult and child binary asthma health outcomes, we used multivariate, generalized marginal models to test the effect of green. All multivariate models were adjusted for season. All analyses were performed using SAS (version 9.4; Cary, NC).

RESULTS AND DISCUSSION

We performed 423 home visits with 235 unique participants in three Boston public housing developments; 188 (80%) participated in two visits over the successive years. The percentage of participants reporting each SBS symptom across every visit was lower in 11 of 14 symptoms in the green homes and is presented in Figure 1. Only reports of headaches were significantly lower using the Bonferroni correction ($p = 0.0012$). Over the two visits, the participants in control homes experienced a mean (SD) of 4.2 (3.4) symptoms as compared to 2.9 (2.9) symptoms in the green homes ($p = 0.0002$).

There were 50 adults that reported having asthma. The percentage of respondents self-reporting any asthma symptoms in the last month, or an asthma attack, hospital visit, and overnight hospital visit in the last year are detailed in Figure 2. In models adjusted for season, the adults reported 0.61 (95% CI: 0.24, 1.51), 0.46 (0.17, 1.2), 0.98 (0.35, 2.67) and 0.40 (0.08, 1.91) times the odds of experiencing asthma symptoms in the last month, or having an asthma attack hospital visit or overnight hospital visit as a results of asthma in the last year, respectively. While adult asthma exacerbations were lower among green homes, none of these findings were statistically significantly lower.

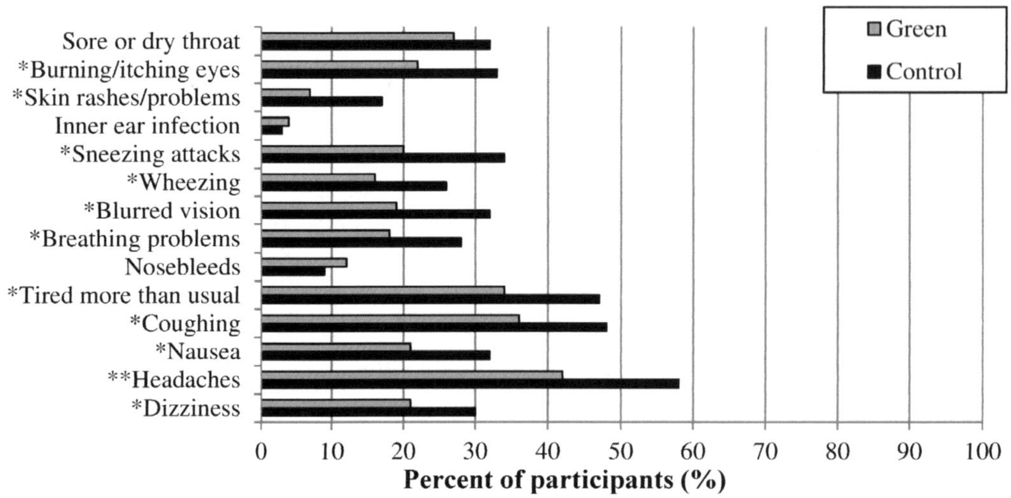

Figure 1. Percent of participants reporting each of 14 SBS symptoms in the last 30 days among participants living in green and control homes.
*$p < 0.05$ test for difference
**$p < 0.0036$ (Bonferroni corrected for multiple tests)

The study population was comprised of 44 children with asthma. The percentage of adult-reported occurrences of asthma symptoms and asthma medication use in the last month, or an asthma attack, hospital visit, overnight hospital visit, and missed day of school in the last year in children are presented in Figure 2. Children's asthma exacerbations were also all lower in the green homes, and in models adjusted for season, this resulted in 0.34 (95% CI: 0.12, 1.00), 0.31 (0.11, 0.88), 0.24 (0.06, 0.88), and 0.21 (0.06, 0.74) times the odds of children experiencing asthma symptoms, an asthma attack, a hospital visit, or missed school (respectively).

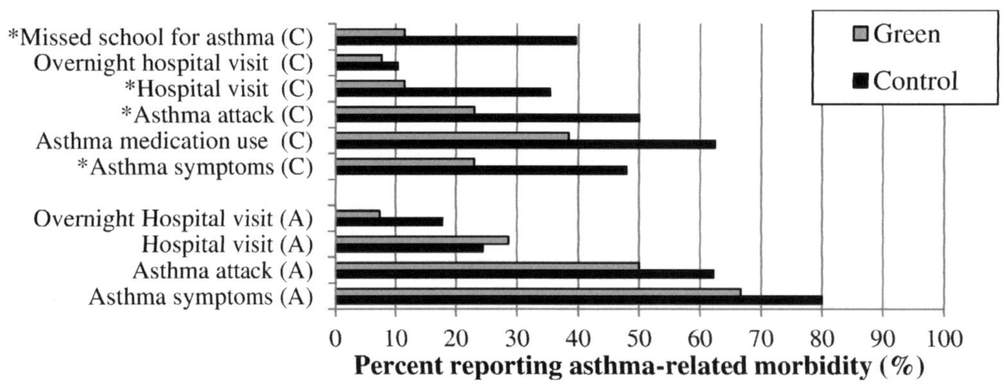

Figure 2. Percentage of adults (A) and children (C) reporting asthma-related morbidities from all asthmatic adults and children respondents living in green and control homes.
*$p < 0.05$ test for difference

We observed a reduction in several key health outcomes among residents living in green buildings compared to conventional, low-income multi-family housing. There are several limitations to this study. First, the predominantly case control design dictates that there may be unmeasured confounders, which explain the observed differences in health outcomes between study populations. However, this study greatly minimized the potential for unmeasured confounding by comparing extremely similar populations within Boston public housing, the majority of whom resided within a 20-acre site of a single housing development. In addition, participants could not be blinded to their building type status, and it is possible that residents were exposed to information about green building and improved health, which influenced their responses. However, we used objective questions on SBS symptoms and asthma exacerbation that have been used in previous studies (CDC, 2003, Brightman et. al. , 2008)

It will be important to ensure that these outcomes are maintained over time. Some argue that multi-component interventions are necessary to comprehensively improve indoor air quality and resident health (Brugge et. al. , 2003, Sandel et. al. , 2004). While green construction is often considered a luxury for middle- or high-income communities, these approaches may provide the greatest value in resource-poor settings where green construction has the potential to simultaneously reduce harmful indoor exposures, promote resident health, and reduce operational costs.

ACKNOWLEDGEMENT

This study was funded by the U.S. Department of Housing and Urban Development (Grant No. MALHH0229-10). The authors would like to acknowledge the Boston Housing Authority, Trinity Management Company, and Beacon Communities, the Committee for Boston Public Housing, and our gracious participants.

REFERENCES

Adamkiewicz, G., Zota, A.R., Fabian, M.P., Chahine, T., Julien, R., Spengler, J.D. and Levy, J.I. (2011) Moving environmental justice indoors: understanding structural influences on residential exposure patterns in low-income communities, *American journal of public health*, **101 Suppl 1**, S238-245.

Brightman, H.S., Milton, D.K., Wypij, D., Burge, H.A. and Spengler, J.D. (2008) Evaluating building-related symptoms using the US EPA BASE study results, *Indoor Air*, **18**, 335-345.

Brugge, D., Vallarino, J., Ascolillo, L., Osgood, N.D., Steinbach, S. and Spengler, J. (2003) Comparison of multiple environmental factors for asthmatic children in public housing, *Indoor Air*, **13**, 18-27.

Centers for Disease Control and Prevention (2003) SLAITS National Asthma Survey, Vol. 2015.

Clausen, G., Beko, G., Corsi, R.L., Gunnarsen, L., Nazaroff, W.W., Olesen, B.W., Sigsgaard, T., Sundell, J., Toftum, J. and Weschler, C.J. (2011) Reflections on the state of research: indoor environmental quality, *Indoor Air*, **21**, 219-230.

Iom (2000) *Clearing the Air: asthma and indoor air exposures,* Washington, DC, National Academy Press.

Noris, F., Delp, W.W., Vermeer, K., Adamkiewicz, G., Singer, B.C. and Fisk, W.J. (2013) Protocol for maximizing energy savings and indoor environmental quality improvements when retrofitting apartments, *Energy and Buildings*, **61**, 378-386.

Sandel, M., Phelan, K., Wright, R., Hynes, H.P. and Lanphear, B.P. (2004) The effects of housing interventions on child health, *Pediatric annals*, **33**, 474-481.

Spengler, J.D. and Chen, Q. (2000) Indoor air quality factors in designing a healthy building, *Annual Review of Energy and the Environment*, **25**, 567-600.

Sundell, J., Levin, H., Nazaroff, W.W., Cain, W.S., Fisk, W.J., Grimsrud, D.T., Gyntelberg, F., Li, Y., Persily, A. and Pickering, A. (2011) Ventilation rates and health: multidisciplinary review of the scientific literature, *Indoor Air*, **21**, 191-204.

VENTILATION PROVISION AND OUTCOMES IN MAINSTREAM CONTEMPORARY NEW-BUILDING FLATS IN LONDON, UK

Tim SHARPE[1] and Antoinette CHARLES[1]

[1]Mackintosh Environmental Architecture Research Unit (MEARU), Glasgow School of Art, UK

[*]Corresponding email: t.sharpe@gsa.ac.uk

Keywords: Housing, Ventilation, MVHR, Natural Ventilation, Occupancy

SUMMARY

Increasingly stringent statutory regulation intended to reduce energy consumption is leading to the use of new approaches, for example greater airtightness and mechanical ventilation. The impacts of these, particularly when they become mainstream, are not yet well understood. This is an important problem for social housing providers who retain a contractual obligation to their tenants and are therefore concerned about possible energy and health impacts. The aim of this research was to examine the environmental performance of new-build housing through the monitoring and evaluation of 20 typical new-build contemporary flats in London. Problems with ventilation provision were endemic – in winter 90% of bedrooms had CO^2 levels over 1000ppm overnight, due to a variety of design, construction and operation issues. This paper describes the nature and causes of these.

INTRODUCTION

The need to address climate change by reducing the significant energy use and consequent carbon emissions of housing (Palmer and Cooper 2011) is leading to increased performance requirements of Building Regulations in the UK (HMG 2010, SG 2010) and this in turn has led to the rapid adoption of new technologies for buildings that seek to reduce energy consumption and carbon emissions. However there is now clear evidence of performance gaps between design intentions and performance (Green Construction Board, 2012; Zero Carbon Hub 2014), and there is also emerging evidence that some measures (for example increasing building air tightness levels and mechanical ventilation) can have detrimental impacts in indoor air quality (Crump et al, 2009; Davies M, Oreszczyn T. 2012). There is increasingly widespread use of Mechanical Ventilation with Heat Recovery (MVHR) and decentralised Mechanical Extract Ventilation (dMEV), but there is very little field information on how these perform in mainstream housing.

METHODOLOGIES

The research was undertaken as part of a Knowledge Transfer Partnership (KTP) between MEARU and Cartwright Pickard Architects (CPA), in collaboration with 5 Registered Social Landlords (RSLs) in London. The project aimed to develop techniques for 'light touch' Building Performance Evaluation (BPE) and undertook a series of BPE studies of 4 houses in 5 new-build housing schemes owned by the RSLs. In each case a series of 2-week monitoring periods were undertaken during 3 different seasonal conditions (spring, summer and winter) during which internal temperature (°C), relative humidity (%) and CO_2 concentration (ppm) was monitored in all apartments. Measurement of these parameters were made at 1-minute intervals using Eltek GD-47 transmitters and recorded as a 5-minute mean value on Eltek RX250AL data loggers. Information was gathered on the construction of the flats, and testing was undertaken including thermography and flow testing of mechanical ventilation systems. Evidence was also gathered from the occupants through surveys and interviews. The general characteristics of the dwellings are described in the Table 1 below:

Table 1. Characteristics of dwellings involved in the study

REF		1	2	3	4	5	6	7	8	9	10	11	12	13	14	15	16	17	18	19	20
Certification		CSH 4				CSH 3				EH 'very good'				BR - 2009				CSH 3			
Dwelling type		MT (H)	GF M)	MT (H)	TF M)	MF F)	TF (F)	MF F)	TF (F)	GF (F)	GF (F)	MF F)	MF F)	GF (F)	MF F)	MF F)	MF F)	MF F)	MF F)	MF F)	MF F)
Floor area (m^2)		104	77	71	76	70	70	104	104	77	113	72	56	92	47	46	46	79	86	70	47
Bedrooms		3	2	2	2	2	2	3	3	2	4	2	1	3	1	1	1	2	3	2	1
Occupants	A	2	2	1	1	1	2	2	2	2	1	1	2	3	1	1	2	2	2	1	1
	C	2	-	2	-	1	2	4	3	2	4	-	-	1	-	-	-	2	3	1	-
Occupancy		24	16	24	24	24	24	24	24	16	16	16	16	24	24	24	24	24	16	24	24
Heating system		HW+R	GB+R	HW+R	GB+R	GB+R	GB+R	GB+R	GB+R	HE+R	HE+R	HE+R	HE+R	GB+R	GB+R	GB+R	GB+R	GB+R	GB+R	GB+R	GB+R
Ventilation		MVHR				MVHR + kit extract				MEV	MVHR	MEV	MEV	MEV				MEV			
Extract flow rates (l/s)	T	-	13	25	23	13	-	11	13	14	17	5	9	9	10	11	-	17	-	11	-
	B	19	17	26	25	19	-	22	26	16	17	13	15	20	12	14	-	21	34	24	-
Supply flow rates (l/s)	T	-	19	22	22	14	-	14	17	-	16	-	-	-	-	-	-	-	-	-	-
	B	33	26	22	22	21	-	27	32	-	19	-	-	-	-	-	-	-	-	-	-
Air permeability in $m^3/h.m^2$		2.9	2.9	2.9	2.9	2.7	2.8	2.3	2.6	7.0	2.6	7.0	7.0	5.0	5.0	5.0	8.0	3.0	3.2	5.0	4.3

CSH 3 / 4 – Code for Sustainable Homes level 3 / 4
EH - Eco homes
BR – Building Regulations
A – Adult C- Child(ren) D – Dog
T – Trickle (low rate)
B – Boost (high rate)

MT (H) – Mid-terrace house
GF (M) – Ground floor maisonette
TF (M) – Top floor maisonette
GF (F) – Ground floor flat
MF (F) – Mid floor flat
TF (F) – Top floor flat

HW+R – hot water cylinder + radiator
GB+R – Gas boiler + radiator
HE+R – heat exchanger + radiator
MVHR – Mechanical Ventilation and Heat Recovery system
MEV – Mechanical Extract Ventilation

These houses are typical of mainstream new-build dwellings currently being constructed in London and the UK. The sample houses are representative of approximately 1158 dwellings built in the 5 schemes, which in turn are typical of approximately 24,000 homes built in London in the same period.

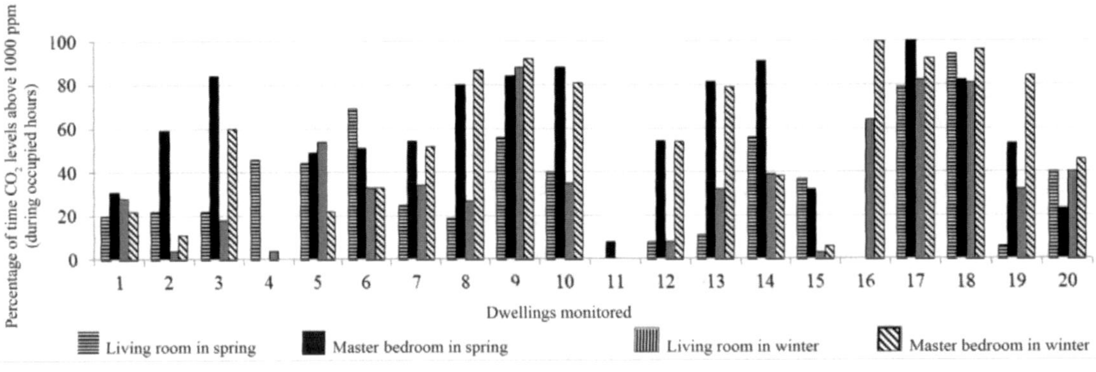

Figure 1. Percentage of time > 1000ppm for occupied spaces in living rooms and master bedrooms during monitoring periods in spring and winter.

RESULTS AND DISCUSSION

In this study CO_2 is being used as a proxy indicator for ventilation. There is a general acceptance that CO_2 keeps 'bad company' and that levels above 1000 ppm are indicative of poor ventilation rates (Porteous 2011) and corresponds well with a ventilation rate of 8 l/s per person (Appleby 1990). A study by Batterman and Peng (1995) identified associations between CO_2 levels and total volatile organic compounds (TVOCs). A more recent paper paper (Wargoki 2013) identified associations between CO_2 levels and health and concluded "The ventilation rates above 0.4 h-1 or CO_2 below 900 ppm in homes seem to be the minimum level to protect against health risks based on the studies reported in the scientific literature".

Figure 1 shows the percentage of occupied time (using the occupancy information) that CO_2 levels exceed the 1000ppm threshold. It is apparent from the results that high levels of CO_2 are endemic in the properties, despite the use of MVHR and MEV ventilation systems. The examination of the houses and interviews with occupants identified a series of common problems and these are summarized below:

Design	**Construction**	**Operation**
• Individual flow rates to rooms inadequate for a delivery of 8 l/s per person in in 96% of room wuth MVHR; • External Inlet and outlet vents are located too close together and near boiler flue, leading to a risk of gases entering the system in in 50% of the MVHR flats • Complaints of smells from the flat below entering the flat above causing the system to be turned off in 2 flats. • Duct runs are complex, duct sizes are small (100 – 125mm dia.) and flexible ducting is used in all the flats. • The system are not sized or designed for actual loads in terms of IAQ (e.g. room occupancy, peak occupancy, household size) in all the flats	• Unbalanced flows in 100% of the MVHR systems. • 2 MVHR bedrooms with missing supply vents • Poorly placed MVHR vents, e.g. too close to the bedroom door 50% of the flats. • Use of flexible ductwork (where rigid ducting has been specified) and poor installation leading to constricted ducts in 100% of the flats. • Evidence in 2 flats of lack of sealing of the system during construction and lack of cleaning before handover leading to construction dust in the system. • Extract rates below Building Standards requirements in 90% of the flats.	• Units located above bedrooms, lack of acoustic separation and insolation causing noise leading to units being turned off in 55% of MVHR flts • Unit located in the loft leading to lack of maintenance and clogging of filters in 55% of MVHR flats. • Lack of occupant guidance on the use of the system – system being turned off, with 77% of MVHR system being disabled or used sub-optimally. • User maintenance of the filters – lack of access and understanding • 'Cold' air perceived as draught • Use of windows over MVHR system. • 63% of flats with dMEV had trickle vents closed.

Figure 2a), close location of inlet, outlet and boiler flue; 2b) dirty filter; 2c) constricted flexible duct

High CO_2 levels were also observed in the houses that had dMEV and are the result of a number of problems. Extract rates were below regulatory standards and trickle vents are frequently closed, and there is lack of clear air-flow paths due to curtains and blinds. Windows and trickle vents are closed at night due to issues of noise and security. The window design in one scheme incorporates a trickle vent, which is occluded by the window frame itself. The boost switch for one of the dMEV systems is not mains connected and does not operate when the battery is flat. Use of windows is preferred to trickle vents – they are more accessible and the effects are more immediate. However this adaptive behavior does not occur at night when occupants are asleep. There are some incidences of high occupancy (2 adults and 2 children in a bedroom).

CONCLUSIONS

The fact that such poor conditions were so common is a cause for concern. Taking this sample as being broadly representative of contemporary construction in London suggests that the proportion of housing with poor ventilation may be very high indeed and this could have important implications for occupants' health. The study found gaps in knowledge at design, construction and occupancy. Designers are generally designing to the regulation, rather than performance standards and are not thinking holistically about what levels of ventilation may be required, and how they may be achieved. There was evidence of poor construction and workmanship on site, with a lack of appreciation of the implications of this. There was little handover or occupant guidance on how the ventilation should be used. Further work is needed to examine the actual impacts of poor ventilation and to develop systems and approaches that can deliver good IAQ in the context of reduced energy consumption.

REFERENCES

Palmer J, Cooper I. (2011). Great Britain's housing energy fact file. DECC 2011. https://www.gov.uk/government/uploads/system/uploads/attachment_data/file/48195/3224-great-britains-housing-energy-fact-file-2011.pdf (accessed 21/03/2014)

HM Government, (2010). The Building Regulations 2010: Conservation of fuel and power in new dwelling. Approved Document L1A. London: NBS, part of RIBA Enterprise Ltd., 2010. ISBN 978 1 85946 324 6, 2010

Scottish Government, (2010). Scottish Building Regulations, Edinburgh: Scottish Government, www.scotland.gov.uk/Topics/BuiltEnvironment/Building/Building-standards (accessed 21/3/14)

Green Construction Board Buildings Working Group, (2012). The Performance Gap: Causes & Solutions: www.greenconstructionboard.org/index.php/resources/performance-gap [accessed May 2014]

Zero Carbon Hub, (2014). Closing the Gap between Design and As-Built Performance – End of Term Report accessed at: www.zerocarbonhub.org [accessed May 2014]

Crump D, Dengel A, Swainson, M. (2009). Indoor Air Quality in Highly Energy Efficient Homes—A Review. IHS BRE Press: London, UK.

Davies M, Oreszczyn T. (2012). The unintended consequences of decarbonising the built environment: A UK case study. Energy Build 2012;46:80-5.

Porteous CDA. (2011). Sensing a Historic Low- CO2 Future, Chemistry, Emission Control, Radioactive Pollution and Indoor Air Quality, Dr. Nicolas Mazzeo (Ed.), ISBN: 978-953-307-316-3, pp 213-246 InTech, DOI: 10.5772/16918

Appleby A. (1990) Indoor Air Quality and Ventilation Requirements. In Buildings and Health, the Rosehaugh Guide to the Design, Construction, Use and Management of Buildings, London: RIBA Publications Ltd., 176 in pp.167-193.

Batterman S, Peng C. (1995). TVOC and CO2 concentrations as indicators in indoor air quality studies. American Industrial Hygiene Association 1995; 56(1): 55-65.

Wargocki, P. (2013). The Effects of Ventilation in Homes on Health. Int. J. Vent. 2013; 12, 101–118.

COMPARISON OF INDOOR AND OUTDOOR AIR QUALITY IN LATINO COMMUNITIES IN PERU, CHILE, AND USA

Wyatt M. CHAMPION[1*], Francisco BARRAZA[2], Héctor JORQUERA[2], and Lupita D. MONTOYA[1]

[1]Civil, Environmental & Architectural Engineering Department, University of Colorado Boulder, Colorado USA
[2]Chemical and Bioprocesses Engineering Department, Pontificia Universidad Católica de Chile, Santiago, Chile

[*]Corresponding email: wyattchampion@gmail.com

Keywords: AIR SAMPLING, particulate matter, biomass combustion, Latino communities, $PM_{2.5}$ exposure

SUMMARY

The air quality inside and outside the homes of three Latino communities (rural Peru, urban Chile and suburban USA) was compared in this study. Filter samples were collected and analysed for fine mass ($PM_{2.5}$) and elemental and organic carbon (ECOC). An optical particle sensor was used in lieu of filter samples for $PM_{2.5}$ mass in Peru. Results indicated that rural Peruvian homes burning dung for heating and cooking had the highest average level of $PM_{2.5}$ indoors (140 $\mu g/m^3$) and relatively clean outdoor air. Urban Chilean homes had a lower average indoor $PM_{2.5}$ concentration (22 $\mu g/m^3$), while suburban USA homes showed elevated concentrations of $PM_{2.5}$ indoors (7.0 $\mu g/m^3$) compared to outdoors, but lower indoor levels than the other two communities studied.

INTRODUCTION

The World Health Organization recognizes that inequality in energy access drives poor air quality in the developing world, and that air pollution from the household combustion of fuels is the most important global environmental health risk (WHO, 2014). In the USA, Latino populations are often disproportionately exposed to environmental hazards (Metzger et al., 1995) including elevated concentrations of indoor air pollutants (Hun et al., 2009; Miller et al., 2009). In developing communities, solid fuel use for heating and cooking as well as cultural practices (e.g. raising animals indoors) can contribute to high levels of particulate matter (PM) indoors and increased exposure of residents to air contaminants. The objective of this study was to compare the indoor and outdoor concentrations and composition of particulate matter (PM) for three distinct Latino communities: Langui Peru (Gomez et al., 2011), Santiago Chile (Barraza et al., 2014), and Boulder USA (Escobedo et al., 2014).

Figure 1 shows sample pictures of the communities compared in this paper. Langui, Peru is a high-altitude (13,000 ft) Andean community where residents rely on dried animal dung as a fuel source for heating and cooking. Though outdoor air in this community is clean, indoor PM levels were expected to be high. Santiago, Chile is a metropolis of over 6 million inhabitants which has experienced severe air quality issues in the past, but has made substantial progress in the last two decades, including halving the annual average outdoor $PM_{2.5}$ concentration (Díaz-Robles et al., 2011). The suburban Boulder, USA community experiences outdoor sources including vehicle emissions, and residents commonly use gaseous fuels indoors for cooking.

METHODOLOGIES

In these studies, indoor and outdoor $PM_{2.5}$ sampling was conducted using TAS MiniVol air samplers with $PM_{2.5}$ impactors (Airmetrics, Eugene OR) for a period of 24 hours in Peru and USA and for 48 hours in Chile. In Peru, samples were collected onto 47mm Tissuquartz filters. In Chile and USA, samples were collected onto 47mm Teflo and Tissuquartz filters. Indoor sampling was conducted in the kitchen/living room area at a height of approximately one meter and field blanks were collected for every tenth sample.

For the USA and Chile studies, $PM_{2.5}$ concentrations were determined from the Teflo filters collected over 24 and 48 hours, respectively. Conditioned Teflo filters were analyzed gravimetrically using a five-digit laboratory scale (Sartorius, Goettingen, Germany). For the Peru study, $PM_{2.5}$ concentrations were determined using UCB Particle Monitors (Berkeley Air, Berkeley, CA) over 24 hours and following published protocols (Chowdhury et al., 2007). For all three studies, Tissuquartz filters were collected to assess ECOC and analysed using a Sun Laboratories Dual Optics Lab Instrument and the NIOSH 5040 thermal optical transmittance (TOT) method.

Figure 1. Study communities

Residents of all three communities were also surveyed. Questionnaires were presented in their native language and pertained to current and past respiratory health, potential exposure to PM at work, home characteristics, and heating and cooking practices.

RESULTS AND DISCUSSION

The average levels of indoor and outdoor $PM_{2.5}$, OC, and EC measured in these studies are shown in Table 1. Peruvian homes showed consistently higher indoor levels for all contaminants studied, and the average indoor $PM_{2.5}$ concentration (140 $\mu g/m^3$) was more than five times greater than the WHO recommended indoor 24-hr $PM_{2.5}$ standard of 25 $\mu g/m^3$. This value was also 20 times higher than the average outdoor $PM_{2.5}$ concentration (7.0 $\mu g/m^3$), suggesting significant contribution of pollutants from indoor sources. All of the Peruvian households used dried animal dung as fuel for heating and cooking and 47% of the homes kept guinea pigs in the kitchen. Chilean homes had a lower average indoor $PM_{2.5}$ concentration (22 $\mu g/m^3$), which was close to the average outdoor $PM_{2.5}$ concentration but still indicated indoor sources. Since the sampling campaign in Chile was conducted during the spring, windows and doors were regularly left open, as reported by the participants. Chile also registered the highest average outdoor $PM_{2.5}$ levels (19 $\mu g/m^3$) in this comparison, as expected, given the urban location.

In both Chile and USA, natural gas and electrical furnaces were the most common means of residential heating; therefore, emissions from these inside sources were likely low. USA homes had a higher average indoor $PM_{2.5}$ concentration (7.0 μg/m^3) compared to outdoor (4.1 μg/m^3), potentially due to cooking (Abdullahi et al., 2013) and cleaning (Kwon et al., 2007). One home in the USA study reported an indoor smoking event prior to the sampling period and it registered the highest average indoor $PM_{2.5}$ level in that study.

Peruvian homes also had the highest carbon concentrations in the $PM_{2.5}$ collected, with average indoor OC and EC levels of 46 and 3.7 μg/m^3, respectively. Chilean homes had the lowest average indoor OC (7.1 μg/m^3). Detailed evaluation of the major sources of pollutants inside the Chilean homes can be found in Barraza et al., (2014) since there was source apportionment conducted in that study. Infiltration of vehicular emissions, high in EC (He et al., 2006), have been reported in other urban settings. USA homes had approximately double the indoor OC concentration (15 μg/m^3) compared to Chilean homes, due potentially to cleaning and cooking. They also showed the lowest indoor average EC (0.2 μg/m^3) of the three studies.

Table 1. Average Indoor and Outdoor $PM_{2.5}$, OC, and EC Concentrations (μg/m^3)

Country	Indoor			Outdoor		
	$PM_{2.5}$	OC	EC	$PM_{2.5}$	OC	EC
Peru	140	46	3.7	7.0	6.4	0.4
Chile	22	7.1	1.2	19	4.3	1.6
USA	7.0	15	0.2	4.1	6.8	0.3

Indoor to outdoor (I/O) ratios can be useful in comparing concentrations inside and outside of the homes, and I/O ratios greater than one suggest sources of air pollution indoors. Peruvian homes exhibited the highest average I/O ratios for $PM_{2.5}$, OC, and EC (20, 7.2, and 10, respectively) due likely to the combustion of dung, a very dirty solid fuel, in simple stoves inside the home. Chilean homes had average I/O ratios for $PM_{2.5}$, OC, and EC of 1.2, 1.7, and 0.75. Elemental carbon likely originated from motor vehicles outside. USA homes had average I/O ratios for $PM_{2.5}$, OC, and EC of 1.7, 2.2, and 0.67, respectively.

Survey results found that 77% of the Peruvian women reported irritation to their eyes, nose, and throat during cooking, and that 51% of the women and 73% of the children reported coughing. Health effects were also observed in the USA study, where 24% of respondents experienced difficulty breathing while walking fast or uphill. These health impacts are due potentially to current and/or past exposures of the population to poor air quality. The study found that 87% of Peruvian homes used traditional stoves with improper exhaust ventilation to the outdoors, while 49% of the USA respondents reported previous exposure to wood smoke for heating and cooking abroad.

CONCLUSIONS

The WHO (2014) has stated that new approaches to problem solving ("practical tools") should be developed to best identify heating and cooking alternatives which decrease exposure to air pollutants inside the home. These alternatives should lower emissions while providing affordable and culturally appropriate heating and cooking alternatives. In comparing three distinct Latino communities, results suggest that common practices, born of tradition or necessity, can affect indoor air quality.

This study found significant differences in indoor air quality between the three distinct Latino communities. Peruvian homes where solid fuel use and pet keeping is common had average indoor $PM_{2.5}$ levels twenty times higher than USA homes. Chilean homes with high infiltration exhibited indoor $PM_{2.5}$ levels similar to outdoor values, and these homes also had the highest average outdoor EC concentration, due likely to vehicular emissions in the urban setting. Suburban USA homes experienced elevated indoor OC concentrations, likely due to cleaning and cooking.

ACKNOWLEDGEMENT

We thank Ernie Escobedo, Mike Hannigan and his lab, and the study participants.

REFERENCES

Abdullahi KL, Delgado-Saborit JM, Harrison RM. (2013). Emissions and Indoor Concentrations of Particulate Matter and Its Species Chemical Components from Cooking: A Review. *Atmospheric Environment,* **71**, 260-294.

Barraza F, Jorquera H, Valdivia G, Montoya LD. (2014). Indoor $PM_{2.5}$ in Santiago, Chile, Spring 2012: Source Apportionment and Outdoor Contributions. *Atmospheric Environment,* **94**, 692–700.

Chowdhury Z, Edwards RD, Johnson M, Shields KN, Allen T, Canuz E, Smith KR. (2007). An Inexpensive Light-Scattering Particle Monitor: Field Validation. *Journal of Environmental Monitoring,* **9**, 1099–1106.

Díaz-Robles L, Saavedra H, Schiappacasse L. (2011). The Air Quality in Chile : 20 Years of Challenge. *Environmental Management (EM), Journal of the Air & Waste Management Association.* 28–33.

Escobedo LE, Champion WM, Li N, Montoya LD. (2014). Indoor Air Quality in Latino Homes in Boulder, Colorado. *Atmospheric. Environment,* **92**, 69–75.

Gomez OM, Handorean AM, Brandt EL, Lopez, AA, Hernandez MT, Montoya LD. (2011). Indoor Air Pollution in an Indigenous Highlands Community in Peru. In: Proceedings of Indoor Air 2011 International Congress, Austin, Texas, USA, 5-10 June 2011.

He LY, Hu M, Huang XF, Zhang YH, Yu BD, Liu DQ. (2006). Chemical Characterization of Fine Particles from on-Road Vehicles in the Wutong Tunnel in Shenzhen, China. *Chemosphere*, **62**, 1565–1573.

Hun DE, Siegel JA, Morandi MT, Stock TH, Corsi RL. (2009). Cancer Risk Disparities between Hispanic and Non-Hispanic White Populations: The Role of Exposure to Indoor Air Pollution. *Environmental Health Perspectives*, **117**, 1925–1931.

Kwon KD, Jo WK, Lim HJ, Jeong WS. (2007). Characterization of Emissions Composition for Selected Household Products Available in Korea. *Journal of Hazardous Materials,* **148**, 192–198.

Metzger R, Delgado JL, Herrell R. (1995). Environmental Health and Hispanic Children. *Environmental Health Perspectives,* ***103****,* 25–32.

Miller SL, Scaramella P, Campe J, Goss CW, Diaz-Castillo S, Hendrikson E, DiGuiseppi C, Litt J. (2009) An assessment of indoor air quality in recent Mexican immigrant housing in Commerce City, Colorado. *Atmospheric Environment*, **43**, 5661-5667.

World Health Organization. (2014). WHO Guidelines for Indoor Air Quality: Household Fuel Combustion.

CARPET DUST CULTURABLE FUNGI AND BACTERIA RESULTS USED TO EVALUATE CARPET MAINTENANCE

Kent A CANDEE[1,*], Krista J SCOTT[1], Craig S BLACK[1], and Dave W HAVICK[1]

[1]Home Office Risk Improvement, EMC Insurance, Des Moines, Iowa USA

[*]Corresponding email: Kent.A.Candee@EMCIns.com

Keywords: Carpet, dust, fungi, bacteria, database

SUMMARY

A systematic method was developed to evaluate carpet maintenance practices by matching the analytical results of a carpet dust sample to a carpet maintenance recommendation. Carpet dust culturable fungi and bacteria results are entered into a database. Each fungi and bacteria identified within the sample is classified as acceptable, borderline, or unacceptable and a respective value assigned. The average value for both fungi and bacteria correlates with a carpet maintenance recommendation. The amount and composition of carpet dust should be a reasonable predictor of the adequacy of carpet maintenance practices.

INTRODUCTION

Many indoor environments within the Midwest (Iowa, Minnesota, Nebraska, and Wisconsin) use carpet as a floorcovering for purposes of aesthetics, noise-reduction, and comfort. Building owners, facility managers, and custodians have limited methods for evaluating carpet maintenance practices and their impact on indoor air quality. Inadequate or improper carpet maintenance may result in the accumulation of particulate matter within the fibers. Building maintenance and housekeeping practices should be designed in a manner to minimize the volume of particulate matter that is available to be embedded into carpeting.

When sufficient moisture and other environmental factors are present, the carpet dust may start to grow fungi and bacteria. Poorly maintained carpets may allow allergens and microorganisms to become airborne, which could negatively impact the indoor air quality. The age or visual condition of the carpet cannot be equated to carpet cleanliness. Professional experience has demonstrated that regardless of appearance or age, carpet may be laden with unacceptable fungi and/or bacteria levels.

Multiple sampling devices (i.e., vacuum sock, DUSTREAM® collector, DustChek™, filter cassettes) are available to collect carpet dust samples (AIHA, 2005). The collected carpet dust can be analyzed for select allergens and microorganisms (AIHA, 2005 and ACGIH, 1999). Allergens are not helpful in evaluating the current carpet maintenance practices as they may not be detected within the carpet dust. Microorganisms are always present at some level; in particular, culturable fungi and bacteria.

Few numerical standards exist for data comparison purposes, leaving the professional to rely on past professional experience and judgement (AIHA, 2005 and ACGIH, 1999). Control samples are not useful since carpet dust within and between carpets is not homogenous and results can vary by several orders of magnitude (AIHA, 2005). Developing culturable fungi and bacteria databases provides a systematic methodology to evaluate and interpret the carpet dust results and make a corresponding carpet maintenance recommendation. Available recommendations are: 1) continue current carpet maintenance practices, 2) improve daily maintenance activities, 3) deep-clean carpet, or 4) remove carpet.

METHODOLOGIES

The carpet dust samples are collected using a small shop vac with the DUSTREAM® collector attached. This device incorporates a nylon filter and slips on the end of the vacuum cleaner hose.

A sampling technique was developed, which covered two square feet of carpet using a Plexiglas template with a 6 inch by 6 inch (0.25 square feet) square opening. Eight separate areas of the carpet are selected from heavy foot traffic areas. Samples are collected using the DUSTREAM® collector by completing 12 horizontal and 12 vertical passes within the opening. After the initial eight sampling locations have been completed, the dust collector is visually checked to estimate whether enough mass (>0.5 grams) has been collected. If not, additional sampling is completed. Typically, rooms larger than 900 square feet either have additional surface area covered during sample collection or multiple samples are collected (Tanner, 2009 and New York University, 2013). The type of room or setting is documented.

The carpet sample analysis identifies the top three taxa for both fungi and bacteria. The results are reported in colony-forming units (CFU) per mass (grams). Analytical data is entered into culturable fungi and bacteria databases. Multiple calculations are performed. The primary calculation is dividing the CFUs/gram by the total sample area (square feet).

Each individual fungi and bacteria is evaluated. A given fungi or bacteria result is deemed acceptable when it is below the quotient of geometric mean (GM) and geometric standard deviation (GSD) or GM/GSD, borderline when within one GSD of the GM, and unacceptable when above the product of GM and GSD or GM x GSD. Numerical values of 3, 2, and 1 are assigned to acceptable, borderline, and unacceptable ratings, respectively. Two overall values are obtained separately by averaging the individual values for the sample's fungi and bacteria levels. The fungi value and bacteria value are then cross-referenced to one of the four carpet maintenance recommendations shown in Table 1.

Table 1. Values, Ratings and Carpet Maintenance Recommendations

Value	Rating	Carpet Maintenance Recommendation
≤1.5	Unacceptable	Remove carpet; install preferably new hard-surface flooring
1.6 to 2.0	Borderline	Deep-cleaning carpet; preferably steam-extraction
2.1 to 2.5	Borderline	Vacuuming practices improved (i.e., frequency, vacuum type)
≥2.6	Acceptable	Maintain current carpet maintenance practices

The results of the carpet dust samples are reported along with the corresponding carpet maintenance recommendation.

RESULTS AND DISCUSSION

Variability in sample collection analysis was minimized by using the standardized DUSTREAM® collector and the same vacuum (when possible), sample collection technique, and laboratory accredited by the American Industrial Hygiene Association's Environmental Microbiology Proficiency Analytical Testing (EMPAT) program.

The carpet dust sample results are reported in CFUs/gram/square foot (CFUs/g/ft^2) to allow for comparison between areas. The database (initiated in 2007) currently has 93 carpet dust samples collected from commercial grade, low-cut pile carpeting. School environments, either a classroom, music room, computer lab, library, or office setting account for 93.5% of the samples collected. The use of the Shapiro-Wilk Expanded Test (W-test) was used to verify the log-normal distribution of the results.

One commonly identified culturable fungi is *Cladosporium sp.* The concentrations within the database range from 143 to 1.52×10^6 CFUs/g/ft^2 with a corresponding GM of 7.40×10^4 and GSD of 8.24. The calculated range of 898 to 6.10×10^4 CFUs/g/ft^2 would be assigned as borderline, with any value above 6.10×10^4 CFUs/g/ft^2 rated as unacceptable and any concentration below 898 CFUs/g/ft^2 rated as acceptable.

The most common culturable bacteria is *Bacillus*. Database values ranged from 714 to 6.00×10^8 CFUs/g/ft^2. The GM is 3.86×10^5 with GSD of 68.2. Calculated concentrations below 5.66×10^3 would be rated acceptable, any concentration greater than 2.63×10^7 as unacceptable, and any concentrations in between as borderline.

The system can be demonstrated using sample data from the database containing only one fungi (*Cladosporium sp.* result of 2.00×10^4 CFUs/g/ft^2) and bacteria (*Bacillus* result of 4.00×10^5 CFUs/g/ft^2) result. Both fall into the borderline range using the aforementioned concentration ranges with a value of 2, as shown in Tables 2 and 3. From Table 1, a recommendation is made to deep-clean the carpet using steam extraction.

Table 2. Fungi Sample Results

Fungal ID	Media	Dilu-tion	Sample Count	%	CFU/gram	CFU/g/ft^2	A,B, U,ISD	Value
Cladosporium	MEA	1:100	1,000	100%	40,000	20,000	B	2
TOTALS			1,000	100%	40,000	20,000	B	2.0

Table 3. Bacteria Sample Results

Bacterial ID	Media	Dilu-tion	Colony Count	%	CFU/gram	CFU/g/ft^2	A,B, U,ISD	Value
Bacillus	TSA	1:1000	20,000	100%	800,000	400,000	B	2
TOTALS			20,000	100%	800,000	400,000	B	2.0

There can be cases where the average value for fungi and bacteria will differ, recommending different carpet maintenance recommendations. In such situations the more stringent recommendation is provided. Since several of the carpet maintenance recommendations require a financial commitment, it is imperative building owners fully understand this prior to collecting a carpet dust sample.

A rating of acceptable, borderline, or unacceptable cannot be determined if a sampling result contains a fungi or bacteria that does not have at least three data points within the database. In these cases, the rating would be reported as insufficient data (ISD) and assigned a value of zero. This reported fungi or bacteria is generally not included in the overall average calculation for the sample.

As more data is collected and entered into the databases a typical range for a given fungi or bacteria would be expected to emerge. A given fungi or bacteria result can shift from one rating to another depending on when the statistics are performed relative to the number of data points and range of data values.

CONCLUSIONS

The culturable fungi and bacteria contained in carpet dust can serve as a tool to evaluate carpet maintenance. Statistical analysis of the fungi and bacteria levels contained in a carpet dust sample provides a systematic methodology for determining a carpet maintenance recommendation. Although, basing carpet maintenance decisions on one sample may or may not be representative of actual conditions. The addition of data points will improve future statistical analysis. Future research could include more detailed statistical analysis (e.g., hypothesis testing, z or t test, P-value, etc.) of the data set. Additional applications include cloud mapping by zip code and taxa as well as the evaluation of health risk through the collection of fungi and bacteria air samples at the time of carpet dust sample collection.

ACKNOWLEDGEMENT

A special thanks to EMC Insurance for their financial, computer programming, and administrative support and to EMC industrial hygienists in their contributions to the ongoing success of this project.

REFERENCES

AIHA (2005). Field Guide for the Determination of Biological Contaminants in Environmental Samples. American Industrial Hygiene Association, Fairfax, VA.

ACGIH (1999). Bioaerosols Assessment and Control. American Conference of Governmental Industrial Hygienists, Cincinnati, OH.

New York University (2013), Space Planning Working Group, presentation to University Space Priorities Working Group, website accessed December 11, 2014 (http://www.nyu.edu/content/dam/nyu/spacePriorities/documents/13-1008%20USPWG%20Classrooms%20FINAL.pdf)

Tanner K (Revised 2009), Minimum Classroom Size and Number of Students Per Classroom University of Georgia, School Design and Planning Laboratory, website accessed December 11, 2014 (http://sdpl.coe.uga.edu/research/territoriality.html)

Emission of Phthalates and Phthalate Alternatives from Vinyl Flooring and Crib Mattress Covers: The Influence of Temperature

Yirui LIANG and Ying XU*

Department of Civil, Architectural and Environmental Engineering, the University of Texas at Austin, Austin, Texas, USA

*Corresponding email: xuying@mail.utexas.edu

Keywords: Temperature, Phthalates, Phthalate alternatives, Chamber test, Model, SVOCs

SUMMARY

Emissions of phthalates and phthalate alternatives from vinyl flooring and crib mattress covers were measured in a specially-designed chamber. The gas-phase concentrations versus time were measured at four different temperatures, i.e., 25, 36, 45, and 55 °C. The key parameter that controls the emissions (y_0, gas-phase concentration in equilibrium with the material phase) was determined, and the emissions were found to increase significantly with increasing temperature. Both the material-phase concentration (C_0) and the chemical vapor pressure (V_p) were found to have great influence on the value of y_0. The measured ratios of C_0 to y_0 were exponentially proportional to the reciprocal of temperature, in agreement with the van't Hoff equation. An emission model was validated at different temperatures, with excellent agreement between model calculations and chamber observations. In residential homes, an increase in the temperature from 25 to 35 °C can elevate the gas-phase concentration of phthalates by more than a factor of 10, but the total airborne concentration may not increase that much for less volatile compounds. In infant sleep microenvironments, an increase in the temperature of mattress can cause a significant increase in emission of phthalates from the mattress cover and make the concentration in breathing zone about four times higher than that in the room, resulting in potentially high exposure.

INTRODUCTION

Phthalates have been used as plasticizers to enhance the flexibility of polyvinyl chloride (PVC) products. These semi-volatile organic compounds (SVCOS) are found in a wide range of building materials and consumer products (Bornehag et al., 2005), and they are present at various levels, with some as high as 10% level or even greater. Because phthalate additives are not chemically bound to the polymer matrix, they usually are emitted slowly from the products into the air or other media (Xu et al., 2012). As a result, phthalates are ubiquitous and among the most abundant SVOCs in indoor environments. Recent studies have suggested that exposure to some phthalates may cause irreversible changes in the development of the human reproductive tract; increase the risk of asthma, rhinitis, and allergies; and affect endogenous hormones. As a result, rapid changes are occurring in phthalates used in PVC products, with a trend toward using phthalates that have higher molecular weight and lower volatility. Also, alternative plasticizers, such as diisononyl cyclohexane-1,2-dicarboxylate (DINCH), di(2-ethylhexyl) adipate (DEHA), and di(2-ethylhexyl) terephthalate (DEHT), have emerged very recently, but currently there is a lack of toxicological information for these compounds. Given that these alternatives have properties that are similar to those of phthalates, similar levels of emissions and environmental fates may be expected.

We must understand the mechanisms by which phthalates and their alternatives are emitted from various sources in order to characterize their fate and develop suitable approaches for reducing their concentrations in indoor environments. However, very few emission studies have targeted low volatility plasticizers (Xu et al., 2012; Clausen et al., 2012), possibly due to the substantial difficulties associated with chamber tests for SVOCs, including the long duration of tests (months to years), the low gas-phase concentrations in test chambers, strong sorption onto surfaces, and ubiquitous contamination in laboratory facilities. Using data collected in a specially-designed stainless steel chamber, Xu et al. (2012) showed that the emission rate of di(2-ethylhexyl) phthalate (DEHP) from vinyl flooring can be predicted based on a priori knowledge of y_0, the gas-phase concentration of DEHP in equilibrium with the material phase, the material to air and air to stainless steel-surface mass-transfer coefficients (h_m), and the stainless steel/air equilibrium relationship. Liang and Xu (2014) recently improved the design of the earlier emission chamber and developed a novel, rapid method to measure y_0 for a range of phthalate compounds released from building materials. The mechanisms that govern emissions of phthalates from polymeric materials were also elucidated further by their study.

Temperature may have a strong influence on the emissions of phthalates in indoor environments, because y_0 is related to vapor pressure, which depends strongly on temperature. Despite its importance for exposure and risk assessment, the influence of temperature on the emission and transport of indoor phthalates have yet to be studied comprehensively. The aim of this study is to investigate the influence of temperature on emission of phthalates and their alternatives from building materials and consumer products. The specific objectives are to: (1) conduct a series of controlled tests in specially designed stainless steel chambers to characterize phthalate emissions from vinyl flooring and crib mattress cover at four different temperatures; (2) measure y_0 (the gas-phase concentration in equilibrium with material phase) and C_0 (the concentration of phthalates present in materials), and investigate the nature of their relationship as a function of temperature; (3) validate the SVOC emission model in chamber experiments over a range of temperatures; and (4) use the obtained data to demonstrate the impact of temperature on phthalate concentrations in residential indoor environments and infant sleep microenvironments.

METHODOLOGIES

Test Materials. Sixteen types of vinyl floorings and ten crib mattresses were purchased primarily in the United States. The contents of phthalates and their alternatives in each sample were quantified experimentally. Among them, we selected four floorings and two mattress covers that contained different phthalates at relatively high concentrations. Each material was cut into two 0.45 m × 0.25 m sheets, wrapped in aluminum foil, and stored at room temperature. Before a measurement, the test pieces were unwrapped and placed in the emission chamber.

Emission Experiments. A specially-designed, stainless steel chamber used in previous experiments (Liang and Xu, 2014) was employed in this study. The chamber that was used for SVOC emission testing maximized the surface area of emission source and minimized the surface area of internal sink thereby substantially reducing the time required to reach steady state. In addition, three precision-ground stainless steel rods, matched to the interior stainless steel chamber surface, were inserted into the chamber and then periodically removed so that the concentration of SVOCs sorbed on the surface could be measured. This method allowed us to relate gas-phase concentration to the concentration sorbed on the surface of the chamber and quickly establish the sorption equilibrium relationship. The specific design of the chamber was described in greater detail by Liang and Xu (2014).

RESULTS AND DISCUSSION

Impact of Temperature on Emissions. Figure 1 shows that temperature had a strong influence on phthalate emissions. When the temperature was increased from 25 to 36 °C, the steady-state air concentration of phthalates and their alternatives in the chamber increased up to an order of magnitude. The specific emission rate (SER) can be estimated based on the gas-phase concentration, chamber air exchange rate, the surface area of emission source, and the chamber's volume. On average, a 30 °C increase in temperature (from 25 to 55 °C) resulted in more than a 300-fold increase in SER. Emission of DnBP from sample 4 showed the highest SER at each temperature. In the case of DnBP emission at 36 °C (SER~ 130 $\mu g/m^2/h$), calculations show that after five years, only 1% of the total mass would have been emitted from the vinyl flooring sample. Although the SER measured in the chamber might be somewhat different from that in actual indoor environments, the results indicate that PVC products may behave as permanent indoor sources of phthalates during the product's lifetime. But, temperature increase has a strong impact on their emissions from these sources, accelerating off-gassing, elevating their air concentrations, and thereby resulting in significantly greater indoor exposures.

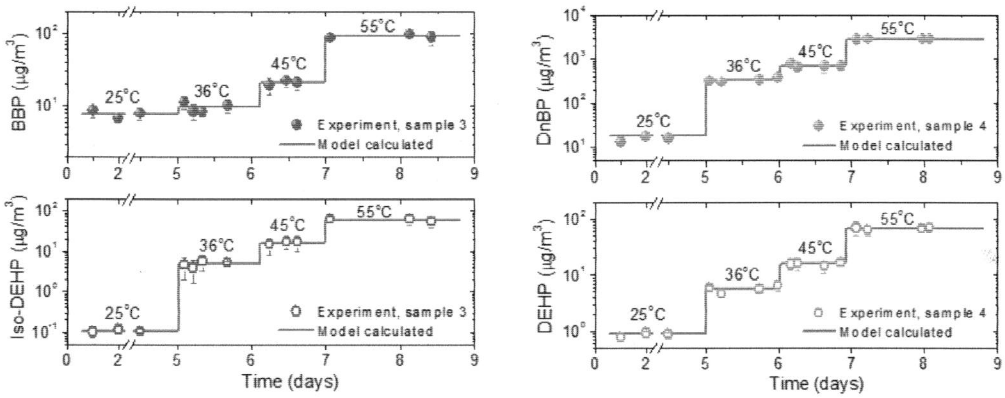

a) b)

Figure 1. Gas-phase chamber concentration of phthalates and phthalate alternatives emitted from tested samples at different temperatures. a) sample 3, b) sample 4.

Exposure to Phthalates by Infants in Their Sleep Microenvironment. Phthalate plasticizers and their alternatives were found in crib mattress covers. As shown in Figure 2, the breathing zone (BZ) consists of the air space surrounding a sleeping infant, with a bottom cross-sectional area that covers the mattress. Using the crib mattress cover that we tested (sample 5, which contains DEHA) as an example, Figure 2 shows the gas-phase concentrations in the BZ and RA zones versus time. The concentrations in the BZ and RA were higher during the sleeping period than when the child was not on the mattress. This finding indicates that the emission of DEHA from the crib mattress cover increased significantly due to the increase in the temperature of the mattress surface caused by heat transfer from the sleeping infant. When the infant was on the mattress, the DEHA concentration in the infant's BZ was several times higher than that in the bulk room air. It suggests that, due to their close contact with the mattress, infants are likely to be exposed to higher phthalate concentrations than other occupants of the room.

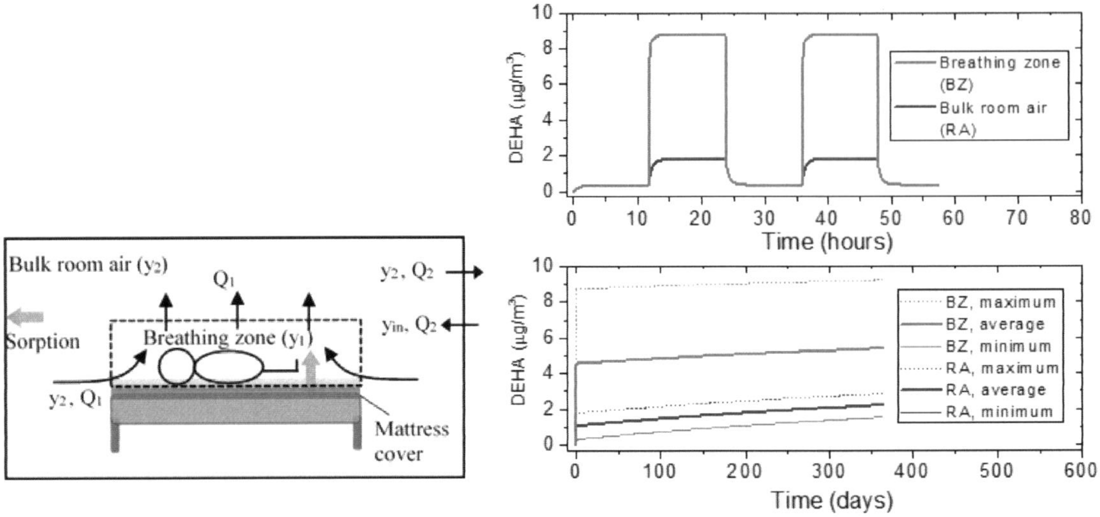

a) b)

Figure 2. a) Schematic representation of the two-zone model, and b) Model prediction of DEHA concentrations in BZ and RA zones.

CONCLUSIONS

This study investigated the influence of temperature on the emission of phthalate from vinyl flooring and crib mattress covers. The gas-phase concentrations versus time were measured at four different temperatures, i.e., 25, 36, 45, and 55 °C. The emissions were found to increase significantly with increasing temperature. In infant sleep microenvironments, an increase in the temperature of mattress can cause a significant increase in emission of phthalates from the mattress cover and make the concentration in breathing zone about four times higher than that in the room, resulting in potentially high exposure.

ACKNOWLEDGEMENT

Financial support was provided by the National Science Foundation (NSF) (CBET-1150713 and CBET- 1066642).

REFERENCES

Bornehag, C. G.; Lundgren, B.; Weschler, C. J.; Sigsgaard, T.; Hagerhed-Engman, L.; Sundell, J., Phthalates in Indoor Dust and Their Association with Building Characteristics. *Environmental Health Perspectives* **2005,** *113*, (10), 1399-1404.

Xu, Y.; Liu, Z.; Park, J.; Clausen, P. A.; Benning, J. L.; Little, J. C., Measuring and Predicting the Emission Rate of Phthalate Plasticizer from Vinyl Flooring in a Specially-Designed Chamber. *Environmental Science & Technology* **2012,** *46*, (22), 12534-12541.

Clausen, P. A.; Liu, Z.; Kofoed-Sorensen, V.; Little, J.; Wolkoff, P., Influence of Temperature on the Emission of Di-(2-ethylhexyl)phthalate (DEHP) from PVC Flooring in the Emission Cell FLEC. *Environmental Science & Technology* **2012,** *46*, (2), 909-915.

Liang, Y.; Xu, Y., Improved Method for Measuring and Characterizing Phthalate Emissions from Building Materials and Its Application to Exposure Assessment. *Environmental Science & Technology* **2014,** *48*, (8), 4475-4484.

LABORATORY AND FIELD PERFORMANCE OF INTERIOR FINISH MATERIALS USING FORMALDEHYDE SCAVENGING TECHNOLOGY

Stanley D. GATLAND II[1*], Douglas GEHRING[1], and Davy VENNEKENS[2]

[1]CertainTeed Corporation, Valley Forge, Pennsylvania, USA
[2]Saint-Gobain Group, Gypsum Activity, Paris, France

[*]Corresponding email: stanley.d.gatland@saint-gobain.com

Keywords: Formaldehyde reduction, Measurement, Interior finishing, Chemical absorption

SUMMARY

Formaldehyde gas is a common indoor air pollutant known to impact human health. Both gaseous dilution by outdoor air ventilation and source control through the specification of low-emitting materials have been the primary remediation strategies. More recently building materials and systems have been designed to capture formaldehyde gas using passive or active building technologies.

The purpose of the study is to describe the methods used to evaluate interior finish materials for their ability to reduce indoor air formaldehyde concentrations. Gypsum board specimens were evaluated in a small environmental chamber for their ability to reduce the concentration of formaldehyde in indoor air. In addition, two classrooms were monitored for formaldehyde concentration over a 16 week period. One classroom was renovated with a formaldehyde scavenging gypsum panel ceiling. Results will be presented and discussed.

INTRODUCTION

Volatile organic compounds (VOCs) are detrimental to human health, and formaldehyde has been identified as a human carcinogen by the International Agency for Research on Cancer (IARC 2004). Increased ventilation and source control are the most common techniques utilized to minimize exposure to VOCs and formaldehyde (Herberger et al. 2009). Offerman (2009) concluded in a large residential housing study that ventilation may not be the most effective solution to reducing formaldehyde concentrations in indoor air. In addition, Srebric (2010) described the conflict between reducing building energy consumption and using increased ventilation rates to improve indoor air quality. Many solutions that actively remove formaldehyde and other VOCs from indoor air are described by Salthammer et al. (2010). Both active and passive formaldehyde sampling techniques are available for laboratory and field investigations (Salthammer et al. 2010, Mullen et al. 2013, Godish et al. 2014). A formaldehyde scavenging gypsum board using an agent derived from ethylene urea and its derivatives, compounds comprising active methylene(s), sulphites, tannins and their mixtures (Sahay-Turner 2013) was evaluated using both small scale chamber and in situ performance testing methodologies. The gypsum material had an average permanent formaldehyde scavenging capacity of 0.4 g/m^2 of surface area (Air Quality Sciences 2012).

ISO 16000-23 TESTING METHODOLOGY AND RESULTS

Formaldehyde emission tests were performed on standard and formaldehyde scavenging gypsum board specimens in accordance with ISO 16000-23 (2009). Environmental chamber inlet sampling flows were calibrated at 450 ml/min to give an air change rate of 0.5. Background aldehyde samples were collected prior to loading each sample into a chamber. Gypsum samples were quickly removed from sealed aluminium foil bags and prepared for testing. One 254 by 254 mm specimen was cut from each board and covered on four edges and one face with aluminium foil and placed in each chamber for testing.

Formaldehyde (HCHO) was allowed to flow from a standard gas cylinder into each chamber at a rate calculated to yield a theoretical chamber concentration of 150 µg HCHO/m^3. The formaldehyde was allowed to flow into the chambers for seven consecutive days at which point the flow was stopped and clean zero-grade humidified air was introduced. Each sample was maintained in the chamber at 23 °C and 50% RH with an effective air change rate of 0.5 for 28 days. Periodic monitoring of formaldehyde concentration was conducted using active dinitrophenolhydrazine (DNPH) Waters Sep-Pak cartridges conforming to ISO 16000-3 (2001). The sorbents were analysed using high performance liquid chromatography (HPLC) in accordance with ISO 16000-9 (2007). Measurements were reported to a quantifiable level of 2 µg/m^3. Results are presented in Figure 1.

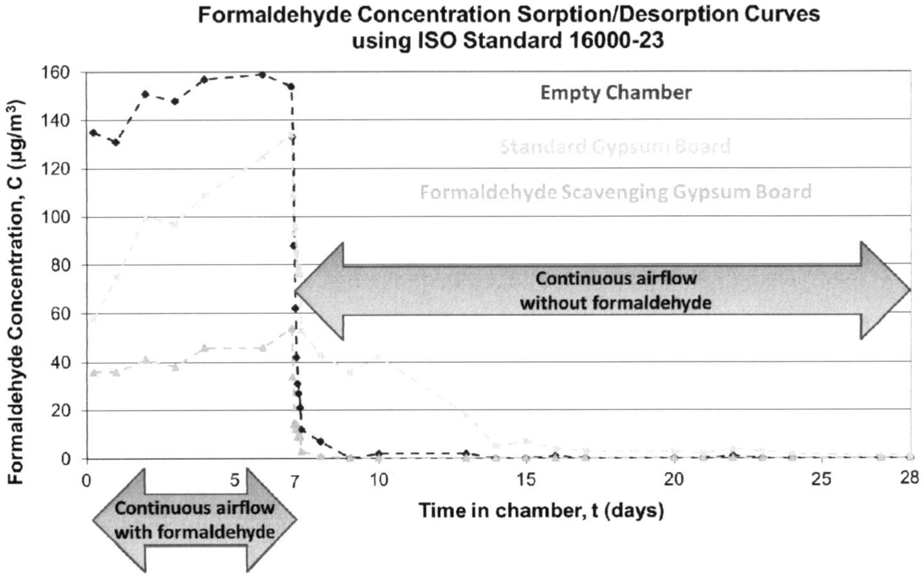

Figure 1. Comparison of standard and formaldehyde scavenging gypsum board performance.

IN SITU SCHOOL PERFORMANCE TESTS AND RESULTS – KALLO, BELGIUM

Two similar elementary school classrooms in Kallo, Belgium were simultaneously monitored for formaldehyde concentration over sixteen weeks (VITO NV 2013). The volume of each room was approximately 140 cubic meters. Prior to collecting samples one classroom's acoustical ceiling was retrofitted. The existing mineral fiber panel ceiling was removed and replaced with a perforated, formaldehyde scavenging gypsum panel ceiling, see Figure 2.

Formaldehyde was collected in each classroom over a 16 week period between February 3rd and May 29th in 2012. The classrooms were populated with approximately 20 children between the ages of six and eight. No new formaldehyde sources were introduced into either classroom during the investigation.

Passive (diffusive) samplers conforming to ISO 16000-4 (2004), strategically placed using ISO 16000-2 (2004), were used to collect formaldehyde in each classroom. The single-use samplers, validated in laboratory and field experiments, contained tape treated with 2, 4 - dinitrophenolhydrazine (DNPH). Sampling periods varied based on building accessibility between 3, 4 and 7 days. The average formaldehyde concentration was calculated for each time period assuming a constant passive sampling rate. The samplers were packaged in individual aluminized pouches that were used to transport the sampler to an accredited laboratory for extraction and analysis using high performance liquid chromatography (HPLC). For sampling periods of 24 hours, the applicable concentration range is 0.003 mg/m^3 to 1 mg/m^3. Classroom formaldehyde concentrations are compared in Figure 3.

Figure 2. Mineral fiber panel ceiling (left) and scavenging gypsum panel ceiling (right).

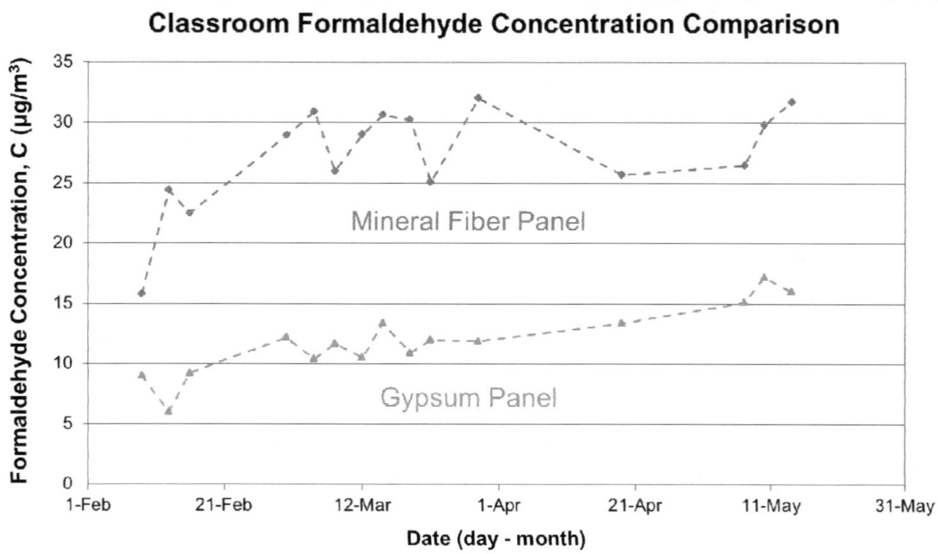

Figure 3. Formaldehyde concentration comparison of two classrooms in Kallo, Belgium

DISCUSSION

The results in Figure 1 clearly demonstrate the difference between standard and formaldehyde scavenging gypsum board under similar environmental chamber conditions. The active material reduced the formaldehyde concentration in the chamber up to 70 percent and completely removed the remaining gas after the seven day exposure. The standard gypsum board initially reduced the formaldehyde concentration by 60 percent, but desorbed the gas within the seven day exposure and throughout the 28 day test period.

The results in Figure 3 clearly demonstrate the difference in formaldehyde concentrations between a classroom using a standard mineral fiber panel and a classroom using a perforated, formaldehyde scavenging gypsum panel as the acoustical ceiling. The formaldehyde concentration in the room using the active gypsum panels was on average 60 percent lower than the standard room.

CONCLUSIONS

The small environmental chamber and in situ performance tests clearly demonstrate the effectiveness of formaldehyde scavenging interior finish materials. The technology combined with good source control and outdoor air ventilation has the ability to reduce formaldehyde concentrations to acceptable levels in field applications.

REFERENCES

Air Quality Sciences (2012). Formaldehyde reduction evaluation testing of a gypsum drywall product. AQS Report: 17284-03, UL Environment, Atlanta, GA.

Godish T, Davis W, and Fu J (2014). *Air Quality, Fifth Edition*. CRC Press. Boca Raton, FL.

Herberger S, Herold M and Ulmer H (2009). Demand controlled ventilation for improved perceived air quality. Proceedings of Healthy Buildings 2009, Syracuse, NY.

IARC (2006). Monographs on the evaluation of carcinogenic risks to humans. International Agency for Research on Cancer, Lyon, France.

ISO (2004). *ISO 16000-2*. Indoor air – Part 2: Sampling strategy for formaldehyde. International Standard Organization, Geneva Switzerland.

ISO (2001). *ISO* 16000-3. Indoor air – Part 3: Determination of formaldehyde and other carbonyl compounds – Active sampling method. International Standard Organization, Geneva, Switzerland.

ISO (2004). *ISO* 16000-4. Indoor air – Part 4: Determination of formaldehyde – Diffusive sampling method. International Standard Organization, Geneva, Switzerland.

ISO (2007). *ISO* 16000-9. Indoor air – Part 9: Determination of the emission of volatile organic compounds from building products and furnishing – Emission test chamber method. International Standard Organization, Geneva, Switzerland.

ISO (2009). *ISO 16000-23*. Indoor Air – Part 23: Performance test for evaluating the reduction of formaldehyde concentrations by sorptive building materials. International Standard Organization, Geneva, Switzerland.

Mullen N, Russell M, Lunden M, and Singer B (2013). Investigation of formaldehyde and acetaldehyde sampling rate and ozone interference for passive deployment of Waters Sep-Pak XPoSure samplers. Lawrence Berkeley National Laboratory, Report LBNL-6386E, Berkeley, CA.

Offermann B (2009). Ventilation and indoor air quality in new homes. California Energy Commission and California Air Resources Board, Document CEC 2009-085.

Sahay-Turner P (2013). Plaster-based material including an agent capable of trapping formaldehyde. BPB Limited, United States Patent - US 8,500,903, Issued August 6, 2013.

Salthammer T, Mentese T, and Marutzky R (2010). Formaldehyde in the indoor environment. *Chemistry Review,* Volume 110, No. 4, American Chemical Society, Washington, DC.

Srebric J (2010). Draft contractor report: Opportunities for green building (GB) rating systems to improve indoor air quality credits and to address changing climatic conditions. US Environmental Protection Agency, Washington, DC.

VITO NV (2013). Analysis of formaldehyde samples. Contractor report 2013/MRG/R/46, Mol, Belgium.

MASS-BASED SIZE DISTRIBUTION OF HOUSE DUST LOADING ON RESIDENTIAL FLOORINGS

Yan MA[1], Yilin TIAN[2], and Andrea R. FERRO[1]

[1]Department of Civil and Environmental Engineering, Clarkson University, Potsdam, New York, USA
[2]Institute for a Sustainable Environment, Clarkson University, Potsdam, New York, USA

Keywords: HVS3-MOUDI, Dust loading, Settled Dust, Collection efficiency, Flooring

SUMMARY

Determination of house dust loading in residences is essential for estimating human exposure to indoor pollutants; however, few studies have investigated surface dust loading for particles with aerodynamic diameters less than 10 µm (inhalable particles). To fill this gap, a new method, HVS3-MOUDI, for determining the mass and size distribution for settled dust was developed and tested in the laboratory with seeded dust on hard wood, cut-pile carpet, and loop carpet flooring. Size-resolved diffusion, bend, and sedimentation losses of the particles in the system were estimated and applied to the sampling efficiency estimates. Sampling efficiency for particles greater than 5 µm were ~90% and higher, depending on the flooring type. However, even after applying the estimated losses in the system, the removal efficiency of the HVS3-MOUDI is very low (<2%) for particles less than 5 µm. The HVS3-MOUDI method was used to collect house dust from 20 houses in St. Lawrence County, NY. The median values of total dust loading on hard floor, cut-pile and loop carpet analyzed for field work were 0.24, 1.1, 3.3 g/m^2, respectively.

INTRODUCTION

House dust contaminants can be associated with a particular size fraction (Fortune et al. 2000). Resuspension of settled dust resuspension from floorings from human activity is a significant factor influencing the inhalation route of human exposure and comprises a large portion of the particle mass inhaled by humans (Ferro et al. 1999; Yakovleva and Hopke 1999). Within the inhalable range, the detachment of larger particles is easier than smaller particles due to the differences in adhesive and external forces on the particles. Thus, the determination of house dust size fraction will contribute to the human exposure to airborne particles in the residential environment, where people spend most of their time (Klepeis et al. 2001). Few studies have investigated surface dust loading in residences for particles with aerodynamic diameters less than 10 µm (inhalable particles), and the size distribution for settled particles in this range have not been reported. In addition, no standard methods are available for determining surface loading and size distribution for settled inhalable particles.

DESIGN OF HVS3-MOUDI METHOD

A vacuum-based method was developed for sampling size-segregated house dust using a combination of the HVS3 sampler and a MSP (Shoreview, MN) MOUDI Model 100-R cascade impactor. A "Y" connector and valve were mounted between the exhaust air duct and the vacuum bag on the HVS3 downstream of the 5 µm size cut cyclone and collector bottle. The MOUDI was used in the non-rotating mode with only 6 stages (size cuts 0.56, 1.0, 1.8, 3.2, 5.6 and 10 µm). An additional vacuum pump was attached to the outlet of MOUDI to maintain the MOUDI's design flow rate of 30 liters per minute (LPM).

METHOD VERIFICATION

The size-resolved sampling efficiency was determined in the laboratory by seeding a known quantity of particles on flooring samples and comparing the seeded amount with the collected amount for each particle size range. Three flooring types were included in the method verification experiments: (a) hard wood, (b) cut-pile carpet, and (c) loop carpet. Two experiments were conducted for each flooring type. Sampling methods followed the standard operation procedure of both the HVS3 and MOUDI (ASTM D5438-05 ; Marple 1991).

The test dust (<45 μm) was collected from homes in the Potsdam, NY area by collecting vacuum bags from standard home vacuum cleaners and pooled as bulk dust. The bulk dust was sieved through a series of mesh screens of 250, 149, 75, 63 and 45 μm on two sieve shakers (Lab-line instruments Inc., Melrose Park, IL and W.S. Tyler, Mentor, OH). The house dust was adulterated with Cab-0-Sil fumed silica (1% w/w) to reduce agglomeration. A Malvern Mastersizer 2000 (Worcestershire, UK) was employed to characterize the size distribution of the test dust. Dust was seeded onto dust-free flooring in a 53 cm × 48 cm × 61 cm acrylic chamber via a steel funnel and mini-vacuum operated under a pressure of 40 psi (Tian et al. 2013). The seeding system was fed to generate a surface dust loading of 2.4 ± 0.26 g/m^2. All flooring pieces were placed in the center of seeding chamber and remaining for five hours after feeding for dust deposition. Before each experiment, the hard wood was wiped using 75% ethyl alcohol, while all carpets were vacuumed by the HVS3 according to adapted method (ASTM D5438-05) to remove the loose fibers. All flooring pieces were stored in a HEPA-filtered conditioning chamber 24-hour before seeding and 40-hour after seeding. The dust on the cut-pile and loop carpets was embedded into the fibers by dragging a roller back and forward on floorings exactly 30 strokes for the rolled cut-pile and rolled loop carpet samples (adapted ASTM F608-89 1989).

As shown in Table 1, size segregated surface dust loadings measured by the HVS3-MOUDI, performed in duplicate, are compared with seeded dust loadings. The sampling efficiency results from high to low are: wood > cut-pile carpet > loop carpet. The reproducibility of the results is good, with relative percent difference < 29%. Compared with previous HVS3 results (Svendsen et al. 2006), which was conducted on worn carpet (86.8% on cut-pile carpet and 89.9% on loop carpet), the sampling efficiency is slightly higher on cut-pile carpet and lower on loop. The sampling efficiency is much higher for particle size larger than 5 μm than particle size less than 5 μm. These results are expected since smaller particles have relatively high adhesive forces. However, the efficiencies have decreased from greater size to smaller size within the range of 1.0 – 5.0 μm.

Table 1: Average Value of Size-Segregated Sampling Efficiency on Different Floorings

Particle Size (μm)	Hard Wood	Cut-pile Carpet	Loop Carpet
0.56 – 1.0	5.6% (18%)	12.3% (3.6%)	5.9% (18%)
1.0 – 1.8	7.1% (4.9%)	15.1% (0%)	7.2% (3%)
1.8 – 3.2	3.4% (26%)	7.8% (3.3%)	6.1% (21%)
3.2 – 5.0	0.7% (0%)	4.1% (14%)	2.5% (25%)
>5.0	96.1% (1.5%)	94.8% (3.8%)	75.6% (1%)

There are several error sources in the sampling efficiency calculations that can be accounted for. First, the seeded dust amount was determined by Mastersizer, detected in optical size, while the HVS3-MOUDI measures the aerodynamic particle size. Thus, a conversion needs to be applied: $D_{ae} = \sqrt{\frac{1}{X}} \times \rho_P{}^{0.5} \times Dp$ (1)

Second, the shape of house dust, which is not spherical, has an error factor (X) on the particle size: $X = 4\pi \times A/P^2$ where A is the surface area and P is perimeter. The shape factor cannot simply be determined through SEM images because of the diverse shape of house dust. The particle density ρ, assumed to be 2.5 g/cm^3, is another error source since the dust is a composite of mineral and biological particles with a range of particle densities.

Particle losses in the HVS3-MOUDI system include diffusion losses, bend losses, and sedimentation losses (Hinds 1999). Losses for the particle size, flow rate, pipe radius, bend angle, horizontal and tubing length were calculated (Table 2). To better demonstrate the performance of HVS3-MOUDI after considering the particle losses, the "removal efficiency" is introduced (

Table 3) and represents the particle dislodged by HVS3-MOUDI vs. total dust on the flooring.

Table 2: Transport Efficiency (Particle Losses) in HVS3-MOUDI System on Three Types of Floorings

Size range (μm)	Hard floor	Cut-pile carpet	Loop carpet
0.56 – 1.0	98.3% (1.7%)	98.4% (1.6%)	98.6% (1.4%)
1.0 – 1.8	94.9% (5.1%)	95.2% (4.8%)	95.8% (4.2%)
1.8 – 3.2	82.5% (17.5%)	84.4% (15.6%)	86.9% (3.1%)
3.2 – 5.0	41.9% (58.1%)	49.6% (50.4%)	59.8% (40.2%)

Table 3: Removal Efficiency of HVS3-MOUDI for Particles from 0.56 to 5.0 μm

Size range (μm)	Wood floor	Cut-pile carpet	Loop carpet
0.56 – 1.0	5.7%	12.5%	6.0%
1.0 – 1.8	7.5%	15.9%	7.5%
1.8 – 3.2	4.2%	9.2%	7.0%
3.2 – 5.0	1.6%	8.3%	4.3%

FIELD SAMPLING USING HVS3-MOUDI METHOD

Twenty homes in St. Lawrence County, NY were selected to collect samples for size-segregated dust loading. During each visit, two or three sampling areas (1-2 m^2) were selected and marked in the high traffic area. The occupants were informed not to clean the floor at least four days ahead of sampling schedule. Replicate field samples accounted for 20% of all field sample quantity. After sampling completion, the substrates were conditioned for 24 hours and weighed to calculate the size-segregated dust loading for size range of 0.56 – 1.0, 1.0 – 1.8, 1.8 – 3.2, 3.2 – 5.0 μm. The dust in the bottle (>5 μm) was sieved through a series of screens: 250, 149, 75 and 45 μm on the sieve shaker. The mass collected from each sieve was measured to the nearest 0.1 mg in order to calculate the dust loading of 5.0 – 45, 45 – 75, 75 – 149 and 149 – 250 μm. The bulk dust from 5 to 45 μm was analyzed by Mastersizer to estimate the dust loading of 5 – 10 μm and 10 – 45 μm.

The median value of total surface dust loading recovered is: loop carpet > cut-pile carpet >

hard floor. Surface dust on carpet is approximately an order of magnitude higher than on hard floor loading for particle less than 149 μm in diameter (Figure 1). The log transformation (base of 10) of original total dust loading passed the Anderson-Darling normality test with a P-value > α-value (0.05). Consequently, the total dust loadings on both hard floor and carpets (including cut-pile and loop carpet) are considered as log-normal distribution. To statistically analyze the reproducibility of HVS-MOUDI method, a two-sample Kolmogorov-Smirnov Test was employed to compare the distribution of dust loading between samples and replicates. There was no statistically significant difference between sample and replicate distributions which indicates that the HVS3-MOUDI method was reproducible during the field work.

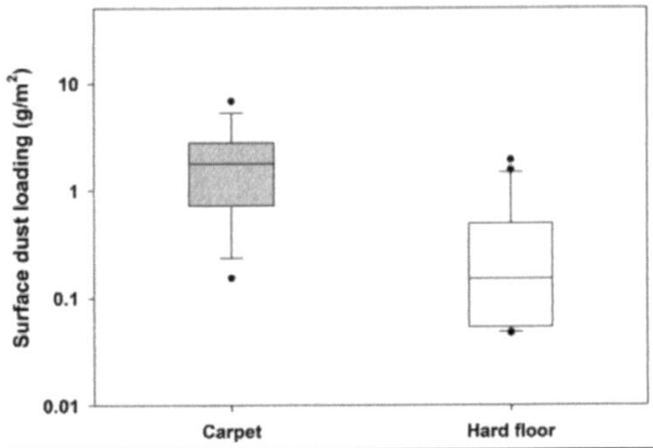

Figure 1: Total surface dust loading <149 μm on carpets and hard floor (n=15).

CONCLUSIONS AND IMPLICATIONS

The collection efficiency of HVS3-MOUDI method on various floorings from high to low is: wood > cut-pile carpet > loop carpet. However, the collection efficiency was very low for particles smaller than 5 μm in diameter. While several design changes are recommended to improve the efficiency of the HVS3-MOUDI system (Ma, 2013), the removal efficiency for a vacuum method is limited by the external force on the particle from the air flow required to overcome the adhesive forces of the particle on the surface. The flooring type significantly affects the dust loading in residential house with carpeted floors having higher floor loading values. To improve accuracy of floor loading estimates, efficiencies specific to flooring type should to be taken into account.

REFERENCES

ASTM D5438-05 (2005) Standard Practice for Collection of Floor Dust for Chemical Analysis.

ASTM F608-89 (1989) Standard Test Method for Evaluation of Carpet Embedded Dirt Removal Effectiveness of Household/Commercial Vacuum Cleaners.

Chuang JC, Gordon SM, Roberts JW, Han W, Ruby MG (1994) Evaluation of HVS3 Sampler for Sampling Polycyclic Aromatic Hydrocarbons and Polychlorinated Biphenyls. EPA/600/R-94/188.

Ferro A, Hildemann L, McBride S, Ott W, P. S (1999) Human exposure to particles due to indoor cleaning activities. Air Pollution VII(487-496)

Fortune CR, Blanchard FT, Ellenson WD (2000) Analysis of Aged In-Home Carpeting to Determine the Distribution of Pesticide Residues between Dust, Carpet, and Pad Compartments. EPA/600/R-00/030

Klepeis NE, Nelson WC, Ott WR, et al. (2001) The National Human Activity Pattern Survey (NHAPS): a resource for assessing exposure to environmental pollutants. Journal of Exposure Analysis and Environmental Epidemiology 11(3):231-252

Tian Y, Sul K, Qian J,Mondal S, and Ferro A.R. (2014) A comparative study of human-induced dust resuspension using a consistent test mechanism.Indoor Air, 24, 592-603 .

Ma, Y. (2013). Mass-based Size Distribution of House Dust Loading on Residential Floorings. Master's thesis, Clarkson University, Potsdam, NY.

Marple VA and Behm SM (1991) A Microorifice Unifom Deposit Impactor (MOUDI): Description, Calibration, and Use. Aerosol Science and Technology 14:434-446

Yakovleva E, Hopke PK (1999) Receptor Modeling Assessment of Particle Total Exposure Assessment Methodology Data. Environmental Science and Technology 33(20)

SAFEGUARDING OUR RESOURCES AND INDOOR AIR – GYPSUM WALLBOARD CLOSED-LOOP RECYCLING IN AN ERA OF INCREASED CHEMICAL TRANSPARENCY

Danielle WAGNER[1], Zhongqi (Joshua) CHENG[1*] Amanda KAMINSKY[2], Sydney MAINSTER[2], Geoffrey Brock[3], Ronald SPALTER[4], and Catherine BOBENHAUSEN[5]

[1]Department of Earth and Environmental Sciences, Brooklyn College of The City University of New York, Brooklyn, New York USA
[2]The Durst Organization, New York, New York USA
[3]Lend Lease (US) Construction LMB Inc., New York, New York USA
[4]Sustainable CUNY, The City University of New York, New York USA
[5]Vidaris, Inc., New York, New York USA

[*]Corresponding email: ZCheng@brooklyn.cuny.edu

Keywords: Wallboard, closed-loop, chemical emissions, stewardship

SUMMARY

New York City generates about 165,000 tons/year of wallboard waste; an estimated 10% of the total installed. In 2014, a pilot project was initiated by The Durst Organization and the Building Product Ecosystems collaborative for closed-loop recycling, engaging other owners/developers, wallboard manufacturers, construction managers, NYC Department of Buildings, recyclers, and academic institutions including the City University of New York in working group meetings and research. At the outset, chemicals of health concern were analyzed to characterize product composition; establish third-party QA/QC for source separation and recycling feedstock; and develop an industry-wide, transparent screening process to safeguard the indoor environment and optimize closed loop recycling. Target analytes include metals with mercury speciated, asbestos, octamethylcyclotetrasiloxane, boric acid, sulfur, volatile organics (including formaldehyde), semi-volatile organics, silica, mineral fibers, and radiation. The metal results are comparable to those reported in one previous EPA study. A comparison with two internal industry limits shows several samples above acceptability limits for arsenic, chromium and copper, while median results were below the acceptability screening limits.

INTRODUCTION

Gypsum wallboard in NYC is a high-volume landfilled waste, on the order of 165,000 tons/year (NYS DEC, 2010), disposed as is (i.e. clean scrap), size-reduced (pulverized or fine), or as recovered screen material (RSM) for landfill cover/grading/shaping. Closed-loop recycling of wallboard scrap into new wallboard manufacture is impeded in this region; as there are not yet economic advantages, no regulations or mandates to recycle gypsum waste, no incentives to source-separate wallboard, and comparatively low cost existing feedstock (especially, flue gas desulfurization (FGD) gypsum). As a result, none of the wallboard manufacturers serving the NYC market currently accept wallboard scrap for reprocessing. Recyclability of waste wallboard (either from demolition or new trim scrap) diminishes upon moisture contamination, compaction and subsequent segregation from other waste, as it falls apart into crumbs and fines. Therefore source separating the wallboard waste stream is required, posing logistical challenges. Staging areas and ample loading dock space is needed to manage multiple containers (as opposed to one commingled container), and the scheduling of pickups must be managed. This has not traditionally been pursued on dense urban construction projects, however the pilot project associated with this study is meant to demonstrate viability and document best practices, resulting in a "clean" stream of wallboard waste for the manufacturer.

Closed-loop recycling is recognized as an ultimate resource management goal. Feeding post-consumer recycled wallboard scrap back into new wallboard manufacturing can diminish new gypsum mining needs, and alleviate dependencies on FGD byproduct from coal-fired power plants, which are increasingly being supplanted by alternatives like natural gas in some locations. Also, in addition to depleting scarce landfill capacity, wallboard disposal is often attributed to emissions of hydrogen sulfide, which can result in bronchial constriction in asthmatic individuals (Wang, 2011). It is hoped that eventual streamlined transportation and processing infrastructure for this closed loop recycling can minimize greenhouse gas emissions. For these and overriding product stewardship reasons, a pilot closed-loop recycling project was initiated in New York City by The Durst Organization and key wallboard manufacturers. Ultimately, the objective is to establish an efficient, standardized screening process that the gypsum manufacturing industry can agree upon, which meets public health objectives and safeguards the indoor environment.

The first critical step is to characterize and verify product composition to begin to evaluate the feasibility of recycling gypsum wallboard. An extensive list of potential contaminants of concern was developed based on input from the major wallboard manufacturers and gypsum processors, a review of relevant studies and reports, an investigation of current gypsum recycling initiatives elsewhere, and the tabulation of regulatory and internal industry limits. The intent is to chemically characterize wallboard used in New York City, share this data among all stakeholders, consider the impacts, if any, of those chemicals on health and material integrity during installation, use, and ultimate re-manufacturing in closed-loop recycling. Then, standardized, published protocols and procedures can be established to support the closed-loop recycling process.

METHODOLOGIES

The wallboard samples tested in this study include new wallboard specified for several large upcoming construction projects, as received directly from the manufacturer or jobsite, and wallboard removed from demolition sites in New York City. The new wallboard analyzed in this study thus far is mined natural gypsum, produced by two domestic gypsum wallboard manufacturers. Additional new gypsum products are being added for study. The source of gypsum for the demolition wallboard samples in this study may be variable, but it is likely to be mined (natural) gypsum as the age of the wallboard corresponds to a time when natural gypsum was the main feedstock for the industry.

Upon receipt, the samples have been logged in, dried and pulverized in preparation for gypsum core analysis. Trace Metal Analysis was completed using EPA Method 3052 (microwave-assisted acid digestion) followed by ICP-MS (EPA Method 6020) at Brooklyn College. Reference materials for soils SRM-2585, SRM-2587 and SRM-2710a (from the National Institute of Standards and References) were used for external quality control. The detection limits are generally 0.1 mg/kg for most metals. Twenty-five samples (with duplicates for each) were analyzed, consisting of 17 from demolition wallboard and 8 from new wallboard.

RESULTS AND DISCUSSION

In Table 1, the maximum and median values of trace metals are compared to results from one published EPA study, and to internal acceptability limits established by a recycling facility and two domestic gypsum wallboard manufacturers. Total trace metal concentrations are below the internal acceptability limits with the exception of the highest report values for arsenic, chromium, and copper. The median values are below limits for all the metals.

Compared to the wallboard manufacturer internal acceptability limits, one of the new domestic-sourced wallboard samples exceeded the copper limit, and two exceeded the chromium limit. One demolition wallboard sample exceeded both the arsenic and copper limit (the same wallboard sample) and an additional demolition wallboard sample exceeded the copper limit.

Table 1
Priority Metals Analysis of Gypsum Wallboard and Comparison with Measured and Internal Acceptability Limits
(mg/kg or ppm)

	GYPSUM WALLBOARD SAMPLES						SPECIFICATIONS	
	New Domestic-Sourced		Demolition		EPA Study[1]		Internal Acceptability Limits for Post-Consumer Recycled Gypsum Wallboard	
Analyte	Maximum	Median	Maximum	Median	Maximum	Median	Recycling Facility[2]	Wallboard Manufacturer[3]
Arsenic	2.6	1.6	<u>10.5</u>	1.3	N/A	N/A	120	<u>10</u>
Beryllium	0.1	0.1	0.1	0.1	N/A	N/A	1.06	5
Cadmium	0.2	0.2	<0.1	<0.1	N/A	N/A	42.5	1
Chromium	<u>30.8</u>	3.2	4.2	2.0	4.3	2.4	10,625	<u>20</u>
Cobalt	3.4	2.3	4.1	1.3	3.0	0.9	425.0	5
Copper	<u>29</u>	8.6	<u>101</u>	7.3	6.2	1.9	2,125	<u>20</u>
Lead	15.1	2.2	6.3	1.9	3.5	2.6	1,062	30
Mercury	<0.1	<0.1	0.3	0.2	<u>2.1</u>	0.1	<u>1.3</u>	<u>2</u>
Nickel	17	6.1	8.2	4.3	5.4	1.9	836	25
Selenium	7.4	0.5	0.3	0.1	8.9	3.2	2,125	25
Strontium	477	376	908	442	1,130	568	20,000	N/A
Thallium	0.1	0.1	<0.1	<0.1	N/A	N/A	2,125	1
Uranium	0.6	0.2	0.5	0.2	N/A	N/A	2,167	5
Zinc	40	9.3	42.1	8.7	10.1	7.2	50	250

In this Table, numbers are underlined when they exceed an internal acceptability limit. The corresponding limit is also underlined.

N/A: not available
[1]U.S. Environmental Protection Agency (2009), Drywall Sampling Analysis. Correspondence dated May 7, 2009 including elemental analysis of four gypsum wallboard panels purchased from stores in Edison, NJ to compare to products manufactured in China.
[2](Confidential)(2014). Trace Metals Acceptability Speciation for Recycled Gypsum Material Delivered to Processor for Gypsum Wallboard Reuse.
[3]USG Corporation (2014). Guide for Using Post-Consumer Recycled Gypsum Panels in the Manufacture of Gypsum Panels, suggested gypsum guideline based on natural soil levels (also cited by a second manufacturer)..

The results presented in Table 1 are consistent with those routinely measured in U.S.gypsum wallboard waste samples as part of New West Gypsum Recycling's QA/QC program. Note that the maximum mercury level reported in the EPA study is above the referenced internal acceptability limits. Several studies report higher levels of mercury in FGD gypsum compared to natural mined gypsum (EPA, 2014b). Additional testing is underway as part of the current study, to incorporate results for FGD gypsum.

The potential for success in developing a widely-adopted wallboard trim scrap screening process depends in part on the adoption of standard protocols and regulations for gypsum testing. To the best of our knowledge, there is no consensus on maximum allowable chemical concentrations in processed wallboard, hence the variable limits presented in Table 1, which are in general based on typical ranges in natural gypsum and soil.

In order to create a transparent and consistent working model for the wallboard recycling industry, a system of protocols for sampling and analysis of potential chemicals of concern in gypsum board are warranted. The tests developed for analyzing solid materials such as soils and sediments must be modified for gypsum, the main component of wallboard, which is a substantially different matrix. The development of standard tests would then necessitate gypsum standards for calibration and quantification. A forum of certified analytical laboratories would be beneficial for developing these tests and standard materials. Systematic tracking of changes in wallboard chemistry would be warranted, with updates made to the list of analytes, to ensure that the protocol reflects changing composition of the source material, e.g., biocides, nanoparticles, or formaldehyde scavengers. In addition, to make the closed-loop recycling process more complete, a similar analytical framework for the paper core and/or other surface materials would be advantageous.

The chemical characterization continues, with an expanded supply of wallboard from several different manufacturers, and the remainder of the analytical scope to evaluate both the gypsum core and the residual post-consumer paper, including asbestos, octamethylcyclotetrasiloxane (D4), boric acid, elemental sulfur, mercury (total and speciated), volatile and semi-volatile organics, formaldehyde, crystalline silica, mineral fibers, and radiation, being completed.

Simultaneously, the next step in the pilot is beginning on construction sites in NYC. This is a "bench scale" feasibility trial via various project teams to establish logistics. It consists of reclaiming wallboard trim scrap through rare onsite source separation of the trim scrap from a number of large construction sites in NYC, delivering it to a nearby wallboard manufacturer, and performing similar QA/QC chemical composition analysis of this incoming feedstock, before incorporating it into new product. Any form of waste commingling tends to bias recyclability to lower grade uses, e.g., landfill topping or agricultural amendment (with its own set of constraints). In this pilot, trim scrap is strategically segregated in separate containers on-site. Realistically, the scrap output must be substantial and regular enough to justify the added space needed for multiple containers. The result, intact, homogeneous loads of trim scrap, is key to maintaining a high quality aggregated regional stream of recovered 'clean' gypsum for wallboard manufacturers.

CONCLUSIONS

This pilot study is part of an initiative by the Building Product Ecosystems Evolving Wallboard Working Group, convened by The Durst Organization to investigate opportunities for optimizing the health of our very broadly used wallboard systems, and feasibility of closed-loop wallboard recycling in New York City. While further testing is needed to characterize the composition of gypsum wallboard, to date trace metal analysis of the samples yielded low concentrations in most samples. Maximum levels measured for arsenic, chromium and copper were above internal acceptability limits, however, the median levels did not exceed the limits. Through development of a system of standardized protocols appropriate for gypsum wallboard waste, a feasible and health-protective method for screening post-consumer gypsum wallboard can be established, reclaiming this resource while addressing potential public health concerns.

ACKNOWLEDGEMENT

The authors gratefully acknowledge the financial support of The Durst Organization and the cooperation of all the members of the Building Product Ecosystems – Evolving Wallboard Working Group.

REFERENCES

New York State Department of Environmental Conservation (2010). Construction and Demolition (C&D) Debris Detailed Composition Analysis for 2010.

U.S. Environmental Protection Agency (2009). Drywall Sampling Analysis. Correspondence dated May 7, 2009 including elemental analysis of four gypsum wallboard panels purchased from stores in Edison, NJ to compare to products manufactured in China.

U.S. Environmental Protection Agency (2014a). Best Management Practices to Prevent and Control Hydrogen Sulfide and Reduced Sulfur Compound Emissions at Landfills that Dispose of Gypsum Drywall. U.S. EPA Contract Number EP-W-09-004.

U.S. Environmental Protection Agency (2014b). Coal Combustion Residual Beneficial Use Evaluation: Fly Ash Concrete and FGD Gypsum Wallboard.

Wang, P., G. Zhang, T. Wondimu, B. Ross, R. Wang (2011). Hydrogen Sulfide and Asthma. Exp. Physiol. 2011 Sept; 96(9):847-852.

INVESTIGATION OF PARTICLE-MEDIATED GAS-PHASE TRANSPORT OF PHTHALATES FROM VINYL FLOORING

Yaoxing WU[1], Eric VEJERANO[1], Cong LIU[2], Steven S. COX[1], Mingjie XIE[1], Linsey C. MARR[1], and John C. LITTLE[1*]

[1]Department of Civil and Environmental Engineering, Virginia Tech, Blacksburg, VA 24061, USA
[2]Department of Building Science, School of Architecture, Tsinghua University, Beijing 100084, China

[*]Corresponding email: jcl@vt.edu

Keywords: DEHP, phthalates, aerosol, particle

SUMMARY

Semivolatile organic compounds (SVOCs), such as phthalates, are ubiquitous in the indoor environment. Airborne particles play an important role in the distribution of SVOCs. The interactions between airborne particles and SVOCs increase the total airborne SVOC concentrations and consequently human exposure. In this study, a specially-designed chamber with an optimized velocity field was used to investigate the particle-mediated mass transfer and to measure the particle/air partition coefficient (K_P) of di-2-ethylhexyl phthalate (DEHP). The time for DEHP-particle equilibrium was determined to be between 4 and 10 min. As the particles were introduced at concentrations of 50-110 μg/m^3, the gas-phase DEHP concentrations decreased by 30-50%, and the total concentrations increased by a factor of 2 to 4. The observations are consistent with model predictions. The measured K_P was 0.034 ± 0.006 m^3/μg. This work reveals quantifiable influence of particles on overall SVOC dynamics and allows for a more comprehensive exposure assessment.

INTRODUCTION

Phthalates, including di-(2-ethylhexyl) phthalate (DEHP), are semivolatile organic compounds (SVOCs) that are widely used in polyvinyl chloride products as plasticizers and in personal care products as solvents. They can be released from sources as diverse as vinyl flooring, electrical insulation, toys, and nail polish. Potential adverse health effects of phthalates include decreased fertility, birth defects, hormone disruption, and reproductive malformations (Jaakkola et al., 2004, Heudorf et al., 2007). SVOC partitioning onto particles tends to result in higher concentrations than the gas phase concentrations due to low vapor pressure (Liang et al., 1997, Weschler and Nazaroff, 2008). These SVOC-containing particles can remain in the air or settle and resuspend, allowing for inhalation and ingestion (Weschler and Nazaroff, 2010). It is therefore important to understand the mechanisms governing the particle-air interactions in the indoor environment. Mechanistic models have been developed to investigate the influence of particles on the mass transfer of indoor airborne DEHP (Weschler and Nazaroff, 2008, Liu et al., 2010, Liu et al., 2012), and showed that particle-enhanced transport could significantly increase emission rates and human exposure by a factor of 4 to 10. This particle-induced enhancement of emissions has been experimentally validated by Benning et al. (2013) using a small-scale circular chamber. However, one of the substantial difficulties associated with chamber tests resulting from extensive SVOC sorption to chamber surfaces is that an extended amount of time (more than a month) is required for the system to reach steady state. The objective of this study is to develop a new system for determining the impact of particles on SVOC dynamics in indoor environments and to measure the particle/air partition coefficient (K_P) of di-2-ethylhexyl phthalate (DEHP). The methodologies will be a useful tool for the research focusing on particle-SVOC interactions.

METHODOLOGIES

Emission chamber. A special semi-circular chamber (1.2 L) designed and tested by Liang and Xu (2014) was used for this study. With the new chamber design, the velocity field inside the chamber was improved and the impact of the sink effect was reduced, which substantially reduces the time required to reach steady state. Two pieces of homogeneous vinyl flooring (VF) containing 20 wt% of DEHP were placed into the chamber as emission source. The setup was kept in an incubator and maintained at a constant temperature of 25°C.

Particle generation. A solution containing 50 mg/L ammonium sulfate in ultrapure water was prepared for particle generation. The inorganic particles with a median diameter of 50-70 nm were generated using an aerosol atomizer (TSI 3076) at pressure of 35 psi. The particle-containing air stream went through a diffusion dryer and a neutralizer before being introduced into the chamber. The distribution and concentration was measured using a Scanning Mobility Particle Sizer system (SMPS, TSI 3936NL). The total suspended particulate (TSP) mass concentration was between 50-110 μg/m^3.

DEHP sampling. The sampling train has been improved to minimize the sink effect along the sampling path (Figure 1). Tenax TA-packed sorbent tubes were used to collect effluent air samples. Two sampling ports were drilled through the chamber wall and the sorbent tubes were directly attached to the chamber through the sampling port using both straight and male Swagelok fittings. One of the sampling ports was fitted with a PTFE membrane filter (Pall Life Science, 0.2 μm pore size) sandwiched between two PTFE shims. The sorbent tubes downstream of the filter collected only gas-phase DEHP while the other without the filter collected DEHP in both particle and gas phases.

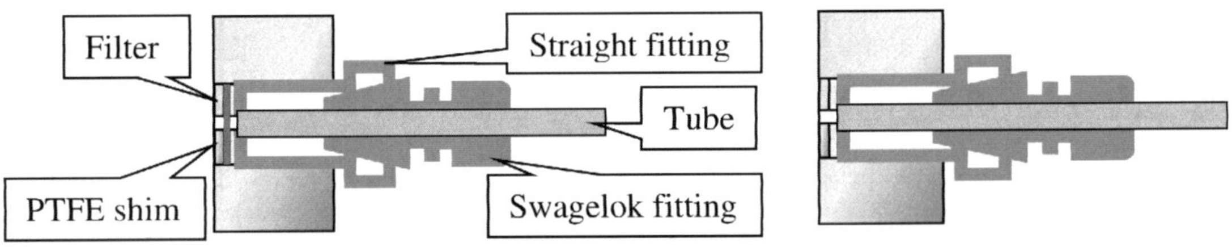

Figure 1. Experimental sampling setup for the collection of gas-phase (left) and gas- plus particle-phase (right) DEHP.

Experiments were conducted at different flow rates to investigate the gas-particle sorption kinetics. The flow rate through the chamber ranged from 120 to 600 mL/min, corresponding to a residence time of 2-10 min. The sampling rate was fixed at 50 mL/min. The system was allowed to reach steady state at the designated flow rate prior to the experiments. Particles were introduced into the chamber for a total of 7 times, each time over a 4 hour period. The particle loss rate in the chamber was monitored downstream of the central ventilation port and was determined to be less than 5%. When each experiment with particles ended, the particle generator was stopped and clean air was reintroduced to the chamber. A new filter was also installed to replace the old one in order to prevent particle resuspension. The gas samples collected before and after particle introduction were compared to ensure that the chamber remained at steady state. All samples were analyzed by thermal desorption coupled with gas chromatography/flame ionization detector (TD-GC/FID). K_P was calculated as q_P / (y × TSP), where q_P is the particle-phase DEHP concentration, and y is the gas-phase concentration,

RESULTS AND DISCUSSION

The new system was confirmed to reach steady state within 48 hrs, which was much faster than the 50 days recorded by Benning et al. (2013). Figure 2a shows the DEHP concentrations measured in gas phase and in both gas and particle phase at a flow rate of 300-600 mL/min (4 experiments) and 120 mL/min (3 experiments). The positive and negative peaks correspond to the introduction of particles in the chamber. It is fairly clear that the DEHP concentrations in particle plus gas phase increase with decreasing flow rates through the chamber, and that the decrease in gas-phase DEHP concentrations become more significant at lower flow rates. The difference might be related to the time scales for DEHP sorption equilibrium with particles. The low residence time (less than 4 min) at flow rate of 300-600 mL/min might not be long enough for the transport or the partition of DEHP to a suspended particle. This finding was consistent with the estimated time (between 1 min to 1h) required to establish equilibrium between gas-phase DEHP and particles (Weschler and Nazaroff, 2008). As the residence time increased to 10 min, the gas-phase DEHP concentrations decreased by 30-50%, and the DEHP concentration in particle plus gas phase increased by a factor of 2 to 4 (Figure 2a). The observation provides direct evidence that DEHP in the gas phase was rapidly depleted as the particles were introduced, which in turn increased the concentration gradient in the boundary layer and resulted in enhanced mass transfer of DEHP from the emission source. In comparison, the decrease in gas-phase concentration in the research by Benning et al. (2013) was not as significant as in this study. This was probably due to the difference in convective mass transfer coefficient (h_m). The value of h_m was calculated as 1.2×10^{-4} m/s in the new chamber, which is lower compared to Benning's case (4×10^{-4} m/s). Thus the resupply of the gas-phase DEHP from the emission source in the new chamber may not be as efficient as in the original chamber. The measured K_P was 0.034 ± 0.006 $m^3/\mu g$ ($n = 3$), in good agreement with Benning et al. (2013).

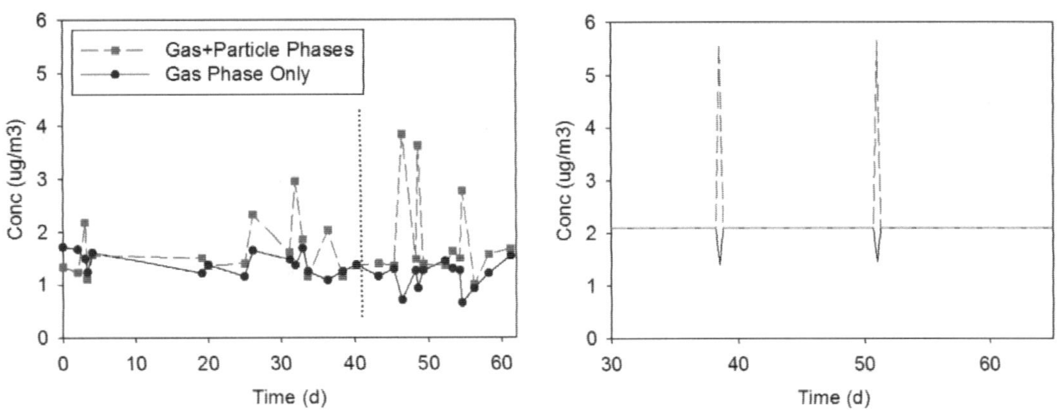

(a) (b)

Figure 2. a) Measured gas-phase DEHP concentrations and DEHP concentration in gas plus particle phases in the chamber exhaust air at 120 – 600 mL/min; b) Model predicted gas-phase DEHP concentrations and DEHP concentration in gas plus particle phases in the chamber exhaust air at 120 mL/min. Each peak coincides with the introduction of particles into the chamber over the period of 4h.

Figure 2b shows the predicted DEHP concentrations in gas phase and in both gas and particle phase at a flow rate of 120 mL/min and particle concentration 100 $\mu g/m^3$ using Liu's model (2012). There was good agreement for the overall tendency, but the concentration levels predicted by the model were higher than those measured in the chamber test. The difference is probably due to the change of convective mass transfer coefficient h_m, as the model predicts that the particle mediation can also increase h_m between gas phase and the surface. The value of h_m is one of key parameters that controls the extent of particle-enhanced mass transfer and this requires further study.

CONCLUSIONS

A specially-designed chamber was used to investigate the particle-air interactions and measure the particle-air partition coefficient of DEHP. The results provide clear evidence of particle mediated gas-phase transport, which can significantly increase human exposure to DEHP in the indoor environment. Further development of appropriate methods for determination of h_m, one of the key parameters that control the extent of particle-enhanced mass transfer, will improve the understanding of the particle-air interaction mechanism.

ACKNOWLEDGEMENT

Financial support was provided by the United States National Science Foundation (CBET 1066802 and CBET 1335722).

REFERENCES

Benning, J. L., Z. Liu, A. Tiwari, et al. (2013). Characterizing Gas-Particle Interactions of Phthalate Plasticizer Emitted from Vinyl Flooring. *Environmental Science & Technology*, **47**, 2696-2703.

Heudorf, U., V. Mersch-Sundermann and J. Angerer (2007). Phthalates: Toxicology and exposure. *International Journal of Hygiene and Environmental Health*, **210**, 623-634.

Jaakkola, J. J., H. Parise, V. Kislitsin, et al. (2004). Asthma, wheezing, and allergies in Russian schoolchildren in relation to new surface materials in the home. *Am J Public Health*, **94**, 560-562.

Liang, C., J. F. Pankow, J. R. Odum, et al. (1997). Gas/Particle Partitioning of Semivolatile Organic Compounds To Model Inorganic, Organic, and Ambient Smog Aerosols. *Environmental Science & Technology*, **31**, 3086-3092.

Liang, Y. and Y. Xu (2014). Improved Method for Measuring and Characterizing Phthalate Emissions from Building Materials and Its Application to Exposure Assessment. *Environmental Science & Technology*, **48**, 4475-4484.

Liu, C., G. C. Morrison and Y. Zhang (2012). Role of aerosols in enhancing SVOC flux between air and indoor surfaces and its influence on exposure. *Atmospheric Environment*, **55**, 347-356.

Liu, C., B. Zhao and Y. Zhang (2010). The influence of aerosol dynamics on indoor exposure to airborne DEHP. *Atmospheric Environment*, **44**, 1952-1959.

Weschler, C. J. and W. W. Nazaroff (2008). Semivolatile organic compounds in indoor environments. *Atmospheric Environment*, **42**, 9018-9040.

Weschler, C. J. and W. W. Nazaroff (2010). SVOC partitioning between the gas phase and settled dust indoors. *Atmospheric Environment*, **44**, 3609-3620.

RESUSPENSION OF DEPOSITED PARTICLES INDUCED BY JETS AT DIFFERENT SURFACE DUST LOADS

Xiong Mei, Tengfei (Tim) Zhang* and Shugang Wang

School of Civil Engineering, Dalian University of Technology, Dalian, China

**Corresponding email: tzhang@dlut.edu.cn*

Keywords: Particles, Resuspension, Jet impingement, Surface dust load, Experiment

SUMMARY

Surface dust load on indoor surfaces can range over several orders of magnitude. At a high surface dust load, particles have more chances of depositing in multiple layer structures and agglomerating together. We experimentally examined particle resuspension of the Arizona test dusts (ATDs) at different surface dust loads after an intensive pulsed-jet impingement. The results reveal that the deposited particles with a low surface dust load (less than 3 g/m^2 in the tested cases) are more difficult to resuspend than those with a high surface dust load. The re-deposition of resuspended dusts soon after a jet impingement is evident at a high surface dust load.

INTRODUCTION

Deposited particles can resuspend from indoor surfaces due to human activities, animal activities (Ferro et al., 2004), or the HVAC system operation (Krauter et al., 2007). The dust loads on surfaces range from 0.1 g/m^2 to 100 g/m^2 (Moritz et al. 2001; Johnson et al. 2009). The broad range of surface dust loads indicates there are two different particle deposition structures. One is the monolayer deposition, in which all particles are directly against the substrate surface; the other is multilayer deposition, which forms a layer-by-layer structure among particles and only one layer is against the substrate surface. It is reported that different surface dust loads have a great impact on particle resuspension from indoor surfaces (Boor et al., 2013).

To examine particle resuspension at different surface dust loads, pulsed jet-induced resuspension tests were conducted. Particle removal mass and airborne particle numbers were measured to evaluate the resuspension strength. In addition, the resuspension process was recorded by a high-speed camera to provide more detailed information of the process.

METHODOLOGIES

Deposited particles were obtained via natural deposition process in a glass enclosure, where the Arizona test dusts (ATDs) were fed to a solid aerosol generator (type:

SAG410; Topas, Germany) and then discharged into the enclosure for natural deposition by the laden gas. The pulsed jet is generated by pressurized nitrogen gas via a nozzle, as shown in Figure 1(a). The test section is connected to a wind tunnel for removing the background particles suspended in the air before each test, as shown in Figure 1(b).

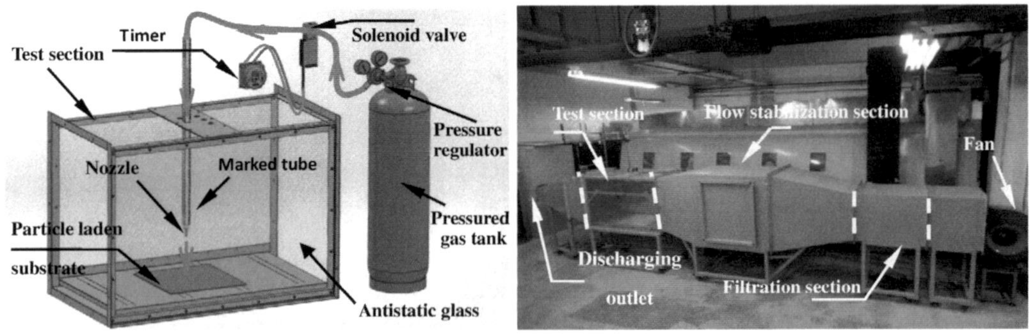

(a) (b)

Figure 1. The pulsed jet device: (a) schematics; (b) picture of the wind tunnel connecting with the test section.

Some major parameters of the experimental tests were listed in Table 1. After a jet impingement, the particle removal mass on the substrate surface was weighed by a digital balance (type: DV215CD; Ohaus, USA) and the airborne particle numbers were counted by a particle counter (type: 9310-02; TSI, USA).

Table 1. Some major parameters of the experimental test.

Item	Value	Item	Value
Particle type	Arizona Test Dust, ATD	**Particle diameters**	1.0-10.0 μm
Jet duration time, t_{jet}	0.2 s	**Jet speed, V_{jet}**	20 m/s
Substrate material	Stainless steel	**Nozzle diameter, D**	6.5 mm
Surface dust load, ρ_s	0.46, 0.71, 1.44, 2.31, 2.98, 3.71, 4.78, 5.34 g/m^2	**Distance of nozzle to substrate, H**	5 mm

Meanwhile, the resuspension process was recorded by a high-speed camera (type: acA2000-340km; Basler, Germany) operated at 200 frames per second. The recording process, which lasts 1 s, was accompanied by a laser sector sheet (wave length: 532 nm, green light) across the central plane of the impingement jet. Basic experimental test procedure is:

(1) Feed ATD particles to the aerosol generator and discharge particles into the glass enclosure for natural deposition of approximately one hour.

(2) Turn on the wind-tunnel fan to remove the background airborne particles. Then close both ends of the test section, shut the fan and load the particle laden plate with deposited particles on it into the test section.

(3) Release the jet, record the particle resuspension process with the digital camera, and count the airborne particles.

(4) Take the particle laden plate out and weigh the remaining mass. Each test was repeated five times to ensure repeatability of the measurement data.

RESULTS AND DISCUSSION

Both the counted total particle number (for size from 1 to 10 μm) and the total removal mass on the substrate surface after a jet impingement increase with the surface dust load, as shown in Figure 2. The errors bars in Figure 2(a) are the standard deviations of counted particles in five repeating tests. Figure 2(b) highlights a dot-dash line for the increase rate of removal mass with surface dust load when the load is smaller than 2.5 g/m^2. Once the surface dust load is greater than approximately 3 g/m^2, the solid line deviates farther away from the dashed line, which means more vigorous particle resuspension. This may be due to more chances of monolayer deposition at a low surface dust load and thus the larger adhesion force between the dusts and substrate surface. As shown in Figure 3, for a surface dust load greater than 3 g/m^2, dusts have large chances to pile together to form multilayer deposition. The multilayer deposition result particle agglomeration with an apparent re-deposition of coarse particles soon after the jet impingement, as shown in Figure 4. However, at a low surface dust load, the particle re-deposition is not evident.

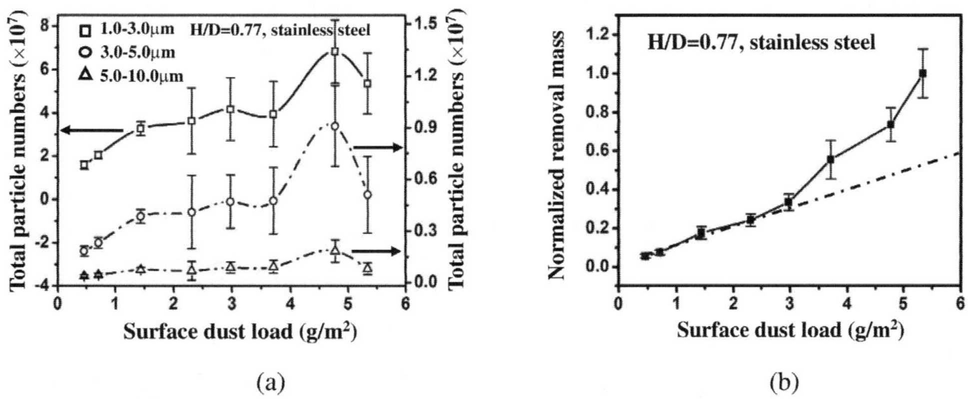

Figure 2. Impacts of surface dust load (g/m^2) to: (a) sampled total particle numbers; (b) total removal mass normalized by 0.01932 mg.

CONCLUSIONS

Experiments on resuspension of deposited particles at different surface dust loads from the stainless steel substrate surface after an intensive pulsed-jet impingement were conducted. Results reveal that particles are more difficult to resuspend at small surface dust loads (more chances in monolayer deposition) due to the outstanding particle-substrate adhesion force. Once deposited dusts are piled together at a large surface dust load, the particle resuspension is easier. The piled particles may agglomerate, which result an evident re-deposition of coarse particles soon after the jet impingement.

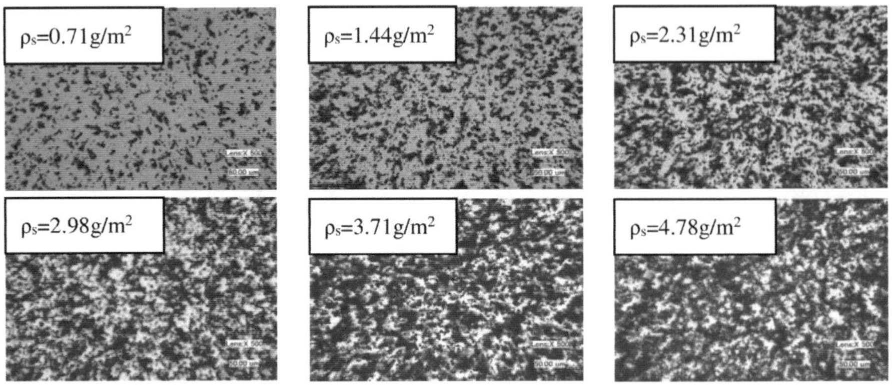

Figure 3. Pictures of deposited particles at different surface dust loads provided by a 3D super depth digital microscope, where dark dots represent particles.

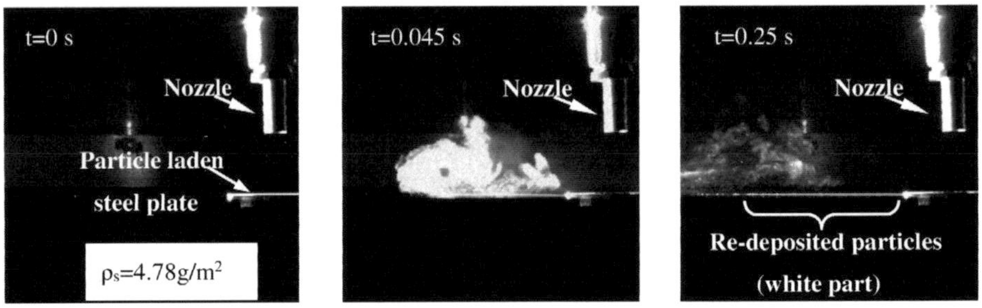

Figure 4. Re-deposition of resuspended particles after a jet impingement.

ACKNOWLEDGEMENT

This work was funded by the National Natural Science Foundation of China through grant 51278077.

REFERENCES

Boor BE, Siegel JA and Novoselac A (2013) Monolayer and Multilayer Particle Deposits on Hard Surfaces: Literature Review and Implications for Particle Resuspension in the Indoor Environment. *Aerosol Science and Technology*, **47**, 831-847

Ferro AR, Kopperud RJ, Hildemann LM (2004) Source Strengths for Indoor Human Activities that Resuspend Particulate Matter. *Environmental Science and Technology*, **38**, 1759-1764.

Johnson DL, Hunt A, Griffith DA, et al. (2009) Geographic Patterns of Non-Carpeted Floor Dust Loading in Syracuse, New York (USA) Homes. *Environ. Geochem. Health*, **31**, 353–363.

Krauter P, Biermann A (2007) Reaerosolization of Fluidized Spores in Ventilation Systems. *Applied and Environmental Microbiology*, **Apr**, 2165-2172

Moritz M, Peters H and Ruden H (2001) Methods to Determine the Dust Loads of Air Ducts in HVAC Systems. *VDI-Berichte*, **1603**, 153–161 (in German).

EXPERIMENTAL STUDY OF ACTIVE FLOWS AND PASSIVE MATERIALS FOR INDOOR AIR QUALITY CONTROL

Denise MAUNEY[1], Christella SUWONGSO[2], Miles ABARR[3], Wil W. SRUBAR[1] and Lupita D. MONTOYA[1] *

[1]Department of Civil, Environmental and Architectural Engineering, University of Colorado, Boulder, Colorado USA
[2]Environmental Engineering Program, University of Colorado, Boulder, Colorado USA
[3]Department of Mechanical Engineering, University of Colorado, Boulder, Colorado USA

*Corresponding email: Lupita.Montoya@colorado.edu

Keywords: Synthetic jets, Airflows, NO_2, Photocatalytic materials, Indoor air quality

SUMMARY

This experimental study investigated the use of active flows and passive materials for the removal of gaseous pollutants indoors. In this study, a commercial electromagnetic synthetic jet actuator (SJA) was characterized to determine its flow patterns and sphere of influence within a room. This SJA was then used in chamber experiments to study the enhanced removal of nitrogen dioxide by a mortar substrate coated with titanium dioxide. Results showed that the SJA uses a small power input to impact the airflow over 1 meter downstream from the jet exit and has significant dynamic pressure per unit energy, which may be useful for flow and air quality control. The removal of nitrogen dioxide by the passive material used was found to be driven by both the surface velocity and the quality of the flow regime generated by the SJA.

INTRODUCTION

Many air pollutants have been found at higher concentrations inside buildings compared to the outside air (Barraza et al., 2014; Escobedo et al., 2014) and sometimes at levels harmful to humans. People in industrialized countries are also spending the majority of their time indoors, exacerbating this exposure. Indoor air quality in buildings is usually controlled through the Heating, ventilating and air conditioning (HVAC) system, which can be energy intensive. In fact, HVAC systems account for about 25% of the energy used by buildings, and about 40% of the energy used in the U.S. is associated with buildings.

Task/ambient conditioning (TAC) systems have been proposed to provide comfortable, healthy indoor air quality while reducing the total building energy consumption (Fanger, 2000; Bauman and Arens, 1996). They range from individually controlled under-floor air supply systems to those where the supply air diffuser is on the desk (Webster et al., 2002). In the study reported here, a commercial electromagnetic SJA was characterized to determine its flow patterns and sphere of influence within a room. A comparison was also made to a computer fan with a power level on the same order as the SJA. This SJA was then used in chamber experiments to study the enhanced removal of nitrogen dioxide by a mortar substrate coated with titanium dioxide. The purpose of this study was to investigate the potential application of this commercial SJA (active flow device) and the mortar substrate coated with titanium dioxide (passive material) for indoor air quality control applications.

METHODOLOGIES

The synthetic jet actuator characterized in this study was a SynJet ZFlow 87 LED Cooler (Nuventix, Austin, TX) and it was operated at the high performance setting to maximize its range. This actuator was designed to be driven by a 12VDC power supply, and it has onboard controls to convert the load to a 100-120Hz sine wave. The current draw from this device was 38mA; at 12 V, this resulted in a power consumption of 0.46 W. Figure 1 shows a diagram of this device.

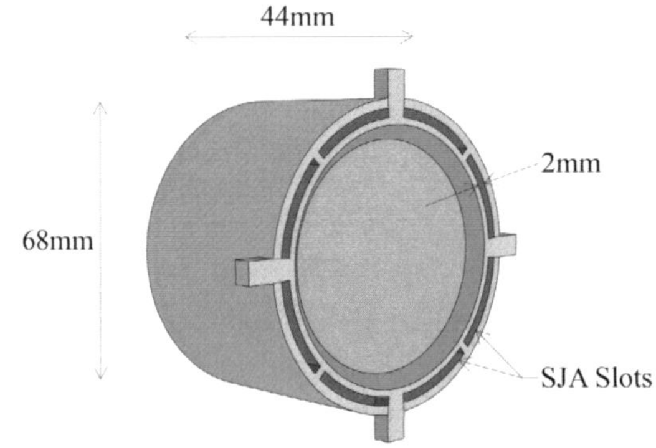

Figure 1. 3D Computer Aided Design (CAD) of Nuventix SynJet Zflow 87 LED Cooler.

A hot-wire anemometer (Climomaster Model 6531, Kanomax, Andover, NJ) was used in these experiments to measure and store air velocities. This model measures air velocities between 0.10 and 30.0 m/s with an accuracy of +/- 2% or 0.0125 m/s, whichever is greater. It has a minimum sampling time of 1 second. The experimental set up is shown in Figure 2.

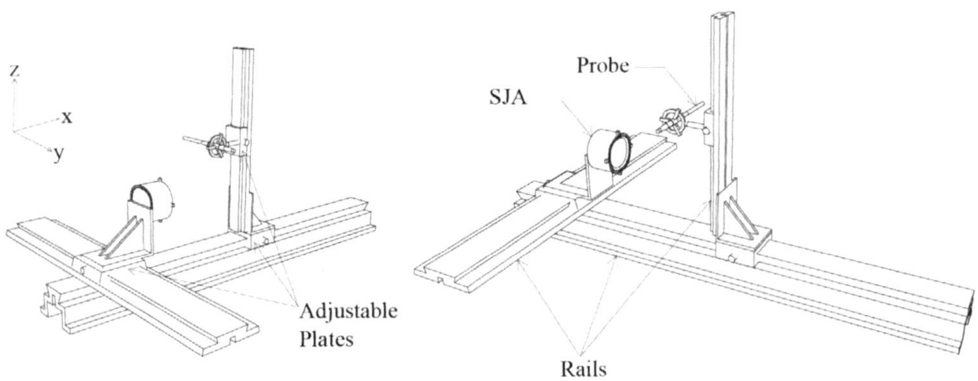

Figure 2. Schematic of experimental setup.

RESULTS AND DISCUSSION

The axisymmetry of the SJA was evaluated at 105 mm, 205 mm, 305 mm, and 500 mm from the jet exit. Figure 3 shows y and z velocity profiles measured at the x=105mm location and both profiles show general agreement. Figure 4 shows the peak velocity of the SJA along the x axis, showing that it can impact the air flow at a distance over 1 meter from the SJA exit.

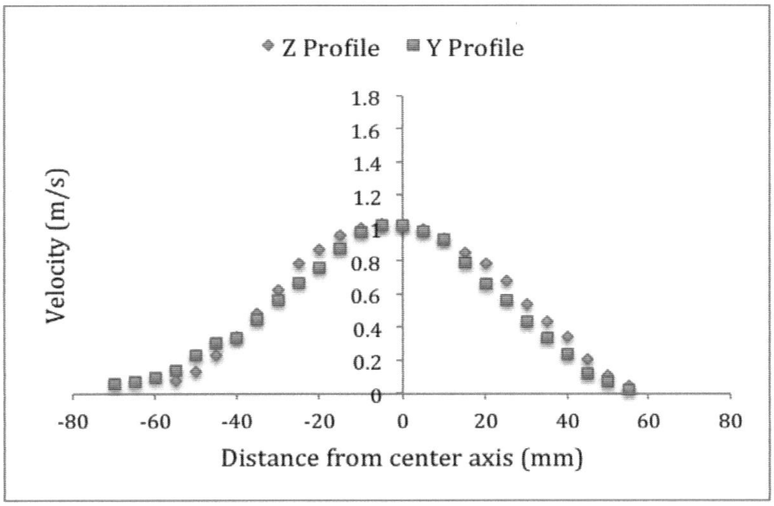

Figure 3. Y and z velocity profiles for SJA at x=105 mm

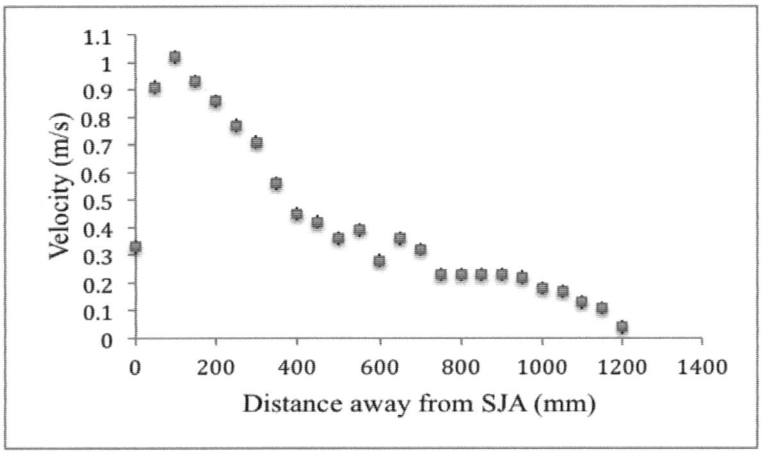

Figure 4. SJA peak velocity measured at various distances from the SJA exit (x direction).

Table 1 shows the removal rates of NO_2 measured at different distances from the active flow device (SJA or the small fan). Results of the NO_2 removal experiments indicated that the optimum distance between the SJA and the surface of the titanium dioxide reactor was approximately 315mm, and a surface velocity of approximately 0.1 m/s. The removal rate due to the small fan was found to be equivalent to the SJA removal rate when the surface velocities were not equal, but the flow regimes had similar shape.

Table 1. Removal rates measured at various distances from the passive material.

Active Flow Device	Distance from passive material (mm)	Removal rate (min-1)
None	-	0.0007
SJA	115	0.0011
SJA	315	0.0013
SJA	510	0.0011
Fan	315	0.0011

CONCLUSIONS

These experiments showed that the SJA uses a small power input to impact the airflow far downstream (over 1m) from the jet exit and this impact could be potentially useful for indoor air quality control applications. This initial study also showed that nitrogen dioxide removal rate by titanium dioxide was driven by both the surface velocity and the quality of the flow regime generated by the SJA.

ACKNOWLEDGEMENT

This work was partially funded by the University of Colorado Innovation Seed Grant program.

REFERENCES

Barraza, F., Jorquera, H., Valdivia, G., and Montoya, L. D. (2014). Indoor PM2.5 in Santiago, Chile, spring 2012: Source apportionment and outdoor contributions. *Atmospheric Environment*, **94**, 692–700.

Bauman, F. S. P. E., and Arens, E. A. P. D. (1996). Task/ambient conditioning systems: Engineering and application guidelines. Cent. Environ. Des. Res. Univ. Calif. Berkeley.

Escobedo, L. E., Champion, W. M., Li, N., and Montoya, L. D. (2014). Indoor air quality in Latino homes in Boulder, Colorado. *Atmospheric Environment*, **92**, 69–75.

Fanger, P. O. (2000). Indoor Air Quality in the 21st Century: Search for Excellence. *Indoor Air*, **10**(2), 68–73.

Webster, T., Bauman, F., and Reese, J. (2002). Underfloor air distribution: thermal stratification. *Cent. Built Environ.* [Online]. Available: http://escholarship.org/uc/item/9145t9gz.pdf.

ASSESSING WILDFIRE EXPOSURE IN HOMES NEAR WILDFIRES

By Ernest R. CRUTCHER III and Heidie K. CRUTCHER-BETTES

Microlab Northwest, LLC, Redmond Washington USA

Corresponding email: russ.c@microlabnw.com

Keywords: WILDFIRE PARTICLES, Microscopy, environmental particles, tapelifts, assemblage analysis

SUMMARY

More than 2000 samples from over 200 homes exposed to 10 different fires in 5 different states representing 9 different biomes were examined to assess the extent of smoke exposure in the home from those fires. There are over 45 common combustion sources impacting most homes that have nothing to do with wildfire. Criteria for exposure has been developed that is capable of detecting smoke from a specific wildfire in the presence of these interferences using tapelifts. Surfaces that have not been cleaned since the alleged exposure are preferred. The surfaces were selected from horizontal surfaces non-systematically. No attempt was made to collect black deposits since much of the debris from wildfires is not black. Tapelifts using frosted 3M Magic tape and assemblage analysis have proven to be very effective.

INTRODCUTION

The smoke from wildfires travels hundreds of miles, often in a concentrated plume. A "smoky" odor is easily detectable even at those distances. Does that mean that the homes at that distance are "damaged" by the smoke from that fire? What are the signatures of an exposure that may be detrimental? How do we know that the combustion products in this home are predominantly from the wildfire in question? These are the questions that need to be answered by the insurance company to justify a claim and by the home owner to place a claim. The purpose of this study was to address the question of exposure.

From an analytical point of view the first question was: what do the particles from a wildfire look like? The answer to that question begins by looking at the fuel. The terrain that has been burned by a wildfire is inhabited by the burnt hulks of trees and shrubs. Most of the bark and leaves of these plants are gone. Grasses and low growing herbs tend to be burned completely. So the particles in the plume tend to be the charred remains of the leaves and bark of the trees and shrubs and the plant fragments of grasses and herbs. The woody portions also burn to some extent but this is different than a fire in a fireplace or pellet stove. The fuel in a fireplace and pellet stove is predominantly the woody portion of the plant and not leaves and bark. Grasses and herbaceous plants are not normally burned in a fireplace. The combustion products from leaves and bark are very different than the combustion products from woody tissue.

The combustion related particles from a wildfire are dominated by ash (the inorganic residue) and char (the coked residue of the fuel). Fire retardant tends to be dropped in large quantities. A significant fraction forms a fine aerosol that drifts with the plume from the fire and other wind currents. The updraft from the fire carries soil with it. This soil is exposed to the temperature of the fire and tends to be oxidized to an orange color by the traces of iron in the soil. This burnt soil and particles of the fire retardant aerosol become additional markers. All of these markers create a signature for the fire that could then be identified and quantified in the home suspected of being impacted by the fire. They have to discriminate these particles from all of the other combustion products present in every home and from all of the other particles that could be mistaken for ash or char. The quantification procedure has to be independent of the time since the last cleaning of the home and provide some information on primary and secondary exposure. The new normal in the environment of the home often included a charred forest or field nearby. How might the primary plume be different than the subsequent "dust" blown in from that direction or tracked in on boots or by pets?

METHODOLOGY

Twenty or more homes were analyzed for each of ten different fires in five different states. That was more than two hundred samples per fire for a total sample set of over two thousand individual samples. Biomes included Arizona chaparral, California pine forest, California chaparral, California steppe, Colorado montane, New Mexico spruce and pine forest, Washington pine forest, and Washington steppe. The analysis was done without the analyst knowing the proximity or direction from the fire for the samples being examined. The time post-fire varied from a few days to as much as eighteen months after the exposure. The time post-fire was not under our control because much of this work was done in response to insurance claims.

Sample Collection

Tapelifts were collected from homes near specific wildfire to assess the amount of combustion products that could be assigned to that wildfire. Typically ten tapelifts were collected from each home. A few of these were collected on an exterior surface. The tapelifts in the home were from windowsills, countertops, book shelves, and other horizontal surfaces. The tape was applied directly to the surface once and removed. Each tapelift was three quarters of an inch wide and about four inches long. The tape used was 3M Scotch Brand Magic Tape, a "frosted" tape. The frosted plastic film was removed by an acetone wash. This left the particles fixed in position on the slide in the adhesive that had a refractive index of about 1.48. A mounting medium with a refractive index of about 1.49 was then place on the particles in the adhesive and a coverslip was added. The result was a high quality optical mount that could be examined in detail under a light microscope.

Microscope Configuration

The microscope was a polarizing compound scope using circular polarized light with a phase condenser and a fiber optic ring light fitted to the barrel of the 10X and 20X objectives to provide simultaneous reflected darkfield illumination. The transmitted or reflected illumination could be easily adjusted or cancelled to optimize specific characteristics of interest in a field of view. The phase turret in the condenser was adjusted to optimize contrast and provide refractive index information using oblique transmitted light.

Analysis

Calcium oxalate phytoliths have a characteristic range of shapes depending on the plant and the part of the plant. These phytoliths tend to retain their shape but change their optical properties depending on the heat to which they were exposed. The phytoliths exposed to fire can be differentiated from those released by the natural degradation of dead plant material by these thermally induced changes. Calcium oxalate phytoliths are concentrated in the leaves and bark of a plant. Since leaves and bark are a predominant fuel in wildfires thermally modified calcium oxalate phytoliths are common in the concentrated plume of a wildfire and are unique (see http://www.microlabgallery.com/PyrolyzedCaOxFile.aspx). Their presence in the samples is a required part of the wildfire signature. These particles are rarely present in homes not exposed to wildfire based on our history of the analysis of tens of thousands of dust samples from homes. In the outdoor environment these particles quickly weather into the soil and so are absent from the track-in or windblown dust within a few weeks of the fire. The structure of the cells in grasses, leaves of various plants, bark, and wood are sufficient to identify the type of source plant. These cell structures are often retained in the inorganic ash and in the char collected by tapelifts in fire exposed homes (see http://www.microlabgallery.com/CharredLeafFile.aspx). Wipe samples destroy this structure. Images of some of these combustion related particles from different types of fires are available on the web (see http://www.microlabgallery.com/FireParticlesFile.aspx).

The degree of exposure is based on the presence of the assemblage of particles characteristic of the specific wildfire. This included char and ash of multiple plant types typical of the burn area, thermally modified calcium oxalate phytoliths from those plants, fire retardant residues, and burnt soil particles. The relative amount of fire retardant and charred soil varys depending on the conditions and activities at the time of exposure. The samples are quantified as trace, low, moderate, or high exposure based on the number of scans across the ¾ inch tape required to detect sufficient members of the assemblage to identify the surface as exposed to wildfire. If the assemblage is not present then the types of black particles are listed and the sample is assigned a "Not Wildfire" designation. If ten or more scans are required before the assemblage could be identified then the tapelift is assigned a "trace" level. If more than two scans but less than ten are required then a "low" level is assigned. If nearly every scan contains the assemblage then a "moderate" level is assigned. If just a few fields of view are required then a "high" level is assigned. This method of quantification is independent of the other particle background.

RESULTS AND DISCUSSION

All of the homes in these samples were within at least fifteen miles of a wildfire. The residents in these homes were acutely aware of the smoke from the fire. The time of sampling ranged from a few months after the fire to as much as eighteen months after the fire. In every biome homes were found that contained the assemblage of particles that marked that wildfire. That included homes that were sampled eighteen months post fire. Not all of the homes suspected of exposure showed evidence of exposure. That included some homes sampled within a month of the fire. Most of samples were collected six months or more after the fire. That is the result of a delay in filing a claim, the insurance company contacting a sampling agent, and the sampling agent getting into the field to collect the sample. Delays of a year or more were primarily due to the late filing of a claim. The filing of claims was often due to the presence of a Public Adjuster in a particular neighborhood so the samples tended to come in groups from different proximities and locations relative to the fire. The table below is an example of the results by home for the Wenatchee Complex fire in Washington State. The first samples were six months after the fire was controlled. Sixty-two homes showed no presence of the particle assemblage related to the fire. Sixty-one homes showed varying levels of exposure. Two showed high levels of exposure. Exposure was not obvious to the inspector at the home, which was why samples were collected.

Months Post Fire	6	7	8	9	10	11	Total
Not Significant	15	18	17	6	4	2	62
Low Exposure	4	8	5	5	0	3	25
Moderate Exposure	4	12	10	3	0	5	34
High Exposure	0	2	0	0	0	0	2

CONCLUSION

Debris from a specific fire can be identified by the assemblage of characteristic particles created by the plants that fuel the fire. These markers are absent in homes not exposed to wildfire. Quantification of exposure based on the set of particles characteristic of that fire provides information independent of total particle loading. Exposure to a primary plume from the wildfire can be assessed months after exposure. Tapelifts using 3M Scotch Brand Magic Tape (frosted tape) are the preferred sampling procedure. It retains the fine structure of the ash and char critical to the identification of the plant sources. Keys to the identification of the plant markers are available on the internet (see http://www.microlabgallery.com/FireParticlesFile.aspx).

ACKNOWLEDGEMENT

I would like to acknowledge Russ Nassof of Risknomics, LLC for most of the samples from the Southwest U.S. and for his support in this study. I would also like to acknowledge Barbara Trenary of Trenary and Associates, LLC and Brian Hunt of Alternative Technology, LLC for most of the samples from Washington State fires.

AIR POLLUTION EXPOSURE MODEL FOR INDIVIDUALS (EMI) IN HEALTH STUDIES: EVALUATION FOR AMBIENT PM2.5

Michael S BREEN,[1,*] Thomas LONG,[1] Bradley SCHULTZ,[1] Ronald WILLIAMS,[1] Jennifer RICHMOND-BRYANT[1]

[1]Office of Research and Development, U.S. Environmental Protection Agency, Research Triangle Park, North Carolina, USA

[*]Corresponding email: breen.michael@epa.gov

Keywords: Exposure modeling, Fine particulate matter (PM), Residential infiltration

SUMMARY

Health studies of fine particulate matter ($PM_{2.5}$) often use outdoor concentrations as exposure surrogates, which fail to account for indoor attenuation of ambient $PM_{2.5}$ and time indoors. To address these limitations, we developed an air pollution exposure model for individuals (EMI) in health studies, which predicts five tiers of exposure metrics from questionnaires and outdoor concentrations. We linked an air exchange rate (AER) model to a mass-balance $PM_{2.5}$ infiltration model to predict residential AER (Tier 1), infiltration factors (Tier 2), indoor concentrations (Tier 3), exposure factors (Tier 4), and exposures (Tier 5). Individual predictions were compared to 591 daily measurements across 31 homes in North Carolina. Relative differences were 39% (median; Tier 1), 18% (Tier 2), 20% (Tier 3), 18% (Tier 4), and 20% (Tier 5). Using outdoor concentrations as surrogates for exposures (Tier 5), differences of 86% (median) were substantially larger. Thus, EMI could improve exposures for health studies.

INTRODUCTION

Health effect studies have shown associations between ambient (i.e., outdoor-generated) fine particulate matter ($PM_{2.5}$) and adverse cardiopulmonary effects. Due to cost and participant burden, many of these health studies have used central-site monitors for exposure assessments. However, there can be a discrepancy between $PM_{2.5}$ concentrations measured at central-site ambient monitors and personal exposures due to house-to-house and temporal variations in indoor infiltration (i.e., attenuation) of ambient $PM_{2.5}$, and variations in the time people spend in different indoor and outdoor locations. The impact of these discrepancies (i.e., exposure errors) can increase uncertainty and introduce bias in health risk estimates. To improve exposure assessments, we developed the Exposure Model for Individuals (EMI) for application in epidemiologic studies.

The EMI predicts personal exposures based on outdoor concentrations, meteorology, questionnaire information (e.g., building characteristics, occupant behavior related to building operation) and time-locations. This study describes a critical aspect of EMI: the development and evaluation of five tiers of individual-level exposure metrics for ambient $PM_{2.5}$, which includes three exposure metrics related to homes (air exchange rates, infiltration factors, indoor concentrations) and two exposure metrics related to personal exposures (personal exposure factors, exposures).

There are several benefits of the EMI. First, EMI is complementary to population-level exposure models (e.g., SHEDS, APEX; Burke 2001). EMI predicts individual exposures for the specific people in the health study by using individual-level input data (e.g., questionnaires, time-locations from diaries or global positioning system devices) from each participant enrolled in the health study. Population models predict distributions of population variability in exposures for demographic groups (e.g., school-age children), not specific individuals, by using population-level input databases from other studies (e.g., U.S. Census, Consolidated Human Activity Database) (Burke 2001). Thus, EMI is designed for epidemiologic studies with individual health outcomes, and population exposure models are appropriate for epidemiologic studies with health outcomes summed by numbers of people across a region. Second, EMI can predict multiple tiers of exposure metrics. Using these various exposure metrics with different levels of complexity in the epidemiologic analysis can help determine the benefit of more complex exposure metrics to optimize the design of health studies.

METHODOLOGIES

PM panel study

For model evaluation, we used daily 24-hour average measurements from a one-year panel study that investigated personal, residential, and central-site $PM_{2.5}$ for participants living in central North Carolina (Williams 2003). Participants and their homes were monitored for seven consecutive days in each of four consecutive seasons (summer 2000 – spring 2001). For $PM_{2.5}$ mass measurements, the Harvard Impactor (HI) was used for residential indoor and outdoor monitoring, and for a central-site monitor. A personal exposure monitor (PEM) was used for personal $PM_{2.5}$ sampling. Filter-based sulfate measurements for the HI and PEM filters were determined using X-ray fluorescence. Residential air exchange rate (AER), which is the airflow into and out of a home, was measured using a perfluorocarbon tracer method. In this paper, we evaluated EMI with 591 daily measurements across 31 detached homes, which includes a low to moderate socioeconomic status cohort of 27 participants with controlled hypertension; and a moderate socioeconomic status cohort of 4 participants with implanted cardiac defibrillators.

For EMI input data, we obtained home building characteristics, weather, and time-locations. Daily questionnaires collected occupant behavior related to building operation (e.g., opening windows). Participants recorded their locations every 15 min. Daily home indoor temperatures, and central-site outdoor temperatures and wind speeds were measured.

Measured tiers of exposure metrics

We determined five tiers of measured exposure metrics (daily 24-hour averages) for each participant. For Tier 1, we used the measured residential AER. For Tier 2, we used indoor/outdoor ratios of sulfate (tracer of ambient $PM_{2.5}$) to determine residential $PM_{2.5}$ infiltration (i.e., attenuation) factors (F_{inf_home}). For Tier 3, we determined residential indoor concentrations of ambient $PM_{2.5}$ (C_{in}) using the steady-state equation defined as

$$C_{in} = F_{inf\_home}\, C_{out\_home} \tag{1}$$

where C_{out_home} is the measured residential outdoor $PM_{2.5}$ concentration. For Tier 4, we used personal/outdoor ratios of sulfate (tracer of ambient $PM_{2.5}$) to determine personal exposure factors of ambient $PM_{2.5}$ (F_{pex}), which accounts for the attenuation of ambient $PM_{2.5}$ and the participant's daily time within different indoor locations (e.g., home, buildings other than home, inside vehicles). For Tier 5, we calculated exposures of ambient $PM_{2.5}$ as defined by

$$E = F_{pex}\, C_{out\_home} \quad (2)$$

Modeled tiers of exposure metrics

We developed five tiers of modeled exposure metrics (daily 24-hour averages) that were time-matched with the five tiers of measured exposure metrics. For Tier 1, residential AER, which is an important mechanism for indoor infiltration of ambient air pollutants, were predicted from building characteristics, indoor-outdoor temperature differences, and wind speeds using a mechanistic AER model, called LBLX (Breen 2010). The LBLX model accounts for leakage, which is the airflow through unintentional openings in a building envelope, and natural ventilation, which is the airflow through controlled openings in a building envelope (e.g., open windows).

For Tier 2, modeled residential $PM_{2.5}$ infiltration factors were predicted with a steady-state mass-balance infiltration model described by

$$\hat{F}_{inf\_home} = P\, AER/(AER + k_r) \quad (3)$$

where P is the penetration coefficient, and k_r is the indoor removal rate of $PM_{2.5}$. We performed a leave-one-home-out jackknife method to estimate P and k_r using the least-squares method. For Tier 3, modeled residential indoor concentrations of ambient $PM_{2.5}$ were predicted from measured outdoor concentrations at central-site ($C_{out\_central}$) defined as

$$\hat{C}_{in,central} = \hat{F}_{inf\_home}\, C_{out\_central} \quad (4)$$

For Tier 4, modeled personal exposure factors of ambient $PM_{2.5}$ were predicted by

$$\hat{F}_{pex} = f_{in\_home}\, \hat{F}_{inf\_home} + (f_{in\_work} + f_{in\_other})\hat{F}_{inf\_other\_bldg} + f_{in\_vehicle}\, \hat{F}_{inf\_vehicle} + (f_{out\_home} + f_{out\_work} + f_{out\_other}) \quad (5)$$

where f is the fraction of time spent in the seven microenvironments (indoors: in_home, in_work, in_other; outdoors: out_home, out_work, out_other; inside vehicles: in_vehicle). The $\hat{F}_{inf\_other\_bldg}$ and $\hat{F}_{inf\_vehicle}$ are the $PM_{2.5}$ infiltration factors for buildings other than homes and for vehicles, respectively. We set $\hat{F}_{inf\_other\_bldg}$ to 0.64 based on the mean value of three literature-reported $PM_{2.5}$ infiltration factors for offices, stores, and restaurants (Burke 2001). We set $\hat{F}_{inf\_vehicle}$ to 0.44 based on the literature-reported $PM_{2.5}$ infiltration factor for cars (Ott, 2008). For Tier 5, modeled exposures were estimated by

$$\hat{E}_{central} = \hat{F}_{pex}\, C_{out\_central} \quad (6)$$

We used a leave-one-home-out cross validation to evaluate each modeled exposure metric.

Exposure metrics based on Central-site PM measurements

Health studies often use measurements from central-site monitors as exposure surrogates without considering the infiltration factors that attenuate ambient $PM_{2.5}$ within indoor microenvironments. To compare this exposure surrogate to our modeled exposure metrics, we also estimated residential indoor concentrations (Tier 3) and exposures (Tier 5) using the central-site measurements without accounting for indoor infiltration (i.e., set all infiltration factors to 1).

RESULTS AND DISCUSSION

For the residential infiltration model (Tier 2), the jackknife estimates for P and k_r were 0.84 and 0.21 h^{-1}, respectively. Based on the cross validation, we compared individual predictions with individual measurements. We first examined the exposure metrics related to the homes (Tiers 1-3). For Tier 1 (AER), relative differences (|measured – modeled|/measured) were 39% (median). For Tier 2 (residential infiltration factors) relative differences were 18% (median). For Tier 3 (residential ambient indoor concentrations), relative differences were 20% (median). Using central-site concentrations as surrogates for residential indoor concentrations, relative differences of 72% (median) were substantially higher. This demonstrates the potentially large exposure error in health studies that use central-site measurements as exposure surrogates without considering the residential infiltration factors that attenuate ambient $PM_{2.5}$ within indoor microenvironments.

We then examined the exposure metrics related to personal exposures (Tiers 4-5). For Tier 4 (personal exposure factors), relative differences were 18% (median). For Tier 5 (exposures), relative differences were 20% (median). Using central-site concentrations as surrogates for exposures, relative differences of 86% (median) were substantially larger. This shows the potential for substantial exposure errors in health studies that use central-site monitoring as exposure surrogates without considering infiltration factors of buildings and vehicles, and participant time in different locations.

CONCLUSIONS

This study demonstrates the ability to predict multiple tiers of individual-level exposure metrics with different levels of complexity and information needs. To improve exposure assessments, we showed the importance of accounting for: (1) daily house-specific infiltration of ambient $PM_{2.5}$ based on a mechanistic AER model linked to a mass balance infiltration model, and (2) time spent in different indoor and outdoor locations. The large differences of the Tier 3 (indoor concentrations) and Tier 5 (exposures) model predictions when compared to the central site measurements suggests that the EMI is substantially more accurate. This implies that use of the EMI can improve the accuracy of risk estimates in health studies that currently rely on central-site monitoring for exposure surrogates.

REFERENCES

Breen MS, Breen M, Williams R, et al. (2010). Predicting residential air exchange rates from questionnaires and meteorology: model evaluation in central North Carolina. *Environ Sci Technol,* **44**, 9349-9356.

Burke J, Zufall M, Ozkaynak H. (2001). A population exposure model for particulate matter: case study results for PM2.5 in Philadelphia, PA. *J Expo Sci Environ Epidemiol,* **11**, 470-489.

Ott W, Klepeis N, Switzer P. (2008). Air change rates of motor vehicles and in-vehicle pollutant concentrations from secondhand smoke. *J Expo Sci Environ Epidemiol,* **18**, 312-325.

Williams R, Suggs J, Rea A, et al. (2003). The Research Triangle Park particulate matter panel study: PM mass concentration relationships. *Atmos Environ,* **37**, 5349-5363.

IMPROVING THE INDOOR AIR QUALITY BY USING A SURFACE EMISSIONS TRAP FOR EXPOSURE REDUCTION

Lennart LARSSON[1,*], Pawel MARKOWICZ[1], and Johan MATTSSON[2]

[1]Department of Laboratory Medicine, Lund University, Sölvegatan 23, 223-62 Lund, Sweden
[2]cTrap Ltd, Scheelevägen 15, 223 62 Lund, Sweden

[*]Corresponding email: lennart.larsson@med.lu.se

Keywords: Indoor air purification, Volatile organic compounds, Building dampness, School environment

SUMMARY

A prototype of a surface emissions trap, a new device for reducing emissions of volatile organic compounds and particulate matter from surfaces while allowing evaporation of moisture, was used to improve the indoor air quality of a school building with elevated air concentrations of 2-ethyl-1-hexanol. A clear improvement of the perceived air quality was noticed a few days after the trap had been attached on the PVC flooring. In parallel, decreased air concentrations of 2-ethyl-1-hexanol were found as well as a linear increase of the amounts of the same compound adsorbed on the installed trap as observed up to 13 months after installation. This study suggests that a surface emissions trap may represent a fast and efficient means of restoring the indoor air quality in a building e.g. after water damage leading to irritating and potentially harmful emissions from building material surfaces indoors.

INTRODUCTION

Unsatisfactory indoor air quality may be due to emissions e.g. of volatile organic compounds (VOCs) from moist building construction parts. Such emissions are formed as a result of the effect of water on a particular material (Andersen et al., 2011, Claeson et al., 2007). For example, 2-ethylhexanol and n-butanol may be formed due to alkaline hydrolysis of PVC flooring and glue applied on a concrete floor (Sjoberg and Ramnas, 2007). Both long- and short-term exposure to VOCs emissions can result in eye, nose, and throat irritation, allergic reactions, headaches, fatigue (Fiedler et al., 2005) etc.

Here we describe the performance of a new device, a prototype surface emissions trap, developed for reducing emissions from building material surfaces indoors such as VOCs and particle-bound mycotoxins, the goal being to prevent such emissions from reaching individuals residing inside the building. The trap is a cloth, a laminate with two protective sheets of nonwoven polyester fabric surrounding an adsorption layer and a hydrophilic polymer sheet. The device is applied directly at the source of the emissions (floor, walls, ceiling, over cavities etc).

METHODOLOGIES

A school built in the 1970:s, with a long history of complaints on air quality among the pupils and the school staff, was studied. Increased ventilation and use of air purifiers in the rooms had not resulted in any major improvements in the perceived air quality. A surface emissions trap prototype was attached on the existing PVC flooring, by using a double sided adhesive tape, in a small office room (room A, 9 m^2). The trap, a laminate with one adsorption and one polymer layer, adsorbs only from the adsorption layer side; hence the device was applied with the adsorption layer facing the floor. After a few hours the clerk reported a considerably improved air quality. It was then decided to evaluate in more detail the result of a trap installation a larger room (room B, 30 m^2); over the device was laid a laminate flooring. Because of the unsatisfactory perceived air quality room B was not in use since several months. The ventilation in both rooms was 2-2.5 air exchanges per hour as measured by the HVAC system. Air samples as well as samples of the trap cloth were taken from the floor (immediately replaced with new pieces of the device) at different time periods for measuring the amounts of 2-ethyl-1-hexanol in the air and adsorbed on the cloth, respectively. Tenax TA tubes were used for passive air samplings, as recommended by Sunesson et al. (Sunesson et al., 2002), for 1 week and sent to IVL (Stockholm, Sweden) for thermal desorption and GC-MS analysis. The trap cloth pieces (approximately 3 cm^2, n=4) were extracted by using dichloromethane following GC-MS, as described elsewhere (Markowicz and Larsson, 2012). In total, approximately 500 m^2 of the device was installed in the school building directly on the PVC flooring and subsequently covered by a laminate flooring.

RESULTS AND DISCUSSION

The results are illustrated in Table 1. Decreased air concentrations of 2-ethyl-1-hexanol, from 6-7 $\mu g/m^3$ to 2 $\mu g/m^3$, were found two months after the trap had been applied; the concentrations of 2-ethyl-1-hexanol in the installed cloth rose from 0 (unused) to 280.3 $\mu g/g$ after 13 months of use. Air concentrations of total VOCs (TVOCs) were 58-127 $\mu g/m^3$.

Table 1. Air concentrations of total VOCs (TVOCs) and 2-ethyl-1-hexanol in room B, and amounts of 2-ethyl-1-hexanol extracted from the trap cloth, at different time periods after installation of the device.

Time	TVOCs [$\mu g/m^3$]	2-Ethyl-1-hexanol	
		Air concentrations [$\mu g/m^3$]	Adsorbed amounts [$\mu g/g$]
before applying cTrap	127	6	0
1 week after application	95	7	17.0
2 months after application	72	2	68.6
8 months after application	119	2	150.6
13 months after application	58	2	280.3

This study shows that attaching a surface emissions trap to a surface from which moisture-driven emissions are spread can prevent humans from being exposed to such emissions. In the studied room (room B) air concentrations of 2-ethyl-1-hexanol were 6-7 $\mu g/m^3$ before the trap had been installed suggesting emissions from the floor. 2-Ethyl-1-hexanol in indoor air is frequently related to chemical (Björk et al., 2003) or microbial (Nalli et al., 2006) degradation of plasticizers in a PVC flooring and/or in the glue used to attach a PVC flooring on concrete. Such cases, where the concrete has not been enough dry when gluing or when moisture, in the absence of a moisture barrier, has been allowed to diffuse from the ground up to the floor, are common in Sweden (Norbäck et al., 2000). In the air of room B, apart from 2-ethyl-1-hexanol, we also found traces of phenol, n-decane and n-hexanal, as well as 5 unidentified aliphatic hydrocarbons (data not shown), indicating that the bad indoor air quality perceived by the pupils and staff may have resulted from a combination of different VOCs.

The fact that the air concentration of 2-ethyl-1-hexanol did not decrease immediately after the trap had been applied on the floor may be due to absorption of the compound in the ceiling or walls, diffusing from these surfaces back into the air after the floor emissions had been stopped. This so-called sink effect, where building materials act as buffers for VOCs, has been described (Meininghaus et al., 2000, Zhang et al., 2002). It has even been suggested that VOCs may diffuse through walls from one room to another and that different barriers may be used to prevent such diffusion (Meininghaus et al., 2000). The odor problems in the studied school disappeared shortly (a few days) after application of the device and rooms A and B could again be used as before the air quality complaints. Interestingly, this improvement in the perceived air quality was noticed well before the air concentrations of 2-ethyl-1-hexanol had started to decrease indicating that the problems were caused by substances other than 2-ethyl-1-hexanol. The amounts of 2-ethyl-1-hexanol in the trap under the laminate flooring in room B increased from 17.0 (one week after installation) to 280.3 (13 months after installation) µg/g in a linear manner (R^2=0.9975) suggesting a constant emission of 2-ethyl-1-hexanol from the floor.

A surface emissions trap has previously been shown to be able to efficiently reduce a range of small and larger VOCs including alcohols, aldehydes, ketones, terpenes, aromatic hydrocarbons, sulfides (Markowicz and Larsson, 2012); as well as formaldehyde (a common emission product from building materials used indoors) and 2-chloroanisole (from moist impregnated wood) (data not shown). Recently, it has also been found that such a device is efficient against radon and odors from cigarette smoke (data not shown). A surface emissions trap may represent a convenient, health-effective and environment-friendly way of improving indoor air in cases when the problems are due to emissions from surfaces of the building indoors. Further studies should include an unbiased evaluation of the perceived the air quality in buildings following installation of the device. Further studies are also required for being able to evaluate if the device, due to its ability to reduce emissions, also will be useful for saving energy by allowing a lowered ventilation rate or a reduced need for conventional electricity-driven air cleaners while at the same time maintaining a satisfactory indoor air quality.

CONCLUSIONS

Emissions of VOCs (including odors) from a surface may be stopped efficiently by applying a surface emissions trap on the surface. In the present study, attaching a surface emissions trap on a PVC flooring in a school with air complaints led to a dramatic improvement in the perceived air quality and decreased 2-ethyl-1-hexanol air concentrations. The device may constitute a useful means of restoring the indoor air quality after water damage.

ACKNOWLEDGEMENT

This research was supported by the Swedish Research Council for Environment, Agriculture Science and Spatial Planning (FORMAS) and Lund University.

REFERENCES

Andersen, B., Frisvad, J.C., Sondergaard, I., Rasmussen, I.S. and Larsen, L.S. (2011) Associations between Fungal Species and Water-Damaged Building Materials, *Applied and Environmental Microbiology*, **77**, 4180-4188.

Björk, F., Eriksson, C.A., Karlsson, S. and Khabbaz, F. (2003) Degradation of components in flooring systems in humid and alkaline environments, *Construction and Building Materials*, **17**, 213-221.

Claeson, A.S., Sandstrom, M. and Sunesson, A.L. (2007) Volatile organic compounds (VOCs) emitted from materials collected from buildings affected by microorganisms, *Journal of Environmental Monitoring*, **9**, 240-245.

Fiedler, N., Laumbach, R., Kelly-Mcneil, K., Lioy, P., Fan, Z.H., Zhang, J.F., Ottenweller, J., Ohman-Strickland, P. and Kipen, H. (2005) Health effects of a mixture of indoor air volatile organics, their ozone oxidation products, and stress, *Environ Health Perspect*, **113**, 1542-1548.

Markowicz, P. and Larsson, L. (2012) The surface emissions trap: a new approach in indoor air purification, *J Microbiol Methods*, **91**, 290-294.

Meininghaus, R., Gunnarsen, L. and Knudsen, H.N. (2000) Diffusion and sorption of volatile organic compounds in building materials - Impact on indoor air quality, *Environmental Science & Technology*, **34**, 3101-3108.

Nalli, S., Horn, O.J., Grochowalski, A.R., Cooper, D.G. and Nicell, J.A. (2006) Origin of 2-ethylhexanol as a VOC, *Environmental Pollution*, **140**, 181-185.

Norbäck, D., Wieslander, G., Nordström, K. and Wålinder, R. (2000) Asthma symptoms in relation to measured building dampness in upper concrete floor construction, and 2-ethyl-1-hexanol in indoor air, *Int J Tuberc Lung Dis*, **4**, 1016-1025.

Sjoberg, A. and Ramnas, O. (2007) An experimental parametric study of VOC from flooring systems exposed to alkaline solutions, *Indoor Air*, **17**, 450-457.

Sunesson, A.L., Liljelind, I., Sundgren, M., Pettersson-Stromback, A. and Levin, J.O. (2002) Passive sampling in combination with thermal desorption and gas chromatography as a tool for self-assessment of chemical exposure, *Journal of Environmental Monitoring*, **4**, 706-710.

Zhang, J.S., Zhang, J.S., Chen, Q. and Yang, X. (2002) A Critical Review on Studies of Volatile Organic Compound (VOC) Sorption by Building Materials, *ASHRAE Transactions*, **108(1),** 162-174.

STATISTICAL EVALUATION OF HUMAN SENSORY RESPONSE TO ACETONE

Tunga SALTHAMMER[1,*], Nicole SCHULZ[1], Ramona STOLTE[1)] and Erik UHDE[1]

[1]Fraunhofer WKI, Department of Material Analysis and Indoor Chemistry, Braunschweig, Germany

[*]Corresponding email: tunga.salthammer@wki.fraunhofer.de

Keywords: Sensory testing, Odour, Acetone, Π-scale, ISO 16000-28

SUMMARY

The release of volatile organic compounds (VOCs) from building products may influence the perceived air quality in the indoor environment. Consequently, building products need to be assessed for chemical emissions and for the acceptability of emitted odours. Recently, the standard ISO 16000-28, which describes the evaluation of perceived odour in test chambers, has been implemented. The procedure requires a panel of trained probands, being calibrated with diluted acetone. However, the calibration procedure itself can be a substantial source of the method uncertainty.

INTRODUCTION

The assessment of perceived air quality is gaining increasing significance due to the low acceptance by end consumers and customers for unusual or unknown material odours or odour sources in the indoor environment. Recently, an ISO standard has been developed in order to standardize the assessment of odours in interior room materials in test chambers. In ISO 16000-28 (2012), the measurement technique for the acceptance, the perceived intensity and the hedonic effect is described. ISO 16000-28 only considers the standard climate of a test chamber and the possible formation of odorous compounds from indoor chemistry (Uhde and Salthammer, 2007) is completely neglected. The procedure for the odour evaluation of indoor air is described in ISO 16000-30 (2014).
Sensory testing is a criterion in many building product evaluation schemes, usually in combination with the determination of volatile organic compounds in the chamber air. The objective of a current research project is to evaluate the proposed method of perceived intensity and hedonics with respect to ISO 16000-28 for their suitability in practice. This publication deals with the statistical evaluation of proband responses to acetone reference concentrations.

METHODOLOGIES

The performance test according to ISO 16000-28 is as follows: trained subjects determine the perceived intensity Π (unit pi) of an air sample using an acetone benchmark. Before a chamber test, the subjects are offered two different concentrations of an acetone/air-mixture (concentration range: 0 pi = 20 mg/m³ (odour threshold of acetone) to 15 pi = 320 mg/m³). A linear scale with $\Pi = 1/20 \times (C_{acetone}-20)$ is applied. A calibrated FID is used to determine the acetone concentration. The accuracy of one single evaluation shall be within +/- 2 pi. If the difference between the nominal value and the actual value does not meet the required accuracy, a second round of testing shall be done within 30 h after the first assessment.

From our panel of 32 persons, two probands, one male (Proband 1) and one female Proband 2, were selected for further discussion. In the following, A1 describes the results of the first round and A2 describes the results of the second round. If the first round was successful, these data also apply for the data set of A2.

RESULTS AND DISCUSSION

In Figure 1, the responses of two the probands versus acetone-air mixtures of different concentrations are shown as ΔΠ(nominal-actual). The red line represents the results of the first round (A1). In A2, the responses from A1, which did not meet the allowed deviation criteria of +/- 2 pi are substituted by the data from the second round (blue line).

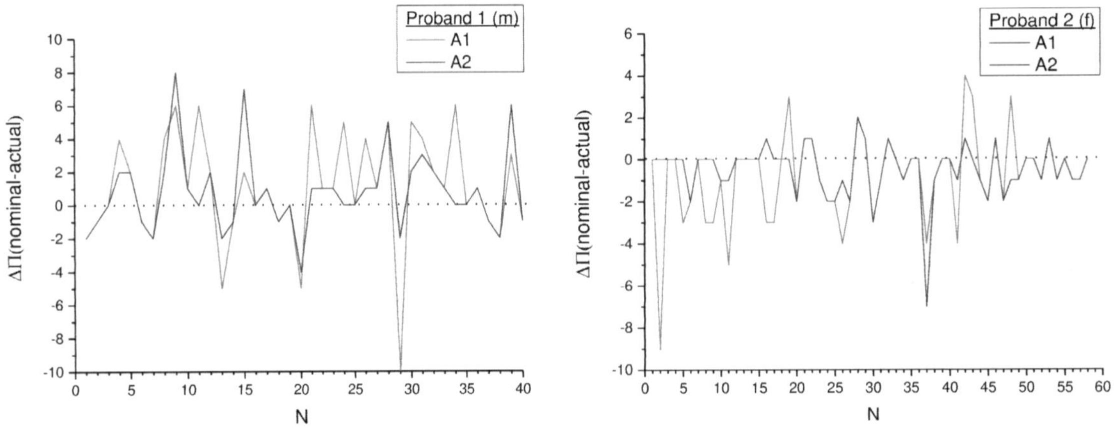

Figure 1. Difference between nominal acetone value and reported (actual) acetone value for the first (A1) and the second (A2) round for Proband 1 (male) and Proband 2 (female). Please note that the data in A1 and A2 only differ if the first round was not successful.

In Table 1, the statistical evaluation of the data sets (A1 and A2) is presented. On average, Proband 1 overestimates the response to the acetone-air mixtures with a mean of +1.03 in A1 and 0.75 in A2. In contrast, Proband 2 underestimates the perceived acetone air concentrations with a mean of -0.83 in A1 and -0.50 in A2. As expected, the standard deviation decreases from A1 to A2.

Table 1. Statistical evaluation of proband responses to acetone-air mixtures (N is the number of performance tests a proband has passed).

Proband	Proband 1 (male)		Proband 2 (female)	
A1 or A2	A1	A2	A1	A2
N	40	40	58	58
Arithmetic mean	1,03	0.75	-0.83	-0.50
Standard deviation	3,36	2.46	2.14	1.31
Median	1	0.5	-0.5	0

ISO 16000-28 (2012) bears one more disadvantage, which might cause a systematic error. As mentioned earlier, each proband is exposed to two different acetone-air concentrations on the same day during the performance test. To avoid saturation of the human odour receptors, the more diluted acetone concentration is always being offered first. Although the absolute acetone concentration is unknown to the proband, he or she always has the information that the second acetone concentration is higher. Moreover, if a proband fails to respond within +/-2 pi, he or she will be told if the assessment was too high or too low. This information will probably influence the second response and may lead to biased results.

CONCLUSIONS

As shown for two probands, the calibration procedure by use of acetone/air-mixtures of a human panel according to ISO 16000-28 might bear a number of shortcomings. Moreover, it can be expected that the variation of parameters like temperature, humidity and air flow conditions will also influence the response. A full statistical evaluation under consideration of 32 probands and different climatic conditions is currently underway (Salthammer et al., 2015).

ACKNOWLEDGEMENT

Financial support by internationaler Verein für Technische Holzfragen (iVTH) is gratefully acknowledged.

REFERENCES

ISO 16000-28 (2012). Indoor Air - Part 28: Determination of odour emissions from building products using test chambers. Beuth Verlag, Berlin.

ISO 16000-30 (2014). Indoor Air - Part 30: Sensory testing of indoor air. Beuth Verlag, Berlin.

Salthammer T, Schulz N, Stolte R, Uhde E (2015). Human sensory response to acetone-air mixtures, manuscript in preparation.

Uhde E, Salthammer T (2007). Impact of reaction products from building materials and furnishings on indoor air quality - a review of recent advances in indoor chemistry. *Atmospheric Environment* **41**, 3111-3128.

CHALLENGES AND OPPORTUNITIES FOR FILTER FORENSICS

Daniel HAALAND[1,*] and Jeffrey A SIEGEL[1,2]

[1]Department of Civil Engineering, University of Toronto, Ontario, Canada
[2]Dalla Lana School of Public Health, University of Toronto, Ontario, Canada

[*]Corresponding email: daniel.haaland@mail.utoronto.ca

Keywords: Filtration, sampling, exposure quantification, particles

SUMMARY

Filter forensics is a promising indoor air investigation technique involving the analysis of dust collected on HVAC filters. In this paper, we summarize past filter forensics research and suggest an approach that quantitatively links contaminants collected on a filter to average indoor concentrations. Although the use of quantitative filter forensics has great potential to measure exposure relevant indoor concentrations of particle-bound contaminants, this paper also identifies potential limitations and challenges to the use of the technique.

INTRODUCTION

Particles are among the most serious indoor air contaminants and their composition includes metals, allergens, semi-volatile organic compounds (SVOCs), bacteria, fungi and viruses. Filter forensics refers to the process of analyzing dust which has collected on heating and air-conditioning filters. Once collected, the dust samples are analyzed for the contaminants of interest. Filter forensics has a long history in industrial hygiene, but has only recently been used as an indoor air research technique. The purpose of this paper is to highlight the opportunities and identify the challenges of filter forensics.

LITERATURE REVIEW

Table 1 presents a summary of prior filter forensic studies. Table 1 shows that most studies have focused on microbiology, with fewer filter forensic investigations of abiotic compounds.

Table 1: Filter forensic papers in the literature

Reference	Contaminant	Building type and number
(Fox and Rosario, 1994)	bacteria	residential (5), institutional (1-5)
(Simmons et al., 1997)	fungi	institutional (7)
(Echavarria et al., 2000)	viruses	institutional (1)
(Moritz et al., 2001)	bacteria, fungi	institutional (2)
(Viladeri et al., 2007)	elem. and org. carbon	commercial (6), institutional (1)
(Ljaljevic-Grbic et al., 2008)	fungi	commercial and institutional (2-15)
(Tringe et al., 2008)	bacteria	commercial (2)
(Stanley et al., 2008)	bacteria	institutional (2)

(Noris et al., 2009)	bacteria, fungi, metals	commercial (1), residential (8)
(Drudge et al., 2011)	bacteria	institutional (4)
(Goyal et. Al., 2011)	viruses	institutional (2)
(Ni et al., 2011)	SVOCs	commercial (1)
(Noris et al., 2011)	bacteria, fungi	residential (5)
(Siddique et al., 2011)	metals	commercial (15)
(Besis et al., 2014)	SVOCs	commercial (18), institutional (2)
(Hoisington et. al., 2014)	bacteria, fungi	commercial (4)
(Barnes et al., 2015)	allergens	residential (56)
(Xu et al., 2015)	SVOCs	commercial (14)

OPPORTUNITIES

The investigations in Table 1 demonstrate the value of the technique. However, there is a gap between identification of a compound on filters and establishing an indoor air concentration. By collecting additional pieces of data during the sampling period there is an opportunity to extend filter forensics to a quantitative measure of indoor air quality. This can be achieved using Equation 1, which relates the collected mass of a contaminant and knowledge of the system operation to the time and space averaged concentrations in indoor air.

$$C = \frac{M \cdot F}{\eta \cdot Q \cdot t} \quad (1)$$

where C is the indoor concentration of a specific particle-bound contaminant, M is the total mass of particles collected on the filter, F is the fraction of the contaminant in the overall dust, Q is the volumetric air flow rate through the filter, t is the total runtime of the system during the period of interest, η is the filter efficiency for the specific particle-bound contaminant. Using Equation 1, it is possible to relate the typical indoor concentrations to the sensitivity of the detection limit for the dust analysis of a contaminant. Figure 1 shows a high and low sensitivity case as well as examples for two example compounds, di-n-butyl phthalate (DnBP) and lead. Along the x-axis, Figure 1 shows upper and lower bounds for typical indoor air concentrations of DnBP and lead, which were obtained from (Xu et al., 2014) and (Paschold et al., 2003) respectively.

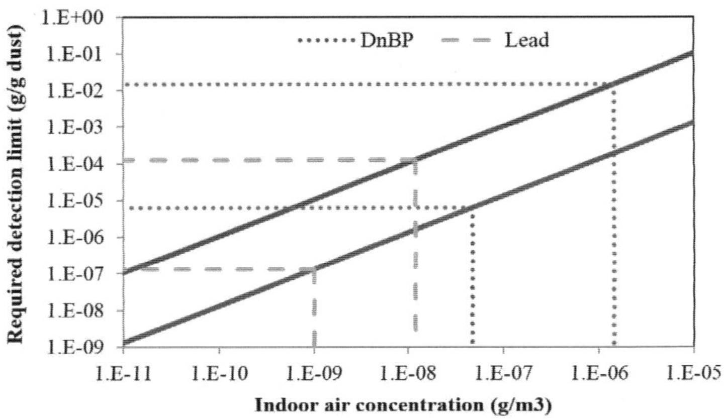

For a 5cm by 5cm piece of filter and 30 days of operation. The low sensitivity (blue) case assumes an air velocity of 1 m/s and filter efficiency of 20%. The high sensitivity (red) case assumes an air velocity of 5 m/s and filter efficiency of 80%.

Figure 1: Connection between indoor air concentration and dust analysis detection limit.

Figure 1 shows that the large air flow rates allow for substantial accumulation of particle mass on the filter. For a compound with relatively high indoor air concentrations (e.g., DnBP) the required detection limits are between 2-5 times greater than those in published studies (e.g., Xu et al., 2015). Even for a contaminant with lower indoor concentrations such as lead, the required detection limits are still 3-6 orders of magnitude above published detection limits of different analytical approaches for extracting lead from dust. These examples suggest that filter forensics shows promise for a wide variety of compounds.

CHALLENGES

Figure 1 shows the potential for quantitative filter forensics, but there are also challenges. The first is that while the technique is readily applicable to buildings equipped with recirculating forced air systems, many buildings do not have forced air systems and when they do, run-times can be short. In cases where filters do not accumulate sufficient mass, portable air cleaners can be used. A second concern is that concentrations may change over the long period that a filter is installed: microorganisms can grow or decay (Maus et al., 2001) and volatile components of dust, such as SVOCs may partition between phases after deposition on the filter. However, there is preliminary evidence that quantitative filter forensics can successfully be used to determine concentrations of SVOCs in filter dust (Xu et al., 2015) and this should be further explored for other compounds. Another consideration is that many HVAC filters have very low efficiency and thus may collect insufficient quantities of some compounds. Noris et al. (2009) suggest that this is less of a concern for culturable fungi and bacteria as well as heavy metals, but it remains unexplored for other compounds. A final concern is uncertainty in the terms of Equation 1. A propagation of error and reasonable uncertainties of the input measurements suggests an approximate uncertainty of 15-20%, which may be insufficient for some uses. Much of the error comes from uncertainty in measuring the filter efficiency (Stephens and Siegel, 2012).

CONCLUSIONS

Filter forensics is a demonstrated and potentially powerful method to assess average indoor air contaminant concentrations. Further exploration and evaluation of the technique is warranted as it could provide considerable insight on indoor particle-bound contaminants

ACKNOWLEDGEMENTS

Funding was provided by the US Department of Housing and Urban Development and the National Science and Engineering Research Council of Canada.

REFERENCES

Barnes CS, Allenbrand R, Mohammed M, et al. (2015). Measurement of aeroallergens from furnace filters. Accepted to *Ann. Allerg. Asthma Im.*

Besis A, Katsoyiannis A, Botsaropoulou E, and Samara C (2014). Concentrations of polybrominated diphenyl ethers in central air-conditioner filter dust and relevance of non-dietary exposure in occupational indoor environments in Greece. *Environ. Poll.*, **188**, 64-70.

Drudge CN, Krajden S, Summmerbell RC and Scott JA (2011). Detection of antibiotic resistance genes associated with methicillin-resistant Staphylococcus aureus (MRSA) and coagulase-negative staphylococci in hospital air filter dust by PCR. *Aerobiologia*, **28**, 285-289.

Echavarria M, Kolavic SA, Cersovsky S, et al. (2000). Detection of adenoviruses (AdV) in culture-negative environmental samples by PCR during an AdV-associated respiratory disease outbreak. *J. Clin. Microbiol.*, **38**, 2982-2984.

Fox A. and Rosario RMT (1994). Quantification of muramic acid, a marker for bacterial peptidoglycan, in dust collected from hospital and home air-conditioning filters using gas-chromatography mass-spectrometry. *Indoor Air*, **4**, 239-247.

Goyal SM, Anantharaman S, Ramakrishnan MA, et al. (2011). Detection of viruses in used ventilation filters from two large public buildings. *Am. J. Infect. Control*, **39**, E30-E38.

Hoisington A, Maestre JP, Siegel JA, and Kinney KA (2014). Exploring the microbiome of the built environment: A primer on four biological methods available to building professionals. *HVAC&R Res.*, **20**, 167-175.

Ljaljevic-Grbic M, Vukojevic J, and Stupar M (2008). Fungal colonization of air-conditioning systems. *Archives of Biological Sciences*, **60**, 201-206.

Maus R, Goppelsroder A and Umhauer H (2001). Survival of bacterial and mold spores in air filter media. *Atmos. Environ.*, **35**, 105-113.

Moritz M, Peters H, Nipko B and Ruden H (2001). Capability of air filters to retain airborne bacteria and molds in heating, ventilating and air-conditioning (HVAC) systems. *Int. J. Hyg. Environ. Health*, **203**, 401-409.

Ni HG, Cao S.P, Chang WJ and Zeng H (2011). Incidence of polybrominated diphenyl ethers in central air conditioner filter dust from a new office building. *Environ. Poll.*, **159**, 1957-1962.

Noris F, Siegel JA and Kinney KA (2011). Evaluation of HVAC filters as a sampling mechanism for indoor microbial communities. *Atmos. Environ.*, **45**, 338-346.

Noris F, Siegel JA, Kinney KA (2009). Biological and metal contaminants in HVAC filter dust. *Ashrae Transactions 2009*, **115**, 484-491.

Paschold H, Li W-W, Morales H, et al. (2003). Elemental analysis of airborne particulate matter and cooling water in west Texas residences. *Atmos. Environ.*, **37**, 2681-2690.

Siddique N, Majid A and Tufail M (2011). Elemental analysis of dust trapped in air conditioner filters for the assessment of Lahore city's air quality. *J. Radioan. Nucl. Ch.*, **290**, 691-699.

Simmons RB, Price DL, Noble JA, et al.(1997). Fungal colonization of air filters from from hospitals. Am. Ind. Hyg. Assoc. J., **58**, 900-904.

Stanley NJ, Kuehn TH, Kim SW, et al. (2008). Background culturable bacteria aerosol in two large public buildings using HVAC filters as long term, passive, high-volume air samplers. *J Environ. Monitor.*, **10**, 474-481.

Stephens, B. and Siegel, J.A. (2012) Comparison of Test Methods for Determining the Particle Removal Efficiency of Filters in Residential and Light-Commercial Central HVAC Systems, *Aerosol Sci Tech*, 46, 504-513.

Tringe SG, Zhang T, Liu X, et al. (2008). The airborne metagenome in an indoor urban environment, *PLOS ONE*, **3**.

Viladeri CV, Piscopou TD and Siskos PA (2007). Chemistry and air-conditioning: Sampling and analysis of particles collected from filters of air conditioning equipment in Athens. *Indoor Built Environ.*, **16**, 563-568.

Xu Y, Liang Y, Urquidi JR and Siegel JA (2014). Phthalates and polybrominated diphenyl ethers in retail stores. *Atmos. Environ.*, **87**, 53-64.

Xu Y, Liang Y, Urquidi JR and Siegel JA (2015). Semi-volatile organic compounds in heating, ventilation, and air-conditioning filter dust in retail stores. Accepted to *Indoor Air*.

Bioaccumulation on HVAC Filters in University Buildings in Singapore

Irvan LUHUNG[1,2], Yan WU[1,2], Victor WC CHANG[1,2,*], and William W NAZAROFF[1,3]

[1]Berkeley Education Alliance for Research in Singapore, Singapore
[2]Civil and Environmental Engineering, Nanyang Technological University, Singapore
[3]Civil and Environmental Engineering, University of California, Berkeley, CA, USA

[*]Corresponding email: wcchang@ntu.edu.sg

Keywords: HVAC filter, DNA-based analysis, Occupancy level, University buildings

SUMMARY

Particle filters are important components of HVAC systems. Biomass accumulation on HVAC filters can potentially have adverse effects on building occupants. The research reported in this paper aims to quantitatively describe the content of biomass on HVAC filters from various indoor office environments in Singapore in relation to key governing processes. In this initial phase, filter samples were collected and analysed using DNA-based methods. The collected DNA was directly extracted from the filters and the final concentration was measured with Qubit and qPCR methods. The DNA concentrations were then related to the cumulative indoor occupancy level. The results show good correlations between bacterial DNA and cumulative occupancy. The proportion of total DNA that belongs to fungal and bacterial species decreases as occupancy level increases, probably because of biomass from other sources such as human skin cells.

INTRODUCTION

More than 90% of office buildings in Singapore are equipped with a centralized ventilation and air conditioning system. In a tropical climate, such systems are commonly referred to as "air conditioning and mechanical ventilation (ACMV), because heating is not needed. However, to be consistent with common international usage, we will refer to these systems using the common HVAC acronym. Current building practices were developed without detailed knowledge about how accumulation of biological material can happen in various parts of the HVAC system. It is feasible that processes such as microbial colonization and metabolism in HVAC systems contribute to adverse consequences for occupants, such as malodors, allergic reactions, asthma or sick building syndrome symptoms (Crook and Burton, 2010; Xu et al., 2011).

Accumulation on the HVAC filter of particulate matter, which includes particles of biological origin, is inevitable as it is the first line of defence in the HVAC system. Studies have reported the existence on filters of opportunistic pathogens (Noris et al., 2011). Research also indicates the possibility of more resilient organisms growing without additional nutrients (Pigeot-Remy et al., 2014). However, to date only a few studies have reported biological characteristics accumulating on building ventilation system filters. To our knowledge, there has not yet been a study using DNA-based analysis methods that *quantitatively* assesses the biological content of used HVAC filters.

The present study utilizes culture-independent DNA-based analysis applied to extracted material from HVAC filters that were collected from office environments in Singapore. The goal is to provide quantitative information about the biological content of the accumulated particulate matter. Total, bacterial and fungal DNA concentration on material directly extracted from the filters was measured. The results are correlated to estimates of cumulative occupancy level in the area serviced by the HVAC system. The results may provide a partial basis to reassess current practices with a view towards improving future building systems.

METHODS

Ventilation system filters from nine different air-handling units (AHUs) at Nanyang Technological University, Singapore, were acquired for this study. The nine AHUs served rooms containing typical office and library facilities at a university. The occupancy levels of the selected indoor environments were determined by conducting hourly population counts by means of direct observation on five weekdays and four weekend days for each of two months. The rooms were also characterized based on activities levels and sizes.

The filter samples (polyester, MERV 8 rating) were obtained from the AHUs' indoor secondary filter bank, which is positioned after the primary outdoor-air filter in the centralized fan-coil HVAC system. This secondary filter processes a mixture of recirculated indoor and make-up outdoor air, as designed to keep the indoor CO_2 level within allowable limits. In accordance with Singapore regulations, these filters are replaced every three months. In this study, all filters were collected after full three-month periods of service.

Pieces were cut from each filter for DNA extraction and analysis. Specifically, a 2×5 cm size was used for filters that were 2.5-5 cm thick and a 1×5 cm size was used for filters with pleats deeper than 5 cm. The filter pieces were directly inserted into the 5-ml bead beating tubes of the MOBIO Power Water (PW) DNA extraction kit. The process of DNA extraction followed the manufacturer's protocol with some modifications to improve DNA yield. Extraction replicates were taken the same filter panel.

The final DNA solutions were analyzed with a Qubit (Invitrogen) for total DNA and with qPCR (LC480, Roche) using universal probes and primers for specific quantification of bacterial and fungal DNA. The DNA concentrations measured by Qubit and qPCR were scaled up using the ratio of the actual area of the AHU's filter panels to the extracted filter area. Weekly average DNA accumulation rates (ng DNA/week) were then determined, by dividing the total accumulation by the weeks of service for the given filter.

RESULTS AND DISCUSSION

Figure 1 displays the weekly fungal and bacterial DNA accumulation rates and also the cumulative occupancy level (person-h/week) of each room studied.

The results show that the bacterial DNA accumulation rate has a stronger relation to occupancy level than its fungal counterpart. It can be observed that filters collected from locations with lower occupancy levels tend to be dominated by fungal DNA (Office 1-6 and Classroom). Conversely, bacterial DNA dominates at locations with higher occupancy, such as the library office with occupancy of 6575 person-h/week and the library reading hall with an occupancy of 15,400 person-h/week. A linear regression of bacterial DNA and cumulative occupancy also displayed a high correlation (R^2=0.69), whereas the correlation between fungal DNA and cumulative occupancy was weak (R^2=0.05). This finding is broadly consistent with previous studies showing that humans are an important source of indoor airborne bacteria and that ventilation-system filter dust is influenced by indoor emissions (Hospodsky et al., 2012; Noris et al., 2011).

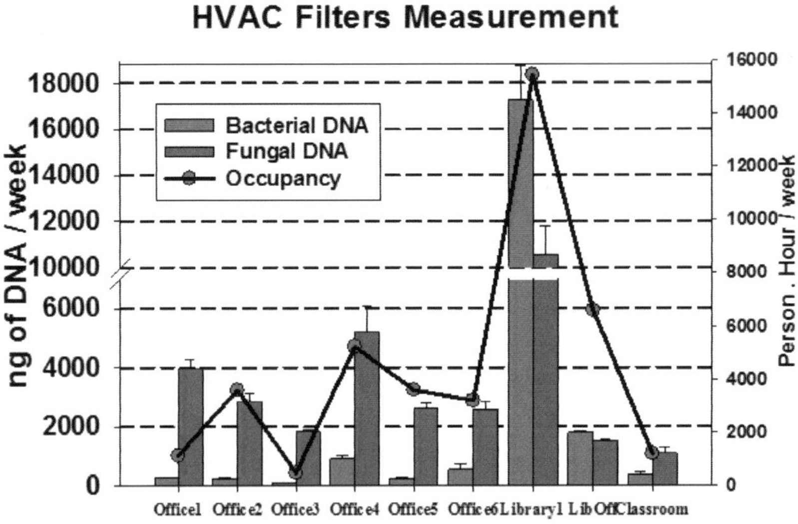

Figure 1. Weekly bacterial and fungal DNA accumulation rates on HVAC filters coupled with occupancy levels (in person-hour/week) for nine AHUs in university buildings.

The main source of fungal DNA on the HVAC filters may be the outdoor air. Outdoor air is known to contain several fungal taxa that may also have reached the filters (Noris et al., 2011). Nevertheless, although weak, some association can be seen between fungal DNA and cumulative occupancy. Occupancy could be a contributor to fungal bioaerosols through several mechanisms, such as resuspension from floors or shedding from clothes. People may be vectors for transport from outdoors, rather than the primary source (Qian et al., 2012).

The amount of non-microbial DNA on the filters also shows a strong relation with occupancy level. The non-microbial DNA was evaluated as the difference between the Qubit determination of total DNA and the qPCR evaluation of fungal plus bacterial DNA. Figure 2 displays the occupancy level and the total DNA accumulation rates on the same filters segmented based on the DNA origin.

Although the accumulation rate of bacterial DNA shows a closer relationship to occupancy level than fungal or total DNA, bacterial DNA only comprises 2-16% of total DNA in these samples. Fungal DNA, on the other hand, comprises a larger share: 5-37% of total DNA. Taken together, bacterial and fungal DNA comprises 12-40% of total DNA on these samples. This evidence suggests that nonmicrobial biomass, which may originate from humans, animals or plants, also are important contributors to total biomass collected on HVAC filters.

CONCLUSIONS

This study presents baseline information for biomass accumulation on HVAC filters that serve typical office environments in buildings in Singapore's tropical climate. Bacterial-based biomass abundance on the filters correlates more closely with occupancy as compared to fungal-based biomass. Other sources of biomass, possibly associated with human, animal or plant cells, contribute the majority of total accumulation on the filters.

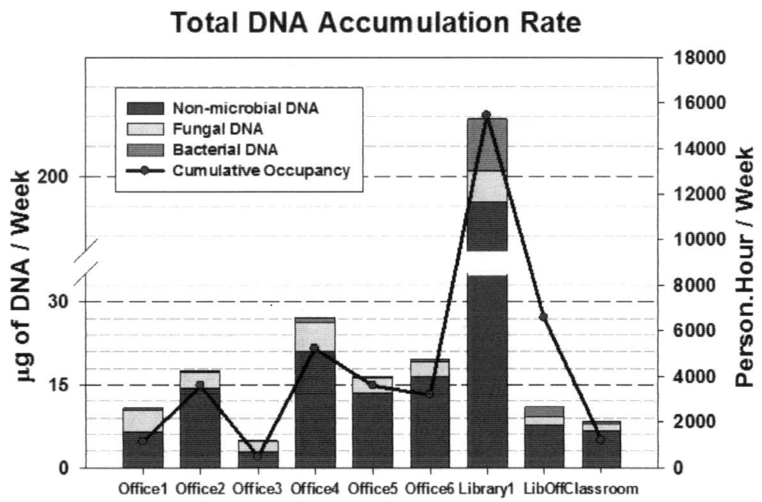

Figure 2. Weekly Total DNA accumulation rates on HVAC filters coupled with occupancy levels (in person-h/week) for nine AHUs in university buildings.

The findings in this study contribute to the foundation needed to further evaluate HVAC filter usage and operational practices, such as filter changing frequency or choice of filter performance rating. Bioaerosol analysis utilizing DNA-based techniques is a promising tool for further studies both to quantitatively estimate concentrations of various biological entities and also to investigate biodiversity on filters by means of sequencing.

ACKNOWLEDGEMENT

This research was funded by the Republic of Singapore's National Research Foundation through a grant to the Berkeley Education Alliance for Research in Singapore (BEARS) for the Singapore-Berkeley Building Efficiency and Sustainability in the Tropics (SinBerBEST) Program. BEARS has been established by the University of California, Berkeley as a center for intellectual excellence in research and education in Singapore.

REFERENCES

Crook, B., and Burton, N. C. (2010). Indoor moulds, Sick Building Syndrome and building related illness. *Fungal Biology Reviews,* **24**, 106-113.

Hospodsky, D., Qian, J., Nazaroff, W. W., et al. (2012). Human occupancy as a source of indoor airborne bacteria. *PLoS ONE,* **7**.

Noris, F., Siegel, J. A., and Kinney, K. A. (2011). Evaluation of HVAC filters as a sampling mechanism for indoor microbial communities. *Atmospheric Environment,* **45**, 338-346.

Pigeot-Remy, S., Lazzaroni, J. C., Simonet, F., et al. (2014). Survival of bioaerosols in HVAC system photocatalytic filters. *Applied Catalysis B: Environmental,* **144**, 654-664.

Qian, J., Hospodsky, D., Yamamoto, N., et al. (2012). Size-resolved emission rates of airborne bacteria and fungi in an occupied classroom. *Indoor Air,* **22**, 339-351.

Xu, Z., Wu, Y., Shen, F., et al. (2011). Bioaerosol science, technology, and engineering: Past, present, and future. *Aerosol Science and Technology,* **45**, 1337-1349.

MODELING THE IMPACT OF RESIDENTIAL HVAC FILTRATION ON INDOOR PARTICLES OF OUTDOOR ORIGIN

Parham AZIMI[1], Dan ZHAO[1], and Brent STEPHENS[1,*]
[1]Department of Civil, Architectural and Environmental Engineering, Illinois Institute of Technology, Chicago, IL, USA

[*]Corresponding email: brent@iit.edu

Keywords: Particulate matter, Infiltration, Human exposure, HVAC filters

SUMMARY

The infiltration of outdoor particles into residential buildings greatly impacts human exposure to particle matter, including the mass of particles smaller than 2.5 μm ($PM_{2.5}$) and ultrafine particles (UFPs: particles smaller than 100 nm). In this project the indoor concentration of outdoor fine and ultrafine particles in three types of typical single-family homes in 22 cities in the U.S. was modeled using a variety of central forced air HVAC filtration scenarios. Metrics of filtration effectiveness and infiltration factors are used to explore the influence of higher efficiency HVAC filters on indoor proportions of outdoor particles. Results suggest that higher efficiency HVAC filtration can meaningfully reduce indoor proportions of outdoor particles inside residences that rely on infiltration for ventilation air and that have central HVAC systems that operate only to meet cooling or heating demands, but both home vintage and climate zone strongly influence the results.

INTRODUCTION

Many studies have shown a variety of adverse health effects associated with elevated outdoor concentrations of airborne particulate matter (e.g., Fann et al. 2012; Stölzel et al. 2006). Because Americans spend most of their time indoors (Klepeis et al. 2001), and airborne particles of outdoor origin can infiltrate into residential buildings with varying efficiencies (Stephens and Siegel 2012), much of human exposure to outdoor particles actually occurs indoors, particularly in residences. High efficiency particle air filters installed in central HVAC systems are increasingly being relied upon to reduce indoor concentrations of particulate matter inside residences (Bräuner et al. 2008). Several previous investigations have explored the impacts of HVAC filters on particle concentrations in residences through a combination of measurement and modeling, but there are some limitations to previous studies, including: they have considered only a narrow range of particle sizes or classes; they have relied on filter classifications other than MERV; or they have investigated only a narrow variety of filters. In this project, we have worked to build and expand upon these previous efforts by simulating the indoor proportion of outdoor fine and ultrafine particles for every hour of the most recent year for which data was available (2012) in three typical vintages of single-family homes in 22 cities in the U.S. under a variety of central forced air HVAC filtration scenarios. The goal is to mechanistically investigate the likely influence of higher efficiency HVAC filtration on indoor concentrations of $PM_{2.5}$ and UFPs of outdoor origin in typical homes with central heating and air-conditioning systems that operate only to meet heating or cooling demands.

METHODOLOGIES

For modeling the indoor proportion of outdoor fine and ultrafine particles, 22 cities were selected to cover all 15 U.S. climate zones as well as the top 15 cities with the highest annual average outdoor $PM_{2.5}$ concentrations summarized in the most recent Integrated Science Assessment for Particulate Matter (US EPA, 2009). Three typical home types were selected in each of 22 cities to represent a range of homes existing across the building stock. First, new, energy efficient homes were designed to have lower outdoor particle infiltration by incorporating well-insulated building envelopes, high airtightness, and properly sized high efficiency heating and air-conditioning systems for each climate zone. Second, typical existing, older, less efficient homes were designed to have higher outdoor particle infiltration by incorporating moderately insulated building envelopes, typical airtightness, and larger and less efficient HVAC systems for each climate zone based on typical existing home characteristics in each area. Third, typical older homes were designed to have the highest outdoor particle infiltration by incorporating poorly insulated building envelopes, low airtightness and larger and less efficient (and often undersized) HVAC systems for each climate zone based on typical older vintage home characteristics in each area. Envelope characteristics were taken from two surveys (Y. J. Huang et al., 1987; J. Huang, et al., 1999).

We utilized a discrete time-varying mass balance on particles of outdoor origin inside a single well-mixed zone where, in the absence of indoor sources, the indoor particle concentration (of $PM_{2.5}$ or UFPs of outdoor origin) at each time step $[C_{i,in}(t_n)]$ is estimated using Equation 1. One-minute intervals were used to improve model stability.

$$C_{i,in}(t_n) = C_{i,in}(t_{n-1}) + \Delta t[P_i\lambda_{inf}C_{i,out}(t_n) - (\lambda_{inf}(t_n) + \beta_i + f(t_n)\eta_{i.HVAC}\lambda_{HVAC})C_{i,in}(t_{n-1})] \quad (1)$$

where P_i is $PM_{2.5}$ or UFP penetration factor of the building envelope (-), λ_{inf} is the air exchange rate due to infiltration (hr^{-1}), $C_{i,out}$ is the outdoor $PM_{2.5}$ or UFP concentration ($\mu g/m^3$ or $\#/m^3$), β_i is the first-order indoor particle deposition rate loss coefficient for $PM_{2.5}$ or UFP (hr^{-1}), $\eta_{i,HVAC}$ is the $PM_{2.5}$ or UFP removal efficiency of the HVAC filter (-), λ_{HVAC} is the HVAC system recirculation rate (HVAC airflow rate divided by volume, hr^{-1}), f is the fractional operation time of the HVAC system (-), t_n is the current time step (hr), and t_{n-1} is the previous time step (hr).

Model geometry was first constructed in BEopt and whole building energy simulations were then performed in EnergyPlus to predict hourly air exchange rates and HVAC system runtimes using historical weather data from each city in the year 2012. Hourly outdoor pollutant data in each location were obtained from the EPA Air Quality System (AQS) Technology Transfer Network (TTN). Eleven HVAC filters were modeled in each home in each location, including: MERV 5, MERV 6, MERV 7 (×2, with low and high removal efficiency), MERV 8, MERV 10, MERV 12 (×2, with low and high removal efficiency), MERV 14, MERV 16, and HEPA. Estimates of HVAC filtration efficiency were based on a recent study that we published mapping a large number of outdoor particle size distributions (194 particle size distributions across the world) to size-resolved removal efficiency curves from typical HVAC filters (Azimi, et al., 2014). Inputs for envelope penetration factors and indoor deposition loss rate coefficients for both $PM_{2.5}$ and UFPs were culled from recent literature (Williams et al. 2003; Stephens and Siegel 2012; Wallace et al. 2013). The model results include indoor $PM_{2.5}$ mass concentrations and indoor ultrafine particle concentrations of outdoor origin for each hour of the year in 2012.

RESULTS AND DISCUSSION

Annual average outdoor $PM_{2.5}$ concentrations ranged from ~5 to ~19 µg/m^3 and the mean outdoor UFP concentration varied from ~4,100 to ~25,000 #/cm^3. Thus, a wide variety of environments are well represented. The median air exchange rate across all locations was just under 0.5/hr in old homes and only ~0.25/hr and ~0.1/hr in existing and new homes, respectively. Average HVAC system recirculation rates multiplied by hourly runtimes for all selected cities were ~1.5/hr, ~0.6/hr, and ~0.4/hr for old, existing, and new homes respectively, reflecting a variety of common operational characteristics. Predicted annual average indoor concentrations of outdoor $PM_{2.5}$ with a MERV 5 filter installed ranged from 2.1 to 6.3 µg/m^3 in old homes, 1.0 to 3.1 µg/m^3 in existing homes, and only 0.09 to 0.28 µg/m^3 in new homes. On average, indoor concentrations of outdoor $PM_{2.5}$ in the old homes with a MERV 5 filter were predicted to be approximately twice that of existing homes and ~25 times that of new homes. Similarly, annual average indoor concentrations of outdoor UFPs with a MERV 5 filter installed ranged from ~850 to 6000 #/cm^3 in old homes, ~370 to 2660 #/cm^3 in existing homes, and only 66 to 544 #/cm^3 in new homes. The modeled data were then used to estimate infiltration factors (F_{inf}), or the indoor proportion of outdoor particles, for both $PM_{2.5}$ and UFPs as a measure of effectiveness for all filters (Figure 1).

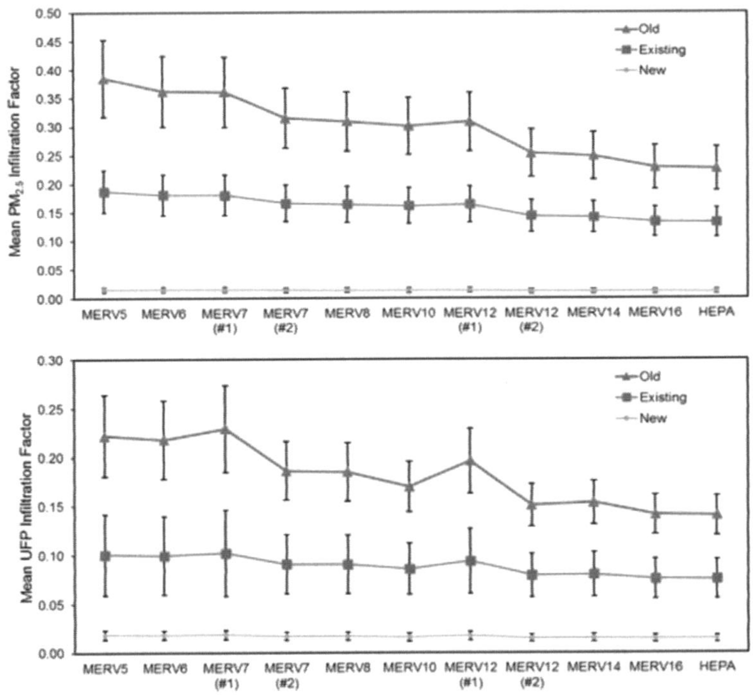

Figure 1. Mean $PM_{2.5}$ and UFP infiltration factors for 3 home types in 22 U.S. locations

The mean infiltration factor for $PM_{2.5}$ was just under 0.40 in the old home with a MERV 5 filter installed, decreasing to under 0.25 with a HEPA filter installed. For existing homes, the mean infiltration factor for $PM_{2.5}$ ranged from just under 0.20 with a MERV 5 filter to around 0.15 with a HEPA filter, suggesting that the impact of even the highest efficiency HVAC filtration is limited by relatively low system runtimes. Finally, the mean infiltration factor for $PM_{2.5}$ was consistently under 0.03 for the new home construction regardless of filter selection. Results for UFP infiltration factors were similar. The mean UFP infiltration factor ranged from ~0.22 in the old home with a MERV 5 filter installed to ~0.14 with a HEPA filter installed, from ~0.10 to ~0.08 in the existing home, and was consistently less than 0.02 with all filters in the new home. These data suggest that high efficiency HVAC filtration is likely to have a much greater influence on indoor $PM_{2.5}$ and UFPs of outdoor origin in older, less efficient homes.

CONCLUSIONS

Results from the simulations herein clearly demonstrate that both home vintage and climate zone strongly influence the impact that HVAC filters can have on indoor proportions of outdoor $PM_{2.5}$ and UFPs inside residences that rely on infiltration for ventilation air and that have HVAC systems that only cycle on and off to meet cooling or heating demands. These data suggest that higher efficiency HVAC filtration can indeed be used to reduce indoor proportions of outdoor particles inside residences, but the impacts are often limited by low HVAC system runtimes that are common in homes that rely on air filtration in central forced air HVAC systems that only operate to meet heating and air conditioning demands.

ACKNOWLEDGEMENT

We acknowledge ASHRAE for funding this research (ASHRAE RP-1691), ASHRAE TC 2.4 membership and leadership, and ASHRAE Project Monitoring Subcommittee members for their input on this work.

REFERENCES

Azimi, Parham, et al. 2014. "Estimates of HVAC Filtration Efficiency for Fine and Ultrafine Particles of Outdoor Origin." *Atmospheric Environment* 98 (December): 337–46.
Bräuner, Elvira Vaclavik, et al. 2008. "Indoor Particles Affect Vascular Function in the Aged: An Air Filtration-Based Intervention Study." *American Journal of Respiratory and Critical Care Medicine* 177 (4): 419–25.
Fann, Neal, et al. 2012. "Estimating the National Public Health Burden Associated with Exposure to Ambient PM2.5 and Ozone." *Risk Analysis* 32 (1): 81–95.
Huang, Joe, et al., 1999. Residential Heating and Cooling Loads Component Analysis. Lawrence Berkeley National Laboratory, University of California.
Huang, Y. J., et al., 1987. Methodology and Assumptions for Evaluating Heating and Cooling Energy Requirements in New Single-Family Residential Buildings. Lawrence Berkeley Lab., CA (USA).
Klepeis, N. E., et al. 2001. "The National Human Activity Pattern Survey (NHAPS): A Resource for Assessing Exposure to Environmental Pollutants." *Journal of Exposure Analysis and Environmental Epidemiology* 11 (3): 231–52.
Stephens, B., and J. A. Siegel. 2012. "Penetration of Ambient Submicron Particles into Single-Family Residences and Associations with Building Characteristics." *Indoor Air* 22 (6): 501–13.
Stölzel, Matthias, et al. 2006. "Daily Mortality and Particulate Matter in Different Size Classes in Erfurt, Germany." *Journal of Exposure Science and Environmental Epidemiology* 17 (5): 458–67.
US EPA. 2009. *Integrated Science Assessment for Particulate Matter*. No. EPA/600/R-08/139F. National Center for Environmental Assessment, Research Triangle Park, NC.
Wallace, Lance, et al. 2013. "Fine and Ultrafine Particle Decay Rates in Multiple Homes." *Environmental Science & Technology* 47 (22): 12929–37.
Williams, Ron, et al. 2003. "The Research Triangle Park Particulate Matter Panel Study: Modeling Ambient Source Contribution to Personal and Residential PM Mass Concentrations." *Atmospheric Environment* 37 (38): 5365–78.

POLLUTANT EXPOSURES AND ASTHMA: WHAT CAN HVAC FILTER DUST TELL US?

Juan P. MAESTRE[1], Sandra DEDESKO[2], Kelly ROYSE[3], Ying, XU[1], Sharon HORNER[3], Jeffrey SIEGEL[2], Kerry A. KINNEY[1]

[1]Department of Architectural, Civil and Environmental Engineering, University of Texas at Austin, Austin, Texas, USA

[2]Department of Civil Engineering, University of Toronto, Ontario, Canada

[3]School of Nursing, University of Texas at Austin, Austin, Texas, USA

[*]Corresponding email: kakinney@mail.utexas.edu

Keywords: Filter forensics, allergens, asthma, microbiome, bacteria, fungi

SUMMARY

The influence of indoor exposures on the development and severity of childhood asthma is an active area of research. In this study, the microorganisms, allergens and chemical contaminants recovered from heating, ventilation and air conditioning (HVAC) filters in 60 rural homes are being analysed to evaluate the relationship between these parameters and asthma severity. The results to date indicate that selected fungal species are enriched homes with self-reported mold problems. Greater than 20% of the bacteria recovered from the filter dust were human associated. Allergens, phthalates and flame retardants were detected in the filter dust collected from the study homes. Interestingly, the phthalate and flame retardant concentrations measured to date were higher in filter dust samples than in settled dust samples collected from the same homes. This study is ongoing and the relationship between the contaminants detected and the severity of asthma for children in these rural homes is being evaluated.

INTRODUCTION

In the US, children spend a significant fraction of their time indoors. Over the past few years, there has been increased interest in exploring the relationship between childhood asthma and chemical and biological exposures in the home. Most previous studies have utilized settled dust samples to investigate these relationships. In the current study, the feasibility of utilizing the dust collected in heating, ventilation and air conditioning (HVAC) filters in homes is being investigated. This filter forensics approach uses the HVAC filters installed in homes as integrated, long-term samplers of particle-bound contaminants, such as phthalates, organophosphate-based flame retardants, microorganisms, endotoxins, and allergens. The potential advantages of using HVAC filter dust are: a) Most central HVAC systems have a filter, b) the filters are in place for long periods of time, and c) they collect particles from a wide spatial area, acting in essence as a high volume sampler. HVAC filter dust samples when combined with HVAC system characterization, offers a controlled way of detecting and assessing indoor air contaminants present at low concentrations in homes. Thus, the objective of this study is to evaluate the use of HVAC filters for assessing housing-related contaminant exposures in homes. In addition, in this work, we evaluate three techniques for filter sampling.

Semivolatile organic contaminants sorb strongly to surfaces such as airborne particles. HVAC filters are in this sense particularly good candidates for investigating phthalate and flame retardant levels in homes. Phthalates are widely used in consumer products as plasticizers and phthalate exposure has been linked to the increased prevalence of asthma and allergies, among other health effects. Similarly, organophosphates are common flame retardants in indoor environments. Even though the potential adverse effects of organophosphates are still unknown, some studies have shown that chronic exposure may cause neurotoxic effects.

METHODOLOGIES

Sixty low-income houses located in rural Texas are being sampled over two seasons, summer and winter. Airborne dust samples are being collected from home HVAC filters as are settled dust samples. Phthalates, organophosphates, endotoxins, and allergens (cat, dog, mite, cockroach, and mouse) concentrations are being determined in dust samples recovered from the HVAC filter samples and compared to the levels recovered from settled dust samples. In addition, microbial DNA is being extracted and total fungal and bacterial loads analyzed by quantitative PCR. Samples are analysed to determine the concentration of the 36 species that comprise the Environmental Relative Moldiness Index (ERMI). Fungal and bacterial communities are being delineated via high-throughput Illumina sequencing, targeting the bacterial 16S region and the fungal ITS-1 region. After quality control, operational taxonomic units, alpha-diversity and beta-diversity indices are calculated to compare the molecular results. Runtimes of the HVAC systems are being estimated through the use of temperature sensors. The HVAC systems are also characterized to estimate the air flowrates through the filters. Questionnaires from the study participants are being used to assess asthma severity (frequency of symptoms, activity limitations) and quality of life. Spirometry tests are performed to evaluate lung function and exhaled breath condensate samples are being collected to analyze for biomarkers of lung inflammation (cytokines) that are critical indicators for asthma.

For the filter sampling techniques evaluation, two HVAC filters from one of the homes were studied for the purpose of investigating sampling repeatability, and to determine differences among three methods with respect to bacterial community characterization. The three sampling techniques investigated were: 1) Cut-out five 1x1 inch squares the HVAC filter, sonication in buffer solution, pre-filtration through 47um, filtration through 2 um filter which is used for DNA extraction; 2) Vacuum five 1x1 inch squares in 40um thimble, the dust cake accumulated was used for DNA extraction; and 3) Swab five 1x1 inch squares with PBST pre-moistened sterile swab, which tip was used for DNA extraction. For each of the techniques, seven independent samples were taken, along with positive and negative controls.

RESULTS AND DISCUSSION

The HVAC runtimes were successfully estimated with temperature sensors. For the sample of homes analysed to date, the runtimes in manufactured homes were found to be longer than those in detached homes. The lower insulation levels in the manufactured homes may explain the longer runtimes observed. The results to date indicate that homes with self-reported mold problems (Fig.1) have high ERMI values (>5) which is reported to be indicative of mold problems. However, further investigation is needed to evaluate homes with high ERMI values but without self-reported mold problems to confirm this relationship. Finally, the results to date indicate that the concentrations of two fungal species, *Cladosporium herbarum* and *Epicoccum nigrum* were statistically different between asthmatic homes and non-asthmatic homes ($p<0.05$; Wilcoxon Rank-sum test).

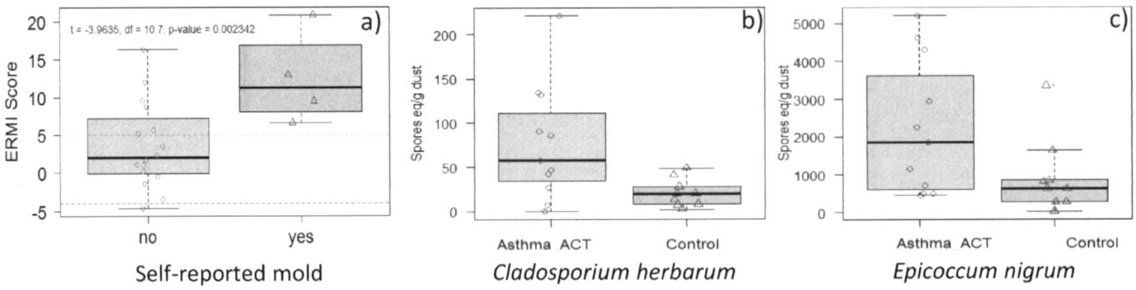

Fig 1. a) ERMI Score in HVAC dust samples collected in homes with and without self-reported mold problems, b and c) Fungal concentrations measured with qPCR.

Microbial community analysis results to date indicate that the bacterial diversity in dust samples obtained from HVAC filters did not correlate with the type of home, the presence of carpet floor, or the presence of pets. More than 20% of the bacterial sequences observed in the filter dust belong to human-associated taxa. *Corynebacterium*, *Acinetobacte*r, *Pseudomonas* and *Staphylococcus* were the predominant genera present in the available dataset.

The bacterial analysis of the community assemblages obtained from the tests performed to evaluate filter dust sampling techniques indicated that even though all three techniques provided similar results, the vacuum technique and the swab technique yielded the most consistent community compositions. The vacuum samples yielded the most diverse communities with the highest level of repeatability (Figure 2).

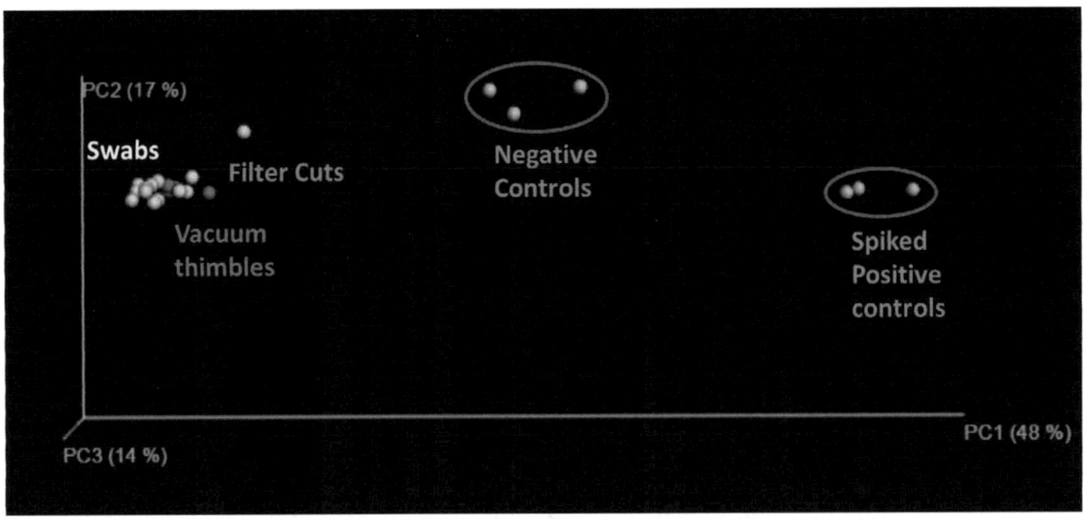

Figure 2. Weighted Unifrac PCoA plot showing the close clustering among swab, cut, and vacuum thimble samples collected from HVAC filter dust samples. Negative and positive controls for all three techniques are also represented. The axes represent the percentage of variance.

Allergens, phthalates, organophosphates, and endotoxins have been detected in the HVAC filter dust samples. Cat and dog allergen were successfully detected in HVAC dust samples collected from homes with pets. The data collected to date suggests that the manufactured homes had higher mouse allergen values than detached homes, potentially due to the lower quality of building materials and the higher potential for animal intrusion into homes. The measured concentrations of phthalates and organophosphates in HVAC filter dust and settled dust are shown in Fig 3 and compared with data in literature. The most frequently detected phthalate compounds are benzyl butyl phthalate (BBzP) and diethylhexyl phthalate (DEHP). Their concentrations in filter dust range from 100 to 2000 ug/g while most organophosphates are present within a lower range, from 20-100 ug/g. The levels of these SVOCs in filter dust were found higher than those detected in settled dust, which might be associated with the different dust sampling methods used or the more frequent use of vinyl based building materials as well as plastic hardware in low income homes.

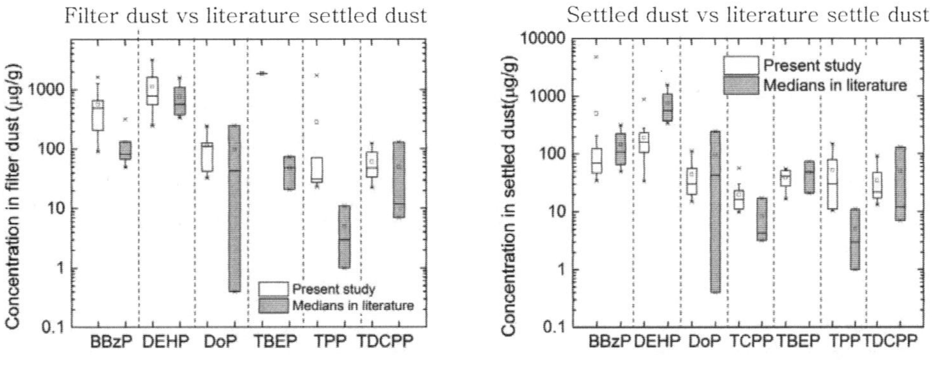

Fig 3. Concentrations of phthalates and flame retardants in HVAC filter dust

CONCLUSIONS

Unveiling the exposures occurring in homes is an essential step toward understanding how they affect the prevalence and severity of asthma and allergy in children. In this ongoing study, the microbial communities, allergens and chemical contaminants recovered HVAC filters in 60 rural homes are evaluated. The HVAC systems were characterized and their runtimes could be estimated with temperature sensors, revealing that manufactured homes HVAC systems were operating longer than those in detached homes. Regarding fungal communities, ERMI performed in the filter dust samples captured successfully self-reported mold problems. Additionally, bacterial communities recovered from filter dust samples revealed a significant presence of human-associated taxa. Also, the vacuum sampling technique yielded the greatest microbial diversity and the highest level of repeatability as compared to two destructive sampling methods that were evaluated. Results to date also show that phthalates and flame retardants concentrations detected in the HVAC filter dust were higher than in settled dust samples. The relationship between the contaminants studied and the severity of asthma for children is currently being evaluated.

ACKNOWLEDGEMENT

This project is funded by the U.S. Department of Housing and Urban Development (HUD: TXHHU0023-13). We want to acknowledge Fernando Almada, Wiley Jennins, Shuyi Yin, Andrew Romano and Shahana Kurshid for their help during the sampling campaigns.

RECOMMISSIONING A ZERO-CARBON MULTIFAMILY RESIDENTIAL BUILDING'S VENTILATION SYSTEM: A NECESSITY

Jelle LAVERGE[*], Eline HIMPE[1], Marc DELGHUST[1], Stijn VAN DE PUTTE[1] and Julio Vaillant REBOLLAR[1]

[1]Department of Architecture and Urban Planning, Ghent University, Ghent, Belgium

[*]Corresponding email: jelle.laverge@ugent.be

Keywords: Commissioning, ventilation, heat exchanger, flowrates

SUMMARY

With increasing airtightness, we rely more and more on ventilation systems to bring fresh and/or filtered air in for the occupants of the 'healthy buildings' we design. By applying heat recovery we hope to do this in a very energy efficient way. These systems, however, are rather complex and do not always operate the way we have planned them.
We recommissioned a ventilation system in a brand new, zero carbon multi family residential building in Belgium. In this paper, we present an overview of the issues we found: low actual ventilation rates, malfunctioning heat recovery, noise and leaks, demonstrating that, especially in situation were we aim for state of the art performance, commissioning is an absolute necessity.

INTRODUCTION

There is a tight trade off between achievable indoor air quality and the ventilation heat loss that it entails (J. Laverge and Janssens 7). Air to air heat recovery ventilation systems, when designed carefully (J. Laverge and Janssens 2012), allow to reduce the heat losses without impacting the indoor air quality. Therefore, they are an integral part of the popular 'passive house' concept (Feist et al. 11) that is often used as the basis to create zero carbon buildings. The drawback of these systems is that they are prone to lots of 'misuse' by all actors involved in their creation and operation. From literature, we know for example that flow rates rarely match the design flow rates (Balvers et al. 2012; Caillou and Van Den Bossche 2011) and that the occupants operate their ventilation systems at very low flow rates (Jelle Laverge, Delghust, and Janssens), achieving, in the end, comparable indoor air quality as in very leaky dwellings without any ventilation system.

Due to concerns about carbon induced climate change (IPCC 2007) and energy shortages, the European commission has issued a decree that requires every new building in 2018 to be nearly zero energy (EU 2010). In several research projects leading up to this decision, zero-energy and zero-carbon demonstration buildings were erected as living labs. The multi-family social housing project 'Venning' was is such a demonstration building within the EU-CONCERTO project 'Eco-Life'. It was completed in 2013. Consisting of 4 parallel blocks of 3-4 floors, it implements the passive house principles combining a very well insulated and extremely airtight envelope with rotary heat exchangers in collective ventilation units for each block (blocks 2 and 3 are served by the same unit). The air flowrates of all the ventholes were measured and adjusted by the contractor upon completion of the project. We recommissioned the ventilation system after 6 months of occupation as part of the monitoring scheme that was included in the project to demonstrate the performance.

METHODOLOGIES

The leakage level of the envelope was measured in accordance with the pressurization method (CEN 2001)(Minneapolis Blowerdoor 4, calibrated 2009). An interlaboratory test with this equipment showed that repeatability and reproducibility standard deviations of the ACH50 values are 1.5 % and 2.5 % respectively (Delmotte and Laverge 2011). The flow rates at each vent hole were measured (Acin Flowfinder Mk 2, ± 3 m3/h or 3%, calibrated 2011). Temperature (± 0.35 °C) and relative humidity (± 2.5 pp) in each of the indoor spaces, in the supply duct of each dwelling, at 3 point in the ventilation unit and the outdoor environment were logged with one logger per room (Onset Hobo U12) on a 5-minute interval. Additionally, in each of the dwellings, the carbon dioxide concentration in the living room and the master bedroom was monitored using a non-dispersive infrared carbon dioxide logger with an accuracy of ± 30 ppm ± 3 % (CO2meter K-33 ELG).

From the temperatures measured in the ventilation unit, the supply and exhaust heat recovery effectiveness of the rotary heat exchangers were calculated (CEN 1997).

RESULTS AND DISCUSSION

As can be seen in figure 1, repeated flow rate measurements revealed actual flow rates well below the design flow rates specified by the Belgian standard (NBN 1991). Measurements 1 and 2 were done to confirm the findings. Before the last measurement, the contractor intervened to improve the situation, but the situation was not fully remediated due to an unsustainable noise level in the dwellings if ventilation flow rates were at the required level. After the recommissioning, additional sound dampers were installed so that the ventilation flow rates could be improved further.

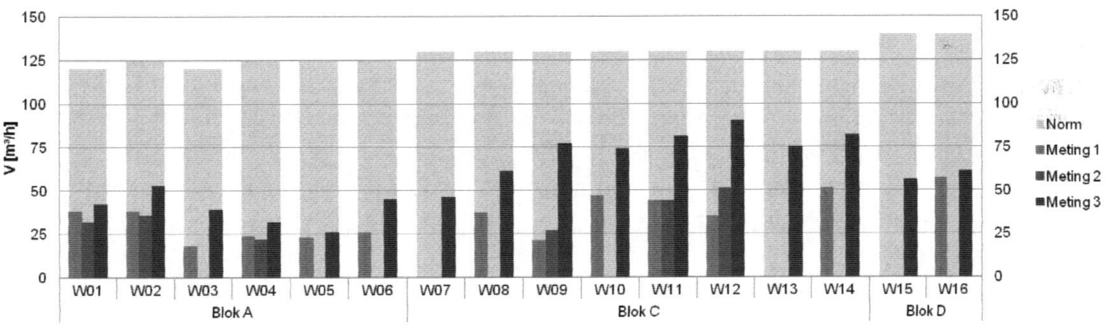

Figure 1. Measured flow rates in the living rooms of a number of dwellings across the different blocks of the project on 3 consecutive occasions and the required flow rate by the Belgian standard (light blue background).

The large inter-dwelling variation in flow rates, even within the same blocks, can be explained by user behaviour: the occupants reported to have 'adjusted' the position of the damper in the vent holes on several occasions. Since this is a collective system without automatic pressure control in the main ducts (still standard practise virtually everywhere), such actions can seriously alter the operation point of the fans. Residential systems are especially prone to this type of 'misuse' since occupants feel a much stronger will to be 'in control' of the indoor environment of their home.

Although the flow rates were in some cases very low, the number of complaints about low indoor air quality was not excessive due to very low occupancy levels in these initial 6 months.

Figure 2 shows the correlations between fresh air, reject air, supply and exhaust temperatures in the ventilation unit of blocks B and C during a 10 day measuring period as a function of outdoor temperature, as well as the supply and exhaust effectiveness calculated based on those measurements.

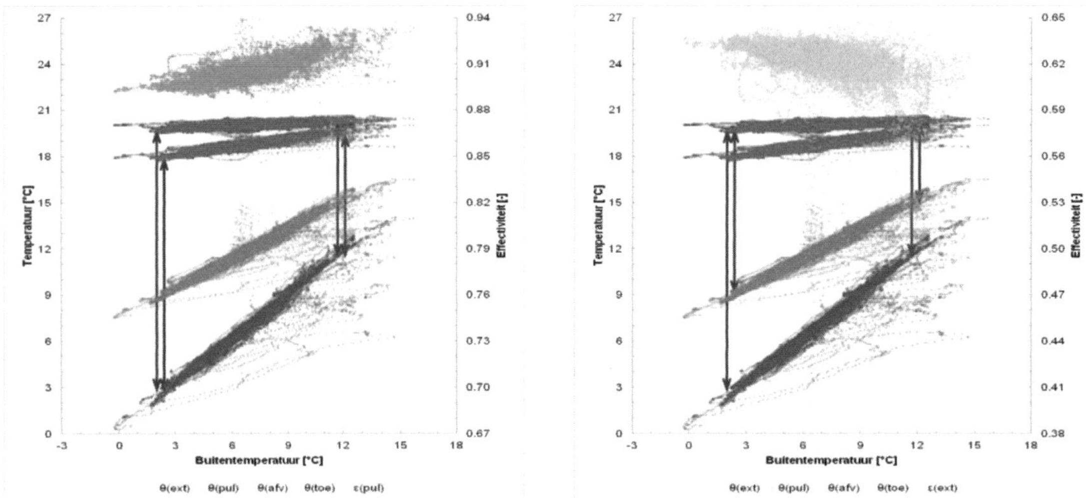

Figure 2. Measured supply, exhaust, fresh air and reject air temperatures for the ventilation unit of block B and C over a 10 day measuring period, as well as supply (left) and exhaust (right) effectiveness as a function of outdoor temperature.

The thermodynamic definition of effectiveness entails that the effectiveness should be constant and independent of the temperature differences across the heat exchanger if the supply and exhaust flow rates are balanced. A closer analysis of the effects seen in the supply and exhaust flow rates indicated that important internal leakages (up to 30% of the flow rate) between the supply and the exhaust flows, short circuiting the rotary heat exchanger.

It also showed that the rotary heat exchangers in Blocks A and D had broken down (unnoticed due to backup heaters), logically leading to almost 0 effectiveness (Figure 3.)

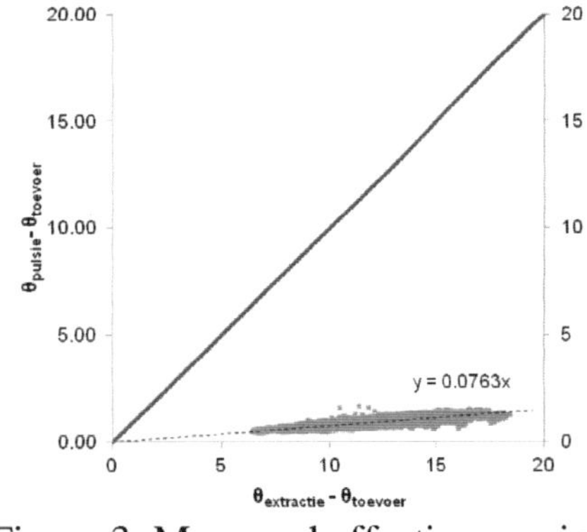

Figure 3. Measured effectiveness in the rotary heat exchanger of block A.

CONCLUSIONS

When heat recovery ventilation units are used to achieve a good trade-off between indoor air quality and required ventilation heat loss in multi-family residential buildings, special care is required to ensure that they are operated and maintained in their design conditions. Showing that, only six months after initial commissioning and completion, a collective heat recovery ventilation system in a zero-carbon demonstration project suffered from seriously lowered flow rates, unacceptable noise levels, leaky heat recovery units and broken down rotary heat exchangers without anyone noticing, the risk of supposedly 'healthy' and 'low energy' buildings to be that only in name and conception is disturbingly high.

ACKNOWLEDGEMENT

We would like to thank the occupants of the project who have participated in the recommissioning effort.

REFERENCES

Balvers, Jaap, Rik Bogers, Rob Jongeneel, Irene van Kamp, Atze Boerstra, and Froukje van Dijken. 2012. "Mechanical Ventilation in Recently Built Dutch Homes: Technical Shortcomings, Possibilities for Improvement, Perceived Indoor Environment and Health Effects." *Architectural Science Review* 55 (February): 4–14. doi:10.1080/00038628.2011.641736.

Caillou, S., and P. Van Den Bossche. 2011. "Optivent Project (IWT-VIS-CO-95076)."

CEN. 1997. *Heat Exchangers - Test Procedures for Establishing Performance of Air to Air and Flue Gases Heat Recovery Devices*. CEN.

———. 2001. *Thermal Performance of Buildings - Determination of Air Permeability of Buildings - Fan Pressurization Method (ISO 9972:1996, Modified)*. CEN.

Delmotte, Christophe, and J. Laverge. 2011. "Interlaboratory Tests for the Determination of Repeatability and Reproducibility of Buildings Airtightness Measuremetnts." In , 9.

EU. 2010. "Directive 2010/31/EU of the European Parliament and of the Council of 19 May 2010 on the Energy Performance of Buildings."

Feist, Wolfgang, Jürgen Schnieders, Viktor Dorer, and Anne Haas. 11. "Re-Inventing Air Heating: Convenient and Comfortable within the Frame of the Passive House Concept." *Energy and Buildings* 37: 1186–1203. doi:10.1016/j.enbuild.2005.06.020.

IPCC. 2007. *Climate Change 2007, Fourth Assessment Report*. Geneva: IPCC.

Laverge, Jelle, Marc Delghust, and Arnold Janssens. "Carbon Dioxide Concentrations in Bedrooms and Living Spaces and Humidity Levels Measured in Belgian Standard and Low Energy Dwellings with Common Ventilation Strategies." *International Journal of Ventilation* under review.

Laverge, J., and A. Janssens. 7. "Optimization of Design Flow Rates and Component Sizing for Residential Ventilation." *Building and Environment* 65: 81–89. doi:10.1016/j.buildenv.2013.03.019.

———. 2012. "Heat Recovery Ventilation Operation Traded off against Natural and Simple Exhaust Ventilation in Europe by Primary Energy Factor, Carbon Dioxide Emission, Household Consumer Price and Exergy." *Energy and Buildings* 50: 315–23. doi:10.1016/j.enbuild.2012.04.005.

NBN. 1991. *Ventilatievoorzieningen in Woongebouwen*. Brussels.

COLLEGE STUDENTS' PERCEPTIONS OF HOUSEHOLD ENERGY EFFICIENCY

Suchismita Bhattacharjee[1], Sandeep Langar[2], and Maggie Hogan[3]

[1]College of Architecture, University of Oklahoma, Norman, OK
[2]School of Construction, The University of Southern Mississippi, Hattiesburg, MS
[3]Student, University of Oklahoma, Norman, OK

Keywords: Energy Efficiency, Energy Behavior, Behavioral Psychology

SUMMARY

Commencing college-life as a student marks an important stage in an adolescent's life as they start to acknowledge responsibility for daily activities and exhibit individuality. Utilizing energy resources to execute daily life is a part of their overall behavioral psychology supported to an extent by the level of knowledge about the topic. The study identifies association between college students' personal beliefs, attitudes, knowledge, psychosocial correlates and energy conservation behavior. Energy conservation correlates were assessed with respect to knowledge about different energy efficiency measures, individual energy savings efforts, awareness about energy efficiency incentives and polices (state and municipality), and perceptions about importance of energy efficiency. The study utilized cross-sectional observation study to assess the energy conservation behavior among students (graduate and undergraduate) enrolled in a university geographically located in Southern US. Perceptions varied as per race, gender, age, and employment status.

INTRODUCTION AND BACKGROUND

Energy conservation and the use of renewable energy sources have gained paramount significance since the oil crisis of 1973, especially within the US. The country has been finding ways to limit consumption and utilize alternative energy sources. A fundamental imbalance between supply and demand defines the energy crisis in the country. According to Demeo and Taylor (1984), social and psychological correlates of human behavior needs to be scientifically explored and employed to achieve Energy Efficiency (EE). Following this, several researchers explored the relationship between energy consumption, human behavior, and race. A review of research since 1975 found that some of the major factors influencing energy consumption are age, income, education, and home-ownership (Bhattacharjee & Reichard, 2011; Guerin, Yust, & Coopet, 2000).

The cumulative impact of choices made by humans have the potential to determine the future of the planet (Shove, 2005). Largely, the adverse environmental impact perceived in modern times and anticipated in the future has its causal roots embedded in human behavior (Nickerson, 2003). Considering the future of the planet, one cannot ignore the looming threat of soaring energy usage by humans. A comprehensive review of existing literature suggests that psychological, demographic, and background characteristics of individuals does have an influence on energy consumption (Bhattacharjee, Reichard, McCoy, Pearce, & Beliveau, 2014). The goal of this study is to identify the association between college students demographic and background characteristics, personal beliefs, attitudes, knowledge and selected energy consumption behaviors. Energy consumption correlates were assessed with respect to knowledge about different energy efficiency measures, energy savings efforts taken by individuals, study participants' awareness about the state government or local municipal energy efficiency incentives and polices, and their perceptions about the importance of energy efficiency.

METHODOLOGY

The objectives of this study were achieved by conducting a cross sectional survey among undergraduate and graduate students at the University of Oklahoma for the academic year 2014-2015. The survey questionnaire was divided into three sub-sections, which included: 1) demographic and background data; 2) energy efficiency knowledge, attitudes, and beliefs data; and 3) personal energy consumption behaviors data.

RESULTS AND DISCUSSION

Approximately, 63.0% of the responding students were female and 90.0% of the respondents were in the age group of 18-20 or 21-30 years old. Analyzing the responses from the perspective of year of enrollment, at the time of the study, 30.9% were enrolled in freshman followed by 24.1% who were at the graduate level. Reviewing the participants from the perspective of educational major, a majority of respondents (4.4 %) were majoring in biology or premedical followed by 2.8% in Chemical Engineering. In addition, a vast majority (68.6%) of respondents resided off-campus. Among the respondents who resided off-campus, 55.5% stayed with roommates as compared to 87.5% of respondents on-campus living with roommates. Thus, a greater number of students staying on-campus stayed with roommates.

The authors have assigned number 1 for 'Strongly Agree' and 5 for 'Strongly Disagree' (2 = Agree; 3 = Not sure; 4 = Disagree). While the numbers only indicate the order, those numbers have been used to compute the descriptive statistics (Table 1).

Table 1. Respondents' Knowledge about Energy Efficiency

Respondents Energy Efficiency Knowledge	Mean	Mode	SD
US can meet future energy needs with the current rate of consumption	3.62	4	1.08
US is a net importer of energy	2.50	2	0.95
EE measures can help save money	1.63	1	0.72
Implementing EE measures can lower environmental damage	1.72	1	0.86
EE incentives/loans/rebates information available through utility companies or state/local government website are useful	3.00	2.98	0.86
Each and every householder can influence policies and decisions regarding the US's energy future	2.00	2.21	1.10

When respondents were asked to rank the US based on global ranking for energy consumption on a scale of 1 to 5 (1= lowest and 5= highest), the mode was 4 and mean 3.71. This is not in accordance with the International Energy Statistics data provided by Energy Information Administration which rated the US as the largest consumer of energy in 2012 (EIA, 2012). Further, the respondents also provided their opinion on the distribution of energy usage by the different equipment used at home as shown in Figure 1. Figure 2 shows the actual distribution of household energy consumption in 2009 by EIA. When performing a cross tabulation of energy usage distribution with the highest level of education completed, no major differences in responses were noted (Table 2).

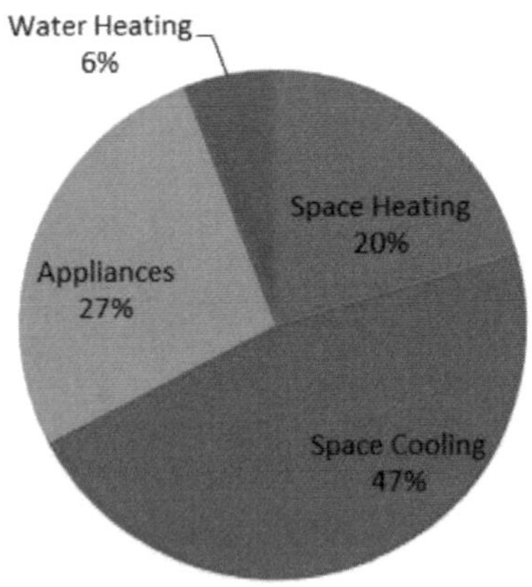

Figure 1. Respondents opinion on household energy consumption distribution

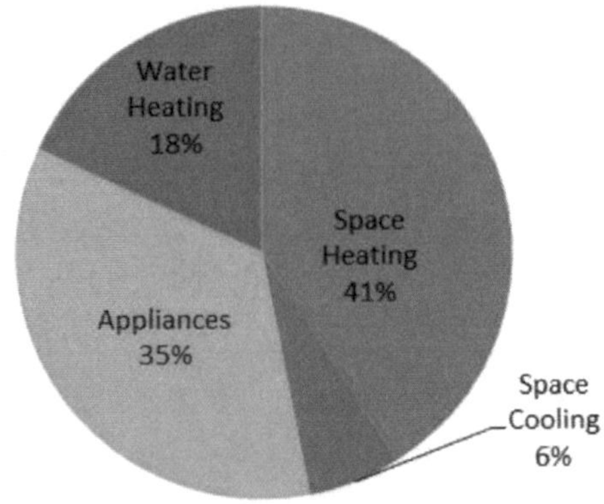

Figure 2. US household energy distribution in 2009 ((EIA, 2013)

Table 3. Crosstab of Energy Consumption distribution with level of Education

	Space Heating	Space Cooling	Appliances	Water Heating	Lighting	Other
Freshman Count (%)	35 (20.0%)	71(40.6%)	30 (17.1%)	13 (7.4%)	23 (13.1%)	3 (1.7%)
Sophomore Count (%)	17 (22.1%)	35 (45.5%)	12 (15.6%)	2 (2.6%)	9 (11.7%)	2 (2.6%)
Junior Count (%)	16 (18.4%)	39 (44.8%)	19 (21.8%)	6 (6.9%)	5 (5.7%)	2 (2.3%)
Senior Count (%)	20 (21.3%)	50 (53.2%)	13 (13.8%)	2 (2.1%)	7 (7.4%)	2 (2.1%)
Graduate Count (%)	28 (20.6%)	73 (53.7%)	15 (11.0%)	10 (7.4%)	5 (3.7%)	4 (2.9%)

When asked if the respondents have implemented EE measures in their home, approximately 68.0% responded yes. When asked about the benefits of implementing EE measures, a majority of the respondents selected reduced energy bill, followed by reduced environmental damage, health, comfort, and wellbeing of inhabitants and community and social benefits as shown in Figure3. Figure 4 depicts barriers to the process of adoption of EE measures as selected by the respondents. Cost is identified as a major barrier by 76.5% respondents, followed by not easily available, and time consumed to implement the measures.

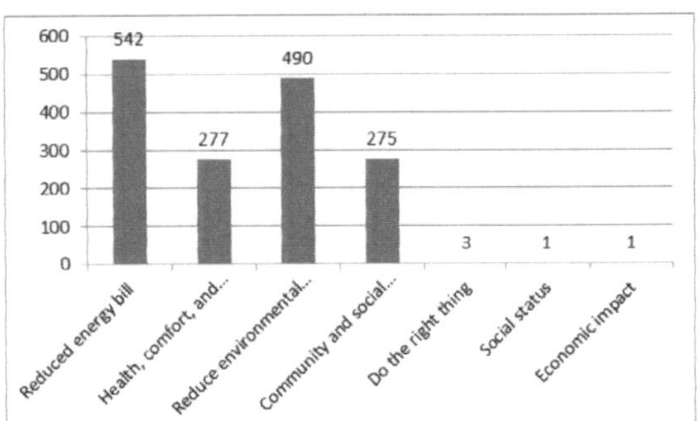

Figure 3. Respondent selection of benefits from Energy Efficiency (EE) measures

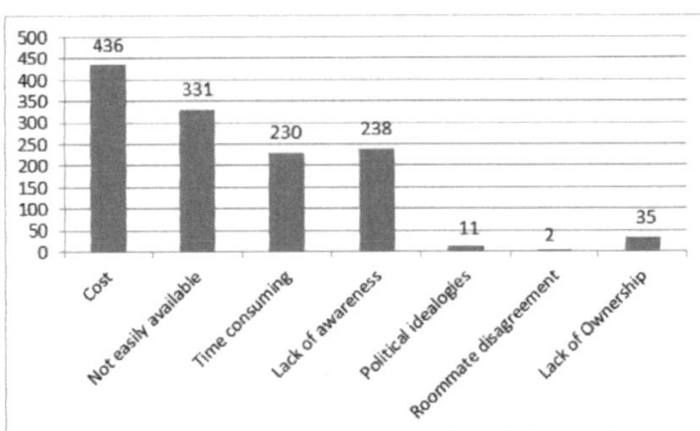

Figure 4: Respondents selection of barriers towards implementation of EE measures

Approximately 42.0% of the respondents mentioned that they have heard about EE incentives/ loans/ rebate programs as promoted through utility companies or state/government local website.

Additionally, several respondents selected different sources from which they have received EE information. As shown in Figure 5, a majority of students (75.8% respondents) identified the internet as a major source of information followed by books/magazines and utility company bills and pamphlets.

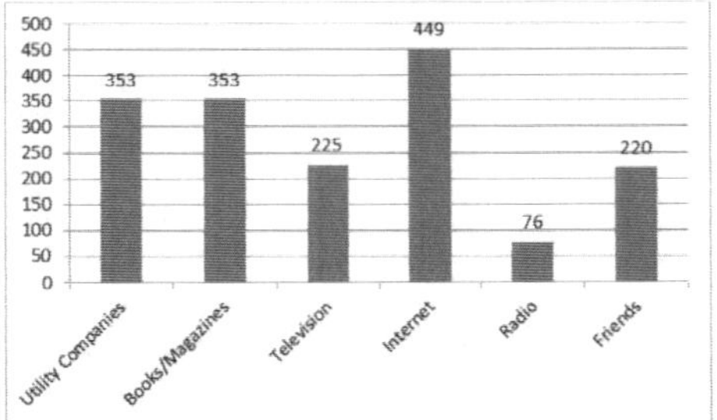

Figure 5: Respondent identification for source of EE information

CONCLUSION

Although respondents of the survey are aware of the existing imbalance within the US energy sector, they are unaware that current consumption patterns within the country were higher than the global norm. Respondents also possessed the knowledge to realize that the country is a net importer of energy and consumption patterns were an obstacle towards achieving independence from foreign energy. The study identified that the students lacked awareness about: incentive programs towards EE and comprehensive understanding of energy conservation, and the impact of subjectivity on such. Thus, one can conclude that even though students are aware of the problem at hand and want to respond appropriately, however they lack information/knowledge about resolving the issue at hand. Hence, the students need to be educated in this regard.

REFERENCES

Bhattacharjee, S., & Reichard, G. (2011). *Socio-Economic Factors Affecting Individual Household Energy Consumption: A Systematic Review.* Paper presented at the ASME 2011 5th International Conference on Energy Sustainability.

Bhattacharjee, S., Reichard, G., McCoy, A., Pearce, A., & Beliveau, Y. (2014). Identification of elements to control and regulate residential energy consumption. *Advances in Building Energy Research*(ahead-of-print), 1-22.

DeMeo, E. A., & Taylor, R. W. (1984). Solar photovoltaic power systems: An electric utility R & D perspective. *Science, 224*(4646), 245-251.

EIA. (2012). *Total Primary Energy Consumption*.

EIA. (2013). *State fact sheets on household energy use*.

Guerin, D. A., Yust, B. L., & Coopet, J. G. (2000). Occupant predictors of household energy behavior and consumption change as found in energy studies since 1975. *Family and Consumer Sciences Research Journal, 29*(1), 48-80.

Shove, E. (2005). Changing human behaviour and lifestyle: a challenge for sustainable consumption?

INDOOR ENVIRONMENTAL RESILIENCE: A REVIEW AND DISCUSSION

Andrew PERSILY[1,*] and Steven EMMERICH[1]

[1] National Institute of Standards and Technology, Gaithersburg, Maryland USA

[*]Corresponding email: andyp@nist.gov

Keywords: Community resilience, disasters, floods, heat waves, indoor air quality, power failure, resilience, wildfires

SUMMARY

NIST is developing a planning guide to define strategies to increase community-based resilience in the face of a broad range of natural disasters and other extreme events. Many of these events will affect indoor environmental quality (IEQ) either through increases in airborne contaminant levels or challenges in maintaining acceptable indoor environmental conditions during the event and afterwards in the recovery phase. However, the elements of indoor environmental resilience (IER) have not been identified and discussed in a systematic fashion, which needs to be done to determine the role of these factors in community resilience programs. This paper summarizes a longer review of existing information, standards, programs and other technical resources related to disasters that are likely to impact IER. That report describes the scope and potential impacts, current activities, important gaps requiring research, and needs for standards and guidance.

INTRODUCTION

The NIST Community Disaster Resilience Planning Guide (NIST 2015) addresses resilience of buildings and infrastructure systems at the community scale and provides guidance on establishing long-term goals and plans for recovery of the built environment following a disaster with consideration of social needs. In addition to developing the planning guide, the NIST resilience program is pursuing other activities in support of the overall program goals. This paper summarizes a longer report describing one such activity, an effort to explore the role of IEQ in the context of community resilience (Persily and Emmerich 2015). This work is motivated by the fact that many of the disasters being considered under the program will also affect IEQ. These effects include both increased airborne contaminant concentrations associated with the disaster or its aftermath and challenges in providing acceptable environmental conditions (such as cooling) during an event or afterwards during recovery.

In considering resilience in the context of the indoor environment, it is important to note what the indoor built environment is expected to provide to occupants. Two key objectives are to maintain thermally comfortable conditions and to limit the concentrations of airborne contaminants to safe and comfortable levels. The indoor environment, primarily via the building enclosure, is also intended to isolate the building occupants from the exterior environment, specifically precipitation, pests and threats to physical security. Finally, the indoor environment is expected to provide amenities such as light, power and food storage to support the intended activities of the occupants including working, learning or residing. In considering IER, relevant features of the indoor environment include those that impact occupant health, comfort and productivity, i.e., IEQ. The four primary elements of IEQ are indoor air quality (IAQ), thermal comfort, acoustics and illumination (ASHRAE 2011).

The role of IEQ issues in the context of disasters has been identified in two prominent documents. The National Climate Assessment summarizes the impacts of climate change on the U.S., now and in the future, and highlights several anticipated changes that are relevant to IEQ (Melillo et al. 2014). Under extreme weather, the increased frequency of heat waves, heavy downpours, floods and hurricanes are all noted. The discussion of human health mentions vulnerable people and communities, wildfire smoke, increased levels of pollen, and impacts on asthma and allergies. Disruptions in energy production and delivery are noted under infrastructure. The Institute of Medicine published a study on climate change and indoor environmental health in 2011, which contained several key findings: poor IEQ is creating health problems today, impairing the ability of occupants to work and learn; climate change may worsen existing IEQ problems and introduce new ones; and, there are opportunities to improve public health while mitigating or adapting to alterations in IEQ induced by climate change (IOM 2011). It also noted several problematic indoor exposures including: indoor contaminants; dampness, moisture and flooding; infectious agents and pests; thermal stress; and, building ventilation, weatherization, and energy use.

More recently, Fisk (2015) reviewed the potential health consequences of climate changes affecting indoor environments including consideration of the IOM report, recent efforts under the Intergovernmental Panel on Climate Change (IPCC 2013) and other resources. Fisk discusses the potential health impacts of increases in urban airborne ozone concentrations (not considered in this report) as well as increases in the frequency and severity of heat waves, flooding associated with severe storms and sea level rise, and wildfires.

The report summarized in this paper identifies and discusses the elements of what is referred to here as indoor environmental resilience through a review of existing information, programs and other technical resources related to events that are likely to impact IER. For each event, existing standards and guidelines are described, as well as other programs and activities to support planning and response strategies.

SCOPE OF INDOOR ENVIRONMENTAL RESILENCE

The first step in this effort was to consider the extreme events with the potential to impact IEQ and which may merit planning in support of increased community resilience. Table 1 lists the events considered, along with the associated indoor environmental exposures.

Table 1 Events relevant to indoor environmental resilience

Type of event	Indoor environmental exposure
Heat waves	High indoor temperatures/heat stress High levels of outdoor pollution
Storms causing power failure	Lack of heating, cooling, ventilation leading to heat/cold stress and elevated indoor contaminant levels CO exposure from portable generators
Floods and mold exposure	Microbial growth affecting occupants and remediation workers
Wildfires	Particulate and other contaminant exposure
Airborne releases of chemical, biological or radiological agents	Exposure to agent

Each of the events listed in Table 1 was analyzed in terms of what is known about the scenarios of interest and their impacts in the U.S. This effort involved examining available information on these events and their impacts on IEQ, technical gaps in understanding these impacts, existing standards, and how the events are being addressed by various guidance documents. Following this discussion of each of the topics, the report discusses two other issues that are relevant to IER: pandemics and the role of healthcare facilities, and indoor environmental conditions in safe rooms and shelter-in-place facilities. These discussions are followed by a review of existing standards and guidelines relevant to IER.

RESULTS AND DISCUSSION

While the amount of available technical information varies across the events considered, it is clear that these topics have received growing interest in recent years due to high profile events, such as heat waves, hurricanes and major storms, wildfires and terrorist attacks, as well as government initiatives in response to these events. Despite the attention given to these issues, the impact of IER issues is not fully appreciated. For example, during the period of 1979 to 2003, more people in the U.S. died from extreme heat than from hurricanes, lightning, tornadoes, floods, and earthquakes combined (CDC 2012).

This review identified important knowledge gaps meriting research, as well as the need for improved and more relevant standards and guidance. Additionally, much of the existing knowledge needs to be better integrated into a more comprehensive community resilience approach to maximize its impact. The important research gaps that were identified include the following, organized by the type of event:

- Heat waves: Development and evaluation of passive building design approaches and retrofit measures to avoid overheating during heat waves. Such research needs to consider a variety of building types (e.g., single family and high-rise residential, institutional) and building occupants (i.e., beyond healthy adults).
- Power outages: Definition of short term acceptable ventilation and IAQ conditions (beyond thermal comfort) for living and working in buildings temporarily during power outages that impact HVAC system function.
- Floods: Coupled thermal/airflow/moisture simulation tools to better predict conditions that will lead to the potential for mold growth.
- Wildfires: Infiltration of smoke into buildings and the use of air cleaning to create clean air shelters in buildings.
- Airborne CBR releases: Building protection approaches based on design and system operation in new and existing buildings. Tools to identify buildings most likely to be impacted by outdoor releases. Determination of how clean is clean enough after decontamination. Tools to support deciding between evacuation and sheltering in place.
- Pandemic response in healthcare facilities: Evaluation and comparison of options to create surge airborne isolation space and temporary negative pressure isolation space, and the impacts on overall building operation.
- Sheltering in place: Development of guidance for community-wide sheltering in response to events such as heat waves, CBR releases, wildfires, and power outages.

Several topics for potential standards development were also identified during this effort and are summarized below. Some are motivated by the fact that most published standards and guidance relevant to IEQ consider only normal operating conditions for buildings and healthy adult occupants, e.g., ASHRAE Standards 55, 62.1 and 62.2, which cover thermal comfort and ventilation (ASHRAE 2010, 2013a and 2013b). A need exists to develop standards and guidance to address these requirements during or following a disaster, when indoor conditions may not be consistent with normal operation and building use. Specific standard and guidance needs identified in this effort include the following:

- Thermal "comfort" standards or guidelines that define conditions that are safe for occupants (including other than healthy adults) during heat waves and power outages.
- Ventilation standards or guidelines that covers extreme conditions, which are beyond the scopes of Standards 62.1 and 62.2 (ASHRAE 2013a and 2013b). Such documents might include separate requirements for safe rooms and shelter-in-place facilities.
- Current efforts underway at the U.S. Consumer Product Safety Commission and UL to address CO emission limits from portable generators should be continued.
- Guidance should be developed and provided for homeowners and volunteers engaged in mold/wet building cleanup following large scale flooding events.
- Guidance to support deciding between evacuation and sheltering-in-place in response to wildfires and CBR releases.
- Standards for portable air cleaner performance to reduce indoor particulate exposure during wildfires, and guidance on system selection.

CONCLUSIONS

As efforts to increase community disaster resilience continue, the indoor environmental impacts need to be considers and their proper role identified. The report summarized in this paper provides the background to support these discussions.

REFERENCES

ASHRAE (2010). Standard 55, Thermal Environmental Conditions for Human Occupancy. American Society of Heating, Refrigerating and Air-Conditioning Engineers, Inc., Atlanta, GA.

ASHRAE (2011). Guideline 10, Interactions Affecting the Achievement of Acceptable Indoor Environments. American Society of Heating, Refrigerating and Air-Conditioning Engineers, Inc., Atlanta, GA.

ASHRAE (2013a). Standard 62.1, Ventilation for Acceptable Indoor Air Quality in Low-Rise Residential Buildings. American Society of Heating, Refrigerating and Air-Conditioning Engineers, Inc., Atlanta, GA.

ASHRAE (2013b). Standard 62.1, Ventilation for Acceptable Indoor Air Quality. American Society of Heating, Refrigerating and Air-Conditioning Engineers, Inc., Atlanta, GA.

CDC (2012). Emergency Preparedness and Response, Extreme Heat. http://emergency.cdc.gov/disasters/extremeheat/

Fisk, WJ (2015). Review of Some Effects of Climate Change on Indoor Environmental Quality and Health and Associated No-Regrets Mitigation Measures, Building and Environment. **86**, 70-80.

IOM (Institute of Medicine) (2011). *Climate Change, the Indoor Environment, and Health.* The National Academies Press.

IPCC (2013). Climate change 2013, the physical science basis, Working Group 1 contribution to the IPCC 5th Assessment Report, Final draft underlying scientific-technical assessment. Geneva, Switzerland: Intergovernmental Panel on Climate Change.

Melillo JM, Richmond TC, and Yohe GW, Eds. (2014). *Climate Change Impacts in the United States: The Third National Climate Assessment.* U.S. Global Change Research Program.

NIST (2015). Disaster Resilience Planning Guide, http://www.nist.gov/el/building_materials/resilience/guide.cfm.

Persily AK and Emmerich SJ. (2015). Indoor Environmental Issues in Disaster Resilience, National Institute of Standards and Technology, NIST Technical Note 1882, Gaithersburg, MD.

AIR QUALITY IN DATA CENTERS: PEOPLE VS. THE MACHINES

Chris MULLER[1,*], Chuck ARENT[1]

[1]Purafil, Inc., Doraville, Georgia, USA

[*]Corresponding email: cmuller@purafil.com

Keywords: Air quality, corrosion, data center, electronic equipment reliability, free cooling.

SUMMARY

When one hears the phrase "indoor air quality" or IAQ, most associate this with the health, well-being, and comfort of humans in an occupied space. However, in mission critical facilities such as data centers, IAQ is being scrutinized less for the human occupants and more for the "health" of the critical informational technology (IT) and datacom equipment.

Regulatory changes in place since 2006 resulted in much higher failure rates for IT and datacom equipment in facilities located in regions with high air pollution levels. The use of outdoor air for free cooling as a way to reduce energy costs has reached the mainstream of data center design and for many companies it is now the standard design approach for all new facilities. However, as the use of free cooling expands many locations are experiencing higher equipment failure rates due to the effects of gaseous pollutants, higher temperatures, and fluctuating humidity inside the data center.

This doesn't mean that free cooling should not be considered where feasible; it is just that a few additional steps are required to assure reliable operation of datacom equipment.

This paper will present:

- Air quality standards for datacom environments.
- Updates on ongoing environmental concerns.
- An overview of free cooling with respect to issues affecting electronic equipment reliability.
- Free cooling case studies with and without application of contamination assessment, control, and monitoring programs.

Key words: Air quality, corrosion, data center, electronic equipment reliability, free cooling.

INTRODUCTION

Indoor air quality (IAQ) is a term that resonates throughout the commercial and residential building sectors and refers to the air quality within and around buildings and structures, especially as it relates to the health and comfort of building occupants. IAQ can be affected by airborne contaminants such as fine particulate matter and gaseous contaminants whose source can be either indoors or outside the building. Poor IAQ can result in occupant complaints due to odors, irritation of eyes and mucous membranes, allergic reactions, and other acute and chronic symptoms.

Particulate and gaseous contaminants cannot only be detrimental to the health of the human occupants of a building, they can also be responsible for damage to other building "occupants" such as the sensitive electronic and electronic devices used in IT and datacom equipment. Whereas IAQ for humans can be affected by indoor or outdoor pollutants, the failure of information technology (IT) equipment is almost exclusively caused by pollutants with sources outside the building. These include sulfur and nitrogen oxides (SOx, NOx), hydrogen sulfide (H_2S), ozone (O_3), chlorine (Cl_2), diesel particulate matter (DPM), and fine and ultrafine particulate matter ($PM_{2.5}$, $PM_{0.1}$) generated from motor vehicle exhaust.

Mass produced commercial computing started in the early 1950s in response to the Korean War. Tom Watson, Sr. asked the president of the U.S., "What can IBM do to help?" The response was "build some computers that can be used by defense contractors to design build ships, aircraft, weapons, etc… things that take years to manually calculate." IBM's response was the 701 Electronic Data Processing Machines in which the central processor alone had 4000 vacuum tubes. It was the first mass produced processor, with 13 of them built between 1952 and 1954. Follow on processors were developed and in increasing quantity. A major technological breakthrough came later that decade with transistors replacing the vacuum tubes.

There have been major advancements through the decades. It the 1980s circuit heat density reached the point where water was needed to transfer heat from the then dominant bipolar transistor technology. Complementary metal-oxide semiconductor (CMOS) technology advancements greatly reduced power consumption and conventional air cooling returned. Currently, we are again faced with the downside of microelectronics and high power densities and thus the need to reintroduce liquid cooling and investigate new methods of data center cooling.

Many things have changed over the past 50+ years. All the support systems and hardware have advanced as well as construction techniques and building materials. Not the least of the changes is the dependency on high availability systems and resources and the worldwide demand and expectations. The basic elements remain the same, just more of it. While the computer room was a "showcase" at a point in time, now clients feel the need to hide them. In the end there is still an air-conditioned, controlled access space just like in the first data center design.

DATA CENTER EQUIPMENT RELIABILITY

The physical environment surrounding a printed circuit board (PCB) is defined by the temperature, humidity and gaseous and particulate contamination in the air. Environmental factors can cause PCBs to fail in two ways: First, electrical open circuits can result from corrosion, such as the corrosion of silver terminations in surface mount components. Second, electrical short circuits can be caused by (a) copper creep corrosion, (b) electrochemical reactions such as ion migration and cathodic-anodic filamentation or (c) settled, hygroscopic particulate matter contamination reducing the surface insulation resistance between closely spaced features on PCBs.

In 2006, the European Union's RoHS directive banning the use of lead in solders led to changes in PCB finishes and the elimination of lead from solders (European Commission 2003). These changes dramatically increased the PCB failure rates due to sulfur creep corrosion. Another common failure mode during this period was that of surface mount resistors suffering open circuits due to the corrosion of their silver terminations. IT equipment manufacturers have since learned to make their hardware more robust against these two failure modes, which used to occur predominantly in geographies with high levels of sulfur bearing gaseous contamination (Muller, et al. 2014). However, even as these IT equipment manufacturing changes were being widely implemented, a change in the design and operational practices used for the data centers in which this equipment was being used has caused a renewed concern with regards to reliability and uptime.

FREE COOLING FOR ENERGY CONSERVATION

It has been reported that data centers account for up to 1.5% of global electricity use and that in the U.S.A. it may be as high as 2.5%. Of the electricity consumed in a data center, as much as 50% or more goes towards cooling the IT equipment. And with the total number of data centers worldwide surpassing 3 million, energy consumption is only be expected to increase. This has not gone unnoticed and significant efforts are being made to reduce the overall energy consumption of data centers – specifically the energy required for cooling. One data center design approach that can offer tremendous energy savings is the use of economizers.

In most climates, data center cooling can be satisfied when the ambient air temperature is cold enough to supplement or replace the refrigeration equipment. This process is known as an "economizer" or "free cooling" cycle because the refrigeration equipment can be shut off and cooling can instead be provided from the outdoor air with lower energy consumption.

ASHRAE Technical Committee 9.9 for Mission Critical Facilities, Data Centers, Technology Spaces and Electronic Equipment created the first edition of the "Thermal Guidelines for Data Processing Environments" in 2004 (ASHRAE 2004). In 2008 and 2011 the temperature and humidity ranges for data centers were expanded so that an increasing number of locations throughout the world would be able to operate with more hours of economizer usage.

In certain geographical locations, economizers can be run using 100% outdoor air for part or all of the year to meet both the cooling and ventilation requirements for data centers. Although this addresses the goal of overall lower energy use in data centers, it can come with a hidden penalty. This being an increase in the amount of particulate and gaseous contamination coming into the data center along with the outdoor air and a concurrent increase in IT equipment reliability concerns.

The rapid expansion of the IT equipment market in the polluted geographies of Asia that have high levels of gaseous and particulate contaminants in the ambient air and the increasing use of free cooling has resulted in an uptick in corrosion related IT equipment failure rates that had been declining in previous years.

AIR QUALITY AND THE EFFECTS OF CONTAMINATION IN DATA CENTERS

Three types of gases are the prime suspects in the corrosion of electronics: acidic gases, such as hydrogen sulfide, sulfur and nitrogen oxides, chlorine, and hydrogen fluoride; caustic gases, such as ammonia and amines; and oxidizing gases, such as ozone. Of the gases that can cause damage to electronic devices, acidic gases are typically most harmful.

Although less of a problem than gaseous contaminants, particulate matter has come under increased scrutiny (Singh et al. 2015). Most attention is being paid to the fine particles

(<2.5µm), of which there are of two types. Primary particles are directly emitted from a source. Secondary particles, which comprise the bulk of the fine particulate pollution, are those formed as a result of chemical reactions in the atmosphere. SO_2 and NO_2 can interact with <0.1 µm carbonaceous material seed particles in a complex, multi-step photochemical process to produce sulfuric and nitric acids. The presence of ozone, a contaminant increasing in both outdoor levels as well in regional coverage, serves to catalyze these reactions.

Each site may have different combinations and concentration levels of corrosive gaseous and particulate contaminants and IT equipment performance degradation can occur rapidly in weeks or months or over many years, depending on the concentration levels and combinations of contaminants present at a site. The currently prescribed level of contaminant control in data centers is listed in Table 1.

Table 1. Particulate and gaseous contamination guidelines for data centers (ASHRAE 2011)

Data centers must be kept clean to ISO 14644-1 Class 8 (ISO 2014). This level of cleanliness can generally be achieved by an appropriate filtration scheme as outlined here: 1. The room air may be continuously filtered with MERV 8 filters (G4/F5, 25-30% dust spot). 2. Air entering a data center may be filtered with MERV 11 (F6, 60-65% dust spot) or MERV 13 (F7, 80-90% dust spot) filters. Sources of dust inside data centers should be reduced. Every effort should be made to filter out dust that has deliquescent relative humidity greater than the maximum allowable relative humidity in the data center.
Gaseous contamination should be within the ISA-71.04-2013 severity level of G1-Mild that meets: 1. A copper reactivity rate of less than 300 angstroms (Å) per month, ***and*** 2. A silver reactivity rate of less than 200 Å per month.
For data centers with higher gaseous contamination levels, gas-phase filtration of the inlet air and the air in the data center is highly recommended.

THE COST OF FREE COOLING

Where free cooling is being employed the number and types of data center equipment failures have increased dramatically in locations with high pollution levels. Because most failures occur in the most common components, there is no simple solution. Faced with this reality, the world's leading IT and datacom equipment manufacturers changed have their warranties to include requirements for the control of corrosion due to gaseous contamination.

Another concern with free cooling is the degree of temperature and relative humidity (RH) control possible. Whereas temperature appears to be generally well controlled, RH is subject to wide fluctuations; with rates of change as high as 10-15% per hour being observed. Couple this with upper boundaries for the recommended and acceptable level of RH being expanded and the potential for equipment failures increases even in areas with moderate pollution – especially for corrosion-related failures due to gaseous contamination.

Corrosion of metals due to gaseous contaminants is a chemical reaction that is accelerated by heat and moisture. Rapid shifts in either temperature or humidity can cause small portions of circuits to fall below the dewpoint temperature, facilitating the condensation of contaminants. RH above 50% accelerates corrosion by forming conductive solutions on electronic components. These microscopic pools of condensation can absorb contaminant gases to become electrolytes where crystal growth and electroplating occur. Above 80% RH, electronic corrosive damage will occur regardless of the levels of contamination.

Case Study: ENI Green Data Center

Located in Ferrera Erbognone (Pavia) the ENI Green Data Centre will host all of Eni's central computer processing systems, both for information management and seismic simulation processing (High Performance Computing). It will provide 5,200 m^2 of usable space, up to 30MW of IT power and up to 50kW/m^2 of energy density. Construction began in 2011 and the data center was officially opened in October 2013

The data center cools its computers using air directly from outside 80% of the year. This direct free-cooling technique means the air conditioners are switched on less than 25% of the time. It consists of 6 individual modules each with 1.4MM m^3/hr (825,000 cfm) of air being delivered to the data center when in free cooling mode.

The plant was built in the immediate vicinity of the Enipower Ferrera Erbognone power station, which provides the data center's energy requirements. The readily available power is produced by gas turbines with natural gas; a clean-burning fuel that produces little or no pollution.

This location provided readily available power; however, it was also adjacent to the Eni Sannazzaro de 'Burgondi refinery (Figure 1). This presented a significant concern due to the potential for high levels of corrosive gases – especially hydrogen sulfide (H_2S), being introduced directly into the data center.

Figure 1. ENI Green Data Center (Trifoglio Nord and Trifoglio Sud) showing location of the Enipower Ferrera Erbognone power station and the Eni Sannazzaro de 'Burgondi refinery .

Corrosion Classification Coupons (CCCs) were placed on the greenfield site in 2010 prior to the start of construction and the results indicated pollutant levels would not meet the environmental specifications and warranty requirements for the IT equipment. Adding to these air quality concerns was the construction project underway to expand the refinery's production capacity.

The data center design provide for filtration of the outdoor air through two stages of particulate filters; medium efficiency prefilters and high efficiency V-bank final filters. Given the volume of air that was required for cooling, almost 7,000 filters were installed. This provided for the control of dust and other particulate matter, however, no allowances were made for the use of chemical filters in the original design – even after the results of the CCC analyses were made available.

As the data center construction was nearing completion (June 2014) additional CCCs were placed inside four of the completed modules. One coupon was placed in front of the prefilters in each module to check outdoor air quality and several others placed inside of one of the modules where IT equipment had been installed. As expected, the CCC analysis indicated that the outdoor air **and** the air being delivered to data center exceeded manufacturers' allowable pollutant levels with all locations showing high levels of sulfur corrosion.

Ongoing monitoring will be used to develop baseline corrosion data over the course of one year to determine if chemical filtration should be installed. Given the location of the data center along with all of the air monitoring data collected to this point, it would seem a clear-cut decision to do so. Of course this would increase operating costs, but would also guarantee the ongoing operation and reliability of the IT equipment housed inside the data center.

Case Study: Idea Cellular

This company had experienced continuing problems with corrosion-related failures of switchgear and network cards in one of their mobile switching centers (MSC). A chemical filtration system was installed and the number of failures per month dropped from an average of 36 per month to ~20 after five months of operation. Although this represented a significant improvement in the operational status of the MSC, the owner wanted to improve this number even further.

The chemical filtration system had been designed to deliver the specified amount of cleaned air to the MSC when running at 70% capacity. The system was adjusted to increase the amount of air being delivered to the MSC. The system is now operating at 90% capacity and since then the number of failures has dropped below 10 per month (Figure 2).

CONCLUSIONS

IAQ is a term that not only has relevance to humans but also when considering the health of the nonhuman "occupants" of mission critical environments such as data centers.

Miniaturization of electronic components combined with reductions in feature spacing on PCBs and the loosening of the data center temperature and humidity envelope to save energy is making electronic hardware more prone to failure due to exposure to ambient pollutants.

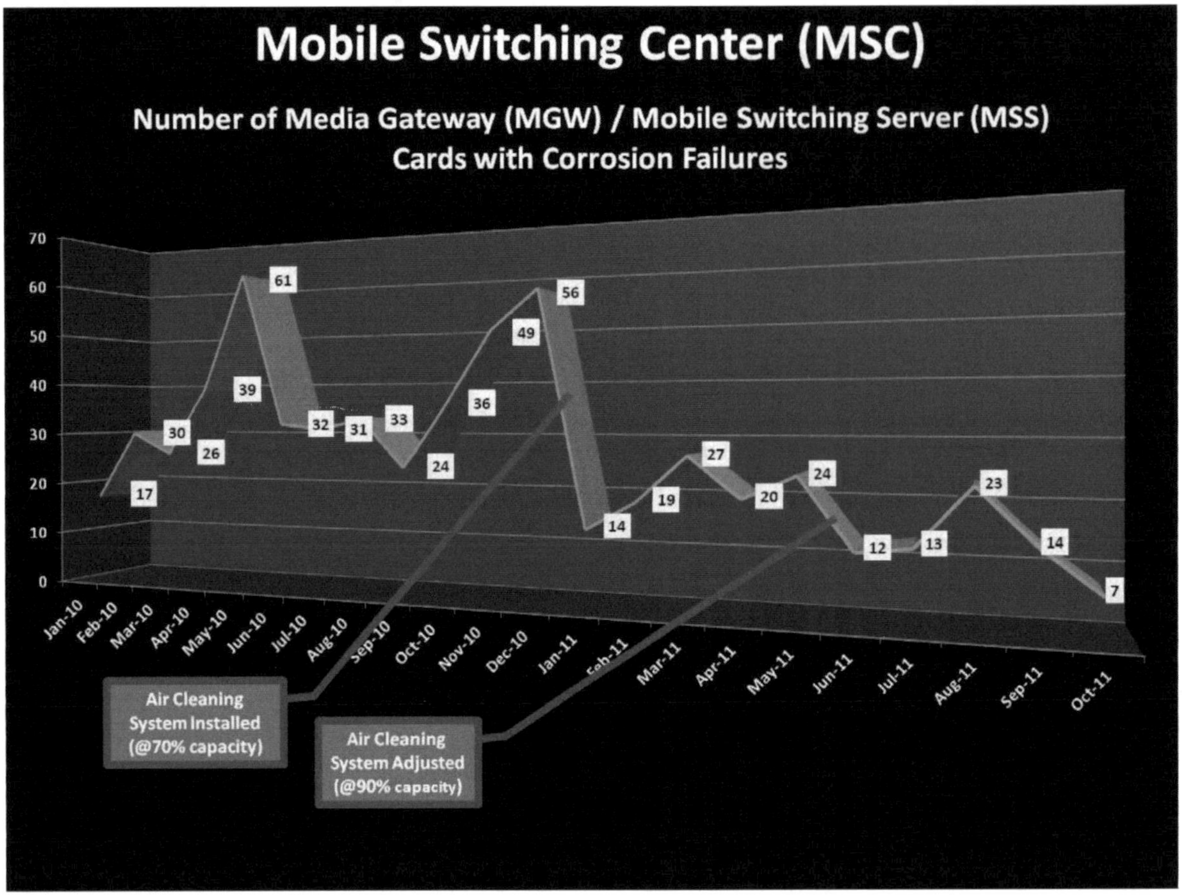

Figure 2. Number of MGW and MSC cards with corrosion failures.

Increasing the maximum allowable temperature and RH ranges for IT equipment means free cooling can be used in more locations than ever before. While this has led to dramatic energy savings and overall lower operational costs, in many locations this has come at the cost of equipment reliability. Although climatic conditions may allow for the use fee cooling, other factors now have to be considered. Primary among these are local and regional air quality.

This does not mean that free cooling should not be considered where feasible; just that additional steps are required to assure reliable operation of datacom equipment. These steps include a determination of the types and levels of particulate and gaseous contamination, establishing a monitoring program inside the data center to check against standard levels and equipment warranty requirements, and lastly upgrading or adding the required type of filtration to remove and reduce contamination to manufacturers' requirements.

REFERENCES

ASHRAE (2011). 2011 Gaseous and Particulate Contamination Guidelines for Data Centers. Atlanta: American Society of Heating, Refrigerating, and Air-Conditioning Engineers, Inc.

ASHRAE (2012). ANSI/ASHRAE Standard 52.2-2012: Method of Testing General Ventilation Air-Cleaning Devices for Removal Efficiency by Particle Size. Atlanta: American Society of Heating, Refrigerating, and Air-Conditioning Engineers, Inc.

Directive 2002/95/EC of the European Parliament and of the Council of 27 January 2003 on the Restriction of the use of Certain Hazardous Substances on Electrical and Electronic Equipment Official Journal L 037, February 13, 2003, 19-23.

Muller C, Singh PJ, White GH, and Finch P, "Solving Air Contaminant Problems in Data Centers: RoHS-Compliant Products Exacerbate the Problem," *The Uptime Institute Journal*, May 2014, Vol. 3: 24-33.

Singh, PJ, Ruch P, Saliba S, and Muller C (2015). Characterization, Prevention and Removal of Particulate Matter on Printed Circuit Boards. Proceedings of IPC APEX EXPO, San Diego, CA, 24-26 February 2015.

A GENERAL METHOD TO PREDICT VENTILATION REQUIREMENTS WHILE INCORPORATING BUILDING MATERIAL EMISSION TESTING

Wei YE[1, 2, *], Doyun WON[3], Xu ZHANG[2], and Qi ZHOU[1]

[1]College of Environmental Science and Engineering, Tongji University, Shanghai, China
[2]School of Mechanical Engineering, Tongji University, Shanghai, China
[3]NRC Construction, National Research Council Canada, Ottawa, Ontario, Canada

[*]Corresponding email: weiye@tongji.edu.cn

Keywords: Ventilation rate, Building material, Volatile organic compounds, Testing

SUMMARY

Building material emissions could be the major source for indoor air pollutants in low-occupant-density spaces, e.g., a private office. Lowest Concentrations of Interest (LCI) references could be used to predict ventilation requirement and the idea and testing procedure could be utilized to build a database that consist of material emission data. Thus, the determined ventilation rate based on an emission database could be generally applied for diverse usage. This paper proposed a general method to predict ventilation requirement while incorporating such database. Three LCI references currently used in EU were adopted to demonstrate the procedure. The results showed that standard ventilation requirement (N_i) could be pre-determined for each material in a database, the ultimate ventilation rate could be determined by adding up all the N_is, after revising all the material loadings. And only two parameters (i.e., diffusion coefficient, D; initial emittable concentration, C_0) were required to build such a database.

INTRODUCTION

Building material emissions could be the major source for indoor air pollutants in low-occupant-density spaces, which are believed to have negative effects on occupants' comfort, health, and productivity. In addition to source control and air purification, ventilation is commonly applied as a basic means to remove pollutants emitted from sources, thereby reducing concentrations as well as human exposure to the pollution.

However, ventilation rates were prevalently determined by an IAQ surrogate mainly emitted from occupant-related sources (e.g., body odor and CO_2) based on the common knowledge that human was the major source of pollution in indoor environment since Yaglou's experiments (Yaglou et al., 1936). Occupants' well-beings may be compromised in indoor environments that the material emissions are the leading pollution, e.g., a newly-refurbished private office.

Recently, methods to determine the ventilation rate based on Lowest Concentrations of Interest (LCI) references were proposed for occupant-density spaces (Ye et al., 2014b). LCIs are health-based values that could be used to evaluate material emissions after 28 days from a single product in a laboratory chamber test and, thus, serve as part of the labeling scheme to enforce source control (ECA, 2013). At present, LCIs are commonly applied in product safety assessment in European Union (EU). Although LCIs were not particularly designed for determining ventilation rate standard, the idea and the testing procedure could be utilized to predict indoor emissions. Imaging that, once a databased built on the LCI concept and hundreds or thousands of commonly-used materials' emission data were included in the database, the determined ventilation rates could be generally applied. At present, there is no available emission database that is used to determine ventilation rate. The aim of this paper is to propose a general method to predict ventilation requirement while incorporating building material emission testing.

METHODOLOGIES

To demonstrate the application, 28 emission sources was adopted from the NRC database (Won et al., 2005), for the following reasons: 1) the selected materials represent a variety of building material usage, including ceiling, wall, floor and furniture materials; 2) all selected materials could be considered approximately as single-layer, uniform and solid materials with known dimensions. Therefore, screening-level emission estimation method based on internal diffusion coefficient and initial material-phase concentration could be applied to determine emission rates (Ye et al., 2014a).

Lowest Concentrations of Interest (LCI) reference values can be adopted to predict required ventilation as a composite index (Ye et al., 2014b). Examples of such composite index, e.g. AFSSET list published in France, that were adopted in this research are summarized in Table 1.

Table 1. Selection and description of indoor air quality references.

No.	LCIs	References	Published year	Brief descriptions	Number of compounds common to NRC database
1	AFSSET	(ANSES, 2009)	2009	165 substances, France	63
2	AgBB	(Däumling, 2012)	2012	176 substances, Germany	50
3	EU-LCI	(ECA, 2013)	2013	82 substances and another 95 substances to be derived, EU	33

A simplified characteristic ventilation rate method, which doesn't take multiple chemical reactions and material coexisting into account, was adopted to determine the ventilation rate and was shown in Eq. (1) (Ye et al., 2014c)

$$N_i = \frac{1}{V}\sum_{j=1}^{m} \frac{F_{i,j}A_i}{I_j} = \frac{1}{V}\sum_{j=1}^{m} \frac{E_{i,j}}{I_j}, i = 1,2,\cdots,n \qquad (1)$$

where $F_{i,j}$ and $E_{i,j}$ are the emission factor and emission rate of the *j*-th compound emitted from the *i*-th material, $\mu g \cdot m^{-2} \cdot h^{-1}$ and $\mu g \cdot h^{-1}$, respectively; A_i is the emission area of the *i*-th material, m^2; N_i is the required ventilation rate for the *i*-th material, h^{-1}; and V is the volume of the room, m^3; I_j is the Lowest Concentration of Interest (LCI) value for the *j*-th compound, $\mu g \cdot m^{-3}$; n and m are the number of the materials and all selected compounds, respectively.

RESULTS AND DISCUSSION

The selected materials from the NRC database were subjected to emission modelling and the required ventilation rates were determined by Eq. (1). The results are shown in Figure 1.

Two major observations can be made from Figure 1 as illustrations of using LCI concept to determine ventilation rates. First, the determined ventilation rates predicted by using the AFSSET list (see Table 1) were the greatest for 22 (out of 28) scenarios. This is likely the results of the more stringent values by the AFSSET list. The most noticeable difference is that the AFSSET list is the only scheme that includes VVOCs (e.g., formaldehyde). Second, the determined ventilation rates vary in a range of six orders of magnitude for all available materials in the NRC database. The ventilation rates for most of carpets (CRP1 – CRP7) and gypsum wallboards (GB1 – GB3) are less than 01 h^{-1}, while ventilation rates for oriented strand boards (OSB1 – OSB5b) are great than 1 h^{-1}. Among all the scenarios, MDF2 (medium density fiberboard) requires the highest ventilation rate (~22 h^{-1} based on the AFSSET list). The main reason is that the estimated formaldehyde emissions was at a ~200 $\mu g \cdot m^{-3}$ level, suggesting that engineered wood materials could result in high ventilation demand for its formaldehyde emissions. However, engineered wood materials may not be the only issue as natural wood materials can also lead to a high ventilation rate due to the emissions of terpenes (see PIN1 (pine) in Figure 1).

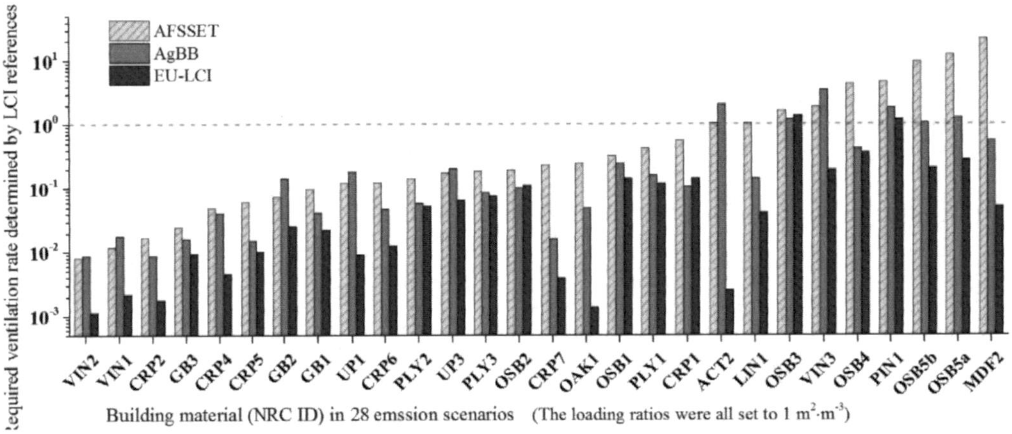

Figure 1. Ventilation requirements determined for 28 scenarios. Some of the material descriptions can be found in the references (Won et al., 2005; Ye et al., 2014c).

The ventilation rate can also be modified according to various loadings as shown in Eq. (2). For example, based on the emissions of MDF2 in Figure 1, the loading ratio should be reduced to ~0.05 $m^2 \cdot m^{-3}$ to produce a required ventilation rate of 1 h^{-1}.

$$N_i' = \frac{A_i'}{A_i} \cdot \frac{V}{V'} \cdot N_i = \frac{LR'}{1} \cdot N_i = LR' \cdot N_i \qquad (2)$$

where N_i' is the modified ventilation rate, h^{-1}; A_i', V' and LR' are the emission area, room volume and loading ratio in a different indoor environment, m^2, m^3 and $m^2 \cdot m^{-3}$.

Furthermore, and most importantly, the combined effect of emissions from multiple materials on the ventilation rate can be determined from a "combined form" as shown in Eq. (3)

$$N = \sum_{j=1}^{m} \left(\frac{A_i'/A_i}{V'/V} \cdot \frac{F_{i,j} A_i}{I_j V} \right) = \frac{1}{V'} \sum_{j=1}^{m} A_i' N_i \qquad (3)$$

Eq. (3) indicates that, a standard ventilation requirement (N_i) could be pre-determined for each material when building the database to predict ventilation rate based on material emissions. The ultimate ventilation requirement could be simply added up using the standard ventilation requirements pre-determined for different materials, after revising each material loading. Therefore, the ventilation rate could even be DIYed once the standard emission characteristics of each material were known. And only two parameters (i.e., diffusion coefficient, D; initial emittable concentration, C_0) were required to predict the standard emission characteristics (Ye et al., 2014b).

CONCLUSIONS

A general method to predict ventilation requirement while incorporating an emission database was proposed. Three LCI references were adopted to demonstrate the procedure. The results showed that standard ventilation requirement (N_i) could be pre-determined for each material in a database, the ultimate ventilation rate could be determined by adding up all the N_is, after revising all the material loadings. And only two parameters were required to predict the standard emission characteristics, as well as to build such a database.

ACKNOWLEDGEMENT

The material emissions database by National Research Council Canada (NRC) is the outcome of two consortium projects, i.e., CMEIAQ I and II. The authors would like to thank the members of the consortiums and the technical advisory committees.

REFERENCES

ANSES, 2009. Emissions de compos´es organiques volatils (COV) par les produits de construction et de d´ecoration, https://www.anses.fr/fr/content/emissions-de-compos%C3%A9s-organiques-volatils-cov-par-les-produits-de-construction-et-de, last accessed 10 November 2014.

Däumling, C., 2012. Product Evaluation for the Control of Chemical Emissions to Indoor Air-10 Years of Experience with the AgBB Scheme in Germany. CLEAN – Soil, Air, Water 40, 779-789.

ECA, 2013. Report No. 29. Harmonisation framework for health based evaluation of indoor emissions from construction products in the European Union using the EU-LCI concept. EUR 26168/EN, ISBN 978-92-79-33194-7. Publications Office of the European Union, Luxembourg.

Won, D., Magee, R., Yang, W., Lusztyk, E., Nong, G., Shaw, C., 2005. A material emission database for 90 target VOCs, In: Proceedings of Indoor Air 2005, Beijing, China, pp. 2070-2075.

Yaglou, C.P., Riley, E.C., Coggins, D.I., 1936. Ventilation requirements. ASHVE Transactions 42, 133-162.

Ye, W., Little, J.C., Won, D., Zhang, X., 2014a. Screening-level estimates of indoor exposure to volatile organic compounds emitted from building materials. Building and Environment 75, 58-66.

Ye, W., Won, D., Zhang, X., 2014b. Practical approaches to determine ventilation rate for offices while considering physical and chemical variables for building material emissions. Building and Environment 82, 490-501.

Ye, W., Won, D., Zhang, X., 2014c. A preliminary ventilation rate determination methods study for residential buildings and offices based on VOC emission database. Building and Environment 79, 168-180.

ASSESSMENT OF THE IMPLICATIONS OF NATURAL AND MECHANICAL VENTILATION ON HUMAN HEALTH IN THE RESIDENTIAL SECTOR

Sandra M BERNAL[1*]

[1] PhD student in Arid Lands Resource Science
[*]sbernal@email.arizona.edu

Keywords: Natural ventilation, mechanical, ventilation, particles, indoor air quality, human health.

SUMMARY

From the alternatives to solve indoor environmental quality in the residential sector, the ventilation system used as source of air exchange and temperature control makes a difference in particle deposition. Using a case study that compares the 3 most common ventilation systems, the difference in particle deposition and temperature regulation was assessed. The ultimate goal of the assessment is to have enough information to define the real impact of natural ventilation and mechanical systems in the inhabitants. The discussion has been focused on health with an important stress on energy consumption and indoor environmental quality. The analysis was done using Energy 2d, passive particle catchers, and digital thermometer.

INTRODUCTION

The environmental and economic circumstances in Mexico City promote the use of natural ventilation as the main source of air exchange. Its population growth define it as a Megacity and therefore, the negative implications of the air supply at home might have extended consequences as more environmental stress is expected. Accordingly to census data, 85% Mexico City has an average income of 9.8 dollars a Day. The environmental conditions have changed experiencing higher surface temperatures since 1987, historical minimum records of precipitations between 1999 and now, as well as the minor percentages of surface relative humidity since 1984. Data produced at NOAA/ESRL PSD at http://www.esrl.noaa.gov

Current policy and building codes do not have regulations regarding indoor air quality and its consideration into the land use and zoning plans is not of main importance yet. Therefore, the ability of inhabitants to adapt to higher temperatures and, at the same time, to stop the negative effects of pollutants in air is an accessory for those who can afford it.

From the alternatives to solve indoor environmental quality in residential sector, the ventilation systems are the sources of air exchange and temperature control. Using a case study that compares the 3 most common ventilation systems, the difference in particle deposition and temperature regulation has been analyzed.

METHODOLOGIES

The property was divided in areas that represent its usage: Private (PR), Public (PA), Semi Private (SP), Kitchen (KT), and Bathroom (BA).

For natural ventilation, temperature was registered during 4 scenarios: having all opening closed, having two openings across the floor plant, closing every room's door and opening one window per room, and having all openings (including doors) opened (figure 1).

	closed openings	cross ventilation	single sided vent	tottally open ventilation
	temperature reached in 10 min in farenhait degrees			
public area	80	78	80	79
semi-public	80	80	80	80
private	80	79	80	79
bath	86	85	86	86
kitchen	80	80	80	81
temp-return	80	80	80	80

Figure 1 Natural ventilation and resulting temperature changes

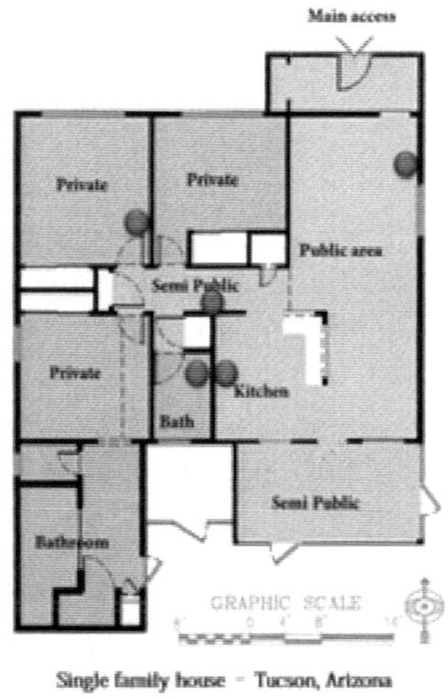

Figure 2 Case study home

The house assessed for the case study has a conventional supply and exhaust design that supplies air from top and intake from the bottom. It is known that this conventional design promotes excessive energy consumption and deposition of polluted air in the house. Temperature was measured to consider the energy consumed when using evaporative cooler and air conditioner. The first step was to turn of the mechanical systems and checking the temperature. It is important to remark that each mechanical system was tested in different periods. In both cases the procedure was the same. Once a steady temperature was reached, the mechanical systems were turned on and the time to reach comfort temperature was checked. The test was ran when the filters were new and after one month of using them (figures 3 and 4).

Collection dishes for sampling were placed in each area to observe the deposition of particles in a one week period in two periods: when the filter was new and repeating the collection during another week period after a month when the filter was dirty. This collection relies only in the comparison of the general characteristics of the outcomes.

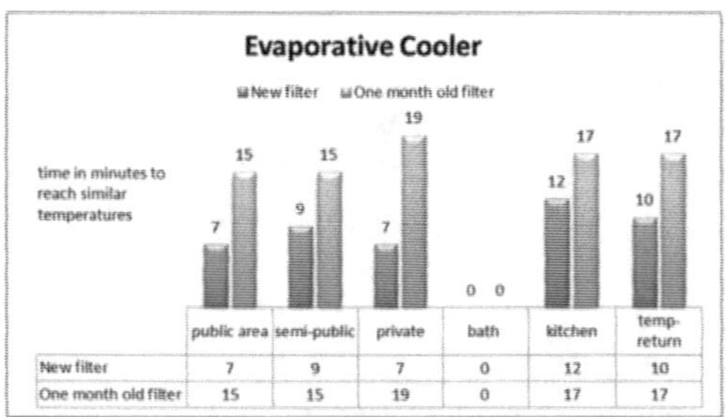

Figure 2 Time to reach comfort temperature with evaporative cooler system

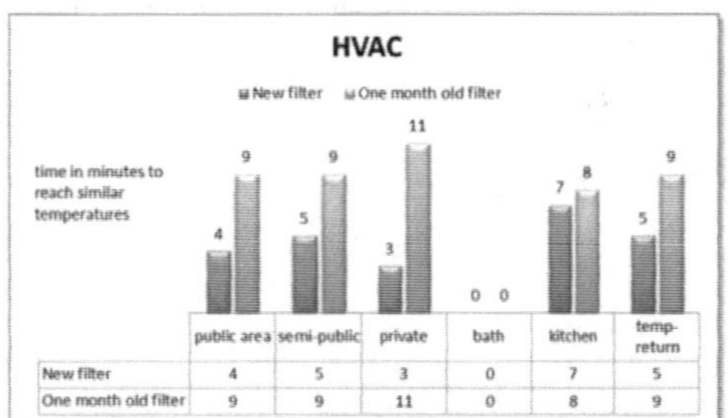

Figure 1Time to reach comfort temperature with HVAC system

The models crated in Energy 2d software were used for testing the two most used modes of natural ventilation, proving the effectivity of cross ventilation. The program was also used to model a room with a mechanic ventilation system.

RESULTS AND DISCUSSION

Natural ventilation registered higher deposition of particles on public areas. The private area in the bedroom appears to have higher content of textile fiber and dust. During the use of mechanical systems the higher collection of particles was in the semi-private area, meaning that the corridor is dirtier than the public area.

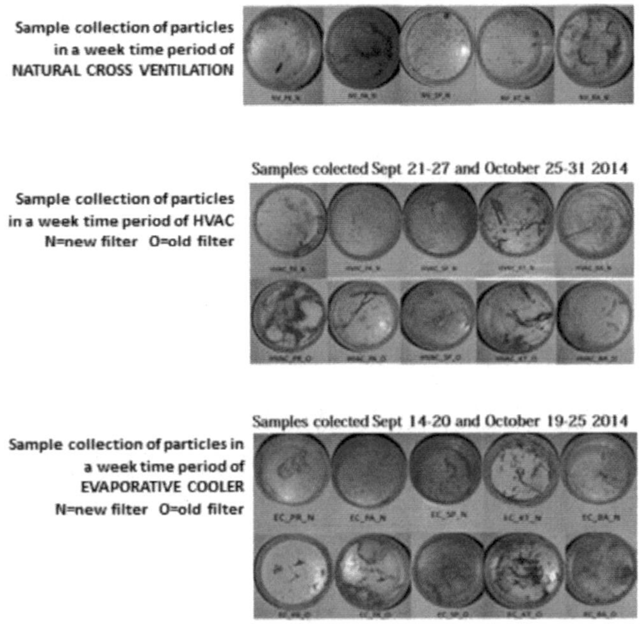

Figure 3 Particle deposition by area

The deposition of the particles is consistent with the location of the intake of the mechanical system from where the air is extracted. It is noted that one of the benefits of the mechanical systems is the particle retention by filtering. As it is seen in the analysis, it is a difference between the deposition when using air conditioner and evaporative cooler, being the evaporative cooler and the use of old filters the promoters of cluster formation. One important consideration is that the more humidity the higher the risk of mold and/or bacteria to grow, whereas the old filters prevent particles for moving, staying longer periods without re-suspension.

Energy 2d test demonstrates that the room using mechanic ventilation system forces the air and temperature to follow a straight forward route to the intake, the power of the fan used for extracting air from the room into de system carries particles to the house. As is seen in the optical analysis of the samples and in figure 7, the particles are deposited in a semi-public area close to the intimate places where people spend at least 8 hours sleeping.

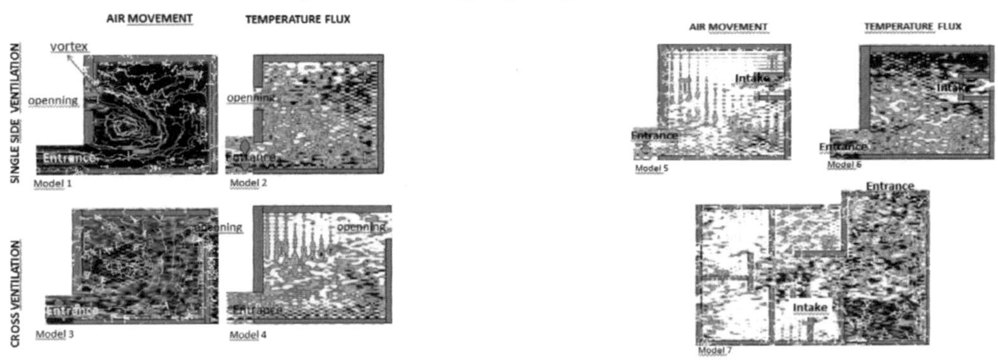

Figure 4 Effectivity of natural ventilation modes

It is important to consider that the breathing when sleeping is a profound breath that allows particles to reach deeper in human respiratory system.

CONCLUSIONS

The deficiency of indoor air quality caused by the intrusion of outdoor pollution into the houses is linked to the ventilation system used to supply air and regulate temperature.
As it is observed in the analysis of the performance of the three systems, natural ventilation prevents bigger particles to settle in semi-private and private areas whereas it allows ultrafine and fine particles to access to the entire space. Ultrafine and fine particles are responsible of chronic diseases due its tracheobronchial and pulmonary deposition and are highly present in Mexico City due the high usage of transportation based on fossil fuels. Nevertheless, the

particles in general are less clustered and clear. These characteristics might indicate that no humidity and grease particles area attached, minimizing the chance of mold and bacteria to grow.

In the case of Mexico City, the article of Escobedo overcrowding homes and the tendency of cooking at home constantly, both promoters of indoor produced particles with higher concentrations that outdoors. This is a relevant remark since the samples from the kitchen (KT) are, in the case of those under mechanical ventilation, dirtier that the sample taken when natural ventilation. It is noted that cooking is a source of indoor produced particles. Using natural ventilation, the particles are dispersed with the flow of air and runs out instead of being re-suspended as in the mechanical systems. Notice that the KT sample after a month is even dirtier and clustered.

In order to meet the need of a healthy indoor environment with the possibility of energy responsible ventilation strategy, the natural ventilation has to include passive and mechanical techniques to avoid medical expenses on chronic diseases due the intrusion of ultrafine and fine particles. Passive techniques, such as shading, will help to avoid the use of mechanical ventilation systems. If needed, air conditioner with a higher SEER (Seasonal Energy Efficiency Ratio) plus an accurate air flow management will be an affordable technique. Physical obstacles and vegetation can be placed in the main access (most used) of the house as well rough surfaces before going to the semi-private and private areas will help to avoid dirt. These techniques can be used to increase the efficiency of air-conditioner too.

ACKNOWLEDGEMENTS

This research could not being done without the orientation of Dr. Nader Chalfoun, Professor of Architecture and Environmental Sciences. To the faculty and students of the House Energy Doctor and the Master of Science in Architecture Program, The Arid Lands Resource Sciences Graduated Interdisciplinary Program of the University of Arizona, the Global Change Ph.D. Minor. Also, Dr. Paloma Beamer and Dr Miranda Loh from the College of Public Health, University of Arizona.

REFERENCES

- Dhalluin, A., Liam, K.. *Comparison Of Natural And Hybrid Ventilation Strategies Used In Classrooms In Terms Of Indoor Environmental Quality, Comfort And Energy Savings,* Indoor And Built Environment, 2012.
- Escobedo, L., Champion, W., Li, N., Montoya, L. *Indoor Air Quality In Latino Homes In Boulder, Colorado,* Atmospheric Environment, 2014.
- Pegas, P., Alves, C., Nunes, E., Bate-Epey, E., Evtyugina, M., Pio, C. *Could Houseplants Improve Indoor Air Quality In Schools?* Journal Of Technology And Environmental Health, 2012.
- Tan, C, Et Al. *Experimental Investigation Of Indoor Air Pollutants In Residential Buildings*, Indoor Built Environment, 2012.
- Thomas, L. *Evaluation Design Strategies, Performance And Occupant Satisfaction: A Low Carbon Office Refurbishment,* Building Research And Information, 2010.
- Faggianelli, G., Brun, A., Wartz,E., Muselli, M., *Natual Cross Ventilation In Buildings On Mediterranean Coastal Zones,* Energy And Buildings, 2014.
- Yassin, M., Althaqeb, B., Al-Mutri, E., *Assessment Of Indoor Pm2.5 In Different Residential Environment,* Atmospheric Environment, 2011.

Time-resolved Aerosol and Fluorescent Bioaerosol Concentrations in an Air-Conditioned and Mechanically Ventilated Office in Singapore

Jin ZHOU[1,2], Victor W.-C. CHANG[1,2,*], and William W NAZAROFF[1,3]

[1]Berkeley Education Alliance for Research in Singapore, Singapore 138602
[2]School of Civil and Environmental Engineering, Nanyang Technological University, Singapore 639798
[3]Department of Civil and Environmental Engineering, University of California, Berkeley, California 94720-1710, USA

[*]Corresponding email: wcchang@ntu.edu.sg

Keywords: Bioaerosol, real time, indoor-to-outdoor ratio, human occupancy

SUMMARY

Using an ultraviolet-light induced fluorescence (UV-LIF) technique, we measured number concentrations of total aerosol particulate matter (tPM) and fluorescent biological aerosol particles (bioPM) (1.0-3.0 μm and 3.0-5.0 μm diameter) in an office and outdoors, sampling with 1-min resolution. The air-conditioning and mechanical ventilation (ACMV) system equipped with high-grade filters was effective in controlling both tPM and bioPM indoors. As expected, removal efficiencies were found to be size dependent. One human subject walking on the carpet was found to be a strong contributor to bioPM, resulting in 2-3 times higher concentration than that outdoors. Compared to the times when the room is vacant, the biological proportion of total airborne particles increased by an order of magnitude during the light walking period. Consequently, indoor-to-outdoor ratios depend on the ACMV operating conditions and on human activities. This pilot study provides preliminary data concerning the bioPM levels in an indoor environment equipped with an ACMV system. Ongoing investigations using this approach promise to improve our understanding of the processes that influence indoor bioaerosol levels and the effectiveness of control alternatives.

INTRODUCTION

Conventional indoor bioaerosol studies have relied heavily on culture-based sampling, with well-documented limitations. Alternatively, filter-based or impinger-based sampling can be used, in combination with DNA-based quantification methods. However, neither approach is well suited for studying short-term dynamic processes that might influence the concentrations and fates of bioaerosols indoors. New developments in biosensing technology utilize ultraviolet-light induced fluorescence (UV-LIF) measurement techniques for real-time detection of biological aerosol particles. This approach provides opportunities in various applications, such as in laboratory studies (Agranovski et al., 2003) and in semiurban outdoor areas. Nevertheless, few studies have used UV-LIF measurement to investigate bioaerosols in indoor environments. One example, reported by Bhangar et al. (2014), characterized fluorescent biological aerosol particle levels and occupant emissions in a mechanical ventilated classroom in the United States. To date, there are no studies utilizing UV-LIF to investigate indoor bioaerosols in tropical regions. Among other considerations, the indoor environment in tropical climates is regularly served by air conditioning and mechanical ventilation (ACMV) systems with a high propensity for condensed water, which might influence indoor levels of biological aerosol particles.

The specific aim of this preliminary investigation is to characterize the impact of an ACMV system and human occupancy on biological aerosol particles in a meeting room in an urban environment in Singapore. We report on the acquisition and analysis of simultaneously measured indoor and outdoor particle number concentrations (PNC) for total aerosol particulate matter (tPM) and for fluorescent biological aerosol particles (bioPM). Information on ACMV operation status and human activities was also obtained, all with 1-min time resolution.

METHODS

A meeting room with carpeted flooring and a central air conditioning and mechanical ventilation (ACMV) system in a university building in western Singapore was selected as the study site. The room volume was 150 m^3 based on physical dimensions. It shared ACMV ducts with nearby offices and labs. The ACMV system was a central forced-air system with integration of outdoor air intake and indoor air recirculation. The ACMV system operated from 8:00 to 22:00 every day. The room air-exchange rate (AER) was 2.6±0.3 per hour when the mechanical ventilation system was on. The supply air to the room, which comprised about 90% recirculated air and 10% outdoor air (according to ACMV design), was filtered with MERV 13 filters that were replaced every 3-6 months. When the mechanical ventilation system was off, the infiltration and leakage caused the room AER to be 0.7±0.03 per hour. The AER was evaluated through sulphur hexafluoride (SF_6) tracer decay tests. The mean value and standard deviation of AER was determined from analysing three SF_6 decay tests (5 hours of data for each test when the ACMV system was on, 12 hours of data for each test when the ACMV system was off).

Observational monitoring was conducted for 6 days (1-6 Jan 2015, labelled D1-D6). The room was unoccupied during the monitoring period except 13:00 to 16:00 on D3-D6. One human subject performed light walking during the occupied period. Biotrak (Model 9510-BD; TSI, Inc.) was used to measure time-resolved number concentrations of tPM and bioPM in two particle-size ranges (1.0-3.0 μm and 3.0-5.0 μm). The device was configured to sample air during 40 s out of every 1 min. The sampling flow rate was 28 L/min. We took measurements indoors and outdoors by using tubing and an auto-switch device. The frequency of switching between indoor and outdoor samples was once per four minutes. The indoor sampling inlet (S1) was placed at the center of the study room, at a height of 1.2 m. Outdoor air (S2) was sampled via an inlet protruding from a window to the building corridor.

Calibration testing of aerosol losses in the tubing and auto-switch device was conducted before the monitoring period. For this purpose, the sampling inlets were collocated for 5 hours in an outdoor environment at a height of 1.2 m. We added a third sampling inlet (S3), which didn't connect with any tubing or switch device. Sample S3 was designated as the reference case. The objective was to adjust data from S1 and S2 to match as closely as possible the response of the S3, so as to offset the aerosol losses during transport from the point of sampling and the point of measurement. The calibration factors specific to S1 and S2 are 0.95±0.07 and 0.94±0.07 for $tPM_{1.0\text{-}3.0}$, 0.54±0.09 and 0.53±0.10 for $tPM_{3.0\text{-}5.0}$, 0.80±0.15 and 0.79±0.11 for $bioPM_{1.0\text{-}3.0}$, 0.52±0.12 and 0.52±0.09 for $bioPM_{3.0\text{-}5.0}$.

RESULTS AND DISCUSSION

Figure 1 illustrates the time series plots of the $tPM_{1.0\text{-}3.0}$ and $bioPM_{1.0\text{-}3.0}$ concentrations measured indoors and outdoors. Comparing outdoor tPM with outdoor bioPM (Figure 1a), the concentration profile for bioPM didn't always follow the broad trends of the tPM, implying that their sources were time dependent and the composition of tPM was changing at different times of the day. The biological proportion of total airborne particles (BPTP) of outdoor particles in the 1.0-3.0 μm and 3.0-5.0 μm diameter range were 0.14±0.12% (mean±sd) and 0.82±0.68%, respectively, during the monitoring period.

The ACMV system served as an important indoor/outdoor air exchange pathway. The ACMV system equipped with high-grade filters effectively reduced the indoor aerosol concentrations. As shown in the concentrations plots for indoor environment (Figures 1b and 1c), the concentrations for both tPM and bioPM dropped significantly as soon as the ACMV system was turned on at 8:00, indicating the effectiveness of this removal mechanism.

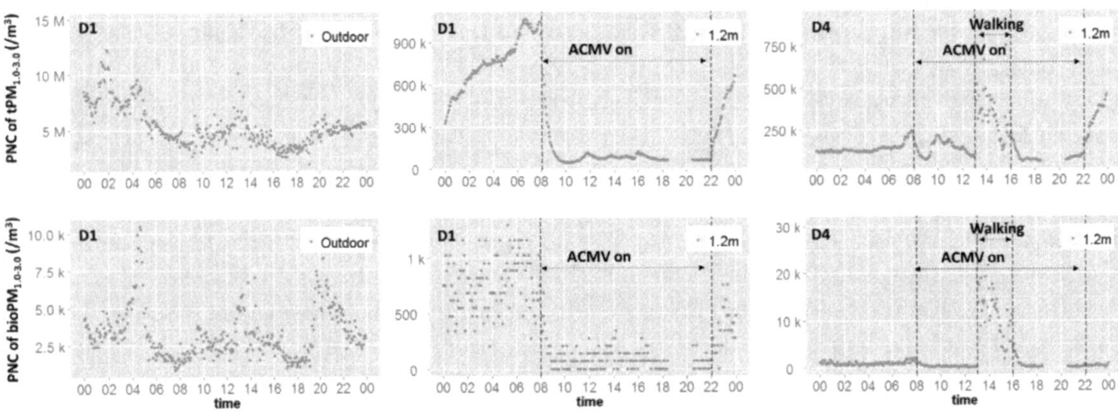

(a) Outdoor concentrations on day D1

(b) Indoor concentrations on day D1

(c) Indoor concentrations on day D4

Figure 1. Time series data for the number concentrations of total particles (tPM, top 3 frames) and biological aerosol particles (bioPM, bottom 3 frames) in the diameter range 1.0-3.0 μm measured (a) outdoors on day D1, (b) indoors on day D1, and (c) indoors on day D4.

The presence of a human occupant is an important factor for both indoor tPM and bioPM. Figure 1(c) shows a significantly higher indoor concentration occurring at 13:00-16:00 on D4, which coincides with a designated light walking period. The probable causes of the observed peaks are resuspension of tPM and bioPM from the carpet and shedding from the clothing, hair or skin of the occupant. The indoor/outdoor (I/O) ratios for bioPM (2.1 for 1.0-3.0 μm particles, 3.3 for 3.0-5.0 μm particles) were significantly higher than the values for tPM (0.10 for 1.0-3.0 μm particles, 0.33 for 3.0-5.0 μm particles) (Table 1), which indicates that walking on the carpet makes a stronger contribution to bioPM than to tPM. The time-averaged BPTP indoors was 1.9% and 5.6% for the particles in the size range 1.0-3.0 μm and 3.0-5.0 μm, respectively, higher than the BPTP estimated for unoccupied periods by one order of magnitude. This result further substantiates the important role of human activities contributing to bioaerosols indoors.

Overall, across the 6 monitoring days, the time-averaged outdoor concentration of total airborne particles in the 1.0-3.0 μm (and 3.0-5.0 μm) diameter range was 4000k/m³ (150k /m³); the value for biological aerosol particles was 3.2k/m³ (1.1k/m³) (Table 1). The indoor concentrations were less than a quarter of the outdoor concentrations when the room was vacant. During occupied times, the average indoor-outdoor (I/O) ratios for bioPM were higher than the value for tPM by a factor of 20 for 1.0-3.0 μm particles and by a factor of 10 for 3.0-5.0 μm particles. These higher ratios indicate that the particles resuspended from carpet and/or shed from occupants were strong determinants of indoor bioPM levels.

Table 1 Time-averaged and size-resolved particle number concentrations (PNC) of total airborne particles (tPM, 1000 particles per m^3), biological aerosol particles (bioPM, particles per m^3), and indoor-to-outdoor concentration ratios (I/O).

Condition	**ACMV on, unoccupied**				**ACMV on, occupied**			
Particle size	**1.0-3.0 μm**		**3.0-5.0 μm**		**1.0-3.0 μm**		**3.0-5.0 μm**	
	tPM (×1000)	bioPM	tPM (×1000)	bioPM	tPM (×1000)	bioPM	tPM (×1000)	bioPM
Indoor								
PNC	106	241	3	79	228	6490	44	2989
I/O	0.03	0.08	0.02	0.10	0.10	2.1	0.33	3.3
t(h)	75	75	75	75	9	9	9	9
Outdoor								
PNC	3420	2829	145	811	2364	3130	134	901
t(h)	75	75	75	75	9	9	9	9

CONCLUSIONS

Real-time data from monitoring tPM and bioPM provides interesting and potentially important insights linking indoor environments with building operation. The number concentrations of tPM and bioPM of the particles in size range 1.0-3.0 μm and 3.0-5.0 μm in an unoccupied meeting room were lower than the outdoor levels by one to two orders of magnitude. It is also worth noting that the I/O ratios in the unoccupied room diminished significantly when the ACMV system was operating. The reduction can be attributed to particle removal provided by the high grade filters and, probably, reduced outdoor particle infiltration owing to slight pressurization of the indoor environment. Activity of human occupants was an important source for tPM and bioPM. Light walking resulted in a 2-3 times higher indoor bioPM concentration than measured outdoors. Data indicate that the biological proportion of total airborne particles was one order-of-magnitude higher during light walking, compared to times when the room was vacant. Further investigations with real-time UV-LIF instruments have the potential to inform deeper understanding of indoor bioaerosol dynamics.

ACKNOWLEDGEMENT

Financial support was provided through a grant to the Berkeley Education Alliance for Research in Singapore (BEARS) for the Singapore-Berkeley Building Efficiency and Sustainability in the Tropics (SinBerBEST) Program. BEARS has been established by the University of California, Berkeley as a center for intellectual excellence in research and education in Singapore.

REFERENCES

Agranovski V, Ristovski Z, Hargreaves M, et al. (2003). Real-time measurement of bacterial aerosols with the UVAPS: Performance evaluation. *J Aerosol Science*, **34**, 301-317.

Bhangar S, Huffman J A, Nazaroff W W (2014). Size-resolved fluorescent biological aerosol particle concentrations and occupant emissions in a university classroom. *Indoor Air*, **24**, 604-617.

Testing of a Downflow Ventilation System for High Risk Infectious Disease Isolation Rooms.

Blanca Beato Arribas[1,2], Catherine J Noakes[1], P. Andrew Sleigh[1]

[1]Pathogen Control Engineering Institute, University of Leeds, UK.
[2]BSRIA, Old Bracknell Lane West, Bracknell, RG12 7AH Berkshire UK.

[*]*Corresponding email: blancab@bsria.co.uk*

Keywords: isolation rooms, ventilation, downflow, airborne infection

Summary

Isolation room airflows for infectious diseases are designed to minimise the risk of transmission of airborne pathogens to those outside the room and to protect healthcare workers who tend to the patient. This study considers the risk in the vicinity of the patient and conducts an experimental investigation into a downflow ventilation design to evaluate whether it is capable of providing protection to a healthcare worker. Anemometry and smoke tests are conducted in a mock up room to assess influence of ventilation rate, extract design, heat loads and flow local to a healthcare worker. Results show a good downward flow can be established, but a fan speed capable of delivering 0.35m/s and central extract are required to create a uniform flow. Heat loads and a healthcare worker leaning over the bed both compromise downflow effectiveness; local flow acceleration and exhaust can mitigate to some extent.

1 Introduction

Design of isolation rooms for the prevention of airborne transmitted diseases depend on the risk of infection and the hazard category of the pathogen. To contain infectious patients, the traditional design has been for negatively pressurised isolation rooms, whilst, to prevent infection entering a space, the space has been positively pressurised. Another alternative is the neutral pressure room with a positively pressurised ventilated lobby (PPVL), advocated by the UK Department of Health[1,2]. In all cases these rooms provide a well-mixed airflow and are recommended for up to hazard category 3 pathogens.

Isolation rooms for hazard category 4 pathogens require a very different approach. Although there is little evidence for airborne transmission, the severity of such infections is such that additional precautions are taken, including when designing airflow. Such rooms are typically are designed for very high negative pressure with the patient located inside a ventilated tent within the room. However, this minimises the ability to provide care to the patient. This study investigates whether a new design based on a downflow ventilation could eliminate the tent, in order to facilitate care to the patient[3]. This paper describes how this design conce

2 Methodology

A reduced size mock-up of the patient room (Figure 1) was constructed with dimensions 3.1 x 2.5 x 3 m height to explore the airflow patterns under a range of design configurations. The room was built within a BSRIA environmental chamber to ensure stability of the room conditions. The ventilation strategy was downward displacement; the air in the room was intended to flow uniformly

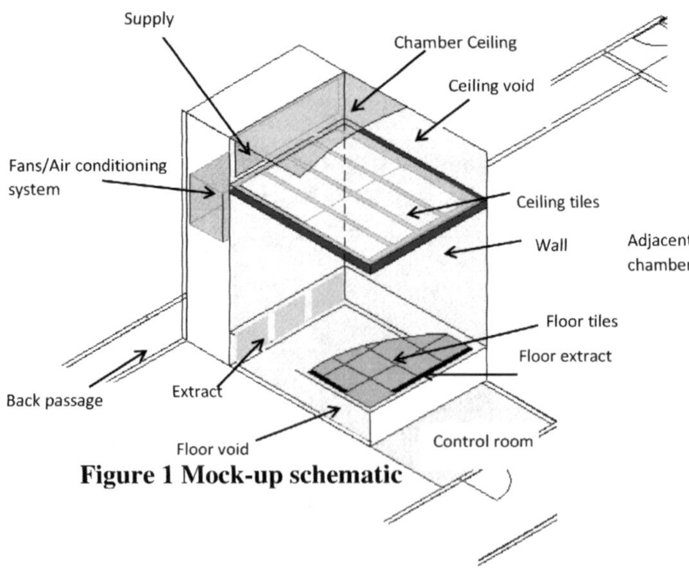

Figure 1 Mock-up schematic

downwards with air supplied into a void in the ceiling and entering the room through eight membrane tiles. Air was extracted at floor level, via grilles and a perimeter extract. The design of the room was based on operating theatres; it was intended to achieve a speed of 0.2 m.s^{-1} at the patient's bed. To enable the required flow rates in the test facility the extract air was recirculated.This would not happen in reality; in a CL4 room none of the air should be recirculated as the design intention is not to dilute the dose the staff might receive, but to make it as near as possible to zero.

Several tests were carried out to determine the airflow patterns in the room including airtightness, anemometry and smoke visualisation. The room was tested empty, with a bed and with heatloads provided by DIN men to simulate staff and patient. The experiments considered different floor extract configurations, with an extract underneath the bed and with different supply fan speeds as well as local control of the flow close to the patient.

3 Results and discussion

Prior to conducting experiments the room leakage and pressure drop over the ceiling supply tiles measured at different fan speeds. This was used to calculate the theoretical downflow speed as shown in Table 1.

Fan speed	ΔP	Flow-leakage	Theoretical speed
Hz	(Pa)	l.s^{-1}	m.s^{-1}
19.8	22.49	1410	0.25
33.0	32.55	2013	0.35
45.0	40.75	2498	0.44

Table 1 Fan speeds-Air speed in room

Initial tests were carried out with a perimeter extract and a fan speed of 19.8 Hz (approximately 0.25 m.s-1). Anemometry results (Figure 2) showed that at the lowest fan speed (enough to achieve a theoretical 0.20 m.s^{-1} at bed level) the air was not flowing uniformly downwards, but coalescing, leaving areas in the room with an insufficient air velocity (<0.05 m.s^{-1}). The air did not achieve a speed of 0.2 m.s^{-1} anywhere in the room at occupied level (0-1.8 m) except where it accelerated near the extract. As shown in Table 1, to achieve uniform down flow, it was necessary to increase the fan speed to 33Hz, although the theoretical fan speed would have been 19.8 Hz.

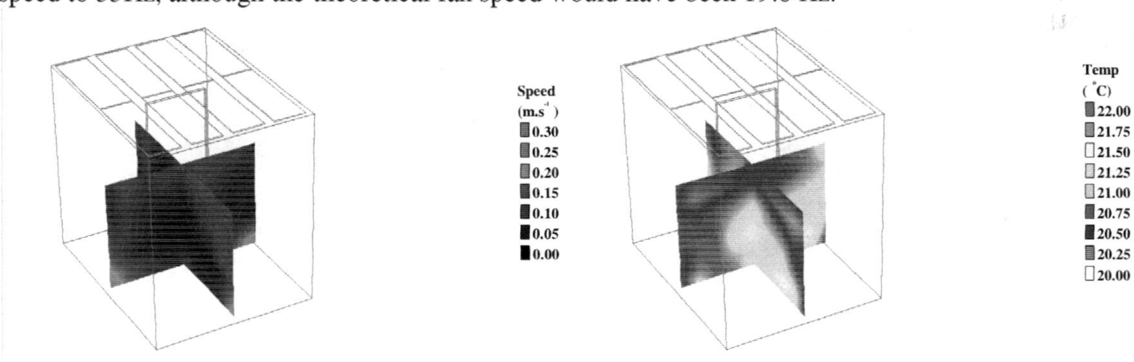

(a) Speed (b) Temperature

Figure 2 Test 02, empty room anemometry results at 19.8Hz

Results showed that it was not sufficient with the *push* effect of the supply from the ceiling, it was also necessary to *pull* the air from the extract (floor) side. With this intention, several extract configurations were tested: a perimeter extract round the room, tartan floor (the floor tiles were extract grilles themselves -with full floor tiles and 50% open tiles set up in a chess board pattern) central extract (with four 50% tiles in the center of the room). Additional tests also examined a pedestal extract (with a perforated box/pedestal underneath the bed), and a bed extract which pulled air from the sides of the bed into the floor void. Figure 3 shows results from anemometory tests to compare tartan and central extract locations. It can be seen that central extract configuration improves the flow field in the centre of the room, minimizing areas of low flow. Subsequent tests showed this worked very well in accelerating the flow above the bed, and smoke tests used to visualize the air coming from the patient showed that any "breath" (not coughs) released by the patient was effectively removed by the extract and not reaching the staff.

	Height	**Test 10 19.8 Hz**				**Test 11 33 Hz**				**Test 13 45 Hz**			
	(m)	ave	max	min	stdev	ave	max	min	stdev	ave	max	min	stdev
Average speed ($m.s^{-1}$)	2.70	0.07	0.11	0.02	0.034	0.23	0.44	0.07	0.066	0.37	0.64	0.11	0.095
	2.55	0.05	0.08	0.02	0.016	0.20	0.41	0.05	0.064	0.35	0.61	0.10	0.099
	2.40	0.03	0.05	0.02	0.008	0.18	0.35	0.06	0.055	0.33	0.48	0.12	0.089
	2.25	0.01	0.05	0.00	0.011	0.16	0.28	0.08	0.043	0.32	0.43	0.14	0.087
	2.10	0.03	0.07	0.02	0.015	0.16	0.26	0.10	0.036	0.29	0.40	0.15	0.059
	1.95	0.03	0.09	0.01	0.018	0.15	0.28	0.09	0.038	0.28	0.36	0.15	0.050
	1.80	0.03	0.10	0.01	0.020	0.15	0.29	0.06	0.039	0.26	0.35	0.15	0.042
	1.65	0.03	0.12	-0.01	0.028	0.16	0.29	0.02	0.038	0.26	0.36	0.14	0.041
Average temperature (°C)	2.70	20.7	21.0	20.4	0.195	20.4	20.7	20.3	0.096	20.6	20.8	20.4	0.075
	2.55	20.8	21.2	20.6	0.171	20.6	20.8	20.4	0.087	20.8	21.0	20.6	0.131
	2.40	20.8	21.6	20.4	0.263	20.3	20.8	20.2	0.108	20.5	20.8	20.3	0.057
	2.25	20.9	21.2	20.6	0.196	20.5	21.1	20.3	0.116	20.7	20.9	20.4	0.086
	2.10	21.0	21.2	20.6	0.171	20.6	20.9	20.3	0.105	20.7	20.9	20.5	0.074
	1.95	20.9	21.2	20.5	0.197	20.5	20.7	20.3	0.120	20.7	20.9	20.5	0.092
	1.80	21.0	21.3	20.7	0.156	20.6	20.8	20.4	0.107	20.8	20.9	20.5	0.071
	1.65	20.9	21.2	20.5	0.168	20.6	20.9	20.4	0.116	20.8	20.9	20.6	0.084

Table 2 Speed and temperature by height in the empty room at three fan speeds.

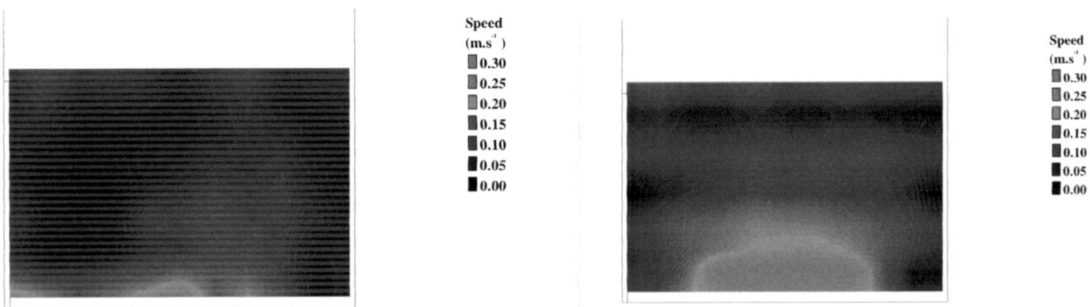

(a) Tartan extract (b) Central extract

Figure 3 Tests 08 and 09, empty room, investigating extract location.

The introduction of a bed effectively reduced the distance the air had to travel downwards, therefore increasing the air speed above the bed. Despite this, smoke tests showed that when heat loads were introduced (DIN men –to simulate staff, patient and equipment) the plume was making the air flow upwards. Anemometry results in Figure 4 also show the low flow above the bed due to the thermal plume from the DIN man counteracting the downflow.

The final set of experiments explored approaches for enhancing the downflow. A local supply ventilation ring was installed above the bed above the occupied space. Despite the higher-speed curtain of air provided by the ring, the downflow at 19.8 Hz still was not enough to overcome the plume due to the heat loads. A fan speed of 33Hz (0.35m.s^{-1}) was still necessary to counteract the effects of the plume in a patient and nurse. Finally the ring was moved to the side of the bed (see figure 5). Smoke tests demonstrated that the ring in this position aided the plume from the patient downwards. Results also showed that even at the highest supply fan speed, a person blocking the air pattern could block the downflow and breathe in contaminated air from the patient. Therefore, a halo system was envisaged, consisting of a mask and a ventilation method over the nurse's head which will blow air downwards the face of the nurse. This system proved effective in preventing contaminated air from reaching the face of the mannequin-nurse when it was lying over the patient.

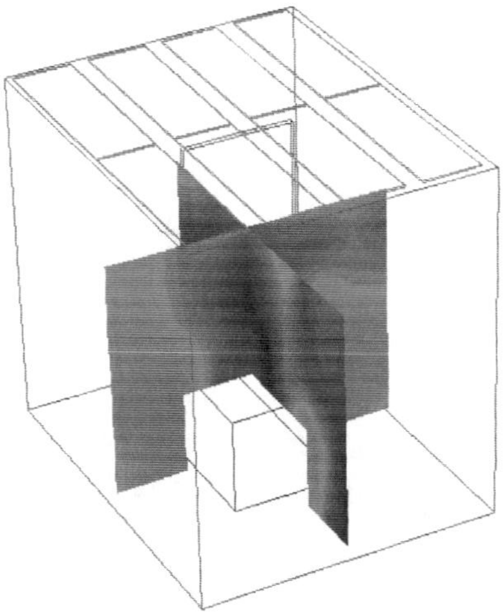

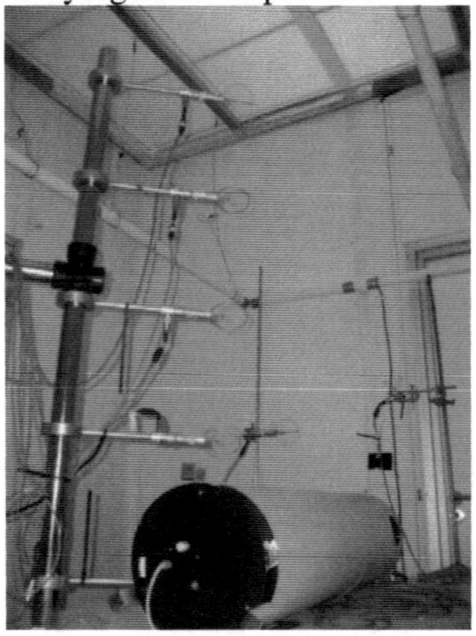

Figure 4 Anemometry results with patient and nurse heatloads, central extract 33 Hz supply fan. Cantilever anemometry pole measuring above DIN-man

4 Conclusions

This design for a CL4 isolation room was intended so that the tent (enclosure over and around the patient's bed) could be eliminated to provide easier access to the patient from the hospital staff. To achieve a uniform downflow and an air speed of 0.20 m.s^{-1} at bed level, and to overcome the plume (upwards) of heatloads in the room (patient, staff, equipment) it was necessary to increase the speed of the fans to a supply design of 0.35 m.s^{-1} (both set ups included the losses from airtightness). It was also necessary to modify the floor extract to create a push (supply) pull (extract) effect on the room, to avoid very low speed areas (or air going upwards). A pedestal extract and a ventilation ring proved to increase the air speed around and above the bed, reducing the risk of infection. The ability of the downflow system to overcome thermal plumes arising from simulated people and equipment was demonstrated. Heatloads remain a challenge and future work should consider ways of reducing or minimizing the effects of people and equipment within a CL4 protected zone.

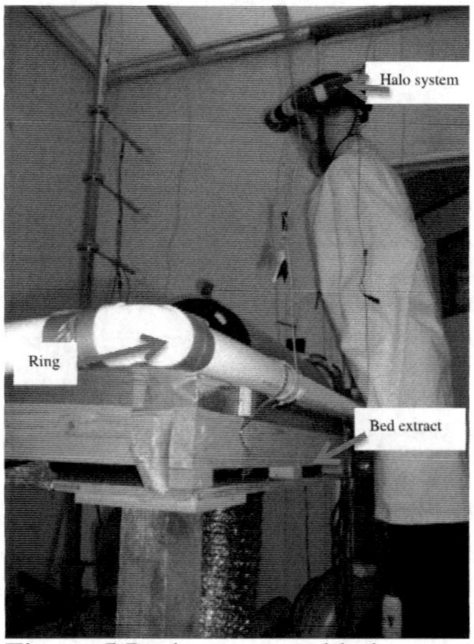

Figure 5 Bed extract and halo system

5 References

1. Department of Health, 2005 HBN4 In-patient accommodation Supplement 1 Isolation facilities in acute settings
2. B. Beato Arribas, W.B. Booth 2009. Ventilation of Isolation rooms. The implications of pressure regimes and ventilation strategy on the prevention of transmission of airborne transmitted infectious diseases. Ventilation 2009, Zurich, Switzerland
3. W.B. Booth B. Beato Arribas, 2012. BSRIA Reports 18914W and 18914-1.

OPEN SOURCE BUILDING SCIENCE SENSORS (OSBSS): AN OPEN SOURCE SENSOR NETWORK FOR INDOOR ENVIRONMENTAL DATA COLLECTION

Akram S ALI[1], Torkan FAZLI[1], Joseph CP HUAN[2], Deion DEBOSE[1], Boyang DONG[2], and Brent STEPHENS[1,*]

[1]Department of Civil, Architectural and Environmental Engineering, Illinois Institute of Technology, Chicago, IL USA
[2]Department of Electrical and Computer Engineering, Illinois Institute of Technology, Chicago, IL USA

[*]Corresponding email: brent@iit.edu

Keywords: Sensors, Measurement, Human occupancy, Data collection, Microbiology

SUMMARY

Accurate characterization of building characteristics that influence indoor air quality (IAQ), thermal comfort, and energy consumption are often limited to the use of proprietary hardware and software. The Open Source Building Science Sensors (OSBSS) project is working to design and develop a network of inexpensive open source devices based on the Arduino platform for measuring and recording long-term indoor environmental and building operational data. Custom OSBSS devices include sensors and data loggers to assess human occupancy, air and surface environmental conditions, various metrics of IAQ including CO_2, HVAC system runtimes, surface water activity, and other important building environmental parameters. The development, calibration, and performance of the sensor network is being documented and made available freely online along with detailed tutorials allowing others to use in future investigations of energy, IAQ, and comfort in the built environment.

INTRODUCTION

The collection of indoor environmental and building operational data is becoming ubiquitous for use in energy, thermal comfort, and indoor air quality research and practice. However, accurate characterization of influential parameters in the built environment is often limited to the use of proprietary hardware and software, which can adversely affect both functionality and costs (Tse and Chan, 2008). In response, the Open Source Building Science Sensors (OSBSS) project was developed to facilitate data collection for a wide variety of building parameters that influence physics, chemistry, and microbial ecology in buildings. We are developing open source sensors based on the Arduino platform to assess human occupancy, air and surface temperatures, relative humidity, various metrics of indoor air quality including CO_2, HVAC system runtimes, surface water activity, and several other important parameters for energy, comfort, and air quality in buildings (Ramos and Stephens, 2014). This development has been made possible by the availability of a large number of high-grade sensors and devices with custom Arduino libraries and active support from the manufacturers themselves. The Arduino platform has already been shown to enable similar efforts such as embedding wireless sensor networks for temperature and humidity monitoring within concrete structures (Barroca et al., 2013). In OSBSS, the development, calibration, and performance of the sensors and data loggers is being documented in its entirety and made available freely online along with detailed tutorials designed to allow other researchers to incorporate the sensors in both ongoing and future investigations of energy, air quality, and comfort in the built environment. Here we report on the development of the OSBSS platform and successful demonstration of several sensors and data loggers, including: (1) air temperature and relative humidity (T/RH), (2) CO_2 concentrations, and (3) dual infrared doorway beam break occupancy counters.

METHODOLOGIES

The core OSBSS data logger platform uses the user-friendly Arduino Pro Mini 328 3.3V/8 MHz as the main controller and a microSD card for long-term data storage. (Other wireless versions are also under development). A Maxim DS3234 real-time clock (RTC) is used for accurate time keeping. The Pro Mini uses an ATmega328 microcontroller, which we have modified in several ways to operate in a very-low-power draw mode. Modifications include: (1) operating the chip in a "power-down" sleep mode during periods in which the sensor is not measuring or logging data; (2) shutting down the microSD card reader and SHT15 when not sensing or logging; (3) using the DS3234 RTC as the master controller to wake and sleep; and various other power-saving techniques in programming.

Electrical circuits are constructed on solderable breadboards of various sizes depending on application, which allows for clear demonstration of how to build the sensors in online tutorials (www.osbss.com). All that is required is off-the-shelf parts and a soldering iron. For temperature and relative humidity (T/RH) measurements we use a U.S. Sensors 'Ultra Precision' thermistor ($8) combined with a Sensiron SHT15 humidity sensor ($42). For CO_2 measurements, we are using a K-30 1% CO_2 sensor ($85) from CO2meter, Inc. Examples of a completed temperature and relative humidity data logger and a prototype CO2 data logger are shown in Figures 1a and 1b, respectively. With low power draw techniques in place, the T/RH logger is expected to have a battery life of approximately 13 continuous months logging at 1-minute intervals with a 3.7V 2500 mAh lithium ion polymer battery. The CO_2 logger requires more power to operate and is best used with a DC power supply (which is not uncommon for most commercial CO_2 sensors). Both of these devices were setup as proof of concepts against lab-grade equipment, including an Onset HOBO U12 for T/RH and a PP Systems SBA-5 analyzer for CO_2, in order to evaluate the performance of the custom sensors.

a) T/RH **b) CO_2**

Figure 1. Examples of OSBSS (a) temperature and relative humidity (T/RH) data logger and (b) CO_2 concentration data logger.

As another example of OSBSS devices, a custom dual infrared doorway beam break sensor and data logger was also designed and built (Figure 2). The logger uses two infrared LED emission sources to direct a beam towards an LED receiver on the opposite side of a doorway. Because the LED emitters have high power draw that would lead to very short battery life, two low-power passive infrared sensors are installed on either side of the emitter to detect occupant presence prior to entering the door. When presence is detected on either side of the doorway, the passive infrared detectors "wake" the sleeping emitters to emit until a beam is broken or a short period of time before sleeping again. The use of two emitters and two occupancy detectors allows for an accurate measure of whether a person is entering or exiting through the doorway. The data logger then records a time stamped entrance or exit and keeps a running tally of the total occupancy in the room.

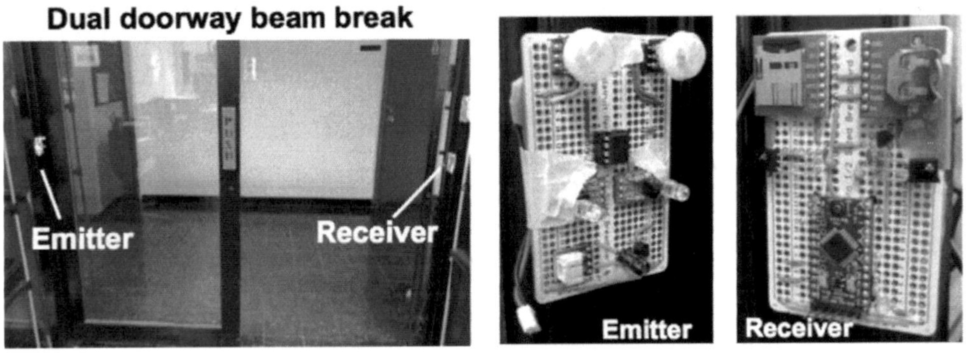

Figure 2. Example demonstration of a dual infrared doorway beam break sensor and data logger, including an emitter and receiver.

RESULTS AND DISCUSSION

A prototype of the OSBSS T/RH logger was co-located alongside a commercially available T/RH logger (Onset HOBO U12) for a period of approximately one week. Time series data and linear regressions of these data are shown in Figure 3.

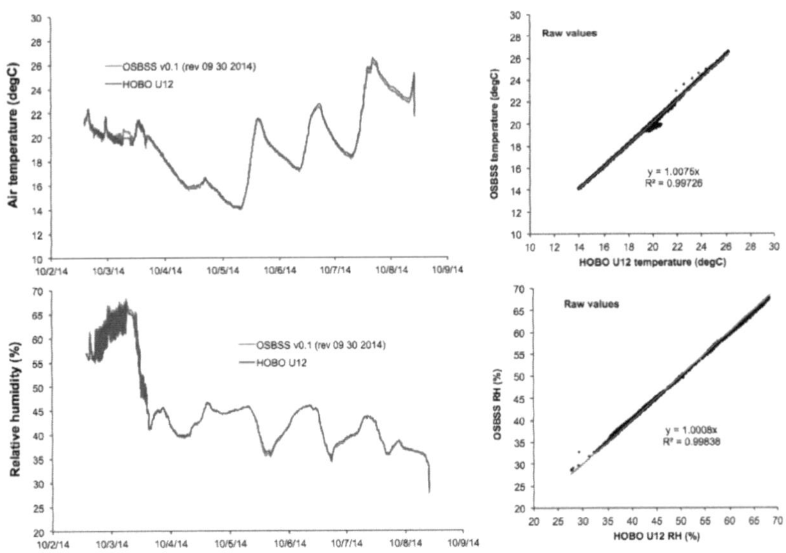

Figure 3. Time series data and linear regressions of temperature and relative humidity data from an OSBSS device and an Onset HOBO U12.

The OSBSS T/RH sensors performed extremely well when compared against the HOBO. For both temperature and relative humidity, regression slopes between OSBSS and HOBO were approximately 1.0 with R^2 values greater than 0.99. Further testing has revealed that multiple custom T/RH sensors yield similar results, demonstrating that this sensor and data logger combination is ready for application in real environments. Moreover, one T/RH logger has been running continuously logging at one-minute intervals for over four months without losing power.

Figure 4a shows data from an example test with the custom OSBSS CO_2 sensor and data logger co-located with an SBA-5 analyzer and data logger. CO_2 was briefly injected from a tank into a small chamber and allowed to decay for a period of several hours. The data reveal strong correlations between the two monitors ($R^2 > 0.99$). Figure 4b shows occupancy counts recorded by the dual infrared beam break sensor installed at the doorway of a student office and laboratory. The sensor was able to accurately count occupancy in the room over a period of several hours. While similar commercially available single and dual beam break products cost ~$400 and ~$900, respectively, this device was made with less than ~$150 in parts.

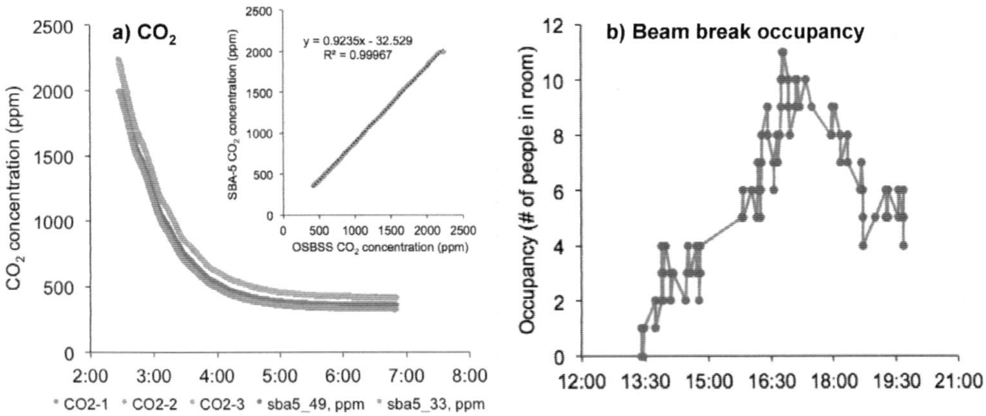

Figure 4. Example data from (a) OSBSS CO_2 sensors compared to PP Systems SBA-5 sensors and (b) occupancy counts in a student office made with the dual beam break sensor.

CONCLUSIONS

The OSBSS network of devices is designed to allow for more flexibility in synchronizing a large number of measurements with high spatial and temporal resolution in a more cost effective manner for use in research projects and building automation and control. Early prototypes have proven successful.

ACKNOWLEDGEMENT

OSBSS is funded by the Alfred P. Sloan Foundation's program on the Microbiology of the Built Environment (Grant No. 2013-10-07). OSBSS is also supported by the PURE and MIND programs in the Armour College of Engineering at Illinois Institute of Technology.

REFERENCES

Barroca, N., Borges, L.M., Velez, F.J., Monteiro, F., Górski, M., Castro-Gomes, J., (2013). Wireless sensor networks for temperature and humidity monitoring within concrete structures. *Constr. Build. Mater., Special Section on Recycling Wastes for Use as Construction Materials* 40, 1156–1166.

Ramos, T., Stephens, B., (2014). Tools to improve built environment data collection for indoor microbial ecology investigations. *Build. Environ.* 81, 243–257.

Tse, W.L., Chan, W.L., (2008). A distributed sensor network for measurement of human thermal comfort feelings. *Sens. Actuators Phys.* 144, 394–402.

DEVELOPMENT OF AN EXPERIMENTAL SYSTEM FOR ASSESSING INDOOR BIOAEROSOL TRANSPORT AND CONTROL

Stephanie KUNKEL*[1], Parham AZIMI[1], Brent STEPHENS[1]

[1]Department of Civil, Architectural and Environmental Engineering, Illinois Institute of Technology, Chicago IL 60616, USA

*Corresponding email: brent@iit.edu

Keywords: Microbial, Aerosol, Measurement, Infectious disease transmission, Pathogen

SUMMARY

Many microbial pathogens are transmitted via airborne routes in indoor environments. Breathing, sneezing, and coughing are important sources of many of these viral, bacterial, and fungal species, with microbes being aerosolized and dispersed in droplets that settle to nearby surfaces or rapidly evaporate into smaller droplet nuclei and disperse throughout a room. Knowledge of how this dispersion is affected by outdoor air ventilation, particle deposition, and central HVAC or stand-alone particle filtration is critical to understanding how the diseases caused by these microbes are spread and controlled. Here we report on the development of a custom nebulizer-based respiratory activity simulator that, when combined with size-resolved bioaerosol sampling and DNA extraction, can be used to aerosolize model organisms from solution and measure their transport and control indoors. Initial findings demonstrate that various model organisms can be successfully aerosolized and collected in a manner that realistically reflects human coughing and breathing.

INTRODUCTION

Communicable respiratory illnesses lead to large excesses in expenses associated with healthcare, absence from work, and lost worker productivity (Fisk, 2000), but the control of airborne infectious disease transmission in indoor environments is not yet entirely understood (ASHRAE, 2014). Several complex physical and biological processes govern the transmission of respiratory pathogens and the associated risk of infection to occupants in indoor environments. Human pathogens that are transmitted via the airborne route include a variety of viral, bacterial, and fungal species. Among the viruses are those that cause measles, mumps, chicken pox, influenza, and the common cold; biologically these include those with DNA genomes as well as both plus and minus strand RNA genomes. Airborne bacterial pathogens include both Gram positive (e.g., *Mycobacterium* (tuberculosis); *Cornybacterium* (diphtheria); *Streptococcus* (respiratory disease) and Gram negative (e.g., *Bordella* (pertussis); *Neisseria* (meningitis)) species. Fungi, which are known to be transmitted via airborne spores, include *Aspergillus* species and *Cryptococcus*, both of which can infect the lungs of immunocompromised individuals. When an individual coughs, sneezes, speaks, or breathes, droplets consisting of liquid water, proteins, salts, and various other organic and inorganic matter are expelled into the air. If the emitter is infected with a particular respiratory infection, these larger droplets will also contain the smaller infectious particles themselves, which may be viruses, bacteria, or fungi depending on the type of infection. With this information in mind, we have developed a custom nebulizer-based respiratory activity simulator that, when combined with size-resolved bioaerosol sampling and DNA extraction, can be used to aerosolize model organisms from solution and measure their transport and control indoors under different ventilation and particle filtration conditions. The experimental system utilizes model (non-pathogenic) organisms that mimic the airborne transmission of pathogenic viruses, Gram positive and Gram negative bacteria, and fungi via aerosols produced by human respiratory activities such as breathing and coughing.

METHODOLOGIES

We have designed and built a custom human respiratory activity simulator using a High-output Extended Aerosol Respiratory Therapy (HEART) Lo-Flo nebulizer (Mini-HEART, Westmed, Inc.) combined with a low-cost air compressor (California Air Tools CAT-1610A) (Figure 1a). The nebulizer is used to mimic aerosol emissions from the human lung and the air compressor is used to mimic lung volume flow. The simulator was first tested in a laboratory setting to evaluate its performance in terms of airflow rates and aerosol emissions. The simulator was then used to aerosolize model organisms in conjunction with simultaneous size-resolved air sampling in an unoccupied test facility on the main campus of Illinois Institute of Technology (Figure 1b).

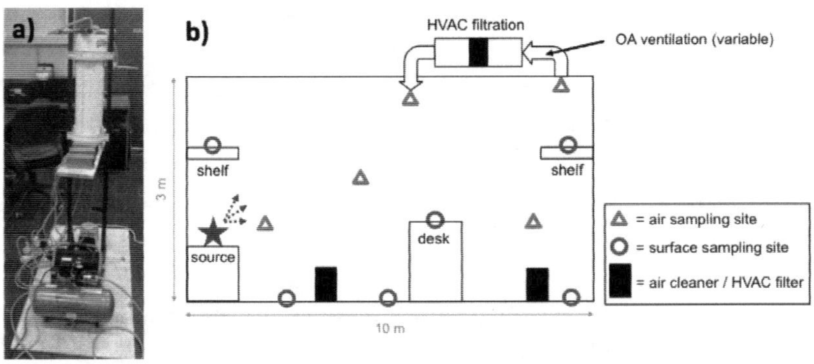

Figure 1. (a) The human respiratory activity simulator, and (b) layout of the experimental facility

In the laboratory, we first measured airflow rates through the nebulizer and the cough apparatus. The lung volume airflow rate was measured using a Fluke 975 AirMeter anemometer installed at the end of a smooth 1-meter long pipe connected directly to the simulator's output orifice. Measured cross-sectional velocities were multiplied by the area of the pipe to estimate the volumetric airflow rate. The airflow rate through the lower flow nebulizer was measured using a TSI Model 4043 mass flow meter. The respiratory activity simulator was then used to aerosolize three model organisms in an unoccupied test facility: *Escherichia coli, Bacillus subtilis*, and *Neurospora crassa. E. coli K12*, a gram negative bacterium (size ~1 μm), is used as a model for *Bordella, Neisseria. B. subtilis*, a gram positive bacterium (size ~3 μm), is used as a model for *Cornybacterium, Mycobacterium*, and *Streptococcus. N. crassa*, a ~4 μm fungus, is used as a model for *Aspergillus* and *Cryptococcus.* Each of the model organisms is either a BSL 1 or BSL exempt organism as designated by the NIH Office of Biotechnology guidelines.

In the test facility, size-resolved airborne particle concentrations were measured using a TSI NanoScan SMPS combined with a TSI Optical Particle Sizer (OPS) while the simulator operated in a small room over a period of approximately 4 hours. Active bioaerosol sampling was then conducted in multiple locations throughout the test facility. Bioaerosol sampling utilized five-stage Sioutas cascade impactors operating at 9 L/min for 6-8 hours in order to collect sufficient biomass, following procedures described in Yang et al. (2011). This allowed for particle collection within 5 size ranges: >2.5 μm, 1.0-2.5 μm, 0.5-1.0 μm and 0.25-0.5 μm, all captured on 25 mm PTFE filters, as well as <0.25 μm captured on 37 mm after-filters.

The total genomic DNA was extracted from the PTFE filters by first cutting the filters into thirds, placing them into PowerBead tubes supplied with the PowerSoil DNA kit (MoBio Laboratories, Solana Beach, CA) and then continued with the manufacturer's instructions. The template DNA concentration from each filter was measured using a Nanodrop2000c spectrophotometer (Thermo Scientific, Wilmington, DE), and determined to be approximately 100 ng/μL. For PCR amplification, specific primers for each organism were ordered (IDT, Coralville, IA) and utilized according to published protocols (Peccia and Hernandez, 2006). This was performed in order to check for efficiencies in sampling time, extraction, and amplification.

RESULTS AND DISCUSSION

Measurements of airflow rates from the compressor main airflow (lung flow) and the nebulizer (aerosol flow) were conducted for two different simulated lung volume settings. Airflow rates ranged from ~2.7 L/s to ~7.5 L/s depending on compressor setting, demonstrating accurate reproductions of human lung flows similar to those observed in previous investigations of human respiratory activities, including coughing (e.g., Lindsley et al., 2013). The resulting particle size distributions (Figure 2) are similar to one of the only studies of which we are aware that utilized instrumentation to measure particle emissions from human respiratory activities of particles as small as 10 nm (Holmgren et al., 2010).

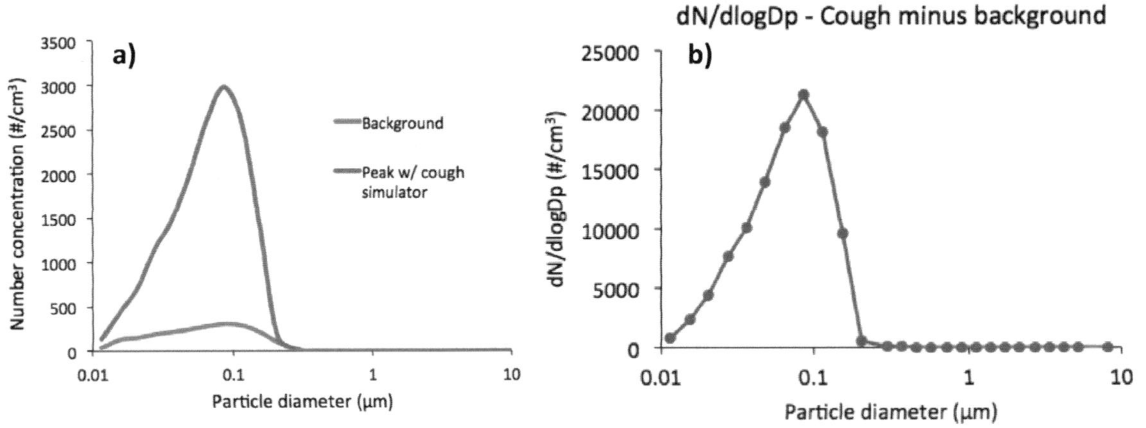

Figure 2. Particle size distributions resulting from operation of the respiratory activity simulator.

Biomass collected on the filters of the air samplers yielded sufficient template DNA after extraction to perform PCR from the first two stages for *N. crassa* and from the first three stages for *E. coli, B. subtilis.* Figure 3 shows PCR amplification (~100 bp) from all three organisms. These size resolutions correspond to particle collection in the first stage which is > 2.5 μm, the second stage is 1.0-2.5 μm, and the third stage is 0.5-1.0 μm. Detection in these stages is reasonable given the characteristic dimensions of the bacteria and fungi used (viruses are expected to appear in the smaller stages).

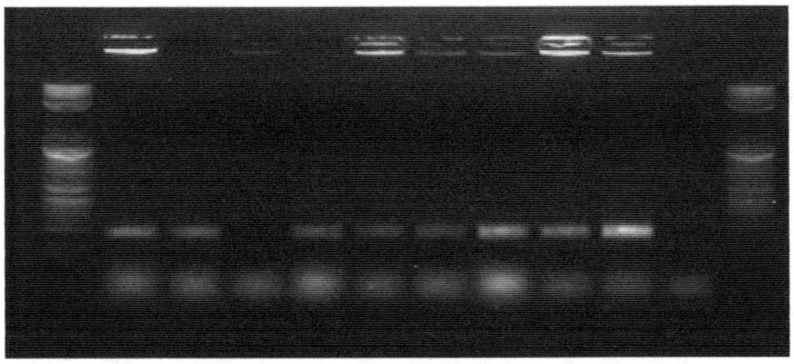

Figure 3. Agarose gel of PCR amplification. Lanes on both left and right, 2-log ladder (New England Biolabs). Lanes in between from left to right, Lane 2-4; *N. crassa* sizes (μm) <2.5, 1.0 - 2.0 and 0.5-1.0; Lane 5-7; *B. subtilis* sizes (μm) <2.5, 1.0-2.0 and 0.5-1.0; Lane 8-10; *E. coli* sizes (μm) <2.5, 1.0-2.0 and 0.5-1.0.

CONCLUSION

Results demonstrate that the model organisms could be successfully aerosolized by the custom respiratory activity stimulator in a manner that realistically reflects human coughing and tidal breathing. Resulting particle size distributions were similar to those reported in the literature. The methods for particle collection, filter extraction, and PCR amplification were tested and found to be sufficient for species-specific identification. Using this method we will next test bacteriophages MS2 and PR722 to model viral particles, and all model organisms will then be evaluated under a variety of ventilation and filtration conditions.

ACKNOWLEDGEMENTS

This work is funded by the Alfred P. Sloan Foundation's program on the Microbiology of the Built Environment.

REFERENCES

ASHRAE, (2014). ASHRAE Position Document on Airborne Infectious Diseases.

Fisk, W.J., (2000). Health and productivity gains from better indoor environments and their relationship with building energy efficiency. *Annu. Rev. Energy Environ.* **25**, 537–566.

Holmgren, H., Ljungström, E., Almstrand, A.C., et al., (2010). Size distribution of exhaled particles in the range from 0.01 to 2.0μm. *J. Aerosol Sci.* **41**, 439–446.

Lindsley, W.G., Reynolds, J.S., Szalajda, J.V., et al., (2013). A Cough Aerosol Simulator for the Study of Disease Transmission by Human Cough-Generated Aerosols. *Aerosol Sci. Technol.* **47**, 937–944.

Peccia, J., Hernandez, M., (2006). Incorporating polymerase chain reaction-based identification, population characterization, and quantification of microorganisms into aerosol science: A review. *Atmos. Environ.* **40**, 3941-3961.

Yang, W., Elankumaran, S., Marr, L.C., (2011). Concentrations and size distributions of airborne influenza A viruses measured indoors at a health centre, a day-care centre and on aeroplanes. *J. R. Soc. Interface* **8**, 1176–1184.

A SIMPLE METHOD FOR MEASURING GAS-PHASE SVOC CONCENTRATION IN EQUILIBRIUM WITH THE MATERIAL PHASE

Yaoxing WU[1], Steven S. COX[1], Mingjie XIE[1], Linsey C. MARR[1], and John C. LITTLE[1*]

[1]Department of Civil and Environmental Engineering, Virginia Tech, Blacksburg, VA 24061, USA

[*]Corresponding email: jcl@vt.edu

Keywords: SVOCs, Phthalates, Diffusion, Exposure, Additives

SUMMARY

Assessing human exposure to semivolatile organic compounds (SVOCs) emitted from consumer products in indoor environmental is difficult because methods are not available to easily measure the key emission parameters. A simple and rapid method using standard sorbent tubes was developed to measure the gas-phase concentration in equilibrium with the material-phase concentration of an SVOC in a product (y_0) and the surface/air partition coefficient (K_s). Phthalates in two types of polyvinyl chloride (PVC) flooring were selected to test the method. The values of K_s and y_0 was obtained by fitting a diffusion model to the experimental data. This novel approach should be useful for assessing potential exposure to SVOCs in consumer products.

INTRODUCTION

Semivolatile organic compounds (SVOCs) are widely used in building materials and consumer products (Weschler and Nazaroff, 2008). These chemicals tend to redistribute from their original sources via indoor air to other solid phases and are ubiquitous in the indoor environment. Exposure to SVOCs has been associated with adverse health effects, including endocrine disruption, asthma, and allergies (Matsumoto et al., 2008). Recently Little et al. (2012) developed a fundamental approach that can be used to obtain screening-level estimates of potential indoor exposure to SVOCs in certain classes of products. Based on this approach, a critical parameter is the gas-phase concentration in equilibrium with material-phase concentration of SVOC in the product (y_0). A few researchers have used small-scale chambers to measure y_0 (Liu et al., 2013, Liang and Xu, 2014), however, the complexity of chamber tests limits the practicality of the approach. The sink effect of SVOCs to chamber surfaces makes chamber test challenging (Clausen et al., 2004, Xu et al., 2012). The heterogeneity of the velocity field in the chamber may compromise the accuracy of y_0. Therefore, in this paper we focus on developing a simple and rapid method to measure y_0 without using a chamber. We first develop a mechanistic model to predict the mass flux of SVOCs from a product through a hollow tube. Then we validate the model using a simple experimental method that employs standard sorbent tubes. Finally values of y_0 and an additional unknown parameter, K_s, the surface/air partition coefficient inside the sorbent tube, are obtained by fitting the diffusion model to the experimental data.

METHODOLOGIES

SVOC diffusion model. Figure 1a provides a schematic representation of the diffusion model that predicts mass flux of SVOCs from the surface of a source.

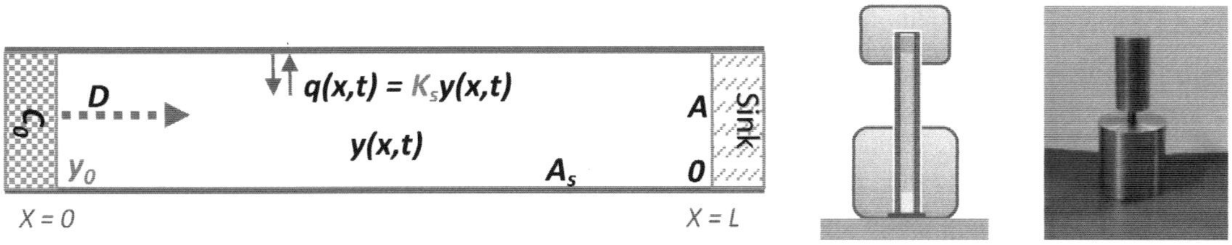

(a) (b) (c)

Figure 1. a) Schematic representation of the diffusion model; b) Schematic of experimental apparatus; c) Photo of the experimental apparatus.

Consider the case of diffusion through a tube of air with length L and diffusion coefficient D. Assume that one end of the tube at x=0 (immediately adjacent to the material surface) is kept at a constant SVOC concentration of y_0 and that the concentration at the other end of the tube at x=L is effectively zero. Then, if the air in the tube is initially free of SVOC, the concentration in the air in the tube changes according to Crank (1975):

$$y(x,t) = y_0\left(1-\frac{x}{L}\right)+\frac{2}{\pi}\sum_{n=1}^{\infty}\left(-\frac{y_0}{n}\right)\cdot\sin\left(\frac{n\pi x}{L}\right)\cdot\exp\left(-\frac{D\,n^2\pi^2 t}{RL^2}\right) \tag{1}$$

where $R= 1+K_S \cdot 4/d$ is retardation factor, and d is the internal diameter of the tube. $y(x,t)$ is the gas-phase concentration at position x at time t, and L is the total length. We assume the gas-phase concentration is distributed uniformly over the cross section of the tube because of its small diameter (mm scale) and that a linear equilibrium relationship exists between the concentration in the gas phase and on the interior surface of the tube, such that

$$q(x,t) = K_s \cdot y(x,t) \tag{2}$$

where $q(x,t)$ is the surface concentration at position x and time t, and K_s is the surface/air partition coefficient. The diffusive flux at x=L can then be obtained as:

$$J(t)\big|_{x=L}=\frac{D\,y_0}{L}\left(1+2\sum_{n=1}^{\infty}\cos(n\pi)\cdot\exp\left(-\frac{D\,n^2\pi^2 t}{RL^2}\right)\right) \tag{3}$$

On integrating eq 2 with respect to x we can also calculate the mass adsorbed on the tube surface at any time t.

Selection of compounds and materials. Three phthalates, diisobutyl phthalate (DiBP), di-n-butyl phthalate (DnBP), and di-2-ethylhexyl phthalate (DEHP) were selected for study. Two homogeneous PVC flooring materials, one (green) containing 15 wt % DEHP and the other (red) containing approximately 5 wt % of DiBP, DnBP and DEHP, were selected.

SVOC diffusion setup. As shown in Figure 1b and 1c, the apparatus consists of two stainless steel cylinders, one standard sorbent tube and one stainless steel shim. The sorbent tube (5mm in diameter) is inserted through the larger cylinder (5cm OD × 5cm L) and has a shim (0.14mm thickness, McMaster) at its sampling end to prevent direct contact between the tube and the PVC flooring. The cap-like smaller cylinder (2.5cm OD× 5cm L) rests upon the top of the sorbent tube, and holds the tube and shim firmly against the material surface. The 15mm long sorbent-retaining restriction sleeve within the tube forms a micro diffusion chamber of the same inner diameter of the tube, where the SVOCs are only transported by random motion of the phthalate molecules. The Tenax sorbent positioned at 15mm from the tube inlet serves as an infinite sink in this case.

Diffusion test. The tests were carried out in a thermally-controlled cabinet maintained at a constant temperature of 25°C. The clean thermal desorption tubes were placed on the PVC samples as shown in Figures 1b and 1c. Tubes were removed and immediately analyzed after 6, 12, 24, 48, 72, 100, 140, and 180 h for the green PVC flooring and after 2, 4, 6, 8, 12, 18, 24, 48, and 72 h for the red PVC flooring. Two diffusion tests were conducted with the green PVC under identical conditions using the same sample of PVC. All samples were extracted and analyzed by thermal desorption coupled with gas chromatography/flame ionization detector (TD-GC/FID).

RESULTS AND DISCUSSION

During the test, the phthalates migrate from the gas-phase adjacent to the PVC surface, diffuse through the gas in the tube and adsorb to the stainless surfaces (Crank, 1975). The extraction of samples should yield the total mass of SVOCs trapped in the tube (M) including the mass adsorbed on the tube surfaces (M_q) plus the mass flux into the Tenax sorbent (M_j). The diffusion test data with model fitting for the green PVC is shown in Figure 2. There is good agreement between the duplicate diffusion tests, and the data compare well to the model simulations. The values of y_0 and K_S were found to be 2.4 µg/m^3 and 1800 m, which agreed very well with literature data for the exact same material (Xu et al., 2012, Liang and Xu, 2014). Based on the model predictions, the mass flux to the sorbent is minimal compared to that on the interior surface of the tube.

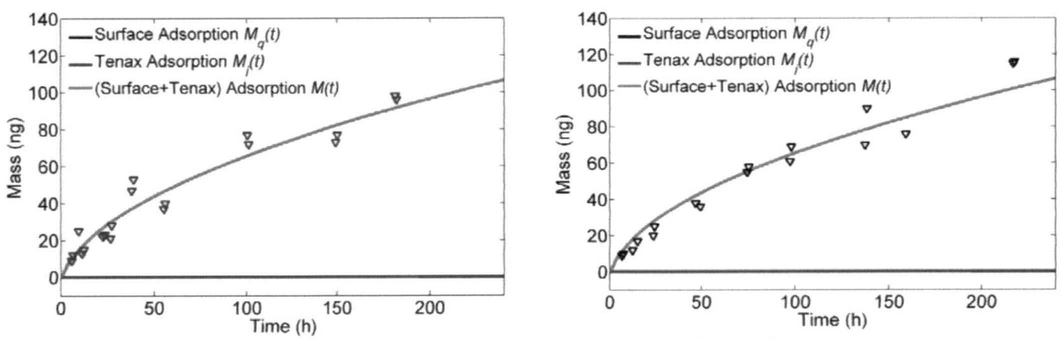

Figure 2. Measured accumulated mass of DEHP for green PVC flooring (duplicate tests).

The applicability of the diffusion model and experimental setup was further tested using SVOC with higher vapor pressures, which in this case are DiBP and DnBP. Figure 3 shows the experimental data and model predictions for the red PVC flooring. For the low concentration of DEHP in the red PVC material, the diffusive mass transport was too low to be detected by our current analytical system in 72 hrs suggesting y_0 is closely correlated with the material phase concentrations and may not be approximated by its vapor pressure. This finding is consistent with recent results (Liang and Xu, 2014). For DiBP and DnBP, the diffusion is much faster than DEHP, due to their higher vapor pressures. The fitted y_0 and K_S are 50 µg/m^3, 130m, and 27 µg/m^3, 150 m respectively, which are also reasonably consistent with the values measured by Liang and Xu using the same material (2014). Overall, the results suggested initial success in developing a simple approach for determining two of the key parameters that influence human exposure to SVOCs.

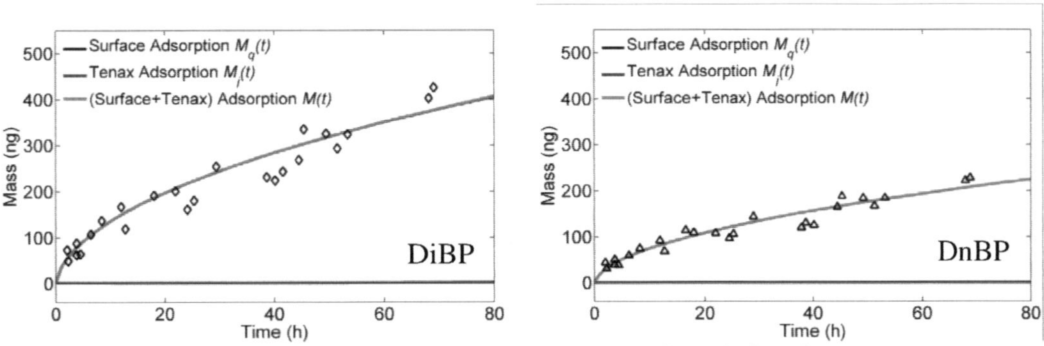

Figure 3. Diffusion curve for DiBP and DnBP from red PVC flooring.

CONCLUSIONS

In this study, we developed a novel, simple and efficient approach based on a diffusion model to determine y_0 and K_S, two of the key parameters that control SVOC emissions and influence human exposure. More tests focusing on a wider range of SVOCs other than phthalates are needed. Further development of the experimental setup and overall approach including additional types of surfaces, such as wood, glass, and artificial skin would extend the utility of this approach.

ACKNOWLEDGEMENT

Financial support was provided by the United States Environmental Protection Agency (Grant Number RD – 83560601 – 0) and National Science Foundation (CBET 1066802).

REFERENCES

Clausen, P. A., V. Hansen, L. Gunnarsen, et al. (2004). Emission of Di-2-ethylhexyl Phthalate from PVC Flooring into Air and Uptake in Dust: Emission and Sorption Experiments in FLEC and CLIMPAQ. *Environmental Science & Technology*, **38**, 2531-2537.

Crank, J. (1975). *The mathematics of diffusion / by J. Crank*. Oxford [England], Clarendon Press.

Liang, Y. and Y. Xu (2014). Improved Method for Measuring and Characterizing Phthalate Emissions from Building Materials and Its Application to Exposure Assessment. *Environmental Science & Technology*, **48**, 4475-4484.

Little, J. C., C. J. Weschler, W. W. Nazaroff, et al. (2012). Rapid Methods to Estimate Potential Exposure to Semivolatile Organic Compounds in the Indoor Environment. *Environmental Science & Technology*, **46**, 11171-11178.

Liu, C., Z. Liu, J. C. Little, et al. (2013). Convenient, Rapid and Accurate Measurement of SVOC Emission Characteristics in Experimental Chambers. *PLoS ONE*, **8**, e72445.

Matsumoto, M., M. Hirata-Koizumi and M. Ema (2008). Potential adverse effects of phthalic acid esters on human health: A review of recent studies on reproduction. *Regulatory Toxicology and Pharmacology*, **50**, 37-49.

Weschler, C. J. and W. W. Nazaroff (2008). Semivolatile organic compounds in indoor environments. *Atmospheric Environment*, **42**, 9018-9040.

Xu, Y., Z. Liu, J. Park, et al. (2012). Measuring and Predicting the Emission Rate of Phthalate Plasticizer from Vinyl Flooring in a Specially-Designed Chamber. *Environmental Science & Technology*, **46**, 12534-12541.

WIRELESS MONITORING OF THE INDOOR AIR ENVIRONMENT IN A WINERY DURING HARVEST

Nicolas M MADRID[1], David J KILLEEN[1], André KNOESEN[1,] David MILLS[2] and Roger B BOULTON[3*]

[1]Department of Electrical and Computer Engineering, University of California, Davis, California USA
[2]Department of Food Science and Technology, University of California, Davis, California USA
[3]Department of Viticulture and Enology, University of California, Davis, California USA

[*]Corresponding email: rbboulton@ucdavis.edu

Keywords: Winery, wireless, carbon dioxide, particles, harvest

SUMMARY

The purpose of this study was to establish a wireless monitoring system for air environment in the fermentation hall of a winery during harvest. Sensor boards with wireless communication were developed for the measurement of carbon dioxide, temperature, humidity, volatile organic carbon and particle counts during the 2014 harvest in the LEED Platinum Teaching and Research Winery at UC Davis. The study confirmed the effectiveness of the vapour capture system of the building design to keep the ambient carbon dioxide levels well below exposure limits and that more studies are needed to understand the source and fate of the volatile carbon compounds and particles and the role of people and equipment movement on the particle patterns.

INTRODUCTION

Wineries, breweries and many food facilities differ from standard office and occupied buildings in that they are unconditioned spaces with significant movement of materials during normal operation. The seasonal nature of the grape harvest and the higher levels of fermentable sugars distinguish wineries from breweries in terms of the levels of carbon dioxide and ethanol vapour released during fermentation. In wineries, grapes are received and fermentations are conducted only during an 8 to 10 week period. During this period outside doors will be open for activities such as grape bin delivery, crushing and pressing with daily cleaning and washing of most process equipment and periodic cleaning of a fermentation tank when fermentation is completed.

The evolution of carbon dioxide and ethanol vapour as emissions from the ethanol fermentation pose significant worker exposure issues and there is essentially no information about the ambient levels of these components in winery buildings during harvest. Carbon dioxide in workplace air has an OSHA permissible exposure level (PEL) of 5000 ppm (Crowl and Louvar, 1990) and when released in a concentrated form such as from a wine fermentation, it is denser that air. Ethanol vapor which is a natural part of the fermentation emission has a PEL of 1000 ppm with most federal and state agencies, and this seems to have been established only recently in 2009 (OSHA, 2014). Current EPA daily particle exposure limits for PM2.5 and PM10 are 35 and 150 ug/m^3 (EPA, 2014) and particles are of concern in agricultural areas where significant natural air intake is common in unconditioned spaces,

such as wineries. As a result, winery fermentation spaces can present a hazardous environment if the rate of removal by ventilation does not exceed the rate of carbon dioxide release by fermentation and if there is poor air distribution within the building or fermentation hall. They can also be spaces in which outside particle can be introduced during working periods and remain and be redistributed with internal air movements.
The design of fermentation spaces is further complicated by the dynamic emission characteristics of batch ethanol fermentations and by the staggered beginning of fermentation that is usually determined by the grape delivery pattern. The number of active fermentations changes as additional grapes are delivered throughout the harvest period. White wine fermentations normally take 14 to 21 days to complete, while red wines fermenting at higher temperatures are completed within 7 to 10 days. Wine fermentations typically release between 60 to 65 L of carbon dioxide per litre of juice, depending on the initial sugar content (Boulton et al. 1996). The rate of carbon dioxide release is proportional to the fermentation rate and this begins slowly then accelerates to a peak at mid-fermentation and then declines back to zero at the end of the fermentation. In order to remain below the permissible exposure limits for carbon dioxide, the rate of fresh air intake needs to be about 200 times the rate of carbon dioxide release and this poses the major challenge in conventional building design approaches. Most indoor winery fermentation spaces are monitored for carbon dioxide levels but often have only one sensor and some sensors are not installed in optimal air flow locations, even worse is that some fan ventilation systems become ineffective once the room is closed up at the end of the working day.

This project aimed to evaluate the ability of wireless sensor board to monitor the air environment in a space that is densely-populated with stainless steel fermenters in the midst of an existing wireless data network, inside a steel structure. This is the first report of specific information about carbon dioxide, ethanol and particle concentrations in a winery during harvest period and hopefully, it will lead to improved design and control of the ventilation patterns in these unconditioned spaces.

METHODOLOGIES

The winery building chosen for this installation was the LEED Platinum Teaching and Research Winery at UC Davis. Like many commercial wineries, it is an unconditioned fermentation space that relies on open doors for most of its air ventilation during working hours but is closed up during the non-working period. A unique feature of this winery is the capturing of fermentation emissions from all of the fermenters rather than releasing them into the room. The building is a steel-framed, steel-roof structure with one hundred and fifty 200 litre, stainless steel, research fermenters and fourteen 2000 litre, stainless steel, teaching fermenters. Each of the research fermenters sends temperature information and juice density every 15 minutes on a separate wireless network. The sensor board used in this project was located in the centre of the fermentation hall underneath a two meter high catwalk in the middle of two rows of the large fermenters.

A wireless point to point connection using IEEE 802.15.4 communication protocol was implemented. The sensor board contained sensors for volatile organic compounds (Applied Sensors, iAQ-2000), carbon dioxide, temperature and humidity (COZIR, GC-0011) with an adjacent particle counter (Dylos, DC1700) to measure two grades of particles greater than 10 um and smaller than 2.5 um and by difference a measure of the 2.5 to 10 um fraction. The board also controlled the data acquisition and transfer using a PSoC5LP module (CY8C5268AXI-LP047) and XBee transmitter (XBee Pro S1, XBP24-AUI-001). The particle counts were measured every minute and accumulated into a data packet along with air measurements taken every 15 minutes. Data packets are transmitted every 15 minutes to a centrally located receiver (ConnectPort X2) that places it on the cloud. During the experimental period two wireless sensor networks were operational simultaneously in the 2.4 GHz spectral region that operated without causing data loss. In addition to IEEE 802.15.4 network, a wireless network configured in a point to multipoint topology using propriety protocol (Cypress CyFi™ Low-Power RF) supports information collection from the research fermenters.

The carbon dioxide and volatile organic carbon sensors were direct reading in parts per million. The particle counts were converted from the Dylos output of counts per $1/100^{th}$ of a cubic foot to counts per cubic meter by multiplying by 3531.5. The mass of 2.5 and 10 micron particles was calculated to be 5.8×10^{-7} ug and 1.2×10^{-4} ug based on the radii reported by Lee et al., 2008 and a particle density of 1.65 g/cm^3. The particle counts were converted into the units of microgram per cubic meter using equation 1.

$$\text{Particle Concentration } (ug/m^3) = \text{Dylos Number} * 3531.5 * \text{Particle Mass } (ug) \qquad (1)$$

RESULTS AND DISCUSSION

A typical data for the pattern of temperature, carbon dioxide, volatile organic carbon is presented in Figure 1. The daily temperature and humidity values in the hall approach those of the outside air when the doors are open during the day and return to the night time conditions after the doors are closed at the end of the day. The carbon dioxide levels are generally less than 2500 ppm due to the capture system that is installed in this facility. The volatile organic carbon levels which might be expected to be dominated by ethanol are not related to carbon dioxide levels and tend to spike after the building is closed. This may be due to emissions from cleaning solutions (peroxyacetate) or residual wine in drains after tank cleaning. The levels of small and large particles show periodic spikes, Figure 2, and these may be due to night air intake fan activity since it is more pronounced when the building is closed. This pattern is the same over longer periods. Almost all values are below the daily permissible exposure levels and would be acceptable on a time-weighted average basis. The two sizes of particles appear to be independent of each other, suggesting difference sources. This needs to be investigated further.

CONCLUSIONS

The results demonstrate the effectiveness of a wireless sensor system to record the air composition in a winery during harvest. The vapor capture system in this winery keeps the ambient carbon dioxide levels to well below permissible exposure limits. The volatile organic carbon measurements were not correlated with the carbon dioxide levels and show peaks that seem to relate to other winery activities rather than fermentation emissions. The airborne particles can be distinguished into two types a fine particle set and a large particle set that were unrelated to each other.

ACKNOWLEDGEMENT

We gratefully acknowledge the support of two TJ Rodgers Fellowships, an Alfred P. Sloan Foundation grant from the Microbiology of the Built Environment Program and the Stephen Sinclair Scott Endowment.

REFERENCES

Boulton R, Singleton V L, Bisson L B and Kunkee R E (1996). *Principles and Practices of Winemaking*. Springer, Berlin, Germany.

Crowl D A and Louvar J F (1990). *Chemical Process Safety: Fundamentals with Applications*. Prentice Hall, Englewood Cliffs, New Jersey, USA.

EPA (2014) http://www.epa.gov/ttn/naaqs/standards/pm/s_pm_history.html

Lee J Y, Shin H J, Bae S Y, et al. (2008). Seasonal variation of particle size distributions of PAHs at Seoul, Korea. *Air Quality Atmospheric Health*, **1**, 57-68.

OSHA (2014) https://www.osha.gov/dts/chemicalsampling/data/CH_239700.html

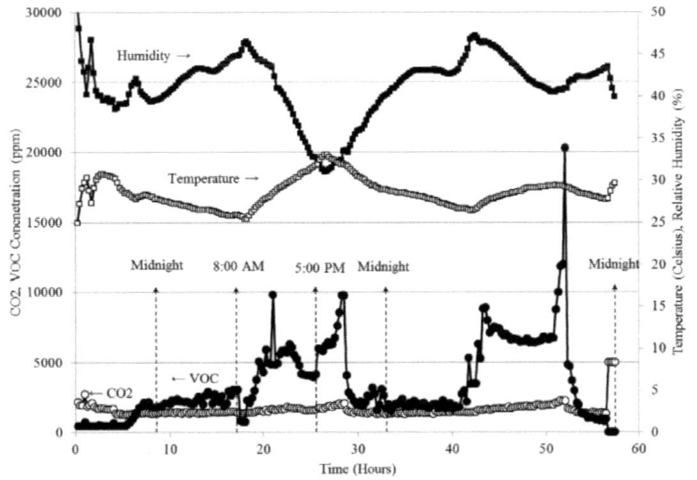

Figure 1. The air temperature, humidity, and concentrations of volatile organic carbon and carbon dioxide during a two day period, during the 2014 harvest.

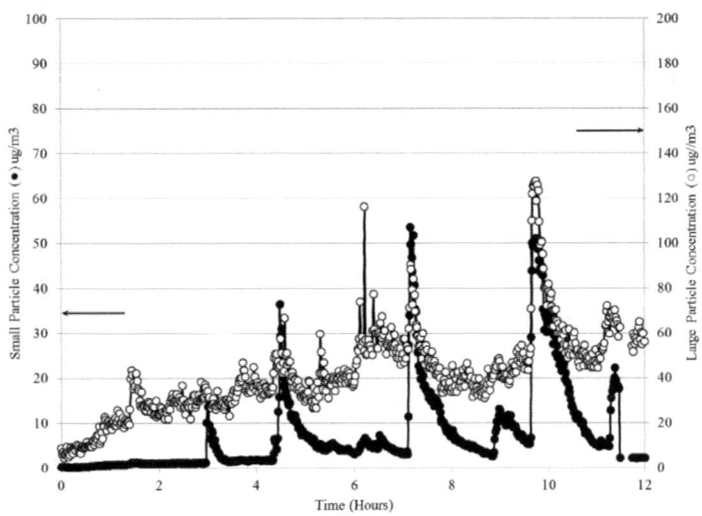

Figure 2. Small and large particle counts during a midday to midnight period.

SOURCES AND MITIGATION OF DIOXIN AND DIOXIN-LIKE COMPOUNDS IN INDOOR DUST

Patricia V CLINE[1]

[1]Protect Gainesville's Citizens

*Corresponding email: pcline@ufl.edu

Keywords: Brominated flame retardants, Brominated furans

SUMMARY

In addition to the chlorinated dioxins/furans regulated by the Environmental Protection Agency (EPA), *in vitro* assays of house dust samples indicate unregulated or unidentified compounds show dioxin like activity contributing to human background exposures to total dioxin toxic equivalents. As part of a study to characterize the contribution of these unregulated compounds to residents impacted by an industrial source, the brominated analogues of the chlorinated compounds were analyzed in dust samples collected from thirty homes. The dust samples were collected in 2012 by hepa vacuuming accessible floor areas and analyzed by a modification of method 1613B. Brominated furan concentrations and patterns related to the specific brominated flame retardant (BFR) source materials in the homes. Two homes resampled and tested in 2014 showed significant decreases corresponding to the phase out of those specific BFRs, however, increases on one of the homologues suggest alternative BFRs may be in use.

INTRODUCTION

Consensus protocols are established for estimating the toxicity equivalents (TEQs) of well studied and regulated mixtures of chlorinated dioxin and furan compounds (USEPA, 2013). Bioassays have been developed that measure toxicity equivalents for a compound or mixture of compounds. Using this assay, high responses have been reported in indoor dust samples that are from unregulated and/or unidentified compounds from sources within the homes. ((Suzuki et al., 2010; Tue et al., 2013)). Where residential soil impacted by chlorinated dioxin compounds are being remediated, reporting of these high bioassay results lead residents to assume all toxicity equivalents are attributed to the industrial source. This lead to the need to better characterize sources of these compounds within homes.

Among compounds responding in these assays are the brominated analogs of the regulated chlorinated dioxins and furans, primarily the brominated dibenzofurans (BDFs). BFRs have been identified as a source of these brominated compounds, in particular two commercial mixtures of polybrominated diphenylethers (PBDEs). EPA has developed an action plan for these compounds and have summarized information on toxicity and exposures (http://www.epa.gov/oppt/existingchemicals/pubs/actionplans/pbde.html). One commercial mixture, pentaBDE, contains a mixture of congeners containing 3-6 bromines but predominantly congeners with five bromines. This was used in textiles and as an additive in polyurethane foams found in materials like foam furniture, mattresses, and automobile seats. The use of pentaBDE was phased out in the US in 2004. Approximately 97% of decaBDE commercial products are fully brominated BDEs containing 10 bromines, but may include other congeners. DecaBDE are plastics found in electronics, wire/cable insulation or in textiles. EPA (2010) reviewed studies of concentrations of PBDEs in house dust and found results to be highly variable. Mean concentrations of pentaBDE and decaBDE were in the range of 2400 parts per billion.

These flame retardants can have traces of brominated dioxins and furans ((Brown, Overmeire, Goeyens, Denison, & Clark, 2001; Ren et al., 2011)). These are predominantly brominated dibenzofurans (BDFs). In addition, furans can form from PBDEs through intra-molecular elimination of Br2 or HBr. For example, the elimination of Br2 from decaBDE (10 bromines) would form OBDF (8 bromines). Photolysis of decaBDE in textiles exposed to natural sunlight have been shown to increase the concentrations of furans.(Kajiwara, Desborough, Harrad, & Takigami, 2013) In addition, debromination of the furans is a photochemical pathway leading to lower molecular weight congeners (Watanabe, 2003).

METHODOLOGIES

Indoor dust from thirty homes were sampled in 2012. Samples were collected following the EPA standard operating procedure (SOP) #2040 "Collection of Indoor Dust Samples from Carpeted Surfaces for Chemical Analysis Using a Nilfisk GS-80 Vacuum Cleaner". The sampling consisted of one composite dust sample per household. Each dust sample was collected from high traffic floor areas that were easily accessible including carpets, rugs, tile, and wood floors. The fine dust that passed the 100-mesh sieve was sent to Vista Analytical Laboratories to be analysed for chlorinated dioxins and furans using High Resolution Gas Chromatography/ High Resolution Mass Spectrometry (Method 1613b) and a modification of 1613b was used to analyse the brominated dioxins and furans. In 2014, comparable protocols were used to resample and analyse the brominated compounds the indoor dust in two of these homes.

RESULTS AND DISCUSSION

Brominated dioxins represented less than one percent of the brominated compounds detected. Brominated dibenzofurans (BDFs) were detected in dust samples from each of the 30 homes, with the highest concentrations found for homologues with four (TBDF) or eight (OBDF) bromines (Table 1)

Table 1. Summary of BDF homologue concentrations (ppt) for 30 home dust samples.

Homologue	Count	Min	Max	Median	Average
Total TBDF	30	267	379,000	11,200	53,917
Total PentaBDF	22	329	30,200	5,485	7,103
Total HexaBDF	17	77	13,550	502	1,508
Total HeptaBDF	30	132	52,300	3,285	6,740
OBDF	29	403	570,000	46,200	88,569

While the absolute concentrations varied, the data show that TBDF and OBDF represent the highest percentage of the BDF compounds. Many homes had nearly 100 percent OBDF, others nearly 100 percent TBDF, while others had some mixture of these. However, as shown on Figure 1, the percentages are not correlated with the total concentration of all the furan homologues.

OBDF is a contaminant in, or formed from, decaBDE which was phased out in 2013. DecaBDE was used in plastics and/or textiles. While the total TBDF congener concentrations were frequently elevated, the more toxic congener (2,3,7,8-TBDF) was not detected in any of the samples. While TBDFs can form from debromination of the higher molecular weight furans, the low concentrations of furans with five, six or seven bromines and lack of the 2,3,7,8-TBDF congener suggest the TBDFs are associated with the presence of pentaBDE. PentaBDE, phased out in 2004, is likely associated with its use in foam furniture.

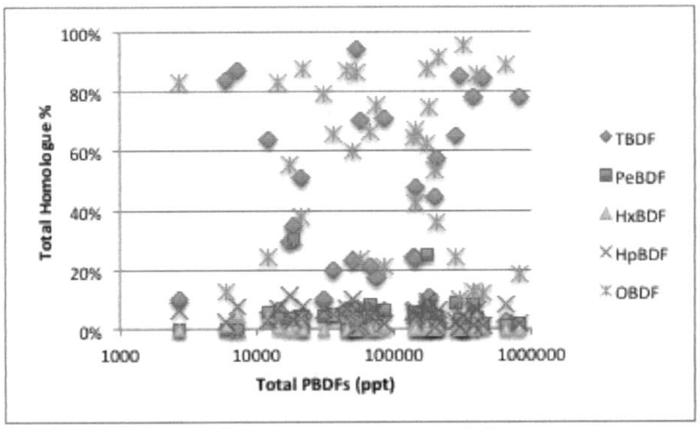

Figure 1. Contributions of BDF homologues to the total furan concentration in dust samples.

Dust from two of the homes were resampled in 2014 (Figure 2). Dust from both showed significant decreases in concentrations of TBDF and OBDF. TBDF decreased 91% in home a, and 40% in home B during that interval. OBDF concentrations decreased about 90%. This suggests removal of sources of pentaBDE and decaBDE resulted in decreased concentrations of the brominated furans.

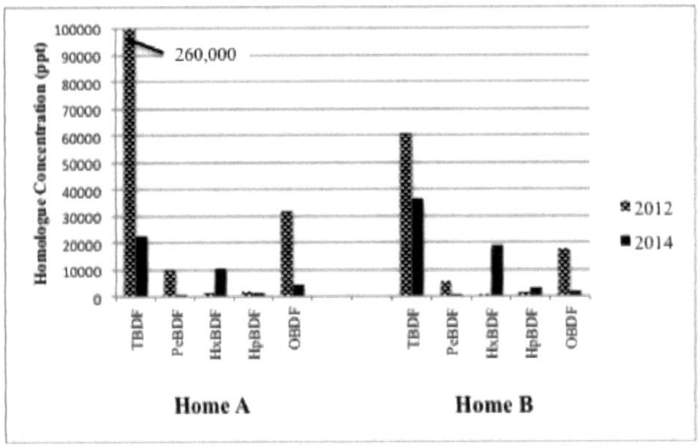

Figure 2. Changes in concentration of BDF homologues in dust from two homes sampled in 2012 and 2014.

However, the HxBDF concentrations in 2014 increased to concentrations of 10,200 and 18,800 ppt and represented 25 to 30 percent of the total furan homologue concentrations, much different from samples collected in 2012. Alternative brominated flame retardant chemicals replacing the PBDEs that have been phased out may be a potential source of HxBDF.

CONCLUSIONS

Brominated flame retardants are known to be sources of brominated furan compounds that can respond in dioxin bioassays. PBDEs have been phased out, but can remain in many materials in our homes. Reductions in concentrations of OBDF and TBDF suggest removal of sources of these compounds will result in decreasing dust concentrations.

In addition to the flame retardant chemicals, the presence or formation of dioxin-like compounds may be a consideration in evaluating potential health effects. Contributions to total background dioxin TEQ exposures from materials within homes should be considered in decision making.

ACKNOWLEDGEMENT

Protect Gainesville's Citizens (PGC) collaborated with Scott Miller, the EPA Region 4 project manager who funded the 2012 dust investigation. Other stakeholders in the dust study include Randy Merchant of the Florida Department of Health and John Mousa of the Alachua County Environmental Protection Department. The PGC research project was funded by an EPA Environmental Justice Small Grant, and our thanks go to our project manager, Daphne Wilson.

REFERENCES

Brown, D. J., Overmeire, I. Van, Goeyens, L., Denison, M. S., & Clark, G. C. (2001). ANALYSIS OF BROMINATED FLAME RETARDANTS AND BROMINATED DIBENZODIOXINS AND BIPHENYLS FOR Ah RECEPTOR ACTIVATION USING THE CALUX ® BIOASSAY. *Organohalogen Compounds*, *54*(12), 3–6.

Kajiwara, N., Desborough, J., Harrad, S., & Takigami, H. (2013). Photolysis of brominated flame retardants in textiles exposed to natural sunlight. *Environmental Science: Processes & Impacts*, *15*(3), 653. doi:10.1039/c3em30887a

Ren, M., Peng, P., Cai, Y., Chen, D., Zhou, L., Chen, P., & Hu, J. (2011). PBDD/F impurities in some commercial deca-BDE. *Environmental Pollution*, *159*(5), 1375–80. doi:10.1016/j.envpol.2011.01.004

Suzuki, G., Someya, M., Takahashi, S., Tanabe, S., Sakai, S., & Takigami, H. (2010). Dioxin-like activity in Japanese indoor dusts evaluated by means of in vitro bioassay and instrumental analysis: brominated dibenzofurans are an important contributor. *Environmental Science & Technology*, *44*(21), 8330–6. doi:10.1021/es102021c

Tue, N. M., Suzuki, G., Takahashi, S., Kannan, K., Takigami, H., & Tanabe, S. (2013). Dioxin-related compounds in house dust from New York State: occurrence, in vitro toxic evaluation and implications for indoor exposure. *Environmental Pollution*, *181*, 75–80. doi:10.1016/j.envpol.2013.06.010

USEPA. (2013). Recommended Toxicity Equivalence Factors (TEFs) for Human Health Risk Assessments of 2,3,7,8-Tetrachlorodibenzo-p-dioxin and Dioxin-Like Compounds.

Watanabe, I. (2003). Environmental release and behavior of brominated flame retardants. *Environment International*, *29*(6), 665–682. doi:10.1016/S0160-4120(03)00123-5

COMPARISON OF ADVECTIVE AND DIFFUSIVE TRANSPORT OF SVOCS THROUGH CLOTH FOR INDOOR CONDITIONS

Alexander KORFF[1], Melissa BUECHLEIN[1], Madison GIBLER[1], Glenn MORRISON[1,*]

[1]Department of Environmental Engineering, Missouri University of Science and Technology, Rolla, MO

[*]Corresponding email: gcm@mst.edu

Keywords: Advection, cloth, dermal uptake, mass transfer

SUMMARY

Clothing fabrics may influence dermal uptake of pollutants by acting as a transport barrier, but also as a reservoir of adsorbed compounds. Transport to and through fabrics is complex, but simplifications for indoor environments would improve exposure estimates by requiring less information about conditions. This study applies a Peclet number analysis to a range of indoor activities to determine when diffusion, advection or both are important for transport of SVOCs through clothing. Our findings suggest that for situations where an occupant is sitting, standing or walking slowly, diffusion dominates. For other indoor conditions, both advection and diffusion must be considered, unless very permeable, thin clothing is worn.

INTRODUCTION

Recent research suggests that dermal uptake to bare skin is an important route of exposure for some organic compounds found in indoor air (Weschler and Nazaroff, 2012). Dermal uptake to bare skin is limited by transport through a boundary layer adjacent to the body. It is unclear how clothing would influence dermal uptake; fabrics could act as a barrier to transport from air, but could also act as a near-skin reservoir of adsorbed or absorbed chemicals.

Fabrics have been recognized as an exposure source according to studies investigating dry-cleaning solvents, pesticides, tobacco smoke, and methamphetamine partitioning coefficients between clothing and air (Morrison et al., 2014). In Morrison et al. (2014), the time for equilibrium to occur for seven different fabrics was determined through analysis of the characteristic time for each fabric. In just a period of days to weeks, methamphetamine can equilibrate with clothing resulting in a highly concentrated source near the skin. These observations suggest that fabrics can be an important source for dermal uptake of SVOCs (Morrison et al., 2014).

Models of chemical transport through clothing have primarily focused on advective transport under outdoor conditions with moderate to high speed wind. In buildings, ventilation, fans or body motion can also result in velocities that will induce significant advection through worn fabrics (Ghaddar et al., 2006). Under these non-quiescent scenarios, advection may be a significant component of transport for VOC's and SVOC's to and into skin. However, for sitting, standing or moving slowly, molecular diffusion through fabrics may be the dominant transport mechanism given the low-velocity air conditions. Under diffusion dominated conditions, Buechlein et al. (2014) found that the flux of methamphetamine from cloth to artificial skin oil was a simple function of the distance between the cloth and the artificial skin oil. If we can identify conditions where advection can be neglected, transport models and exposure analysis can be simplified.

To determine if this is the case, we examine the relative importance of advection and diffusion through cloth by applying a Peclet analysis to typical fabric types and typical indoor conditions. The purpose of this analysis is to identify possible air velocities through clothing under common conditions where diffusion, advection or both play significant roles in overall flux.

METHODOLOGIES

To compare the relative importance of diffusion and advection, we first assume that transport of SVOCs can be described reasonably well by models of transport of contaminants through porous media (Fetter and Fetter Jr, 1999). The one-dimensional, non-steady-state concentration (C) can be described by Equation 1,.

$$\frac{D}{k_p}\frac{\partial^2 C}{\partial x^2} - \frac{V}{k_p}\frac{\partial C}{\partial x} = \frac{\partial C}{\partial t} \qquad (1)$$

where D is molecular diffusivity and V is velocity. When modeling the flux of a compound from cloth, the partitioning of compounds to the cloth must be considered as a possible 'retardation' factor. If the compound favors the cloth over the air phase, a large partitioning coefficient (k_p) could reduce the flux from cloth to skin.

A comparison of advection and diffusion was performed, using the Peclet number (Pe) (Equation 2). If the diffusive component is much larger than the advective component, Pe will be very small, indicating that advection is negligible. For $Pe < 0.1$, diffusion dominates. For Pe greater than 10, advection dominates. For Pe between 0.1 and 10, both advection and diffusion must be considered in transport models.

$$Pe = \frac{V_{face} * L}{D * n} \qquad (2)$$

A diffusivity (D) of 0.05 cm^2/s was used to represent common SVOCs. The face velocity (aka ventilation rate) due to air penetration of outer fabric of a clothed body surface (V_{face}) was determined from empirical correlations found in Ghaddar's (2004) review that included Loten's (1993) dissertation, seen in Equation 3 below, where (v_{eff}) is the effective wind velocity(m/s) and (α) is the air permeability through the outer fabric in L/(m^{2}s) at 200 Pa (Ghaddar et al., 2004).

$$V_{face} = 4.5 * 10^{-5} * \frac{v_{eff}{}^{0.5+0.05*\sqrt{\alpha}}}{0.16} \qquad (3)$$

Values for air permeability ($\boldsymbol{\alpha}$) were determined based on the range of common cloth-air permeabilities found in (ASTM, 1975). This correlation combines a sitting or standing position, wind, and activity. In order to relate this correlation to the velocity (V) through the cloth, the face velocity (V_{face}) was divided by the porosity (n) of the cloth. A linear approximation for the porosity was determined, assuming an average 1 mm thickness of a cloth (L), based on data by Das and Kothari, which related the permeability of various fabrics to porosity Das and Kothari (2012). An average value of 0.8 was used for (***n***) in this analysis.

RESULTS AND DISCUSSION

Figure 1 depicts the range of activity-related velocities and permeabilities used in this analysis to determine the range of Peclet values. Peclet values less than 0.1 are in white for situations where advection can be considered negligible. For these conditions, mass transfer could be described by Fick's first law, equation 4.

$$\frac{\partial C}{\partial t} = -\frac{D}{k_P}\frac{\partial C}{\partial x} \qquad (4)$$

The two-toned gray section depicts the range of Peclet values where a transition from molecular diffusion to advection--dominated flux occurs. The lighter gray region indicates that diffusion slightly dominates (Pe<1). The darker gray region indicates advection is more dominant (Pe>1). The black region represents scenarios where diffusion can be ignored and transport is influenced primarily by the effective gas velocity, as seen in Equation 5. Note however, at higher relative velocities, dispersion may need to be considered in addition to advection.

$$\frac{\partial C}{\partial t} = -\frac{V_{face}}{k_p n}\frac{\partial C}{\partial x} \qquad (5)$$

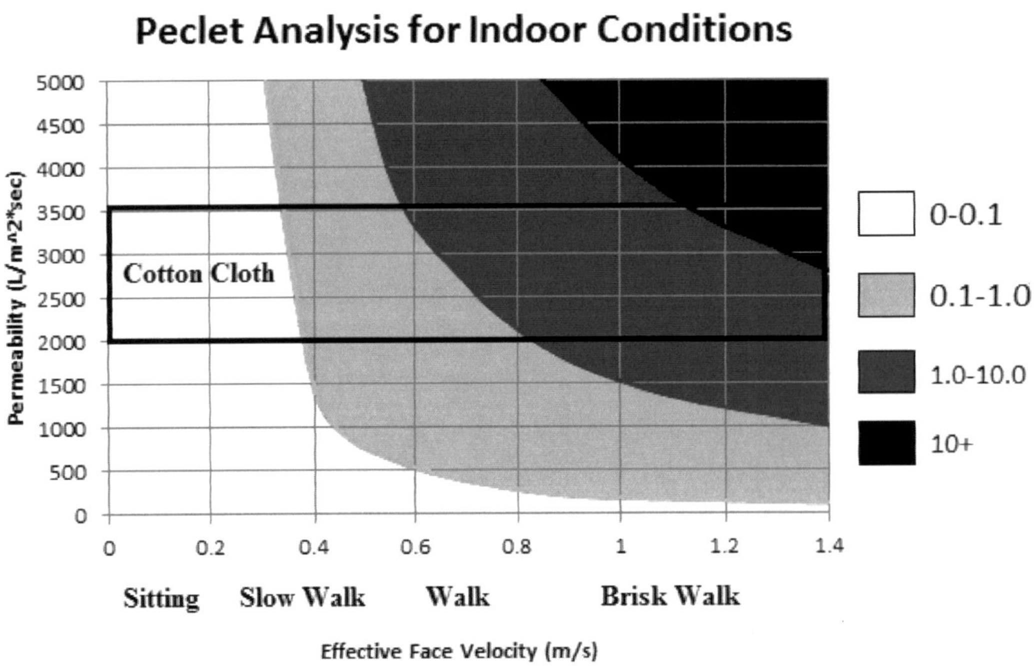

Figure 1: Peclet Analysis for typical indoor conditions ranging from sitting to a brisk walk with varying permeabilities. The permeability region for 1mm thick cotton cloth is denoted by the box.

Given that molecular diffusion dominates for "typical conditions", it is possible to estimate the characteristic time (k_pL^2/D) required for fabric to equilibrate with contaminants while worn. For methamphetamine adsorbing to cotton, the characteristic time is ~ 9 days. Therefore, clean cotton cloth should act as a substantial barrier to dermal uptake of methamphetamine if only worn for one day. On the other hand, if cotton clothing is allowed to equilibrate in a house contaminated with airborne methamphetamine (e.g. hanging in a closet for several weeks) a large amount of this pollutant will be placed directly next to the skin. Further, the flux to skin will remain high while being worn for one day.

CONCLUSIONS

For typical low-velocity conditions indoors, advection can be considered negligible and molecular diffusion is likely the dominant mechanism driving contaminant transfer through clothing to skin. However, advection through the cloth can be a real source of exposure in high wind or high activity conditions. In some cases, such as mild movement or mild wind conditions, advection and dispersion both play non-negligible roles in dermal exposure. Factors such as the compound's ability to partition into and out of cloth, temperature, the effect of relative humidity, and the skin's ability to sorb the compound will influence overall exposure and requires further investigation. These results suggest that models appropriately accounting for advective and diffusive transport are necessary when estimating dermal uptake in exposure models.

REFERENCES

ASTM D737 (2012). Standard method of test for air permeability of textile fabrics.

Buechlein M, K Parker, and G Morrison (2014). Skin Uptake of Gas Phase Methamphetamine: Effect of Clothing. In: The 13th Internation Conference on Indoor Air Quality and Climate. Hong Kong, Hong Kong.

Das S, and V K Kothari (2012). Moisture vapour transmission behaviour of cotton fabrics. *Indian Journal of Fibre and Textile Research,* **37**(2), 151-156.

Fetter C W, and C Fetter Jr (1999). Contaminant hydrogeology. Prentice hall New Jersey.

Ghaddar N, K Ghali, and J Harathani (2004). Inertia effects on the modulated air layer heat and moisture transport from clothing with open aperture. In: Proceedings of the ASME Heat Transfer/Fluids Engineering Summer Conference 2004, HT/FED 2004.

Ghaddar N, K Ghali, and B Jones (2006). Convection and ventilation in fabric layers. *Thermal and Moisture Transport in Fibrous Materials.* 271-307.

Morrison G, N V Shakila, and K Parker (2014). Accumulation of gas-phase methamphetamine on clothing, toy fabrics, and skin oil. *Indoor Air.*

Weschler C J, and W W Nazaroff (2012). SVOC exposure indoors: Fresh look at dermal pathways. *Indoor Air,* **22**(5), 356-377.

POLLUTION TRANSPORT BY THE HUMAN CONVECTION FLOW – IMPACT OF THE ROOM AIR TEMPERATURE AND SEATED BODY INCLINATION

Dusan LICINA[1,2,*], Arsen MELIKOV[2], Chandra SEKHAR[1], Kwok Wai THAM[1]

[1]Department of Building, School of Design and Environment, National University of Singapore, Singapore
[2]International Centre for Indoor Environment and Energy, Department of Civil Engineering, Technical University of Denmark, Kgs. Lyngby, Denmark

[*]Corresponding email: licinadusan@yahoo.com

Keywords: Convective boundary layer, Pollution, Thermal manikin, Personal exposure

SUMMARY

This study investigates the ability of a human convective boundary layer (CBL) to transport the pollution in a quiescent indoor environment. The impact of the room air temperature and seated body inclination is examined in relation to pollution distribution in the breathing zone and the thickness of the pollution boundary layer (PBL). The human body is resembled by a thermal manikin whose body shape, size and surface temperatures correspond to those of a real person. The pollution, simulated by tracer gas Nitrous oxide (N_2O), was located at the chest and groins. The results show that reducing the room air temperature or backward body inclination intensifies the transport of the pollution to the breathing zone which increases personal exposure and reduces the pollution spread across the room.

INTRODUCTION

In many indoor environments air mixing is minimized and the pollution concentration gradients occur, especially in those that operate with low supply air velocity (Cermak and Melikov, 2006). These concentration gradients may occur near a person forming a "personal cloud" that tends to have different pollution concentrations in the breathing zone than in the surroundings of the room (Wallace, 2000). Thus, assuming "well mixed" condition may lead to incorrect exposure prediction. The human convection flow, known as convective boundary layer (CBL) plays two important roles. Firstly, it contributes to the convective heat loss from the body accounting to 29% of the total heat loss (Murakami et al. 2000) and secondly, it transports the gases or particles around the human body (Rim and Novoselac, 2009). Several other studies demonstrated the ability of the CBL to transport the air from the lower room levels to the breathing zone (Brohus and Nielsen, 1996; Melikov, 2004).

Pollution generated by the human body (bioeffluents) is one of the major indoor air quality concerns as numerous volatile compounds can be emitted from different body parts (e.g. feet, groins, armpits, etc.) that are disposed to odor generation (Wysocki and Preti, 2004). The pollution source emitted from different body parts is not likely to spread equally within the CBL. Therefore, it is important to consider to what extent the pollution emitted from different body parts spreads across the CBL and whether it can be transmitted to the breathing zone of other occupants. The influence of room air temperature (Licina et al. 2014) and body posture (Clark and de Galcina-Goff, 2009) on the development of the CBL has been investigated in some detail. The influence of the room air temperature and seated body inclination on the pollutant distribution in the breathing zone is unknown. The objective of this study is to determine pollutant distribution in the breathing zone as a function of the room air temperature and seated body inclination angle.

METHODOLOGIES

The measurements were conducted in a climate chamber with dimensions 4.7 x 6.0 x 2.5 m^3. The chamber was ventilated through the floor with a low velocity upward piston flow (100% outdoor air). Most of the air (85%) was supplied through the floor, while the remaining air was introduced through the space between the wall and the vinyl sheets. This kind of construction ensured that the room air temperature is equal to the mean radiant temperature and the absence of radiant temperature asymmetry.

A calibrated non-breathing thermal manikin with a complex female body shape of 1.23 m height in the sitting posture was used to resemble a realistic human body. The manikin was positioned in the center of the chamber and it was dressed in the tight-fitting attire (t-shirt, trousers, underwear, socks and shoes). The manikin heat output was 65 W/m^2, which is equal to the dry heat loss from a human body in a thermally comfortable state. The air introduced to the room was supplied at such a low velocity so as to minimally disturb the natural convection around the thermal manikin. A horizontal plate (2.0 x 1.54 m^2) was placed below the manikin to prevent the supply airflow from affecting the manikin's natural convection. In this way, the air movement around the manikin was induced solely due to the manikin's body heat. The velocity measured at numerous locations around the unheated manikin was below 0.05 m/s, which indicated that quiescent indoor conditions had been achieved. The air was exhausted from the chamber through the reduced area of the ceiling (2.4 x 2.4 m^2) located above the manikin. This was done to ensure that the rising thermal plume of the manikin would be directly exhausted above, without additional air circulation.

The influence of room air temperature on the pollutant distribution in the breathing zone was examined under three temperatures: 20, 23 and 26 °C. The influence of a seated body inclination angle was examined when the manikin was leaned 25° forward and backwards. The gaseous pollution was supplied isothermally with negligible initial velocity at the chest and groins. Both pollution sources were chosen to simulate pollution emitted from the occupant's body, e.g. human bioeffluents. The room air was kept at a constant temperature of 23 °C. A pollution source was simulated by tracer gas (N_2O) that was injected through a sponge ball (diameter 0.05 m) at a steady emission rate. The tracer gas was simultaneously sampled at 20 locations along the horizontal line in front of the mouth, with a distance between two consecutive points of 0.03 m. The first sampling point was located at the upper lip of the manikin at 0 mm distance from the surface, to represent the sampling of inhaled air. The tracer gas was sampled through the sampling tubes and sent to two calibrated Innova multi-gas samplers (Model 1303) and analyzers (1312) placed outside the chamber. Total sampling time took 3 hours, which corresponded to 45 gas samples in each point and the results were averaged. The thickness of the pollution boundary layer (PBL) was defined to determine the extent to which the pollution spreads horizontally from the mouth. It was assumed that the thickness of the PBL extended to a distance where the concentration of the tracer gas reached 10% of the maximum sampled concentration in front of the mouth.

RESULTS AND DISCUSSION

The impact of the room air temperature on the breathing zone concentrations, personal exposure and thickness of the PBL is shown in Figure 1. For both source locations, reducing the room air temperature from 23 to 20 °C increased the pollutant concentration in the breathing zone. As the lower room air temperature strengthened the CBL and increased its velocity (Licina et al. 2014), it also intensified the transport of the pollution to the breathing zone. The opposite effect can be seen when the room air temperature increased from 23 to 26 °C, as shown in Figure 1 (left). As seen on Figure 1 (right), the pollution emitted at the groins spread more compared to the pollution from the chest, for all three temperatures studied. Elevating the room air temperature from 23 to 26 °C increased the thickness of the PBL for both the source locations; however, it also reduced personal exposure. Furthermore, increasing the room air temperature from 20 to 26 °C reduced personal exposure by 30% for the pollution emitted from the chest. This can be correlated with the peak velocity decrease of 33% in the breathing zone, when the same temperature change was applied, as suggested by Licina et al. (2014). The findings suggest that the room air temperature can be used to control the amount of the transported pollution/clean air to the breathing zone. In practice, however, the room air temperature set point usually aims to satisfy the thermal comfort requirements.

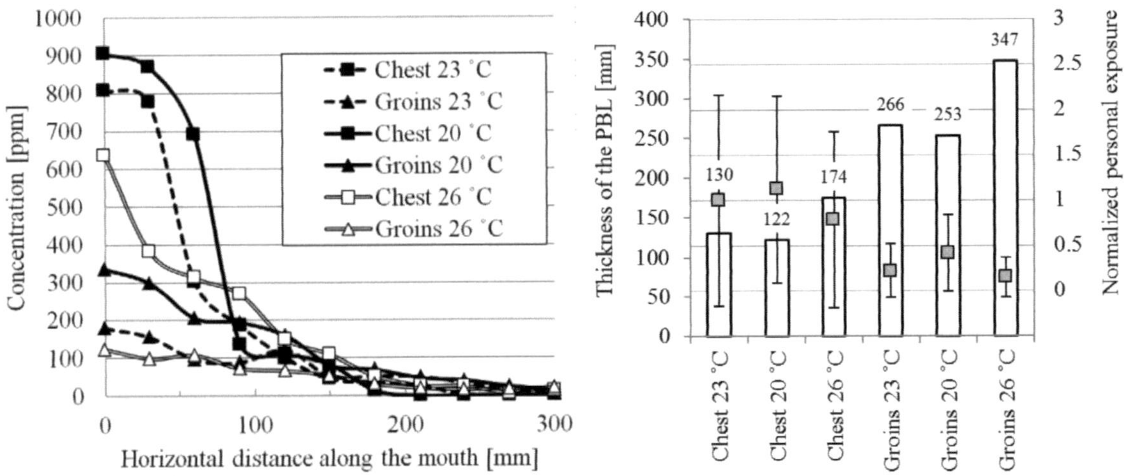

Figure 1. Concentration of tracer gas in the breathing zone (left) and normalized personal exposure and the thickness of the PBL (right) - Impact of room air temperature.

The impact of a seated body inclination angle on the breathing zone concentrations, personal exposure and thickness of the PBL is shown in Figure 2. The results show that forward body inclination decreased the pollution concentration in the breathing zone as much as 82.5% in case of the chest and 74% in case of the groins. The opposite but milder effect was observed in case when the manikin was leaned backwards. As seen, when the manikin was inclined backwards the pollution decay profile was milder compared to the case with no inclination. This occurred because the pollution could not "stick to" the body and follow its contours as easily as it was the case with the straight body posture, causing more detachment from the surface. The manikin leaned forward caused most of the pollution from the groins and chest to avoid the mouth region because of the blocking effect of the chin. In general, the pollution emitted from the groins was more spread, compared to pollution from the chest, as shown in Figure 2 (right). It can be seen that the backward body inclination increased the thickness of the PBL and personal exposure, which was opposite from forward body inclination. The personal exposure changed in the same way as the velocity in the breathing zone, as the occupant leaned backwards increases the velocity, while forward body inclination decreases the velocity (Licina et al. 2014).

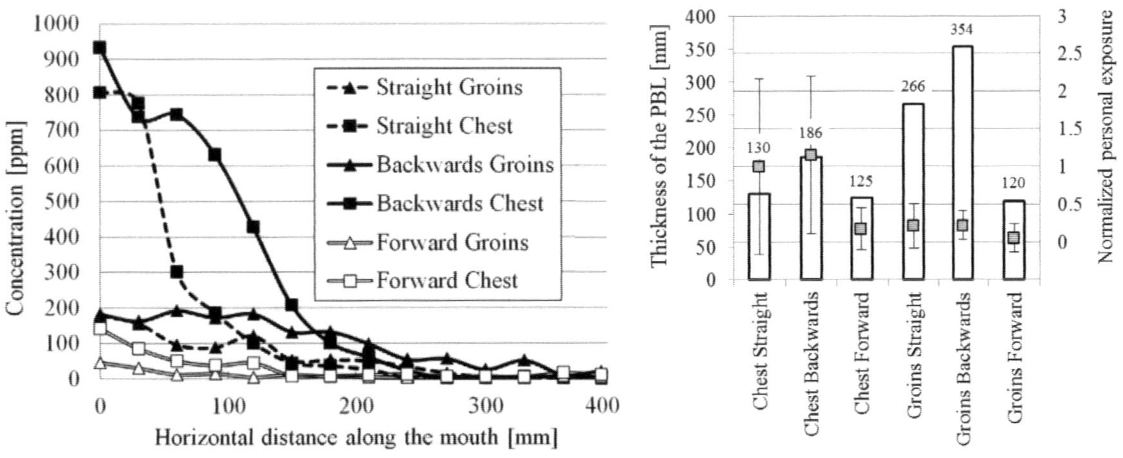

Figure 2. Concentration of tracer gas in the breathing zone (left) and normalized personal exposure and the thickness of the PBL (right) - Impact of a seated body inclination.

CONCLUSIONS

This study investigates the impact of the room air temperature and body inclination on a distribution of gaseous pollution in the breathing zone of a thermal manikin in a quiescent indoor environment. The source emitted at the groins reduces personal exposure, but spreads more across the room, compared to the pollutants from the chest. Reducing the room air temperature intensifies the transport of the pollution to the breathing zone, which increases personal exposure and reduces the pollution spread. The personal exposure increase is observed when the occupant is leaned backwards in the chair. The results suggest that in room with little air movement, a "well-mixed" mass balance model may underestimate the exposure. A detailed understanding of the pollutant distribution in the vicinity of a human body is therefore essential for minimized human exposure.

REFERENCES

Brohus, H. and Nielsen, P.V. (1996) Personal Exposure in Displacement Ventilated Rooms, Indoor Air, 6, 157–167.

Cermak, R. and Melikov, A. (2006) Air quality and thermal comfort in an office with underfloor, mixing and displacement ventilation, Int J Vent, 5, 323-332.

Clark, R.P. and de Galcina-Goff, M.L. (2009) Some aspects of the airborne transmission of infection, J. Royal Soc. Interface; 6: S767-82.

Licina, D., Melikov, A., Pantelic, J., Sekhar, C. and Tham, K.W. (2014) Experimental investigation of the human convective boundary layer in a quiescent indoor environment, Build. Environ., 75, 79-91.

Melikov, A.K. (2004) Personalized ventilation. Indoor Air, 14 (Suppl. 7), 157–167.

Murakami, S., Kato, S. and Zeng, J. (2000) Combined Simulation of Airflow, Radiation and Moisture Transport for Heat Release from a Human Body, Build. Environ., 35, 489–500.

Rim, D. and Novoselac, A. (2009) Transport of particulate and gaseous pollutants in the vicinity of a human body, Build. Environ., 44, 1840-1849.

Wallace, L. (2000) Correlations of Personal Exposure to Particles with Outdoor Air Measurements: A Review of Recent Studies, Aerosol Sci. Tech., 32, 15-25.

Wysocki, C. and Preti, G. (2004) Facts, fallacies, fears, and frustrations with human pheromones, Anat Rec A Discov Mol Cell Evol Biol., 281,1201-11.

THE INFLUENCE OF TEMPERATURE ON THE FATE AND TRANSPORT OF INDOOR PHTHALATES: A CASE STUDY IN A TEST HOUSE

Chenyang BI[1] and Ying XU[1,*]

[1]Department of Civil, Architectural, and Environmental Engineering, The University of Texas at Austin, Austin, Texas, USA

[*]*Corresponding email: xuying@mail.utexas.edu*

Keywords: Emission, Adsorption, Temperature, Phthalates, Field measurements

SUMMARY

In this study, the concentrations of benzyl butyl phthalate (BBzP) and di-2-ethylhexyl phthalate (DEHP) in indoor air, settled dust, and on different interior surfaces including mirror, dish plate, cloth and wood were measured periodically in a test house. The measurements were conducted at temperatures of 21°C, 25°C and 30°C, respectively. In addition, sorption kinetics was also monitored at the temperature of 21 °C. With the temperature increase to 25°C and 30°C, the average airborne concentrations of phthalates increased by about two and four times, respectively, and the surface concentrations on various surfaces also increased correspondingly. This investigation suggests that temperature has an important influence on the fate and transport of phthalates in indoor environments.

INTRODUCTION

Indoor environments contain a wide range of products that can emit a variety of pollutants. Among these chemicals, phthalate esters are particularly important, because they are used extensively as additives in consumer products and are associated with serious health concerns. Phthalates are ubiquitous indoors, redistributing from their original source to indoor air and interior surfaces including airborne particles, dust, and skin. As semi-volatile organic compounds (SVOCs), phthalates partition strongly to surfaces and may persist for long periods, even after the primary source is removed. Environmental conditions such as temperature and humidity may also have great impacts on the fate and transport of phthalates in indoor environments. Clausen et al. (2012) and Liang and Xu (2014) measured di (2-ethylhexyl) phthalate (DEHP) emissions from vinyl flooring in standard emission test chambers and have shown that temperature has a large impact on the emission of phthalates. However, Several studies reported no association between room temperature and phthalate concentrations in indoor air (Fromme et al., 2004; Bergh et al., 2011a, 2011b), while another study found significant correlations for higher vapor pressure phthalates (i.e., dibutyl phthalate [DBP], diethyl phthalate [DEP], and diisobutyl phthalate [DiBP]) (Gaspar et al., 2014). The associations between temperature and phthalate levels are difficult to examine in field campaigns because of the differences of environmental conditions such as humidity, ventilation rate and building characteristics in each site. Therefore, it is necessary to investigate the influence of temperature in a controlled test house.

In this study, we conducted field measurements in a residential test house (UTestHouse) to investigate the influence of temperature on the fate and transport of phthalates in indoor environments. The specific objectives are to: 1) measure the dynamic change of indoor air concentration of phthalates at three different temperatures; 2) determine the equilibrium partition coefficients of phthalates between air and interior surfaces, including dust, windows, dish plates, mirrors, fabric cloth, and wood, at the different temperatures; 3) investigate the sorption kinetics of phthalates onto these surfaces; and 4) interpret the measurement data using a previous developed multimedia indoor fate model for phthalates and conduct a screening-level exposure assessment.

METHODS

UTestHouse. Field measurements were conducted in UTestHouse located at Austin, Texas. The house, which has a floor area of 110 m^2, consists of three bedrooms, two bathrooms, a laundry room, a living room and a kitchen. The experiments were conducted in the living room and the two bedrooms. The vinyl flooring in this house contains 10% (w/w) of BBzP and the house is furnished with necessary furniture and appliances. The temperature was controlled by an air-conditioning system with no mechanical ventilation. The ventilation rate was measured by monitoring the natural decay of a tracer gas (CO_2). The relative humidity was monitored by Onset HOBO Data Loggers.

Sampling protocol. The concentrations of BBzP and DEHP in indoor air, settled dust, and on different interior surfaces, (e.g., mirror, hard wood, window, dish plates and cloth surfaces) were measured periodically in the test house at controlled temperatures of 21 °C, 25 °C and 30°C, respectively. After indoor air concentration reached steady state, sorption kinetics was monitored by measuring BBzP and DEHP accumulation onto the interior surfaces. Temperature, relative humidity and total suspended particle concentration were also monitored in real time. PUF cartridges were used to collect air sample at a flow rate of 3L/min for 48h. Floor dust and non-floor dust were separately collected by a vacuum cleaner connected with a crevice aluminum inlet. Other interior surfaces were wiped three times with pre-cleaned gauze pads. The PUF, dust, fabric cloth and wipe samples were ultrasonically extracted, concentrated, cleaned-up, and analyzed by gas chromatography with flame ionization detector (GC/FID).

RESULTS AND DISCUSSION

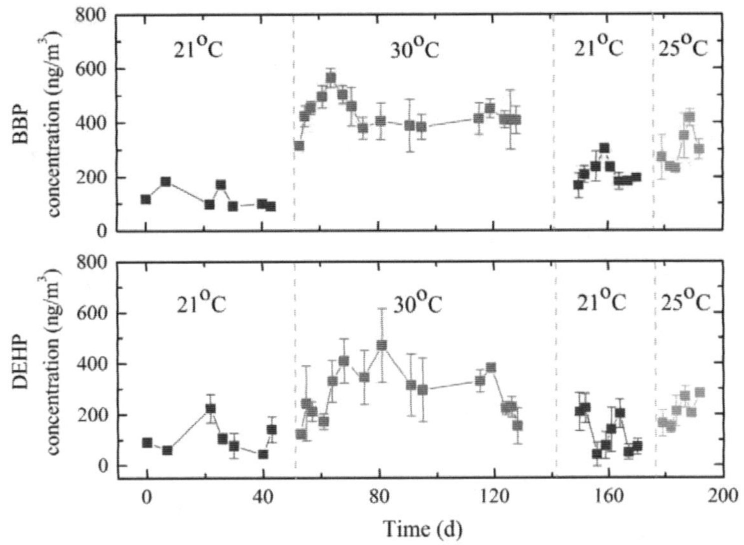

Figure 1. The airborne concentrations of BBzP and DEHP

Figure 1 shows the airborne concentrations of BBzP and DEHP at different temperatures. At 21°C, the gas phase concentrations of BBzP and DEHP range from 141 ng/m^3 to 210 ng/m^3 and 66 ng/ m^3 to 100 ng/ m^3, respectively (Figure 1a). When temperature was raised to 30°C, it took 12 days for the gas-phase phthalates to reach steady state. Although the concentrations of both BBzP and DEHP fluctuated at steady sate, the average levels at 30°C are generally four to five times higher than that at 21°C. It indicates that temperature have a strong influence on the concentration of phthalates in indoor air.

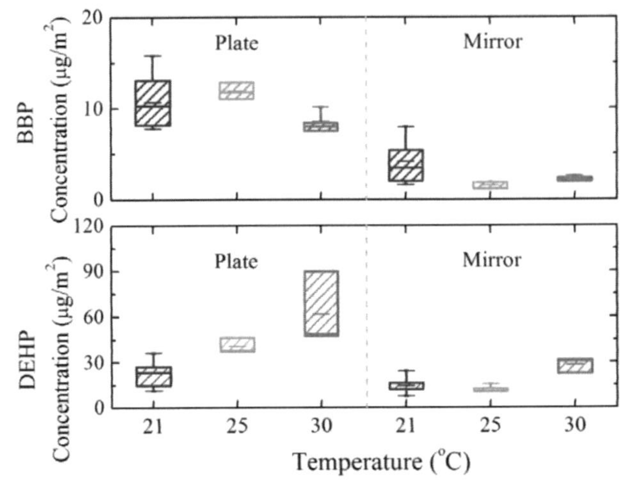

Figure 2. The concentrations of BBzP and DEHP on dish plates and mirrors

The concentrations of BBzP and DEHP on dish plates and mirrors are illustrated in Figure 2. The results suggest that for impervious surfaces such as dish plate and mirror, the concentration of BBzP was not influenced greatly by the temperature. In contrast, for both cotton and polyester cloth shown in Figure 3, the concentrations of BBzP and DEHP increased by about 6 times. In addition, sorption kinetics is much faster for the impervious surfaces (typically within hours), than porous surface like hard wood (>1 week) and dust (> months).

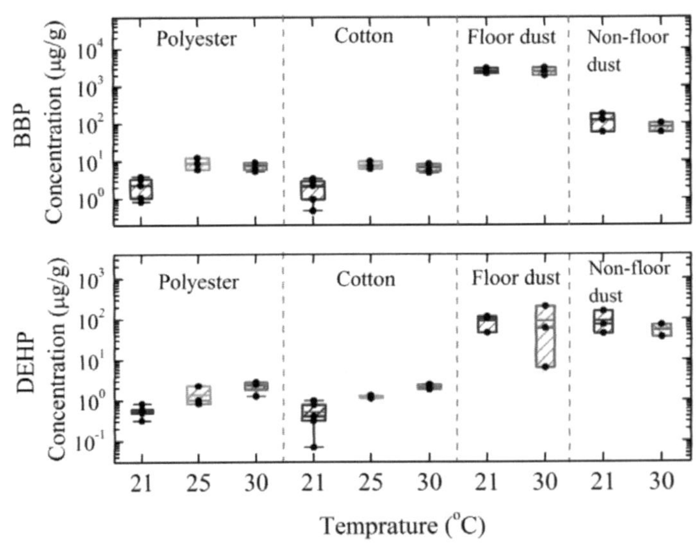

Figure 3. The concentrations of BBzP and DEHP in fabric cloth and dust.

As shown in Figure 3, the level of BBzP in floor dust (range from 2823 to 3518 μg/g) was 17-20 times higher than the non-floor dust (range from 140 to 203 μg/g), while little difference was observed for DEHP. This is because the flooring material installed in the test house used only BBzP as plasticizer. Therefore, particles deposited on source (e.g., vinyl flooring) may contain much higher levels of phthalates than on other types of surfaces. Furthermore, sorption kinetics is very slow (over months) for BBzP and even worse for DEHP. As a result, the concentrations of phthalates in settled dust are not influenced significantly by temperature in a relatively short time period.

CONCLUSIONS

This investigation suggests that temperature have important influences on the fate and transport of phthalates in indoor environments. It possibly implied that differences seen within and between indoor air studies may be, in part, due to differences in room temperature while sampling and vary by season. That is, temperature is an important parameter to record in environmental filed sampling for SVOCs. In additional, the results enable environmental intervention designs to reduce indoor exposures by developing a clear understanding of the factors that govern emissions and sorption of phthalates and their indoor fate and transport.

ACKNOWLEDGEMENT

Financial support was provided by the National Science Foundation (NSF) (CBET-1150713 and CBET- 1066642).

REFERRANCES

Bergh, C.; Torgrip, R.; Emenius, G.; Ostman, C. Organophosphate and Phthalate Esters in Air and Settled Dust - a Multi-Location Indoor Study. *Indoor Air,* **2011a**, *21*, 67–76.

Bergh, C.; Magnus Åberg, K.; Svartengren, M.; Emenius, G.; Östman, C. Organophosphate and Phthalate Esters in Indoor Air: A Comparison between Multi-Storey Buildings with High and Low Prevalence of Sick Building Symptoms. *J. Environ. Monit.* **2011b**, *13*, 2001–2009.

Clausen, P. A.; Liu, Z.; Kofoed-Sorensen, V.; Little, J.; Wolkoff, P. Influence of Temperature on the Emission of Di-(2-Ethylhexyl)phthalate (DEHP) from PVC Flooring in the Emission Cell FLEC. *Environ. Sci. Technol.* **2012**, *46*, 909–915.

Fromme, H.; Lahrz, T.; Piloty, M.; Gebhart, H. Occurrence of Phthalates and Musk Fragrances in Indoor Air and Dust from Apartments and Kindergartens in Berlin (Germany). *Indoor Air* **2004**, *14*, 188–195.

Gaspar, F. W.; Castorina, R.; Maddalena, R. L.; Nishioka, M. G.; McKone, T. E.; Bradman, A. Phthalate Exposure and Risk Assessment in California Child Care Facilities. *Environ. Sci. Technol.* **2014**, *48*, 7593–7601.

Liang, Y.; Xu, Y. Emission of Phthalates and Phthalate Alternatives from Vinyl Flooring and Crib Mattress Covers: The Influence of Temperature. *Environ. Sci. Technol.* **2014**, *48*, 14228–14237.

INVESTIGATING INDOOR CHEMISTRY OF TERPINOLENE THROUGH THE USE OF A NEW DERIVATIZATION AGENT

J.R. WELLS*, Jason E. HAM, Stephen R. JACKSON, and Joel C. HARRISON

Exposure Assessment Branch, Health Effects Laboratory Division, National Institute for Occupational Safety and Health (NIOSH), Morgantown, West Virginia USA

*Corresponding email: ozw0@cdc.gov

Keywords: Ozone, terpinolene, carbonyls, gas-phase

SUMMARY

The gas-phase ozonolysis of terpinolene was investigated using a new carbonyl derivatization agent: O-*tert*-butylhydroxylamine hydrochloride (TBOX). Reaction products from the terpinolene/O_3 system were collected in a water-filled impinger, TBOX was added, and the mixture was heated to 70°C for two hours followed by extraction with 0.5 mL toluene. Using this lower molecular weight derivatization agent resulted in the observation of a tricarbonyl reaction product: 3,6-dioxoheptanal. Additionally, three other carbonyl compounds were observed: methyl glyoxal, 4-methylcyclohex-3-en-1-one, and 6-oxo-3-(propan-2-ylidene)heptanal. The mechanism(s) for formation of these compounds is most likely due to O_3 addition to terpinolene's carbon-carbon double bonds while the tricarbonyl formation may be due to a reaction from secondary OH• radicals. Understanding the formation of di- and tricarbonyl species as a result of terpene ozonolysis is an important element of indoor occupant exposure assessment.

INTRODUCTION

Terpenes are introduced indoors by outdoor ventilation, emissions from building materials, and the use of various cleaning products (Nazaroff and Weschler, 2004). In indoor environments these terpenes can react with oxidants such as ozone (O_3), the nitrate radical (NO_3•) and the hydroxyl radical (OH•) in the gas phase and/or on indoor surfaces and can transform into oxygenated organic and/or nitrated species. An indoor ozone concentration of 1×10^{12} molecules/cm^3 (50 parts per billion) (ppb) has been typically measured or estimated (Sarwar et al., 2002). Using 50 ppb O_3 and a typical terpene/ozone reaction rate constant (2-20 x 10-16 cm^3 molec^{-1} s^{-1}) , terpene removal rates in the range of 0.9 – 9 h^{-1} are calculated which is faster than 0.6 h^{-1} of typical indoor air exchange (Wilson et al., 1996).

The terpene/O_3 reaction products for terpene normally found indoors encompass a wide variety of oxygenated organic compounds. These reactions proceed by addition to carbon-carbon double bonds which leads to carbon ring opening or fragmentation reactions and the secondary formation of OH• (Atkinson and Arey, 2003). The reaction products formed include: aldehydes, ketones, dicarbonyls, carboxylic acids, and organic nitrates (e.g. hydroxynitrates, dinitrates, etc.) (Atkinson and Arey, 2003; Spittler et al., 2006; Wangberg et al., 1997). It is expected that many of these compounds may have harmful health effects and should be characterized. All of these oxygenated organic compounds have the potential to induce a respiratory response, including work related asthma (WRA) (Magnano et al., 2009). Thus, identifying the terpene/O_3 reaction products for terpenes typically present indoors is critical to characterize occupant exposure (Makela et al., 2011; McHugh et al., 2010; Quirce and Barranco, 2010).

Terpinolene (1-methyl-4-(propan-2ylidene)cyclohexene) is a ubiquitous compound in consumer products used in indoor environments. Due to its volatility and two carbon-carbon double bonds, terpinolene reacts with O_3 approximately nine times faster than limonene. In a gas-phase mixture of terpenes such as limonene and α-terpineol, terpinolene is preferentially oxidized by O_3. This highlights the need to identify the oxidation products of terpinolene as a guide for understanding exposures and potential health effects.

METHODOLOGIES

The experimental method used to investigate the gas-phase ozonolysis of terpinolene is similar to a recent investigation of limonene ozonolysis (Wells and Ham, 2014). Briefly, 50 ppb of O_3 was added to 80 L of air in a Teflon chamber containing approximately 2 ppm of terpinolene. This mixture was allowed to react for 30 minutes then 60 L of this mixture was pulled through a 60 mL Teflon impinge containing 25 mL of deionized water. To this collected sample 100 μL of 250mM aqueous O-*tert*-butylhydroxylamine hydrochloride (TBOX) was added. This sample was then heated in a 70 °C water bath for 2 h to initiate the derivatization of carbonyl compounds to oximes (Wells and Ham, 2014). After cooling the sample 0.5mL of toluene was added and the sample was shaken to extract the oximes into the toluene layer. Then 100 uL of the toluene layer was removed and placed into an auto-sampler vial and 1 uL was injected into the analysis system.

All samples were analyzed using a Varian (Palo Alto, CA) 3800/Saturn 2000 GC/MS system operated in the electron impact (EI) mode. Compound separation was achieved by an Agilent (Santa Clara, CA) HP-5MS (0.25 mm I.D., 30 m long, 0.25 μm film thickness) column and the following GC oven parameters: 40 °C for 2 min., then 5 °C min^{-1} to 200 °C, then 25 °C min^{-1} to 280 °C and held for 5 min. One μL of each sample was injected in the splitless mode, and the GC injector was returned to split mode 5 min after sample injection, with the following injector temperature parameters: 130 °C for 2 min then 200 °C min^{-1} to 300 °C and held for 10 min. The Saturn 2000 ion trap mass spectrometer was tuned using perfluorotributylamine (FC-43). Full-scan EI ionization spectra were collected from *m/z* 40-650.

RESULTS AND DISCUSSION

In Figure 1 below the chromatograms of terpinolene only and terpinolene/O_3 are compared. The peaks observed are the derivatized carbonyl compounds formed as a result of the terpinolene/O_3 reaction.

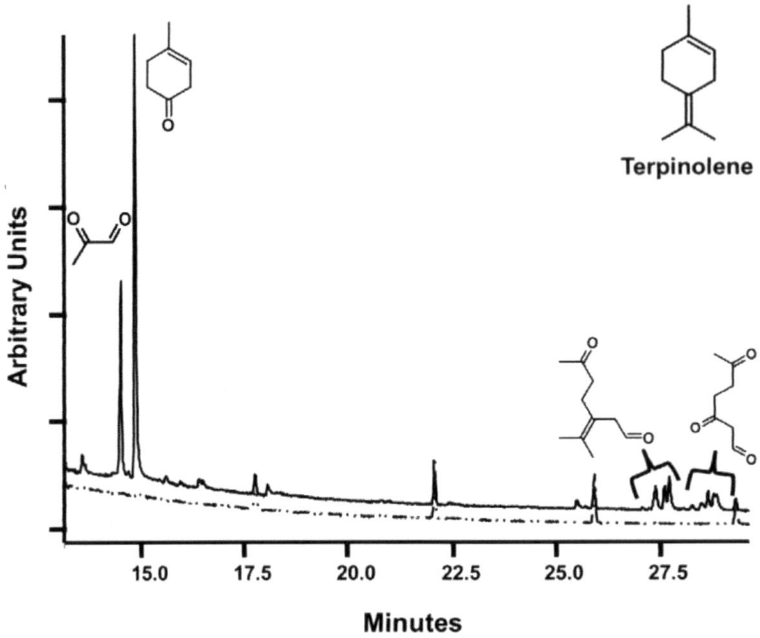

Figure 1. The terpinolene only (dashed line) versus terpinolene/O_3 (solid line) chromatograms are compared. The chromatograms are vertically shifted slightly for clarity. The peaks observed from terpinolene/O_3 system are identified as reaction products and the reaction product structures are shown next to peaks.

The gas-phase oxidation products identified by retention time (in parentheses) are: methyl glyoxal (14.5 min), 4-methylcyclohex-3-en-1-one (14.8 min), 6-oxo-3-(propan-2-ylidene)heptanal (27.4-27.7 min), and 3,6-dioxoheptanal (28.2-28.8 min).

The terpinolene oxidation products observed are similar to previously published results (Harrison and Wells, 2013). The significant difference in the data presented here is the observation of the tricarbonyl species which would not have been possible using *O*-(2,3,4,5,6-pentafluorobenzyl)hydroxylamine (PFBHA) because the resultant PFBHA oxime (727 amu) would be outside the 650 amu mass range of the mass spectrometer. The formation mechanism of the tricarbonyl, 3,6-dioxoheptanal, is most likely the result of secondary OH• addition to the carbon-carbon double bond of the dicarbonyl 6-oxo-3-(propan-2-ylidene)heptanal. The likely OH• dependence of 3,6-dioxoheptanal is similar to recently published results for tricarbonyl formation in the limonene/O_3 reaction system (Wells and Ham, 2014).

The detection of gas-phase tricarbonyl reaction products could be important results for exposure assessment, terpene mechanism modeling, and particulate formation models.

CONCLUSIONS

The detection of a tricarbonyl reaction product from terpinolene ozonolysis was achieved using a lower molecular weight carbonyl derivatization agent – TBOX. The advantages of this aqueous procedure could be utilized by indoor field sampling efforts. Future work will include the incorporation of TBOX with other derivatization agents in order to identify many other oxygenated species such as alcohols, carboxylic acids and organic nitrates simultaneously in a single sample.

DISCLAIMER

The findings and conclusions in this report are those of the author(s) and do not necessarily represent the official position of the Centers for Disease Control and Prevention/the Agency for Toxic Substances and Disease Registry. Mention of any commercial product or trade name does not constitute endorsement by the Centers for Disease Control and Prevention/NIOSH.

REFERENCES

Atkinson, R., Arey, J., 2003. Atmospheric degradation of volatile organic compounds. Chem. Rev. 103, 4605-4638.

Harrison, J.C., Wells, J.R., 2013. Investigation of terpinolene+ozone or terpinolene+nitrate radical reaction products using denuder/filter apparatus. Atmospheric Environment 80, 524-532.

Magnano, M., Silvani, S., Vincenzi, C., Nino, M., Tosti, A., 2009. Contact allergens and irritants in household washing and cleaning products. Contact Dermatitis 61, 337-341.

Makela, R., Kauppi, P., Suuronen, K., Tuppurainen, M., Hannu, T., 2011. Occupational asthma in professional cleaning work: a clinical study. Occup. Med.-Oxf. 61, 121-126.

McHugh, M.K., Symanski, E., Pompeii, L.A., Delclos, G.L., 2010. Prevalence of Asthma by Industry and Occupation in the US Working Population. Am. J. Ind. Med. 53, 463-475.

Nazaroff, W.W., Weschler, C.J., 2004. Cleaning products and air fresheners: exposure to primary and secondary air pollutants. Atmospheric Environment 38, 2841-2865.

Quirce, S., Barranco, P., 2010. Cleaning Agents and Asthma. J. Invest. Allergol. Clin. Immunol. 20, 542-550.

Sarwar, G., Corsi, R., Allen, D., Weschler, C.J., 2002. Production and levels of selected indoor radicals: a modeling assessment, Indoor Air 2002, Monterey, CA, pp. 80-85.

Spittler, M., Barnes, I., Bejan, I., Brockmann, K.J., Benter, T., Wirtz, K., 2006. Reaction of NO3 radicals with limonene and a-pinene: Product and SOA formation. Atmos. Environ. 40, S116-S127.

Wangberg, I., Barnes, I., Becker, K.-H., 1997. Product and Mechanistic Study of the Reaction of NO3 Radicals with a-pinene. Environ. Sci. Technol. 31, 2130-2135.

Wells, J.R., Ham, J.E., 2014. A new agent for derivatizing carbonyl species used to investigate limonene ozonolysis. Atmospheric Environment 99, 519-526.

Wilson, A.L., Colome, S.D., Tian, Y., Becker, E.W., Baker, P.E., Behrens, D.W., Billick, I.H., Garrison, C.A., 1996. California residential air exchange rates and residence volumes. J. Expos. Anal. Environ. Epidem. 6, 311-326.

VALIDATION OF PM CONCENTRATION ESTIMATES USING CONTAM MULTIZONE INDOOR AIR QUALITY MODEL WITH RESUSPENSION DEPOSITION MODULE

Lisa D. BRAMWELL[1], Yilin TIAN[2] and Andrea R. FERRO[2]

[1]Department of Chemical, Biological and Agricultural Engineering, University of Technology, Kingston, Jamaica

[2]Department of Civil and Environmental, Clarkson University, Potsdam, NY

[*]Corresponding email: lbramwell@utech.edu.jm

Keywords: resuspension, residential, particles, CONTAM

SUMMARY

The National Institute of Standard and Technology (NIST) added a particle resuspension and deposition module to their multizone indoor air quality modeling software CONTAM. Using experimental data collected in residences, this study validates the use of CONTAM 3.1 for estimating human exposure to particles resuspended via human activity. The particle size range investigated for this study was 0.56 – 10 μm. CONTAM was able to satisfactorily predict exposure concentration profiles for both fine PM and coarse PM resuspended from residential carpeted flooring. Adjusted R-squared values for the measured versus the modeled concentrations ranged from 89 – 97%. Using CONTAM, the predicted increase in exposures to fine and coarse PM above background for 10 minute walking activities within a residential setting were 0.50 and 18 μg/m^3 respectively for slow walking, and 0.72 and 33 μg/m^3 respectively for fast walking.

INTRODUCTION

Emmerich (2001) reviewed validation studies comparing several IAQ models. The results for CONTAM (Walton and Dols, 2010) were highly favourable based on the statistical markers derived from the study. Zonal and whole house air change rates predicted with CONTAM were well correlated with rates from actual residences (R >0.90) for most studies.

Particle concentrations (0.3 - 5 μm) for a single zone test house also saw good correlation between predicted concentrations and actual events (R >0.94).

This study aims to validate the use of the particle resuspension and deposition module in CONTAM 3.1 by comparing CONTAM's results with those obtained from walking experiments conducted within a residential setting. Following validation, the model can be used to predict and compare exposures for a range of human activity patterns and experimental conditions (e.g., flooring types, dust loading, ventilation settings).

METHODOLOGIES

To validate the resuspension and deposition module in CONTAM, data were used from 2 single family homes in St. Lawrence County, NY, H1 and H2, visited on February 11th and 22nd, 2013, respectively (Ma et al., 2013; Yin et al., 2013). For each home, experiments were carried out with one person walking 10 minutes afterwhich particles were allowed to settle for 50 minutes. The particle number concentrations were measured using a Grimm Technology (Douglasville, GA) Model 1.108 Portable Dust Monitor (OPC). Air exchange rate was estimated for the experimental area of each home by emitting a pulse of CO_2 from dry ice and measuring the concentration decay. The model validation was conducted for each of these homes, comparing the measured and modelled fine and coarse PM concentration time series. The model inputs used for the validation process are given in Bramwell (2013).

RESULTS AND DISCUSSION

The measured concentration variability, shown as somewhat regular wave like patterns in Figure 1, may be due in part to the mixing regimes in the two homes during the 60 minute walking experiment. The adjusted R – squared values comparing the measured fine and coarse PM concentration time series with that predicted by CONTAM for both slow and fast walking (Figure 1) activities ranged from 89 to 97 %. The OPC generated time concentration series for the 60 minute experiment, beginning with the 10 minute walking phase was split up into 10 minutes segments. For each of these segments a linear regression curve was fitted in excel. The sum of the squares of the differences (SSE) between the actual data as measured by the Grimm OPC monitor and the linear regression curve were then calculated (Figure 2).

Mage and Ott (2004) define an indoor air pollution episode as having three distinct chronologically ordered phases; the alpha, beta and gamma phases. Within the alpha phase the source is emitting and is typically characterized by elevated and highly variable concentrations in close proximity to the monitor as the pollutants within the air column are not well mixed. The beta period describes the period after the source is off and lasts until the pollutants are well mixed. Within the gamma period, pollutants are well mixed within the mixing volume, the source is no longer active and concentrations decay exponentially in all locations within the mixing volume (Ott, 1999). Current data indicated that the beta period persisted for at least ten minutes after walking ceased as SSE were consistently highest within this period. For this study SSE values ranged from 5 – 140 for fine PM and 50 – 100,000 for coarse PM.

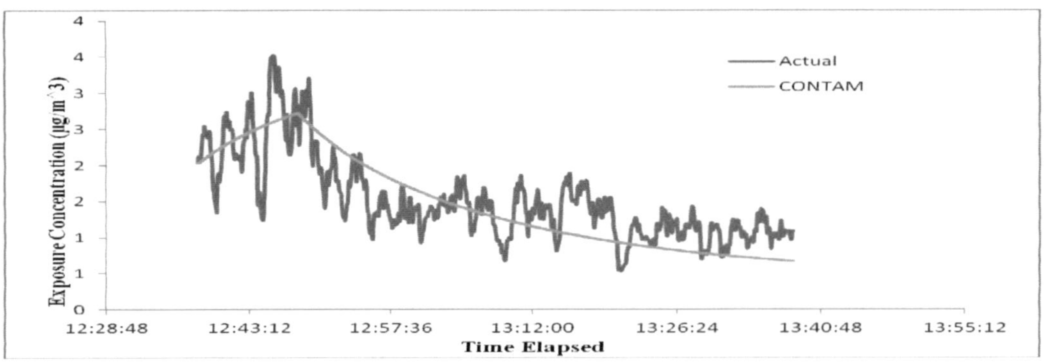

Figure 1 Fine PM concentration time series for slow walking in H1

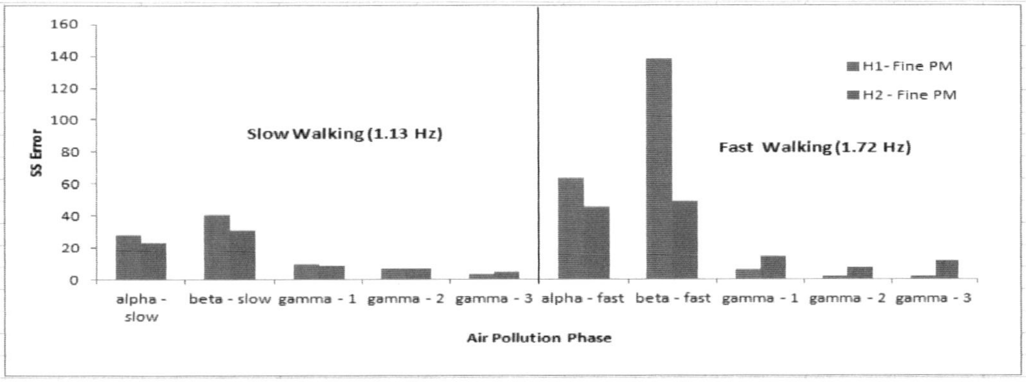

Figure 2 **Variability in OPC measurements for fine PM during each air pollution phase**

CONCLUSIONS

CONTAM can be reliably used to predict exposure concentration profiles for both fine PM and coarse PM resuspended from residential carpeted flooring. While the SS error was relatively high for coarse PM, the converse was true for fine PM (50 - 140). Furthermore adjusted R-squared values, for the measured versus the modeled concentrations ranged from 89 – 97 % for both fine and coarse PM.

ACKNOWLEDGEMENTS

This research material is based upon work supported by the U.S. Department of Housing and Urban Development (HUD) under Grant Number NYLHH0168-08 and the U.S. National Science Foundation (NSF) under Grant Number CBET 0846704. The contents of this article are the views of the authors and do not necessarily reflect the views or policies of HUD, NSF or the U.S. Government.

REFERENCES

Bramwell, L. (2013). *An Evaluation of the Impact of Flooring Types and Ventilation Conditions on Exposures to Fine and Coarse Particles within the Residential Micro-Environment using CONTAM.* Doctoral thesis, Clarkson University, Potsdam, NY.

Emmerich S.J. (2001).Validation of multizone IAQ modeling of residential-scale buildings: A Review, *Transactions-American Society of Heating Refrigeration and Air Conditioning Engineers*: 107 (2): 619-628.

Mage, D. and Ott, W. (1997). Accounting for nonuniform mixing and human exposure in indoor environments. In: Characterizing Sources of Indoor Air Pollution and Related Sink Effects. ASTM STP 1287 (Tichenor BA, ed). West Conshohocken, PA:American Society for Testing and Materials: 263-278.

Ma, Y. (2013). *Mass-based Size Distribution of House Dust Loading on Residential Floorings.* Master's thesis, Clarkson University, Potsdam, NY.

Walton, G. N., and W. S. Dols. CONTAM 3.0 *User Guide and Program Documentation.* National Institute of Standards and Technology Technical Report NISTIR 7251 (2010)

Yin et al., 2013. Personal Communication.

SECONDARY ORGANIC AEROSOL FROM α-TERPINEOL OZONOLYSIS

Yanan YANG[1], Michael S WARING[1,*]

[1]Department of Civil, Architectural and Environmental Engineering, Drexel University, Philadelphia, Pennsylvania USA

[*]Corresponding email: msw59@drexel.edu

Keywords: Indoor chemistry, terpenoids, ozone, aerosol mass fraction (AMF), SOA yield

SUMMARY

Secondary organic aerosol (SOA) generated indoors from terpenoid ozonolysis can add to indoor particle concentrations. α-terpineol is one reactive terpenoid that is very often emitted from cleaning products and air fresheners, though its SOA formation potential is unknown. As such, we studied SOA formation due to α-terpineol ozonolysis in a continuously mixed flow reactor (CMFR) during seven experiments with an air exchange rate (AER) of 0.61 h^{-1}. The SOA maximum mass and number concentrations ranged from 5.23 to 275 $\mu g/m^3$ and 3.36×10^5 to 3.48×10^6 #/cm^3, respectively, for initial α-terpineol concentrations of 10.4 to 208 ppb and very high O_3 (~25 ppm) to ensure complete oxidation. Aerosol mass fractions (AMF) ranged from 0.080 to 0.23, and AMFs were positively correlated to the initial α-terpineol concentration. These results allow the estimation of SOA formation by α-terpineol ozonolysis indoors with existing SOA formation models.

INTRODUCTION

Exposure to particulate matter (PM) can cause adverse health effects, including increased mortality and respiratory diseases. Secondary organic aerosol (SOA) formed indoors due to ozonolysis of reactive organic gases (ROG), e.g. terpenoids, can sometimes be an important PM source (Weschler and Shields, 1999; Weschler, 2000; Waring, 2014; Youssefi and Waring, 2014). α-Terpineol is a ROG that is a component of cleaners and air fresheners (Singer et al., 2006), but its SOA formation potential has been unexplored except in one study (Waring et al., 2011), though not systematically.

Predicting SOA formation with explicit chemical models is challenging, so the aerosol mass fraction (AMF), a.k.a. SOA yield, is often used instead, which is the ratio of formed SOA mass to reacted ROG mass. Though ozone + α-terpineol reaction rates are fast enough to be meaningful indoors (Wells, 2005), AMFs for those SOA forming reactions have not been quantified, so we did so for seven experiments at different α-terpineol concentrations.

METHODOLOGIES

Experiments were conducted in the continuous flow mixed reactor (CMFR) previously described (Youssefi and Waring, 2014). The chamber (stainless steel, 1 m^3 volume, 6 m^2 surface area, well mixed) was flushed with zero air (Environics 7000) until clean. Then a solution of α-terpineol (Sigma-Aldrich, 99.8%) in dichloromethane (Sigma-Aldrich, 96%) was injected into a heated tube flushed by zero air to reach a chamber target concentration. Five minutes later, the α-terpineol concentration was measured by gas chromatograph-flame ionization detection (GC-FID, SRI), and then again every 21 minutes. Seven minutes after the α-terpineol injection, O_3 (~25 ppm) was introduced by a powerful generator (Absolute Ozone Nano) and measured each minute (2B Technologies 205). SOA size distributions were measured with a Fast Mobility Particle Sizer (FMPS, TSI 3091), and mass concentrations were calculated by assuming spherical particles with a density of $1 g/cm^3$. Seven experiments were conducted at varying α-terpineol concentrations (10.4 to 208 ppb) at high O_3 to ensure complete oxidation, at moderate ($0.61 h^{-1}$) air exchange rates (AER). Relative humidity (RH = 0%) and temperature (24.6 ± 0.7 °C) were measured each minute (Onset HOBO U12).

The AMF was calculated using the maximum observed mass of SOA and ROG oxidized, as in Equation 1 (Odum et al., 1996; Presto and Donahue, 2006)

$$\mathrm{AMF} = \frac{\Delta C_{\mathrm{SOA}}}{\Delta \mathrm{ROG}} = \sum_i \alpha_i \left(1 + \frac{c_i^*}{C_{\mathrm{SOA}}}\right)^{-1} \quad (1)$$

where ΔC_{SOA} ($\mu g/m^3$) is the maximum SOA mass concentration; ΔROG ($\mu g/m^3$) is the ROG mass concentration reacted; α_i is the mass-based yield of product i; and c_i^* ($\mu g/m^3$) is the effective gas-phase saturation concentration of product i. Equation 1 shows that the AMF is not constant and increases with C_{SOA}. Equation 1 can fit AMF curves as a function of C_{SOA}, we did so using the 'volatility basis set' (VBS) framework by fitting four α_i at $c_i^* = \{0.1, 1, 10, 100\ \mu g/m^3\}$ (Presto and Donahue, 2006).

RESULTS AND DISCUSSION

Table 1 lists conditions for the seven experiments. Figure 1a illustrates a standard experimental trend, using as an example Experiment M5, which had an initial α-terpineol concentration of 111 ppb. Though we could not measure its actual decay due to the time-resolution of the GC-FID, the α-terpineol lifetime in the chamber was ~5 s when considering the peak average O_3 concentration (24.8 ppm) so it likely reacted away very quickly. For M5, the SOA reached peak mass of 128 $\mu g/m^3$ and peak number of 34.2×10^5 #/cm^3 approximately one minute after O_3 introduction. For all experiments, the peak SOA mass concentrations ranged from 5.23 to 275 $\mu g/m^3$ and peak number from 3.36×10^5 to 34.8×10^5 #/cm^3.

Other researchers have observed that SOA mass formed is linear with ROG consumed, especially at higher values of consumed ROG (Kroll and Seinfeld, 2005). Over our experiments, the linear fit of peak SOA mass concentrations (in $\mu g/m^3$) as a function of reacted α-terpineol (ppb) was: $y = 1.4x - 21$ ($R^2 = 0.99$). Peak SOA number increased linearly up to ~80 ppb of reacted α-terpineol, but after that the peak number concentration decreased. Though not shown, we estimated (*i*) with modeling (Waring and Siegel, 2013) and characterization experiments any formation due to surface reactions with α-terpineol sorbed to chamber walls and determined that it very likely had a negligible impact on the SOA mass concentration; and (*ii*) with modeling that almost all of any generated hydroxyl radicals due to α-terpineol ozonolysis reacted with α-terpineol and not with the dichloromethane solvent.

Figure 1b shows AMF results for each peak SOA mass concentration and the VBS model fit for those AMFs. The AMFs ranged from 0.080 to 0.23 and increased with the SOA concentration as typically observed. Using Equation 1, the VBS framework was applied to fit the AMF results at the different peak SOA mass concentrations for each experiment. Specifically, values of α_i were to fit AMF data for each peak SOA mass concentration by minimizing the square of the residuals for $c_i^* = \{0.1, 1, 10, 100\ \mu g/m^3\}$. The corresponding $\alpha_i = \{0.040, 0.029, 0.023, 0.16\}$. These VBS fits can be used in Equation 1 to recover the AMF for mass balances that predict SOA formation (Hoffmann et al., 1997, Youssefi and Waring, 2012).

Table 1. Results for seven SOA formation experiments due to α-terpineol ozonolysis.

Experiment identification	Initial α-terpineol (ppb)	Peak O_3 (ppm)	Peak SOA mass (μg/m^3)	Peak SOA number (×10^5 #/cm^3)
M1	10.4	26.3	5.23	3.36
M2	21.3	23.5	16.5	13.8
M3	55.3	22.6	49.1	26.5
M4	79.9	25.4	79.8	34.8
M5	111	25.9	128	34.2
M6	163	29.4	233	30.6
M7	208	27.5	275	24.2

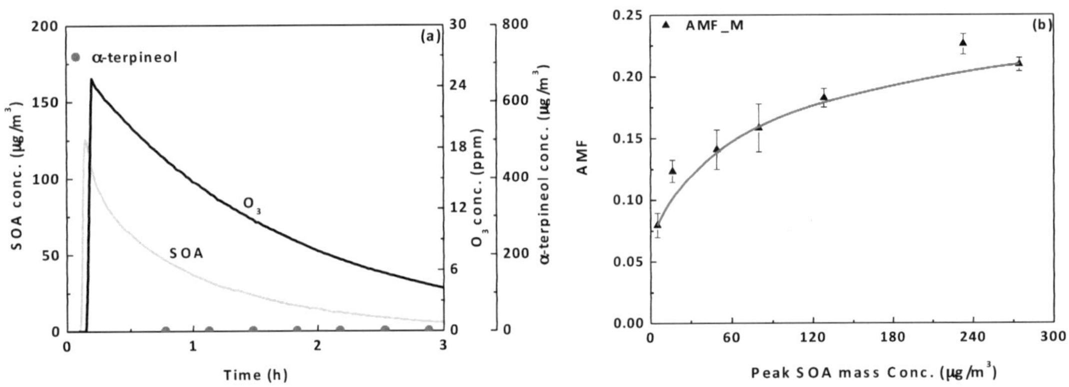

Figure 1. (a) SOA formation from α-terpineol ozonolysis in Experiment M5; (b) measured AMFs and VBS fit of α_i = {0.040, 0.029, 0.023, 0.16}.

Due to its prevalence in consumer products, α-terpineol ozonolysis could contribute substantially to indoor aerosol concentrations. For instance, Singer et al. (2006) quantified terpenoid emissions from various cleaning products and air fresheners. α-Terpineol comprised 67 ± 11 mg/ml of the solution of a common general purpose cleaner (GPC-1), which was 7% of its total measured mass per solution volume. When used in three different counter-cleaning experiments in a mock 50 m^3 room, its 1-hour integrated concentration reached ~41–110 ppb, implying that its peak concentrations were much higher. As an example, using the AMF fit herein and modeling framework in Youssefi and Waring (2012), for peak α-terpineol concentrations of 100 and 500 ppb in the presence of 25 ppb of O_3, we used a Runga Kutta order 4 (RK4) numerical solution to estimate that SOA would reach peak concentrations of 9 and 32 μg/m^3, respectively. (This analysis assumed an AER = 0.61 h^{-1}; that O_3 and SOA mass were lost to surfaces at rates of 2 h^{-1} and 0.1 h^{-1}, respectively; O_3 was replenished by an outdoor concentration; and there was no background organic aerosol present.)

CONCLUSIONS

The experiments in this study investigated indoor SOA formation for ozonolysis of α-terpineol. Parameters quantified in this work can used to recall AMFs that predict SOA formation due to ozonolysis of α-terpineol emitted from consumer products. These reactions have the potential to increase aerosol exposure and impact human health.

ACKNOWLEDGEMENT

This work was funded by the National Science Foundation (Award #1055584).

REFERENCES

Hoffmann, T, Odum, JR, Bowman, F, Collins, D, Klockow, D, Flagan, RC, Seinfeld, JH (1997). Formation of organic aerosols from the oxidation of biogenic hydrocarbons. *Journal of Atmospheric Chemistry*, **26**, 189-222.

Kroll JH, Seinfeld JH (2005). Representation of secondary organic aerosol laboratory chamber data for the interpretation of mechanisms of particle growth. *Environmental Science & Technology*, **39**, 4159-4165.

Odum JR, Hoffmann T, Bowman F, et al. (1996). Gas/particle partitioning and secondary organic aerosol yields. *Environmental Science & Technology*, **30**, 2580-2585.

Presto AA, Donahue NM (2006). Investigation of α-pinene+ ozone secondary organic aerosol formation at low total aerosol mass. *Environmental Science & Technology*, **40**, 3536-3543.

Singer BC, Destaillats H, Hodgson AT, et al. (2006). Cleaning products and air fresheners: emissions and resulting concentrations of glycol ethers and terpenoids. *Indoor Air*, **16**, 179-191.

Waring MS (2014). Secondary organic aerosol in residences: predicting its fraction of fine particle mass and determinants of formation strength. *Indoor Air*.

Waring, MS, Wells, JR, Siegel, JA (2011). Secondary organic aerosol formation from ozone reactions with single terpenoids and terpenoid mixtures. *Atmospheric Environment*, **45**, 4235-4242.

Wells J (2005). Gas-phase chemistry of α-terpineol with ozone and OH radical: Rate constants and products. *Environmental Science & Technology*, **39**, 6937-6943.

Weschler CJ (2000). Ozone in indoor environments: concentration and chemistry. *Indoor Air*, **10**, 269-288.

Weschler CJ, Shields HC (1999). Indoor ozone/terpene reactions as a source of indoor particles. *Atmospheric Environment*, **33**, 2301-2312.

Youssefi S, Waring MS (2012). Predicting secondary organic aerosol formation from terpenoid ozonolysis with varying yields in indoor environments. *Indoor Air*, **22**, 415-426.

Youssefi S, Waring MS (2014). Transient secondary organic aerosol formation from limonene ozonolysis in indoor environments: impacts of air exchange rates and initial concentration ratios. *Environmental Science & Technology*, **48**, 7899-7908.

CHANGES IN OUTDOOR AEROSOL CHEMICAL COMPOSITION UPON TRANSPORT INTO THE INDOOR ENVIRONMENT

Anita M JOHNSON[1], Michael S WARING[1], and Peter F DECARLO[1,2,*]

[1]Department of Civil, Architectural and Environmental Engineering, Drexel University, Philadelphia, Pennsylvania, USA
[2]Department of Chemistry, Drexel University, Philadelphia, Pennsylvania, USA

[*]Corresponding email: pfd33@drexel.edu

Keywords: Volatility, Indoor/outdoor ratio, PM, Composition, Aerosol mass spectrometer

SUMMARY

We explored changes to the chemical composition of outdoor particulate matter (PM) transported to the indoor environment by a building mechanical system. Chemical-specific indoor/outdoor (I/O) concentration ratios of submicron PM were determined using near-simultaneous measurements of indoor and outdoor air in April 2013 in Philadelphia. PM composition was determined using an Aerodyne mini-Aerosol Mass Spectrometer (mAMS) and an aethalometer. Differences in chemical-specific I/O values among components can be attributed largely to volatility. For instance, the median I/O ratio of non-volatile sulfate was 0.60 and for semi-volatile nitrate the median was 0.25. The organic aerosol (OA) component was further analyzed using Positive Matrix Factorization (PMF). The different resulting OA factors exhibited a wide range of I/O values; factors with secondary components had low I/O values, while some like hydrocarbon like-OA, which may have had indoor sources, had much higher I/O values.

INTRODUCTION

Indoor and outdoor air quality are inextricably linked because of transport of gases and aerosols from outdoors to the indoor environment via infiltration and ventilation. Upon transport indoors, outdoor PM is reduced by mechanical losses such as loss in the envelope, filtration, or deposition to surfaces (Riley et al., 2002), as well as by volatility changes for certain types of PM. Physicochemical properties such as volatility are a function of the chemical composition of outdoor PM and determine how the bulk PM will be affected by environmental changes, such as temperature and relative humidity (RH) gradients, that occur upon transport between the two environments.

Organic aerosol (OA) species have been of particular interest, and transformations in OA as a result of changes in temperature have been explored and quantified in ambient aerosol with an AMS coupled with a thermodenuder (e.g. Huffman et al., 2009). Using data from the Relationships of Indoor, Outdoor, and Personal Air (RIOPA) study, Hodas and Turpin (2014) used ambient volatility distributions to estimate outdoor OA phase changes in indoor environments. In this work, we measured indoor/outdoors (I/O) ratios of non-refractory nitrate, sulfate, ammonium, and organics using an mini-aerosol mass spectrometer (mAMS, Aerodyne Research Inc.), as well as black carbon using an aethalometer (AE-33, Magee Scientific).

METHODOLOGIES

Sampling was performed for two weeks in April 2013 in a campus building at Drexel University in Philadelphia. A custom automatic valve system rotated between 3 sampling inlets: indoor, outdoor, and building supply (vent). Each inlet was sampled for 2 minutes using the mAMS, AE-33, and Scanning Electrical Mobility Sizer (SEMS, Brechtel Manufacturing Inc.). During this sampling campaign, the indoor space was always warmer and of lower RH than outdoors during the sampling period. The high time resolution of sampling allowed the determination of the 'lag time' or the time for changes in outdoor PM to be reflected indoors. The calculated lag time, which was consistent throughout the measurement period, was applied to all data before lumping into hour averages and calculating the I/O ratio, and it was also compared to the residence time measured with carbon dioxide tracer gas tests. Positive Matrix Factorization was applied to the OA mass spectral matrix (e.g. Ulbrich et al., 2009).

RESULTS AND DISCUSSION

For the entire two-week sampling period, the mean outdoor PM concentration was 7.4 μg/m^3 and the mean indoor concentration was 3.9 μg/m^3. The OA species were the largest chemical component in both environments, followed by black carbon. Due to volatile losses in the indoor environment, nitrate comprised 11% of the outdoor concentration, but only 4% indoors. We quantified the differences in the composition of the aerosol using the I/O ratio (Table 1). PMF analysis of the OA components resulted in four factors: a hydrocarbon-like (HOA), cooking OA (COA), and two oxygenated OA (OOA) factors, which were combined into a single OOA factor due to spectral similarities. Together, these factors describe the broad range of I/O values observed for the total OA concentration. A full time series of indoor and outdoor concentrations and the I/O ratio for aerosol species and factors is shown in Figure 1.

Table 1. Percentiles (per.) of I/O ratios for each chemical species and PMF factor, indicating the spread of I/O ratios.

Species/Factor	10th per.	25th per.	50th per.	75th per.	90th per.
SO_4	0.44	0.51	0.60	0.71	0.84
NO_3	0.10	0.16	0.25	0.37	0.50
BC	0.25	0.38	0.61	1.07	2.15
Organics	0.35	0.48	0.73	0.98	1.27
COA	0.14	0.23	0.46	0.98	1.54
OOA	0.24	0.41	0.58	0.71	0.87
HOA	0.54	0.93	1.61	2.93	4.49

Outdoor aerosol sulfate (SO_4) was consistent over time with no diurnal cycle, indicating a regional character and little contribution from local sources. Outdoor changes in sulfate are reflected indoors consistently, resulting in a narrow distribution of I/O ratios. Nitrate (NO_3), due to its semi-volatile nature partially evaporated upon indoor transport, and therefore exhibited very low indoor concentrations even during high outdoor concentration periods, resulting in lower overall I/O ratios and a narrower distribution of I/O ratios than sulfate. Black carbon had a median I/O ratio similar to that of sulfate, but with a larger distribution spread, potentially due to the lower signal-to-noise ratio of the aethalometer.

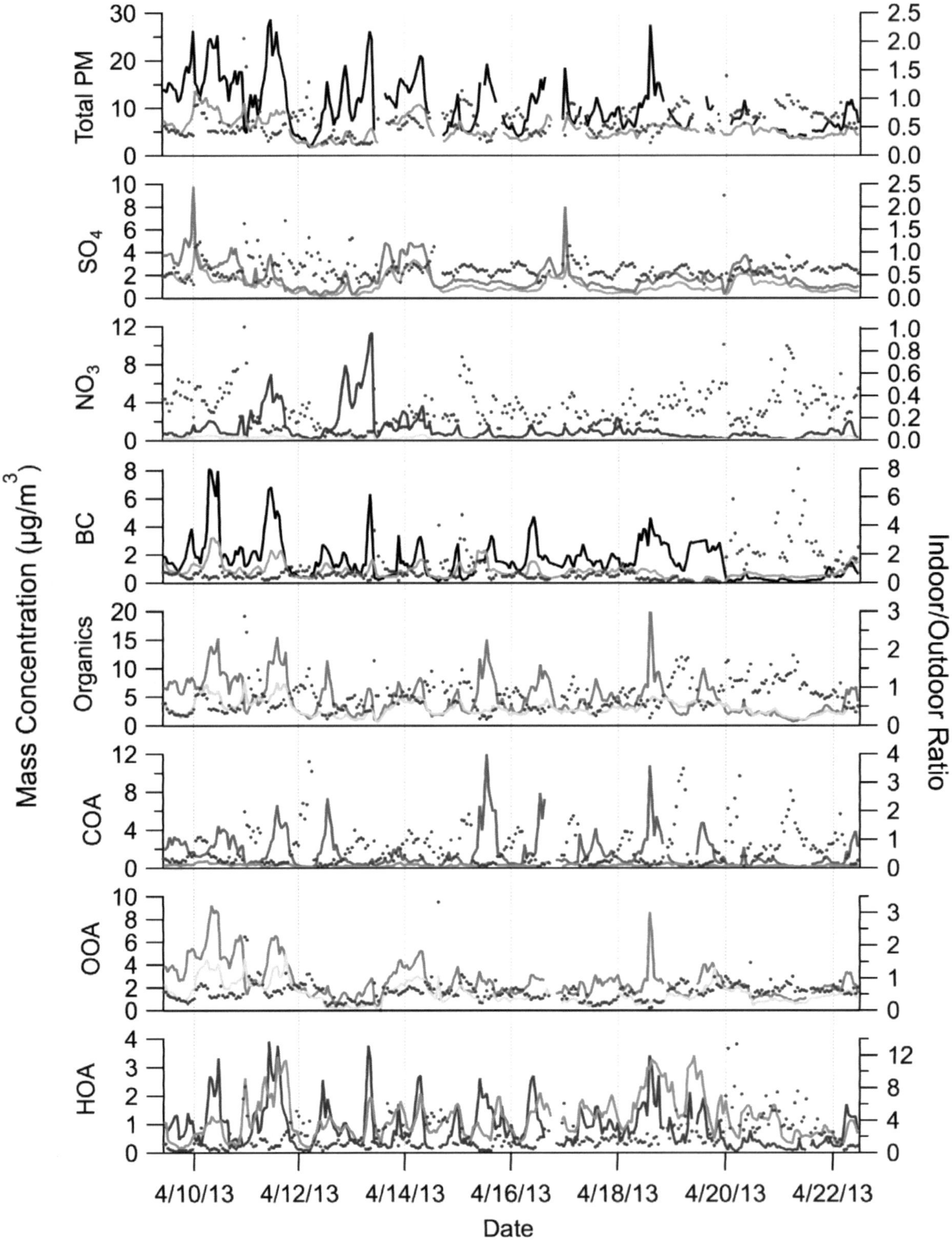
Mass Concentration (µg/m3)
Indoor/Outdoor Ratio
Total PM
SO4
NO3
BC
Organics
COA
OOA
HOA
4/10/13
4/12/13
4/14/13
4/16/13
4/18/13
4/20/13
4/22/13
Date

Figure 1. Time series of concentrations (left) and I/O ratios (right) of each chemical species and PMF factor identified. The outdoor (solid, dark) and indoor (light, dashed) concentrations are distinguished from the I/O ratio (dots).

OA comprised the largest portion of aerosol mass in both environments, but exhibited the largest ranges of concentrations and widest distributions of I/O values. PMF results indicate that different sources of OA account for this variation. The COA had a median I/O ratio of 0.46, possibly due the proximity of the outdoor inlet to a row of food service trucks near the sampling site. The OOA, which was a more aged and regional source, exhibited an I/O value similar to aerosol sulfate. The OOA was the largest observed portion OA in both environments, but increases in COA corresponded to times of known lunchtime cooking activity that appeared in the overall organics series. Finally, the HOA factor was consistently low outdoors (only 8% of total PM), but regularly showed I/O ratios greater than unity, indicating a potential indoor source.

Variation in the I/O ratio due to volatility of different aerosol components is important to understand in the indoor environment as heating and cooling processes create conditions that promote condensation or evaporation of these chemical species. Parameterization of these variations in I/O ratio can inform and improve existing models of the relationship between indoor and outdoor PM.

CONCLUSIONS

Differences in chemical composition coupled with environmental condition gradients can impact the amount of outdoor PM that is transported into the indoor environment. High time resolution sampling has allowed for these mechanisms to be explored, and analysis of the I/O ratio for individual species can be applied to existing indoor exposure models. These results can furthermore help to inform regionally-specific exposure impacts of outdoor PM.

REFERENCES

Hodas, N and Turpin, BJ (2014) Shifts in the Gas-Particle Partitioning of Ambient Organics with Transport into the Indoor Environment, *Aerosol Science and Technology*, **48**, 271–281.

Huffman, JA, Docherty, KS, Aiken, AC et al. (2009) Chemically-resolved aerosol volatility measurements from two megacity field studies, *Atmos. Chem. Phys.*, **9**, 7161–7182.

Paatero, P and Tapper, U (1994) Positive matrix factorization: A non- negative factor model with optimal utilization of error estimates of data values, *Environmetrics*.

Riley, WJ, McKone, TE, Lai, ACK et al. (2002) Indoor Particulate Matter of Outdoor Origin: Importance of Size-Dependent Removal Mechanisms, *Environ. Sci. Technol.*, **36**, 200–207.

Ulbrich, IM, Canagaratna, MR, Zhang, Q., et al. (2009) Interpretation of organic components from Positive Matrix Factorization of aerosol mass spectrometric data, *Atmos. Chem. Phys.*, **9**, 2891–2918.

A PRACTICAL GUIDE FOR ACCEPTABLE INDOOR AIRBORNE MOLD LEVELS (SPORETRAP) WITH EMPHASIS ON ASPERGILLUS-PENICILLIUM

Michael R. MEYER, MS, CMI

Mold Testing Services, LLC, Sioux Falls, SD

Corresponding email: moldymike@moldtestingsd.com

Keywords: Airborne mold, Aspergillus-Penicillium, sporetraps, total (non-viable) mold

SUMMARY

The main concern from mold is airborne exposure. Although this is a serious problem in many thousands of buildings, no widely accepted guidelines have been developed. Because there is no well accepted guide, health departments and others have seen no need for testing or enforcement. The lack of a guide has led to confusion and abuse. The most widely used test method are spore-trap cassettes. Based on testing at >2000 buildings over the last 15 years, this author has developed an easy to use guideline. As is well documented, the key parameter is almost always *Aspergillus-Penicillium* (AP). Normal room AP levels are usually <500-1000 spores/m^3. Problems are often associated with levels >1000-2000 spores/m^3. Common molds, such as *Cladosporium* or Basidiospores, often do not cause problems while problem wet molds, such as *Chaetomium* and *Stachybotrys* can. Lab analysis should be at x600 or x1000. Many commercial labs use x400 and this greatly undercounts the small and clear *Aspergillus-Penicillium* spores.

INTRODUCTION - THE NEED FOR AN AIRBORNE MOLD GUIDE

Dampness, water damage and associated mold problems are widespread in the world. For mold, the key concern is AIRBORNE exposure. Numerous studies have shown that high levels of problem indoor molds associated with health symptoms (Bornehag, 2004; WHO, 2009). However, no widely accepted exposure guide by the US EPA or others exists. This has created a "catch-22" situation. If there are no standards for airborne mold, there is no point in testing for it. This lack of a guide hinders testing and enforcement. In the US, most state agencies refuse to look at airborne mold levels because there are no standards. Lacking a clear guide, consultants and clean-up firms rely on a variety of often unreliable ways to meet their goals (comparing indoor to outside air, etc). Since interpretation of the results is left to the consultants or cleaning firms, the potential for confusion or deliberate misuse is obvious. In most cases, the client is simply given the lab analyses which are incomprehensible to the average person. We have seen many cases where a consultant recommended expensive clean-ups when the mold tests were well within normal levels.

METHODOLOGIES

The most widely used airborne mold testing method on a commercial basis by consultants and many others are "spore-trap "cassettes (such as Allergenco or Zefon) to collect total (also called non-viable) airborne mold (Godish, 2008). They do not require culture and can be analyzed immediately (within a day). They collect virtually all mold spores (both living and dead or non-viable) in the air. Note: both living and dead mold spores and mold fragments are allergens/toxins/irritants. By contrast, viable (culturable) mold tests require culture (often

1-2 weeks) and typically collect only a fraction (living) of the total mold (Yang & Heinsohn, 2007). The molds detected may depends on the selected culture media. Viable tests provide more accurate identification of specific molds and seem preferred by the academic community (and published literature).

However, in terms of speed, accurate mold levels (living and dead) and usefulness, spore traps are preferred by professional consultants. In many cases, clients cannot wait 1-2 weeks for results (evacuated rooms, etc). The cassettes are submitted to a laboratory for microscope identification (preferably x600 to x1000). Note: many commercial labs analyze molds at x400 and this may often undercount the small and transparent *Aspergillus-Penicillium* spores (Godish, 2008). For the last 15 years, this author has collected many thousands of spore trap tests at a few thousand building (Meyer, 2013). Airborne samples are usually collected at a flow rate of 15 liters/minute usually for 5 minutes. These have been mostly houses and apartments but have also includes offices, schools, hospitals, etc. We have been involved in hundreds of cases where high airborne mold levels were associated with health symptoms (and where mold clean-ups reduced the symptoms).

Based on this work, we developed the guideline below. To help develop the guideline, we analyzed data from a few hundred buildings (Meyer, 2009). Since that time, we have continued to use this guideline at many hundreds of sites. As is well documented in the published literature, the key problem indoor molds are almost always *Aspergillus-Penicillium* (these are lumped together when identified using spore trap methods). In a small percentage of cases, some problem wet molds, especially *Chaetomium* and *Stachybotrys*, may be the primary indicator. In non-problem rooms. no *Chaetomium* or *Stachybotrys* are present.

We found a noticeable break in many cases at about 1000-2000 spores/m^3 for *Aspergillus-Penicillium* between problem and non-problem rooms. In fact, non-problem rooms are often<500 spores/m^3.

Common molds, such as *Cladosporium*, or in damp weather, Basidiospores, are typically not associated with health symptoms. Elevated indoor levels of these molds is often due to open windows or doors rather than an indoor mold problem. Note: we prefer the use of a "range" of values rather than a single number. This is easier for the layman to understand and allows for some flexibility. In contrast, the layman assumes a rigid number to "safe" above the limit and "unsafe" below.

MTS Guide – Total (Non-Viable) Indoor Mold Guide. Units in Spores/m^3

Normal (upper Range) Non-Problem Indoor Mold Levels				
	Total Mold	Cladosporium	Asp&Pen	Chaet& Stachy
Total mold = common& problem molds		Common mold	Problem damp mold	Problem wet mold
Upper Range Most Buildings	<2000-4000	<2000-3000	<1000-2000	Zero
Upper Range Good Filtration	<1000-3000	<1000-2000	<500-1000	<50-250

Although not presented in this paper for reasons of space, we offer many examples of using this guide in problem situations. Some of these have been previously provided in Meyer (2013, 2014).

DISCUSSION

High levels of indoor airborne molds, especially *Aspergillus-Penicillium*, may be associated with health symptoms. The lack of a well recognized guide has hindered testing, enforcement and also led to widespread confusion and abuse (expensive clean-ups often required with no justification). In our experience, a "guideline" of about 1000-2000 spores/m^3 *Aspergillus-Penicillium* is usually reliable to distinguish problem from non-problem rooms (and is also a useful clean-up guide).Other investigators have also noted that a level of about 1000 spores/m^3 for *Aspergillus-Penicillium* appears to roughly divide normal from problem airborne mold room levels (Baxter, 2005; Spurgeon, 2014).

In contrast to "problem molds", the most widespread and common "non-problem" mold in the USA is *Cladosporium*. This is usually the dominant mold in non-problem rooms. Very high levels of this mold often occur outside with no noticeable increase in health symptoms. Of course, high levels of any mold should not occur indoors. Another common (and often non-problem) mold are the Basidiospores. These are often dominant in damp weather or just after a rain. Higher levels than normal of these common molds found indoors may often simply indicate an open window or door rather than a serious mold problem.

It is often argued that airborne mold testing is too variable to be relied on. In our experience, total airborne mold testing using lab analysis with at x600 to 1000 is fairly reliable. However, many commercial labs analyze at x400 and this often misses (greatly undercounts) the small and transparent *Aspergillus-Penicillium* spores. Godish (2008) recommends using x1000 for lab analysis. In some cases, large amounts of settled mold spores may be present (such as in vacant buildings) which are not detected in standard (undisturbed) testing. For this reason, sometimes including a few lightly disturbed airborne mold tests may discover problems not detected by the standard tests (Meyer, 2014).

This is a simple and easy to use guide. It is readily understood by the layman and cleaning firms. It is true there is some variability in testing and lab analyses. We emphasize the use of a "range" of values rather than a single rigid number. This helps the layman realize there is some flexibility in assessing results. A single rigid number as a guide is too often misunderstood.

CONCLUSIONS

Elevated airborne mold levels are common in many buildings due to water damage and dampness. There are no well established guidelines for these levels. This has discouraged testing and enforcement as well as often led to confusion and abuse in interpreting tests. Based on many years of experience at thousands of buildings, we developed a simple and practical guide based on normal room levels using spore trap cassettes with the key indicator being *Aspergillus-Penicillium*. Lab analysis should preferably be at x600 to x1000 rather than at x400.

REFERENCES

Baxter,D. et.al., (2005), A regional comparison of mold spore trap concentrations outdoors and inside "clean" and "mold-contaminated" southern California buildings. *Jour. Occup. & Env. Hygiene*, v2.,p8-18

Bornehag, C. et.al. (2004) Dampness in buildings as a risk factor for health effects, *Indoor Air*, v.14,p.243-257

Godish, D, and Godish,T (2008), Total airborne mold particle sampling: Evaluation of sample collection, *Jour. Occup. & Env. Health*, v.5,p100-106

Meyer, M (2009) 2014 What are normal and problem indoor total (non-viable) airborne mold levels? An IAQ consultant's perspective based on 10 years experience at hundreds of houses and offices in the Midwest, *Healthy Buildings 2009*, Syracuse, NY, paper 203

Meyer, M (2013) An argument for developing an airborne total (non-viable) mold guideline for indoor exposure, *Indoor Air Quality Association* Conference, San Antonio, TX

Meyer, M (2014), Are standard (undisturbed) airborne mold tests reliable? The usefulness of disturbed tests for mold assessments with several examples, *IAQA Conference*, Nashville, TN, March

Spurgeon, Joe,(2014), Elements of an exposure assessment strategy, *IAQA conference*, Nashville, TN, March 2014

WHO, (2009) *Dampness and Mould - WHO Guidelines for Indoor Air Quality*, World Health Organization, Copenhagen, Denmark

Yang, C. and Heinsohn,P (2007), *Sampling and Analysis of Indoor Micro-organisms*, John Wiley & Sons, NY

CONTROL OF MOLD CONTAMINATION IN THE BATHROOM

Satomi YANASO*, Hiroshi YAMAGISHI, Takayuki HASEGAWA, and Kosuke TANAKA

Living Care Research Laboratory, LION Corporation, Tokyo, Japan

*Corresponding email: satomi-y@lion.co.jp

Keywords: Mold contamination, ceiling, spore, *Cladosporium*, fumigation

SUMMARY

Mold contamination of the bathroom is a common type of microbial contamination in our living environment. In this study, we examined mold contamination of bathrooms in 25 homes. Mold was detected on ceilings as well on walls and floors. The rate of ceiling mold occurrence as examined by microscopy was 84% (21 homes), and that of spore formation was 68% (17 homes). Mold detected on the ceiling was more likely to produce spores and to spread them throughout the bathroom. Since mold contamination in the bathroom likely originated from ceiling mold, elimination of mold on ceilings via fumigation was an effective method to reduce all bathroom mold.

INTRODUCTION

The area in the home where mold thrives most easily is the bathroom (Moriyama et al., 1992; Lee et al., 2007). A bathroom is often warm and quite humid, and contains an abundance of nutrient sources for mold such as sebum and scurf (Hiroshi et al., 2012). Therefore, a bathroom is the ideal environment for microbes to reproduce and thrive. Usually, people use ventilation systems or chlorine bleach to inhibit mold growth. However, in spite of these efforts, it is difficult to prevent mold completely, as it grow back easily. The aim of this study was to examine the origin of bathroom mold and to investigate how ceiling mold might spread to other areas. In addition, a fumigation system that can deliver the sterilizing agent to the ceiling was evaluated, as a novel anti-fungal treatment method.

METHODOLOGIES

The amount of mold in bathrooms of 25 homes throughout 5 prefectures of Japan was measured from July to August in 2008 (Hiroshi et al., 2014). Mold samples were obtained using transparent dressing tapes attached to various points in the bathroom. One half of each piece of tape was used to observe mold morphology with a microscope, and the other half was cultivated on potato dextrose agar medium supplemented with 0.01% chloramphenicol (PDAcp), and mold species were identified from isolated colonies. To measure the number of viable mold colonies, 10 × 10-cm swabbed samples were obtained using swabs consisting of a shank and a cotton ball impregnated with phosphate buffered saline (PBS). Ten-fold serial dilutions of swabbed samples were cultivated on PDAcp plates for 5 days, and those that formed colonies were counted.

As 1/20-scale models of a bathroom, 55 × 70 × 60-cm plastic containers were prepared to investigate how ceiling mold affected other areas. An agarose plate inoculated with *Cladosporium cladosporioides* was attached to the upper surface of a plastic container and placed on the floor of the other container. Fifty-seven clean PDAcp plates were set on the remaining floor, walls, and upper surfaces, and the containers were cultivated at 25°C for 21 d to observe how the ceiling mold contaminated the other plates. The experimental design is shown in Figure 1.

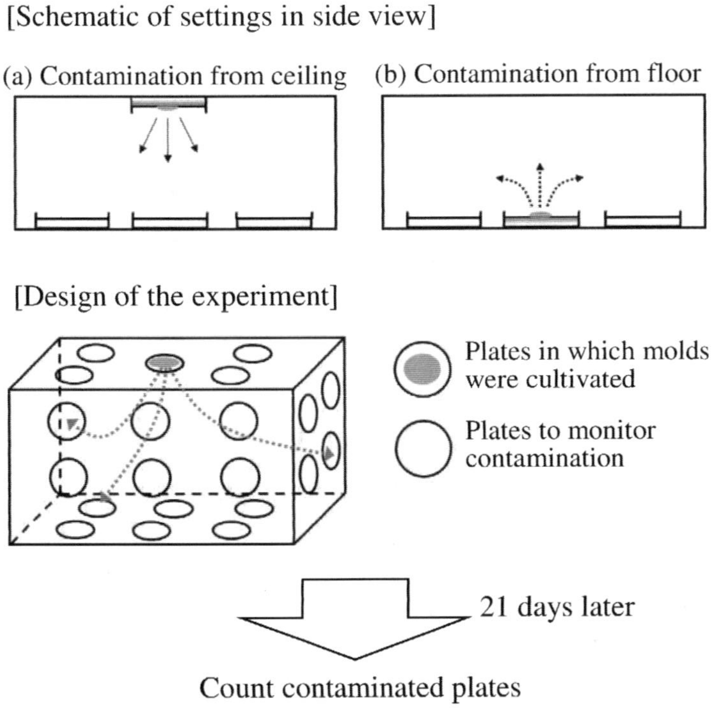

Figure 1. Design of 1/20-scaled bathroom experiment to determine mold spread.

Figure 2. Fumigation agent in a fogging can generate fog.

We also tested the effectiveness of a fumigation system to prevent mold. The fumigation agent was prepared with azodicarbonamide as the blowing agent and silver zeolite as the sterilizing agent. Fiber reinforced plastic plates, on which 2.0E+6 CFU/25 cm^2 of *C. cladosporioides* was inoculated and dried, were set on the ceiling, floor, and walls of a model bathroom. The fumigation agent was placed into fogging cans and then heated to create fog using hydration heat of calcium oxide. The fumigation apparatus is shown in Figure 2. After 90 min, viable mold on the plates was extracted in glucose peptone broth containing lecithin & polysorbate 80 medium, and cultivated on a PDAcp plate at 25°C for 5 d; the colonies obtained were then counted.

The effectiveness of fumigation in bathrooms of 20 homes was also evaluated from June to August in 2011. To measure the number of viable mold colonies, 10 × 10-cm swabbed samples were obtained to compare before and after use of fumigation. In addition, swabbed samples were also acquired after 2 months from fumigation. Ten-fold serial dilutions of swabbed samples were cultivated on PDAcp plates for 5 d, and those that showed colony formation were counted.

RESULTS AND DISCUSSION

Table 1 shows the occurrence rate of mold in the 25 bathrooms sampled. Even though a low amount of mold was visible on ceilings, microscopic detection revealed that a much higher amount of mold was actually present on this surface. Thus, bathroom ceilings may be contaminated with mold that is not visible to the naked eye.

Table 1. Occurrence rate of mold in the bathroom

	Ceiling	Wall	Floor
Occurrence rate of mold examined by visual inspection	36%	88%	60%
Occurrence rate of mold examined by microscope	84%	91%	80%

Regarding the type of fungi isolated from the ceiling, *Cladosporium* was detected most frequently, with the detected amount reaching 10^7-10^8 CFU/100 cm^2 in 40% of homes. In most cases, mold in the home that looks black is an aggregation of thickly grown hyphae. In particular, the spore formation rate of the ceiling mold was high (Table 2). These results suggest that mold on the ceiling consists mostly of spores, so that it does not appear black, but spreads spores throughout the bathroom.

Table 2. Microscopic morphology of mold growth on the ceiling

	Microscopic morphology		
	Only hypha	Spore formation	Total
Number of detected homes	4/25 (16%)	17/25 (68%)	21/25 (84%)

Figure 3 shows an example of the model bathroom experiment and the percentage of plates on which mold grew based on their position. It was confirmed that the mold contamination on the walls and floor was caused by ceiling mold.

(a) (b)

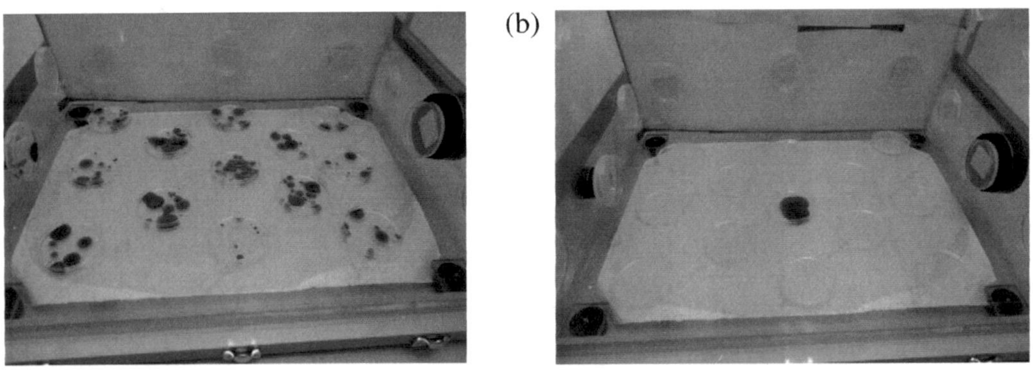

(a) The rate of contaminated plating medium

Ceiling	Wall	Floor
38%	47%	100%

(b) The rate of contaminated plating medium

Ceiling	Wall	Floor
0%	0%	17%

Figure 3. Mold contamination from the ceiling (a) and the floor (b).

Normally, bathroom mold can be eliminated using chlorine bleach or detergent. However, in the case of ceiling mold, it is too difficult and dangerous to apply bleach to the overhead surface. For this reason, we investigated the use of a fumigation system in the bathroom. During fumigation, the fog reaches the entire space, delivering the sterilizing agent onto the ceiling as well as the walls and floor. The effectiveness of fumigation in preventing mold growth was examined using a model bathroom and bathrooms in 20 homes. Before fumigation, viable mold counts reached 2.0E+6 CFU/25 cm^2 in the model bathroom. After 90 min of fumigation, less than 10 viable mold colonies were detected (Table 4). The effectiveness of this treatment in each of the 20 homes is shown in Figure 4. The amount of mold decreased after fumigation, and the sterilizing effect lasted for 2 months. These results suggested that fumigation is an effective way to reduce all mold in a bathroom.

Table 4. Viable colonies of mold in a model bathroom

	ceiling	wall	floor
Before Fumigation	2.0E+06	2.0E+06	2.0E+06
After Fumigation	<10	<10	<10

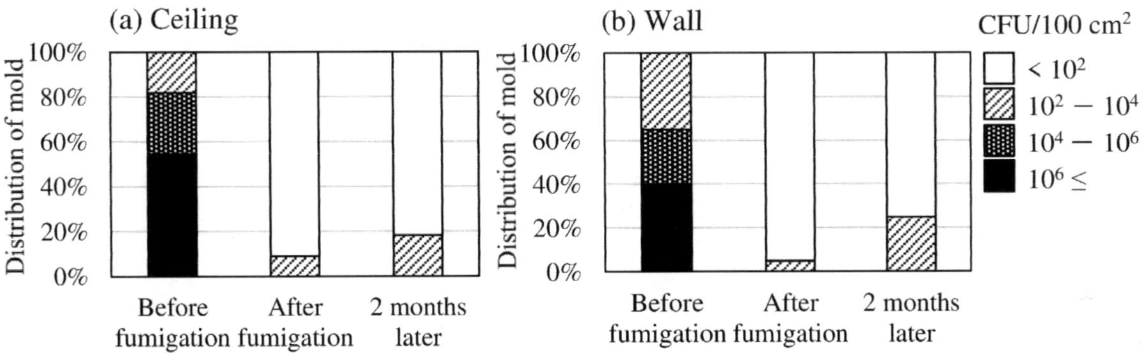

Figure 4. Viable colonies of mold on the ceiling (a) and on the wall (b) in the field study.

CONCLUSIONS

Mold on the bathroom ceiling was detected at a high rate, even when it was not visible to the naked eye. Spore formation rate by the ceiling mold was also high, and it was confirmed that mold contamination on the walls and floor could be caused by ceiling mold. Fumigation is an effective method to reduce mold throughout the bathroom.

ACKNOWLEDGEMENT

We gratefully acknowledge Dr. Hunjun Lee, Director of the Hygiene & Microbiology Research Center Corporation. We also express our gratitude to many others who joined in this survey.

REFERENCES

Lee H, et al. (2007). The effects of soap on the growth of major fungi contaminating the bathroom. *Journal of Antibacterial and Antifungal Agents,* **35**, 197–200.

Hiroshi Y, et al. (2012). Fungus and bacteria. *Lion Science Journal* **6**, 6–11.

Hiroshi Y, Satomi M, Shohei F, et al. (2014). Fungal contamination on the ceiling of bathrooms. *Journal of Antibacterial and Antifungal Agents,* **42**, 295–298.

Moriyama Y, Nawata N, Tsuda T, et al. (1992). Occurrence of moulds in Japanese bathrooms. *International Biodeterioration and Biodegradation*, **30**, 47–55.

INVESTIGATING AND AVOIDING MOISTURE-RELATED PROBLEMS IN EXISTING BUILDINGS

Lewis G. HARRIMAN[1], Terry R. BRENNAN[2]

[1] Mason-Grant Consulting, Portsmouth, NH USA

[2] Camroden Associates, Westmoreland, NY USA

Corresponding eMail: LewHarriman@MasonGrant.com

Keywords: Moisture, Buildings, Investigation, Moisture Problems, Humidity Problems

SUMMARY

Musty or earthy odors provide an important early warning of dampness-related problems. When investigating such odors, it's useful to keep in mind that the really major problems come from water intrusion rather than from high humidity. Locating and stopping rain water and plumbing leaks is generally the most important 1st step. After that, measuring and keeping the indoor air dew point low enough to avoid moisture absorption and condensation is the logical next concern.

INTRODUCTION

Moisture-related problems in buildings are common. Research suggests that nearly all the readers of this document will have experienced a moisture-related problem either personally, or indirectly through the experience of a relative or a close friend.[1] Sometimes, these problems lead to adverse health effects.[2,3.]

At the same time, it's useful to keep in mind that over their long lifetimes, nearly all buildings will get wet from time to time. Moisture problems are common, but moisture accumulation need not become severe enough and persist long enough to pose a health risk.

Building investigations over the last 40 years provide many suggestions for those who would like to understand and avoid such problems in new construction and existing buildings.[1] In brief summary, four suggestions may be especially useful for those who do not intend to make a more detailed study of the complex issues:

1. Musty odors are significant. "Earthy" odors are an early warning of future problems that can sometimes be severe. Locate and fix the long-term dampness and microbial growth that produces such odors, quickly.
2. Damp materials present a bigger risk than humid air. Keep rain water out of the building and fix plumbing leaks quickly.
3. Leaks and spills happen. Dry out surfaces quickly whenever they get wet.
4. High indoor dew points create the potential for condensation and moisture accumulation in all climates. Keep the indoor air dew point low enough to prevent indoor condensation and persistent dampness of materials.

Severe moisture-related problems rarely have a single cause. It's important to locate, understand and fix all of the conditions that contributed to persistent dampness rather than indicting the most visually-apparent problem, which is often a symptom rather than a cause.

INVESTIGATING AND AVOIDING MOISTURE-RELATED PROBLEMS IN EXISTING BUILDINGS

MUSTY ODORS

Investigations of building-related health problems performed by the U.S. Centers For Disease Control's National Institute of Occupational Health (NIOSH) have consistently shown that musty odors are common to buildings that were damp long enough to pose a risk to occupants' health.[2,3.] Further, a meta-analysis by the Indoor Air Division of the California Department of Public Health of worldwide, peer-reviewed building dampness literature also suggests that a pattern of musty or earthy odors is consistently associated with buildings in which health problems have been reported.[4.]

Based on these correlations, it seems prudent to investigate and fix the dampness that leads to musty odors, even if the causal relationship between dampness, microbial amplification and health problems is not yet entirely clear.

RAIN WATER, PIPE LEAKS AND CRAWL SPACES

While a "muggy building" is often rightly perceived as a precursor to dampness-related problems, humid air alone has never been shown to be a problem. It's dampness in materials that allows microbial growth. So generally, the most productive investigation strategies focus first on water intrusion and leaks.

When investigating musty odors, it's useful to focus on joints to and around windows and piping joints to and from plumbing fixtures. Windows and fixtures themselves rarely leak. The problems come at the gaps in connections and flashing around windows, and at the supply and sanitary connections to sinks, tubs and toilets.

DRYING QUICKLY

Even daily heavy wetting is not automatically a problem, as long as the water is drained away and dried out quickly. For example, sinks and shower stalls get very wet, often several times each day. But they are rarely at the root of major moisture problems.

Even moisture-sensitive materials need not be a problem, if they are dried out quickly. Untreated paper-faced gypsum wall board is notorious for supporting mold growth when it stays damp over time. But damp wall board requires days or weeks of persistent dampness before mold growth becomes visible to the unaided eye, and periodic light wetting followed quickly by drying may persist for months without significant microbial growth.[5,6.]

Consequently, a useful guideline for building occupants and managers is to dry out materials and building assemblies that get wet accidentally as quickly as possible, ideally in less than 48 hours. Over the last 30 years, a commercially and technically robust industry has developed in North America and Northern Europe to dry buildings quickly.[7.] So there's no need to allow dampness to persist.

INDOOR DEW POINT

Building owners, managers and occupants often focus on relative humidity of indoor air rather than on its dew point. However, relative humidity measured in the open air is quite different from the relative humidity of that same air in direct contact with a cool surface.

So instead of monitoring relative humidity, it is more useful to monitor and control its dew point. For example, during warm and humid weather when the indoor dew point is above 55°F (12.8°C), the risks of condensation or moisture adsorption into cool surfaces is much greater. In most buildings, there is little risk of excessive condensation or sorption when the indoor dew point is less than 55°F (12.8°C) during the summer, and less than a 40°F (5°C) dew point during the winter[14].

When investigating musty odors in buildings, a robust risk assessment will measure and compare the dew point of the indoor air with the surface temperature of the coldest parts of the HVAC system, the exterior wall and the foundation.

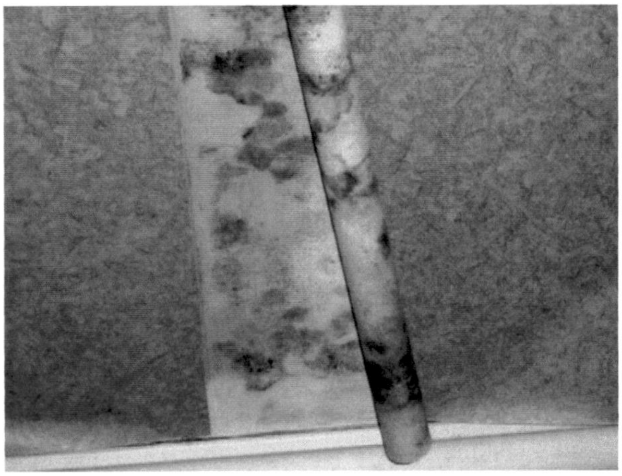

Figure 1. Classic mold growth behind vinyl wall covering

COMPLEXITY IS COMMON

The evidence of past investigations suggests that dampness sufficient to cause problems seldom has a single cause. More often, a series of events including decisions in many areas of professional and personal responsibility combine in complex ways to cause a problem.

Figure 1 shows an example; The classic and problematic practice of installing vinyl wallpaper on the indoor surfaces of exterior walls in mechanically-cooled buildings in hot humid climates.

High dew point outdoor air infiltrates through exterior walls. Its moisture is then absorbed into hidden cool surfaces of interior gypsum wallboard. Because the vinyl wallpaper is relatively impervious to water vapor transport, moisture accumulates in that gypsum board, resulting in mold growth and eventually leads to decay, rot or corrosion of structural members or their fasteners.

Note that the problems illustrated by figure 1 required more than one element; High outdoor dew point for many days or weeks, extensive humid air infiltration into the enclosure, over-chilled indoor surfaces, vinyl wallpaper and untreated paper-faced gypsum board.[6,7]

In this example:

1. The owner or interior designer made a decision to install vinyl wall covering rather than a more permeable wall finish, such as flat-finish latex paint.
2. The architectural designer apparently designed and/or the contractor built a building that allows extensive humid air infiltration and/or or sun-driven water vapor.
3. The HVAC system is designed and/or installed such that it overcools wall surfaces.

4. The toilet exhaust duct connections are not sealed it extracts air from building cavities in addition to extracting air from the bathroom, increasing humid air infiltration. The problem is made worse by the leaky return plenum of the air handler. The fan pulls in air from wall cavities, depressurizing them; which turn in turn pulls more humid outdoor air into the building.

This example illustrates the common circumstance in which risks from not one but several decisions made by different professionals at different times act in combination to produce enough moisture accumulation in wall cavities for a long-enough period to create a microbial growth problem.

SUMMARY

Public health investigations have consistently shown that when there are odors, there is a potential problem. At the moment of musty odor perception, it becomes prudent for all who are in a position of authority to take action to locate and eliminate the sources of the moisture accumulation that allows the microbial growth that generates such odors, and to do so as quickly as resources and circumstances allow.

REFERENCES

1. EPA 2013. Moisture Control for Building Design, Construction and Maintenance.EPA 402-F-13053. U.S. Environmental Protection Agency. http://www.epa.gov/iaq/moisture/

2. Mudari, David and Fisk, William J. "Public health and economic impact of dampness and mold." Indoor Air, June 2007. Volume 17, Issue 3. pp 226-235. Journal of the International Society of Indoor Air Quality, www.blackwellpublishing.com

3. NIOSH Alert 2013. Preventing occupational respiratory disease from exposures caused by dampness in office buildings, schools and other nonindustrial buildings. DHHS NIOSH Publication 2013-102

4. Mendell, M; Mirer, A; Chung, K. et al, 2011. "Respiratory and allergenic heath effects of dampness, mold and dampness-related agents: A review of the epidemiologic evidence." Environmental Health Perspectives, Vol. 119, No. 6, June 2011. National Institute of Environmental Health Sciences NIEHS.gov

5. ASHRAE 2009. ASHRAE/ANSI Standard 160 – 2009, Criteria for Moisture-Control Design Analysis in Buildings. Atlanta: American Society of Heating, Air-Conditioning and Refrigeration Engineers. www.ashrae.org

6. Harriman, Lewis G, Lstiburek, J, Kittler, R. 2000. "Improving Humidity Control in Commercial Buildings" ASHRAE Journal, vol. 42, no. 11, p. 24, 26, 28-32 American Society of Heating, Air-Conditioning and Refrigeration Engineers, Inc. www.ashrae.org

7. ASHRAE 2006. Humidity Control Design Guide. Harriman, Brundrett & Kittler. (ISBN 1-883413-98-2). Atlanta: American Society of Heating, Air-Conditioning and Refrigeration Engineers, Inc. www.ashrae.org.

AIRBORNE BACTERIA AND FUNGI CONCENTRATIONS IN AIRTIGHT CONTEMPORARY DWELLINGS

Gráinne M McGill[1,*], John E MOORE[2], Tim SHARPE[1], Damian G DOWNEY[3] and Lukumon OYEDELE[4]

[1] Mackintosh Environmental Architecture Research Unit, Glasgow School of Art, UK
[2] Northern Ireland Public Health Laboratory, Department of Bacteriology, Belfast City Hospital, UK
[3] Northern Ireland Regional Adult Cystic Fibrosis Centre, Belfast City Hospital, UK
[4] Bristol Enterprise Research and Innovation Centre, University of West of England, UK

[*]Corresponding email: g.mcgill@gsa.ac.uk

Keywords: Microbiology, Housing, Mechanical Ventilation, Fungi, Bacteria

SUMMARY

The contemporary energy reduction agenda for housing is leading to improved levels of air-tightness and increasing adoption of Mechanical Ventilation with Heat Recovery (MVHR). The effect of these measures on indoor microbiology remains significantly under-researched and may have important consequences on the concentrations and diversity of airborne fungi and bacteria. This study investigated indoor microbiology in six contemporary airtight dwellings. Concentrations of fungi and bacteria were measured during the winter season. Mean bedroom fungi concentrations of 18.4 CFU/m^3 and bacteria concentrations of 212.6 CFU/m^3 were recorded. Concentrations of fungi ranged from 3-59 CFU/m^3 and bacteria from 19-607 CFU/m^3. It is suggested that the filtration of air through the MVHR system may have limited the migration indoors of fungi of outdoor origin, explaining the low bedroom concentrations. The findings demonstrate the need for further research to identify the effect of ventilation strategies on fungal and bacterial concentrations indoors, particularly in airtight dwellings.

INTRODUCTION

Choice of materials and ventilation strategies can have a significant influence on indoor microbial concentrations and diversity. As suggested by Kelley and Gilbert (2013), 'the climate, materials and design of artificial environments could have unexpected and interesting consequences for the selection and growth of microbes that might help us to design healthier buildings'. The effect of ventilation strategies on indoor bioaerosol concentrations however is inconclusive. For instance, a study by Reponen et al. (1989) examined the effect of ventilation strategies on indoor concentrations of fungi and bacteria and found reduced levels of fungi concentrations in mechanically ventilated dwellings. The reduction however was not evident for bacteria concentrations. Similarly, in an investigation of design features, Kembel et al. (2012) found less diverse microbial communities in mechanically ventilated health care facility rooms compared to naturally ventilated rooms. However, as explained by Batterman and Burge (1995), ventilation systems can also act as potential sources of indoor bioaerosols, which is exacerbated by poorly designed and inadequately insulated ductwork (DCLG, 2008), contamination of ductwork and filters through inadequate maintenance (Dorer and Breer, 1998), and incorrectly installed condensate drains (Lowe and Johnston, 1997). Despite this, few studies have explored microbiology in a domestic context; particularly in airtight, mechanically ventilated dwellings. The aim of this study therefore was to investigate the concentrations of airborne bacteria and fungi in six dwellings ventilated with Mechanical Ventilation with Heat Recovery (MVHR) systems.

METHODOLOGIES

The Case Study consisted of a row of six, new build (constructed in 2010) timber frame dwellings. The three bedroom homes were built without a central heating system, utilising heat recovery ventilation in combination with an airtight envelope (ranging from 1.4-2.2 m3/h/m2 @ 50 Pa). Household occupancy ranged from 2-5 people, with the presence of mould reported in three of the Case Study dwellings (as illustrated in Table 1).

Table 1. Household Characteristics

House No.	Occupancy A=Adult C=Child	Reported presence of mould	Monitored bedroom conditions	Main heating fuel
No.1	2A, 1C	Yes	Spare room	Natural gas (fire)
No.2	2A, 1C	No	Playroom	Natural gas (fire)
No.3	2A, 2C	Yes	Childs bedroom	Natural gas (fire)
No.4	2A, 1C	No	Childs bedroom	Electric (fire)
No.5	2A	No	Spare room	Electric (fire)
No.6	4A, 1C	Yes	Teenagers bedroom	Natural gas (fire)

Air sampling was conducted to identify the overall concentrations of airborne bacteria and fungi in the six Case Study dwellings. Sampling was conducted in the north facing bedroom of the six dwellings to determine concentrations of airborne bacteria and fungi. This room was selected because of its potential for mould growth. Air samples were collected during the winter season (November- December 2013), using an electrical slit air sampler (CF Casella Ltd., London, UK) in accordance with the manufacturers' instructions. Air samples were collected for 0.5min, 2min and 5min periods at a flow rate of 175 l/min (n=3 replicate runs for each medium). Outside samples were also collected for each home at 2min periods (n=2 replicate runs for each medium). Field blanks were employed for each indoor sampling location. For microbiological analysis, nutrient agar plates and Sabouraud dextrose agar plates (packed in house) were utilised. Plates were incubated at 22°C for 72 hours and mean concentrations of airborne bacteria and fungi (CFU/m^3) were calculated.

RESULTS AND DISCUSSION

Bedroom concentrations of airborne fungi remained well below ≤150 CFU/m^3, with an average value of 18.4 CFU/m^3 (n=6). In comparison, an average outside fungi concentration of 67.5 CFU/m^3 was recorded. Occupant reported presence of mould was identified in house No. 1, No.3 and No.6. Despite this, bedroom concentrations were low (mean of 59 CFU/m^3) in house No.3 and very low (<50 CFU/m^3) in the remaining dwellings. Outside concentrations were higher than bedroom concentrations; with the exception of house No.3 where bedroom concentrations exceeded outside levels, indicating indoor sources of contamination. These results are supported by findings of previous studies, which reported low concentrations of airborne fungi during the winter season (Berry et al., 1996; Frankel et al., 2012; Haas et al., 2007).

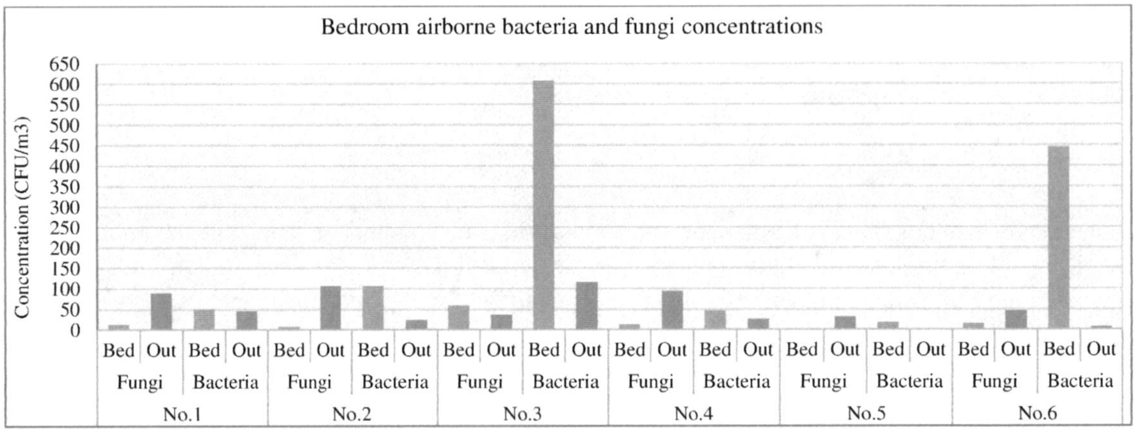

Figure 1. Bedroom (Bed) and outdoor (Out) concentrations of airborne fungi and bacteria in Colony Forming Units per cubic meter (CFU/m^3)

On average, bedroom bacteria concentrations of 212.6 CFU/m^3 and outside concentrations of 36.5 CFU/m^3 were recorded. In house No.3, a mean bedroom airborne bacteria concentration of 607 CFU/m^3 was identified. This bedroom was occupied by a teenage girl who used a hairdryer in the room to dry her hair prior to sampling, which may have affected the results.

In all six monitored dwellings, bedroom bacteria concentrations exceeded outside concentrations, suggesting indoor sources of contamination. A similar study of residential airborne bacteria concentrations found highest concentrations in spring (median, 2,165 CFU/m3) and lowest in summer (median, 240 CFU/m3), with a median concentration in winter of 465 CFU/m3 (Frankel et al., 2012). Correspondingly, the BRE study (Berry et al., 1996) reported a geometric mean concentration of airborne fungi in 163 UK dwellings of 366 CFU/m3.

CONCLUSIONS

This study investigated the concentrations of fungi and bacteria in airtight dwellings with MVHR. The results indicate low bedroom concentrations of airborne fungi and bacteria during the winter season, despite occupant reported mould growth in three of the Case Study dwellings (No. 1, No.3 and No.6). Airborne fungi concentrations were higher outdoors than indoors, with the exception of House No.3. In contrast, concentrations of airborne bacteria were higher in bedrooms compared to outdoors. As explained by the WHO (2009), 'the relationships between dampness, microbial exposure and health effects cannot be quantified precisely'; therefore since an identification of species was not conducted, the risk to occupant health is not clear. However the results do provide valuable insight on the relationship between indoor and outdoor microbial concentrations in airtight mechanically ventilated dwellings.

It is suggested that the filtration of supply air may have resulted in a reduction of indoor fungi concentrations originating from outdoors. In comparison, airborne bacteria concentrations were higher indoors, suggesting human related sources. Future research is required to compare concentrations of fungi and bacteria in naturally and mechanically ventilated dwellings, and the impact of ventilation on concentrations. In addition, the identification of airborne fungal and bacterial species in contemporary domestic environments is required, including a microbiological investigation of domestic MVHR systems.

REFERENCES

Batterman SA and Burge H (1995) HVAC systems as emission sources affecting indoor air quality: a critical review. *HVAC&R Research* 1(1): 61-78.

Berry RW, Brown VM, Coward SKD, Crump DR, Gavin M, Grimes CP, et al. (1996) *Indoor Air Quality in Homes: Part 1, the Buildng Research Establishment Indoor Environment Study*. Bracknell: IHS BRE Press, 1-118.

DCLG (2008) *Performance Testing of Buildings: BD 2535.* London: Demartment of Communities and Local Government.

Dorer V and Breer D (1998) Residential mechanical ventilation systems: performance criteria and evaluations. *Energy and Buildings* 27(3): 247-255.

Frankel M, Bekö G, Timm M, Gustavsen S, Hansen EW, Madsen AM, et al. (2012) Seasonal Variations of Indoor Microbial Exposures and Their Relation to Temperature, Relative Humidity, and Air Exchange Rate. *Applied and Environmental Microbiology* 78(23): 8289-8297.

Haas D, Habib J, Galler H, Buzina W, Schlacher R, Marth E, et al. (2007) Assessment of indoor air in Austrian apartments with and without visible mold growth. *Atmospheric Environment* 41(25): 5192-5201.

Kelley ST and Gilbert JA (2013) Studying the microbiology of the indoor environment. *Genome Biol* 14: 202.

Kembel SW, Jones E, Kline J, Northcutt D, Stenson J, Womack AM, et al. (2012) Architectural design influences the diversity and structure of the built environment microbiome. *The ISME Journal* 6(8): 1469-1479.

Lowe R and Johnston D (1997) Mechanical Ventilation with Heat Recovery in Local Authority, Low-Rise Housing: Final Report on the Derwentside Field Trial. *Leeds: Leeds Metropolitan University*.

Nathanson T (1993) *Indoor Air Quality in Office Buildings: A Technical Guide: A Report of the Federal-Provincial Advisory Committee on Environmental and Occupational Health.* : The Committee.

National Environmental Agency (1996) Indoor air quality guidelines. In: Anonymous *Guidelines for Good Indoor Air Quality in Office Premises (1st Ed.).* Singapore: Institute of Environmental Epidemiology, 1-47.

Reponen T, Nevalainen A and Raunemaa T (1989) Bioaerosol and particle mass levels and ventilation in Finnish homes. *Environment International* 15(1): 203-208.

Wanner H, Verhoeff A, Colombi A, Flannigan B, Gravesen S, Mouilleseaux A, et al. (1993) *Biological Particles in Indoor Environments, European Collaborative Action, Indoor Air Quality and its Impact on Man.* Luxembourg: Office for Official Publications of the European Communities.

World Health Organization (1990) *Indoor Air Quality: Biological Contaminants: Report on a WHO Meeting, Rautavaara, 29 August-2 September 1988.* : World Health Organization, Regional Office for Europe.

MOISTURE MEASUREMENT AND IMPLICATIONS FOR FUNGAL GROWTH IN DRYWALL

Sandra DEDESKO[1,*], Jeffrey SIEGEL[1,2]

[1]Department of Civil Engineering, University of Toronto, Ontario, Canada
[2]Dalla Lana School of Public Health, University of Toronto, Ontario, Canada

[*]Corresponding email: sandra.dedesko@mail.utoronto.ca

Keywords: Gypsum, Wetting, Mold, Relative humidity, Water activity

SUMMARY

Moisture problems in buildings are a regular occurrence, and fungal growth on building materials as well as negative health effects to building occupants are common outcomes. In this paper, the effects of moisture on gypsum's material properties and susceptibility to fungal growth are discussed, as gypsum is ubiquitous in the building stock, and is often subjected to excess moisture. Such moisture has been shown to cause fungal growth on gypsum, however current moisture measurement techniques and standards for acceptable levels of indoor moisture have little to do with mold. Accordingly, we discuss the need for more suitable in-situ and surface moisture measurement devices for gypsum products, which could allow one to better detect moisture-damaged gypsum and ideally allow for remediation to prevent and/or minimize fungal growth.

INTRODUCTION

Unwanted moisture is a common problem in buildings, and can result from a number of issues, such as flooding, plumbing and/or envelope leaks, improper envelope design (e.g., inadequate venting/drainage, lack of moisture barriers), condensation on cold surfaces, and poor climatic control (i.e., indoor temperature and humidity). Such moisture problems can degrade building materials, compromise structural integrity, and pose a threat to occupant health (Mendell et al., 2011). Moisture availability is a fundamental factor in the growth of fungi, which are known to produce spores, mycotoxins, and volatile compounds that can degrade the quality of indoor air and potentially pose a health threat to building occupants when inhaled (e.g., Vittanen et al., 2010). Building materials that are subjected to high levels of moisture can become prime substrates for mold growth. Gypsum is no exception, as findings show that it experiences heavy fungal growth in response to wetting (Andersson et al., 1997), yet it is still a commonly used building material. *Penicillium*, *cladosporium*, *acremonium*, *aspergillus*, *epicoccum*, *alternaria*, and *ulocladium* were all found growing on the surface of drywall that had experienced wetting, starting as early as two weeks after the wetting event (Krause et al., 2006). Even more concerning are the effects of fungal growth on human health, which range from minor problems to more severe conditions, with a critical and rare example being the association of pulmonary hemoraging in small children with toxic fungi in moisture damaged homes. *Stachybotrys* was the predominant fungi, which requires wetted cellulose to grow (Menetrez et al., 2009), and is found more frequently and in higher abundances on wetted gypsum than on other building materials (Hyvärinen et al., 2002).

IMPLICATIONS OF WETTING ON GYPSUM: MATERIAL PROPERTIES AND MOLD GROWTH

Despite evidence that harmful fungi can proliferate on wetted drywall, it is still commonly used as an interior wall finish, or as an exterior sheathing in exterior insulation finishing systems (EIFS) and behind brick veneer walls (CMHC, 1993). However gypsum boards are often exposed to moisture problems when used as either an interior or exterior component. Gypsum can absorb and hold almost 200% of its weight in water and its behaviour and integrity are variable depending on the amount of moisture present (CMHC, 2007). The structural qualities of gypsum boards have been found to rapidly decrease when moisture content (MC) increased from 0-2%, and continued to degrade above 2% MC (but at a slower rate). Gypsum boards have also been found to crumble at approximately 5% MC (CMHC, 2007), indicating that large amounts of excess moisture will likely degrade the structural properties of gypsum products.

Unfortunately, the effects of moisture on gypsum are not limited to the above. Gypsum's high water absorbtance and moderate porosity make it an excellent substrate for mold growth when subjected to certain moisture conditions. Viitanen et al. (2008) conducted experiements (albeit, based on other building materials, such as wood) and found that periods of cyclic high and low relative humidity (RH) may not be enough to facilitate mold growth if the periods of high RH are too brief. Periods of low RH were also found to retard mold growth and also delay the rate of growth once the RH reached higher levels (Viitanen et al., 2008). Another study investigated fungal growth in an office environment by implementing typical building materials and environmental conditions (i.e., daytime temperatures of 21-24 °C and nighttime temperatures of 15-18 °C with the RH ranging from 45-65% and dropping to 35% in the presence of afternoon sun). All gypsum samples that had experienced wetting and successive drying exhibited mold growth, while un-wetted control samples showed no growth (Krause et al., 2006). This effect of moisture on fungal growth is further supported by the finding that mold will not grow on gypsum at a low RH (Nielsen et al., 2004) unless it is exposed to liquid water (Pasanen et al., 1991). Furthermore, a 1.4% MC was found to be sufficient to cause mold and mildew growth (CMHC, 1997), and exposure to high humidities (e.g., during construction) was found to produce MCs of 8-10% in drywall (CMHC, 2007), indicating that mold growth is likely in humid environments. The cumulative findings indicate that gypsum exposed to either a high RH (i.e., close to saturation) or liquid water is likely to grow mold.

MEASUREMENTS OF MOISTURE AND FUNGAL GROWTH

Given the effect of moisture on drywall, it is critical that its moisture can be assessed. Although a variety of sophisticated techniques are available for use in the laboratory, in-situ techniques are much more limited. The most common in-situ moisture assessment techniques are measurements of air RH and of drywall MC. Air RH is simple and inexpensive to measure, but may not provide an accurate environmental characterisation in some instances (e.g., periods of high moisture or large variations in RH may not be captured by short term measurements of a dynamic air RH). An alternate approach is to measure the MC of drywall, often with a pin moisture meter. This also has several challenges, including the fact that most such meters are designed for wood and have produced poor results in the past when used in drywall, and also that the measurement is dependent on the electrical contact made between the pins and the material, which is likely unknown and varying. Furthermore, an evaluation of various in-situ moisture measurement devices (including thermocouples, RH sensors, MC sensors, and gravimetric sensors) revealed limits on accuracy (e.g., tested moisture meters were observed to be accurate up to approximately 6% MC) and discrepencies among devices (e.g., moisture meter values were 3-3.4% higher than gravimetric values above 6% MC) (CMHC, 2007).

More reliable methods of measurement are required, given the effects of mold growth on gypsum in response to excess moisture, as well as the ambiguity surrounding current measurement techniques. Future approaches should explore alternate measurements of moisture in drywall and also on its surface. Much mold growth occurs at the surface of materials and is also dominated by nutrient and moisture availability at the material surface compared to that in the ambient surroundings (Pasanen et al., 1992; Pasanen et al., 1991), and so measurements of microbially-available water at the material surface would likely be a revealing measure. Water activity (a_w) would be the ideal parameter (Nielsen et al., 1999), but current measurement techniques for water activity are limited to destructive testing of samples in a laboratory and are not appropriate for in-situ testing in operating buildings. An alternate approach is to measure equilibrium RH which is equivalent to water activity at equilibrium conditions (Moon, 2005). Although equilibrium RH sensors exist, their microbiological relevance remains untested. Until more sophisticated approaches to measure moisture in drywall become available, simply identifying and replacing, or in certain situations, applying a tested and verified antimicrobial cleaner (e.g., Krause et al., 2006; Price and Ahearn, 1999) on any drywall that has become wet may eliminate the need for measurement and also the potential mycological growth and negative health effects that could result from moisture-damaged gypsum.

CONCLUSIONS

Despite the known susceptibility of drywall to moisture-induced material degradation and fungal growth, it is still commonly used in modern construction. Unfortunately, excess moisture is a frequent occurrence in buildings, and so mold could very likely be present on gypsum components of buildings. Adding to this problem is a lack of reliable moisture measurement techniques that are relevant to mold growth, as many are measures of ambient moisture (e.g., RH) as opposed to surface moisture (e.g., a_w), and also are more suitable for laboratory testing as opposed to monitoring operating buildings. Furthermore, despite much research, there is also no uniform standard on levels of acceptable indoor moisture to prevent mold growth, as most pertain to controlling ambient RH for human comfort (e.g., ASHRAE, 2013). However, serious health concerns have resulted from mycological growth in response to moisture problems in buildings, and so the investigation of mold growth on common building materials under a variety of environmental scenarios, as well as more suitable, mold-related moisture measurement devices deserves attention.

ACKNOWLEDGEMENT

The Alfred P. Sloan foundation provided funding for this research.

REFERENCES

Andersson MA, Nikulin M, Köljalg U, et al. (1997). Bacteria, Molds, and Toxins in Water-Damaged Building Materials. *Applied and Environmental Microbiology,* ***63*(2)**, 387-393.

ASHRAE. (2013). *Standard 55: Thermal Environmental Conditions for Human Occupancy.* American Society of Heating, Refrigerating and Air–Conditioning Engineers, Inc., Atlanta, GA.

CMHC. (1993). *Investigation of the Performance of Gypsum Sheathing.* Canada Mortgage and Housing Corporation, Ottawa, ON.

CMHC. (1997). *Protecting Gypsum Sheathing in Insulated Steel-Stud Walls.* Canada Mortgage and Housing Corporation, Ottawa, ON.

CMHC. (2007). *Relationship Between Moisture Content and Mechanical Properties of Gypsum Sheathing.* Canada Mortgage and Housing Corporation, Ottawa, ON.

Hyvärinen A, Meklin T, Vepsäläinen A, et al. (2002). Fungi and actinobacteria in moisture-damaged building materials - concentrations and diversity. *International Biodeterioration & Biodegradation, **49***, 27-37.

Krause M, Geer W, Swenson L, et al. (2006). Controlled Study of Mold Growth and Cleaning Procedure on Treated and Untreated Wet Gypsum Wallboard in an Indoor Environment. *Jounral of Occupational and Environmental Hygiene, **3(8)***, 435-441.

Mendell MJ, Mirer AG, Cheung K, et al. (2011). Respiratory and Allergic Health Effects of Dampness, Mold, and Dampness-Related Agents: A Review of the Epidemiologic Evidence. *Environmental Health Perspectives, **119*(6)**, 748-756.

Menetrez MY, Foarde KK, Webber TD, et al. (2009). Mold Growth on Gypsum Wallboard - A Summary of Three Techniques. *Journal of Environmental Health, **72*(1)**, 24-28.

Moon, H. J. (2005). *Assessing Mold Risks in Buildings under Uncertainty.* Georgia Institute of Technology.

Nielsen KF, Gravesen S, Nielsen PA, et al. (1999). Production of mycotoxins on artificially and naturally infested building materials. *Mycopathologia, **145***, 43-56.

Nielsen KF, Holm G, Uttrup LP, et al. (2004). Mould growth on building materials under low water activities. Influence of humidity and temperature on fungal growth and secondary metabolism. *International Biodeterioration & Biodegradation, **54***, 325-336.

Pasanen AL, Juutinen T, Jantunen MJ, et al. (1992). Occurrence and Moisture Requirements of Microbial Growth in Building Materials. *International Biodeterioration & Biodegradation, **30*(4)**, 273-283.

Pasanen AL, Kalliokoski P, Pasanen P, et al. (1991). Laboratory studies on the relationship between fungal growth and atmospheric temperature and humidity. *Environmental International, **17*(4)**, 225-228.

Price DL and Ahearn DG. (1999). Sanitation of wallboard colonized with *Stachybotrys chartarum. Current Microbiology, 39*(1), 21-26.

Viitanen H, Vinha J, Peuhkuri R, et al. (2008). Development of an improved model for mould growth: Modelling. *Proceedings of 8th Nordic Symposium on Building Physics.* Copenhagen.

Vittanen H, Vinha J, Salminen K, et al. (2010). Moisture and Bio-deterioration Risk of Building Materials and Structures. *Journal of Building Physics, **33*(3)**, 201-224.

THE IVAIRE STUDY: CORRELATIONS BETWEEN VENTILATION AND MYCOLOGICAL PARAMETERS IN CANADIAN SINGLE FAMILY HOMES WITH ASHTMATIC CHILDREN

Daniel G AUBIN[1], Doyun WON[1*], Hans SCHLEIBINGER[1], Wenping, YANG[1], Pierre LAJOIE[2], Denis GAUVIN[2], Véronique GINGRAS[2], Jean-Marc LECLERC[2] and Don FUGLER[3]

[1]National Research Council of Canada, Ottawa, Canada
[2]Institut national de santé publique du Québec, Québec City, Canada
[3]A researcher formerly with the Canada Mortgage and Housing Corporation, Ottawa, Canada

*Corresponding email: Doyun.Won@nrc.ca

Keywords: field study, ventilation, indoor air quality, mould, asthma

SUMMARY

A randomized controlled intervention study was conducted to measure the impact of a ventilation intervention on indoor air quality and the respiratory symptoms of asthmatic children. During year 1 (pre-intervention) and 2 (post-intervention), the ventilation rates and a number of IAQ relevant chemical, biological and physical parameters were measured in the homes. Following year 1, the participants were randomized into intervention (n=43) and control groups (n=40). The ventilation rates were increased in the homes of the intervention group by optimizing the existing mechanical ventilation systems or by the addition of an Energy or Heat Recovery Ventilator. In all homes prior to the intervention the highest median concentrations of airborne mold spores was measured in the child's playroom (often located in the basement), which were on average 2-3 times higher than those measured in the child's bedroom and/or the living room. Airborne and settling mould spore concentrations were positively correlated with the relative humidity and temperature and negatively correlated with the air exchange rate. After the intervention the geometric mean air exchange rate (h^{-1}) increased from 0.17 to 0.34 in the intervention group compared to 0.18 and 0.21 in the control group. A statistically significant reduction of 58% and 40% was observed in the concentrations for airborne and settling mould spores respectively in the intervention group. This paper focusses on the impact of increased ventilation on the concentration of various mycological parameters and their correlation with certain environmental parameters.

INTRODUCTION

The overall objective of this study was to determine whether increased ventilation will lead to a corresponding decrease in the number of respiratory symptoms in asthmatic children, to correlate venti-lation rates with indoor air quality (IAQ) parameters, and to address questions related to health based ventilation rates. The first phase of the field study involved 111 homes where three home sampling visits were conducted (two during the heating season (October to April) and one in summer). In the second phase, any home with one air exchange rate measurement below 0.25 h^{-1} or two in between 0.25 h^{-1} and 0.30 h^{-1} was considered eligible for the intervention. Of the 111 participants involved in the study, 83 were eligible for the intervention and these were randomly divided into two groups. One group of participants (n=43), had their ventilation rates increased by a combination of: installation of a Heat or Energy Recovery Ventilator; modification of the existing ventilation system; and/or modification of occupant behaviour relating to the use and maintenance of the mechanical ventilation system in their home. The other group (n=40), did not have any modifications done to the ventilation rate and served as a control group. The monitoring in the second phase after the intervention was identical to the first phase. The following sections focus on the impact of the ventilation intervention on the measured mycological parameters and their correlation with various environmental parameters. A detailed description of the impact of the intervention on the respiratory health of the children and the chemical and physical parameters is provided in Lajoie *et al.* (2015).

METHODOLOGIES

The average air exchange rates (AER) were measured in the child's bedroom and living room over a 6 to 8 day period using the perfluorocarbon tracer (PFT) method (Dietz and Cote, 1982). The AER was also measured over a 4-5 hour period in the child's bedroom using SF_6 tracer gas decay according to the ASTM test method E 741-00 (ASTM, 2006) using an Innova 1312 or 1412 photoaccoustic field gas monitor. The relative humidity (RH) and temperature were measured in the child's bedroom and the living room with an Onset HOBO U12-013 data logger. Airborne mould spores were measured in the child's bedroom, the living room and the playroom with a Staplex model MAS-2 two-stage air sampler for 5 min during the heating-season. Settling mold spores were collected for 4–5 h in the child's bedroom onto petri dishes. Tye airborne and settling mould spores samples were collected on both dichloran 18% glycerol (DG-18) and malt extract agar (MEA) covered petri dishes and then incubated at 25 °C (77 °F) for 7 days and then analyzed visually with a microscope to obtain a colony count.

RESULTS

The majority of the homes examined during the pre-intervention phase were under-ventilated and 78% failed to reach our nominal ventilation goal of 0.30 h^{-1} during the heating-season. Following the intervention, the geometric mean AER increased from 0.17 h^{-1} to 0.34 h^{-1} in the intervention group compared to 0.18 h^{-1} and 0.21 h^{-1} in the control group ($p<0.0001$). From Table 1 it can be seen that during the pre-intervention phase the highest median concentrations of airborne mold spores was measured in the child's playroom (often located in the basement), which were on average factor of 2-3 times higher than those measured in the child's bedroom and/or the living room. Following the intervention decreases in the median concentration of airborne mold spores were observed in all three locations. The largest decrease was observed in the child's playroom.

Table 1: Median values of airborne mold spores measured for the pre- and post-intervention period at different locations of the home during the heating-season using MEA and DG-18 growth media.

Location	Pre-Intervention (cfu m^{-3})		Post-Intervention (cfu m^{-3})	
	MEA	DG-18	MEA	DG-18
Child's Bedroom	67 (n=43)	64 (n=43)	60 (n=43)	27 (n=43)
Living Room	60 (n=28)	45 (n=28)	49 (n=42)	33 (n=42)
Playroom	110 (n=36)	174 (n=36)	49 (n=23)	51 (n=23)

Table 2 shows the spectra of the different mold genera: in the child's bedroom airborne *Aspergillus* was the most prominent genus, closely followed by *Penicillium*. In the intervention group airborne *Aspergillus* was reduced by nearly 75%, airborne *Penicillium* by 30 %. Concentrations of airborne *Cladosporium* spores were close to the detection limit, both before and after the intervention. Settling mold spores showed a similar pattern, both regarding the *Asp/Pen* ratio, as well as to the pre/post intervention behaviour. Total airborne and settling mold cfu's showed a moderate correlation (R = 0.37 to 0.76). The total airborne and settling mold spores concentrations both showed a statistically significant (p=0.05) decrease after the intervention.

Table 2: Median values for the genera of the airborne and settling mold spores in the child's bedroom during the heating season for the intervention group (DG-18).

Parameter	Genera	Pre-Intervention (n=43)	Post-Intervention (n=43)
Airborne (cfu m^{-3})	*Aspergillus*	19	5
	Penicilium	16	11
	Cladosporium	2	2
	Total mold spore	64	27
Settling (cfu h^{-1})	*Aspergillus*	0.33	0.14
	Penicilium	0.24	0.17
	Cladosporium	0.10	0.08
	Total mold spore	0.83	0.50

The concentration of the mold spores were found to be correlated with a number of environmental parameters. From Table 3 it can be seen that the concentration of the airborne mold spores and the air exchange rate were negatively correlated, although the correlation is weak.

Table 3: Spearman rank order correlations (R values) between mold spores concentrations and selected environmental parameters during the pre-intervention heating season.

Mold Spores	Medium	AER (SF_6)	AER (PFT)	T_{surf} of the coldest point in the Cbr	RH_{surf} of the coldest point in the Cbr	T_{surf} of the coldest point in the house	RH_{surf} of the coldest point in the house
Total airborne	MEA	-0.04	-0.15*	0.20*	0.43*	0.31*	0.42*
	DG-18	-0.07	-0.14*	0.11*	0.22*	0.15*	0.23*
Total settling	MEA	0.01	-0.10	0.18*	0.39*	0.28*	0.41*
	DG-18	-0.04	-0.16*	0.09	0.30*	0.14*	0.35*

Note: *$p < 0.05$, T_{surf} = surface temperature, and RH_{surf} = surface relative humidity, Cbr = child's bedroom

The negative correlation with the AER suggests that the collected mold spores were from indoor sources. The airborne and settling mold spores were significantly, but moderately positively correlated with the RH measured close to the coldest point in the child's bedroom ($p<0.05$, R = 0.22 – 0.43). The correlation between the airborne and the settling mold spores was high (R = 0.77, $p < 0.001$, n =615) for the combined data sets for the control and intervention groups.

DISCUSSION

The amount of total airborne mold spores, and even more so the concentrations of spores of the genera *Aspergillus* and *Penicillium* could be significantly reduced by increasing the ventilation rate. It was observed that there was no increase in mold concentrations due to higher ventilation rates, which might have resulted in higher linear air speeds, and consequently the re-suspension of already settled mold spores was not observed. The reduction in airborne mold spores is obviously mainly due to dilution of spores by outdoor air being low in terms of airborne mould spores. It can be assumed that outdoor concentrations of mold spores during Quebec's heating season are low. This is also substantiated by the fact, that *Cladosporium* concentrations, typically originating from outdoor sources, are low. The reduction of the airborne mold genera *Aspergillus* and *Penicillium* went in parallel with the concentrations of settling mold spores. Due to the high volume of samples the species under the genera *Aspergillus* and *Penicillium* could not be speciated, which limits partly the value of this aspect of the study; however, this was not a mold study, and mold growth was hardly observed in the homes under investigation. The analysis of the pre-intervention phase showed clear links between low ventilation rates and elevated concentrations mold spores (Lajoie *et al.*, 2015).

CONCLUSION

It was clearly demonstrated that under-ventilation can be corrected through a careful retrofit with Energy of Heat Recovery Ventilators. The concentrations of potentially health relevant mycological parameters can be decreased by increased ventilation, without revealing any negative side effects like re-suspending particles or mold spores due to higher air flows.

ACKNOWLEDGEMENTS

The authors would like to thank Karine Barriault (NRC), Jean-Francois Chouinard (INSPQ), Simon Godbout (INSPQ), Pierre-Luc Lamarre (NRC), Wenping Yang (NRC), Kristina Boros (NRC), and Stephanie So (NRC). This work was funded by the Government of Canada's Clear Air Agenda, Québec's Ministère de la santé et des services sociaux, the Canada Mortgage and Housing Corporation and Health Canada.

REFERENCES

ASTM, 2006. ASTM Test Method E 741-00: Standard test method for determining air change in a single zone by means of a tracer gas dilution. West Conshohocken, American Society for Testing and Materials. 2011.

Dietz R.N. and Cote E.A., 1982. Air infiltration measurements in a home using a convenient perfluorocarbon tracer technique, Environment International, 8(1-6), 419-433

Health Canada, 1987, Exposure Guidelines for Residential Indoor Air Quality, ISBN 0-662-17882-3

Lajoie, P., Aubin, D., Gingras, V., Daigneault, P., Ducharme F., Gauvin, D., Fugler, D., Leclerc, J.-M., Won, D., Courteau, M., Gingras, S., Heroux, M.-E., Yang, W., and Schleibinger, H. The IVAIRE project – a randomized controlled study of the impact of ventilation on indoor air quality and the respiratory symptoms of asthmatic children in single family homes, Indoor Air, 2015 (published online 21 Jan 15)

SYNTHETIC ORIGIN OF MICROBIAL VOLATILE ORGANIC COMPOUNDS

Hyunok CHOI,[1*] Norbert SCHMIDBAUER,[2] Carl-Gustaf BORNEHAG[3,4]

1. Department of Environmental Health Sciences, University at Albany, School of Public Health
2. Norwegian Institute for Air Research, PO Box 100, 2027 Kjeller, Norway, Instituttveien 18, 2007 Kjeller, NORWAY Tel +47 63 89 82 15 Fax +47 63 89 80 50; ns@nilu.no
3. Technical Research Institute of Sweden, Box 857, SE-501 15 Borås, Sweden
4. Department of Public Health Sciences, Karlstad University, SE-651 88 Karlstad, Sweden; Carl-Gustaf.Bornehag@kau.se

*Corresponding email: hchoi@albany.edu

Keywords: indoor, asthma, allergies, dampness, mold

SUMMARY

BACKGROUND: True sources of so-called microbial volatile organic compounds (MVOCs) remain unclear.

OBJECTIVE: To test a hypothesis that the most plausible sources of so-called MVOCs are water-based paint, cleaning fluids, and/or polyvinyl chloride (PVC) material.

METHODS: In a cross-sectional investigation (n=390), the possible sources of the so-called MVOCs was examined using the concentration of propylene glycol and glycol ethers (PGEs $\mu g/m^3$), Texanol®, six phthalates (mg/g dust), including di(2-ethylhexyl) phthalate (DEHP) and benzyl butyl phthalate (BBzP), and qualitative indication of microbial contamination as benchmark of sources.

RESULTS: None of the so-called MVOCs were associated qualitative or semi-quantitative markers of microbial contamination. However, in the homes with lowest microbial contamination and the highest quartile of BBzP ($\geq 0.267\ \mu g/m^3$), the composite sum of the so-called MVOCs were significantly associated with associated with PGEs and Texanol®, respectively.

CONCLUSION: PVC-based flooring and wall material represents the most plausible source of so-called MVOCs.

INTRODUCTION

Microbial volatile organic compounds (MVOCs) were first recognized as markers of microbial putrefaction of food items during the 1970s (Kamiński et al., 1972; Lee et al., 1979; Miller et al., 1973). MVOCs has been proposed as a screening tool for hidden microbial infestation within buildings (Korpi et al., 2009). Yet, MVOCs are inherently limited in their validity of as markers of microbial damage within the indoor settings (Korpi et al., 2009; Sahlberg et al., 2013; Schleibinger et al., 2008). For example, fungal inoculation within a controlled chamber do not necessarily generate clear profile of MVOCs (Schuchardt et al., 2013). Within a number of populations, the indoor concentration MVOCs are generally too low to be toxic to humans (Schleibinger et al., 2008). Even when MVOCs are detected in conjunction with microbial agents, the correlations between MVOCs with the microbial contamination within the indoor setting are null (Korpi et al., 2006; Matysik et al., 2008; Sahlberg et al., 2013; Schuchardt et al., 2013).

To date, the question whether the name, microbial VOCs, for at least some of the compounds described by Korpi et al. (2009) inaccurately attribute their sources has never been directly examined. This is at least partly due to the difficulty in disentangling complex and multiple contributions by both synthetic and biogenic sources of so-called MVOCs within actively occupied buildings (Kim et al., 2007; Newsome et al., 1965). Not only is the emission profile of completed building structure different from the individual components (Järnström et al., 2008), occupant behavior and indoor environmental conditions could influence the indoor VOC concentrations. Furthermore, the extent to which human behavior, the sources, and the home microenvironments contribute to emission and/or retention of the so-called MVOCs might be inherently different from those within the controlled chamber microenvironment. Considering these, the primary aim of the present investigation is to measure the associations between biogenic versus synthetic sources, respectively, with so-called MVOCs. Specifically, we test a hypothesis that the most plausible sources of 16 so-called MVOCs and TXIB of our interest are water-based paint, cleaning fluids, and PVC.

METHODS

The participant enrollment and methods of an on-going Dampness in Buildings and Health (DBH) study have been described (Bornehag et al., 2005a; Bornehag et al., 2005b; Choi et al., 2010b; Holme et al., 2010). The present investigation is based on 390 homes. Six inspectors conducted a walk-through assessment on building construction, materials, ventilation type, and mold and moisture problems. They collected air and dust samples. They also measured water vapor content of both the entire home and the bedroom of the index child. They measured average ventilation rates using validated PFT (perfluorocarbon tracer) technique (Nordtest, 1997; Stymne et al., 1994). The inspector graded each home regarding: 1) first impression of indoor air quality; 2) unpleasant smell along the skirting board; 3) visual signs of water damage (i.e. stains) on walls or ceiling; and 4) damaged structural material (e.g., blackened, bubbly, or loosening flooring material). Each outcome is graded zero (no issue); 1–2 (mild issue); and 3 (severe mold issue). As described in Choi et al. (2010), SKC pocket pumps (model 210-1002, SKC Blandford, Dorset, UK) collected a sample of air in child's bedroom at 80 ml/min for 60 to 90 minutes (5 to 8 liters) through Perkin Elmer adsorption tubes (glass, 300 mg Tenax TA).

We restricted our analysis to 17 of 89 possible MVOCs found in at least 40% of the 381 homes (n=152 homes). The 15 compounds of interest include 1-butanol, 2-butanone, acetic acid, α-pinene, benzaldehyde, β-pinene, decanal, ethylbenzene, hexanal, longifolene, limonene, nonanal, octanal, toluene, and O-Cymene. To preclude possible bias, we restricted our analyses to those above the detection limit (2.22 $\mu g/m^3$) (Choi et al., 2010b). Microbial sources of the 15 compounds were examined using: a) the level of culturable fungi (colony forming units/gram dust); b) building inspector assessment of mold/water damage (Holme et al., 2010); c) parental report of mold/water damage at home (Holme et al., 2010); and d) mold-index (Holme et al., 2010).

On the other hand, non-biogenic sources of 15 VOCs is examined using three proxy groups of the underlying sources; 1) the total concentration of propylene glycol and glycol ether concentration (PGEs, $\mu g/m^3$) as markers for paint and/or water-based cleaning fluids (Choi et al., 2010a); 2) 2,2,4-trimethyl-1,3-pentanediol monoisobutyrate (i.e. Texanol®), as a coalescent in paint; 3) di(2)ethylhexyl phthalate (DEHP) and benzyl butyl phthalate (BBzP) concentrations (*mg*/m^3), respectively, in dust as indicators for PVC building material (Bornehag et al., 2005b). We ran a linear regression analysis for sum of solvent-related related VOCs (i.e. 1-butanol, 2-butanone, acetic acid, benzaldehyde, hexanal, nonanal, and octanal). All compounds were natural log (ln)—transformed, to ensure approximate normal distribution of the compounds (all Komogorov-Smirnov tests > 0.05) with comparable standard deviations.

RESULTS

Figure 1: Correlation between summed total so-called MVOCs with viable fungal concentration in dust sample.

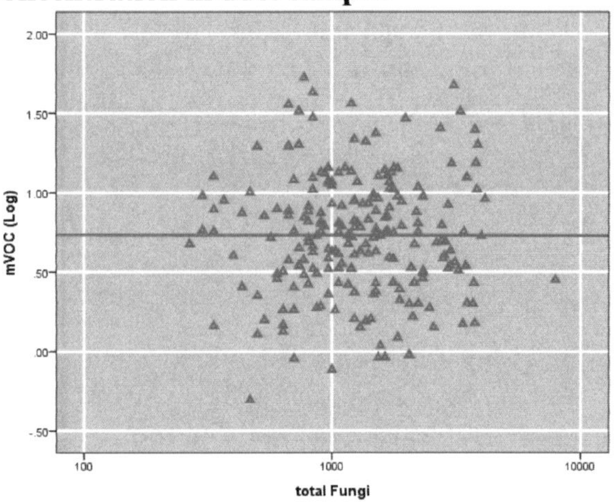

The total concentration of so-called MVOCs is not different between semi-quantitatively determined houses with microbial infestations and those without an infestation. The total MVOCs were weakly but significantly associated with water vapor content (Spearman's rho = 0.289, P< 0.001) as well as the ventilation rate (Spearman's rho = —0.104, P< 0.046). The homes whose dust samples yielded highest quartile of viable culturable fungi (cfu/g dust) are not associated with marked changes in median concentrations of the 15 VOCs, compared to the homes with lower culturable fungi (Figure 1). Inspectors' rating on four signs of mold/water damage (i.e., the moldy odor in the whole home or along the skirting board, respectively; water-damage stain in at least one room, signs of floor moisture in at least one room) are not associated with considerable increase in any of 15 VOCs of interest (all J-T P >0.002). The homes, which are semi-quantitatively determined to be moldy, are associated with 40% greater median concentrations of 1-butanol (3.89 vs. 2.73 $\mu g/m^3$, J-Tp= 0.002). However, the median concentrations of the remaining compounds are not markedly different for the moldy homes, compared to the non-moldy homes.

The homes with the highest quartile of home indoor PGEs ($\geq$15.65 $\mu g/m^3$) in air are associated with 80% higher 1-butanol (4.76 vs. 2.62 $\mu g/m^3$), 60% higher 2-butanone (3.67 vs. 2.31 $\mu g/m^3$), 30% higher benzaldehyde (5.63 vs. 4.42 $\mu g/m^3$), hexanal (6.77 vs. 5.03 $\mu g/m^3$), 60% higher junipene (4.83 vs. 2.95 $\mu g/m^3$), and octanal (4.31 vs. 3.33 $\mu g/m^3$) compared to the median concentrations of the same compound in the remaining homes (all J-T Ps < 0.002). As shown in Figure 2, there is a significant linear correlation between PGE (quartile categorized) and solvent-related related VOCs (J-T P < 0.001).

Figure 2. Distributions of solvent-related VOCs, 2-ethyl-1-hexanol, and TXIB according to quartile categories of Propylene Glycol and Glycol Ethers. Solvent-related VOCs reflect the total summed concentration of 1-butanol, 2-butanone, acetic acid, benzaldehyde, hexanal, nonanal, and octanal in child's bedroom air.

** denotes P < 0.001 using Jonckheere-Terpestra test for linear trend.

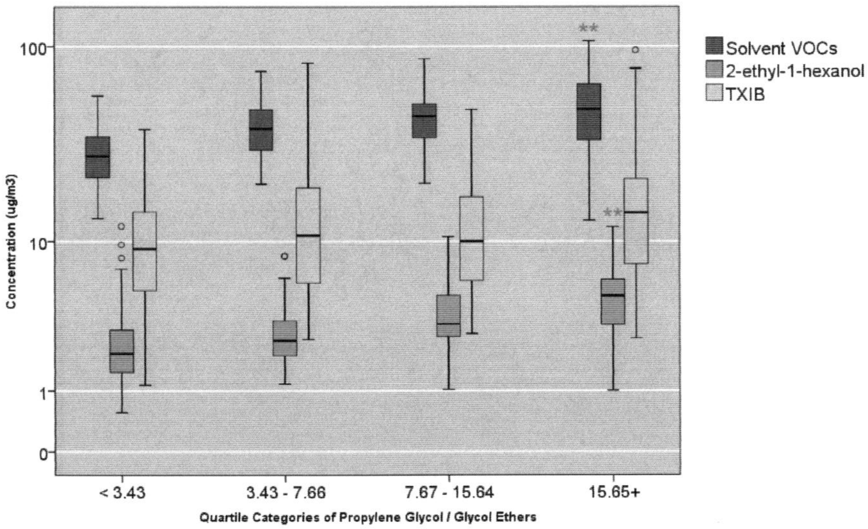

In a forward-selected linear regression model, (ln) MVOCs is explained by (ln) Texanol® (R^2=0.478), water vapor content (R^2=0.143), and (ln) PGEs (R^2=0.029), respectively.

DISCUSSION

Our present findings add to the consistent body of evidence that more plausible sources of the 15 so-called MVOCs are Texanol® and PGE-containing materials (Kim et al., 2007). PVC within completed building structures release a diverse range of VOCs (e.g., plasticizers, solvent residues, unreacted monomers, and secondary degradation products) (Järnström et al., 2008). Within damp building structures, discolored PVC, and 2-ethyl-1-hexanol in particular, have often been associated with alkaline concrete (Norback et al., 2000). Under humid and/or poorly ventilated conditions, rotting PVC emits glycol ethers and their esters, 2-ethyl-1-hexanol, 1-butanol, and TXIB (Järnström et al., 2008; Tuomainen et al., 2004). Furthermore, the hydrolysis of TXIB generates Texanol® (Kempf et al., 2009). In addition, PVC-containing products and cleaning formulations contain glycol ethers and their esters as well as the DEHP (Dodson et al., 2012; Singer et al., 2006). Both 2-ethyl-1-hexanol and butanol are known as traditional markers of microbial contamination (Korpi et al., 2009). Yet, these two compounds could also originate in sterile building materials without microbial contamination (Korpi et al., 1998) or in older buildings with aging building materials (Wieslander et al., 2010). An inverse correlation between a group of so-called MVOCs with viable bacteria and viable mold in air, but their positive correlation with plasticizers, including TXIB within a school building (Kim et al., 2007) supports a synthetic origin of these compounds.

CONCLUSIONS

PVC-based material represents a plausible source of 1-butanol, 2-butanone, acetic acid, benzaldehyde, hexanal, nonanal, and octanal in damp indoor environment. PVC-based materials were also correlated with PGEs and Texanol®.

ACKNOWLEDGMENTS

The study has been supported by the Swedish Research Council for Environment, Agricultural Sciences and Spatial Planning (Formas), Swedish Asthma and Allergy Association's Research Foundation, the Swedish Foundation for Health Care Sciences and Allergy Research, Norwegian Research Council, the U.S. National Research Service Award (T32 ES 07069), the American Scandinavian Foundation (Research Grant for Young Investigators), and County Council of Värmland, Sweden. The funders had no role in study design, data collection and analysis, decision to publish, or preparation of the manuscript.

REFERENCE

Bornehag, C.G., Sundell, J., Lundgren, B., Weschler, C.J., Sigsgaard, T. and Hagerhed-Engman, L. (2005a) Phthalates in indoor dust and their association with building characteristics, *Environmental health perspectives*, **113**, 1399-1404.

Bornehag, C.G., Sundell, J., Lundgren, B., Weschler, C.J., Sigsgaard, T. and Hagerhed-Engman, L. (2005b) Phthalates in indoor dust and their association with building characteristics, *Environ Health Perspect*, **113**, 1399-1404.

Choi, H., Schmidbauer, N., Spengler, J. and Bornehag, C.G. (2010a) Sources of propylene glycol and glycol ethers in air at home, *International journal of environmental research and public health*, **7**, 4213-4237.

Choi, H., Schmidbauer, N., Sundell, J., Hasselgren, M., Spengler, J. and Bornehag, C.G. (2010b) Common household chemicals and the allergy risks in pre-school age children, *PLoS One*, **5**, e13423.

Dodson, R.E., Nishioka, M., Standley, L.J., Perovich, L.J., Brody, J.G. and Rudel, R.A. (2012) Endocrine disruptors and asthma-associated chemicals in consumer products, *Environ Health Perspect*, **120**, 935-943.

Holme, J., Hägerhed-Engman, L., Mattsson, J., Sundell, J. and Bornehag, C.G. (2010) Culturable mold in indoor air and its association with moisture-related problems and asthma and allergy among Swedish children, *Indoor Air*, **20**, 329-340.

Järnström, H., Saarela, K., Kalliokoski, P. and Pasanen, A.L. (2008) Comparison of VOC and ammonia emissions from individual PVC materials, adhesives and from complete structures, *Environ Int*, **34**, 420-427.

Kaminśki, E., Libbey, L., Stawicki, S. and Wasowicz, E. (1972) Identification of the predominant volatile compounds produced by Aspergillus flavus, *Applied microbiology*, **24**, 721-726.

Kempf, M., Ramm, S., Feuerbach, T. and Schreier, P. (2009) Occurrence of 2,2,4-trimethyl-1,3-pentanediol monoisobutyrate (Texanol®) in foods packed in polystyrene and polypropylene cups, *Food Additives & Contaminants. Part A: Chemistry, Analysis, Control, Exposure & Risk Assessment*, **26**, 563-567.

Kim, J.L., Elfman, L., Mi, Y., Wieslander, G., Smedje, G. and Norback, D. (2007) Indoor molds, bacteria, microbial volatile organic compounds and plasticizers in schools--associations with asthma and respiratory symptoms in pupils, *Indoor Air*, **17**, 153-163.

Korpi, A., Järnberg, J. and Pasanen, A.-L. (2006) *The Nordic Expert Group for Criteria Documentation of Health Risks from Chemicals: 138. Microbial volatile organic compounds (MVOCs).*

Korpi, A., Järnberg, J. and Pasanen, A.-L. (2009) Microbial volatile organic compounds, *Critical reviews in toxicology*, **39**, 139-193.

Korpi, A. and Pasanen, A.-L. (1998) Volatile Compounds Originating for Mixed Microbial Cultures on Building Materials under various, *Applied & Environmental Microbiology*, **64**, 2914.

Lee, M.-L., Smith, D. and Freeman, L. (1979) High-resolution gas chromatographic profiles of volatile organic compounds produced by microorganisms at refrigerated temperatures, *Applied and environmental microbiology*, **37**, 85-90.

Matysik, S., Herbarth, O. and Mueller, A. (2008) Determination of volatile metabolites originating from mould growth on wall paper and synthetic media, *Journal of Microbiological Methods*, **75**, 182-187.

Miller, A., Scanlan, R., Lee, J. and Libbey, L. (1973) Volatile compounds produced in sterile fish muscle (Sebastes melanops) by Pseudomonas putrefaciens, Pseudomonas fluorescens, and an Achromobacter species, *Applied microbiology*, **26**, 18-21.

Newsome, J., Norman, V. and Keith, C. (1965) Vapor phase analysis of cigarette smoke, *Tob. Sci*, **9**, 102-110.

Norback, D., Wieslander, G., Nordstr, K., Ouml, R, W., Aring and Linder. (2000) Asthma symptoms in relation to measured building dampness in upper concrete floor construction, and 2-ethyl-1-hexanol in indoor air, *The International Journal of Tuberculosis and Lung Disease*, **4**, 1016-1025.

Nordtest. (1997) Ventilation: Local mean age of air - homogenous emission techniques. Nordtest method NT VVS 118. Nordtest, Finland.

Sahlberg, B., Gunnbjörnsdottir, M., Soon, A., Jogi, R., Gislason, T., Wieslander, G., Janson, C. and Norback, D. (2013) Airborne molds and bacteria, microbial volatile organic compounds (MVOC), plasticizers and formaldehyde in dwellings in three North European cities in relation to sick building syndrome (SBS), *Science of The Total Environment*, **444**, 433-440.

Schleibinger, H., Laussmann, D., Bornehag, C.G., Eis, D. and Rueden, H. (2008) Microbial volatile organic compounds in the air of moldy and mold-free indoor environments, *Indoor Air*, **18**, 113-124.

Schuchardt, S. and Strube, A. (2013) Microbial volatile organic compounds in moldy interiors: A long-term climate chamber study, *Journal of Basic Microbiology*, **53**, 532-538.

Singer, B.C., Destaillats, H., Hodgson, A.T. and Nazaroff, W.W. (2006) Cleaning products and air fresheners: emissions and resulting concentrations of glycol ethers and terpenoids, *Indoor Air*, **16**, 179-191.

Stymne, H., Boman, C. and Kronvall, J. (1994) Measuring ventilation rates in the Swedish building stock, *Building and Environment*, **29**, 373-379.

Tuomainen, A., Seuri, M. and Sieppi, A. (2004) Indoor air quality and health problems associated with damp floor coverings, *Int Arch Occup Environ Health*, **77**, 222-226.

Wieslander, G., Kumlin, A. and Norback, D. (2010) Dampness and 2-ethyl-1-hexanol in floor construction of rehabilitation center: Health effects in staff, *Arch Environ Occup Health*, **65**, 3-11.

SUMMARY OF SLOAN SYMPOSIUM – HEALTHYBUILDINGS 2015-EUROPE

Hal LEVIN[1,4], Martin TÄUBEL[2], and Mark HERNANDEZ[3]

[1]Building Ecology Research Group, Santa Cruz, California
[2]National Institute for Health and Welfare , Kuopio, Finland
[3] Department of Civil, Environmental and Architectural Engineering, University of Colorado, Boulder, Colorado.
[4] International Society for Indoor Air Quality and Climate Santa Cruz, California.

*Corresponding email: HLevin6@gmail.com

Keywords: Microbial ecology, Indoor environment, Practice, Research agenda, Symposium

SUMMARY

The paper summarizes a planned symposium on the microbiology of the indoor environment to be held in Eindhoven at Healthy Buildings 2015-Europe in May, 2015. The workshop has not been held as of the deadline for preparation and submission of this extended abstract, so here we only summarize the plans, the intent and the nature of the expected results. While the paper does not report the results, it provides background for the presentation of results that will occur at the HB2015-America conference.

The Sloan Symposium, embedded in the Healthy Buildings 2015-Europe conference program together with a post-conference Annex Workshop, has been funded by the Alfred P. Sloan Foundation through a grant to International Society for Indoor Air Quality and Climate (ISIAQ). The grant is part of the Sloan Foundation's Microbiology of the Built Environment (MoBE) Program.

The purpose of the planned Sloan Symposium is to improve understanding of the potential uses of molecular methods by building scientists and practitioners. During the symposium, we expect to identify key research questions based on practical needs that can lead to more widespread and contextually appropriate use of molecular methods and collection of useful metadata. This paper provides some background and describes the plans for the symposium.

INTRODUCTION

The main issue addressed by the planned symposium and the reason it is important is that there are numerous barriers to the use of molecular methods by practitioners working to create more healthful indoor environments. Professionals are addressing a variety of potential and existing microbial contamination situations (e.g., post-flooding mold growth, microbial surveys for public health surveillance, and investigations of the indoor microbiome associated with adverse occupant health reports).

While the methods supported by and promoted by the Sloan Foundation's MoBE program have obvious impacts in the Applied Biology and Building Science community, these advances have so far not necessarily provided translational paths to interpret emerging results— either for researchers or practitioners. For lasting impact to the field, practitioners need to be able to interpret and apply cohorts of modern research results, be assured of their reproducibility, and use them in practical contexts. Funders (e.g., clients such as building owners and operators) as well as building occupants and public health officials want to know the "everyday" and immediate potential for short- and long-term implications. Given this translational scenario, this special symposium has been designed to move the understanding of the meaning and implications of microbial evaluation of the indoor environment to the next stage; where scientific effort can be focused on obtaining information to assist in interpretation of results into pragmatic realms of constructive implementation(s).

Until relatively recently, historical investigations of the microbiology of the indoor environment depended largely on the recovery of cultured samples and light microscopic counting and identification, highly operator-dependent. The introduction of molecular based biology methods have increased the number of identified organisms by a factor of ten during the past decade, while the associated costs cost have come done by a factor of almost five hundred. The huge cost reduction in molecular methods, and the increase of their use within and outside the Sloan Foundation's MoBE program, has been accompanied by marked increases in interest from the practicing sectors; however there is a tremendous accompanying need for help in understanding the strengths and limitations of these methods, and the laboratory processing resources, for practical use of results. Today, much of the microbial work supporting building investigations is still done using traditional, cultivation techniques, which clearly displays that the threshold for using advanced molecular methods outside the science applications is high in this field. This situation requests change, as only once this link to practical implementation and interpretation has been made, can there be self-sustainability of the MoBE program.

Among its many goals, Sloan's MoBE program aims for wider adoption of genetic sequencing of microbial samples collected in the built environment. While use by researchers has increased due to Sloan Foundation funding, and a corresponding tremendous cost reduction has been realized in the same time period, appropriate sample collection, genetic processing, and interpretation of results, remains a major challenge, especially for the practitioners. Where microbial ecologists tend to report relative abundance, associations, or distribution of attributable sources as major findings, practitioners have to relate these types of findings in to a network of other indicators in many different environmental contexts.

Given this challenge, the Sloan Symposium at the Healthy Buildings 2015-Europe conference will focus on identifying a new generation of translational practices to facilitate the transfer of modern indoor air science (including molecular microbiology) into meaningful tools for wider application, especially in the (non-research-oriented) practitioner field. This is the original (1988) and on-going goal of the Healthy Buildings conference series.

METHODOLOGIES

The Symposium is planned to begin with a plenary lecture by Dr. Miia Pitkäranta of Finland, followed by nine presentations and a short discussion at the two successive technical sessions, then a workshop within the conference. The presentations are scheduled as follows:

Jordan Peccia (USA): *Revolution/evolution—DNA sequencing to identify indoor microorganisms*
Jeffrey Siegel (USA): *Building science and indoor air determinants of the indoor Microbiome*
Anne Hyvärinen (Finland): *Assessment of moisture and mold problems – the Finnish example*

Kristian Fog Nielsen (Denmark): *Microbial growth and interactions on indoor surfaces - microbial secondary metabolites and mycotoxins*
Maria Nunez (Spain, Norway): *Microbial sampling in building surveys: what and why are we sampling?*
Tiina Reponen (Finland, USA): *Microbial sampling in building surveys: how to choose a sampling method?*
Martin Täubel (Austria, Finland): *Quantitative PCR in microbial assessments of indoor spaces.*
Mark Hernandez (USA): *A perspective on leveraging new generation sequencing for bioaerosol assessments of the built environment: differences and commonalities of processing pipelines and databases*
Darrell Baumgartner (Argentina, USA) *A practical database approach for leveraging catalogues of fluorescence signatures for real-time bioaerosol assessments of the built environment*

The Conference Workshop is planned to follow immediately after the technical sessions. The purpose of the Conference Workshop is to gather input from practitioners on the barriers and challenges they face in adopting molecular methods for characterizing the indoor microbiome. The issues identified in the workshop are intended to establish the agenda for discussion at the post-conference, one-day Annex Workshop

The Annex Workshop will focus on the scientific issues identified as barriers during the Conference Workshop. The participants will include the plenary lecturer, the 10 invited presenters (plenary and technical sessions) plus 5 additional invited participants including Jack Gilbert, David Thrasher, Illka Hanski, Dick Heedrick, and Ulla Haverinen-Shaughnessy. The Annex Workshop was planned to begin with a discussion of biodiversity by Ilkka Hanski and Jordan Peccia.

RESULTS AND DISCUSSION,

The results of the Sloan Symposium will be summarized and presented at the HB2015-America conference.

Expected Products from the Symposium and the grant to support it, are as follows:

- Improved understanding and increased use of molecular methods by practitioners in the field of indoor environmental research and practice.
- Development of a pragmatically focused agenda that can contribute to the translation of Sloan MoBE program priorities into practical use
- Preparation of a manuscript for submission to a peer reviewed journal offering open access, disseminating the results of this Symposium's activities and identifying research needs to develop guidance on interpretation of results.

CONCLUSIONS

Conclusions will be drawn from the analysis of the results after they become available in late May. Expected outcomes are as follows:

- Increased networking between indoor air scientists, practitioners and microbial ecologists.
- Increased understanding of the obstacles to wider use of molecular methods.
- Improved interpretation of the results of indoor microbiome characterization by scientists and practitioners.

- Contribution of the indoor microbiome studies conducted thus far to 'best practice' of practitioners, e.g., in sampling and in interpretation of results.

ACKNOWLEDGEMENT

The authors wish to thank all of the invited participants in the Sloan Symposium at Healthy Buildings 2015-Europe for their willingness to contribute to the symposium. We also thank the organizers of HB2015-Eruope for their cooperation in the incorporation of the Sloan Symposium into the conference program. We wish to acknowledge Paula Olsiewski, Program Manager of the Microbiology of the Built Environment Program at the Alfred P. Sloan Foundation, for her valuable suggestions on the organization of the Symposium. Finally, we thank the Sloan Foundation for the financial support of the Symposium.

REFERENCES

Alfred P. Sloan Foundation Microbiology of the Built Environment Program - http://www.sloan.org/major-program-areas/basic-research/mobe/

Healthy Buildings 2015-Europe conference: http://hb2015-europe.org/

International Society for Indoor Air Quality and Climate (ISIAQ) –http://isiaq.org

EXPERIMENTAL AND CFD INVESTIGATION OF HOSPITAL OPERATING ROOM AIR DISTRIBUTION

Zhiqiang (John) ZHAI[1,*], James MCNEILL[1] and Jean HERTZBERG[2]

[1]Department of Civil, Environmental, and Architectural Engineering, University of Colorado, Boulder, CO, USA
[2]Department of Mechanical Engineering, University of Colorado, Boulder, CO, USA

[*]Corresponding email: john.zhai@colorado.edu

Keywords: Hospital operating room, Air distribution, Experiment, CFD

SUMMARY

Effective ventilation is critical to the successful prevention of surgical site infections (SSI) in hospital operating rooms (ORs). ASHRAE Standard 170-2008 provides specific requirements for the design of hospital OR ventilation system. This research verifies the effectiveness of the system by using full-scale laboratory experimentation and computational fluid dynamics (CFD) simulation. Both the experimental and CFD results show a strong inward contraction of the supply air jet, deviated from the original intent of providing unidirectional flow to protect patient from cross-infection. Further tests point to a strong correlation between the supply air to room air temperature difference and the room air distribution pattern. The buoyancy forces on the downward supply air jet are determined to be the cause of the observed air distribution pattern. The study demonstrates the applicability of using Reynolds-Averaged Navier-Stokes (RANS) CFD with RNG k-e turbulence model for prediction and analysis of the surgical environment air flow.

INTRODUCTION

The hospital operating room (OR) is a unique indoor environment with stringent medical and engineering requirements designed to protect the patient and surgical staff from risk of cross-infection, while accommodating multi-faceted requirements for occupant comfort, facilitation of surgical tasks, and services for the operation of medical equipment. ASHRAE Standard 170-2008 (2008) requires that ventilation is provided from the ceiling in a downward direction with a concentration over the patient and surgical team. The area of the primary ventilation air diffusers must extend at least 305 mm (12 in) beyond each side of the surgical table. It is required that the air is exhausted from at least two grilles on opposing side of the room near the floor. The standard requires the use of non-aspirating, Group E outlets that provide a unidirectional flow pattern in the room. These outlets were previously known as laminar flow diffusers, but are more accurately described as unidirectional, non-aspirating since the air flow leaving the diffuser is technically in the turbulent flow regime.

This project is to verify the effectiveness of the current ventilation systems in protecting patients in ORs by using both lab experiment and computational fluid dynamics (CFD) techniques. The research provides first-hand knowledge of airflow transport characterized in a realistically simulated OR environment.

FIELD EXPERIMENT

The study first measured and collected on-site thermal conditions that may influence the thermal-fluid boundary conditions of operating rooms in lab experiment and CFD simulation. Observations and measurements were conducted during 19 live surgeries at the Department of Orthopaedics at Denver Health Medical Center. Non-contact infrared thermography was used to determine the surface temperatures of heat sources within the operating room, using a Fluke TI-30 thermal imager mounted on a tripod parallel to a photographic camera.

The mean measured surgical site temperature was 78.1 ± 3.7°F (25.6 ± 2.1°C) with a surrounding skin temperature of 85.1 ± 7.3°F (29.5 ± 4.0°C), producing a mean temperature difference of -3.9 ± 2.4°F (-2.1 ± 1.4°C) between the surgical site and surrounding body. All measurements of the surgical site were found to be cooler than the surrounding skin due to evaporation of body fluids from the subcutaneous tissue.

The study examined the thermal conditions of the three ORs by measuring supply air conditions, enclosure surface temperature, and internal heat gains from equipment. The measured supply air ACHs, are, respectively, 26.7, 22.8, 29.3. Equipment in the operating room was grouped into the following categories for thermal measurements: surgical lights, computer monitors, and misc. surgical equipment. Both surface temperatures and heat gains of the equipment (and staff) were measured, which serve as a basis for objects to include in the lab experiment and CFD simulation.

The total room load was calculated using the values that were determined for the heat gains from equipment, lighting and occupants. The total room cooling load for frequently used equipment, excludes the electrosurgical generator, mobile c-arm x-ray system, and blanket warmer is 8663 Btu/h (2539 W). For a typical OR size (22 x 22 x 10 ft = 6.7 x 6.7 x 3.0 m) and an air change rate of 20 ACH a supply air to room air ΔT of approximately 5 °F (3°C) would be required to meet the steady-state cooling load.

FULL-SCALE LAB EXPERIMENT

A full-scale operating room chamber was constructed for conducting experiments to measure the fluid and thermal conditions inside a typical OR. The lab chamber was fit-up with typical furnishings for ORs that act as blockages to indoor air flow. The thermal characteristics of the room were selected based on a combination of design specifications, field testing results, and professional experience. The characteristics are designed to provide a baseline for a typical OR. Figure 1 shows the full-scale lab chamber used for evaluation of air distribution via both quantitative flow visualization and particle image velocimetry (PIV) techniques. Table 1 lists the main laboratory experiment specifications.

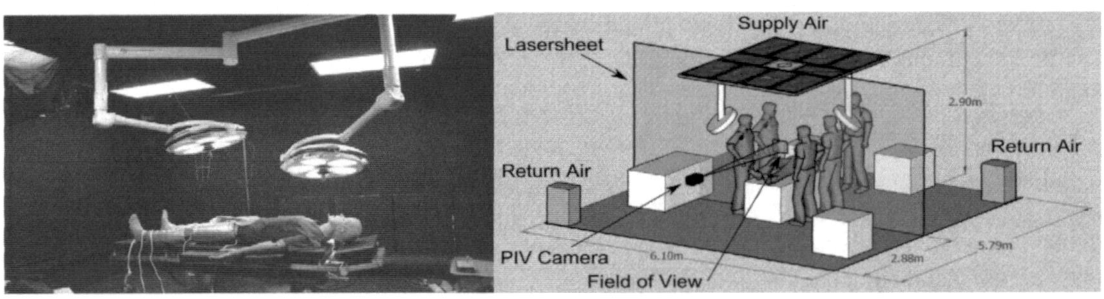

Figure 1. Full-scale laboratory chamber used for evaluation of air distribution via flow visualization and PIV.

Table 1. Laboratory experiment specifications

Parameter Range	Value
Diffuser array dimensions	2.44 x 3.05 m
Diffuser coverage area	7.06 m^2
Air change rate	31.6
Nominal face velocity	0.127 $m^3/s- m^2$
Room air temperature	20°C
Supply air temperature	18.3°C
Room pressurization	+2.5 Pa

This experimental results showed that the air flow pattern in a hospital OR is characterized by complex physics that causes the jet to contract inwards towards the centreline (Figure 2). This location is coincident with the typical location of the patient. The physics of the indoor air distribution contains impingement, recirculation, annular flow, free shear layer flow, and buoyancy. The primary source of turbulence production in the OR is free shear layer at the edge of the diffuser array. The study further found that larger ventilation temperature differences between the supply air and the room air causes more severe supply air jet contraction and higher air velocities near the surgical table. The inlet flow conditions of the supply air appear to be less important to the overall room air distribution than the design of the air distribution system. Pressurization of the room also had little influence on the indoor air distribution in a steady-state case, but may cause significant changes to the indoor air distribution pattern in a transient case (e.g. opening of door).

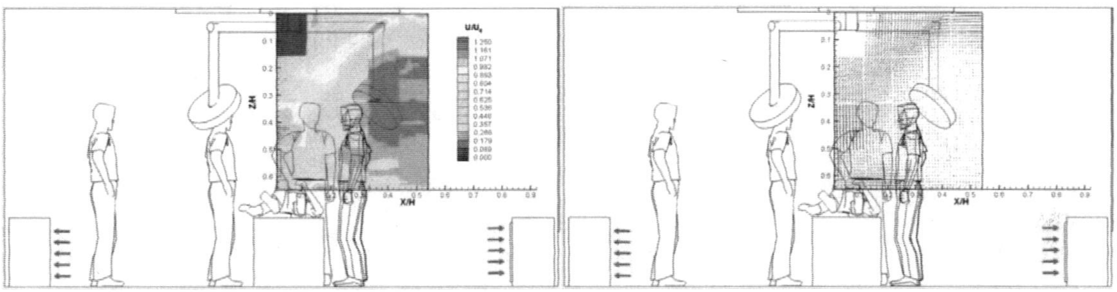

Figure 2. Normalized velocity magnitude contours and velocity vectors in the plane intersecting the center of the room in the sterile region above the patient.

CFD SIMULATION

CFD models were developed employing both the Reynolds-Averaged Navier-Stokes (RANS) and Large Eddy Simulation (LES) methods to simulate the indoor air distribution in a prototypical hospital OR. Figure 3 illustrates the built CFD models, which replicates the lab experiment settings including geometries, system conditions and internal objects. The models were verified using the lab experimental data (Figure 4). In general, good agreement was observed between experiment and simulation. The steady-state RANS CFD with the RNG k-e turbulence model (Yakhot and Orszag 1986) and the mesh of 675K can adequately model the represented physics that were identified in the operating room air distribution. LES may provide some more details but requires much significant computing effort, time and skill. Further parametric studies using the validated CFD model confirm the influence of the temperature difference between the supply air and room air on the distorted airflow patterns in the OR (Figure 5).

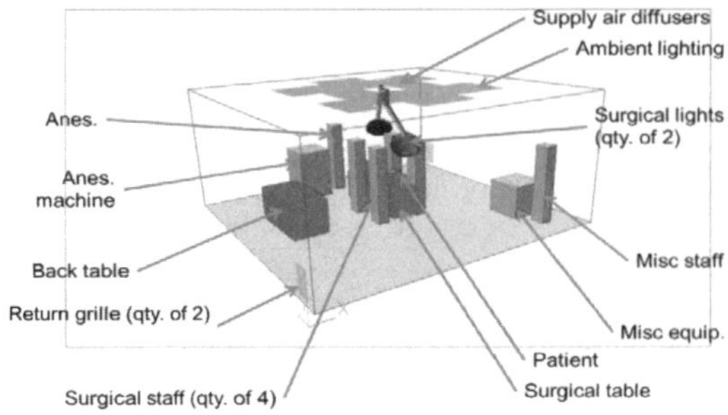

Figure 3. Base CFD model setup

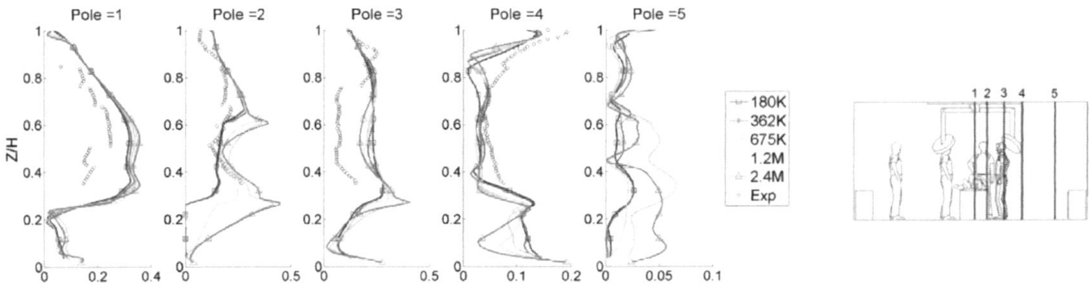

Figure 4. Comparison of velocity magnitude in the u/w plane

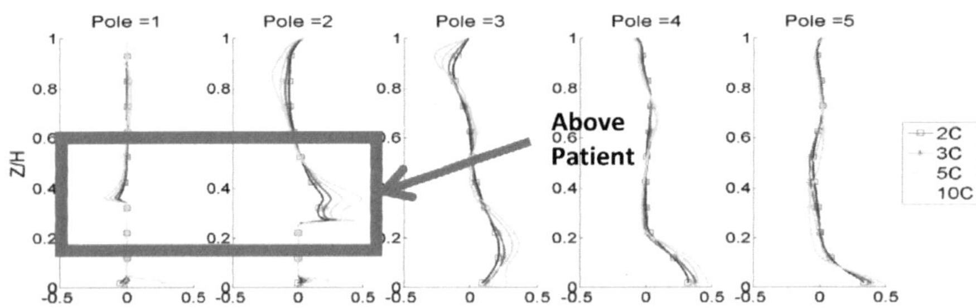

Figure 4. Comparison of U-velocity with varying supply and room temperature differences

CONCLUSIONS

Modern techniques such as flow visualization, PIV, and CFD were employed to evaluate the air flow within a typical OR environment, which all revealed unexpected airflow distortion from the intended unidirectional flows. A strong correlation was verified between the supply air to room air temperature difference and the room air distribution. The current design standard using face air velocity as an indicator may not be appropriate or adequate for ORs.

ACKNOWLEDGEMENT

The authors would acknowledge the funding support for this study from American Society of Heating, Refrigerating, and Air-Conditioning Engineers (ASHRAE) under project TRP-1397.

REFERENCES

ASHRAE. 2008. ANSI/ASHRAE Standard 170-2008.Ventilation of Healthcare Facilities. American Society of Heating, Refrigerating, and Air-Conditioning Engineers, Inc., Atlanta, 2008.

Yakhot, V., & Orszag, S. A. 1986. Renormalization group analysis of turbulence. I. Basic theory. Journal of Scientific Computing, 1, 3-51.

The Role Of Surfaces In The Transmission Of Bioaerosols From Source To Patient In Hospital Rooms

Marco-Felipe. King[1,*], Catherine J. Noakes[1], and P. Andrew Sleigh[1]

Pathogen Control Engineering Institute, School of Civil Engineering, University of Leeds, Leeds, UK.

[*]Corresponding email: m.f.king@leeds.ac.uk

Keywords: Airflow, bioaerosols, hospital infection, modelling

SUMMARY

INTRODUCTION: Aerial dispersion of bioaerosols and subsequent contamination of surfaces is recognised as a potential transmission route for health-care acquired infections. Pathogens accrue on health-care workers' (HCW) hands as they touch surfaces and can subsequently be transmitted to other patients.

METHODS: Computational fluid dynamics (CFD) was used to predict bioaerosol deposition in a single and multi-bed hospital rooms. A Monte-Carlo model was developed using the CFD deposition patterns in conjunction with clinical observation of surface contact sequences to predict the contamination levels of bacteria on HCWs' hands as they perform patient care in the two rooms.

RESULTS: Hand colonisation depends on care type, room layout and in particular on the spatial distribution of pathogens between surfaces, which is influenced by ventilation. During care within multi-bed rooms colonisation levels increase due to the spatial spread of microorganisms contaminating multiple patient surfaces caused by the ventilation strategy. Positioning infectious patients within an unobstructed path between the inlet and outlet diffuser significantly reduces cross contamination to other patients surfaces.

CONCLUSIONS: Colonisation levels of HCWs' hands are likely to be significantly lower after care in single patient rooms than after care in a multi-bed ward and ventilation design is vitally important in curtailing bioaerosol spread.

INTRODUCTION

Risk of acquiring hospital acquired infections (HCAI) is omnipresent in health-care facilities worldwide and understanding transmission routes is key to effective control. Conservative estimates by Harbarth et al. (2003) show that potentially 20% of infections contracted through contact transmission may be preventable. Several recent studies have highlighted the importance of surface contamination and hinted at a causal link to subsequent patient infection (Bhalla et al., 2004). Pathogens have been shown to accrue on health-care workers' (HCW) hands as they touch surfaces (Pittet et al., 1999) and hence can subsequently be transmitted to patients (Hayden et al, 2008). However, there is currently little robust understanding as to how HCW surface contacts and activities in the health care environment result in patient exposure to such pathogens. Differences in benchmarking of surveillance data often make comparisons difficult on any level, however the significance of the problem is undisputed (Smith et al., 2012).

Aerial dispersion of bioaerosols and subsequent contamination of surfaces has been recognised as a potential transmission route for some of these infections (Bhalla et al. 2004). However the combined role of airborne dispersion, pathogen contamination of hospital room surfaces and interaction with human behaviour is still poorly understood and constitutes an area of much controversy and challenging research. Furthermore the influence of airflow patterns and ward design on the risk is not well understood; single bedrooms are widely advocated for their infection control potential, yet there is little data to quantify the benefits. This research considers the question: Are single-bed patient rooms more effective than their multi-bed counterparts at reducing the risk of infection from environmental contamination? The study combines CFD modelling of particle deposition with a contact risk model framework for quantifying the number of colony forming units (cfu) contaminating HCWs' hands following care in two room types.

METHODOLOGIES

Three mechanically ventilated case scenarios were considered: A single-patient room and a two-patient room where the position of the infectious subject was varied with respect to the inlet diffuser, effectively creating two sub-cases (see Table1 and Figure 1).

Table 1: Case-study scenarios

Case Nº	**1**	**2a**	**2b**
Scenario	Single-bedroom	Two-bed room	Two-bed room
Aerosol release	Patient head	Patient 1	Patient 2

Bioaerosol deposition

Airflow and bioaerosol behaviour was modelled through CFD RANS simulations in Fluent (Ansys v13) using the Reynolds' Stress Model turbulence closure which treats Reynolds' stresses independently. Solutions were mesh independent and mesh size was in the region of 4million cells. Simulations were considered converged when there was less than a 5% difference compared to anemometry data. Lagrangian particle tracking of 2.5micron-sized droplets, validated through experiments (King et al. 2013), was used to predict spatial surface distributions of bioaerosol deposition in the single (Figure 1a) and a two-bed room scenarios (Figure 1b). Simulations were based on a comparable experimental set-up. Heated mannequins (DIN-men) were used to represent human patients lying supine on the beds. Ventilation is set to 6 air changes per hour in all cases via wall-mounted inlet and outlet diffusers on opposing façades. Full details can be found in King et al., 2013.

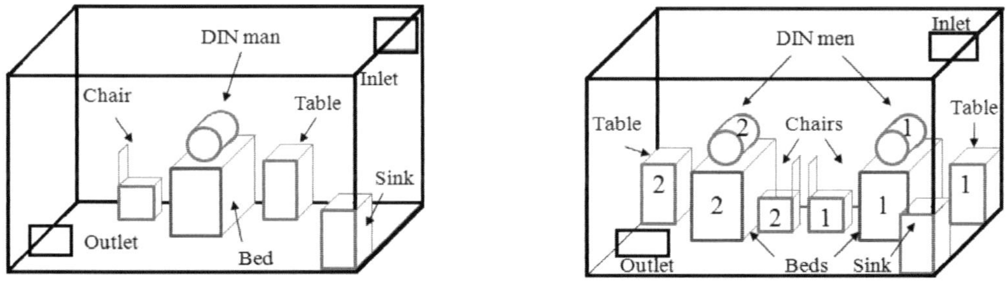

a) Single room set-up b) Two-bed room set-up

Figure 1. Drawings of room set-up representing hospital single and two-bed rooms (both 2.26m x 3.36m x 4.2m)

Surface contact sequences

Sequences of health care worker surface contacts during typical patient care episodes were determined from an observation study in a community hospital. Care types included: *Direct care*, *Housekeeping*, *Mealtimes*, *Medication rounds*, *Personal care* and *Miscellaneous care*. Over 400 observations were conducted and five surface categories were monitored during each care episode, namely: Within patient reach (Near-patient), out of patient reach (Far-patient), the patient themselves, medical equipment, and hygiene products. All surface categories apart from the patient are considered to be made of hard, non-porous material. Hand hygiene was also recorded for all care types. Data collected was used to create contact frequencies for each surface and care type.

Modelling pathogen accretion

A Monte-Carlo simulation of the mechanics of pathogen transfer from surface-to-hands was developed using the CFD predicted deposition patterns in conjunction with the clinical observation of surface contact sequences. The quantity of pathogens accrued on HCWs hands (Y) during patient care was modelled as a function of the number of surface touched (i=1..n), surface contamination levels (V) and the surface area of skin in contact with the surface (A) (Brouwer et al. 1999). However, it is reasonable to assume that not all of the pathogens in contact with the surface area of skin touching the surface are transferred. Therefore a transfer efficiency (λ) is defined to represent the proportion of pathogens that are transferred in the upward direction (Rusin et al. 2002). During hand-to-surface contact it is equally reasonable to assume that some quantity of pathogens already acquired (βY) are deposited from the hand onto the surface during a contact (Rusin et al. 2002). However this quantity deposited will depend on the current hand inoculum level (Y_{i-1}). Therefore this model will consider transfer in both directions or bi-directional transfer. Consequently pathogen accretion (Y) can be modelled by means of a recurrence relationship given in equation 1.

$$\boldsymbol{Y_i = \lambda_i V_i A_i + \beta_i Y_{i-1}}, \qquad \textbf{(1)}$$

This was used to predict the number of pathogens (cfu) on HCWs' hands as they perform the observed routine patient care in the two rooms. Hand hygiene is included by assuming that a certain number of pathogens are removed after care concludes according to observations and experiments by Girou et al. (2002). For each scenario

1,000 simulations were conducted to produce a distribution, and the model validated against available literature (Pittet et al. 1999).

RESULTS AND DISCUSSION

Bioaerosol deposition

Figures 2a and b) show simulated temperature contours and velocity vectors for the single and two-bed patient rooms, plotted on the horizontal and vertical surface through the bed. Complex flow structures can be observed, with the cold inlet air impinging on the opposite wall and multiple recirculation zones at the foot of the bed. A vertical heat plume emanates from the supine mannequin and is depicted in the vertical plane.

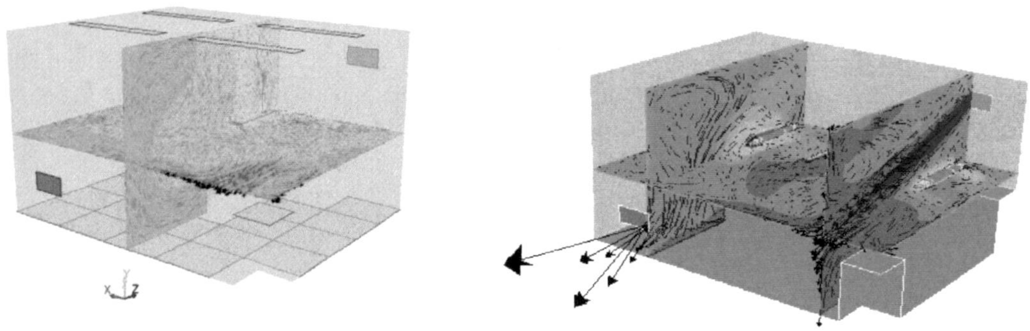

a) Single room   b) Two room

Figure 2. Velocity vectors (0.001-0.1m/s) superimposed over temperature contours (22-37ºC)

Figures 3 a) and b) depict the predicted pathogen concentrations in the three scenarios normalised with respect to the global average. In the two-bed room cases, when patient 1 is the infectious source, bioaerosols have a tendency to disperse, contaminating the adjacent surfaces to patient 2. Conversely no such marked trend is observed when patient 2 is infectious, where the particles are likely extracted by the ventilation rather than deposited on far away surfaces.

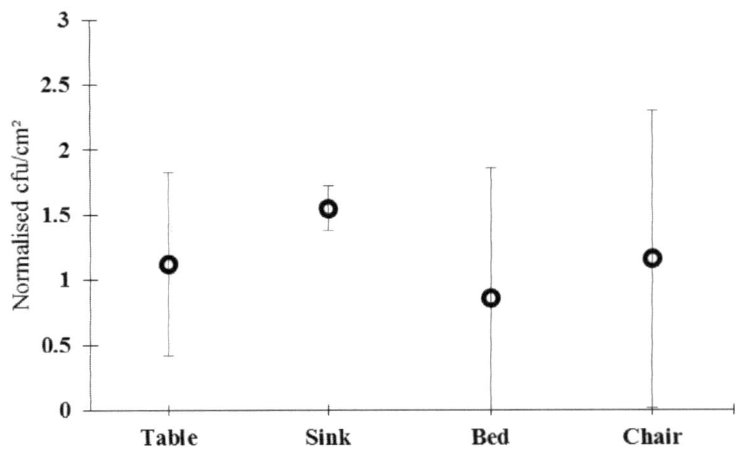

a) Scenario 1: Single-bed room

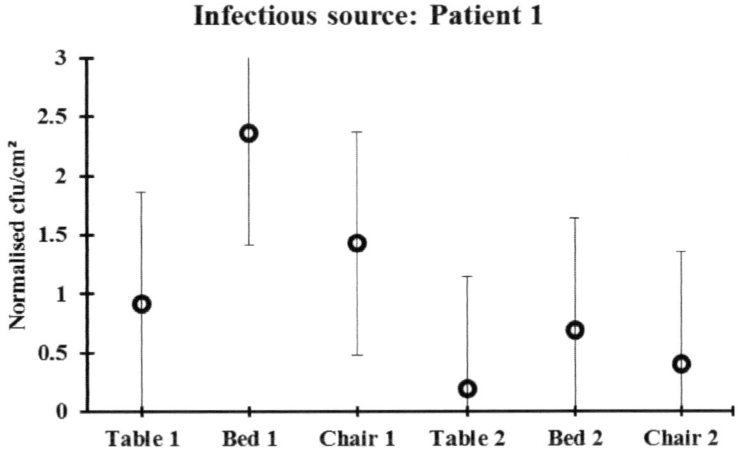

b) Scenario 2a: Two-bed room, infectious patient: 1

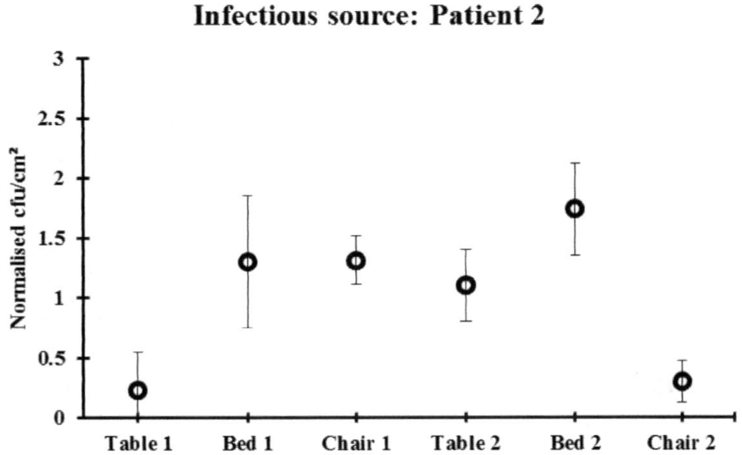

c) Scenario 2a: Two-bed room, infectious patient: 2

Figure 3. Predicted surface colony forming units/cm^2 based on room type, normalised with respect to global average.

Clinical observations

Figure 3 shows surface contact distributions categorised by care type, which exhibit a strong influence on the HCWs' movements. Care types could not be distinguished with respect to the frequency of patient contacts; however environmental surface contacts exhibited a statistically significant variation.

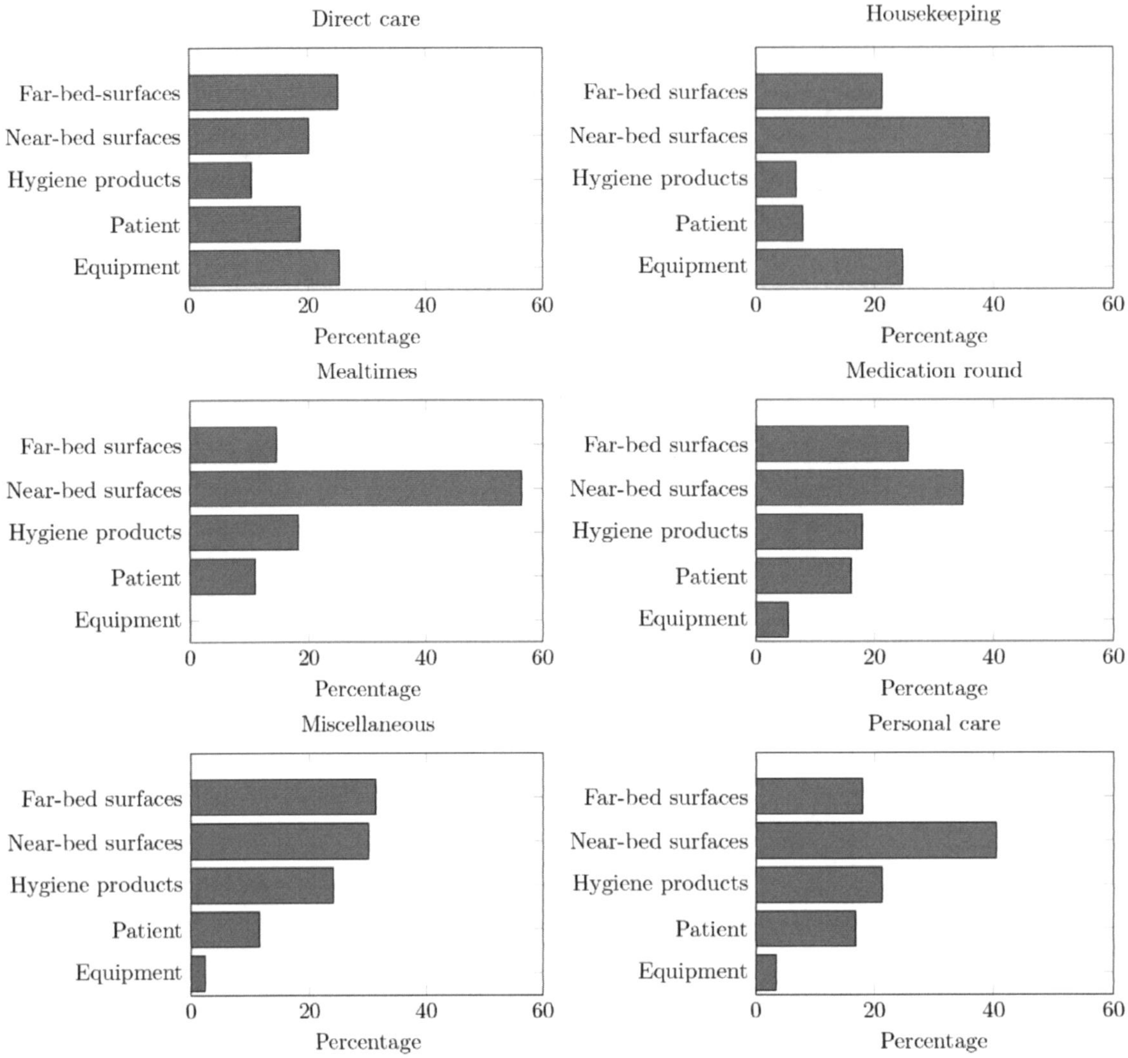

Figure 3. Surface contact distribution subdivided by care type

Pathogen accretion

Figure 4 shows box-plots representing the predicted HCW's hand contamination levels for the six care types for all three room scenarios. Contamination (Y) values have been normalised with respect to the mean contamination levels of the HCWs after *direct care* in the single-bed room to enable comparison between rooms and care. Single rooms results are consistently lower than their two-bed room counterparts. Pick-up of pathogens during *housekeeping* appears to be highest in all scenarios, with *mealtimes* the lowest, reflecting the different likelihood of contact with surfaces during these two activities. The results for the two-bed scenarios show that spatial deposition of particles and subsequent accretion by HCWs is influenced by the location of the ventilation supply inlet relative to the source. Locating a susceptible patient closer to the supply air is likely to reduce the risk of environmental contamination due to bioaerosol release from a neighbouring infectious patient.

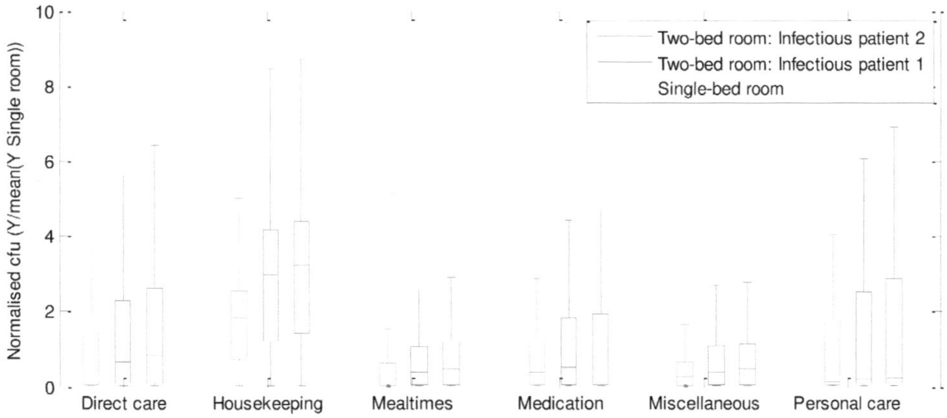

Figure 4. Predicted colony forming units on HCWs hands after each type of patient care. Horizontal lines in the boxplot represent the sample median.

CONCLUSIONS

Results demonstrate that hand colonisation is likely to depend on care type, room layout and in particular on the spatial distribution of pathogens between surfaces, which is influenced by ventilation strategy. Contamination on the HCWs' hands after patient care in a single-bed room, even after hand hygiene, is by no means negligible. However during care within the two-bed room colonisation levels are significantly higher throughout due to the spatial spread of microorganisms into the zone of the neighbouring patient. Positioning infectious patients within an unobstructed path between the inlet and outlet diffuser significantly reduces cross contamination to other patients surfaces (Two-bed room: Infectious patient 1).

Results indicate that colonisation levels of HCWs' hands are likely to be significantly lower after care in single patient rooms than after care in a two-bed room and that patient positioning and ventilation design is important in helping curtail the risk of infection transmission.

ACKNOWLEDGEMENT

This work was carried out as part of a PhD studentship supported by the UK Engineering and Physical Sciences Research Council (EPSRC) and Arup. The authors would like to thank the South East Wales Research Ethics Committee (Ref: 11/WA/0200).

REFERENCES

Bhalla A., Pultz, N.J., Gries, D.M. et al. (2004) Acquisition of nosocomial pathogens on hands after contact with environmental surfaces near hospitalized patients. *Infection control and hospital epidemiology*, 25(2):164–167.

Brouwer, D.H., Kroese, R., Van Hemmen, J.J. (1999). Transfer of contaminants from surface to hands: experimental assessment of linearity of the exposure process, adherence to the skin, and area exposed during fixed pressure and repeated contact with surfaces contaminated with a powder. *Applied Occupational and Environmental Hygiene*, 14:231–239.

Girou, E., Loyeau, S., Legrand, P. et al. (2002) Efficacy of handrubbing with alcohol based solution versus standard handwashing with antiseptic soap: randomised clinical trial. *BMJ British Medical Journal*, 325(7360):362–367.

Hayden, M.K., Blom, D.W., Lyle, E.A. et al. (2008) Risk of hand or glove contamination after contact with patients colonized with vancomycin-resistant enterococcus or the colonized patients' environment. Infection Control and Hospital Epidemiology; 29(2):149-154

Harbarth, S., Sax, H, Gastmeier, P. (2003) The preventable proportion of nosocomial infections: an overview of published reports. Journal of Hospital Infection; 54:258-266.

King M-F, Noakes C.J., Sleigh P.A. et al. (2013) Bioaerosol deposition in single and two-bed hospital rooms: A numerical and experimental study. *Building and Environment,* **59**:436-447

Pittet, D., Dharan, S., Touveneau, S., et al. (1999). Bacterial contamination of the hands of hospital staff during routine patient care. *Arch. Intern. Med*, 159:821–826.

Rusin, P., Maxwell, S., Gerba, C.. (2002) Comparative surface-to-hand and fingertip-to-mouth transfer efficiency of gram-positive bacteria, gram-negative bacteria, and phage. *Journal of Applied Microbiology*, 93(4):585–592.

Smith, S.J., Young, V., Robertson, C. et al. (2012) Where do hands go? An audit of sequential hand-touch events on a hospital ward. *Journal of Hospital Infection,* 80:206–211.

Tian Z.F., Tu J.Y., Yeoh, G.H. et al. (2006) On the numerical study of contaminant particle concentration in indoor airflow. Building and Environment: 41:1504-1514

Tian L, Ahmadi G (2007). Particle deposition in turbulent duct flows- comparisons of different model predictions. Journal of Aerosol Science; 38:377-397.

Zoon, W.A.C., Loomans, M.G.L.C., Hensen J.L.M. (2011). Testing the effectiveness of operating room ventilation with regard to removal of airborne bacteria. Building and Environmen; 46(12):2050-2077

HOSPITAL SURGE CAPACITY: PRACTICAL ASPECTS OF TEMPORARY ISOLATION WARD DESIGN

Nicholas CLEMENTS[1,*], Shelly L MILLER[1,2], Shobha S SUBHASH[2], Aaron E EAGAN[2]

[1]Department of Mechanical Engineering, University of Colorado Boulder, Colorado, USA
[2]National Center for Occupational Health and Infection Control, North Florida/South Georgia Veterans Health System, Gainesville, Florida, USA

[*]Corresponding email: nsclements@gmail.com

Keywords: Surge capacity, Isolation room, Anteroom, Negative pressure, Hospital infection control

SUMMARY

In the event of a pandemic, hospital isolation room capacity must increase to deal with a surge of infectious patients. One method of increasing surge capacity is converting an existing hospital ward into a temporary isolation ward through manipulation of the air handling unit fan speeds and by setting up anterooms for ingress and egress. The Palo Alto VA Surge Capacity Engineering Project (PAVA-SCEP) aims to test the effectiveness of this type of surge capacity response, while collecting vital information about its practical limitations and planning requirements. Here we detail essential planning steps leading up to the final demonstration, which will occur in early 2015. Two planning visits conducted in 2014 provided baseline pressure and flow data for isolation ward design, while highlighting the importance of close coordination between the engineering, infection control, and nursing departments in surge capacity planning.

INTRODUCTION

Recent infectious disease outbreaks (e.g. SARS, H1N1, Ebola) highlight the need for hospitals to prepare for surges of patients with a variety of symptoms and diseases with a range of transmission modes. In the event of a disaster or pandemic, successful medical emergency response requires coordination between hospitals, ambulatory services, private physicians, the U.S. Department of Veterans Affairs (VA), and many others (Lurie et al., 2008; IOM, 2010). At the hospital level, Hick et al. (2010) proposed a three-phased pandemic response plan, which includes conventional, contingency, and crisis levels of care and detailed triggers for each care level. During an infectious disease outbreak causing contingency or crisis conditions, hospital isolation room capacity must be able to expand to meet demand. While it is recognized that increased surge capacity is vital to medical emergency response (Rubinson et al., 2005), little scientific knowledge has been collected to validate proposed surge capacity methods (Subhash and Radonovich, 2011).

Negative pressure airborne infection isolation rooms (AIIRs) are used to prevent the spread of airborne infections. The pressure difference between the corridor and AIIR is recommended to be -2.5 Pa in the U.S., with an air exchange rate of 12 ACH. Through dilution of pollutants and limiting escape volumes, isolation rooms significantly reduce the ability of airborne contamination to leak into adjacent corridors (Adams et al., 2011). The use of an anteroom between the AIIR and hospital corridor is also recommended (Subhash et al. 2013). Door motion and health care worker movement impact AIIR containment effectiveness, and inclusion of an anteroom in AIIR design can limit the escape of motion-dispersed contamination (Adams et al., 2011). In the event of an infectious patient surge, expanding isolation room capacity can be accomplished by either conversion of existing spaces (e.g. Rosenbaum et al., 2004) or by constructing temporary isolation zones (e.g. Johnson et al., 2009) using HEPA-filtered negative air machines. Another ventilation option is to adjust the operating conditions of the hospital's air handling unit (AHU) to establish a negative pressure ward adjacent to the rest of the hospital. To test the effectiveness of this approach, engineers from the National Center for Occupational Health and Infection Control (COHIC) and the University of Colorado Boulder are collaborating with the Palo Alto VA Hospital (PAVA) to perform a demonstration of the establishment of a temporary isolation ward. Here we discuss the planning stages leading up to the Palo Alto VA Surge Capacity Engineering Project (PAVA-SCEP) final demonstration, including two visits to PAVA to conduct pressure and airflow tests in the isolation ward. We also discuss the practical aspects of designing and performing such a demonstration in a functioning hospital and provide recommendations for similar projects in the future.

METHODOLOGIES

A ward on the top floor of the main hospital building at the PAVA campus was chosen as the location of the temporary isolation ward because this part of the hospital has a dedicated AHU, a fire wall separating the ward from the rest of the hospital, and a dedicated exhaust system. The ward has two hallways connecting it to the rest of the hospital, and during the demonstration one hallway will contain an anteroom while fire-doors will be closed in the other to establish the pressure envelope. Upper-room UVGI lamps will be installed in the ceilings of the two stairwells that access the ward. After planning visits described below, it was determined that during the demonstration, the AHU will be operated with supply, return, and exhaust fan speeds set to 60%, 100%, and 100% of their normal operating speeds, respectively. All return air will be exhausted, as isolation room air cannot be recirculated. DG-700 (The Energy Conservatory) pressure sensors placed at anteroom, fire, and stairwell doors will monitor pressure differences between the isolation ward and the rest of the hospital, while T-VER-PXU-X (Veris/Onset) pressure sensors will monitor pressure variability on the ward between bedrooms, bathrooms, and hallways. A balometer (TSI Inc.) will be used to measure register flow rates and estimate air changes per hour. The demonstration will include a 24 hour period of maintaining the temporary negative pressure isolation ward, with pressure and flow data also being collected throughout the day before and after the demonstration to understand baseline conditions. Discussed below are pressure and flow rate data collected during two planning visits conducted at PAVA from 8/4-7/2014 and 12/10-15/2014. During these visits, a DG-700 and balometer were used to collect preliminary data about the ward for designing and planning purposes.

RESULTS AND DISCUSSION

During the August visit to the PAVA campus, pressure and flow rate data from an isolation ward were collected, along with pressure data for the two stairwells with access to the ward and a conference room adjacent to the ward. Figure 1 details the average pressure differences, with green arrows representing the approximate airflow direction across each doorway. It was observed that both stairwells are positively pressurized relative to ward hallways, which aids in preventing the escape of contamination. The isolation room was pressurized slightly below recommended levels and the anteroom was slightly pressurized (-0.5 Pa) relative to the hallway. Balometer measurements of isolation and conference room flow rates differed from building design airflows, highlighting the need for actual measurements when considering isolation ward design. During the visit, it was noted that most of the ward had drop ceilings, while AIIRs had solid ceilings, and all windows were sealed closed, important parameters when considering possible leakage areas.

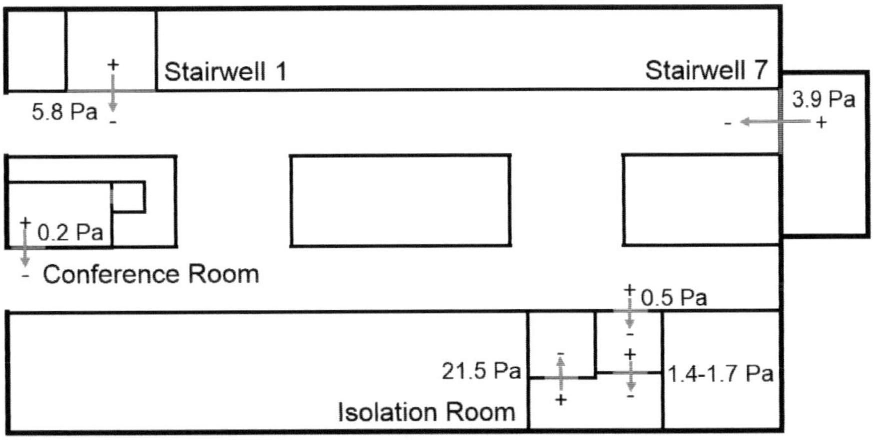

Figure 1. Pressure differences on the test ward.

The goal of the December visit to PAVA was to perform a test of the AHU's ability to create the pressures and flow rates required of an isolation ward. During this negative pressure test, a DG-700 monitored the pressure difference between Stairwell 7 (in Figure 1), and the isolation ward hallway. Initially, supply, return, and exhaust fan speeds were set to 20%, 100%, and 100% of their operating speeds, respectively, and fire doors were left open. As shown in Figure 2, upon closing the fire doors, the ward became highly pressurized, reaching an average pressure of -43.0 Pa. Due to wind noise and difficulty opening the fire doors, the supply fan flow rate was increased until noise diminished to unnoticeable levels. When the supply fan was set to 60% of its normal operating speed, the pressure dropped to -15.7 Pa and noise issues were resolved. With these AHU settings, an average of -25.1 Pa was measured when the DG-700 was moved to measure pressure differences across the fire doors.

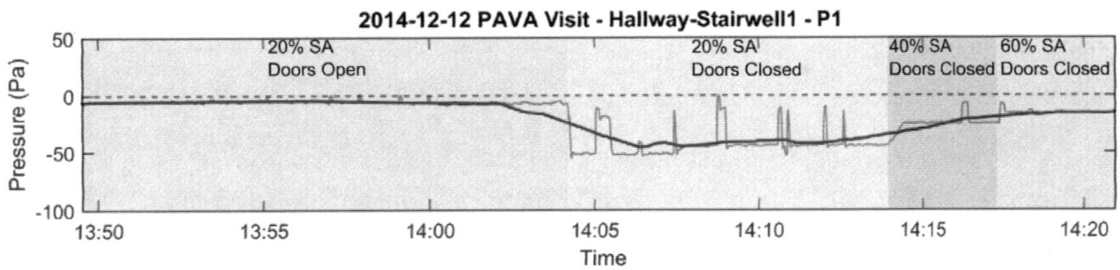

Figure 2. Isolation hallway–stairwell pressure difference during the negative pressure test at various supply air (SA) flow rates (red line – raw data, black line – moving average).

Throughout the planning stages of PAVA-SCEP, properly designing an anteroom to serve the designed isolation ward has been a challenge, especially in light of the difficulties of proper removal and disposal of personal protective equipment (PPE) during the ongoing West African Ebola epidemic (WHO, 2014). After multiple design phases and discussions with hospital staff, it was determined that building and testing a fully-functional isolation ward anteroom was out of the scope of the project, and instead mock-up anterooms consisting of hung plastic sheeting will be used to simulate an anteroom during the demonstration. Ideally, two anterooms for donning and doffing PPE would be established. Only a single anteroom will be used during the PAVA-SCEP demonstration, as nursing requested that we maintain an unaltered entrance for visitors and mobile patients.

CONCLUSIONS

Following two visits to the PAVA facilities in which pressure and flow rate data were collected, many details of the upcoming surge capacity demonstration were finalized. The AHU on the isolation ward is capable of generating -15 to -25 Pa of pressure relative to the stairwell and hallway adjacent to the isolation ward, respectively. It will be important to estimate air changes per hour for many rooms on the ward during the final demonstration, as this is an important aspect of isolation ward design. An additional planning visit in which details of the mock-up anterooms will be finalized is currently scheduled. It is anticipated that the final demonstration will be conducted during the spring of 2015.

ACKNOWLEDGEMENT

The authors wish to thank the following PAVA personnel for their help: Mark Keeth, Laura Markman, Scott Sutorius, Steve Elliott, and Joe Torrano.

REFERENCES

Adams, NJ, Johnson, DL, Lynch, RA (2011). The effect of pressure differential and care provider movement on airborne infectious isolation room containment effectiveness. *American Journal of Infection Control*, **39(2)**, 91-97.

Hick, JL, Christian, MD, Sprung, CL (2010). Chapter 2. Surge capacity and infrastructure considerations for mass critical care. *Intensive Care Medicine*, **36(Suppl 1)**, S11-S20.

Institute of Medicine (IOM, 2010). *Medical Surge Capacity: Workshop Summary*. Washington, DC: The National Academies Press.

Johnson, DL, Lynch, RA, Mead, KR (2009). Containment effectiveness of expedient patient isolation units. *American Journal of Infection Control*, **37(2)**, 94-100.

Lurie, N, Dausey, DJ, Knighton, T, et al. (2008). Community Planning for Pandemic Influenza: Lessons From the VA Health Care System. *Disaster Medicine and Public Health Preparedness*, **2(4)**, 251-257.

Rosenbaum, RA, Benyo, JS, O'Connor, RE, et al. (2004). Use of a Portable Force Air System to Convert Existing Hospital Space Into a Mass Casualty Isolation Area. *Annals of Emergency Medicine*, **44(6)**, 628-634.

Rubinson, L, Nuzzo, JB, Talmor, DS, et al. (2005). Augmentation of hospital critical care capacity after bioterrorist attacks or epidemics: Recommendations of the Working Group on Emergency Mass Critical Care. *Critical Care Medicine*, **33(10)**, 2393-2403.

Subhash, SS, Radonovich, LJ (2011). Hospital Surge Capacity. *ASHRAE Journal*, September 2011, 76-78.

Subhash, SS, Baracco, G, Fennelly, KP, et al. (2013). Isolation anterooms: Important components of airborne infection control. *American Journal of Infection Control*, **41(5)**, 452-455.

WHO (2014). *Interim Infection Prevention and Control Guidance for Care of Patients with Suspected or Confirmed Filovirus Haemorrhagic Fever in Health-Care Settings, with Focus on Ebola.* Ref: WHO/HIS/SDS/2014.4.

EXPLORING SPATIAL PLANNING AND FUNCTIONAL PROGRAM IMPACT ON MICROBIAL DIVERSITY AND DISTRIBUTION IN TWO SOUTH AFRICAN HOSPITAL MICROBIOMES

Jako A NICE[1], Piet VOSLOO[2]

[1]Built Environment, Architectural Engineering (AE), CSIR, Pretoria, South Africa
[2]Department of Architecture, University of Pretoria, Pretoria, South Africa

*Corresponding email: jnice@csir.co.za

Keywords: Architecture, microbial, Healthcare Associated Infection, planning, Hospital

SUMMARY

This paper presents a theoretical and experimental research approach on the impact of spatial planning and functional program on the microbial load, distribution and organism diversity in hospital environments. The investigation aims to identify design markers and define potential risk environments in design and planning of buildings to facilitate appropriate design and administrative interventions. The investigation studies two hospitals in the Western Cape (WC) South Africa (SA), born from the same design brief but with varied typologies and building systems. The study period considers two seasons, and will start in 2015 with four sampling days per season. A three tier experimental methodology is followed: 1) microbial sampling using air samplers, fluorescent particle counter (FPC), and settling plates. Analysis will be done by total count and molecular deoxyribonucleic acid (DNA) techniques, polymerase chain reaction (PCR) and pyrosequencing 2) observational analysis, using space syntactical methods; and 3) static environmental monitoring using data loggers and weather stations. Ethical approval is under way and the initial results are planned for publication in late 2015. The study anticipates conclusive baseline data towards developing a framework for an architectural design microbial risk model (ADMRM) for hospitals.

INTRODUCTION

The complexity in planning healthcare facilities coupled with the diversity of user clients, direct health application and varied disease profiles marks hospital environments as ideal baseline investigation environments for microbial data collection. In SA no microbiome data currently exists. This study will become the baseline dataset for future studies. The impact of *Tuberculosis* (TB) is both a global problem and an epidemic in SA. *Mycobacterium tuberculosis* (Mtb), human immunodeficiency virus (HIV) and healthcare associated infection (HAI) are of concern for developing countries and motivates this microbial investigation. The obligate aerobe nature of Mtb transmitted by airborne means places all with an immune deficiency disease i.e. HIV, unsuspected patients, healthcare workers and healthy people at risk of contagion, with the emphasis on immunosuppressed persons. Furthermore, studies seem to indicate that health care facilities are contributing to the spread of Mtb (Eshun-Wilson et al. 2008). In 2008 the World Health Organisation (WHO) ranked South Africa as the worst infected country in the world (per capita) for Mtb. A WHO assessment in 2008 indicated the epidemiological burden of TB and HIV co-infection in SA at an estimated 70%. WHO estimates that in SA, out of every 100 000 people approximately 768 are Mtb positive and 5.5 million people in SA have HIV Aids. These facts point to alarming risk exposure in hospital environments and the raised potential for HAI.

The spread of infectious bacteria, fungi, viruses and single cell organisms (prokaryotic & eukaryotic) specifically in hospitals are widely known to be first by human contamination (Hospodsky et al. 2012) and secondly dependent on environmental conditions (Basu. et al. 2007). This is exacerbated when microbial favourable environmental conditions are provided (Wolfaardt. et al. draft). HAI is prevalent worldwide in health care facilities but current research indicates that we might have overlooked a critical area in the response to non-tuberculosis bacteria (NTB), Mtb and other invasive pathogenic microbes. That is the microbial environment in which bacteria survive, live, are aerosolised and deposited. What benefit is there in studying the biome of an indoor environment? Does the built environment community (BEC) know the health impact of planning and construction decisions on building users? Both these questions are inherently the same. One needs to understand the ecology of the indoor environment to gauge the impact of design decisions. A microbial built environment study not only illuminates the need for microbial environment surveillance as part and parcel of the full building system planning, but also provides the data to create the tools required to facilitate BEC and building user clients in health design solutions.

Our research hypotheses postulate that the microbial load and diversity of the indoor biome bears a strong relationship to the functional use planning and functional program and that potential amplified microbial load and organism diversity can be correlated to occupancy, functional program and the environmental conditions as a direct result of the nominated building system. This study has fourfold objectives: 1) To analytically compare spatial design layouts and facility specific functional clinical program with the microbial load distribution and community diversity using two case study hospital sites; 2) To determine the variation in microbial load of two healthcare environments born of the same design brief but different building typologies and building systems: full mechanical ventilation and hybrid natural ventilation; 3) To identify risk patterns in design planning (identify and establish universal environmental markers) through microbial investigations; and 4) To establish an architectural design assessment protocol to assist in risk assessment of existing and new buildings.

METHODOLOGIES

The study design investigates two hospital sites in the WC, SA over two seasons (winter and summer) with a sampling period of four days per season. The hospitals were selected based on similar burden of disease and patient load; distinction in building systems and typological responses; but born from a common design brief. The area of investigation is the Accident & Emergency (A&E) department of each facility.

Microbial Investigation

A pre-experimental sampling run has been developed to gauge efficacy and quantity in sampling intervals. Five zones have been selected to study in each facility ranging from high occupancy spaces to low occupancy clinical spaces. 1) For surface sampling the study will utilise settling plates with universal agar base as guided by standard microbiological methods with an exposure duration estimated at 20 min. 2) For air sampling the study intends to use two standard microbiological techniques, including: A) an Andersen air sampler with a flat membrane filter of 5μm, positioned at a height of 1.1 m and 1.5 m away from solid stationary objects. The samples are to be collected at a flow rate of 28.3 L/min for 20 minutes as described by Chunyang et al. B) using a fluorescent particle counter (FPC) as per standard microbiological techniques (Chunyang et al. 2015). A sampling duration of four days with two samples in each zone daily over two seasons. The total planned settling plates and air samples amount to 320 excluding the FPC.

Analysis
Two molecular techniques will be applied using the MXIS BE standard data set. 1) PCR for specific taxa with total DNA gene pool of each sample searching for selected bacteria based on most common United States of America (USA) HAI pathogens, this is done in the absence of an SA standard (Ducel et al. 2002): *Staphylococcus aureus, Pseudomonas aeruginosa,* and airborne pathogens common to South African indoor risk environments: *Pneumocystis carinii pneumonia, Mycobacteria Tuberculosis.* 2) Pyrosequencing of each sample to determine the full community and quantity of organisms. The total planned analysis by PCR: 1280 samples, and by Pyrosequencing: 320samples.

Observational Investigation
The observation methodology used in this study is based on University College London (UCL) Architecture Department, Space Syntax software manuals with author variations where required. The Space Syntax observational manual were developed by Tad Grajewski in 1992 and updated by Laura Vaughan in 2001. A two-task process is to be followed by observers over the period of a day for the total study duration (four days, two seasons). A 12-hour period will be observed at each facility following a mapped route of the study sites.

- Task 1: "Mental snap shot" Requires a floor plan at 1:50 scale of the selected department. Following a mapped tour covering all rooms that are frequently used to ensure that all spaces (excluding storage areas) are observed (based on pre observation test and questionnaire). The observation assessment will be conducted hourly, by manually inscription on the 1:50 scale drawings in each space. Information collected include the number of people and user task using a defined coding system.
- Task 2: "Movement tracer" This technique will collect precise routes taken by people moving through the same spaces. This assessment enables one to determine space use and flow through the space. Following on task 1, the observer will spend 3-5 min in each of the rooms/spaces marking movements of people entering or exiting the space, then manually record and trace all the movement through the space, and notate exit routes. This exercise will be done twice hourly following the same route plan. The collected data will be analysed through Depth Map (DM) software (Al_Sayed et al. 2014). DM software provides graphical representation and analytical prediction of flow and space use.

Static Environmental Monitoring
The study will utilise data loggers and weather stations. 1) Data loggers are preconfigured manufactured units by CO2Meter, Inc. (Ormond Beach, FL, USA, CM0018AA). The unit includes built in sensors collecting CO_2, Temperature and relative humidity (RH) data as well as date and time stamp. A pre-sampling study will be undertaken to check feedback and calibrate the logger units. The study site matches the previous two investigations. The study duration will be continuous over the prescribed four days in the two seasons, utilising a total of five loggers matching the 5 identified zones. The loggers are placed either on a surface at counter height (900 mm) or suspended in the air (2100 mm). The data is recorded via secure digital (SD) card that will be disseminated on completion of the investigation using GasLab® software 2) A weather station will be mounted on each hospital building collecting ambient outdoor weather conditions. The unit will measure outdoor temperature, wind direction, wind speed, CO2 and RH.

RESULTS AND DISCUSSION

By correlating the observational syntactical findings with environmental and microbial data to determine potential microbial spread, which could be linked to risk. This could enable prediction of transmission through flow and user space use modelling. Variation of the floorplan layouts present the variation in the impact of architectural program and user program changes if planning is altered. The results will be used towards developing a theoretical framework of a quantitative model for risk design guidance. This research contributes to known evidence that the indoor biome of an environment does directly relate to its user; but in addition, that dispersal is a direct consequence of functional use and flow patterns in planning. Thereby confirming that spatial configuration does have an impact on disease spread, highlighting the impact of architectural planning on health risk.

CONCLUSIONS

When considering sustainable design, a microbial architectural built environment study illuminates the need for microbial environment surveillance as part and parcel of the full building system planning. Developing world countries and associated health burdens requires sustainable solutions that are designed into buildings to reduce cost, and promote sustainable infrastructure, while providing equitable safe low risk environments for all patients aimed at mitigating HAI. The potential development of intelligent design review tools in current building information modelling software platforms could inform designers on predicted high risk planning decisions. Furthermore such an application provides guidance as to alterations and recommendations to facility staff in considering risk when planning clinical programs within a given building envelope.

ACKNOWLEDGEMENT

This author acknowledges the CSIR and the Alfred P Sloan Foundation for student scholarship funding for financial support to attend the conference.

REFERENCES

Al_Sayed, K., Turner, A., Hillier, ,et al. (eds) 2014, *"Space Syntax Methodology"
*, 2nd Edition edn, Bartlett School of Graduate Studies, UCL, London, London.

Basu, S., Andrews., J.R., Poolman., E.M. ,et al. 2007, "Prevention of nosocomial transmission of extensively drug-resistant tuberculosis in rural South African district hospitals: an epidemiological modelling study", *Lancet,* vol. 370, pp. 1500-150001507.

Chunyang, D., Yan, Z., Xiaoling, M., et al. 2015, "Real-time measurements of airborne biologic particles using fluorescent particle counter to evaluate microbial contamination: Results of a comparative study in an operating theatre", *American Journal for Infection Control,* vol. 43, pp. 78-81.

Ducel, G., Fabry, J., Nicolle, L., et al. 2002, *Prevention of hospital-acquired infections, A practical guide. 2nd Edition.*, WHO, World Health Organization, Geneva, Switzerland.

Eshun-Wilson, I., Zeier, M.D., Barnes, J. ,et al. 2008, "TB infection among staff at Tygerberg Academic Hospital, South Africa", *South African Journal of Epidemiology Infection,* vol. 23, no. 4, pp. 07/03/2013.

Hospodsky, D., Qian, J., Nazaroff, W.W., et al. 2012, "Human Occupancy as source of Indoor Airborne Bacteria", *Plos One,* [Online], vol. 7, no. 4, pp. 11/05/2013-1-10.

Wolfaardt., G., Ghadakpour, M., Bester, E., et al. draft, "Integration and proliferation of Pseudomonas aeruginosa PAO1 in multispecies biofilms", *Microbial Ecology,* vol. tbc

Model and measurement derived estimates of residential $PM_{2.5}$ infiltration in 10 US cities

Jennifer M. Logue[1,*] Neil E. Klepeis[2,3], and Brett C. Singer[1]

[1]Environmental Technologies Area, Lawrence Berkeley National Lab, Berkeley, CA
[2]Department of Civil and Environmental Engineering, Stanford University, Stanford, California, USA;
[3]Center for Behavioral Epidemiology and Community Health (C-BEACH), Graduate School of Public Health, San Diego State University Research Foundation, San Diego State

[*]Corresponding email: jmlogue@lbl.gov

Keywords: $PM_{2.5}$, infiltration, housing stock

SUMMARY

Infiltration of outdoor $PM_{2.5}$ in the housing stock has a significant impact on exposure. Many attempts have been made to estimate $PM_{2.5}$ infiltration, however there are biases to both model and measurement based estimates. In this paper we compare and contrast model and measurement derived estimates of infiltration factor (the ratio of indoor $PM_{2.5}$ of outdoor origin to outdoor $PM_{2.5}$ concentration) in 10 US cities.

INTRODUCTION

Elevated concentrations of $PM_{2.5}$ in the atmosphere have been associated with increased morbidity and mortality in the population (Pope et al. 2009). While most epidemiological studies have examined relationships between health outcomes and measured ambient $PM_{2.5}$ concentrations, exposure to outdoor $PM_{2.5}$ predominantly occurs indoors.

Exposure and health scientists have measured indoor and outdoor $PM_{2.5}$ and its sub-components to assess the infiltration of outdoor $PM_{2.5}$ into homes. Measuring infiltration in homes is challenging because infiltration varies over time for a given home based on weather and occupant behaviour (Kearney et al. 2014). Studies that measured infiltration have predominately done so over the course of one to a few days and reported a time averaged value. A measurement event can be thought of as a snapshot of what is occurring in the building over time and the result from this snapshot may or may not represent mean conditions over time. In addition to variability in a single home over time, there is large variability between homes. The sample of homes selected for measurements may or may not represent the population of homes in the specified geographical area.

Simulation offers an alternative to measurements as a method to estimate $PM_{2.5}$ infiltration distributions for the housing stock of a given region. The Population Impact Assessment Modeling Framework (PIAMF) was developed to quantify the energy and health impacts of possible or actual changes to the U.S. housing stock(Logue et al. 2014). The PIAMF applies physics-based simulation models to calculate one or more environmental or energy performance parameters for each home in a sample developed to represent one or more segments of the population. Results from the individual homes are compiled to provide the statistics for population impacts. The PIAMF has been modified to model indoor concentrations of $PM_{2.5}$ components due to indoor and outdoor sources (Logue et al. 2014).

Discrepancies between actual population infiltration and modeled values can result from differences between modeled values and actual conditions related to (1) magnitude and temporal patterns of window opening, (2) envelope air tightness levels (which has the largest effect when there is little or no window use), and (3) temperature and wind driving forces.

This paper presents PIAMF simulation results estimating $PM_{2.5}$ infiltration distributions and compares these results to measurements reported in the literature for 10 US cities.

METHODOLOGIES

The infiltration factor (F_{inf}) is a common metric used to represent the indoor concentration of outdoor $PM_{2.5}$. As shown in Equation 1, F_{inf} is the ratio of the time-averaged indoor concentration of outdoor particles to the time-averaged outdoor concentration.

$$F_{inf} = \overline{\left(\frac{C_{in}}{C_{out}}\right)} = \frac{1}{T}\int_0^T \frac{C_{in,t}}{C_{out,t}} \qquad (1)$$

We aggregated data from 4 studies that estimated F_{inf} based on measurements in 10 urban locations (Table 1). $PM_{2.5}$ F_{inf} cannot be calculated directly because homes were occupied and likely had indoor sources of PM2.5. All studies except Allen et al. (2003) measured indoor/outdoor concentrations of the sulfur content of PM2.5, a species thought to have only outdoor sources (Sarnat et al. 2002). The ratio of indoor/outdoor sulfur content is then reported as the $PM_{2.5}$ F_{inf}. A large study in Edmonton, Canada found that F_{inf} estimates derived from ratios of sulfur measurements overestimated F_{inf} derived from high-time resolution indoor and outdoor PM_1 measurements (Kearney et al. 2014). Allen et al. (2003) used high-time resolved measurements to estimate the fraction of indoor $PM_{2.5}$ due to outdoor sources.

For each city included in Table 1, we ran the PIAMF for a sample of model homes representing the housing stock located in the same state and climate zone as the city with measurements. For all cities except Seattle, we compare the sulfur derived measured F_{inf} values with the week-average modeled ammonium sulfate F_{inf}. For Seattle we compare the measurement derived $PM_{2.5}$ F_{inf} with the week-average modeled $PM_{2.5}$ F_{inf}.

The PIAMF approximates the expected F_{inf} across a set of homes based on population probabilities of window opening, expected distributions of home properties based on easily determined characteristics (such as age and foundation type), and average seasonal weather and pollutant conditions. The PIAMF is intended to produce the season average distribution across the population of homes in a given geographical area. We compare the modeled, seasonal-average infiltration distributions with those reported from measurement-based studies from the same regions.

Table 1 Studies with reported infiltration factors (F_{inf}) based on a component of $PM_{2.5}$.

Study Name	Reference	Location (number of homes measurement periods[1])
MESA Air (MA)	**(Allen et al. 2012)**	Baltimore, MD (39/48), Chicago, IL (28/40), Los Angeles, CA (53/80), New York, NY (26/24), Rockland County (11/12), St. Paul (23/56), Winston-Salem (39/47)
Allen	**(Allen et al. 2003)**	Seattle, WA (26/29)
Habre	**(Habre et al. 2014)**	New York, NY (61/53)
RIOPA	**(Weisel et al. 2005)**	Los Angeles, CA (4/6/15/14), Elizabeth, NJ (12/7/8/14), Houston, TX (15/14/8/11)

[1] For Allen et al. (2012) the number of homes listed is for warm (>18°C) / cold (≤18°C) weather sampling from Figure 1, for Allen et al. (2003) the number of homes listed is for non-heating (March-September) season and heating (October-February) season. For RIOPA the number of homes listed is for winter/spring/summer/fall.

RESULTS AND DISCUSSION

Figure 1 compares annual F_{inf} distributions. Modeled distributions have equal contributions from all four seasons. Measured distributions are not equally dispersed over the year. The mean IQR for all locations are similar for both modeled (IQR= 0.21) and measured distributions (IQR=0.24). For the 12 locations, mean and median measured F_{inf} values are, on average, 23% and 18% higher than modeled respectively. Measured mean and median values of location specific F_{inf} distributions were within 10% of modeled for 5 and 4 locations and within 25% for 6 and 8 locations. Measured distributions have larger spreads and higher mean/median values than modeled.

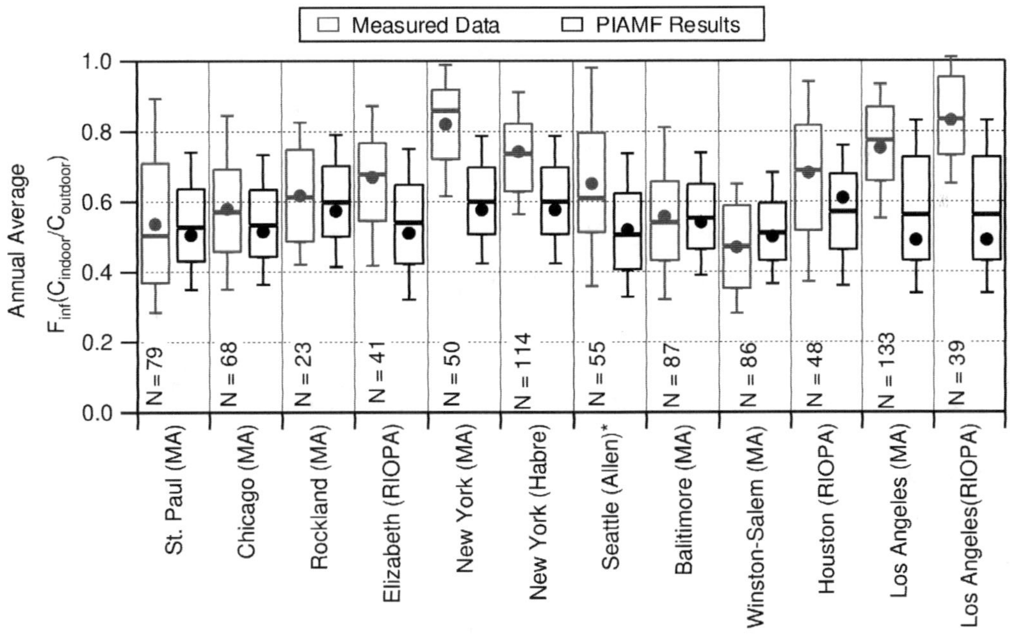

Figure 1. Comparison of measured infiltration factor (ratio of indoor/outdoor sulfur) to modeled ratios of indoor/outdoor ammonium sulfate. Boxes span from 25th to 75th percentiles; the line inside the box is the median and the circle is the mean. Whiskers extend to the 10th and 90th percentiles. Asterisks indicate the mean infiltration factor for each size bin and species.

CONCLUSIONS

Measured distributions have larger spreads and higher mean/median values than modeled. There are several potential causes of differences between the measured and modeled distributions. Homes that volunteer for monitoring may not represent the populations in the area, either in terms of average conditions or of the variability. Half of the homes in the RIOPA dataset reported being low income; these homes tend to have more air leakage and less access to air conditioning than the general population. The MESA air study intentionally selected homes to increase variability in their sample; that may play a factor in those homes having more variability than the representative housing sample. The modeling assumptions made about home occupant behavior (window opening, thermostat settings) also may not reflect the behavior of occupants in the measured homes.

ACKNOWLEDGEMENT

Funding was provided by the U.S. Dept. of Energy Building Technologies Program, Office of Energy Efficiency and Renewable Energy under DOE Contract No. DE-AC02-05CH11231; by the U.S. Dept. of Housing and Urban Development Office of Healthy Homes and Lead Hazard Control through Interagency Agreement I-PHI-01070, and by the California Energy Commission through Contract 500-08-061.

REFERENCES

Allen R, Larson T, Sheppard L, Wallace L and Liu LJS.2003. "Use of Real-Time Light Scattering Data To Estimate the Contribution of Infiltrated and Indoor-Generated Particles to Indoor Air." Environmental Science & Technology **37**(16): 3484-3492.

Allen RW et al.2012. "Modeling the Residential Infiltration of Outdoor PM2.5 in the Multi-Ethnic Study of Atherosclerosis and Air Pollution (MESA Air)." Environmental Health Perspectives **120**(6): 824-830.

Habre R et al.2014. "Sources of indoor air pollution in New York City residences of asthmatic children." Journal of Exposure Science and Environmental Epidemiology **24**(3): 269-278.

Kearney J et al.2014. "Residential infiltration of fine and ultrafine particles in Edmonton." Atmospheric Environment **94**: 793-805.

Logue JM, Lunden MM and Singer BC.2014. Development and application of a physics-based simulation model to investigate residential PM2.5 composition and size distribution across the US. The 13th International Conference on Indoor Air Quality and Climate. Li, Y. Hong Kong, China

Pope CA, III., Ezzati M and Dockery DW.2009. "Fine-Particulate Air Pollution and Life Expectancy in the United States." The New England Journal of Medicine **360**(4): 376-386.

Sarnat JA et al.2002. "Using sulfur as a tracer of outdoor fine particulate matter." Environmental Science & Technology **36**(24): 5305-5314.

Weisel CP et al.2005. Relationships of Indoor, Outdoor, and Personal Air (RIOPA) Part I. Collection Methods, Health Effects Institute, Mickely Leland National Urban Air Toxics Research Center and Descriptive Analysis:HEI Research Report 130 http://pubs.healtheffects.org/getfile.php?u=25

URBAN MICROBIOME PILOT STUDY: PARKS AND PARKING LOTS

Gwynne Á. MHUIREACH[1,2], Josh LADAU[3], James F. MEADOW[2], G.Z. BROWN[4], Bart R. JOHNSON[1], Jessica L. GREEN[2,5]

[1]Department of Landscape Architecture, University of Oregon, Eugene, Oregon
[2]Biology and the Built Environment Center, Institute of Ecology and Evolution, University of Oregon, Eugene, Oregon
[3]Gladstone Institutes, University of California, San Francisco, California
[4]Energy Studies in Buildings Laboratory, Department of Architecture, University of Oregon, Eugene, Oregon
[5]Santa Fe Institute, 1399 Hyde Park Rd, Santa Fe, New Mexico

*Corresponding email: gwynhwyf@uoregon.edu

Keywords: Urban, vegetation, microorganisms, agriculture, land use

SUMMARY

Airborne microorganisms comprise a significant fraction of particulate matter (PM) in urban areas, yet the relative contributions of local, regional, and distant particle sources are poorly understood. The goal of this pilot study was to explore whether and how nearby vegetation influences airborne microbial communities at a particular site. We collected microbial samples from five pairs of parks and parking lots in Eugene, Oregon, using gravitational settling dishes and high-throughput sequencing of the bacterial 16S RNA gene. We found that samples from all ten sites were strikingly similar, suggesting that a homogenized background flow of microbes driven by broad-scale mixing of air currents largely overwhelms subtle local effects. A single taxon, *Sphingomonas* spp., was hyperabundant in all samples, comprising an average of 34% of sequences in each sample. Further work is needed to determine the upwind source of this taxon, but regional agricultural activities are suspected as a likely source.

INTRODUCTION

Humankind evolved under constant exposure to microbes originating from plants, soil, water, and other sources. In recent centuries, we have implemented strategies of industrialized agriculture, impermeable surface construction, and urban inhabitation that have likely altered the composition of microbial communities with which we interact on a daily basis. Although people in developed countries spend the majority of their time inside buildings and cars (Klepeis et al., 2001), the ultimate source of indoor air is the immediate outdoor air. Thus the composition of airborne microbial communities found outdoors and the factors that control their assembly are of concern for both indoor and outdoor air quality.

Outdoor airborne microbial communities have been shown to be highly variable across space and time (Fierer et al., 2007; Brodie et al., 2007). Bowers et al. (2010) found that microbial communities vary by major land uses (forest, agricultural, suburban), and suggest that vegetation and soil are primary sources of microorganisms in the air. However, the rate at which plants emit microbes to the air is not well-understood. Early culture-based investigations concluded that the composition of airborne microbial communities was influenced by nearby vegetation (Kinkel 1997), but a more recent study by Vokou et al.

(2012) found little resemblance between airborne bacterial communities and those on leaf surfaces of several plant species in the Mediterranean.

In additional to natural processes (e.g. wind) that cause microbes to become airborne, human activities such as agricultural crop harvesting can release large loads of dust and microbes to the atmosphere. For example, Lighthart (1984) calculated from microbial plate counts that grass seed harvesting equipment generates a downwind plume of bacteria and fungi that may comprise 41.9% and 35.1%, respectively, of microbial communities in the airshed of the Willamette Valley, Oregon. Human activities may thus contribute to observed seasonal patterning in airborne microbial communities (Bowers et al., 2013). This study was intended, in part, to answer the question, "How much influence do nearby sources, such as vegetation, have on airborne microbial community structure in comparison with the background flow of particles from regional or more distant sources?"

METHODOLOGIES

Air samples were collected from 5 pairs of parks and parking lots in Eugene, Oregon (Figure 1) on July 24th, 2013 during 0800 - 1600. Each parking lot was approximately 400 meters from its nearest park, and each park/parking lot pair was about 2500 meters from its nearest neighboring pair. Samples were collected simultaneously at each site approximately two meters above ground level in a relatively open area (i.e. not directly underneath tree canopy or other obstruction) using a custom tray (sterilized prior to use with 99% isopropyl alcohol) containing 3 passive settling dishes with their lids. Technicians monitored the sampling equipment during collection and performed hourly measurements of wind speed and direction. All samples were frozen at -80°C immediately following sampling and stored frozen until processing. The petri dishes and their lids were swabbed with nylon flocked swabs (copanusa.com; #552C) and whole genomic DNA was extracted directly from the swabs using the MO BIO PowerWater DNA Isolation Kit (MO BIO Laboratories, Carlsbad, CA, USA) according to manufacturer's instructions. The protocols used for DNA extraction, PCR amplification and Illumina library preparation were identical to those described in Meadow et al. (2014). Negative controls were used in all laboratory procedures and used to identify potential contaminants. Samples were sequenced on the Illumina MiSeq platform as paired-end reads at the Dana-Farber/Harvard Cancer Center DNA Resource Core (Boston, MA, USA; dnaseq.med.harvard.edu).

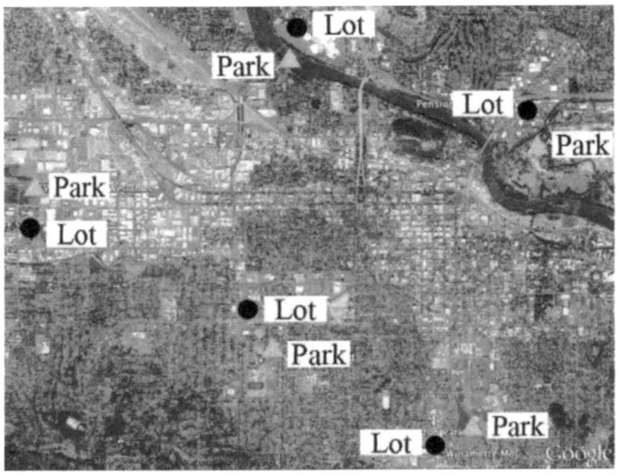

Figure 1. Park and parking lot sampling sites were paired.

We used the FastX Toolkit (http://hannonlab.cshl.edu/fastx_toolkit) and the QIIME pipeline (Caporaso et al., 2010) to process the raw sequences. Sequence read lengths averaged 250 bp, and taxonomy assignment was performed using a 97% similarity clustering. After quality filtering, all plant and mitochondrial sequences were removed, as were single- and doubletons and the top three most abundant potential laboratory contaminants, leaving 36,016 operational taxonomic units (OTUs) for downstream statistical analyses. Statistical analyses were implemented in R (R Development Core Team, 2010). We used the variance-stabilizing transformation function in the DESeq2 package (Love et al., 2014) to account for differences in sample library sizes.

RESULTS AND DISCUSSION

We found that samples from all ten sites were strikingly similar, sharing approximately 74% of taxa (Figure 2a) and 22 out of the top 25 most abundant taxa. In terms of diversity, there was no significant difference between parks and parking lots (lot mean = 5.42, park mean = 5.50, $t = -1.24$, $df = 52.8$, p-value = 0.22). Barplots comparing the relative abundance of taxonomic groups for parks versus parking lots also showed a congruent pattern (Figure 2b). Although all samples were roughly similar in composition, some patterns were observed. Of the 20 OTUs that were significantly more abundant in parks 11 were from the Acidobacteriaceae family, and of the 22 OTUs that were significantly more abundant in parking lots, 10 were from the Acetobacteraceae family. A single taxon, *Sphingomonas* spp., was hyperabundant in all samples, comprising on average about 34% of sequences in each sample. This genus is common on plant leaves, and a BLAST search identified the OTU as either *S. faeni* or *S. aurantiaca*, both previously observed from hay dust. Other genera that were highly represented in both parks and parking lots included *Hymenobacter*, *Pedobacter*, *Agrobacterium* and *Rhodococcus*, all of which are often found in soil.

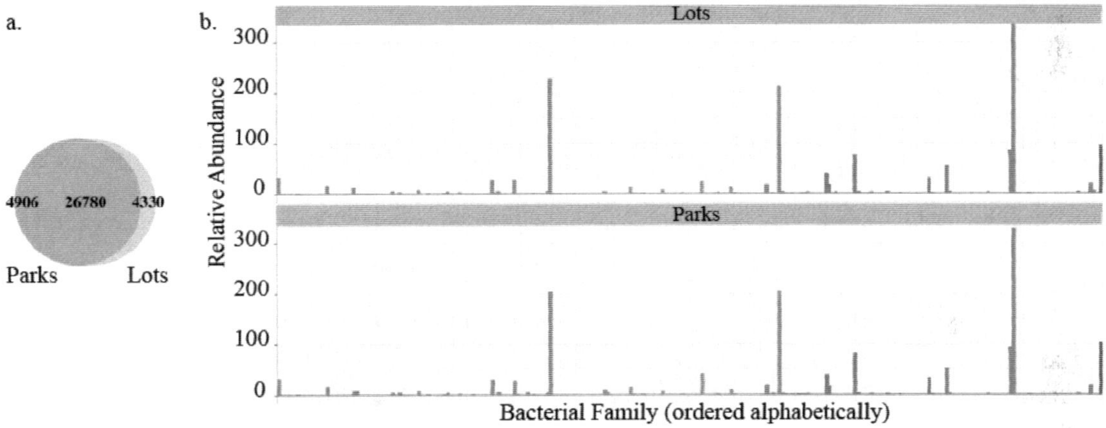

Figure 2. Airborne bacterial communities were remarkably similar whether sampling in parks or parking lots, and this was consistent for occurrence (a) as well as abundance (b).

The results reported here suggest that the well-mixed background flow of airborne microbes from regional and distant sources largely overwhelms subtle local effects, at least for sampling sites in the open air (e.g. not sheltered by tree canopies or other structures). Additional sampling over space and time should be conducted to confirm that the results of this study were not significantly affected by unusual conditions at these particular sites on the day of sampling. Further work is needed to determine the upwind source of the hyperabundant taxon, *Sphingomonas*, but regional agricultural activities are a likely source, since A) sampling took place on July 24th, which is prime harvesting season; B) Linn County, immediately north of the study area, is known as "the grass seed capital of the world"; and C) for the majority of the sampling period, as well as the 72 hours preceding (Figure 3) near-surface wind was blowing from the north. Grass seed harvesting might be expected to generate large amounts of airborne dust carrying soil-associated microbes, which could additionally explain the profusion of *Hymenobacter*, *Pedobacter*, *Agrobacterium* and *Rhodococcus* across all of the samples.

In the Pacific Northwest and other regions with large-scale agricultural production, tilling, livestock operations and crop harvesting can release copious amounts of fine dust and organic particulates, including microorganisms, to the atmosphere, which may then impact downwind air quality. Future research should address the degree to which this regional atmospheric loading is related to health problems in downwind urban areas.

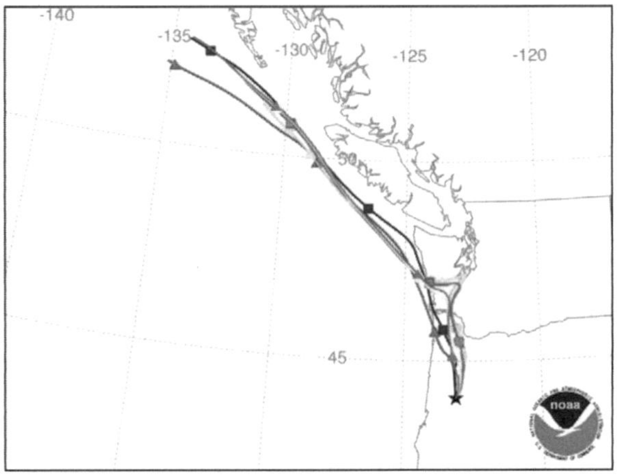

Figure 3. NOAA HYSPLIT backwards air trajectory plot at 300 meters above ground level for the 72 hours preceding sampling.

ACKNOWLEDGMENTS

This work was supported by a grant from the Alfred P. Sloan Foundation to the Biology and the Built Environment Center, and by the Department of Landscape Architecture at the University of Oregon. We would like to thank Adam Altrichter, Ashley Bateman, Ann Womack and Brendan Bohannan for their support and assistance.

REFERENCES

Bowers, R.M. et al., 2013. Seasonal Variability in Bacterial and Fungal Diversity of the Near-Surface Atmosphere. Environmental Science & Technology, 47(21), pp.12097–12106.

Bowers, R.M. et al., 2010. Spatial variability in airborne bacterial communities across land-use types and their relationship to the bacterial communities of potential source environments. The ISME Journal, 5(4), pp.601–612.

Brodie, E.L. et al., 2007. Urban aerosols harbor diverse and dynamic bacterial populations. Proceedings of the National Academy of Sciences, 104(1), p.299.

Caporaso, J.G. et al., 2010. QIIME allows analysis of high-throughput community sequencing data. Nature Methods, 7(5), pp.335–336.

Fierer, N. et al., 2007. Short-Term Temporal Variability in Airborne Bacterial and Fungal Populations. Applied and Environmental Microbiology, 74(1), pp.200–207.

Kinkel, L.L., 1997. Microbial Population Dynamics on Leaves. Annual Review of Phytopathology, 35(1), pp.327–347.

Klepeis, N.E. et al., 2001. The National Human Activity Pattern Survey (NHAPS): a resource for assessing exposure to environmental pollutants. Journal of exposure analysis and environmental epidemiology, 11(3), pp.231–252.

Lighthart, B., 1984. Microbial Aerosols: Estimated Contribution of Combine Harvesting to an Airshed. Applied and Environmental Microbiology, 47(2), pp.430–432.

Love, M., Anders, S. & Huber, W., 2013. Differential analysis of count data-the DESeq2 package.

Meadow, J.F. et al., 2014. Bacterial communities on classroom surfaces vary with human contact. Microbiome, 2(1), p.7.

R Development Core Team, 2010. R: A Language and Environment for Statistical Computing. Available at: http://cran.r-project.org.

Vokou, D. et al., 2012. Exploring Biodiversity in the Bacterial Community of the Mediterranean Phyllosphere and its Relationship with Airborne Bacteria. Microbial Ecology, 64(3), pp.714–724.

ENVIRONMENT PARAMETERS RELEVANT TO MICROBES SPANNING HUMAN URBANIZATION IN SUB-TROPICAL CLIMATE

Humberto CAVALLIN [1], Jean F. RUIZ[2], Atila NOVOSELAC[3,*], Henrique PEREIRA[4], Luciana C. PAULINO[5], Raphael RIOS [6], Jean H. PEREZ. [7], Martin J. BLASER[8], Maria Gloria DOMINGUEZ-BELLO[2,8]

[1]School of Architecture, University of Puerto Rico, USA
[2]Department of Biology, University of Puerto Rico, USA
[3]Depart. of Civil, Architectural and Environmental Eng., University of Texas at Austin, USA
[4]Depart. of Fundamental Sciences and Agriculture, Federal University of Amazonas, Brazil
[5] Center of Natural and Human Sciences, Federal University of ABC, Brazil
[6]College of Natural Sciences, University of Puerto Rico, USA
[7]National University of Amazonia Peruana, Peru
[8]School of Medicine, New York University, USA

[*]Corresponding email: atila@mail.utexas.edu

Keywords: Indoor environment, Microbial communities, Urbanization, Field study

SUMMARY

When compared urban/modern and traditional/rural dwellings urban homes have tighter building envelopes and higher levels of heating, ventilation and air-conditioning. The consequences of these alterations in building environmental conditions are not fully understood, particularly to which extent the environmental and occupants' interactions affect building environment and microbial communities. The goal of this study is to examine for associations between modernization in households and the impact it has on the microbial communities of houses and their occupants. The study is based on a large filed campaign where environmental data and biological samples are collected for four levels of urbanization, ranging from dwellings in a village in the jungle to a major metropolitan area. Results show alterations in building design, surface materials, house appliances, use of space, and level of environmental control were considered when analyzing household modernization. Also, the study discusses building parameters relevant to changes in microbial communities.

INTRODUCTION

Urban development and modern living standards are followed by changes in habitations and buildings. Most often urbanization and modernization of buildings means tighter building envelopes and higher levels of heating, ventilation, and air-conditioning which together result in greater separation between indoor and outdoor environments. The ways in which the occupants use and maintain these buildings also alters with modernization. Generally, occupant density per home decreases, while the use of personal and house cleaning products significantly increases. This means that the urban household environments are converging towards habitats that are isolated from outdoor environments and in which human microorganisms may dominate. This alteration in building environmental conditions and occupancy is not fully understood, especially the extent to which (1) the environment shapes the building's and occupants' microbiome, and (2) the occupants' behavior affects the microbial transmission between humans and indoor surfaces.

During the last decade, advancement in bio sampling enabled studies on building microorganisms (Martini et al., 2006; Su et al., 2012). Since people spend most of their time in indoor environment many reserchers started investigating microbial diversity within the built environment (Humphries, 2012; Kelley and Gilbert, 2013; Konya and Scott, 2014). This includes various buildings including office and educational buildings, healthcare facilities, and homes (Stephens, 2014). The number of studies related to different type of buildings and environments is rapidly increasing in the last few years, however very few studies tied to identify how these microbial communities in indoor environment evolved with the change of environmental conditions caused by modernization of homes and more urban living stile of building occupants. The aim of this study is to examine for associations between modernization in households and the impact it has on the microbial communities of houses and their occupants. Alterations in building design, surface materials, house appliances, use of space, and level of environmental control were considered when analyzing household modernization and their impact on microbial communities.

METHODOLOGY

Environmental data suitable for adequate interpretation of the microbiological data of buildings and their occupants were collected for levels of urbanization, ranging from dwellings in a village in the jungle to a major metropolitan area. Specifically, our study was conducted in the Amazon Basin area of Peru and Brazil, at a similar latitude, in a gradient from a remote jungle village with indigenous populations (Checherta) to rural settings (Puerto Almendras), mid-size cities (Iquitos) to a modern metropolis (Manaus). Parameters that characterize houses and indoor and outdoor environments were collected in 10 houses from all four communities, including indoor and outdoor temperatures and relative humidity, light intensity, particle concentrations, and several parameters that characterize buildings such as air exchange rate, surface temperatures and moisture content in wood and stucco walls. Parameters that more precisely characterize the occupants, including personal temperature and humidity were used to evaluate the conditions of human skin over the daily activity cycle.

Figure 1. Examples of typical homes at the four sampled locations that illustrate different level of urbanization; from left to right: Checherta, Puerto Almendras, Iquitos, and Manaus

Samples of microbes from building occupants, home pets, building materials and home objects were collected to assess the influence that urbanization level and corresponding environmental conditions have on microbial communities. Bacterial communities are being assessed by internal transcribed spacer (ITS) amplifying and sequencing *16S rRNA* gene; then, site-specific communities were compared using QIIME (Quantitative Insights Into Microbial Ecology tool). The active sampling process was one moths long (December 2013), and this paper presents some of the process data. Figure 2 illustrated the data collection and analysis process.

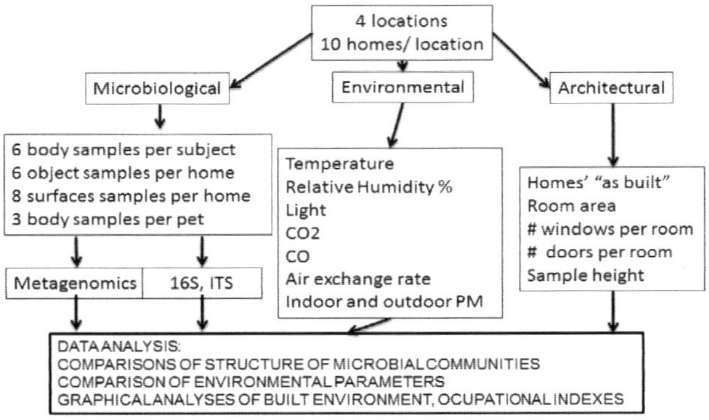

Figure 2. Data collection and analysis process

RESULTS AND DISCUSSION

Results show that changes in the use of home space and architecture lead to altered environmental parameters, ventilation rate, and building materials. One of the most significant change is in the ventilation flow rate. The air exchange rate (ACH) decreases as the level of urbanization increases. However, even with the order of magnitude decrease in ACH, when compared rural and urban homes, all studied houses had high ventilation rate (ACH > 3). This large ventilation rate exceeded 10 ACH in most homes, resulting that environmental parameters such T and RH was relatively similar in most studded indoor environments, because the outdoor climate conditions were very similar, too. Also, levels of CO_2 and CO were very low due to the large ventilation rate. This large ventilation also caused that measured indoor and outdoor PM concentrations were relatively similar.

The architectural properties, building design solutions and occupancy level provide information on privacy and personal space in studies homes (Figure 3). Data show that the mean values for the floor area consistently increase with the urbanization. As a natural consequence of the increment of surface area by household, the density of the households decreases; furthermore, number of occupant per household decreases too, causing more dramatic decrease in occupancy density. Also, architectural solutions such as increase of number of rooms combined with decrease in occupancy results in smaller fraction of space sharing.

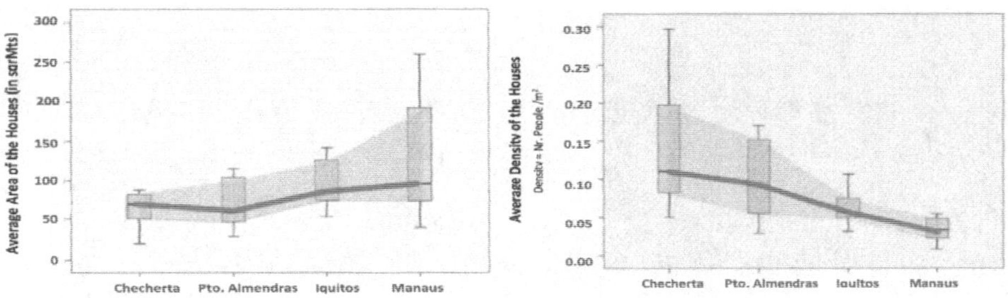

Figure 3. Correlation between density and privacy

Initial analyses of the human microbiome show differences across the gradient of transculturation. Figure 4 show initial results in attempt to examine biodiversity in home microbes, while considering environmental parameters, architectural features and level of occupancy. Results show that floor samples have the greatest bacterial diversity, which is significantly higher than walls for all level of urbanization. This indicates occupants' activity has a significant role in generating biodiversity. Also, when compared the cups sampled in Manaus to cups in other homes the biodiversity is lower which may be attributed to the higher privacy and/or cleaning frequency.

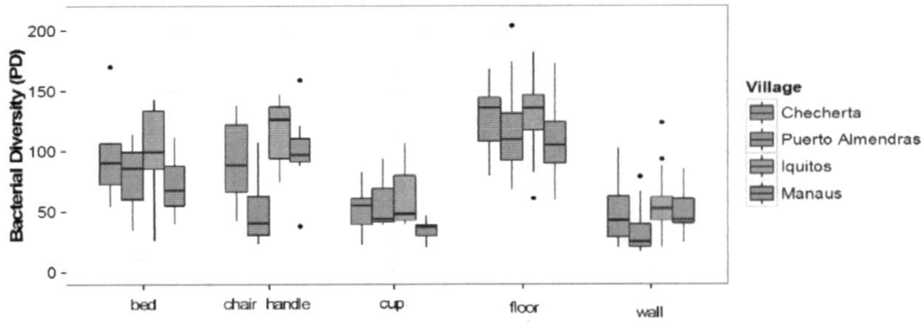

Figure 4. Phylogenetic microbial diversity by sampled surface and home location

One of the challenging result in Figure 4 is that surfaces in homes in Puerto Almendras have smallest biodiversity for almost all sampled surfaces. This cannot be explained nether with measured environmental nor architectural parameters. This indicates that biodiversity may be driven with other factors, such as local diet and/or other factors. Also, it may be the case that analyses that go beyond bio diversity are needed to analyze how microbial communities affect buildings and occupants.

CONCLUSIONS

This study show that urbanization results in more private space in homes and higher isolation from outdoor environment. However, climate condition in Amazonian basin and exclusive use of natural ventilation in homes for all level of urbanization did not caused a dramatic differences in thermal environmental parameters they could in more air tight homes in colder climates. The initial results provide insights on how modernization in households affect colonized surfaces. For example, the biodiversity is lower in urban homes, which may be attributed to the larger privacy and cleaning frequency. Further work on microbial quantification which includes metrics relevant to health of occupants should provide further details on how indoor microbial communities affect occupants.

ACKNOWLEDGEMENT

The study presented in this paper is sponsored by Alfred P. Sloan Foundation.

REFERENCES

Humphries, C., (2012). Indoor Ecosystems. *Science* 335, 648–650.

Kelley, S.T., Gilbert, J.A., (2013). Studying the microbiology of the indoor environment. *Genome Biol.* 14, 202.

Klepeis NE, Nelson WC, Ott WR, Robinson JP, Tsang a M, Switzer P, et al. The National Human Activity Pattern Survey (NHAPS): a resource for assessing exposure to environmental pollutants. *J Expo Anal Environ Epidemiol* 2001;11:231–52.

Konya, T., Scott, J.A., (2014). Recent Advances in the Microbiology of the Built Environment. *Curr. Sustain. Energy Rep.* 1, 35–42.

Martiny JBH, Bohannan BJM, Brown JH, et al.(2006). Microbial biogeography: putting microorganisms on the map. *Nat Rev Microbiol*;4:102–12.

Stephens, B. (2014). Building science to advance research in the microbiology of the built environment (MoBE), Workshop Report and Meeting Transcript. Prepared for: The Alfred P. Sloan Foundation. 630 Fifth Avenue. Suite 2550. New York, NY 10111.

Su, C., Lei, L., Duan, Y., et al. (2012). Culture-independent methods for studying environmental microorganisms: methods, application, and perspective. *Appl. Microbiol. Biotechnol.* 93, 993–1003.

Controlling Pollutants During New Construction To Reduce Exposure Upon Occupancy By Sensitive Individuals

Carl Grimes[1]

[1]Hayward Healthy Home Institute, Monterey, California, 93940 USA

*Corresponding email: grimes@habitats.com

Keywords: Construction, chemical, particulate, sensitive

INTRODUCTION

If reduction of exposure to pollutants – by removal – from existing construction is more effective than isolation, reduction, or filtration, then it is logically true that avoidance by removal during construction will be more effective than isolation, reduction, or filtration after construction; and certainly more so than during occupancy. A custom home in Carmel, CA, USA, was built according to this principle in 2003 and 2004 using only non-toxic or the least-toxic building materials available. Control of "dust" during construction was attempted by specifying work practices. Two similar methods of measurements were conducted prior to occupancy to determine the relative success of the preventive actions. The objective was to observe whether these practices could actually be implemented on the job-site, and if so, if results were measurable.

METHODOLOGY

Multiple techniques were used to construct the house to be as "healthy" as possible upon occupancy. This was in consideration of all seven members of the family who had become variously reactive to a previous house that had a variety of construction defects, as professionally identified, resulting in water damage and chronic dampness.

The house was built in modules in a covered warehouse to prevent exposure to the weather, especially to moisture. The modules were then transported by truck to the site where the modules were lifted by crane onto the foundation and assembled. Plywood was used instead of oriented strand board (OSB). All materials were screened for "toxic" ingredients by evaluating the manufacturer's product label, technical specifications, and Material Safety Data Sheet (MSDS). On-site work practices were specified to ensure that only the approved materials were installed, and to prevent generation of dust inside the building envelope.

After all work was completed the house was thoroughly cleaned with HEPA filtered air "scrubbers" and HEPA filtered vacuums, air flushed for 24 hours, and then closed and rested for 48 hours. An experienced professional then collected air samples in three locations for a panel of Volatile Organic Compounds (VOCs), U.S. Environmental Protection Agency (EPA) panel of Hazardous Air Pollutants (HAP), particulate mass, temperature, relative humidity, carbon monoxide, carbon dioxide, nitrous oxide, and ozone. Samples were sent to the appropriate laboratories for analysis by standard industry methodologies. Simultaneously, and for another 18 hours, data was logged for the interior of the building envelope.[1]

Occupant questionnaires including the SF-12® Health Survey for functionality[2], Grimes Personal Impact Rating[3], and the Miller QEESI[4] will be used in the 2nd phase of follow-up in early February, 2015.

LIMITATIONS

There were a number of challenges during construction and prior to testing that limited the ideal success of the project: Some are now known to be avoidable with hind-sight, such as training of architects and contractors plus providing materials information in advance; some were failures of compliance with work practices, such as contractors turning off "air scrubbers" so the filters didn't clog or because the equipment was too noisy, power sawing wood inside instead of outside to save time; others limitations included product information that could not be obtained until after installation or not at all, such as incomplete or incorrect information from labels, specification sheets, and the MSDS.

A follow-up round of testing 90 days after the first testing will provide an informative comparison of presence and levels of airborne pollutants with the initial results. That testing is scheduled for early February, therefore this report does not include comparative results.

Despite these rather significant limitations and the current lack of adequate controls, there are several surprising findings that warrant revision of construction protocols, training of on-site contractors, and identify further research questions.

RESULTS

Table 1 shows the particulate mass measured for $PM_{2.5}$ and PM_{10} in three locations plus the outside. Our hypothesis was that particulate mass would be insignificant compared to standard levels of 15 ug/L and 50 ug/L, respectively. The significance it that the dust control construction work practices and the thorough cleaning with HEPA filtered equipment by professionals just prior to the sampling did not meet expectations and was not sufficient for the purposes of this family.

Location	Time	Particle Mass $PM_{2.5}$ µg/m³	Particle Mass PM_{10} µg/m³
Living Room	1520	14.30	86.31
	1730	15.29	116.04
Master Bedroom	1527	9.69	37.28
	1739	8.29	31.91
Basment	1542	18.44	117.96
	1745	17.27	92.40
Exterior Front	1627	5.81	27.77

Table 1: Particulate mass

Table 2 shows the measurements for VOCs. Total VOCs is greater than the 500 ug/m3 of the Well Building Standard[5] by a factor of 26. Most of the others, except for formaldehyde and acetone, did not appear on any labels, specification sheets, or the MSDS.

Compound	Basement ng/l	Living Room ng/l	Master Bedroom ng/l
Total VOC (TVOC)	13,000	9,500	10,000
Formidehyde	290	240	290
Acetone	1,400	980	990
2-Butoxyehtanol	2,300	750	500
a-Pinene	470	530	570
DEGEE	180	340	320
Hexanal	140	280	220
trans 1,2-Dichloroehtene	120	190	210
Ethanol	140	180	170
m,p-Xylene	110	170	150
Butane (C 4)	100	160	220
Propylene Glycol	BDL	160	280

BDL = Below Detection Limit

Highest of 3 Locations

Lowest of 3 locations

Middle value of 3 locations

Table 2: Volatile Organic Compounds

Table 3 shows the detection and measurements of the EPA list of Hazardous Air Pollutants (HAP). As with the VOCs above, none of these were identified on any labels, specification sheets, or the MSDS of any materials used within the air sealed building envelope or on the exterior near windows, doors, or other penetrations. Another consideration is that although no outside measurements were taken, this house is located two blocks downwind from the coast of the Pacific Ocean, with no industry or other pollutions sources upwind.

EPA HAP (Hazardous Air Pollutant) Compound	Basement ng/l	Living Room ng/l	Master Bedroom ng/l
Carbonyl sulfide	< 1	< 1	< 1
Carbon disulfide	< 1	< 1	1
Methylene Chloride	< 1	< 1	1
Hexane (C 6)	2	2	2
1,1,1-Trichloroethane	< 1	< 1	< 1
Benzene	< 1	< 1	< 1
1,2-Dichloroethene	< 1	< 1	< 1
Trichloroethene	< 1	< 1	< 1
Methyl methacrylate	2	2	2
Toluene	51	89	66
Tetrachloroethene	9	16	10
Ethylbenzene	33	51	44
m,p-Xylene	110	170	150
o-Xylene	40	57	52
Styrene	11	15	17
1,4-Dichlorobenzene	< 1	< 1	< 1

Table 3: EPA Hazardous Air Pollutants

Table 4 is a composite of the data logging for TVOC, CO_2, CO, and Total Particulate Mass. There are two findings of significance: 1. The total VOCs, while barely within the limits of the Well Building Standard, increased with time but then abruptly dropped. This was when the ERV ventilation system was turned on. 2. The particulate mass increased after about three hours with activity by the inspectors, and again the next day when several people walked through the house as observers. The near 240 ug/m3 is nearly 5 times the maximum recommendation of 50 ug/m3 of the Well Building Standard.

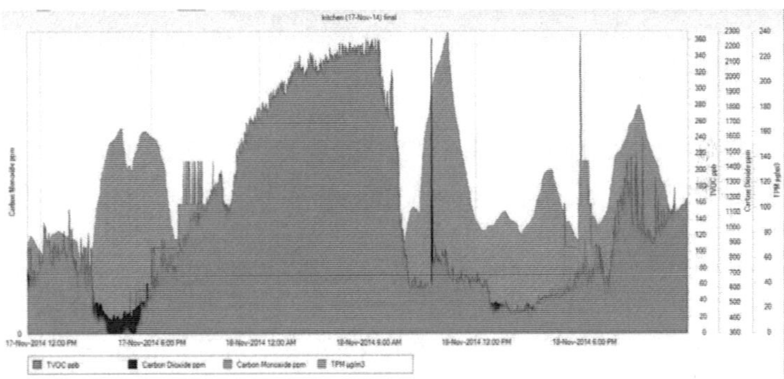

Table 3: Data logging for approximately 24 hours.

Initial results from the occupant questionnaires are incomplete at this time, but verbal reports by the adults were not encouraging. At the beginning of the pre-occupancy testing there were reports of eye, nose, and throat irritation, headaches, brain fog, and a return of fatigue. However, after the ERV ventilation equipment had run for about 24 hours, the entire family verbally reported that they were mostly complaint free.

CONCLUSIONS

Despite the lack of sufficient controls for an academic study, this on-site field experience generated considerable descriptive information useful for demonstrating that the non-routine, unusual, even extraordinary efforts to reduce, limit, and prevent sources of exposure to pollutants from new construction failed to meet expectations of a house ready for occupancy by the more sensitive members of the family. Dust control was often non-existent and cleaning was insufficient to attain acceptable levels of particulate. VOCs and HAPs were detected, some at exposure levels of concern, which were not predictable by the available information on building products. A significant number of which were not indicated as being present in the available literature of building materials. Despite the failure of the intended practices to achieve the expected results, the continuous outside air through the ERV reduced the reported complaints to near zero. Finally, an implication warranting further study is how to protect occupants from sources of pollution associated with the outside the building envelope, when many of those same pollutants appear to originate from the construction materials themselves; and, subsequently, their potential additive and synergistic participation with other sources such as contents and occupant use.

REFERENCES

[1] GrayWolf AdvancedSense IAQ Plus Indoor Air Quality Monitor

[2] Ware, SF-12®Health Survey, Rand Health, available through: http://www.outcomes-trust.org/instruments.htm#SF-36

[3] Grimes, Starting Points for a Healthy Habitat. GMC Media, 1999.

[4] Miller, Prihoda, The Environmental Exposure and Sensitivity Inventory (EESI): a standardized approach for measuring chemical intolerances for research and clinical applications.ToxicolIndHealth
1999 Jun;15(3-4):370-385.

[5] International WELL Building Institute, Well Building Standard, 2014, downloadable at: http://wellcertified.com/well-building-standard/download-the-standard/

Understanding the challenges of assessing "good" air quality in the indoor environment

Jennifer M. Logue[1,*], Max H. Sherman[1], and Brett C. Singer[1]

[1]Energy Technologies Area, Lawrence Berkeley National Lab, Berkeley, CA

[*]Corresponding email: jmlogue@lbl.gov

Keywords: Indoor air quality, standards, guidelines,

SUMMARY

There is some benefit to being colloquial. I would consider reframing as follows: The objective of ensuring "good" or "acceptable" indoor air quality for building occupants is complicated by ambiguity about appropriate standards for air pollutant concentration limits. Hundreds of chemicals have been measured in the indoor environment at varying concentrations. For many pollutants, there is large uncertainty about the levels at which healthy and/or health-compromised individuals can be exposed without adverse effects. There is also large inter-individual variability. Existing standards and guidelines incorporate uncertainty based on the objective of the standard. Not considering the intent of the standard/guideline put forth can lead to erroneous assessments. This paper explains briefly how uncertainty impacts the setting of standards and guidelines and explores the overall variability in standards and guidelines for non-cancer chronic inhalation.

INTRODUCTION

Identifying a "safe" level of exposure for all inhaled constitutes of indoor air is a daunting task. There is limited epidemiological data on the health impacts of exposure at the levels that occur in homes. Epidemiological data that do exist overwhelming link outdoor concentrations with changes in population health outcomes. Most health impact relationships are derived via toxicological studies that expose animals to pollutants at a limited number of high concentration levels over short periods of time. Correction factors are then used to extrapolate those results to human exposures over acute (short term) to chronic (lifetime) durations of exposure. Extrapolating from animal to human and acute to chronic or sub chronic time frames provides an uncertain estimate of the potential or likely impact of a given exposure on the human population. There is also variability across the population in the human response to a given exposure. Additionally, the idea of designating a "safe" level is based on the idea that there is a threshold below which there are no adverse effects, which is not always the case (Schwartz et al. 2002). For pollutants without thresholds, safety can be designated based on a tolerable level of disease risk.

For pollutants with sufficient toxicological or epidemiological research, governmental and non-governmental bodies have developed standards and guidelines that are thought to represent safe levels of exposure or concentration limits that result in safe levels of exposure for the general population. Organizations that publish standards or guidelines could have the goal of identifying a threshold concentration that is extremely unlikely to cause harm or that carries an acceptable risk of harm when humans are exposed at the threshold level over an associated duration. Standards and guidelines can be set to protect the general population, more vulnerable subpopulations such as the elderly, or less vulnerable subpopulations such as working age, healthy adults. The structure of a standard or guideline can vary considerably.

A standard or guideline can be designed to deal with acute or chronic exposures and therefore must specify both concentration threshold and an averaging time. The standard/guideline can stipulate that this threshold is never to be exceeded or that there is a limit to the number of exceedences.

Standards and guidelines have different objectives that can affect the selection of threshold values. For this publication we use the term standard to refer to a legally enforceable limit or a target developed with the attention that it could be legally enforced. Guidelines are limits set for the purpose of risk assessment or hazard avoidance and are not designed to be legally enforceable. Meeting a standard in all locations means that the average concentration will be lower than the standard. In other words a standard is, in most cases, legislating the upper tail of the desired distribution of concentrations. Guidelines tend to be set to represent a level that would be considered safe for a specific population. If there is uncertainty about the concentration at which a certain adverse effect occurs, guidelines would be set at the lower end of that uncertainty. The larger the perceived uncertainty in the health effects of a given standard, the lower the guideline threshold concentration is likely to be.

Policy-guided decision-making is necessary in standard and guideline setting. A decision maker is required to draw a line in the uncertainty of what concentration results in the desired outcome based on the underlying goals of their organization. The science is not currently sufficient in most cases to identify an exposure concentration that is definitely safe or to define the exact health impact of a given exposure. Different standard and guideline promulgating organizations may develop different values using the same information based on the operating philosophies and legal requirements imposed on those organizations. The existence of a specific value for a standard or guideline does not mean that there are no uncertainties in what concentration would be considered low or no risk of adverse outcomes. Since there is varying information about different pollutants, the standards and guidelines for different pollutants do not represent equivalent levels of risk of an adverse outcome. The concentration-response function for many chemicals is non-linear (Huijbregts et al. 2005). Therefore the amount by which you exceed a standard or guideline cannot be directly linked to a comparable increase in risk. Exceeding a guideline indicates that there is potential for harm, but does not indicate that there is harm. Guidelines can be used to exclude pollutants from a list of potential pollutants of concern but cannot be used to say that a pollutant will have a harmful effect.

Standards and guidelines do not incorporate the severity of the outcome of a given exposure. If a pollutant has multiple adverse outcomes, the standard can be set for the most severe outcome. However, if you have two pollutants, one that can potentially cause a mild health outcome and the other that can cause a severe health outcome, their individual standards and guidelines will only reflect the concentration that is expected to avoid that outcome. There are few established methodologies to assess whether exceeding one standard or guideline may be worse than exceeding another. The standard or guideline chosen for the assessment may significantly impact the conclusions.

METHODOLOGIES

In order to illustrate their variability, we aggregated standards and guidelines promulgated by US entities to protect against chronic non-cancer inhalation effects of air pollutant exposures. The organizations included are: **EPA** (US Environmental Protection Agency, publishes criteria pollutant standards and non-criteria pollutant exposure guidelines through the IRIS database), **CA** (Criteria pollutant standards set by the California Air Resources Board and exposure guidelines set by the California Office of Environmental Health Hazard Assessment), **ATDSR** the U.S. CDC (Centers for Disease Control) Agency for Toxic Substances and Disease Registry, and **TCEQ** the Texas Commission on Environmental Quality. Data for CA were obtained from state websites (http://oehha.ca.gov/air/allrels.html, http://www.arb.ca.gov/research/aaqs/caaqs/caaqs.htm). The remaining standards and guidelines were obtained from the International Toxicity Estimates for Risk Assessment (ITER) database maintained by Toxicological Excellence for Risk Assessment (TERA) (www.tera.org).

We assess the organizational variability in promulgated guidelines and standards. For each pair of organizations, *i* and *j*, for each pollutant *k*, we calculated the normalized root mean square deviation (RSMD) between the two organizations. Only pollutants that standards/guidelines promulgated by both organizations were included. For each pollutant *k*, we normalized by the product of the values from the 2 organizations. The values are normalized because there is large variability in toxicity between pollutants.

$$CV(BIAS_{i,j}) = \sqrt{\frac{\sum_{k=1}^{n} \frac{(standard_{i,k} - standard_{j,k})^2}{(standard_{i,k} * standard_{j,k})}}{n}} \quad (1)$$

RESULTS AND DISCUSSION

In total, the organizations noted above promulgate standards and/or guidelines for 122 pollutants. The ratio of the highest to lowest proposed standard/guideline ranged from 1 to 300 across the chemicals. The mean ratio was 13.3 and the median was 3.0. The diagonal of Table 1 indicates the number of pollutants for which each organization provides a standard or guideline. The non-diagonals indicate the number of pollutants each pair of organizations both promulgate standards or guidelines for. EPA and CA have the most overlap. TCEQ has the least overlap with other organizations. TCEQ guidelines/standards were constantly higher than ATDSR. For all the remaining organizations there was no consistent trend where one organization had consistently lower or higher standards/guidelines.

Table 1 also reports the CV (RMSD) calculated for each pair of standard groups. Values ranged from 1.0 to 3.5. The lower the value, the closer the two standards are to one each other. EPA and CA have a CV (RSMD) of 3.5 indicating that there is, on average, greater than a factor of 3 difference between the values promulgated by the two organizations. Both EPA and CA had similar CV (RSMD) values when compared with ATDSR potentially indicating that ATDSR sets standards/guidelines for pollutants with less uncertainty regarding outcomes.

Table 1. Number of pollutants both organizations promulgate standards for and CV(BIAS) value (number of pollutants, CV(BIAS)). When the row and column of the table are the same, the number is the standards/guidelines promulgate by that organization.

	EPA	CA	ATDSR	TCEQ
EPA	88, ---			
CA	47, 3.5	78, ---		
ATDSR	26, 2.2	25, 2.2	35, ---	
TCEQ	0, ---	1, 3.1	3, 1.0	3, ---

CONCLUSIONS

Standards and guidelines provide a valuable tool for IAQ assessment. But it is important to understand how uncertainty is incorporated into the standard/guideline and what information can be derived from a standard or guideline. Standards and guidelines do not always represent a transition point from "safe" to "unsafe"; rather, they present a value thought to be safe incorporating the uncertainty of the health impacts of pollutants based on the goals of the organization promulgating them.

If "safe" levels of exposure can be determined for all chemicals, there is still a question of what resources should be expended to keep concentrations below that threshold for all pollutants. With current technology we could likely achieve any "safe" levels determined but it may require significant cost or large reductions in occupant comfort. Potentially there should be a "good enough" IAQ level that designates an acceptable level of air quality based on the required resources need to achieve it. In a resource-limited world, there is also a need to prioritize pollutants for intervention. Determining a "safe" level is insufficient for this task because it provides no information on the impacts of exceeding that "safe" level. Standards/guidelines can be used to eliminate pollutants

ACKNOWLEDGEMENT

Funding was provided by the U.S. Dept. of Energy Building Technologies Program, Office of Energy Efficiency and Renewable Energy under DOE Contract No. DE-AC02-05CH11231; by the U.S. Dept. of Housing and Urban Development Office of Healthy Homes and Lead Hazard Control through Interagency Agreement I-PHI-01070, and by the California Energy Commission through Contract 500-08-061.

REFERENCES

Huijbregts MAJ, Rombouts LJA, Ragas AMJ and van de Meent D.2005. "Human-toxicological effect and damage factors of carcinogenic and noncarcinogenic chemicals for life cycle impact assessment." Integrated Environmental Assessment and Management **1**(3): 181-244.

Schwartz J, Laden F and Zanobetti A.2002. "The concentration-response relation between PM2.5 and daily deaths." Environ Health Perspect **110**(10).

PM2.5: AN EMERGING ISSUE FOR STANDARD OF CARE IN INDOOR SPACES

Terry BRENNAN[1*]

[1]Camroden Associates, Inc.

[*]Corresponding email: terry@camroden.com

Keywords: particulate matter, PM 2.5, PM 10, health, air cleaning, filtration

SUMMARY

Indoor particulate matter poses significant health risks and existing interventions are available. Risks associated with outdoor particulate matter (PM) exposures, including cardiovascular and respiratory effects are known to be significant. PM concentrations in indoor air may be higher than those in outdoor air. Personal indoor exposures monitored in subjects breathing zone are frequently higher than indoor exposures. A recent analysis of risk posed by indoor air contaminants in US homes found PM2.5 poses the greatest health burden. PM2.5 in indoor spaces may be emitted from indoor sources or may be transported into buildings by infiltration and ventilation. Studies have shown that the use of improved air filtration in buildings reduces levels of PM 2.5 and the use of kitchen range hood exhaust can dramatically reduce a major source if indoor PM2.5. Some studies have found improvements in microvascular function or reductions in cardio-vascular disease blood markers. Cognizant bodies that establish guidance for the control of indoor air contaminants have an opportunity to impact exposure to PM2.5 by adopting requirements or incentives for improved filter efficiency in recirculating and ventilating air handlers and minimum capture efficiency for kitchen range exhaust.

INTRODUCTION

Particulate Matter has long been identified as an outdoor air contaminant with significant public health impacts. In the US Total Suspended Particulate (TSP) was identified and regulated as a criteria pollutant in 1971. In 1987 the regulation changed from TSP to respirable particulate matter less than 10 microns (PM10). In 1997 the regulation separated into a standard for PM10 and a standard for PM2.5. In 2012 driven by an increasing depth of health effects studies that distinguished between PM10 and PM2.5, the PM2.5 standard was lowered from an annual average concentration of 15 $\mu g/m^3$ to 12 $\mu g/m^3$.

Health effects resulting from exposure to PM2.5 include both premature mortality and morbidity. Health effects include lung cancer, chronic obstructive pulmonary disease (COPD), cardiovascular disease, decreased lung function and wheeze, cough and shortness of breath in astmatic individuals (Karottki 2014, Fann et al 2011, Pope et al 2009). Fann et al (2011, Habre 2014) predict 130,000 premature fatalities annually as a result of exposure to 2005 levels of outdoor PM 2.5. Most of the health studies use outdoor air PM10 and PM2.5 data as the basis for analysis. However people spend around 90% of their time indoors (Klepeis et al 2001). Exposure to indoor PM 2.5 must also be considered. Less is known about the health effects of exposure to PM2.5 particles generated within buildings. The best information applies to allergens and inflammatory agents of biological origin (IOM 2000, IOM 2004).

PM2.5 in indoor spaces consists of particles that are generated indoors plus particles that enter from outside (Wallace 2003, Turpin et al 2007, Pope 2009). Indoor sources are often associated with occupant activities and include tobacco smoke, cooking, cleaning activities, burning candles and incense.

Particles may enter with outdoor air drawn in through unintentional air leakage sites and open windows in the building enclosure by stack, wind or mechanically induced pressures. They may also enter through outdoor air intakes or through air leaks in return ductwork located in vented attics and crawlspaces. Studies have reported particle penetration factors to be significant sources of indoor PM2.5. Wallace et al (2003) estimated that outdoor particles penetrated indoors with an efficiency of 48% and were responsible for 25% of the mean indoor concentration, and the major indoor source was smoking. Bhangar et al. (2011) found that indoor sources contributed more to indoor residential PM exposures than particles originating outdoors, and that episodic indoor sources, especially cooking, caused the highest peak exposures to ultrafine PM. Stephens and Siegel (2012) conducted careful particle penetration studies in 19 residential buildings in Austin, Texas. The mean particle penetration factor was 47%, but the range was from 17% to 78%.

The Relationship of Indoor, Outdoor and Personal Air (RIOPA) study included 219 nonsmoking residences located in three cities. Indoor, outdoor and personal exposure data was collected for several classes of air pollutants (VOCs, carbonyls and PM2.5). The RIOPA study found that the median indoor PM2.5 concentrations were about equal to the median outdoor concentrations and that the personal air concentrations were about twice as high as outdoor PM2.5 (Turpin 2007). Particles originating in the outdoor air represented around 57% of those in the indoor air. However, PM 2.5 particles originating in outdoor air compose a smaller fraction of the particles in personal air (25% to 33%). The RIOPA study identified home cleaning (sweeping) and heating and combustion activities (e.g., oil furnace, oven, fireplace) were the largest contributors to the indoor PM2.5 concentrations. Although the RIOPA subjects were non-smokers, ETS emissions were reported in a very small subset of homes, and the median indoor PM2.5 concentrations for these homes was almost double that of homes without ETS.

Logue et al (2010) conducted a hazard assessment of indoor air contaminants. The assessment included results from 77 studies reporting measurements of indoor air pollutants that are representative of US residences. Mid and upper concentrations were calculated for and compared to 97 health standards for chronic and acute health effects. Based on a review of 16 studies PM 2.5 was identified as one of 15 indoor pollutants of high concern in the majority of US homes.

The health impact of the 15 indoor air pollutants identified in the Logue et al (2010) study plus two additional pollutants was conducted so relative risk could be compared (Logue at al 2011). Chronic health impacts were estimated in Disability Adjusted Life Years (DALYs) for each pollutant. DALYs are a combined estimate of years of life lost (YLL) and years living with disability (YLD) providing an estimate of the total health burden. PM2.5 was the pollutant with the greatest DALY (700 DALYs lost per 100,000), followed by second hand smoke (100 DALYs), radon-smokers (80 DALYs) and formaldehyde (35 DALYs). The DALY for PM 2.5 is several times larger than the second hand smoke, the indoor pollutant with the next highest ranking. PM 2.5 is identified by the Logue study as the indoor air contaminant with the greatest health impact in US buildings.

Indoor PM 2.5 concentrations may be reduced in following ways.

1. The elimination of avoidable sources (e.g. avoid burning candles). Many indoor sources of PM2.5 are associated with occupant activity. Elimination of indoor sources (if possible) may have the largest potential for reduction in both general indoor air and personal air. For example, although the RIOPA subjects were non-smokers, ETS emissions were reported in a very small subset of homes, and the median indoor PM2.5 concentrations for these homes was almost double that of homes without ETS.
2. Air sealing the building enclosure to reduce entry of PM2.5 in outdoor air. There is some evidence in the literature that the tighter a building enclosure is the lower the particle penetration from outdoor air. Stephens and Siegel (2012) found there was a significant positive relationship between outdoor particle penetration and building air leakage rates measured by blower door tests – that is that particle penetration was lower for airtight buildings than for leaky buildings.
3. Containing and removing PM2.5 with exhaust ventilation by exhaust ventilation (e.g. kitchen range hoods). Recent research on measuring the effectiveness of residential kitchen range hoods finds that capture efficiency (CE) ranges from 46% to >90% depending on fan speed and whether cooking is being conducted on the back or front burners (Singer et al 2011) and from 31% to 94% (Rim et al 2012).
4. Filtering incoming ventilation air (e.g. using high efficiency filters on outdoor air intakes). Filtering fan powered outdoor air with greater than MERV 13 filters provides a dilution effect for indoor PM2.5 that cannot otherwise be achieved with ventilation. Combined with airtightening measures so that the ventilation is able to pressurize the building by several Pascals would greatly reduce the amount of PM2.5 leaking in through the building enclosure.
5. Filtering PM2.5 from recirculated indoor air (e.g. improved filtration in central air handlers or local use of free standing high efficiency particle filters). (Polidori et al 2012) report that the replacing MERV 7 filters with MERV 16 filters in the air handlers in three classrooms resulted in nearly 90% reductions in indoor PM2.5 concentrations. Bhangar et al. (2011) showed that active particle removal (filtration) can significantly reduce indoor particle levels. Bräuner et al. (2007) investigated the effects of controlled exposure to indoor air particles on microvascular function in a group of healthy elderly adults in Denmark, which showed significant improvement after using filtration to reduce particle exposure.

A number of studies have examined efforts to reduce indoor air PM2.5 concentrations and health outcomes. The evidence that indoor concentrations of PM2.5 can be reduced by local exhaust of cooking equipment is strong. The evidence that indoor PM2.5 concentrations can be reduced by using high efficiency filters on recirculating central air handlers and air handler with outdoor air capability is strong. Fisk (2014) conducted a review of particle filtration studies that measured health outcomes and concluded that particle filtration can be effective at reducing allergic and asthma outcomes and the greatest potential benefit may be reducing the morbidity and mortality that is associated with exposure in indoor spaces to particles that originated in outdoor air.

CONCLUSIONS

Evidence that exposure to PM2.5 results in a significant health burden is strong. Methods for intervening in indoor PM2.5 exposures have been demonstrated to be effective. This evidence is compelling enough for the building community to raise the standard of care in regard to reducing concentrations of PM2.5 in indoor air. The author cognizant bodies that establish guidance for the control of indoor air contaminants to consider improved filtration filter efficiency requirement in recirculating and ventilating air handlers and to adopt capture effectiveness for kitchen range exhaust.

REFERENCES

Bhangar et al (2011). Ultrafine particle concentrations and exposures in seven residences in California, *Indoor Air*, **21**, 132-144

Bräuner, E.V., et al., "Indoor particles affect vascular function in the aged". American *Journal of Respiratory and Critical Care Medicine*, 2007. **177**, p. 419-425.

EPA (2010). Quantitative Health Risk Assessment for Particulate Matter. US Environmental Protection Agency. Research Triangle Park, NC

Fann at al (2011). Estimating the National Public Health Burden Associated with Exposure to Ambient PM2.5 and Ozone. *Risk Analysis*. **10.1111/j**

Fisk W (2014). Health Benefits of Particle Filtration. *Indoor Air* **23**, 357

Habre et al (2014). The effects of PM2.5 and its components from indoor and outdoor sources on cough and wheeze symptoms in asthmatic children. *Journal of Exposure Science and Environmental Epidemiology* **24**, 30-387

IOM (2000). Clearing the air: asthma and indoor air exposures. Washington, D.C., Institute of Medicine, National Academy of Sciences, National Academy Press.

IOM (2004). Damp indoor spaces and health, Institute of Medicine, National Academy of Sciences, National Academy Press.

Karottki et al (2014). Cardiovascular and lung function in relation to outdoor and indoor exposure to fine and ultrafine particulate matter in middle-aged subjects, *Environment International*, **73**, 372-381

Klepeis et al (2001). The National Human Activity Pattern Survey (NHAPS): a resource for assessing exposure to environmental pollutants. *Journal of Exposure Analysis and Environmental Epidemiology* 2001;**11**:231–252.

Logue et al (2010). Hazard assessment of chemical air contaminants measured in residences Lawrence Berkeley National Laboratory Report Number 3650-E

Logue el al (2011). A method to estimate the chronic health impact air pollutants in U.S. residences. *Environmental health perspectives* doi:**10**.1289/ehp

Polidori et al (2012). Pilot study of high-performance air filtration for classroom applications *Indoor Air 2012*doi:**10:**1111

Pope et al (2009). Fine-Particulate Air Pollution and Life Expectancy in the United States. *The New England Journal of Medicine*. **360;4**, 376-386

Rim et al (2012). Reduction of exposure to ultrafine particles by kitchen exhaust hoods: The effects of exhaust flow rates, particle size, and burner position *Science of the Total Environment* **432** (2012) 350–356.

Stephens B, Siegel JA (2012). Penetration of ambient submicron particles into single-family residences and associations with building characteristics. *Indoor Air* **22:** 501-513.

Turpin, B.J., et al. (2007). "Relationships of Indoor, Outdoor, and Personal Air (RIOPA), Part II. Analyses of Concentrations of Particulate Matter Species". Health Effects Institute, 2007. Research report 130, part II.

Wallace (2003). Particle concentrations in inner city homes of children with asthma: the effect of smoking, cooking, and outdoor pollution, *Environmental Health Perspectives*, **111(9)**, 1265-1272

High Performance IAQ Specification for Net Zero Energy Homes

Anthony BERNHEIM[1], Alfred HODGSON[2], and Andrew PERSILY[3,*]

[1]Bernheim + Dean, Inc., San Francisco, California, USA
[2]Berkeley Analytical Associates, LLC, Richmond, California, USA
[3]National Institute of Standards and Technology, Gaithersburg, Maryland USA

[*]Corresponding email: andyp@nist.gov

Keywords: Indoor air quality, high performance buildings, netzero energy, residential, ventilation

SUMMARY

The National Institute of Standards and Technology (NIST) constructed a Net Zero Energy Residential Test Facility (NZERTF) to support the development and adoption of cost-effective Net Zero Energy (NZE) designs and technologies. One key design objective was to provide for occupant health and comfort through adequate ventilation and reduced indoor contaminant sources. To improve source control, guidelines were implemented to utilize products with relatively low volatile organic compound (VOC) emissions, focusing on toxicity, sensory irritation and odor. Emphasis was placed on reducing formaldehyde emissions as well as VOCs from wet-applied materials and acetic acid. These guidelines were based largely on published research and product information, but were not part of the NZERTF architectural specifications. From the lessons learned in their application to the NZERTF, the guidelines were updated and formalized in the form of architectural specifications for use by architects and contractors interested in addressing high performance IAQ in netzero and other residential buildings.

INTRODUCTION

In 2009, the NIST Energy and Environment Division received funding to design, construct and operate a net zero energy residence in Gaithersburg, MD. The 250 m^2 two-story single-family unoccupied NZERTF residence was completed in 2012 (Pettit et al., 2014). It functions as a laboratory to support the development and adoption of cost-effective NZE designs, technologies, construction methods, and building codes. The primary design goal was to meet the comfort and functional needs of the presumed occupants. Other important goals include siting to maximize renewable energy potential, establishing an airtight and highly insulated building enclosure designed for water and moisture control, providing controlled ventilation, and installing highly efficient mechanical equipment, lighting and appliances. The NZERTF achieved its goal of generating more energy than it consumed during its first year of simulated occupancy, despite a severe winter (NIST, 2014).

Indoor air quality (IAQ) directly impacts the comfort, health and well-being of building occupants, making the achievement of acceptable IAQ in newly constructed and renovated buildings an important design objective. IAQ in homes is particularly important because Americans, on average, spend about 90 % of their time indoors with the majority of this time at home (EPA, 2011). Indoor concentrations of many air pollutants are often elevated in homes relative to outdoors because many materials and products used indoors contain and release a variety of air pollutants (Hodgson et al., 2002; Offermann and Hodgson, 2013).

With respect to indoor air contaminants for which inhalation is the primary route of exposure, the critical design and construction parameters are the provision of adequate ventilation and the reduction of indoor contaminant sources.

Both of these pollutant control techniques were incorporated into the design of the NZERTF. Ventilation system design and specifications are described in Pettit et al. (2014). To control indoor contaminant sources, guidelines were developed for the selection of relatively low VOC emitting products used inside of the airflow control layer (or air barrier membrane), thereby reducing potential toxicity, sensory irritation, and odor annoyance. Particular emphasis was placed on reducing formaldehyde emission sources including composite woods. Acetic acid, an odorant and suspected chemical irritant found in numerous interior products, and VOCs emitted from wet-applied materials were also targeted.

These source control guidelines were largely prescriptive, based on published research and product information available at the time. The design team implemented the guidelines to the best of their ability, but found that it was generally difficult for product manufacturers and suppliers to verify the emission requirements except for the relatively simple product credits described in the LEED® for Homes building standard (Pettit et al., 2014). Despite these challenges, the IAQ objectives for the NZERTF largely were accomplished. The efforts to limit airborne concentrations of formaldehyde and other VOCs through ventilation and source control were investigated through a year-long IAQ monitoring study conducted to simulate occupancy (Poppendieck et al., 2014). This study concluded that the measures taken, such as reducing use of medium density fiberboard and particleboard in the cabinetry and other finished products were effective in reducing indoor formaldehyde concentrations. The results additionally suggested that specifications for low-emitting interior products along with ventilation rates consistent with ASHRAE Standard 62.2 (ASHRAE, 2013) could result in residences with lower indoor VOC emission rates and VOC concentrations than typical new homes that did not employ such specifications.

IAQ SPECIFICATIONS

Based on the lessons learned from the NZERTF effort, the guidelines for low-emitting interior sources were updated and formalized into a detailed architectural specification intended specifically for residential new construction and major renovations. This specification differs substantially from the original guidelines and emphasizes a more performance-based approach. This change was made due to the availability of considerably more VOC emissions product data generated by manufacturers interested in demonstrating compliance with LEED® low-emitting materials credits and the requirements of other high-performance building codes and standards. This specification is now available for download in Word format at the NZERTF web page on the NIST website (www.nist.gov).

Specification Section 01 81 13.01, Sustainability Requirements – Indoor Air Quality, aims to provide guidance for building professionals on IAQ topics for the design and construction of a "healthy" NZE home. It is intended for use by architects, contractors, and design-build contractors and provides guidance for the review, selection and specification of systems and products for residential projects, and is structured to assist in developing integrated specification documents governed by this Division 01 section.

The practices contained in the specifications are focused on interior finishes and other known indoor pollutant sources. Some of the main practices are summarized as follows:

1. Paints and Coatings: Shall not contain formaldehyde; shall be formulated with water-based technologies; VOC content (g/L) shall meet South Coast Air Quality Management District (SCAQMD) rule 1113 and/or California Air Resources Board (CARB) Suggested Control Measure (SCM).
2. Adhesives and Sealants: Shall not contain formaldehyde; VOC content (g/L) shall meet SCAQMD rule 1168; only limited quantities of silicone rubber caulks or sealants containing acetic acid shall be used.
3. Thermal and Acoustic Insulation: Shall not contain formaldehyde-based binder as an ingredient; shall be fire retardant free unless required by local building code.
4. Carpet and Carpet Cushion: Shall comply with Carpet & Rug Institute "Green Label Plus" program testing requirements for carpet and "Green Label" program testing requirements for carpet cushion and shall comply with VOC emission requirements of CA Dept. of Public Health Standard Method V1.1.
5. Resilient and Tile Flooring: Shall comply with Resilient Floor Covering Institute (RFCI) "FloorScore®" program testing requirements in accordance with VOC emission requirements of CA Dept. of Public Health Standard Method V1.1.
6. Wood, Composite Wood, Agrifiber Products and Components: Shall comply with VOC emission requirements of CARB Airborne Toxic Control Measure (ATCM); interior moldings and trim shall be solid wood; shelving and panels shall be hardwood plywood with no-added formaldehyde veneer core.
7. Gypsum Board Walls and Ceilings: Gypsum wallboard shall be applied with conventional mechanical fasteners; solvent-containing adhesives shall not be used; panels shall comply with VOC emission requirements of CA Dept. of Public Health Standard Method V1.1.
8. Moisture and Mold Control: Drywall, other porous materials and components, and items with high organic content shall not be loaded into or stored in partially enclosed buildings; visible signs of mold and mildew that appear during construction shall be reported to Owner and Architect; damaged products shall not be used.
9. Attached Garage Pollutant Protection: Garage shall be enclosed with air barrier to effectively separate the habitable area of the home from the garage; garage shall use an exhaust fan with automatic timer controls linked to an occupant sensor, light switch, garage door opening mechanism, and a carbon monoxide sensor.

It is recognized that each residential project will have different objectives and requirements based on Owner's programmatic requirements, site conditions, climatic conditions, project size, and complexity. Accordingly, the specification is intended to be modified by the user. Editor's notes are provided throughout the Specification to guide the user in selecting appropriate requirements and in editing the document consistent with project requirements. An IAQ Compliance Table is provided to assist the specification user in tracking the IAQ compliance requirements for systems and products and for recording submittal documentation such as VOC emission test reports, product technical data sheets, product certifications and manufacturer's self-declared claims, as applicable.

Control measures for pollutants that that may expose occupants by pathways other than, or in addition to, inhalation are not as well established as control measures for VOCs. The specification contains optional requirements that may be used to expand the scope of pollutants by limiting products based on information about chemical ingredients.

The specification also addresses building commissioning, a process of verifying that the building is built and operates as designed and as intended. It recommends that users develop a process that is appropriate for the type, size, and complexity of their project in accordance with industry guidelines, building code requirements, and LEED certification requirements.

CONCLUSIONS

The IAQ specification is intended to support architects and contractors who pursue a higher level of IAQ performance in net zero and other residential buildings. The formal, but flexible, architectural specification language is intended to facilitate use by building professionals.

DISCLAIMER

Certain commercial programs and documents are identified in this paper in order to fully describe the concepts therein. Such identification is not intended to imply recommendation or endorsement by NIST, nor is it intended to imply that the programs and documents identified are necessarily the best available for the purpose.

ACKNOWLEDGEMENT

The efforts of Anthony Bernheim, FAIA, LEED® Fellow, and Alfred Hodgson were performed under contract to the National Institute of Standards and Technology. The authors gratefully acknowledge the contributions of Peggy White of White + GreenSpec, as well as Dustin Poppendieck and Lisa Ng of NIST in preparing the specifications.

REFERENCES

ASHRAE (2013). Standard 62.2-2013: Ventilation and Acceptable Indoor Air Quality in Low-Rise Residential Buildings. American Society of Heating, Refrigerating and Air-Conditioning Engineers, Inc., Atlanta, GA.

Berkeley Analytical Associates, Bernheim + Dean, White + GreenSpec (2015). High Performance Indoor Air Quality Specification for Net Zero Energy Homes. NIST GCR 14-980, National Institute of Standards and Technology. Gaithersburg, MD.

EPA (2011). Exposure Factors Handbook: 2011 Edition, Chapter 16 – Activity Factors. Report EPA/600/R-09/052F, September 2011. U.S. Environmental Protection Agency, Washington, D.C.

Hodgson AT, Beal D, and McIlvaine JER (2002). Sources of formaldehyde, other aldehydes and terpenes in a new manufactured house. *Indoor Air*, **12**: 235–242.

NIST (2014). NIST Test House Exceeds Goal; Ends Year with Energy to Spare. NIST Tech Beat, July 1, 2014. Accessed at http://www.nist.gov/el/building_environment/netzero-070114.cfm, September 23, 2014.

Offermann FJ and Hodgson AT (2011). Emission Rates of Volatile Organic Compounds in New Homes. In: Proceedings Indoor Air 2011, 12th International Conference on Indoor Air Quality and Climate, Austin, TX USA, June 5-10, 2011.

Pettit B, Gates C, Fanney H and Healy WM (2014). Design Challenges of the NIST Net Zero Energy Residential Test Facility. NIST Technical Note 1847, National Institute of Standards and Technology. Gaithersburg, MD.

Poppendieck D, Ng L, Schlegel M, Persily A, and Hodgson A (2014). Long-term air quality monitoring in a Net-Zero Energy Residential Test Facility designed with specifications for low emitting interior products. Paper HP0087, In: Proceedings Indoor Air 2014, 13th International Conference on Indoor Air Quality and Climate, Hong Kong, July 7-12, 2014.

WEATHERIZATION VENTILATION STRATEGY AS A FACTOR IN INDOOR HUMIDITY

William ROSE[1], Stacy GLOSS[1,*] Paul FRANCISCO[1], Zachary MERRIN[1], Salvatore CALI [2], and David JACOBS [2,3]

[1]Indoor Climate Research and Training, University of Illinois, Champaign IL
[2] School of Public Health, University of Illinois at Chicago IL
[3]National Center for Healthy Housing, Columbia MD

[*]Corresponding email: sgloss@illinois.edu

Keywords: Humidity, moisture balance, vapor pressure, ventilation, ASHRAE 62.2, housing, weatherization

SUMMARY

The HEALTH-V study evaluated health and environmental outcomes associated with two ventilation approaches taken in existing homes. The number of homes studied was 52 in the Chicago area and 35 in Indiana. The control homes were weatherized based on ASHRAE 62-1989 and the treatment homes were ventilated to ASHRAE 62.2-2010, both in the course of weatherization (wx) energy retrofit.

One of the outcomes studied was humidity using moisture balance—the comparison of vapor pressure excess indoors with outdoor temperature. Excessive moisture has been linked to mold, asthma, pests and other healthy housing deficits. The house moisture balance was evaluated pre-wx and post-wx. Low moisture balance represents drier conditions indoors.

This study found that there was a significant lowering of moisture balance following weatherization with both ventilation approaches. The study also found that the 62.2 treatment showed a significant lowering of moisture balance compared to control (62-1989).

INTRODUCTION

Weatherization agencies for the Chicago area and for Indiana conducted weatherization and field data collection using a study protocol developed by the investigators. Agency assessors visited the client homes, solicited recruitment and enrollment in the Housing and Environmental Aspects Linked To Health through Ventilation (HEALTH-V study), and placed and retrieved the instrumentation and sampling devices. Devices were placed in living spaces, and in basements solely for homes with basements. The specific weatherization modifications differed slightly from house to house, and may have included addition of a ground cover in the crawl space, attic insulation, wall insulation, equipment modification or change-out, duct tightening, and/or weatherstripping, depending on baseline housing conditions.

Ventilation for each house was randomly selected to comply with either ASHRAE 62-1989 or ASHRAE 62.2-2010. Under ASHRAE 62-1989, a lower limit of airtightness is established and the air-tightening is conducted down to the tightness limit, but no further. As a result, the homes complying with ASHRAE 62-1989 did not, in actuality, receive whole-building ventilation, but relied on passive infiltration. Under ASHRAE 62.2-2010, the air tightening is conducted to achieve the maximum airtightness achievable, and mechanical ventilation is introduced to meet ventilation requirements, relying on active infiltration.

The study homes were instrumented for temperature and relative humidity. Hourly data were collected and compared to outdoor conditions to determine the moisture balance (see below). As a result, the homes complying with ASHRAE 62-1989 did not, in actuality, receive whole-building ventilation. The changes in humidity from pre-wx to post-wx are attributed to the total weatherization effort, and cannot be attributed solely to the introduction of a ventilation approach.

METHODOLOGIES

Battery operated self-contained temperature (T) and relative humidity (RH) loggers were placed in basements and living spaces where they would be unaffected by sunlight or proximity to sources of cold or heat, and where they would be undisturbed by occupants. Loggers were placed on the initial visit (following enrolment and informed consent) and retrieved prior to wx (Visit 3). Following weatherization (Visit 4), loggers were placed again and retrieved at the final visit. The aim was for at least one week of hourly readings prior to and following wx. The data were downloaded by the field data collector and sent to the research team.

Hourly temperature and relative humidity data contained a stamp for date and time. Vapor pressure was calculated from T and RH using ASHRAE psychrometric coefficients. Outdoor temperature and humidity data were downloaded from Midway Airport in Chicago. Outdoor temperature and relative humidity were used to calculate an outdoor vapor pressure, as above. The difference between indoor and outdoor vapor pressure is the vapor pressure excess (vpe).

Humidity in the homes was characterized using the moisture balance approach (Francisco and Rose, 2010). During heating season, houses typically have higher vapor pressure indoors than outdoors. This vapor pressure excess is usually greater at lower outdoor temperatures; at 20°C, the vapor pressure excess may be zero if the house is open to the outside. This analysis does not apply during air-conditioning season, since the humidity is artificially modified by air conditioning.

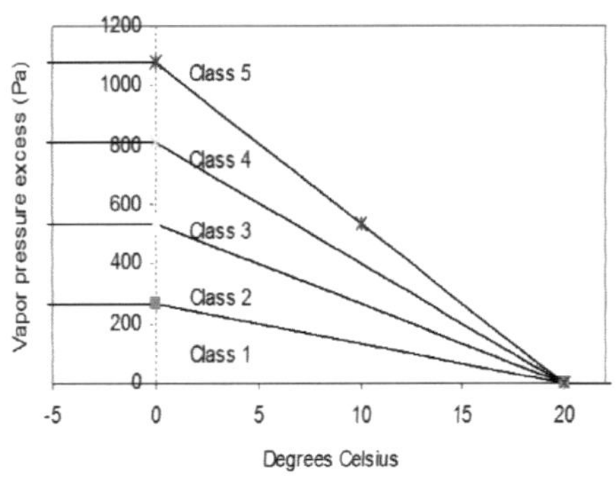

Figure 2. Climate classes as defined under ISO 13788.

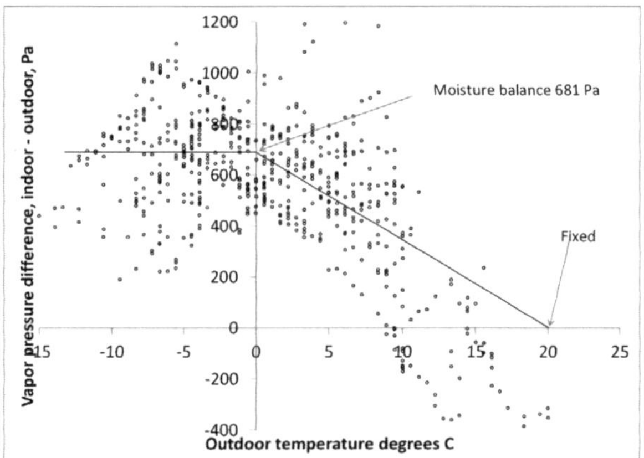

Figure 1. Vapor pressure excess data plotted against outdoor temperature, with regression line passing through 20C. Intersection of regression with 0C is the moisture balance value. Data for IL-49 Living, pre-wx.

IEA Annex 41 developed ISO 13788, which defined climate classes as inputs for hygrothermal modelling. See Figure 1. Higher humidity buildings were presumed to have higher vapor pressure excess as a function of outdoor temperature. In the present study, the same approach is turned around.

Vapor pressure excess is calculated from indoor and outdoor vapor pressure, and a "dogleg" regression is calculated to fit the data. The value of vpe where the regression line intersects the 0°C axis is taken as the value for moisture balance. See Figure 2. This means of analysis has the advantage of allowing comparison of vapor pressure excess at different temperatures during heating season. The primary disadvantage is that cooling season data are ignored.

Hourly data were excluded under two conditions:

1. If outdoor hourly temperature was higher than 20°C, or
2. If the hourly data occurred on a day when 16 or more hours in that day had outdoor temperature in excess of 20°C

The second condition was predicated on the assumption that during warm days, the cooler night-time hours would be affected by dehumidification offered by (daytime) air conditioning.

RESULTS AND DISCUSSION

The individual house results are shown in Tables 1 and 2. MB pre- and post- represent the moisture balance results pre-wx and post-wx. The number of hours in the sample for pre- and post- is shown in columns 5 and 7 of the two tables. The percentages in columns 6 and 8 represent the percent of all measured values that are included in the sample after applying the exclusion criteria. Of the 87 homes in the study, 39 had incomplete data, and data from 16 homes fell within the exclusion criteria.

Table 1. Moisture balance results 62-1989

House	control/ treat.	MB pre- (Pa)	MB post- (Pa)	no. hours pre-	pct of all hours	no. hours post	pct of all hours
IL_01-L	control	663	165	125	96%	97	72%
IL_05_L	control	935	1752	151	97%	15	1%
IL_05_F	control	792	2376	151	97%	15	1%
IL_09_L	control	775	656	672	82%	4120	63%
IL_13_L	control	1210	945	69	47%	163	98%
IL_15_L	control	1478	1426	90	72%	21	17%
IL_15_F	control	2509	1382	78	62%	21	14%
IL_21_L	control	898	488	93	54%	87	73%
IL_23_F	control	948	1366	94	56%	44	26%
IL_29_L	control	2216	716	105	64%	167	100%
IL_31_L	control	2360	1527	51	26%	152	89%
IL_31_F	control	2021	1512	38	22%	152	89%
IL_41_L	control	2086	1836	79	85%	139	83%
IL_41_F	control	861	1184	79	85%	139	83%
IL_43_L	control	458	363	167	100%	193	100%
IL_43_F	control	295	279	167	100%	193	100%
IL_49_L	cont.	681	405	629	100%	169	100%
IN_27_L	control	702	202	26	31%	145	100%
IN_27_F	control	2446	230	26	31%	145	100%
IN_29_F	control	1309	2709	52	50%	85	50%
IN_30_L	control	2401	791	46	43%	121	100%
IN_30_F	control	2054	696	46	43%	121	100%

Table 1. Moisture balance results 62.2 2010

House	control/ treat.	MB pre- (Pa)	MB post- (Pa)	no. hours pre-	pct of all hours	no. hours post	pct of all hours
IL_04_L	treat.	1753	507	106	82%	219	19%
IL_04_F	treat.	1370	529	108	82%	219	19%
IL_08_L	treat.	663	438	155	94%	110	100%
IL_14_L	treat.	-82	1454	46	64%	103	53%
IL_14_F	treat.	912	1370	46	64%	103	53%
IL_20_L	treat.	119	812	44	56%	90	73%
IL_22_F	treat.	2671	1310	43	37%	91	78%
IL_26_L	treat.	1149	291	19	2%	163	100%
IL_26_F	treat.	411	272	19	2%	163	100%
IL_32_L	treat.	2435	884	39	23%	147	100%
IL_32_F	treat.	2812	826	39	23%	147	100%
IL-34_L	treat.	1296	873	39	22%	172	100%
IL_34_F	treat.	1663	647	39	22%	172	100%
IL-36-L	treat.	1803	299	105	64%	190	100%
IL_36_F	treat.	1388	274	105	74%	190	100%
IL_38_L	treat.	978	412	22	7%	169	100%
IL_38_F	treat.	947	164	40	8%	169	100%
IL_40_L	treat.	1814	636	131	37%	166	100%
IL_40_F	treat.	1347	451	131	37%	166	100%
IL_52_L	treat.	582	205	167	100%	148	100%
IN_01_L	treat.	490	268	1611	40%	145	100%
IN_01_F	treat.	687	273	1611	40%	145	100%
IN_06_L	treat.	1245	806	32	21%	97	100%
IN_11_L	treat.	748	200	132	100%	45	62%
IN_11_F	treat.	517	-259	134	100%	45	62%
IN_13_L	treat.	882	3	156	100%	45	62%
IN_13_F	treat.	619	570	156	100%	45	62%
IN_14_L	treat.	1026	576	155	100%	76	78%
IN_14_F	treat.	1182	1893	155	100%	76	78%

A comparison of the pre-wx to post-wx moisture balance is shown in Table 3. The mean value for moisture balance is lower following weatherization. The p-value from a two- tailed paired t-test shows that this reduction is significant ($p<0.05$). The overall aim of the research was to compare the results of complying with ASHRAE 62-1989 (defining and achieving tightness limits: controls) to complying with ASHRAE 62.2 -2010 (tightening to the extent

Table 3. Comparison pre-/post-

All cases	pre-	post-	*p pre-/post-*
mean	1246	784	*2E-04*

Table 4. Comparison control/treatment for pre-/post

mean	pre-	post-	*p pre-/post-*
control	1368	1046	*0.116*
treatment	1153	586	*3E-04*
p ctrl/treat	*0.591*	*0.008*	

Table 5. Comparison living level/foundation for pre-/post

mean	pre-	post-	*p pre-/post-*
living	1164	687	*0.001*
foundation	1353	912	*0.048*
p liv/found	*0.36*	*0.209*	

possible and meeting ventilation needs with mechanical ventilation: treatment). Table 4 shows that prior to weatherization the control and treatment homes had similar moisture balance with no significant difference (similar mean, $p > 0.05$). For the control homes, the reduction was not significant, but for the treatment homes the reduction was significant ($p = 0.0003$). Table 5 shows the results of comparing moisture balance in the living level to the foundation. There were significant declines following weatherization in both locations; the magnitude of decline was greater in the living space than in the foundation.

CONCLUSIONS

The characterization of house humidity using the moisture balance method has been presented. This method provides a single-value humidity characterization for non-air conditioning periods. The method was applied to 32 homes in Illinois and Indiana as part of the HEALTH-V program.

The results show a significant decrease in house humidity as a consequence of weatherization.

Homes that were ventilated during weatherization to ASHRAE 62.2-2010 show significantly lower indoor humidity than the homes ventilated to ASHRAE 62-1989.

ACKNOWLEDGEMENT

This study was part of the Health and Environmental Aspects Linked to Home Ventilation (HEALTH-V) study. Funding for this project was provided by the U.S. Department of Housing and Urban Development (HUD), grant ILLHH0230-10. This study received UIC Internal Review Board approval for research involving human subjects under expedited review, Research Protocol # 2011-0813. The authors thank Eugene Pinzer at HUD and Harold Dawson at Community and Economic Development Association of Cook County, Incorporated in Chicago, Illinois and Steve Nall and Dan Phillips at Indiana Community Action Association in Indianapolis, Indiana, who collected all field data. We also thank Jill Breysse and Sherry Dixon at the National Center for Healthy Housing for their help on this project.

REFERENCES

ASHRAE (2013). Handbook–Fundamentals. American Society of Heating, Refrigerating and Air–Conditioning Engineers, Inc., Atlanta, GA.

Francisco P and Rose W (2010). Temperature and humidity measurements in 71 homes participating in an IAQ improvement program. In: Proceedings of Thermal Performance of Exterior Envelopes of Whole Buildings XI, Florida, USA, December.

IEA ECBCS Annex 41, 2008. Proceedings of the Closing Seminar, Copenhagen, June 19, 2008 (C. Rode, editor), Dept. of Civil Engineering, BYG R-191, Technical University of Denmark, Kgs. Lyngby, Denmark, 2008

Midwestern Regional Climate Center, University of Illinois at Urbana–Champaign, Champaign, Illinois

A Study on TVOC Concentration Emitted by Plywood: Finish, Age and Environment

E.V.H. Dia[1] and L.M.T. Bo-ot[1,2,*]
[1] College of Architecture, University of the Philippines, Diliman, Quezon City, Philippines 1101
[2] National Institute of Physics, University of the Philippines, Diliman, Quezon City, Philippines 1101

In this paper, actual concentrations of total volatile organic compounds (TVOCs) of Philippine plywood were measured. The measurements were taken with the plywood having different finishes during the time period of May to October. The finishing considered were unpainted panels of plywood, newly painted with regular and green-labelled paint, and the readings were conducted in a controlled room with accompanying readings of temperature and relative humidity. The measurements were taken in an enclosed space and each set of test samples was measured every hour for seven (7) straight days and compared to guideline emission and behaviour of TVOC in plywood in other countries. This study can contribute as a baseline data for the TVOC emissions of a Philippine building material.

Keywords: Total volatile organic compound (TVOC), concentration, Philippine plywood.

1. INTRODUCTION

Indoor environmental quality (IEQ) refers to the overall environment that affects the health, comfort and productivity in an enclosed space. IEQ can be influenced by temperature, relative humidity, air movement, MVAC, lighting, odor, noise and vibration (Godish, 1995). A research conducted in Canada reported that there already have 545 Volatile Organic Compounds identified and already have set Threshold Level Values by different agencies and labelling schemes from different countries (Charles et. al., 2005). All of these VOCs are the components of TVOC.

Molhave(1990) suggested in his study that there are four ranges of TVOC exposures: (1) <200μg/m^3 comfort range, (2)200-3000 μg/m^3 multi-factorial exposure range, (3) 3000-25,000 μg/m^3 discomfort range, and >25,000 μg/m^3 toxic range. The TLVs of TVOC emission in Japan and Hong Kong are 400μg/m3 (0.103 ppm) and 600μg/m^3 (0.155 ppm) respectively. In the Philippines, awareness and studies on TVOC impact on public health and the environment is still in its developmental stage and this study can contribute to setting up an IAQ standard in the Philippines.

Plywood is a popular construction material because of its flexibility, strength and economy. In 2011 the production of plywood in the Philippines had an 8% increase over the previous year from 276,000 cubic meters to 299,860 cubic meters (Philippine Forestry Statistics, 2011).

The more commonly used parameter used in IEQ is the emission rate. This however requires the key element of initial concentration (Yang et.al. 2001; Rudd et.al. 2008) which this study addresses. The goal of this study is thus to find how much of TVOCs can initially be emitted from plywood.

2. RELATED STUDIES

In most studies of emissions, a small-scale chamber is used to measure the individual VOCs and TVOCs of buildings materials. The values for the concentrations required are usually those at the chamber inlet and those after the collected air sample. GC/MS are used on those samples to ascertain which specific VOC is read. Results show that emission rates of the new materials were relatively high from 1st-3rd day but decreased and became stable 1 -2 weeks after (Funaki and Tanabe, 2002).

Actual emissions of VOCs of 75 apartments in urban areas in China were measured using GC/MS. Results show that higher concentrations of VOCs were found in

*Corresponding author: ltboot1@up.edu.ph

apartments with more interior decorations and apartments where windows are always closed because of the low outdoor temperature. However the relative humidity was not found a significant factor for formaldehyde. (Yoshino, et. al, 2005). Haghighat, et al. (1998) conducted a study on effects of different indoor air conditions; temperature, RH and surface air velocity on paint and varnish and reported temperature and RH have significantly affected the emission of the individual VOCs and TVOC of the materials. The TVOC emission increased as the temperature increased although individual VOCs did not reflect the same result. Some VOCs increased emission at lower temperatures. No trend can be found on the effect of RH on the materials because emission rates showed different results. Although increase in RH reflected higher TVOC emission rates, individual VOCs showed increase in emission rates at 32% RH.

3. METHODOLOGY

The site is located in an empty lot in Paranaque City. Figure 1 shows the exterior and interior images of the house. The structure was built particularly for the experiment. Interior room dimensions are 2.4m x 2.4m with a total floor area of 5.76 m^2 and ceiling height of 2.4m. Plain plywood was installed on the floor and ceiling. Fibre cement board was used for the exterior wall and nailed to 2"x2" wood frame. The structure is ensured to be sealed from the outside.

One of the main features in the experimental set-up is the fact that the space is kept closed during the reading of measurements. We didn't want to include ventilation because we wanted to quantify the initial concentration of TVOCs. We decided however to study the concentrations with different paint finishes and ages of the plywood.

A multifunction IAQ monitor (Tongdy Model DPM 1084) was used. The instrument can measure temperature, relative humidity and TVOC concentration of a room every minute for 30 straight days. The IAQ monitor can measure TVOC emission up to 30 sq.m area. It can only read the TVOC and not any of a specific VOC.

Six (6) pieces of panels of plywood with the dimensions of x 4'-0" x 4'-0" x ¼" (2400mm x 1200mm x 6mm) are installed in each sample set. A sample set is composed of either one of (a) new plain plywood, (b) newly painted (regular paint) plywood, (c) newly painted (green labelled) plywood, (d) 6 month-1 year old plywood, or (e) 2 years and older plywood.

All set samples have a testing period of 10 days. The room was cleaned and left open on the first day of the experiment. On the 2nd day the room was cleaned before it was closed and sealed for another two days to register a reference reading. The room was again opened but carefully, on the 4th day to install the test samples. A 24-7 observation period was implemented on all data sets for all the samples. No human intervention or contact was permitted during the tests. At the end 7 days, the test sample was removed and the data retrieved. Then the whole cycle is started again for the next sample. The values for the temperature, RH, and TVOC levels were then collected and tabulated after each observation period.

Fig. 1. Exterior and Interior Images

All set samples were measured for 10 days using the IAQ Monitor. The room was cleaned and left open on the first day of the experiment. On the 2nd day the room was cleaned before it was closed and sealed for another two days to register a reference reading. The room was again opened but carefully, on the 4th day to install the test samples. A 24-7 observation period was implemented on all data sets for all the samples. No human intervention or contact was permitted during the tests. At the end of two weeks, the test sample was removed and the data retrieved. Then the whole cycle is started again for the next sample. The values for the temperature, RH, and TVOC levels were then collected and tabulated after each observation period. The measurements were conducted during the period of May to October 2014 to cover both wet and dry seasons of the year. Temperature, RH and rainfall data were collected from National Weather Bureau for comparison of actual exterior weather conditions during the experiment period.

4. RESULTS AND DISCUSSION

Table 1 shows the test results of the TVOC emission for all sample set average per day for a 10 day period. Results show that the new plywood with regular has the highest average concentration at 26.59 ppm during its 7 day period and the lowest average emission rate of 14.36 ppm was found at the 1 month old plywood with green labelled paint sample set. However, the new plywood with green labelled paint was found to also have a high concentration at 26.39 ppm during its first measurement. It has also reflected that TVOC concentration decreased for both sets of 1 month old plywood with regular and green labelled paint.

Set	Description	1 (ppm)	2 (ppm)	3 (ppm)	4 (ppm)	5 (ppm)	6 (ppm)	7 (ppm)	8 (ppm)	9 (ppm)	10 (ppm)	Ave. (ppm)	Temp .°C	RH %
1	Plain Plywood	4.19	8.91	10.36	14.72	15.87	14.68	13.45	13.94	13.1	14.42	**14.41**	**36.36**	**42.68**
2	New Plywood Regular Paint	7.23	12.95	15.92	27.51	27.23	26.72	26.92	27.09	27.03	26.36	**26.31**	**34.83**	**46.90**
3	New Plywood Regular Paint	-	17.44	17.96	29.37	29.54	29.43	29.2	28.74	28.5	28.72	**29.06**	**33.96**	**49.33**
4	New Plywood GL Paint	19.73	22.57	22.31	26.77	26.72	26.67	26.77	26.95	26.01	24.86	**26.39**	**32.83**	**56.71**
5	6 Mos. old Plywood Regular Paint	21.88	22.99	22.81	22.68	22.47	22.81	22.27	22.03	22.52	22.25	**22.2**	**32.99**	**55.06**
6	2+ years Plywood Regular Paint	17.31	19.72	20.47	20.36	19.79	19.45	19.85	19.96	20.39	19.91	**19.92**	**31.85**	**58.13**
7	1Mo. Old Plywood Regular Paint	6.84	13.37	15.49	19.92	19.65	19.63	20.46	21.54	21.48	21.48	**20.6**	**31.37**	**60.48**
8	1Mo. Old Plywood GL Paint	13.93	16.66	17.35	19.81	20.09	20.57	20.93	20.6	21	21.08	**20.58**	**30.85**	**62.65**
9	New Plywood Regular Paint	5.47	9.53	11.56	29.02	27.86	26.91	27.26	26.13	25.96	24.63	**26.59**	**33.14**	**53.34**
10	6 Mos. old Plywood Regular Paint	6.3	17.85	18.73	18.13	18.06	17.68	17.26	18.32	18.17	17.83	**17.92**	**31.55**	**56.57**
11	2+ years Plywood Regular Paint	11.95	16.76	16.95	16.28	15.98	16.1	16.11	15.79	15.35	16.15	**15.97**	**31.56**	**59.24**
12	New Plywood GL Paint	-	12.73	12.84	26.87	25.04	23.85	23.47	23.27	22.61	20.17	**23.61**	**33.27**	**50.62**
13	1Mo. Old Plywood Regular Paint	8.07	15.01	14.65	19.89	20.15	20.08	20.27	20.69	20.24	19.02	**20.05**	**32.52**	**55.30**
14	1Mo. Old Plywood GL Paint	0.19	4.16	4.93	13.76	13.72	13.78	13.06	14.38	16.05	15.79	**14.36**	**31.94**	**53.73**

Table 1. TVOC emission results for the 14 sample set per day.

Results also show that the new plywood with regular paint has the highest concentration than all other sample se which continued until its last day of testing. New plywood with green labelled paint was also recorded to have a high concentration. However, decrease on the concentration during the 7 day experiment period was more visible than the new plywood with regular paint. A significant decrease in TVOC concentration was observed in both 1 month old plywood with regular and green labelled paint. This decline was continued to be observed at the 2 years old plywood applied with regular paint.

Results also show that the new plywood with regular paint has the highest concentration than all other sample set. High concentration continued until its last day of sample set. New plywood with green labelled paint was also recorded to have a high concentration. However, decrease on the concentration during the 7 day experiment period was more visible than the new plywood with regular paint. A significant decrease in TVOC emission rates were observed in both 1 month old plywood with regular and green labelled paint. This decline was continued to be observed at the 2 years old plywood applied with regular paint.

The measurements of new plywood with regular paint and 1 month old plywood with regular paint during different periods and exterior weather conditions were found to have a good correlation. Both sample sets have almost the same average for both measurnent when compared with each other despite having different exterior environmental conditions.

5. REFERENCES

Department of Environment and Natural Resources, Forest Management Bureau http://www.forestry.denr.gov.ph

Funaki, R. and Tanabe, S., Chemical Emission Rates from Building Materials Measured by a Small Chamber, JAABE, 1(2), 2002.

Godish, T. 1995, Sick Building Syndrome: Definition, Diagnosis and Mitigation, Lewis Publication-CRC Press Inc.

Guidance Notes for the Management of Indoor Air Quality in Offices and Public Spaces, Indoor Air Quality Management Group, The Government of the Hong Kong Special Administrative Region, September 2003, Retrieved from http://www.iaq.gov.hk/cert/doc/GN-eng.pdf

Haghighat, F., De Bellis, L., 1997, Material Emission Rates: Literature Review, and the Impact of Indoor Air Temperature and Relative Humidity, Building and Environment, Vol. 33, No. 5, pp. 261-277, 1998

Molhave, L., Volatile organic compounds, indoor air quality and health, Walkinshaw, D.S (ed.) Proceedings of Indoor Air 1990, Ottawa, Canada Mortgage and Housing Corporation, Vol.5, p.15-33.

A. Rudd, A.T. Hodgson and S. Chandra, Volatile Organic Compound Concentrations and Emission Rates in New manufactured and Sire-Built Houses, Research Report 0803, building science.com

X. Yang, Q Chen, J, Zeng, and C.Y. Shaw, Mass transfer model for simulating volatile organic compound emission from wet coating materials applied to absorptive substrates, Intl. J. Heat Mass Trans., 44 (2001), 1803.

H. Yoshino, J Zhao, Y .Yushino, and K. Kumagai, A Study on Indoor Air Quality of Urban residential Buildings in China, JAABE, 4 (2), 2005.

A NEW APPROACH TO ESTIMATE EMISSION FACTORS OF SEMI-VOLATILE ORGANIC COMPOUNDS BASED ON THEIR PHYSICAL PROPERTIES

Jiping ZHU*, Peter MOCHUNGONG, and Julie S. O'BRIEN

Exposure and Biomonitoring Division, Health Canada, Ottawa, Ontario, Canada

*Corresponding author's email: jiping.zhu@hc-sc.gc.ca

Keywords: Emission, SVOC, Empirical Model, Indoor Air

SUMMARY

Knowledge of emissions of semi-volatile organic compounds (SVOCs) from indoor materials is important for assessing SVOC levels in indoor environments. Current measurement methods for determining SVOC emissions from products are chamber-based. One of the challenges of chamber-based methods is the strong absorption of SVOCs to the interior surface of the testing system. In this paper, we introduce a non-chamber approach for estimating SVOC emissions. A linear relationship describing the emission factor as a function of the chemicals' air diffusivity and vapour pressure is established using six aliphatic hydrocarbons (C11 to C16). A constant value (k) by the linear relationship was measured to be 1.04, 6.89, and 12.6 under surface air velocities of 0.01, 0.25, and 0.53 m/s, respectively. Estimated convective mass transfer coefficients of several phthalates are in good agreement with values reported by others. This approach has the potential to be used in estimating SVOC emissions from products.

INTRODUCTION

Semi-volatile organic compounds (SVOCs) are chemicals that have a saturated vapour pressure of 10^{-4} to 10^{-14} atm (1 atm = 760 mmHg) at an ambient temperature of 25 °C (Weschler and Nazaroff, 2008). Many SVOCs added to consumer products, especially plasticisers and certain flame retardants, are in additive form in the material, meaning that they are not chemically bound to the product matrix. Additive SVOCs in these products can emit, often at a slow rate, into the surrounding environment. Once emitted into the indoor environment, the fate of these SVOCs is influenced by their physical properties, especially their ability to partition between the gas and solid phases. Therefore, knowledge of SVOC emissions is important to assessing human exposure to these chemicals in indoor environments (Xu et al. 2010).

Compared to volatile organic compounds (VOCs), emission testing of SVOCs from products is still a relatively new research area. So far, determination of SVOC emissions has been largely carried out using chamber-based methods. However, due to their low vapour pressure, SVOCs have large sorption potentials to surfaces, a phenomenon that is called the sink effect, which poses a serious challenge in chamber-based emission tests when accurately determining the emission rates of SVOCs (Liang and Xu, 2014a).

A new method for estimating SVOC emissions is required, by creating a testing system that eliminates the impact of the sink effect. In this study, we attempt to generate a non-chamber-based approach to estimate SVOC emissions through the development of an empirical prediction model. This model was based on the measured emission factor (EF) of individual SVOCs and their physical properties.

METHODOLOGIES

Micro-balance system set-up. An in-house built micro-balance testing system was used to measure the weight loss of chemicals over time. The system is comprised of a 0.1 m^3 housing made of aluminum panels with inner dimensions of 60 x 40 x 50 cm for length, height, and width, respectively. The interior surface of the housing was electronically polished. A micro-balance (WXSS205, Mettler-Toledo AG, Switzerland) with a readability of 10 μg was placed on an anti-vibration platform (Vibrasorb® Vibration Damping Mount, 45 x 56 cm) that sits on the floor of the chamber. The left side of the housing is fitted with two probes. The first probe delivers moisturized air into the housing at a nominal flow of 200 ml/min. The moisturized air is a mixture of 150 ml/min of ultra-zero air passed through a water column and 50 ml/min of ultra-zero air, controlled by a FLEC Air Control/1000 device. Air exits from the housing through an exhaust opening located on the left side of the chamber's roof. The second is a telescoping probe (TSI9545-A, TSI Incorporated, USA) that records temperature, relative humidity, and air velocity inside the chamber. The right side of the chamber has a round hatch with a diameter of 0.17 m^3 for introducing samples onto the microbalance. For the experiment involving the investigation the impact of surface air velocity on emission rates, a small fan was placed in a corner inside the housing. The speed of the fan was controlled by a L501 Variable Control Transformer Power Source (Staco Energy Products Co., Dayton, OH, USA). The correlation between the power source's voltage output and surface air velocity was calibrated.

Weight loss measurement. Chemicals (undecane (≥99%), dodecane (≥99%), tridecane (≥99%), tetradecane (≥99%), pentadecane (≥99%), hexadecane (≥98.0%)) were purchase from Sigma-Aldrich Canada. At the beginning of each experiment, about 4 g of sample was added into an aluminum weighing dish that had been previously fitted with a filter paper (which covered the flat bottom surface of this weighing dish). The filter paper acted as the substrate on which the sample spread uniformly. The evaporation surface area of each sample was 0.00196 m^2. After the balance display unit was zeroed, the sample dish was placed on the micro-balance through the hatch on the right side. The hatch was then sealed, and the weight loss measurements began. The micro balance transmitted minute-by-minute changes in weight to a laptop, where the data was saved in an excel file. Minute-by-minute changes in temperature, relative humidity, and air velocity inside the chamber were recorded and saved simultaneously (on the air velocity meter), and were later paired together with the emission data during analysis.

RESULTS AND DISCUSSION

The temperature inside the housing depended on the temperature of the room, which was typically between 20 °C and 25 °C. Although the desired nominal relative humidity (RH) value was 50 ± 10% inside the housing, recorded RH values varied between 30% and 50% depending on the weather conditions of the day and the RH of the room where the testing system was located. The surface air velocity was found to be very low at less than 0.01 m/s when no fan was operated inside the housing. The other desired air velocities were achieved with the aid of a small mixing fan operated inside the housing.

The weight of the test chemicals was recorded over time. The slope describes the rate of weight loss, and hence the emission rate (ER, g/min). The emission rate was converted to the emission factor (EF, μg/(m²s)). The emission factor is calculated as follows:

$$EF = h_m\, y_0 = k\, D_{air}^{2/3}\, P_{sat} \qquad (1)$$

where h_m is the convective mass transfer coefficient (m/s) and y_0 is the chemical concentration (μg/m³) in the boundary layer of the evaporation surface (Liang and Xu, 2014a). Both air diffusivity (D_{air}, m²/s) and saturated vapour pressure (P_{sat}, μg/m³) values were obtained online from "EPI Suite" available at the U.S. Environmental Protection Agency's website. The values and significance of k will be discussed below.

Table 1. Air diffusivity (D_{air}), vapor pressure (V.P.), saturated vapor pressure (P_{sat}), and emission factor (EF) of the tested aliphatic hydrocarbons (C11 to C16) at the air velocities (v), as indicated.

Name	D_{air} (m²/s)	V.P. (25 °C) (mmHg)	P_{sat} (μg/m³)	EF v ≤0.01 m/s (μg/(m²s))	EF v =0.25 m/s (μg/(m²s))	EF v =0.53 m/s (μg/(m²s))
Undecane	5.63×10^{-6}	4.12×10^{-1}	3.47×10^{6}	1.12×10^{3}	7.52×10^{3}	1.38×10^{4}
Dodecane	5.36×10^{-6}	1.35×10^{-1}	1.24×10^{6}	4.33×10^{2}	2.70×10^{3}	4.94×10^{3}
Tridecane	5.13×10^{-6}	5.58×10^{-2}	5.54×10^{5}	1.86×10^{2}	1.14×10^{3}	1.82×10^{3}
Tetradecane	4.92×10^{-6}	1.16×10^{-1}	1.24×10^{5}	4.80×10^{1}	3.20×10^{2}	5.65×10^{2}
Pentadecane	4.74×10^{-6}	3.43×10^{-3}	3.92×10^{4}	1.19×10^{1}	9.61×10^{1}	2.17×10^{2}
Hexadecane	4.56×10^{-6}	1.43×10^{-1}	1.74×10^{4}	4.70×10^{0}	3.84×10^{1}	5.38×10^{1}

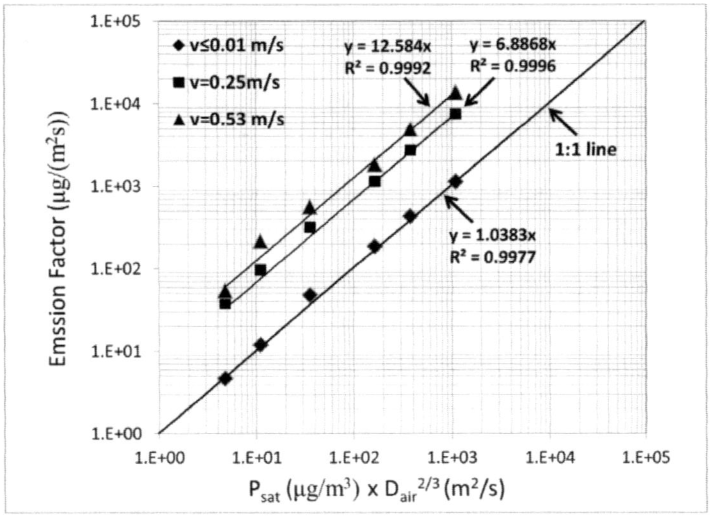

Figure 1. Linear correlations between emission factor and the product of the saturated vapor concentration (P_{sat}) multiplied by the air diffusivity raised to two-thirds power ($D_{air}^{2/3}$), for three different surface air velocities (v). The slopes are the k values that can be used in equation (1).

EFs of the six aliphatic hydrocarbons, each at three different surface air velocities (v), were determined (Table 1). The measured EFs were compared with the predicted EFs based on $D_{air}^{2/3} \times P_{sat}$ according to Equation 1 (Figure 1). The slope in Figure 1 represents the k value; they were 1.04, 6.89, and 12.6 for v = 0.01, 0.25, and 0.53 m/s, respectively.

The term $k \times D_{air}^{2/3}$ is considered to be equal to h_m, a critical parameter in estimating SVOC emissions (Liang and Xu, 2014a). The dependency of the k value on air velocity is clearly demonstrated in Figure 1. Interestingly, when air velocity approaches zero, the k value is close to 1. This implies that in the absence of bulk air movement, where only diffusion (random thermal motion) occurs, the EF of an SVOC can be predicted by its D_{air} and P_{sat} values (k = 1, equation 1). Table 2 lists the estimated h_m values calculated by equation 1 and those reported by Liang and Xu, 2014a for three common phthalates, both at 25 °C and low air velocity ($v \leq 0.01$ m/s). Good agreement between the h_m values is present.

Table 2. Comparison of estimated and reported h_m values for three phthalates.

Phthalate	D_{air} (m^2/s)	V.P. (mmHg)	P_{sat} ($\mu g/m^3$)	h_m, (m/s), $v \leq 0.01$ m/s	
				This Study	Lit[(a)]
Dimethyl phthalate	5.92E-06	3.08E-03	3.22E+04	3.40E-04	3.12E-04
Di-n-butyl phthalate	4.55E-06	2.01E-05	3.01E+02	2.85E-04	2.54E-04
Di-(2-ethylhexyl) phthalate	3.66E-06	1.42E-07	2.98E+00	2.46E-04	2.10E-04

(a): Liang and Xu, 2014a

CONCLUSIONS

Herein, we have reported a novel approach to estimate the emissions of individual SVOCs. Future research will focus on (1) estimating SVOC emission factors from products, where concentrations of additive SVOCs at the boundary layer (y_0) will be used (Liang and Xu, 2014a), (2) exploring y_0 and temperature dependence (Liang and Xu, 2014b), and (3) developing a prediction model to estimate y_0 based on SVOC concentrations and product types.

ACKNOWLEDGEMENTS

This work is financially supported by the Canadian government's Clean Air Regulatory Agenda.

REFERENCES

Weschler CJ, Nazaroff WW. (2008). Semivolatile organic compounds in indoor environments. *Atmospheric Environment*, **42**, 9018 – 9040.

Xu Y, Cohen Hubal EA, Little JC. (2010). Predicting residential exposure to phthalate plasticizer emitted from vinyl flooring: Sensitivity, uncertainty, and implications for biomonitoring. *Environmental Health Perspectives*, **118**, 253 – 258.

Liang Y, Xu Y. (2014a). Improved method for measuring and characterizing phthalate emissions from building materials and its application to exposure assessment. *Environmental Science & Technology*, **48**, 4475 – 4484.

Liang Y, Xu Y. (2014b). Emission of phthalates and phthalate alternatives from vinyl flooring and crib mattress covers: The influence of temperature. *Environmental Science & Technology*, dx.doi.org/10.1021/es504801x.

NEW VOC COMPOUNDS IN THE INDOOR ENVIRONMENT FROM THERMOPLASTIC FILAMENTS FOR 3D PRINTING

Olaf WILKE*, Stefan SEEGER

BAM Federal Institute for Materials Research and Testing, Berlin, Germany

*Corresponding email: olaf.wilke@bam.de

Keywords: VOC, Emission, 3D printer

SUMMARY

A screening test for potential emissions of volatile organic compounds (VOC) was run on different thermoplastic filaments used for 3D printing. The method of direct thermal desorption was used to simulate the high temperatures during the 3D printing process and to identify the main compounds emitted from the filaments. A large number of unexpected compounds were detected that might affect the user's health and have an impact on indoor air chemistry.

INTRODUCTION

The use of desktop 3D printers is increasing. Compared to other devices with known emissions, e.g. laser printers, there is still a lack of information on possible emissions of VOC and ultrafine particles during operation and the effect on indoor air quality. Most of the commercially available desktop 3D printers operate with a molten polymer deposition. For this process a solid thermoplastic filament is heated in an extrusion nozzle. Most filaments for desktop 3D printers use either acrylonitrile butadiene styrene (ABS) or polylactic acid (PLA) as filament. Alternatives are polyvinyl alcohol (PVA) or polycarbonate (PC).

This study is the first step of an investigation of potential indoor air pollution caused by 3D printing.

METHODOLOGIES

Eight different thermoplastic filaments for 3D printers were analysed by direct thermal desorption followed by GC-MS identification of the emitted substances. Direct thermal desorption was done by desorbing 5 mg of the feedstock for 1 minute at a temperature of 210°C. This is an average temperature for 3D printing with thermoplastic filaments.

RESULTS AND DISCUSSION

The main identified compounds for each filament are listed in Table 1. Most compounds are completely new with regard to VOC emissions into the indoor environment.

The comparison of the 4 different filament groups showed the highest overall emissions from ABS, followed by PLA, PC and PVA. Filament ABS 2 emitted mainly SVOCs and triphenyl phosphate, the latter has the highest emission for a single compound from all evaluated filaments.

Table 1. Compounds (main emission) from thermoplastic filaments when heated to 210 °C

Filament	Compound 1	Compound 2	Compound 3	Compound 4
ABS 1	Styrene	Not identified		
ABS 2	Triphenyl phosphate	Phenol	BHT	Hexadecanoic acid
PLA 1	Lactide	Lactic acid		
PLA 2	Lactide	1,6-Dioxacyclododecane-7,12-dione	Lactid acid	Not identified
PLA 3	Lactide	Lactid acid	n-Hexadecane	n-Heptadecane
PLA 4	Lactide	Lactic acid	1-Dodecanol	2,6-Di-tert-butyl-p-quinone
PVA	Glycerine	Triethanolamine		
PC	Chlorobenzene	Phenol	Bisphenol A	Diethyl phthalate

CONCLUSIONS

Thermoplastic filaments are a new source of VOC emissions due to the high temperatures associated with 3D printing, that can reach up to 270°C. Some of the detected compounds like lactic acid, lactide and bisphenol A have never been described before in the indoor environment. Additionally some of the main substances could not be identified and some others might have the potential to affect the indoor air chemistry.

The appearance of some newly detected compounds raises concerns about potential health effects for the users of 3D printers at home.

Test chamber measurements will be set up to measure the emission rates of these VOC as well as particulate emissions under realistic conditions.

ACKNOWLEDGEMENT

We thank Doris Brödner, Sabine Kalus, Ria Juritsch, Monika Rothhardt and Maximilian Ebisch for the experimental work.

REFERENCES

Brent Stephens et al. (2013). Ultrafine particle emissions from desktop 3D printers. *Atmospheric Environment*, 79, 334-339.

CHARACTERIZING UNCERTAINTY OF A FORMALDEHYDE REFERENCE STANDARD

Dustin POPPENDIECK[1,*]

[1]Indoor Air Quality and Ventilation Group, Engineering Laboratory, 100 Bureau Drive, Mail Stop 8633, Gaithersburg, MD 20899-8633

[*]Corresponding email: dustin.poppendieck@nist.gov

Keywords: Chamber testing, formaldehyde, reference standard

SUMMARY

High performance buildings need to be energy efficient and provide adequate ventilation for indoor air quality, which can be made easier with low emitting building materials. Based in part on these motivations, the United States Environmental Protection Agency (EPA) issued a proposed rule to implement statutory formaldehyde emission standards for hardwood plywood, medium density fiberboard, and particle board products. A Formaldehyde Reference Standard (FRS) is being developed by the National Institute of Standards and Technology (NIST) to assist the implementation of this rule. The goal of this effort is to produce a FRS with small, stable level of uncertainty that can be used in emission chamber system evaluation and troubleshooting. This presentation will highlight uncertainty quantification from preliminary testing of the proposed NIST FRS to begin in February 2015.

INTRODUCTION

High performance buildings need to balance energy efficiency and providing adequate ventilation for indoor air quality. The level of required ventilation can potentially be reduced when low emitting building materials are used. The United States Environmental Protection Agency (EPA) issued a proposed rule to implement statutory formaldehyde emission standards for hardwood plywood, medium density fiberboard, and particle board products. The proposed rule requires formaldehyde emission chamber testing of various wood products and third-party certification of emission testing. A Formaldehyde Reference Standard (FRS) is being developed by the National Institute of Standards and Technology (NIST) to assist implementation of this rule. The intent of the FRS is to assist in evaluating chamber testing performance and support the small chamber/large chamber equivalency requirements in the proposed EPA rule and the existing California Air Resources Board (CARB) Airborne Toxic Control Measure (ATCM) to Reduce Formaldehyde Emissions from Composite Wood Products. Previous interlaboratory testing of standard wood products has shown large variability (e.g. 20 % standard deviations, Yrieix 2010). The goal of this effort is to produce a FRS with small, stable level of uncertainty that can be used in emission chamber system evaluation and troubleshooting.

This presentation will highlight uncertainty quantification from preliminary testing of the proposed NIST FRS. All measurements related to chamber testing (formaldehyde concentration, temperature, relative humidity, flow and pressure) will be traceable to NIST primary standards. The experiments will be conducted in the spring of 2015; this abstract reviews the methods to be used.

METHODOLOGIES

Experiments will be conducted following ASTM D6007-02 (2008) (Standard Test Method for Determining Formaldehyde Concentrations in Air from Wood Products Using a Small-Scale Chamber) as closely as possible. In this method, small samples of wood are introduced into a climate-controlled chamber (0.02 m^3 to 1 m^3 volume, 0.5/h air change rate, 25 °C ± 1 °C temperature and 50 % ± 4 % relative humidity) at a standard loading ratio. Formaldehyde concentrations are measured once the chamber has reached equilibrium and emission rates are calculated. This work will deviate from ASTM D6007 in the following ways:

- The tested material is a formaldehyde-water solution contained in a small bottle (as described below) instead of a wood sample.
- The samples will be analyzed using a quantum cascade laser trace gas monitor. This instrument provides real time data with standard uncertainties of the measurements less than 0.1 ppb_v (0.13 $\mu g/m^3$), making it advantageous compared to the chromotropic acid test procedure described in D6007.

Experiments to test the formaldehyde reference standard will be conducted in the NIST small chamber system. Only one 50 L chamber will be used in this research. A zero air generator will supply formaldehyde free air to the chamber. To achieve the desired relative humidity (50 %), humidified and dry air streams will be mixed in a controlled fashion using two mass flow controllers (MFC).

Each experiment will consist of placing a formalin-containing Teflon and stainless steel bottle in the chamber (Figure 1). This design is based on work done by Wei et. al. (2013). The 1 mL ampules of 16 % formalin solution (mass of formaldehyde/volume of distilled deionized water) will be acquired from an outside manufacturer.

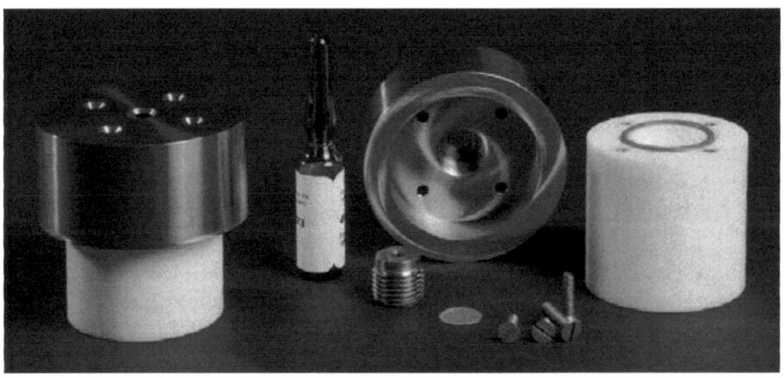

Figure 1: NIST formaldehyde reference standard (FRS). The FRS consists of a formalin ampule, a stainless steel lid, a compression set screw, a polydimethylsiloxane (PDMS) membrane, four stainless steel screws, a Teflon base with 3 mL formalin well, and an o-ring. The lid is sealed to the bottle with the screws.

Formaldehyde from the FRS will diffuse through a thin polydimethylsiloxane (PDMS, 1.0 mm) membrane, eventually reaching a steady state formaldehyde concentration in the chamber. The target steady state formaldehyde concentration is between 20 and 50 ppb_v (25 $\mu g/m^3$ and 63 $\mu g/m^3$). The key component of the FRS is the replaceable membrane. Each time the FRS is used a new membrane and new formalin solution will be placed in the bottle. Formaldehyde concentrations will be measured using a quantum cascade laser trace gas monitor. Given the known absorption spectra for each of the known constituents, the measured spectra can be deconvoluted to determine the concentration of the formaldehyde, formic acid and water in the sample. Currently experiments are planned to be run until the formaldehyde concentration reaches a value that does not change more than 2 % over six hours. Ten bottles have been manufactured in this first phase of the project. Initially a random selection of five of these bottles will be tested twice. Each replicate experiment of each bottle will use a new membrane and new formalin ampule. These initial data will be used to confirm that the uncertainty of the emission rate is within acceptable bounds.

Measurement Equation

The chamber system's small mixing fan produces a uniform contaminant concentration. Hence, the emission rate of the FRS will be determined using a single-zone mass balance approach. The formalin concentration in the bottle should remain relatively constant for the duration of the experiment, resulting a steady-state diffusion across the membrane. This constant emission rate will result in the chamber reaching a steady state formaldehyde concentration. The emission rate can be determined as follows:

$$E = Q(C - C_{in}) \tag{1}$$

Where E is the emission rate (μg/h), Q is the total flow rate (m^3/h), C is the formaldehyde concentration in the chamber (μg/m^3) and C_{in} is the formaldehyde concentration entering the chamber from sources other than the formaldehyde diffusing out of the bottle (μg/m^3, which should be zero).

Uncertainty in the emission rate will be quantified using the "Evaluation of measurement data — Guide to the expression of uncertainty in measurement" (JCGM 2008). This approach assumes that measurement uncertainty "reflects the lack of exact knowledge of the value" of the emission rate. Uncertainty is "best described by means of a probability distribution over the set of possible values" for the emission rate. To determine the probability distribution for the value of the emission rate, a range of factors that could influence the measured value must be considered. The emission rate is expected to depend on a range of factors including temperature, relative humidity, pressure, flow, exposed membrane area, membrane thickness, and formalin concentration. The first four factors are set points for the experimental system and may have a systematic impact on the emission rate. The last three factors will impact the emission rate in a random manner. Random variation in exposed membrane area, membrane thickness, and formalin concentration will be captured in the standard uncertainty of the emission rate. (Any instability in the values of the set point factors that lead to variations in the data will be included in these standard uncertainties as well.) Systematic errors will not be captured in the standard uncertainty values for flow and concentration. Hence, systematic errors need to be included in the uncertainty in another manner. The measurement equation used to determine the expanded uncertainty is:

$$E = Q(C - C_{in}) * f_{temperature} * f_{relative\ humidity} * f_{pressure} * f_{HCHO\ monitor} * f_{flow\ rate\ \mathrm{dry}} * f_{flow\ rate\ \mathrm{wet}} \tag{2}$$

Where f values are the influence quantities, which have a value of 1 for computation of the emission rate and a standard uncertainty determined via calibration of the measurement instruments to NIST primary standard reference values. Influence quantities are not required in the definition of the measurand (E in this case) but do have an effect on the measurement result through their contributions to the uncertainty.

The standard uncertainty in flow (Q) and concentration (C-C_{in}) will be quantified using standard deviations of the flow and concentration data, respectively. The relative contributions to the total variation seen between bottles, between membranes and ampules, and within each bottle, membrane, and ampule will all be considered as part of the assessment of the standard uncertainty of the emission rate. The standard uncertainties of the values of the influence factors will be calculated from the uncertainty comparison between the measured parameter and the relevant NIST primary standard. For example the uncertainty for the influence factor for the formaldehyde monitor will be determined from the variation seen in comparisons between the concentration values from the formaldehyde monitor and the NIST primary gravimetric standard formaldehyde reference source.

The uncertainty in the emission rate will be quantified using the measurement equation (2). A first-order Taylor series expansion with n=1 can be used to determine a linear approximation for the measurement equation near the value of the measurand. The Taylor series then can be used to determine the propagation of uncertainties to determine the combined standard uncertainty. The expanded uncertainty will be calculated from the combined standard uncertainty using a coverage factor of two.

RESULTS AND DISCUSSION

Experiments will commence in February 2015 and complete by June 2015. A summary of the data will be presented at the conference.

In order to develop an acceptable NIST standard reference material based on the FRS approach, additional tests will need to be conducted if the uncertainty from these preliminary experiments is deemed acceptable. Further testing will include testing of a subset of production bottles, membranes and formalin. In addition, a subset will be tested to determine stability of the system after 6 and 12 months. After stability has been confirmed and the final uncertainty deemed acceptable, untested production bottles may be certified and sold.

CONCLUSIONS

Confidence in testing results is needed if emission data are to be submitted to regulatory agencies or is used to meet voluntary product labelling standards. Verification of formaldehyde emission testing data are of particular interest, and therefore merit a reference standard with small, stable level uncertainty. This research will determine if the proposed NIST FRS will have acceptable uncertainty to serve as a verification tool.

ACKNOWLEDGEMENT

Funded in part by EPA Interagency Agreement Number DW-13-92388501-1.

REFERENCES

Evaluation of measurement data — Guide to the expression of uncertainty in measurement. JCGM 100:2008 GUM 1995 with minor corrections: Joint Committee for Guides in Metrology; 2008.

Yrieix, C.; A. Dulaurent; C. Laffargue; F.; Maupetit; T. Pacary, and E. Uhde. (2010). Characterization of VOC and formaldehyde emissions from a wood based panel: Results from an inter-laboratory comparison. Chemosphere 79: 414–419. doi:10.1016/j.chemosphere.2010.01.062

Wei, W.; C. Howard-Reed; A. Persily; Y. Zhang. (2013). "Standard formaldehyde source for chamber testing of material emissions: model development, experimental evaluation, and impacts of environmental factors." Environ Sci Technol 47(14): 7848-7854.

INDOOR MICROBIOMES CHANGE WHEN WE TRY TO CONTROL THEM

Alexander MAHNERT[1*], Christine MOISSL-EICHINGER[2,3], Henry MÜLLER[1], Alexander J. PROBST[4], Roscel Amor ORTEGA[1], Parag VAISHAMPAYAN[5], Kasthuri VENKATESWARAN[5], Gabriele BERG[1]

[1]Institute of Environmental Biotechnology, Graz University of Technology, Petersgasse 12/I, 8010 Graz, Austria
[2]Medical University Graz, Department of Internal Medicine, Auenbruggerplatz 15, 8036 Graz, Austria
[3]BioTechMed Graz, Krenngasse 37, 8010 Graz, Austria
[4]Lawrence Berkeley National Laboratory, Earth Sciences Division, 1 Cyclotron Rd., Berkeley, CA 94720, USA
[5]Biotechnology and Planetary Protection Group, Jet Propulsion Laboratory, 4800 Oak Grove Drive, Pasadena, CA 91109, USA

[*]Corresponding email: a.mahnert@tugraz.at

Keywords: Viable controlled and uncontrolled indoor microbiomes, cleanrooms, intensive care units, wildlife park, greenhouse

SUMMARY

The excessive removal of microbes from the built environment is daily routine. However effects of these stringent maintenance procedures on our surrounding microbiomes are less known. Deep sequencing technologies combined with viability assays and quantitative measures revealed high reduction of viable microbial abundance in cleanrooms, higher similarities of viable microbiomes between controlled and uncontrolled areas compared to the total microbial fraction, and distinct profiles of microbial communities from floors, medical devices and workplaces in intensive care units. Surprisingly, we found not only an overlap of the indoor microbiome with humans, but also the plant microbiome. Hence plant leaves could be identified as a main source for microbial distribution on the surrounding indoor environment.

Knowledge of certain key species and their ecological key functions in controlled as well as uncontrolled built environments will help to control indoor environments in a more sophisticated way for our health inside buildings in the future.

INTRODUCTION

Since humans built homes, people establish new artificial environments for microbes. Nowadays, in many developed areas of the world, this environment is our prime habitat for living. However modern buildings are completely different in respect of their microbiomes compared to the natural outdoor environments where mankind evolved. Especially when human try to control an indoor habitat by fast removal or eradication of microbes the microbiome of the built environment is changed. Recently many studies showed that most microbes are beneficial and how important the human microbiome is towards people's health. Hence, our hypothesis is that a well-balanced microbiome with a higher diversity will result in more healthy indoor environments (Berg *et al.*, 2014; see Fig. 1 for an overview).

The objective is to better understand differences of controlled and uncontrolled built environments in respect towards human health and develop new strategies to manage them in the future.

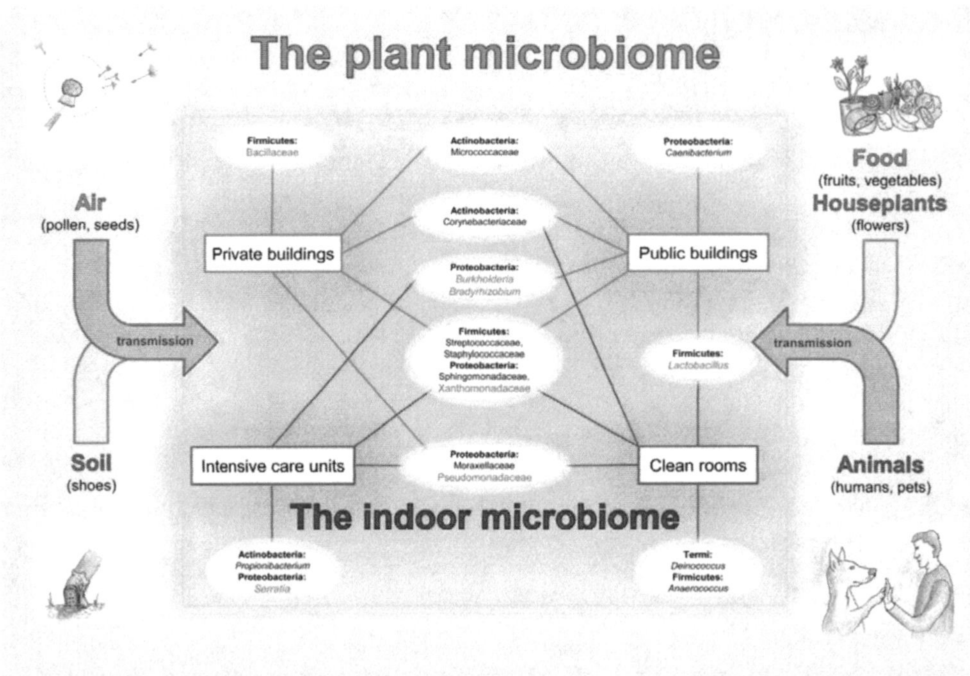

Figure 1. Supposed relationships between the plant and the indoor microbiome (Berg *et al.*, 2014)

METHODOLOGIES

Molecular samples from floor, wall and plant leave surfaces were obtained with Alpha Wipes® (TX1009, VWR International GmbH, Vienna, Austria) and the BiSKit (biological sampling kit, QuickSilver Analytics, Abingdon, MD). The SKC BioSampler® (SKC Inc., PA, USA) was applied to sample the indoor air.

Shotgun metagenomes 16S/18S rRNA and ITS gene amplicons of Archaea, Bacteria, Eukaryota and fungi were investigated via high throughput deep sequencing technologies (454 pyrosequencing and Illumina HiSeq and MiSeq). In addition microarray technology (PhyloChip G3™) and traditional 16S cloning were applied to round out molecular analysis. A special focus was set on potential viable microbiomes. Hence, viability assays using ATP (adenosine tri-phosphate) (Venkateswaran *et al.*, 2003) and PMA (propidium monoazide) were deployed (Vaishampayan *et al.*, 2013) and supported by qPCR to investigate microbial abundance. Finally, VOC's (volatile organic compounds) assays of culturable indoor bacteria (Cernava *et al.*) were performed to elucidate the biotechnological potential of microbes living indoors.

RESULTS AND DISCUSSION

Influences from human and plants manifest themselves in a characteristic composition of microbiomes in controlled and uncontrolled built environments. Microbial compositions could be further correlated to different room maintenance:

The common indoor plant *Chlorophytum comosum* significantly increased microbial abundance on room surfaces and changed microbial diversity. Plant leaves could be identified as a main driver on shifts in abundance and diversity (Mahnert *et al.*). In the setting of a greenhouse the phyllosphere microbiome correlated to a certain magnitude with the leave morphology and room climate.
In more controlled built environments like an intensive care unit distinct profiles were determined for samples from floors (dominated by *Acinetobacter*) and medical devices as well as workplaces (e.g. *Burkholderia* and *Bradyrhizobium*) (Oberauner *et al.*, 2013). Highly controlled built environments like cleanrooms showed a tremendous reduction of their viable microbiomes from 45 to 1% (relative abundance) (Moissl-Eichinger *et al.*, 2015). The viable microbiome of an uncontrolled gowning room and a controlled cleanroom had a higher similarity than their total microbial fractions as revealed by PMA treatment. Masking the dominant amount of DNA from dead microbes helped to detect 16S rRNA gene sequences of 33 underrepresented genera. The viable microbiome of a cleanroom harbored several survival specialists such as *Haloferax* or *Sporosarcina*, which established themselves independently in the cleanroom from the surrounding indoor environment (Mahnert *et al.*). Investigations of a whole cleanroom complex (see Fig. 2) exhibit additional reductions of food and soil associated microbes inside the high controlled areas and the changing room could be identified as the main route of contamination into the cleanrooms (Moissl-Eichinger *et al.*, 2015).

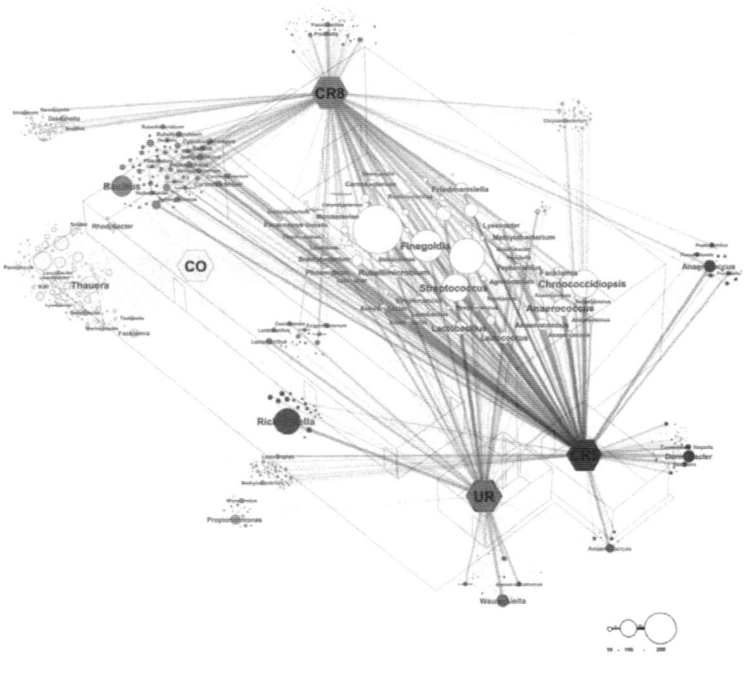

Fig. 2: Network analysis of OTUs of a cleanroom facility visualized with the Cytoscape 2.8.3 layout edge-weighted spring embedded eweights. Samples from four different rooms are displayed as hexagonal nodes on an abstracted map of the clean room facility in colors yellow (CO – check out room), red (UR – changing room), light blue (CR8 – ISO8 clean room) and dark blue (CR5 – ISO5 clean room). pOTUs are displayed as nodes. Node color is defined by sample source and respective color mixtures (according to Itten's color circle) are used for shared pOTUs. pOTUs resolved till genus level are labelled. Node and font size

represents abdundance of pOTUs (score is indicated in the legend below). Edge width and opacity correlates with respective eweights (edge weights represent sequence abundance in respective pOTUs). Network layout (pOTUs distance) was slightly modified to fit on the abstracted map (Moissl-Eichinger *et al.*, 2015).

CONCLUSIONS

The excessive removal and eradication of mainly beneficial microbes from the indoor environment could have adverse effects towards human health, since human well-being is dependent on the human microbiota and its diversity. However, stringent cleaning and maintenance procedures in intensive care units, operation theaters or cleanrooms is daily routine to achieve an impossible goal - microbial-free environments. Hence this unsophisticated removal of microbes should be reconsidered and new techniques for beneficial microbiome control should be developed to stabilize beneficial microbial communities inside buildings.

ACKNOWLEDGEMENT

Lisa Oberauner-Wappis, Tobija Glawogger, Jessica Cisneros, Taguhi Arakelian, Myron T. La Duc, Anna Auerbach, Simon Barczyk, Gerhard Kminek, Wolf-Gunthram Freiherr von Schenck, Isabelle and Franziska Mahnert for support during sampling and the following laboratory assays. KUWI scholarship for short time academic research at Caltech and the JPL Visiting Student Researchers Program.

REFERENCES

Berg, G., Mahnert, A. & Moissl-Eichinger, C. (2014). Beneficial effects of plant - associated microbes on indoor microbiomes and human healt. *Frontiers in Microbiology*. doi:10.3389/fmicb.2014.00015

Cernava, T., Aschenbrenner, I.A., Grube, M., Liebminger, S. and Berg G. A novel high-throughput assay for the detection of bioactive volatiles: evaluation by screening of lichen-associated bacteria. *Submitted*

Mahnert, A., Moissl-Eichinger, C., and Berg, G. Plants Substantially Shape the Microbiome within the Built Environment *Submitted*

Mahnert, A., Vaishampayan, P., Probst, A.J., Auerbach, A., Moissl-Eichinger, C., Venkateswaran· K., and Berg, G. An unexpected and unique microbial diversity is alive in cleanroom facilities. *Submitted*

Oberauner, L., Zachow, C., Lackner, S., Högenauer, C., Smolle, K.-H., and Berg, G. (2013) The ignored diversity: complex bacterial communities in intensive care units revealed by 16S pyrosequencing. *Sci. Rep.* **3**: 1413.

Moissl-Eichinger C, Auerbach AK, Probst AJ, Mahnert A, Tom L, Piceno Y, *et al.* (2015). Quo vadis? Microbial profiling revealed strong effects of cleanroom maintenance and routes of contaminantion in indoor environments. *Sci Rep* **5**: 9156.

Vaishampayan, P., Probst, A.J., La Duc, M.T., Bargoma, E., Benardini, J.N., Andersen, G.L. and Venkateswaran, K.. (2013). New perspectives on viable microbial communities in low-biomass cleanroom environments. *ISME J.* **7,** 312–24.

Venkateswaran, K., Hattori, N., La, M. T. & Kern, R. (2003). ATP as a biomarker of viable microorganisms in clean-room facilities. *Journal of Microbiological Methods.* **52,** 367–377

BUILDING CHARACTERISTICS AS DETERMINANTS OF MOISTURE IN WATER-DAMAGED HOMES – A STUDY DESIGN

Michal P SPILAK[1,*], Iman SYLVAIN[2], Rachel ADAMS[2] and Michael S Waring[1]

[1]Department of Civil, Architectural and Environmental Engineering, Drexel University, USA
[2]Department of Plant and Microbial Biology, University of California Berkeley, USA

[*]Corresponding email: mps95@drexel.edu

Keywords: building characteristics, dampness, indoor air, moisture in materials

SUMMARY

Moisture-related problems in residential buildings have been associated with increased risk of negative health outcomes such as allergic and respiratory problems. The presence of dampness in indoor environment increases the risk of fungi contamination and presence of mould which has likewise been associated with high risk factor for respiratory discomfort.

This study was designed as a replicated survey of 60 residential units. Half of the residences experienced exacerbated water damage surrounding the Superstorm Sandy in 2012. During three-week period continuous measurements of indoor air parameters and concentrations of particles were performed. In the end of each period moisture content was measured. House characteristics and occupancy behaviours were collected by means of researchers' observation and occupants' diary, respectively.

The expected outcomes of the study are to elucidate building characteristics and occupancy behaviour that increase risk of dampness in indoor environment.

INTRODUCTION

In October 2012 south east coast of the USA was struck by the Superstorm Sandy. The extensive flooding in the aftermath of Sandy resulted in total damages over $68 billion USD (either direct or indirect). Aside from large flooding, many buildings in NYC area suffered from severe water intrusion. Both, the flooding and the water intrusion created ideal conditions for increased risk of dampness.

A recent reviews report that living or working in buildings with dampness and mould is consistently associated with negative health outcomes, including allergic and respiratory problems (Bornehag et al. 2004a; Bornehag et al. 2004b; Quansah et al. 2012).

Presence of dampness has been generally associated with older homes, building with poor housing conditions and lack of ventilation (Hägerd-Engman et al. 2009; Salo et al. 2008; Howden-Chapman et al. 2005). Aside from building characteristics different climate conditions are likely explanations of high number of moisture-related problems in building studies. There is also a wide variation how is dampness defined (Bornehag et al. 2001). Studies focusing on dampness in buildings have reported visible mould, damp stains, condensation on walls and windows and mouldy odours as common indicators of dampness indoors (Suna and Sundell 2013). Results from these studies are based on self-reported dampness. This brings up a risk of recall bias on the actual dampness presence indoors (Quansah et al. 2012; Bornehag et al. 2001).

This project aims to elucidate building characteristics and occupants' behaviour as a factor significantly influencing risk of indoor dampness and compare self-reported dampness with measured values.

METHODOLOGIES

In total 60 units in Brooklyn and Manhattan area were selected for this study. The design aimed for 30 residences with varying severity of water damage as a result of either water intrusion during the Superstorm Sandy or caused by pipe leaks. Shown in Figure 1 is the scheme of investigated houses. The study was designed as a set of two three-week measurement periods. The first period was completed within three-month period starting January 2015. The second period is scheduled for summer 2015.

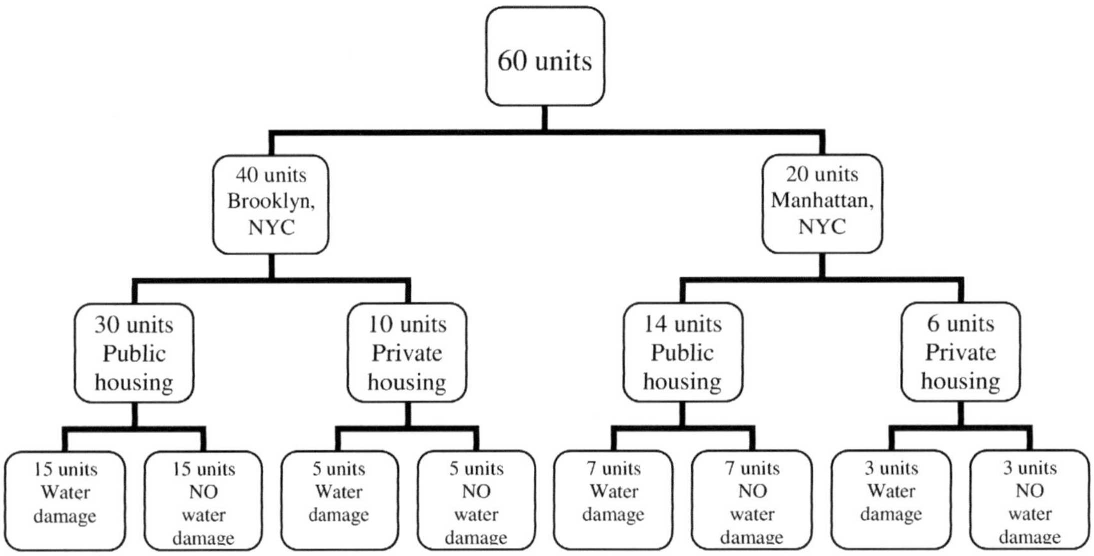

Figure 1. Design of the house selection.

Prior the measurement information about the building and the unit was collected by means of researchers' observation and occupants' survey. Collected information included: location of the building; design and age of the building; square footage of each room; flooring material; wall type for each room; number of windows; pet type and number; houseplant size and number; HVAC presence and type; and use of any portable air cleaners, humidifiers etc. Occupants participating the study were surveyed regarding their number at home and approximate time per day each occupant spend home. Recorded were occupants' cooking and cleaning habits had due to mold or water damage in last 12 months. The questionnaire took into account whether the known water damage occurred within the individual unit or in the whole building.

Throughout the measurement period we collected real-time data of indoor air temperature (°C), indoor air relative humidity (%) and levels of CO_2 (ppm). Additionally, the real-time use of the heating or cooling systems and window opening in each home was captured with another data logger.

During the last day of the measurement period moisture content in exterior walls and floors (%) was determined using invasive and non-invasive measurement technique. Type of exterior-wall material was recorded together with surface temperature of measured spot (ºC). Any visible mold present in unit was photographed and size of the mold area was documented. Additional indicators of inadequately ventilated environment, e.g. steam condensation on walls or windows, presence of moldy odor were recorded.

RESULTS AND DISCUSSION

The dampness assessment will be based on detailed information about the building and dwelling characteristics, occupants' survey, occupants' diary, measurements of moisture content in building materials and measurements of indoor air conditions. In general, sources of dampness can be divided into three groups (i) building characteristics; (ii) indoor sources-mainly from occupants' activities and behavior and (iii) outdoor sources that permeate the building due to leaks or intrusion. Building characteristics, together with the impact of geographical location and season of the year, are considered as significant determinants of moisture-related problems. In many studies high risk of dampness is associated with older houses, poor housing conditions with inadequate or no insulation and houses with absence of an HVAC system (Haas et al. 2007; Hägerd-Engman et al. 2009). We expect similar results in our study due to differences in quality of housing between private- and public-housing groups. We hypothesize that partial reconstructions after the Superstorm Sandy took minimal effect in reducing moisture-related problems in the public-housing group. Measurements of indoor air conditions (CO_2 levels, indoor air RH and T) will reflect short term variations of indoor sources (occupants' activities), whereas moisture-content measurements will reflect long term variations of indoor sources (e.g. bad window-opening habits, frequent hang drying of clothes). Information about accidental leaks from outdoors, for example occasional leaks during heavy rains, will help to eliminate the influence of outdoor sources.

Moisture in building materials and dampness in indoor air have been studied independently in the past (Nielsen et al 2004; Viitanen et al. 2010; Nguyen et al. 2014). In the present study we will perform both measurements in order to estimate which specific building characteristics and occupants' behavior contributes dominantly to increased risk of dampness in residential buildings.

CONCLUSIONS

We expect that results from this study will help to identify determinants of moisture in residential buildings. This may aid in avoiding or limiting dampness in homes as an effective way to avoid fungal growth and resulting health effects.

ACKNOWLEDGEMENT

The authors would like to acknowledge the help of Red Hook Initiative for recruiting participants. The study was initiated and supported by Alfred P. Sloan foundation.

REFERENCES

Bornehag, C.G., Sundell, J., Bonini, S. et al. 2004a. Dampness in buildings as a risk factor for health effects, EUROEXPO: a multidisciplinary review of the literature (1998-2000) on dampness and mite exposure in buildings and health effects. Indoor Air 14:243-257.

Bornehag, C.G., Sundell, J. and Sigsgaard, T. 2004b. Dampness in buildings and health (DBH): Report from an ongoing epidemiological investigation on the association between indoor environmental factors and health effects among children in Sweden. Indoor Air 14:59-66.

Bornehag, C.G., Blomquist, G., Gyntelberg, F., et al. 2001. Dampness in buildings and health. Nordinc interdisciplinary review of the scientific evidence on associations between exposure to "dampness" in buildings and health effects (NORDDAMP). Indoor Air 11:72-86.

Haas, D., J. Habib, H. Galler, W. Buzina, R. Schlacher, E. Marth, and F. F. Reinthaler. 2007. Assessment of indoor air in Austrian apartments with and without visible mold growth. Atmospheric Environment **41**:5192-5201.

Hägerd-Engman, L., Bornehag, C.G. and Sundell, J. 2009. Building characteristics associated with moisture related problems in 8,918 Swedish dwellings. International journal of environmental health research. 19:251-256.

Howden-Chapman, P., Saville-Smith, K., Crane, J. and Wilson, N. 2005. Risk factors for mold in housing: a national survey. Indoor Air 15:469-476.

Nguyen, J.L., Schwartz, J. and Dockery D.W. 2014. The relationship between indoor and outdoor temperature, apparent temperature, relative humidity, and absolute humidity. Indoor Air 24:103-112.

Nielsen, K.F., Holm, g., Uttrup, L.P et al. 2004. Mould growth on building materials under low water activities. Influence of humidity and temperature on fungal growth and secondary metabolism. Int. Biodeteor. Biodegr. 108:1023-1028.

Quansah, R., M. S. Jaakkola, T. T. Hugg, S. A. Heikkinen, and J. J. Jaakkola. 2012. Residential dampness and molds and the risk of developing asthma: a systematic review and metaanalysis. PLoS ONE **7**:e47526.

Salo, P.M., Arbes, S.J., Crockett, P.W., Thorne, P.S., Cohn, R.D. and Zeldin, D.C. 2008. Exposure to multiple indoor allergens in US homes and its relationship to asthma. Journal of Allergy and Clinical Immunology. 121:678-684.

Suna, Y and Sundell, J. 2013. On Associations between Housing Characteristics, Dampness and Asthma and Allergies among Children in Northeast Texas. Indoor and Built Environment 22:678-684.

Viitanen, H., Vinha, J., Salminen, K. et al. 2010. Moisture and bio-deterioration risk of building materials and structures. Journal of Building Physics 33:201-224.

MICROBIAL COMMUNITIES IN HOUSE DUST AND ASTHMA: A BIRTH COHORT STUDY

Rachel I ADAMS[1*], Martin TÄUBEL[2], Pirkka KIRJAVAINEN[2,3], Balamuralikrishna JAYAPRAKASH[2], Anne KARVONEN[2,4], Anne HYVÄRINEN[2], and Juha PEKKANEN[2,4]

[1] Plant & Microbial Biology, University of California, Berkeley, California USA
[2] Department of Health Protection, National Institute for Health and Welfare, Kuopio Finland
[3] Institute of Public Health and Clinical Nutrition, University of Eastern Finland, Kuopio, Finland
[4] Department of Public Health, University of Helsinki, Helsinki, Finland

*Corresponding email: adamsri@berkeley.edu

Keywords: Atopy, Diversity, Farming, Moisture damage, Mold, Protective

SUMMARY

The overall objective of this cohort study is to identify risk factors for asthma and allergy development. The study includes over 400 Finnish children and their homes. The children have been followed since the third trimester of pregnancy, and they are now about 10 years of age. House dust, collected while the children were on average five months of age, are being analyzed for fungal and bacterial communities using high-throughput sequencing of molecular markers. Residences include aspects of farming, in that approximately one quarter of the cohort resides on farms, and moisture damage, as standardized building inspections were undertaken at each home at two times points. This paper describes the microbial characteristics of the house dust samples with respect to the macroenvironment, including the presence of moisture damage, and lays the groundwork for further studies identifying the protective and risk quality of house dust microbial communities.

INTRODUCTION

Mold and moisture damage in the home are known to negatively affect respiratory health among both children and adults, including the exacerbation and development of asthma. (Mendell et al., 2011; Kanchongkittiphon et al., 2015). With roughly half of the houses in US affected with dampness or mold, as many as one in five asthma cases may be attributable to these exposures (Mudarri and Fisk, 2007). On the other hand, exposure to high levels of microbes and overall biodiversity, e.g. on farms, of which microbial communities are part, may actually be protective of these same respiratory ailments (Ege et al., 2011; Hanski, 2014; Ruokolainen et al., 2015). The characteristics of microbial exposure that determine if high microbial exposure indoors is associated with protective or adverse respiratory effects are not known.

The LUKAS study has followed a cohort of over 400 children from birth and targeted children living in both urban and rural settings. Previous work within this long-term study has revealed that doctor-diagnosed wheezing at 18 months was associated with building engineer-diagnosed moisture damage and mold in the home, particularly when the water damage was found in the main living area, kitchen, and the child's bedroom (Karvonen et al.,

2009). Associations persisted when the study subjects reached 6 years of age (Karvonen et al., *in press*).

The long-term goal of the study is to uncover links between early life microbial exposure and respiratory health outcomes, including possible protective effects of farming-related factors. To begin to tackle this, we conducted a detailed description of the fungal and bacterial communities in the house dust of the cohort children, specifically exploring the effects of geolocation, environmental setting, and moisture damage of the building on indoor microbes.

METHODOLOGIES

The LUKAS cohort comprises the Finnish arm of the PASTURE study (Protection against Allergy STUdy in Rural Environments; von Mutius et al., 2006) and its extended cohort, consisting of an on-going birth cohort of 442 children in Eastern Finland. The study subjects have been followed up from third trimester of pregnancy. 123 children live on farms with livestock and the rest of the children are from rural areas or small towns. The children have been followed up with repeated questionnaires, health visits, and home inspections by trained civil engineers to assess moisture damage up to 6 years of age. Details on the environmental sampling including house dust, collected on average five months after birth, from the living room floors – as used in this current study – have been provided elsewhere (Karvonen et al., 2014).

Genomic DNA was extracted from 20mg of dust using bead beating method and chemagic DNA plant kit (Perkin Elmer) on the KingFisher DNA extraction robot. The extracted DNA was used to amplify a targeted region of the multi-copy ribosomal DNA from both bacteria and fungi. For bacteria, the V4 region of the 16S marker was amplified using 515F/806R primers (Caporaso et al., 2012). For fungi, the ITS1 region of the Internal Transcribed Spacer (ITS) was amplified using ITS1F/ITS2 primers (Smith and Peay, 2014). Amplicon were sequenced on the Illumia MiSeq Platform (350bp). Taxonomic identification for bacteria was determined using the Greengenes reference database (DeSantis et al., 2006), version released August, 2013, and for fungi the UNITE database (Abarenkov et al., 2010), version released December, 2014.

RESULTS AND DISCUSSION

Analysis is ongoing. Preliminary results corroborate a rich microbial presence of fungi and bacteria in indoor environments. Richness of both fungal and bacterial taxa is elevated on farms, residences with pets, and older residences; bacterial richness is also greater in homes with engineer-diagnosed moisture damage. Farming is the measured factor with the greatest effect on the community composition of the microorganisms indoors. Nevertheless, there is great overlap in the most abundant fungal and bacteria taxa between farms and non-farms, indicating that differences between the two settings are predominately driven by rare taxa. Further determination of the environmental and building factors that predict the richness and composition of indoor microbes is continuing.

CONCLUSIONS

Combining information from environmental factors, building inspections and concomitant high-throughput molecular assessment of bacteria and fungi in dust from this well-characterized birth cohort study offers a promising approach for evaluating the existence of microbial signatures and characteristics distinguishing protective and adverse human microbial exposure.

REFERENCES

Abarenkov, K., Nilsson, R.H., Larsson, K.H., *et al.* (2010) The UNITE database for molecular identification of fungi — recent updates and future perspectives, *New Phytol.*, **186**, 281-285.

Caporaso, J.G., Lauber, C.L., Walters, W.A., *et al.* (2012) Ultra-high-throughput microbial community analysis on the Illumina HiSeq and MiSeq platforms, *ISME J.*, **6**, 1621-1624.

Desantis, T.Z., Hugenholtz, P., Larsen, N., *et al.* (2006) Greengenes, a chimera-checked 16S rRNA gene database and workbench compatible with ARB, *Appl. Environ. Microbiol.*, **72**, 5069-5072.

Ege, M.J., Mayer, M., Normand, A.-C., *et al.* (2011) Exposure to Environmental Microorganisms and Childhood Asthma, *New England Journal of Medicine*, **364**, 701-709.

Hanski, I. (2014) Biodiversity, microbes and human well-being, *Ethics in Science and Environmental Politics*, **14**, 19-25.

Kanchongkittiphon, W., Mendell, M.J., Gaffin, J.M., *et al.* (2015) Indoor Environmental Exposures and Exacerbation of Asthma: An Update to the 2000 Review by the Institute of Medicine, *Environ Health Perspect*, **123**, 6-20.

Karvonen, A.M., Hyvärinen, A., Rintala, H., *et al.* (2014) Quantity and diversity of environmental microbial exposure and development of asthma: a birth cohort study, *Allergy*, **69**, 1092-1101.

Karvonen, A.M., Hyvärinen, A., Roponen, M., *et al.* (2009) Confirmed Moisture Damage at Home, Respiratory Symptoms and Atopy in Early Life: A Birth-Cohort Study, *Pediatrics*, **124**, e329.

Mendell, M.J., Mirer, A.G., Cheung, K., *et al.* (2011) Respiratory and allergic health effects of dampness, mold, and dampness-related agents: a review of the epidemiologic evidence, *Environ Health Perspect*, **119**, 748-756.

Mudarri, D. and Fisk, W.J. (2007) Public health and economic impact of dampness and mold, *Indoor Air*, **17**, 226-235.

Ruokolainen, L., Von Hertzen, L., Fyhrquist, N., *et al.* (2015) Green areas around homes reduce atopic sensitization in children, *Allergy*, **70**, 195-202.

Smith, D.P. and Peay, K.G. (2014) Sequence depth, not PCR replication, improves ecological inference from next generation DNA sequencing, *PLoS ONE*, **9**, e90234.

Von Mutius, E., Schmid, S. and The, P.S.G. (2006) The PASTURE project: EU support for the improvement of knowledge about risk factors and preventive factors for atopy in Europe, *Allergy*, **61**, 407-413.

EFFECT OF ULTRAVIOLET GERMICIDAL IRRADIATION ON MICROBIAL LOADING OF HVAC HEAT EXCHANGERS IN HUMID VERSUS DRY ENTERING CONDITIONS

Julia C LUONGO[1], Shelly L MILLER[1,*]

[1]Department of Mechanical Engineering, University of Colorado, Boulder, Colorado

[*]Corresponding email: Shelly.Miller@colorado.edu

Keywords: Coil fouling, Biofilm, Cooling coil, UVGI, UVC

SUMMARY

One reason for degraded HVAC (Heating, Ventilation, & Air Conditioning) performance is biofilm growth on cooling coils. Anecdotal evidence exists suggesting that Ultraviolet Germicidal Irradiation of HVAC heat exchangers can save operational and maintenance costs, but few experimental studies demonstrate the effect on heat exchanger microbiology. An HVAC test apparatus was built consisting of two parallel ducts, each with its own cooling coil. One coil was exposed to UVC radiation while the other was the control to test the efficacy of UV coil cleaning technology. This investigation aimed to measure changes in microbial loading resulting from coil irradiation in dry and humid conditions. Biological samples were taken from the coil surfaces and surrounding air to assess the impact of UVC radiation on biofilm growth and microbial impaction/resuspension. Much higher concentrations of microorganisms were found on the downstream side of the cooling coils under wetted conditions and, conversely, microbial concentrations were higher on the upstream surfaces under dry conditions. Coil irradiation reduced total surface microbial counts and increased downstream air counts due to suspension of inactivated biofilms off of the coil. In dry conditions, desiccation caused the same effect as irradiation but to a lesser degree, causing suspension of biological material from the coil.

INTRODUCTION

Ultraviolet Germicidal Irradiation (UVGI) has a long history of being used for the disinfection of air streams, primarily in environments with higher risk of airborne pathogen transmission such as healthcare facilities, schools, and prisons (Reed, 2010). Using UVGI as a coil cleaning technology in air-handling units (AHUs) has recently gained popularity. Heat exchanger surfaces are an ideal site for biofilms due to the presence of adequate nutrients (debris inherent on coil surfaces) and moisture (Morey, 1988). High bacterial and fungal concentrations have been documented within AHUs, specifically on cooling coils and drain pans (Hugenholtz & Fuerst, 1992; Levetin et al., 2001; Menzies et al., 2003). The objective of this study is to report detailed measurements of total microbial counts on surfaces and in air for an irradiated coil versus a non-irradiated coil in both dry and humid conditions. Biological samples were analysed via epifluorescent microscopy.

METHODOLOGIES

An HVAC test apparatus was built in the Air Quality Laboratory at the University of Colorado, consisting of two parallel ducts, each with its own cooling coil, but supplied by the same temperature and relative humidity controlled airstream (Figure 1). The test apparatus is equipped with sensors to measure duct velocities, static pressure drops, entering and exiting water temperatures, and entering and exiting air temperatures and relative humidities for each branch.

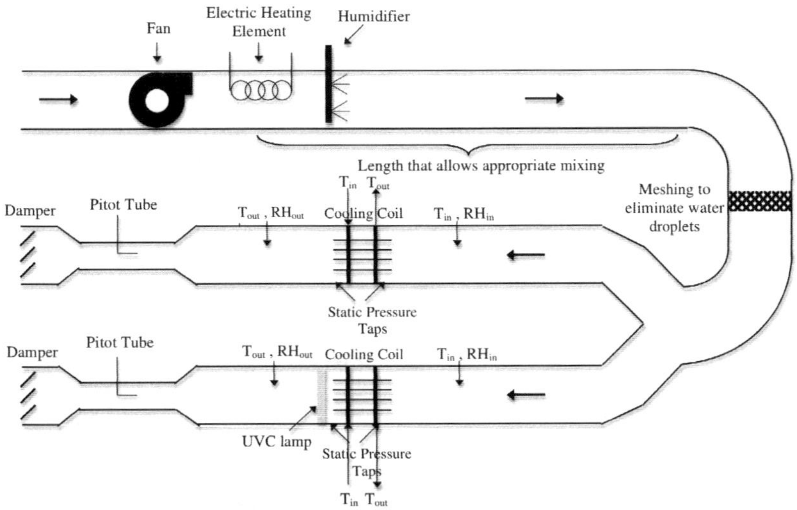

Figure 1. Schematic of HVAC test apparatus.

Air enters each cooling coil, on average, at 72 °F and 45% relative humidity and chilled water enters at 45 °F, satisfying conditions for condensation onto the coils. These conditions, however, are mild compared to the condensing conditions of cooling coils in very humid climates. The system mimics a constant volume HVAC system, meaning the volumetric flow rate is held constant. The flow rates through each coil are held equal to one another using dampers since the static pressure drop across the coils may not be equal given equivalent flow rates. Air and water inlet temperatures, inlet relative humidity, and water flow rate are all held constant.

The system ran undisturbed for 10 months without UV on either coil to ensure that both coils "fouled" at an equivalent rate and to establish a robust baseline dataset. After 10 months of operation, the UV lamp was turned on, irradiating the downstream side of one of the cooling coils (labeled "Top" coil). The control coil is labeled "Bottom" coil. The irradiance at the surface of the Top coil was 200 $\mu W/cm^2$ at the center and 150 $\mu W/cm^2$ at the corners.

Coil surface samples were taken with sterile BBL CultureSwabs (BD, Sparks, MD). A 10-cm^2 area of the coil surface was swabbed and extracted into HPCL water. Both the upstream and downstream side of each coil was swabbed. Air samples were collected isokinetically with 0.45-μm cellulose nitrate membrane filters (Thermo Scientific, Waltham, MA) at 10 L/min for 18 hours and extracted into HPLC water. All samples were stained using SYTO BC Green Fluorescence stain (Life Technologies, Carlsbad, CA) and deposited on a 0.2-μm black polycarbonate membrane (Millipore, Billerica, MA). SYTO BC is a nucleic acid stain that penetrates both Gram-negative and Gram-positive bacteria, yielding total cell counts. Samples were directly counted using an epifluorescent microscope (Nikon E600).

RESULTS AND DISCUSSION

Results from epifluorescent microscopy show higher total microbial counts on the downstream surface samples versus upstream surface samples on both coils prior to being exposed to UV under wetted conditions ("Baseline" in Figure 2). This result is consistent with evidence of higher viable microbial loads on the downstream side of cooling coils when culturing surface samples (Hugenholtz & Fuerst, 1992). Total surface cell counts for the irradiated coil remained unchanged until 3 weeks later (10/3) even though the coil appeared visually cleaner within one week.

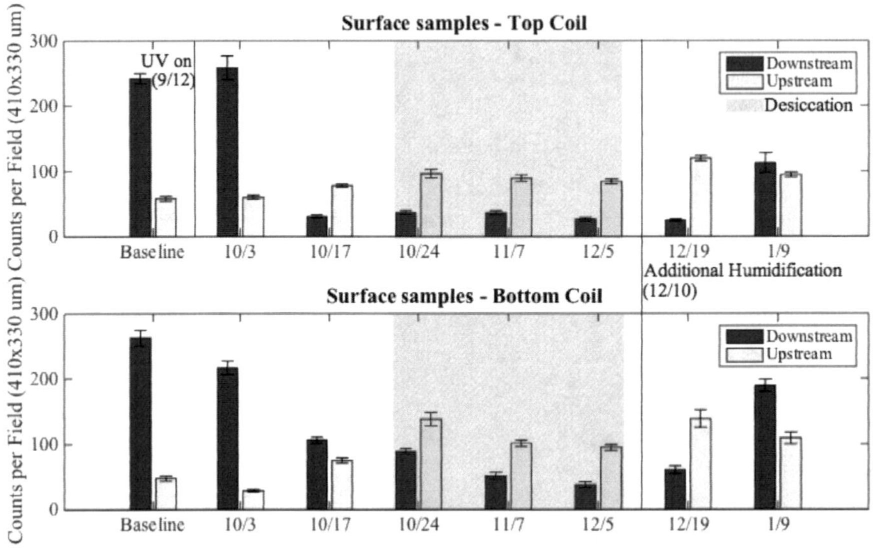

Figure 2. Total microbial surface sample cell counts per field of view using epifluorescent microscopy. Error bars represent the standard error between triplicate samples.

In conjunction with reduced surface counts, total counts in air samples downstream of the irradiated coil increased the same week (Figure 3). Two weeks later (10/17), the UV-irradiated coil saw a large reduction in downstream surface counts as well as the appearance of large conglomerates of biomass with many cells embedded in air samples, likely large chunks of biofilm that were inactivated and later detached from the coil surface (asterisks in Figure 3).

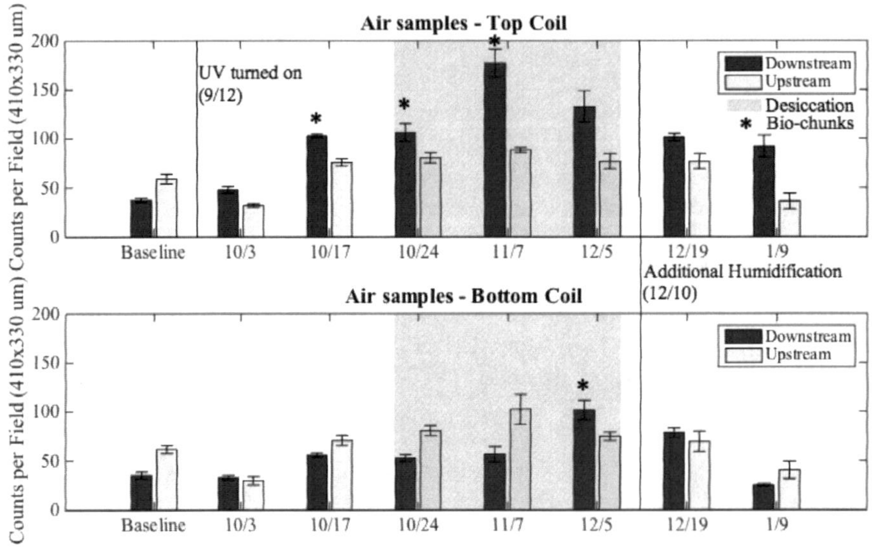

Figure 3. Total microbial air sample cell counts per field of view using epifluorescent microscopy. Error bars represent the standard error between triplicate samples.

The following week (10/24), both coils began to dry due to decreased entering humidity. At this time, surface total counts on the non-UV coil shifted from being higher downstream to being higher upstream. We hypothesize that the lack of moisture on the downstream side of the coil drove the microbes to the next most favorable location, the upstream side where the majority of particulate matter deposits. Siegel & Nazaroff (2003) modelled particle deposition on HVAC heat exchangers and found that the majority of particles smaller than 8 μm in diameter deposit on the leading fin edge via impaction. Both cooling coils began to desiccate over the following 6 weeks (10/24-12/5). After 6 weeks of dry conditions, large chunks of biofilm appeared in air samples downstream of the non-UV coil. This is likely due to inactivation from desiccation causing chunks of biofilm to detach from the fin surfaces. The following week, additional humidification was installed to return to wetted conditions. Within two weeks, higher microbial loading returned to the downstream side of both coils.

CONCLUSIONS

Microbial surface sampling indicates a higher degree of biofouling on the downstream side of the cooling coils in wetted conditions and a higher degree of biofouling on the upstream side in dry conditions in the absence of UV irradiation. While both UV irradiation and desiccation serve as methods for biofilm detachment, the UV-irradiated coil had a higher degree of detached biofilm chunks (pointing towards a higher degree of biofilm inactivation), began detaching faster than with desiccation, and was visibly cleaner than the desiccated coil. UV also reduced total surface cell counts more quickly than desiccation. The preliminary results presented here imply that UV coil cleaning technology is most effective in humid, condensing conditions and encourages our recommendation that lamps be placed downstream of the coil in wetted conditions when microbial loading is greatest downstream. However, lamps may be more effective on the upstream side in dry conditions due to higher microbial loading upstream.

ACKNOWLEDGEMENT

We gratefully acknowledge financial support for this project from the University of Colorado (CU) Boulder Innovative Seed Grant Program, The Department of Mechanical Engineering CU Boulder, and an Industry Consortium consisting of American Ultraviolet, Lumalier, Sanuvox Technologies, and UltraViolet Devices, Inc.

REFERENCES

Hugenholtz, P. & Fuerst, J. a, "Heterotrophic bacteria in an air-handling system" *Applied and environmental microbiology*, 58(12):3914–20 (1992).

Levetin, E., Shaughnessy, R., Rogers, C.A., et al., "Effectiveness of Germicidal UV Radiation for Reducing Fungal Contamination within Air-Handling Units" *Applied and Environmental Microbiology*, 67(8):3712–3715 (2001).

Menzies, D., Popa, J., Hanley, J. a, et al., "Effect of ultraviolet germicidal lights installed in office ventilation systems on workers' health and wellbeing: double-blind multiple crossover trial" *Lancet*, 362(9398):1785–91 (2003).

Morey, P.R., "Microorganisms in buildings and HVAC systems: A summary of 21 environmental studies" In *Proceedings of the ASHRAE Conference on Indoor Air Quality, American Society of Heating, Refrigeration, and Air-Conditioning Engineers*. Atlanta, GA, pp. 10–24 (1988).

Moyer, E.S., Miller, W.E., Commodore, M. a., et al., "Aerosol and Biological Sampling of a Ventilation Fan-bank Modified with Ultraviolet Germicidal Irradiation and Improved Filter Holders" *Indoor and Built Environment*, 19(2):230–238 (2010).

Reed, N.G., "The history of ultraviolet germicidal irradiation for air disinfection" *Public Health Reports*, 125:15–27 (2010).

Siegel, J. A. & Nazaroff, W.W., 2003. Predicting particle deposition on HVAC heat exchangers. *Atmospheric Environment*, 37(39-40), pp.5587–5596.

SURVEY OF GREEN BUILDING WATER SYSTEMS REVEALS ELEVATED WATER AGE AND MICROBIAL CONCERNS

William J. RHOADS[1,*], Amy PRUDEN[1], Marc A. EDWARDS[1]
[1]Charles E. Via Department of Civil and Environmental Engineering, Virginia Polytechnic Institute and State University, Blacksburg, VA

[*]Corresponding email: wrhoads@vt.edu

Keywords: Green buildings, Water quality, Water age, Disinfectant decay, Opportunistic pathogens

SUMMARY

Widespread adoption of water conservation strategies has potential for unintended consequences for water quality. Three green buildings and a conventional building were surveyed, focusing on quantitative metrics of water plumbing system retention time (i.e., water age), water chemistry, and levels of opportunistic pathogen genetic markers. Water age was estimated to be 2-6.7 months in an off-grid office, 8 days in a healthcare suite, and was increased by 1.7 days due to installation of a solar "pre-heat" water tank in a net-zero energy house. Disinfectant residuals were often completely absent in the building plumbing system and decayed up to 144 times faster than water in the utility distribution systems. Concentration of total bacteria and opportunistic pathogen genetic markers were 1-4 orders of magnitude higher compared to conventional buildings. This study raises concerns with respect to current green building practices and brings to light the importance of considering potential public health impacts in the design of sustainable water systems.

INTRODUCTION

Although the indoor environment is a primary focus of modern buildings, public health risks that may result from increased water stagnation in building plumbing systems (i.e., increased water age) due to reduced water demand is a generally overlooked aspect of water conservation (Rhoads et al., 2015; Edwards et al., 2014). Water age is likely to increase in both distribution and premise plumbing systems as water conservation practices become more commonplace.

It is well-established that high water age is problematic in main distribution systems, contributing to problems with corrosion, stability of disinfectant residuals, taste and odors, and microbial regrowth (EPA, 2002). High water age in building plumbing can further amplify these problems and the proliferation of human pathogens (Pruden et al., 2012; NRC, 2006). Opportunistic pathogens in premise plumbing (OPPPs), including *Legionella* spp. (especially *L. pneumophila*) and *Mycobacterium avium* complex (Non-tuberculosis mycobacteria), are now the leading cause of waterborne disease in the U.S. (NRC, 2006). OPPPs are native to natural water sources and have complex life cycles, including amplification and enhanced virulence when hosted by free-living amoebae such as *Vermamoeba vermiformis*.

Few prior studies have touched upon the potential for unintended consequences in green water systems (Elfland et al., 2010; Nguyen et al., 2012). Although OPPPs have been documented to preferentially colonize some water efficient devices (Sydnor et al., 2012) and have been implicated in disease associated with rainwater harvesting systems (Ahmed et al., 2008), prior studies have failed to document basic system features such as storage methods, plumbing materials, water chemistry, water demand, or water treatment in such facilities.

The purpose of this study was to survey a cross-section of cutting-edge green buildings in order to establish a baseline understanding of water age and water quality characteristics relative to conventional buildings. Water systems were characterized in terms of disinfectant residual, disinfectant decay rates, and levels of total bacteria, *Legionella* spp., *Mycobacterium* spp., and *V. vermiformis* using quantitative polymerase chain reaction (qPCR).

METHODOLOGIES

Description of field sites. Four buildings with a cross-section of energy and water conservation strategies were surveyed and sampled to explore effects of green plumbing systems on water quality. The buildings included 1) a net-zero energy house with a solar water heater, 2) a conventional house with no conservation features located near the net-zero energy house (for comparison), 3) a Leadership in Environmental Engineering Design (LEED) certified healthcare suite with very low water demand, and 4) a net-zero energy and net-zero water office building with 3,000 gallons of rainwater storage for all potable and non-potable needs.

Sampling routine. A sampling plan was developed for each field site based on building performance data, investigation of building plumbing blueprints, and field observations. Samples were designed to capture profiles of water quality as it was flushed from taps. Starting with the first flush, grab samples were collected until a disinfectant residual consistent with the main distribution system was obtained (where applicable). Afterwards, the taps were shut off and water was allowed to sit stagnant. Small volume samples (~35 mL) were collected at regular intervals during stagnation. At the net-zero office, only stagnant and moderately flushed (3 minutes) water samples were collected because the facility was not connected to a water main (i.e., had no disinfectant residual) and flushing had no observable impact on water quality.

Water-quality and biological sampling. Temperature, pH, ammonia, nitrite, and total chlorine were measured in the field at the time of collection using a pH 110-Series meter (Oakton Research, Vernon Hills, Il) or a DR2700 spectrophotometer (Hach, Loveland, CO) according to Standard Methods. EPA Method 300.1 was used to measure anions. Cations were measured by inductively coupled plasma mass spectrometry. Biological activity reaction tests (BARTs) indicating the presence/absence of nitrifying bacteria were used for select samples.

Water sample collection methodology, DNA extraction protocols, and quantitative polymerase chain reaction assays are described elsewhere (Wang et al., 2012). If samples had detectable gene copies, but were below the quantification limit, they were recorded as 3 gene copies/mL (well below the quantification limit) for graphing and statistical purposes, which included non-parametric Mann-Whitney U-tests performed on non-transformed qPCR data to identify statistically significant differences in the concentrations of bacterial genes (p-value $<$ 0.05).

RESULTS AND DISCUSSION

This study documented key design factors in water systems of three green buildings. After reviewing the characteristics of each water system and closely examining trends in disinfectant residuals within the buildings, a final section provides an overview of microbial sampling from each building.

Water age. A survey of three green buildings revealed extremely high water age associated with innovative features intended to conserve water and energy. The net-zero office building had an estimated 2-6.7 month total water age. The healthcare suite uses only 0.26 gallons/ft^2/year, which is 60 times lower than typical commercial buildings (BOMA, 2014), resulting in an average water age of 8 days. At the net-zero energy house, the solar water heater increases the hot water storage and hot water age from < 1 day to ~2.7 days. The conventional home used more than 4 times more water than the net-zero energy house, but it is important to note that the net-zero energy house only simulates hot water demand as part of an on-going experiment evaluating the feasibility of achieving net-zero energy in a residential setting. Because cold water demand draws no energy, cold water use is eliminated in this house.

Temperature profiles. In the healthcare facility, 70 minutes of water flushing was required for temperatures to drop below cold water recommendations for OPPP control (20° C). At the net-zero house, temperature in the solar tank did not recover after flushing (on a partly cloudy day), whereas the backup electric water heater recovered to its set point of 49 °C within one hour. The net-zero office building was characterized by warm temperatures, even in the cold water (>23 °C). Flushing for this building did not lower the temperature because the sole water source was the rainwater cistern located outdoors.

Water sources in the net-zero office. Water quality analysis revealed that alkalinity, calcium, and magnesium (~110 ppm as $CaCO_3$, ~12 ppm Mg^{2+}, and ~30 ppm Ca^{2+}) were much higher than expected for rainwater. This validated the estimate that groundwater used for routine maintenance of the system, rather than rainwater, made up 38-58% of total water use in the facility.

Disinfectant residual decay. Maintaining a disinfectant residual was generally a challenge in the green buildings. At the healthcare suite and net-zero house, >80 minutes and 15 minutes of flushing, respectively, was necessary to establish a residual similar to that present in the main distribution system. Since taps generally are not flushed to this extent during normal use, it is unlikely that residual is ever present in portions of these systems. In contrast, the conventional home had a residual of 0.71 mg/L as Cl_2 in stagnant cold water after overnight stagnation.

Disinfectant residuals also decayed rapidly. During a second site visit at the net-zero house, despite an influent residual of 0.95 mg/L as Cl_2, concentrations never increased at the point of use during a 30 minute warm water draw in a shower. Therefore, the chlorine residual was either decaying as it flowed through the pipes to the shower (via reactions with the pipe material or biofilms) or an instantaneous chlorine demand was created within the hot water system. At the healthcare facility, chloramine decayed 20-144 times faster than in well-flushed water collected from the same tap and placed into a glass container with no headspace.

Regrowth in green buildings. At all field sites, there were more total bacteria in stagnant samples than flushed samples. The median increase in 16S rRNA (an indicator of total bacteria) gene copies/mL was 2.8 log (p-value 0.007; $n_{stagnant}=18$, $n_{flushed}=10$). At the healthcare suite, there was also a median increase of 1.64 log in 16S rRNA gene copies/mL in hot water sample taps compared to water collected directly from the water heater. The net-zero house yielded >20X more 16S rRNA gene copies/mL in stagnant cold water than the conventional house receiving the same municipal water and using approximately four times more water than the net-zero house yielded. In addition, levels of *Legionella* spp., *M. avium* and *V. vermiformis* gene copies in the conventional house were below the quantification limit in stagnant samples and below the detection limit in flushed samples, whereas levels of these organisms were frequently present in high concentrations (up to 4 log higher than levels reported in environmental samples in the literature; Wang et al., 2012) in hot and cold stagnant and flushed samples in the green buildings. Median bacterial concentrations between stagnant and flushed samples at the net-zero office were relatively small in magnitude (0.41 log gene copies/mL), indicating that the majority of the growth occurred in the cistern with very high water age. This suggests that in such buildings, storage tanks may be the primary concern for growth of OPPPs, rather than pipes leading to taps.

CONCLUSIONS

The results of this study provided the following insight: 1) the green buildings sampled have exceptionally high water age, 2) the conservation strategies employed at each green building created hot and/or cold water temperature profiles that are more conducive to OPPP growth, 3) truly achieving net-zero status may be difficult in practice, 4) disinfectant residuals were lower in the green buildings with high water age than typically encountered in conventional buildings, 5) disinfectants decayed rapidly in the systems that routinely had high water age, 6) stagnation can increase total bacterial loads and OPPPs, and 7) storage inherent to green building design may contribute to increased OPPP growth.

ACKNOWLEDGEMENT

This research was partially funded by the Water Research Foundation on project 4383 *Green Building Design: Water Quality Considerations,* as well as by the Alfred P. Sloan Foundation Microbiology of the Built Environment program. Any opinions, findings, conclusions, or recommendations expressed here are those of the authors and do not necessarily reflect the views of the sponsors.

REFERENCES

Pruden, A.; Edwards, M.A.; Falkinham, III, J.O. Research Needs for Opportunistic Pathogens in Premise Plumbing: Experimental Methodology, Microbial Ecology and Epidemiology. Water Research Foundation Project 4379, Denver, CO, 2012.

Drinking water distribution systems: assessing and reducing risk. National Research Council, Washington, DC, 2006.

EPA (US Environmental Protection Agency), 2002. Effects of Water Age on Distribution System Water Quality. Office of Water and Office of Ground Water and Drinking Water. Cincinnati, OH.

Sydnor, E.R.; Bova, G.; Gimburg, A.; Cosgrove, S E.; Perl, T M.; Maragakis, L.L. Electronic-eye faucets: *Legionella* species contamination in healthcare settings. Infect. Cont. Hosp. Ep. 2012, 33(3), 235-240.

Ahmed, W.; Huygens, F.; Goonetilleke, A.; Gardner, T. Real-time PCR detection of pathogenic microorganisms in roof-harvested rainwater in Southeast Queensland, Australia. *Appl. Environ. Microb.* **2008**, *74(17), 5490-5496.*

Wang, H.; Edwards, M.A.; Falkinham, III, J.O.; Pruden, A. Molecular survey of the occurrence of *Legionella* spp., *Mycobacterium* spp., *Pseudomonas aeruginosa*, and amoeba hosts in two chloraminated drinking water distribution systems. *Appl Environ Microb.* **2012**, *78(17), 6285-6294.*

BOMA BEst Energy and Environment Report – The National Green Building Report. Building Owners and Managers Association Building Environment Standards (BOMA BEst), Toronto, ON, 2014. http://www.bomabest.com/wp-content/uploads/BBEER-2014-Full-Report.pdf (accessed December 19,2014).

Rhoads, W.J.; Pearce, A.; Pruden, A.; Edwards, M.A. Anticipating Water Quality and Infrastructure Impacts due to Green Buildings. *J. Am. Water Works Ass.* (**2015**-in press).

Elfland, C.; Scardina, R.P.; Edwards, M.A. Lead-contaminated water from brass plumbing devices in new buildings. *J. Am. Water Works Ass.*, **2010**, 102(11):66-76.

Nguyen, C.; Elfland, C.; & Edwards, M.A. Impact of advanced water conservation features and new copper pipe on rapid chloramine decay and microbial regrowth. Water Research. **2012**, 46(3):611-621.

Edwards, M.A.; Rhoads, W.J.; Pruden, A.; & Pearce, A. Green Water Systems and Opportunistic Premise Plumbing Pathogens. Plumbing Engineer. **2014**. Available: http://www.plumbingengineer.com/content/green-water-systems-and-opportunistic-premise-plumbing-pathogens

TEMPORAL AIR EXCHANGE RATES AND RELATIVE HUMIDITY MAY AFFECT AIRBORNE MICROBIOMES MORE THAN WEATHERIZATION EFFORTS

Mytien T NGUYEN, Denina HOSPODSKY, Catherine SPIRITO, and Largus T ANGENENT

Department of Biological and Environmental Engineering, Cornell University, Ithaca, NY

*Corresponding email: la249@cornell.edu

Keywords: indoor microbiome, weatherization, bioaerosols, air exchange rate, humidity

SUMMARY

The Department of Energy estimates that approximately 7 million homes have been weatherized in the U.S. since 1976. Weatherization is a nationwide effort to improve home energy efficiency through insulation and infrastructure repairs. Infrastructure patterns, such as ventilation rate and the nature of ventilation, alter the composition and diversity of indoor airborne microbiomes. However, it is not clear whether weatherization results in a significantly changed airborne microbiome. In this ongoing study, we aim to characterize the airborne microbiomes before and after weatherization in ten single-family homes. In addition, microbiome dynamics in weatherized homes are being compared to control homes that did not undergo structural changes during the same study period. Even though this study is currently ongoing, our preliminary results with two homes shows that air exchange rate and relative humidity are significant drivers of indoor airborne microbiome composition.

INTRODUCTION

In 2012, nearly 40% of energy consumption in the U.S. occurred in residential and commercial buildings (EIS 2014). In a nationwide effort to lower energy consumption, the U.S. Department of Energy established the Weatherization Assistance Program to aid single-family home-owners in achieving a higher home energy efficiency by a combination of procedures including insulation, fixture repairs, and crack sealing. However, a major concern is a reduced air exchange rate that accompanies these procedures, which could adversely affect the indoor air quality. Decreased air exchange rates may lead to the accumulation of, and prolonged exposure to, harmful microbes that can cause asthma- and allergy- related symptoms such as hypersensitivity and wheezing.

Infrastructure patterns are parameters that affect microbial load and diversity in indoor air. Air exchange rate has been implicated as a major driver for indoor exposure to fungi (Frankel et al. 2012). In addition, how a building is ventilated can impact the microbial composition, regardless of ventilation rate. Recent studies of mechanical *versus* window ventilation in a hospital (Kembel et al. 2012) and a university (Meadow et al. 2014) showed that different ventilation strategies alter the indoor airborne microbiomes. The openness of a house, its air exchange rate, the type of ventilation present, and the presence or absence of a basement are all important considerations in understanding the dynamics of the indoor airborne microbiome.

In our ongoing study, the objective is to examine and compare the microbiomes on surfaces and in air before and after weatherization efforts. We are monitoring ten homes with weatherization and five without infrastructural changes. Here, we are showing preliminary results from two homes (one weatherized and one not) that we monitored five times throughout a one-year time period. We are analyzing microbiomes using culture-independent methods, including quantitative PCR and high-throughput DNA sequencing. We aim to identify parameters and home characteristics that drive microbiome dynamics to inform architectural design considerations and decision-making in weatherization efforts.

METHODOLOGIES

Study homes: Two single-family homes with a basement and without mold or water problems were sampled throughout a year during five sampling events. One of the houses was located in Ithaca, NY and one in Pittsburgh, PA. The Ithaca house was not weatherized (control), while the Pittsburgh house was considerably altered, including a weatherization effort. Sampling occurred before, during, and after weatherization of the Pittsburgh house. **Sampling and sample collection**: Sampling is conducted for 36-h in a continuous mode. Air samples from these two homes are collected on a one-stage PM10 impactor (SKC Inc., Eighty Four, PA). Our continuous home sampling campaign also includes surface sampling, air sampling, air exchange rate estimation *via* CO_2 decay (LI-COR Biosciences, Lincoln, NE), blower door testing to determine air exchange rate, and continuously monitoring of temperature, humidity (Onset Computer Co., Cape Cod, MA), water activity (AquaLab, Pullman, CA), and radon concentration (AccuStar, Medway, MA). In addition, occupant health and house characteristics are being collected in a questionnaire. Sterile swabs and wipes are used to collect samples from surfaces. Indoor and outdoor air samples are being collected through impaction onto 0.2-µm pore sized PCTE filters at 70 lpm in a 4-stage Anderson cascade impactor (Tisch Environment Inc., Cleves, OH), with size bins (µm): <2.1, 2.1-4.7, 4.7-9.0, and >9.0. Filters are weighed before and after sampling with a Mettler Toledo Microbalance (Columbus, OH). Laboratory blanks are weighed alongside field samples and one field blank per set of three samples is also weighed. All samples are stored at -20°C prior to further processing. **Sample processing and DNA extraction**: Surface samples from wipes are eluted using a method described in Yamamoto, Shendell, and Peccia (2011). In detail, wipes are submerged in 100 ml of PBS with 0.1% Tween-80, then shaken at 250 rpm for 5 h. The eluted liquid is filtered through a 0.2-µm pore sized Micropore filter. DNA extraction methods for all samples were described previously (Boreson, Dillner, and Peccia 2004; Hospodsky, Yamamoto, and Peccia 2010). Indoor airborne bacterial load is quantified using *B. subtilis* standards in a qPCR method targeting universal 16S rRNA genes as described by Nadkarni et al. (2002). **PCR and sequencing**: Samples were sequenced with the paired-end 250 bp Illumina MiSeq sequencing approach using 515F and barcoded 806R primers targeting the V4 hypervariable region of the 16S rRNA gene (Caporaso et al. 2012). Sequences were analyzed using the standard QIIME v1.8.0 pipeline in (Caporaso et al. 2010). Rare taxa (singletons) were removed and OTU counts were normalized *via* the Cumulative Sum Scaling (CSS) method (Paulson et al. 2013) prior to analysis. Relationship between samples were visualized using Principal Coordinates Analysis (PCoA) of weighted and unweighted UniFrac distance matrices (Lozupone and Knight 2005). To explore the relationship between the airborne microbiome and environmental variables, we performed constrained ordination analysis using the capscale method of the vegan R package (Okansen et al. 2013).

RESULTS AND DISCUSSION

Relationships between samples were visualized with weighted and unweighted UniFrac distances, and showed no clustering between homes (**Figure 1**) or time of sampling (not shown).

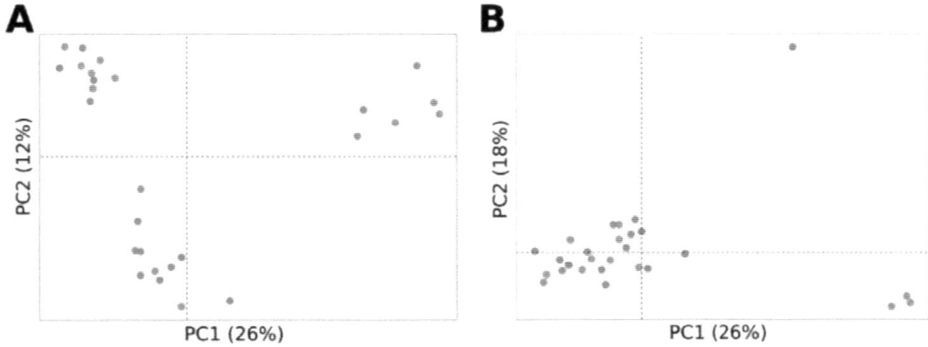

Figure 1: Normalized principal coordinate analysis of UniFrac distances between air samples. Each point represents one sample, colored by house (red: Ithaca; blue: Pittsburgh). Shown are the first and second principal coordinates based on UniFrac distance matrices, calculated from CSS-normalized OTU counts (**A**: weighted UniFrac; **B**: unweighted UniFrac).

Constrained ordination analysis was performed with temperature, relative humidity (RH), and building air exchange rate (AER) as constraints. This statistical analysis determines the degree to which variability between sample microbiomes is explained by the variability in the constraining variables (environmental factors). Both relative humidity and air exchange rate were found to individually correlate (low co-linearity: VIFs < 1.5) with microbial (bacterial) variations between air samples (**Figure 2**).

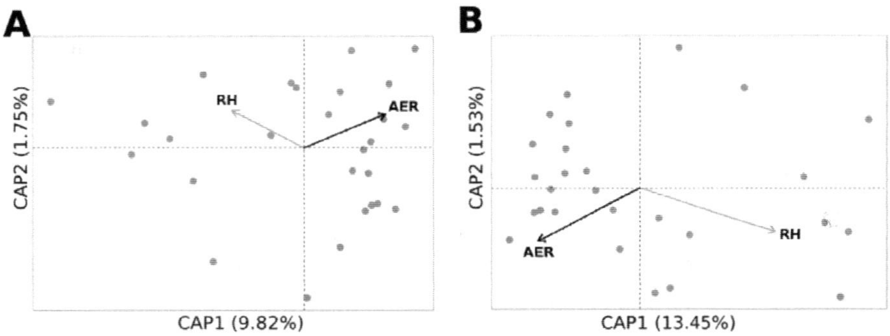

Figure 2: Normalized constrained ordination analysis of UniFrac distances between air samples. Each point represents one sample, colored by house (red: Ithaca; blue: Pittsburgh). The first two constrained coordinates for unweighted UniFrac (**A**) and weighted UniFrac (**B**) distance matrices explaining 30.4% and 34.0% of the variance displayed in the unconstrained PCoA plot above, respectively. Arrows of significant environmental factors (ANOVA $p<0.05$; VIF < 1.5) are in the direction of increasing magnitude. Arrow length indicates the degree of correlation between the variable and the constrained axis. In both unweighted and weighted UniFrac analyses, relative humidity (**RH**-grey) and air exchange rate (**AER**-black) were correlated with the constrained axes.

CONCLUSIONS

Homes that undergo weatherization efforts are characteristically more insulated and sealed, with lower air exchange rates. In our preliminary study of two homes, we found that both temporal relative humidity and air exchange rate drove diversity and composition of airborne microbiomes. However, inter-sampling comparisons could not distinguish between homes, which were located in two different Northeastern states of the U.S., or between sampling times, which for the Pittsburgh home included a change in the building envelope as part of a weatherization effort. Therefore, the effect of weatherization in older homes may either be: 1) not severe enough to change the airborne microbiome; or 2) secondary to significant effects of the temporal changes in relative humidity and air exchange rates on the microbiome. Upcoming analysis of ten weatherized and five non-weatherized homes will provide more data to elucidate whether these relationships from the first two houses will hold.

ACKNOWLEDGEMENTS

This study is funded by the Alfred P. Sloan Foundation. The authors would like to thank the homeowners for welcoming the scientists into their homes for extended periods of times. We would also like to thank Joseph Laquatra Jr, Ph.D for his advise in building sciences, Siddarth Govindarajan for his help during sampling, and Howard Chong, Ph.D for the use of his blower door equipment.

REFERENCES

Boreson, Justin, Ann M. Dillner, and Jordan Peccia. 2004. 'Correlating bioaerosol load with PM2.5 and PM10cf concentrations: a comparison between natural desert and urban-fringe aerosols', *Atmospheric Environment*, 38: 6029-41.

Caporaso, J. Gregory, Justin Kuczynski, Jesse Stombaugh, Kyle Bittinger, Frederic D. Bushman, Elizabeth K. Costello, Noah Fierer, Antonio Gonzalez Pena, Julia K. Goodrich, Jeffrey I. Gordon, Gavin A. Huttley, Scott T. Kelley, Dan Knights, Jeremy E. Koenig, Ruth E. Ley, Catherine A. Lozupone, Daniel McDonald, Brian D. Muegge, Meg Pirrung, Jens Reeder, Joel R. Sevinsky, Peter J. Turnbaugh, William A. Walters, Jeremy Widmann, Tanya Yatsunenko, Jesse Zaneveld, and Rob Knight. 2010. 'QIIME allows analysis of high-throughput community sequencing data', *Nat Meth*, 7: 335-36.

Caporaso, J. Gregory, Christian L. Lauber, William A. Walters, Donna Berg-Lyons, James Huntley, Noah Fierer, Sarah M. Owens, Jason Betley, Louise Fraser, Markus Bauer, Niall Gormley, Jack A. Gilbert, Geoff Smith, and Rob Knight. 2012. 'Ultra-high-throughput microbial community analysis on the Illumina HiSeq and MiSeq platforms', *The ISME Journal*, 6: 1621-24.

EIS. 2014. 'How Much Energy Is Consumed in Residential and Commercial Buildings in the United States?'.

Frankel, Mika, Gabriel Bekö, Michael Timm, Sine Gustavsen, Erik Wind Hansen, and Anne Mette Madsen. 2012. 'Seasonal Variations of Indoor Microbial Exposures and Their Relation to Temperature, Relative Humidity, and Air Exchange Rate', *Applied and Environmental Microbiology*, 78: 8289-97.

Hospodsky, Denina, Naomichi Yamamoto, and Jordan Peccia. 2010. 'Accuracy, Precision, and Method Detection Limits of Quantitative PCR for Airborne Bacteria and Fungi', *Applied and Environmental Microbiology*, 76: 7004-12.

Kembel, Steven W., Evan Jones, Jeff Kline, Dale Northcutt, Jason Stenson, Ann M. Womack, Brendan J. M. Bohannan, G. Z. Brown, and Jessica L. Green. 2012. 'Architectural design influences the diversity and structure of the built environment microbiome', *ISME J*, 6: 1469-79.

Lozupone, Catherine, and Rob Knight. 2005. 'UniFrac: a New Phylogenetic Method for Comparing Microbial Communities', *Applied and Environmental Microbiology*, 71: 8228-35.

Meadow, J. F., A. E. Altrichter, S. W. Kembel, J. Kline, G. Mhuireach, M. Moriyama, D. Northcutt, T. K. O'Connor, A. M. Womack, G. Z. Brown, J. L Green, and B. J. M. Bohannan. 2014. 'Indoor airborne bacterial communities are influenced by ventilation, occupancy, and outdoor air source', *Indoor Air*, 24: 41-48.

Nadkarni, Mangala A., F. Elizabeth Martin, Nicholas A. Jacques, and Neil Hunter. 2002. 'Determination of bacterial load by real-time PCR using a broad-range (universal) probe and primers set', *Microbiology*, 148: 257-66.

Okansen, J, FG Blanchet, R Kindt, P Legendre, PR Minchin, RB O'Hara, GL Simpson, P Solymos, MH Stevens, and H Wagner. 2013. 'vegan: Community Ecology Package', *R package version 2.0-10*.

Paulson, Joseph N., O. Colin Stine, Hector Corrada Bravo, and Mihai Pop. 2013. 'Differential abundance analysis for microbial marker-gene surveys', *Nat Meth*, 10: 1200-02.

Yamamoto, N., D. G. Shendell, and J. Peccia. 2011. 'Assessing allergenic fungi in house dust by floor wipe sampling and quantitative PCR', *Indoor Air*, 21: 521-30.

AN EXPERIMENTAL STUDY ON HEAT GAIN IN WINDOW WITH AN EXTERNAL SHUTTER

Esam M ALAWADHI
Department of Mechanical Engineering, Kuwait University, Kuwait
Corresponding email: esam.alawadhi@ku.edu.kw

Keywords: External window shutter, solar radiation, heat gain through window.

SUMMARY

Windows are account for the majority of the heat gain in buildings, and to reduce the heat gain, exterior shutters are commonly installed in residential buildings in hot climates. The shutter is typically incompletely close during the daytime to have indoor natural lighting, which potentially reducing the thermal effectiveness of the shutter. Since there is temperature difference between the window glass and shutter, natural convection flow in induced in the space between the window and shutter. Experimental measurements are employed to study the effect of the shatter on the heat gain through the window during the month of June. The results indicate that when the shutter is incompletely closed, the heat gain through window can be increased by 44.5%, depending on the shutter opening distance.

NOMENCLATURE

h	heat transfer coefficient (W/m^2-^{o}C)
H	height (m)
k	thermal conductivity (W/m-^{o}C)
Q	heat flow (Watts)
t	time (sec.)
T	temperature (^{o}C)
u	velocity in x-direction (m/s)
v	velocity in y-direction (m/s)
W	width (m)

Greek

α	Solar absorptivity

Subscripts

a	air gap
i	indoor
is	indoor surface
o	outdoor
os	outdoor surface
net	net
s	shutter
sh	shutter
w	window

INTRODUCTION

The building's envelope is the barrier that separates the indoor controlled environment from outdoor unstable environment, and it has a significant impact on the building's comfort and energy consumption. For roof and external walls, thermal insulation materials are extensively used, and have a significant impact on heat flow through them. Windows are considered as a weak link for heat loss, because they account for 40% of the cooling load (Maccari and Zinzi, 2001). In this regards, a thermal analysis of simple glass window models was addressed in literature with the aim of understanding the effect of glass properties of the heat gain (Ismail and Henriquez, 2003). The direct solar radiation account most of heat gain through windows, and the most effective technique to reduce it is to install the window shutter. The window shutters are typically made of foam filled aluminum rolling shutter slat, and Figure 1 shows a residential building in Kuwait with shutters installed on the windows. A comparative study of thermal effectiveness of a wooden and aluminum shutters was accomplished by F. Yazicioglu (2013). For central Europe and during heating season, C. Oleskowicz-Popiel and M. Sobczak (2014) indicated that the external roller blinds could save about 45% of heat loss of the double-glazing window, the spacing between window glass and roller blind surfaces must be larger than 2 cm. Energy saving using dynamics external shutter in an office building was presented by F. Hammad and B. Abu-Hijleh (2010). The proposed dynamic shutter system with light dimming can save up to 34.02% of energy used in lighting.

Practically, the shutters are incompletely closed during the daytime to reduce the solar heat gain, and to provide the natural illumination. The air in the spacing between the shutter and window is heated during the daytime. The objective of the present research is to study the thermal characteristics of an incompletely closed shutter with experimental measurements. The results of the incompletely closed shutter are compared to window with completely close one. The experiments are conducted for a window facing west and month of June

Figure 1. A residential building in Kuwait with shutters installed on the windows.

METHODOLOGIES

The outdoor surface of the shutter is subjected to time dependent solar radiation and convective boundary conditions, while the indoor surface of the glass is subjected to time independent convection boundary condition. The thickness of the shutter and window are (L_s) and (L_w), respectively, and they are spaced with distance (L_a). The height and width of the window are (H) and (W), respectively. Figure 2 shows the measured average hourly variations of the ambient air temperature, and global radiation, for the month of June.

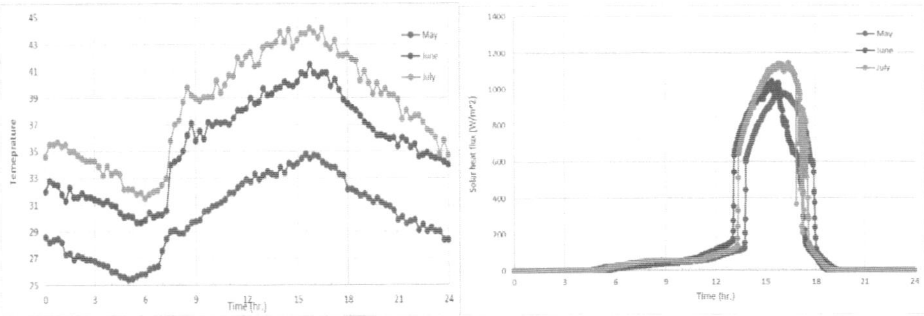

(a) (b)

Figure 2. (a) The measured average hourly variations of the ambient air temperature, and (b) global radiation, for month of June.

The heat flux at the indoor glass surface at any instant is calculated using the following expression:

$$Q_{is} = \int_0^H h_i(T_i - T_{is})\, dy \qquad (1)$$

The heat flux at indoor glass surface can be integrated to obtain the net heat flux during a specific period:

$$(Q_{is})_{net} = \frac{1}{t}\int_0^t Q_{is}\, dt \qquad (2)$$

RESULTS AND DISCUSSION

Experimental measurements were performed for the window with shutter. The system consists of a double plate glass with total thickness of 2 cm, and a metallic shutter with a thickness of 0.75 cm, apart from the glass by a distance of 20 cm. The experiment was conducted in the month of May. An incompletely closed shutter by a distance of 10 cm, 20 cm, and 30 cm. These cases are compared with a completely close one. The indoor air temperature is maintained at 23.5±0.5 °C. Heat flux is measured at the indoor and outdoor surfaces of the window using heat flux sensors, and the temperature is measured along the indoor and outdoor surfaces of the window and shutter, indoor and outdoor air. Figure 3 shows infrared image of the incompletely closed shutter with 40 cm. The image indicates that the surface temperature of the shutter can reach as high as 36.1°C during the daytime.

Figure 3, infrared image of the incompletely closed shutter with 40 cm.

Temperature and heat flux measurements

Heat flux at the indoor surface is calculated using equation (1) for window facing west direction, and during the month of June. Figure 4 shows measured heat flux at the indoor surface of the window. The presented results are a completely closed shutter, and for an opened shutter by a distance of 10 cm, 20 cm, and 30 cm. The noticeable heat flux increase is due to direct sun radiation, as well as increasing the outdoor air temperature. Increasing the opened part increases the temperature and the heat flux at the indoor surface of the window. The increase of the temperature and heat flux indicates the heat gain through the window is increased.

For a completely closed shutter case, solar radiation is completely blocked and prevented from entering the indoor space. The general behavior of the heat flux is similar to the completely closed shutter, but the magnitude of the heat flux is significantly less than the completely closed one. The natural convection flow has an important contribution in transporting the thermal energy from the shutter to the glass by reducing the thermal resistance in the enclosure. Figure 6 shows an infrared image of the incompletely closed shutter with 40 cm

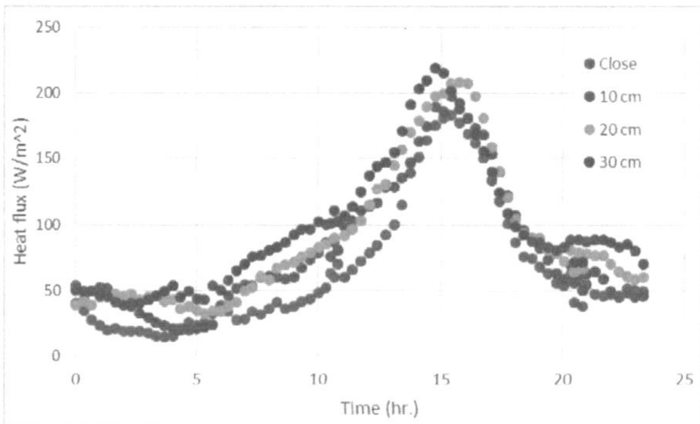

Figure 4. Heat flux at the indoor surface of the window.

Net heat gain calculations

The instantaneous heat flux in Figure 4a is integrated using equation (2) to obtain the net heat flux during a day. As the opened distance increased, the net heat gain increases significantly. As On the other hand, when the shutter is completely closed, the solar radiation is blocked, and the natural convection flow is induced in the spacing, but it has insignificant effect on the heat gain. The figure indicates that when the shutter is completely close, the net heat flux gain is equal 7610 W/m^2 and less by 40% than a shutter opened by a distance of 30 cm. As the opening distance increases, the heat gain can increases to reach as high as 11003.32 W/m^2 for shutter opened by 30 cm. Notice that the net heat flux for window with just opened by 10 cm, the heat flux is increased by 10% compared to a completely closed shutter.

CONCLUSIONS

Windows are account for the majority of the heat gain in buildings, and to reduce the heat gain, exterior shutters are installed. Keeping the shutters incompletely closed is common during the daytime, and since there is temperature difference between the window and shutter surfaces, natural convection flow in induced. The thermal characteristics of a completely closed shutter are investigated with experimental measurement, and the characteristics are compared with window with a completely closed shutter. The results indicate that the heat gain through window can be increased by 44.5% if the shutter is incompletely closed by distance 30 cm.

ACKNOWLEDGEMENT

The work presented was carried out with the financial and technical support of the research administration at Kuwait University under the grant EM02/12.

REFERENCES

Hammad F and Abu-Hijleh B (2010). The energy saving potential of using dynamic external louvers in an office building. *Energy and Buildings*, **42**, 1888-1895.

Ismail K and Henriquez J (2003). Modeling and simulation of a simple glass window. *Solar Energy Materials and Solar Cells*, **80**, 355-374.

Maccari A and Zinzi M (2001). Simplified algorithms for the Italian energy rating scheme for fenestration in residential buildings. *Solar Energy*, **69**, 75-92

Oleskowicz-Popiel C and Sobczak M (2014). Effect of the roller blinds on heat losses through a double-glazing window during heating season in Central Europe. *Energy and Buildings*, **73**, 48-58.

Yazicioglu F (2013). A comparative analysis of the energy performance of traditional wooden shutters and contemporary aluminum roller shutters Istanbul, a case study. *Energy Procedia*, **42**, 483-492.

THE ENERGY CONSEQUENCES OF EXCESS STATIC PRESSURE IN CENTRAL RESIDENTIAL HEATING AND AIR-CONDITIONING SYSTEMS

Torkan FAZLI[1], Rou Yi YEAP[2], Brent STEPHENS[1*]

[1]Department of Civil, Architectural and Environmental Engineering, Illinois Institute of Technology, Chicago, IL, USA
[2]Department of Chemical Engineering, Illinois Institute of Technology, Chicago, IL, USA

[*]Corresponding email: brent@iit.edu

Keywords: HVAC systems, Energy simulation, Static pressures, EnergyPlus

SUMMARY

The use of higher efficiency particle filters in central residential forced air heating and cooling systems have become more widespread recently. Improved filtration products often introduce excess static pressure to systems, which can have significant energy and economic impacts. However, the overall energy consequences of excess static pressure drop have not been explored in depth across a large number of climates, homes, or system characteristics. In this study, we model the energy impacts of a range of realistic static pressure conditions for both newly constructed energy efficient and existing single-family homes in all climate zones in the U.S. using building energy simulations in BEopt combined with EnergyPlus. Results suggest that severe excess external static pressure can increase annual energy consumption by up to ~14% in homes with a permanent split capacitor (PSC) blower installed but only ~2% in homes with an electronically commutated motor (ECM) installed.

INTRODUCTION

Heating, ventilating, and air-conditioning (HVAC) systems use almost 50% of the energy in buildings, or ~10-20% of the total energy consumption in developed countries (Pérez-Lombard, Ortiz, and Pout, 2008). Many central residential heating and air-conditioning systems contain high-pressure drop elements such as high-efficiency or dust-loaded filters, dirty coils, constructed or undersized ductwork, and/or closed registers or grilles. In particular to this study, many occupants are relying on the use of higher efficiency particle air filters for improved filtration; however, high efficiency filters also yield excess static pressure that the systems must overcome (Walker, 2014). Excess static pressures can have considerable energy and economic impacts, but the impacts are depending on the type of blower motor used in the air handling unit and the amount of excess static pressure in the system (Rodriguez et al., 1996; Parker et al., 1997). Although magnitudes of excess static pressure drops have been reported in detail in a number of studies (e.g., Stephens, 2014), the overall energy consequences have not been explored in depth across a large number of climates, homes, or system characteristics. In this work, the energy consequences of excess static pressures on central residential HVAC systems are investigated by using whole building energy simulation of two home vintages with different fan blower motors across 15 U.S. climate zones.

METHODOLOGIES

Two types of homes were selected for simulating in each of 15 cities located in different climate zones according to ASHRAE 62.2 (ASHRAE, 2010): modern high-efficiency homes and typical existing homes. BEopt software was used to model the same home geometry with different envelope characteristics for all climate zones. Modern high-efficiency homes were designed to meet or exceed most minimum energy code requirements in all locations according to the 2009 International Energy Conservation Code (IECC, 2009). They have a tight building envelope with low air infiltration (3 ACH_{50}), supplemental mechanical ventilation systems (supply or energy recovery ventilation systems), and properly sized high efficiency heating and air-conditioning systems for each climate zone. 'Typical' homes were tailored to represent existing, older, and less efficient homes with moderate building envelope insulation, moderate airtightness (10 ACH_{50}), and larger and less efficient heating and air-conditioning systems for each climate zone based on typical existing home characteristics in each area. There were no mechanical ventilation systems modeled in the existing homes. Envelope characteristics were taken from two primary surveys of existing housing characteristics for homes built after 1979 (Y. J. Huang et al., 1987; J. Huang, Hanford, and Yang, 1999).

A range of target total external static pressures were identified as realistic values from the literature (e.g., Proctor, et al., 2011). Thirteen external static pressures were chosen to model, increasing from a low of 50 Pa as the baseline static pressure drop to as high as 350 Pa. Two types of blower motors were examined in this study because the airflow rate and overall fan and motor efficiency will vary according to fan type and thus have different effects on fan power draw and system runtimes: permanent split capacitor (PSC) motors and electronically commutated motors (ECM). Representative fan curves (airflow rate versus static pressure and fan efficiency versus static pressure) from the U.S. Department of Energy for single-stage virtual model furnaces (for PSCs) and two-stage virtual model furnaces (for ECMs) were used to define blower characteristics (DOE, 2011). Rated and maximum airflow rates were set to 400 cfm/ton (a commonly used assumption in residential system sizing). These airflow rates were assumed to be the same during both cooling and heating operation for simplicity. Rated airflow rates for HVAC equipment and duct sizes were kept at the maximum (nominal, or lowest pressure) value for each simulation case, but the design and specified airflow rates were adjusted in each case.

EnergyPlus input (IDF) files were first generated by running the BEopt simulations. The required parameters such as airflow rate, fan pressure, fan efficiency, and nominal cooling and heating capacities were changed for each scenario. Heating and cooling equipment capacities and HVAC airflow rates were changed from 'auto-size,' which is calculated by BEopt, to proper sizes based on realistic existing HVAC systems on the market. A total of 780 annual energy simulations were then performed in EnergyPlus using the appropriate typical meteorological year (TMY3) data and all of the combinations of input scenarios. Outputs of EnergyPlus simulations are consisted of annual electric use for the AHU fan and condenser–compressor unit, as well as annual natural gas usage for the furnace. The annual heating and cooling outputs were used to explore impacts of blower types in different home types and in various climate zones on total HVAC energy use and national and regional costs on an annual basis were estimated.

RESULTS AND DISCUSSION

Table 1 shows the average heating, cooling, fan, and total HVAC energy consumption for the baseline static pressure conditions (50 Pa) across all 15 climate zones. The annual total site energy consumption for the baseline static pressure ranged from 14431 to 51297 kWh, depending on climate. The baseline annual heating and cooling costs ranged from $272 to $3060, again depending on climate. The electricity used for cooling and HVAC fans ranged from 97 to 8872 kWh and gas consumption for heating ranged from 11 to 35242 kWh, for which all minimum values refer to new efficient homes with ECM blowers and the maximum values refer to existing homes with PSC blowers.

Table 1. Average heating, cooling, fan and total HVAC energy consumption of all 15 cities for baseline static pressure (50 Pa).

Blower Types	PSC				ECM			
Energy Use (kWh)	**Heating**	**Cooling**	**Fan**	**Total**	**Heating**	**Cooling**	**Fan**	**Total**
New Homes	8431	906	461	9797	8134	881	221	9237
Existing Homes	12931	2380	720	16032	12378	2299	342	15019

Figure 1 shows the percentage increase in average HVAC energy consumption across all 15 cities for two types of fan blowers (PSC and ECM) and home types (new and existing).

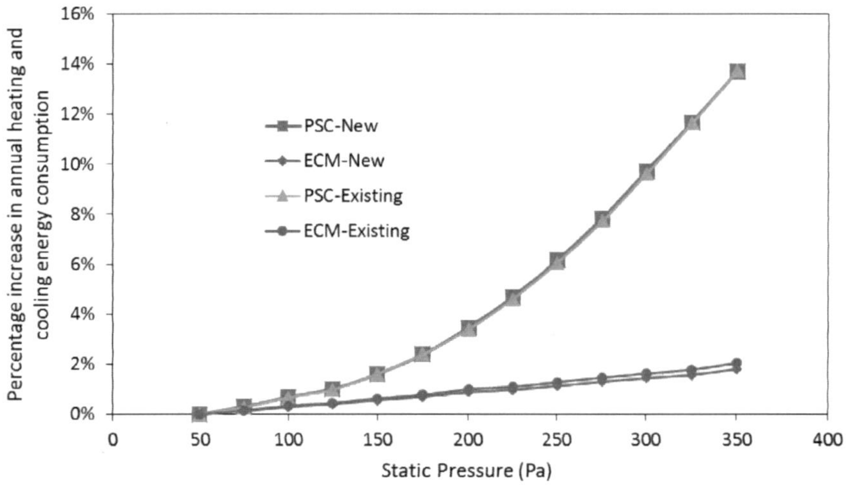

Figure 1. Percentage increase in average total HVAC energy consumption across all 15 climate zones in response to increasing static pressure

The general trends for total annual energy consumption were very different for PSC and ECM blowers. The largest increases in annual energy use with higher static pressures occurred with PSC blowers installed, particularly in the climate zones with higher heating loads. For PSC blowers, the average cooling energy increase was ~36% at the highest static pressure (350 Pa) compared to the lowest static pressure (50 Pa); the average heating energy increase was ~12%; and the average fan energy increase was ~35%. All increases in static pressure reflect an increase in system runtimes with a PSC blower caused by as much as a 56% decrease in airflow rates at the highest pressure. For ECM blowers, the average increase in cooling energy use at the highest pressure (relative to the lowest pressure) was only 2%;

the average heating energy use actually decreased by ~3%; and the average fan energy increase was ~159% when moving from lowest pressure to the highest pressure drop. These effects are because in ECM systems higher pressure drops only minimally impact airflow rates while increasing fan power draw, but fan power is only a small fraction of overall energy use. Moreover, results of this study also show that PSC blowers consume ~330% energy more than ECM blowers and annual energy cost of PSC blowers are ~80% more than ECM Blowers, on average, suggesting that ECMs are more efficient than PSCs across a wide range of pressure drops.

CONCLUSIONS

Results from the 780 annual building energy simulations demonstrate that home vintage, climate zone, and fan blower motor type all influence the energy impacts of excess static pressures on central residential HVAC systems. Extreme values of excess static pressure (moving from 50 to 350 Pa) are predicted to yield average increases in annual energy consumption across all modelled homes of ~14% with PSC blowers installed and only ~2% with ECM blowers installed. Moreover, although the absolute magnitude of incremental changes in energy consumption in existing homes was larger than in new energy efficient homes, relative increases were similar.

REFERENCES

ASHRAE, Standard. 2010. "Standard 62.1-2010." *Ventilation for Acceptable Indoor Air Quality, ASHRAE.*

DOE. 2011. "Technical Support Document: Energy Efficiency Program for Consumer Products: Residential Central Air Conditioners, Heat Pumps, and Furnaces."

Huang, Joe, et al., 1999. *Residential Heating and Cooling Loads Component Analysis.* Building Technologies Department, Environmental Energy Technologies Division, Lawrence Berkeley National Laboratory, University of California.

Huang, Y. J., et al., 1987. *Methodology and Assumptions for Evaluating Heating and Cooling Energy Requirements in New Single-Family Residential Buildings: Technical Support Document for the Pear (program for Energy Analysis of Residences) Microcomputer Program.* Lawrence Berkeley Lab., CA (USA).

IECC. 2009. "International Energy Conservation Code." http://energytaxincentives.org/uploaded_files/iecc09.pdf.

Parker, D. S., et al., 1997. "Impact of Evaporator Coil Airflow in Residential Air-Conditioning Systems," December.

Pérez-Lombard, et al., 2008. "A Review on Buildings Energy Consumption Information." *Energy and Buildings* 40 (3): 394–98.

Proctor, J., et al., 2011. "Efficiency Characteristics and Opportunities for New California Homes." *Final Report of Project Number PIR-08-019 California Energy Commission–PIER Program.*

Rodriguez, et al., 1996. "Effect of Reduced Evaporator Airflow on the High Temperature Performance of Air Conditioners." *Energy and Buildings* 24 (3): 195–201.

Stephens, Brent. 2014. "The Impacts of Duct Design on Life Cycle Costs of Central Residential Heating and Air-Conditioning Systems." *Energy and Buildings* 82 (October): 563–79.

Walker, Iain S. 2014. "Energy Implications of In-Line Filtration in California Homes." *ASHRAE Transactions*, April.

A SURVEY ON EVALUATION OF WORKERS AND THEIR THERMAL ENVIRONMENT IN OFFICES

Koichi IKEDA[1,*], Hikaru ENOMOTO[2], and Hitoshi TAKAONO[3]

[1]Department of Architecture, College of Science & Tech., Nihon University, Japan
[2]Department of Education, Tokyo University of Social Warfare, Japan
[3]DAI-DAN Co., Ltd., Japan

[*]Corresponding email: k.ikeda@arch.cst.nihon-u.ac.jp

Keywords: Office Environment, Thermal Sensation, On-sight Survey, Clo Value, Predicted Mean Vote (PMV)

SUMMARY

To investigate the relationship of thermal environment and sensation, onsite measurements and questionnaire surveys were conducted in 9 office floors in Tokyo and Gifu. Total number of 646 office workers was responded to the questionnaire about thermal sensation. Main findings were (1) most of floors' environment were within the criterion of the Law for Maintenance of Sanitary Environment in Building in Japan, and most of average PMV values were within a comfort range of ISO7730. (2) Although 20% office workers responded as uncomfortable, half of the workers did not satisfy in total thermal environment. Eighty percent of workers answered "Satisfied" with their clothes and about 70 percent answered "Satisfied" with the room temperature, and (3) females felt dryness more strongly than males and the counter measures for protecting the room air humidity were considered important.

INTRODUCTION

Several years have passed since cool biz and warm biz policy had been introduced in Japanese offices. The policy was aiming both energy conservation and comfort by controlling air conditioning temperature and changing clothes as measures for inhibiting carbon dioxide emission increase. Purpose of this research was to conduct a thermal investigation of the office spaces unable to control their own environment in order to understand the actual status and to investigate validity and effectiveness of the current standards.

INVESTIGATION METHOD

PMV meter set on the site

Office room surveyed

Photo. 1 Experimental settings

Outlines of Investigation and Schedule

The investigation was conducted in 2013. The investigation in the summer period was conducted from the middle of July to the beginning of August, that in the intermediate period was conducted in October, and that in the winter period was conducted in December. The buildings surveyed were six floors of five companies in the Tokyo metropolitan area and three floors of one company in Gifu City, with nine floors of six companies in total (Photo. 1). Table 1 shows the building details.

Table 1. Details of Investigated Floors

Office	A	B	C	D	E1	E2	F1	F2	F3
Place	Aarakawa, Tokyo	Chiyoda, Tokyo	Shinjuku, Tokyo	Shinjuku, Tokyo	Chuo, Tokyo		Gifu city, Gifu Pref.		
Construction	RC	RC	SRC, Steel frame	Steel frame, RC, SRC	RC		RC	RC	SRC
Construction year	2004	1973	1999	1994	2009		1995		1970
Total floor space	208.68 ㎡	375.61 ㎡	1576.23 ㎡	955.66 ㎡	224.41 ㎡		370 ㎡	262 ㎡	480 ㎡
Storeys / Number of floors	5F/5F	3F/7F	9F/27F	19F/31F	4F/8F	7F/8F	1F/6F	2F/6F	6F/7F

Indoor environmental measurement

The air temperature and its relative humidity, wind velocity, and average radiant temperatures in the offices were measure using a measurement device (Kyoto Electronics Manufacturing Co.'s product: AM101) every minute. The values at the time when the subjects filled in the questionnaire form were used for the PMV calculation.

Questionnaires conducted on the subjects (office workers)

1) Thermal environment evaluation: A questionnaire survey was conducted in order to investigate how the subjects evaluate in their present work environments. They were asked to respond to warm and cool sensations by selecting from a 9-point scale (very cold, cold cool, slightly cool, no opinion, slightly warm, warm, hot and very hot), and to thermal comfort sensations by selecting from a 7-point scale (very uncomfortable, uncomfortable, slightly uncomfortable, no opinion, slight comfortable, comfortable and very comfortable). In addition, they were asked to select whether or not they were sensitive to the cold and heat, their desired clothes and temperature, and whether or not they were satisfied in their present working environment.

2) Clothing worn during office hours: A questionnaire was conducted for the status of clothes and the amount of clothing, and was calculated using the equation by Hanada.

3) In addition, a questionnaire was conducted concerning body height, body weight, age and cooling and warming items used on the day of investigation.

RESEARCH RESULT AND DISCUSSION

Result of the Measurement of Thermal Environment

Although the temperatures and relative humidity during the summer and intermediate periods were almost same, the temperature and relative humidity during the winter period were approximately 1.0 degree and 12% lower than those during the summer and intermediate periods, respectively (Fig. 1).

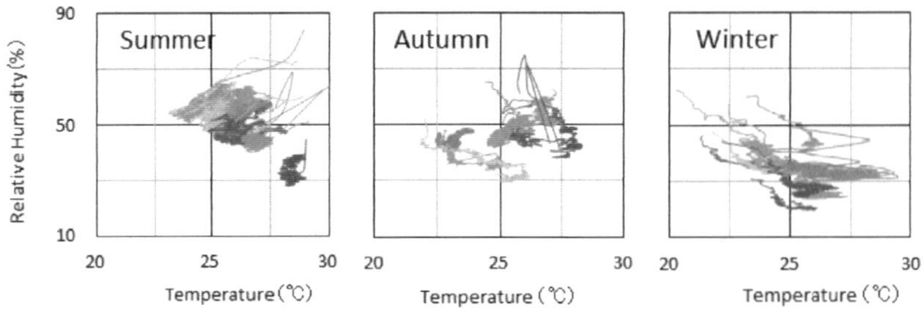

Fig. 1 Room temperature and its relative humidity in each floor

Although the temperature values were within the standard range of the Japanese Law for Maintenance of Sanitary Environment in Buildings in most of the cases, there were many cases where humidity in the winter period was lower than the standard level.

Attributes of the Subjects

Responses to the questionnaire were received from 646 subjects (505 males and 141 females). Fig. 2 shows the age distribution of the subjects. The average age±S.D. was 45.2±11.6 for males, 40.3±10.9 for females, and 44.1±11.6 as a whole. Fig. 3 shows the distribution of BMI of the subjects. The average BMI±S.D, was 23.8±3.9 for males, 20.3±2.4 for females, and 23.0±3.9 as a whole. There was a significant difference in BMI between genders at $p<0.001$ by an χ^2test.

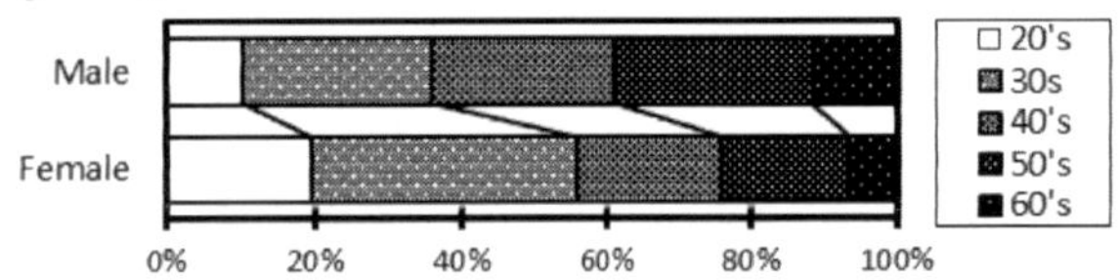

Fig. 2 Age structure of the subjects

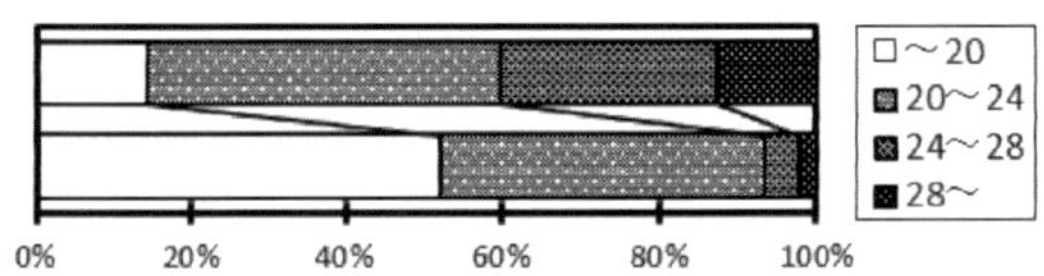

Fig. 3 BMI of the subjects

Clothing status of the subjects

Fig. 4 shows the distribution of the mount of clothing. The average clo value±SD was 0.56±0.08 clo for males and 0.47±0.10 clo for females in the summer period, 0.62±0.16 clo for males and 0.58±0.15 clo for females in the intermediate period, and 0.79±0.18 clo for males and 0.75±0.14 clo for females in the winter period. The result shows that the clo value for females was lower than that of males in all seasons. The difference in clo value between males and females was the largest at 0.9 clo in the summer period.

Males in the summer period, Females in the summer period, Males in the intermediate period, Females in the intermediate period, Males in the winter period, Females in the winter period

Distribution of Individual PMV values

Fig. 5shows the average values of individual PMV of the subjects. The average PMV value±S.D. was 0.41±0.38 for males and 0.12±0.43 for females in the summer period, 0.57±0.35 for males, 0.22±0.57 for females in the intermediate period, and 0.14±0.39 for males and 0.01±0.30 for females in the winter period. Although most PMV values were within the comfort range of ISO7730 (from -0.5 to 0.5), the average value of males in the intermediate period exceeded 0.5. This is considered to be because the amount of clothing was large compared to room temperature.

Distribution of Individual PPD (Predicted Percentage of Dissatisfaction)

Fig. 6 shows the distribution of PPD of the subjects. The PPD value of females at the intermediate period was large compared to other periods. Although the average PMV value was within the range of comfort, the result showed that the number of subjects whose PPD

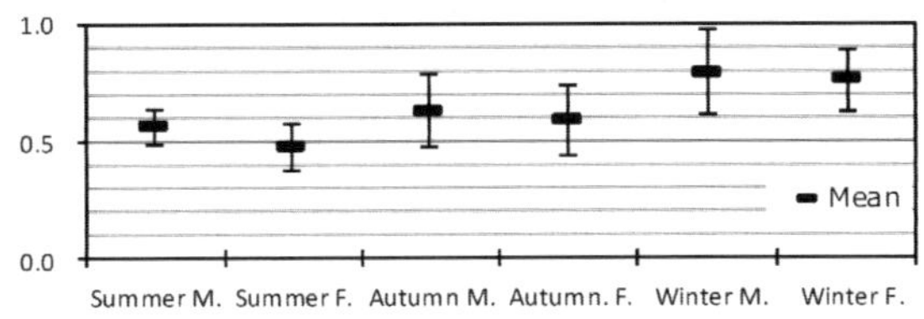

Fig. 4 Amount of clothing of the subjects (±S.D)

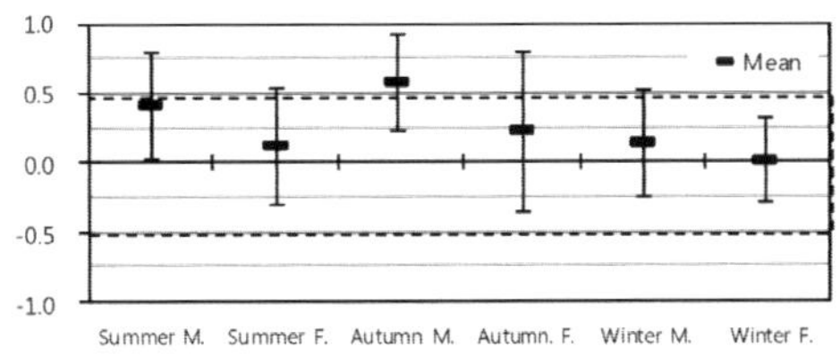

Fig. 5 PMV values (±S.D.) of the subjects

value was more than 10 counted for 58%. It is considered to be caused by the differences in temperature and the amount of clothing worn by subjects on each floor, and there were many cases where PMV exceeded 0.5 or was lower than -0.5.

Response of Thermal Comfort

Fig. 7 shows the responses to comfort, which shows that a quarter of subjects responded uncomfortable. The proportion of uncomfortable sensation in intermediate period showed no difference with other periods. Regarding the difference between male and female, more females responded at the side of comfortable than males in all seasons, which corresponds with the trend of PMV. Significant difference in comfortable sensation between male and female was observed by χ^2 test at $p<0.01$.

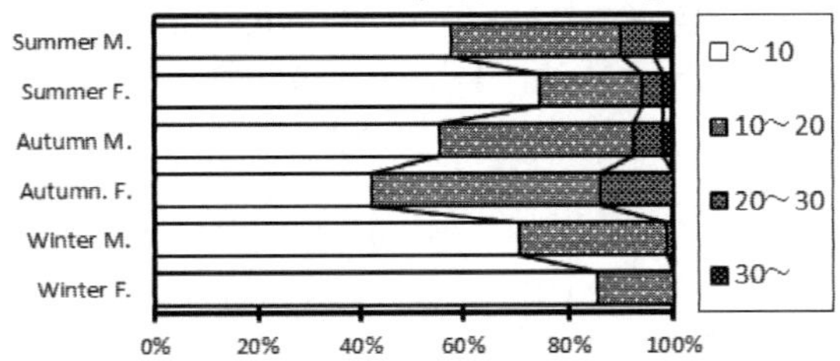

Fig. 6 Distribution of Individual PPD values

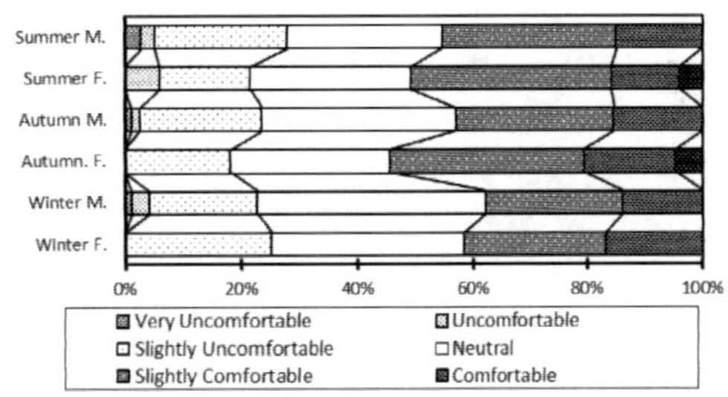

Fig. 7 Distribution of the responses for Thermal comfort

CONCLUSION

1) There were approximately 20% of the responses at the side of uncomfortable. Although 20% of the office workers were not satisfied, approximately half responses answered not satisfied with their total thermal environments. Eighty percent of responses approximately about 70 percent of responses answered as " Satisfied" with clothes and room temperature.
3) For the purposes of further comfortable environment, females felt dryness more strongly than males and the counter measures for humidity protection were considered important.

REFERENCES

Enomoto H, et al.(2009), Thermal Environment Evaluation by the Workers of the Offices that Introduced "Cool Biz" and Occupational Health and Safety Research pp. 5-10.
Haneda H, et al.,(2009) Actual Measurement of Thermal Environment in Cool Biz Office and the Evaluation on Intellectual Productivity by a Questionnaire Survey, Journal of Environmental Engineering, Vo. 635, pp.389-396.

ACKNOWLEDGEMENT

We would like to express our sincere appreciation for the cooperation extended by Professor Ryoichi Inaba of Gifu University School of Medicine, the staff of the Japan Building Maintenance Association and the offices subjected to the research.

Development of an Automatic Thermal Control System Using Human Facial Skin Temperature

Bo YI[1,*], Joon-Ho CHOI[1], Douglas NOBLE[2], Kyle KONIS[3]

[1]School of Architecture, University of Southern California, Los Angeles, California USA

[*]Corresponding email: yib@usc.edu

Keywords: Individual control, skin temperature, thermoregulation, chamber

SUMMARY

Since physiological conditions and satisfaction standards vary considerably from person to person, each individual normally has a different acceptable range of air temperature in a certain thermal condition. It is possible to create an automatic system by utilizing physiological signals to help control and optimize buildings' indoor air temperature to help maintain thermal neutrality in a one-occupant room. Skin temperatures can signal appropriate thermoregulatory behaviour in response to specific conditions, since skin temperature changes with different thermal sensations. The main goal of this study is to better understand the relationship between facial skin temperature and human thermal sensation in order to achieve the appropriate conditioned air temperature dynamic settings, thus helping to achieve thermal comfort by reading facial temperature signal patterns. A series of human subject experiments were conducted in an environmental chamber at USC. Facial skin temperatures were tested continuously by six electronic thermocouples at six locations and sensation surveys were provided by subjects. During the tests, the air temperature was regulated between 20°C and 30°C, while the other five thermal parameters were kept constant. A thermal sensation diagnostic model and control logic with decision tree was built by using data mining and machine learning technology.

INTRODUCTION

Currently in industry, HVAC systems are generally designed according to a certain industrial criterion. However, since physiological conditions and satisfactory conditions vary considerably from person to person, each individual normally has a different thermal preference. However, it is possible to create an autonomic system by utilizing physiological signals to help control and optimize building indoor thermal conditions. Facial skin temperature can play a signalling role to indicate human individual thermal sensation based on thermoregulation.

Thermoregulation is a function of the human body to keep thermal homeostasis. In order to achieve health, a small range of core temperature, 36.8° ± 0.4 °C, must be maintained. To maintain a constant body core temperature, the heat produced by the body must be transported to skin surface through blood, and release into ambient environment (Carroll, 2007). Skin temperature is controlled by blood flow; and blood flow is controlled by the sympathetic neural system, which includes the noradrenergic vasoconstrictor system and a sympathetic active vasodilator system. Facial skin presents higher temperature with warm sensation and lower temperature with cold sensation.

To approach the proposed goal, the objectives are as follows:

Object 1: to complete experimental system with sensors and controlled devices.
Object 2: to investigate the relationship between human facial skin temperature, ambient thermal conditions, and thermal sensation with thermal conditions.
Object 3: to determine the optimal sensing and control intervals to minimize any error rate of a thermal sensation estimation while enhancing the comfort condition.

METHODOLOGIES

In the study, 25 human subject experiments were conducted in a thermal chamber on the basement of Watt Hall, University of Southern California. A total of 25 subjects, aged from 20-30, were recruited in this study. For each experiment, one subject was asked to sit in the chamber doing computer-based work. The subject's facial temperature and ambient environmental conditions were measured with sensing interval of 10 seconds. Each experiment with contains one-hour heating process and one hour cooling process. The test started with heating process, and the indoor dry ball temperature was changed by controlled HVAC system from 20℃ to 30℃. Then, the indoor temperature was changed from 30℃to 20℃. When the indoor temperature was 20°C, 22°C, 24°C, 26°C, 28°C, 30°C during both heating and cooling process, a seven-point scale thermal sensation vote (TSV) survey developed for the PMV model (ASHRAE-55, 2013) was given to the subject to collect human thermal sensation.

Two methods were employed for measuring facial skin temperature. The first method is using 5 thermistors attached on human facial skin at six points in the region of forehead, upper rim of eye, bottom rim of eye, upper ROI area, bottom ROI area, and jaw. Another method is using an infrared temperature sensor, which is fixed on the monitor targeting the subject's face to sense overall facial temperature. Indoor air dry ball temperature was sensed at the height of 0.1m, 0.6m, 1.1m. Relatively humidity, mean radiant temperature (MRT), and CO2 level were also measured as ambient environmental condition.

Based on the collected data from the sample #1, the time-series plot of air temperature, infrared facial temperature and the mean value of facial surface temperature are shown in Figure 1.

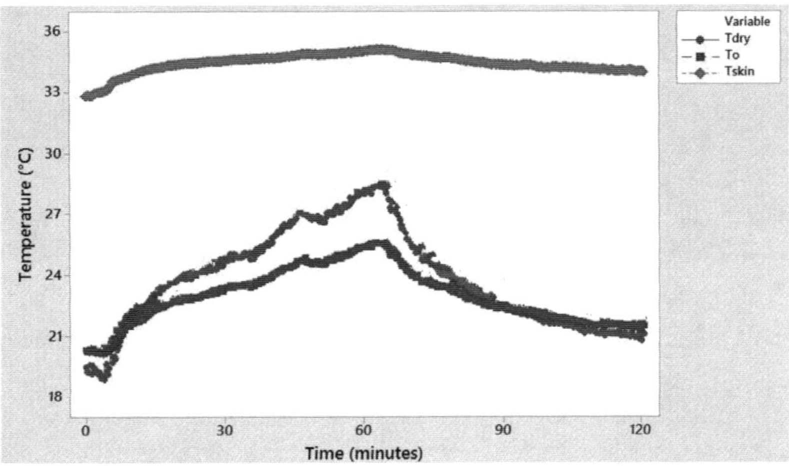

Figure 1. The time series plot of air temperature, infrared facial temperature and the mean value of facial surface temperature.

The temperature change rate (TCR) of both overall facial skin temperature and indoor air temperature were calculated using the following equation:

$$TCR = (T_2 - T_1) / 2 \quad (1)$$

where *TCR* the temperature change rate (℃/min), T_2 is the current temperature, T_1 is the previous temperature measured 2 minutes ago.

The correlation between overall facial skin temperature and indoor air temperature is shown as below in Figure 2. The result of human questionnaire answers is shown in Figure 3.

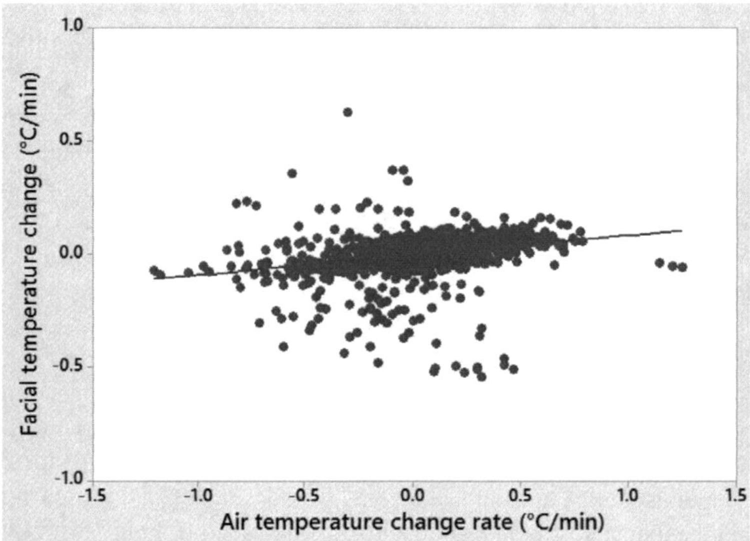

Figure 2. The correlation between overall facial skin temperature and indoor air temperature.

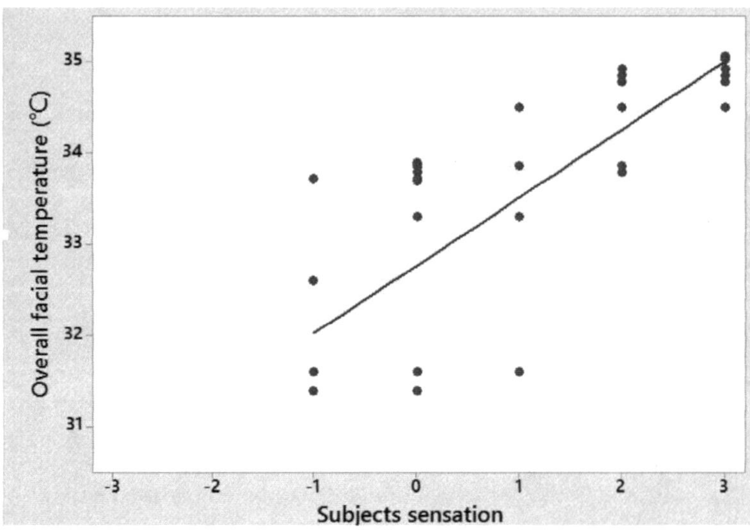

Figure 3. The correlation between thermal sensation votes (-3 very cool, 3 is very wall) and human overall temperature.

RESULTS AND DISCUSSION

According to the data collected, human overall facial skin temperature increased with ambient air temperature increasing and decreased with the reduction of air temperature. It was proved that human facial skin temperature could response accordingly to the ambient environment.

Overall facial skin temperature change rate has linear regression to air temperature change rate. According to estimation, the regression complies with the following equation:

$$\Delta T_{facial} = -0.004573 + 0.08753\Delta T_{Air} \quad (2)$$

According to the TSV survey, subjects claimed warmer when the facial temperature is higher and claimed cooled when a facial temperature is lower. The regression equation was estimated as below:

$$T_{facial} = 32.76 + 0.7469TSV \quad (3)$$

CONCLUSIONS

Through a series of human subject tests in an environmental chamber, this research established a regression model to estimate a user's thermal sensation based on reading the facial skin temperature and its gradient. Human face, as a bare skin area without being covered by any clothing, is a most sensitive area to the ambient condition. This study identified the human facial skin temperature as a high-potential parameter to indicate a human thermal sensation. There were some signal variations found depending on the subjects' physiological conditions, including gender, age, and body mass index. These additional parameters will be considered in the future study to establish a more accurate sensation prediction model with using an additional number of human subjects in a chamber study.

REFERENCES

Carroll, R. (2007). Elsevier's integrated physiology. Mosby Elsevier, Philadelphia, Pennsylvania, USA.

ASHRAE. (2013). ANSI/ASHRAE Standard 55-2013. Thermal environmental conditions for human occupancy. American Society of Heating, Refrigerating, and Air-Conditioning Engineers, Inc.: Atlanta, GA.

de Dear, R.J. & Brager, G.S. (2002) Thermal comfort in naturally ventilated buildings: Revisions to ASHRAE Standard 55. Energy & Buildings, 34(6), 549-561.

McIntyre, D.A. & Griffiths, I.D. (1972) Subjective response to radiant and convective environments. Environmental Research, 5(4), 471-482.

Fanger, O. (1972) Thermal Comfort: Analysis and Applications in Environmental Engineering. McGraw-Hill, New York, USA.

Energy Efficiency Improvements and Indoor Environmental Quality through Climate-Responsive Architecture

Shahryar Habibi[1*]

[1]Department of Architecture, University of Ferrara, Ferrara, Italy

[*]Corresponding email: hbbsh@unife.it

Keywords: Smart materials, Indoor Environmental Quality (IEQ), smart architecture.

SUMMARY

This paper explores about retrofitting of the existing buildings through smart and innovative ways which can adapt the buildings by means of responsive architecture to improve Indoor Environmental Quality (IEQ). In order to meet the requirements of performance for a built environment, optimum values of the indoor environment should be considered as the most important factor. Responsive Building Elements (RBEs) are one of the most important design parameters for improving the efficiency of energy and IEQ. This paper reviews examples of responsive and smart architecture and its building elements. In a first step, the paper gives an overview of the research responsive architecture focusing on aspects that are most relevant for the development of the indoor environment in terms of environmental factors.

INTRODUCTION

At present, the architectural world with outstanding improvements in science and technology is changing our life as smart systems in the form of smart cities, smart buildings and smart space. These changes in architecture are not limited to a set system and sensor but also have significant progress in the building components such as materials, structures and etc. For instance, façades as an important building component which can control environmental factors by means of its components are being developed to be improvement in terms of light, heating, cooling, ventilation, humidity, acoustic and other factors. Responsive Façade is one of the design strategies that including smart building components such as smart elements and smart materials that be able to adjust the indoor environmental conditions with respect to the outdoor environment.

Figure 1. Smart envelope comprised of ETFE encased solar-activated (left) Flare Façade – The modular façade (right)

The term responsive is often used interchangeably with adaptive, but most simply it is used to describe how natural and artificial systems can interact and adapt (P. Beesley et al., 2006).Adaptive as one other of the terms in architecture and smart systems can sense, actuate and adapt efficiently (Figure 1). For instance, WHITEvoid has developed a modular system for building facades that also communicates information, albeit more subtly than through the application of buttons and patches. "Acting like a living skin, it allows a building to express, communicate and interact with its environment. FLARE turns the building facade into a penetrable kinetic membrane, breaking with all conventions of the building surface as a static skin" (WHITEvoid.com). FLARE-façade (Figure1, right side) is composed of metal flake elements that are controlled by a series of pneumatic cylinders. These elements reflect sunlight in a way that casts a shadow on some of the faces of each element, giving them darker colours. The façade does not optimize the energy performance of the indoor spaces, but socially interacts with people outside the building (M. Mansour El Sheikh , 2011).

SMART MATERIALS

The importance of development technology in relation to materials in the buildings is emerging significant. It is entitled to the innovative materials which include smart materials and nanomaterials. Innovative can also be divided into new materials such as aerogels, thermoelectric and conventional materials which include phase change materials, concrete and vegetables.

In the context of the design material, the term smart is frequently transparent, lightweight, malleable and responsive. Although, in some cases these materials are not fully compatible with environmental, but they can easily be adapted to the built environment. Smart materials can adapt itself easily to the external environment in terms of moist, heat, light, pressure and other fundamental factors. Adaptation of materials improved by a smart process. In this process, the material shows self-initiative and adapts according to its own discretion. In general, it is debatable whether a material can be referred to as being smart or intelligent since materials do not have cognitive capabilities (Srinivasan, A. V. et al. 2001). Though, in the practice smart materials have shown their performance in the structural performance, climate and energy performance and architectural performance. The following examples indicate examples of smart materials used in the architecture. For instance, Architect Doris Kim Sung developed her research at University of Southern California by focus on new smart material which requires neither control and nor energy. This material is set to temperature and is a self-ventilating building skin with smart thermo bimetal which deforms and responds automatically to changes when heated or cooled (Figure 2).

Figure 2: One example of smart thermo bimetal self-ventilation skin, (Doris Kim Sung, 2014)

-Advanced glazing and smart windows

The glazing is installed on the facade of buildings to transform light and prevent heat loss and noise. Glazing as a bridge which separate the outside and inside through façade plays a constructive role in the exchange of heat and light and sound. Advanced glazing systems as new smart systems can significantly influence to reduce conventional glazing in terms of energy usage, visual comfort and acoustic performance. It is estimated that the buildings are losing plenty of energy through windows. For this reason, efforts to reduce energy consumption on the windows can be considered as an important control strategy. Smart windows as new technologies in the buildings are able to adjust condition of indoor to be a desired level in terms of lighting, heating and acoustic. For instance, smart windows when outside temperature increases become dark and also become mostly transparent to light in cold temperature. So, smart windows aim to energy-related performance challenges in the buildings to control heat loss and provide daylight with minimal solar heat. Smart windows have also potential to reduce glare and improve the quality of natural light.

Today, responsive materials and intelligent control methods, promise to add new dynamics to windows, including the real-time, adaptation of their aesthetic presence and their performance, based on given conditions. Smart windows aim to take an active role in the dynamic optimization of building performance with low operational cost. For example, by actively controlling the state of electrochromic glass panes, the solar transmittance of windows becomes a programmable feature, by which it is possible to regulate the interior illuminance and temperature (E. S. Lee, et al. 2006) (T. Hausler et al.2003)(S. D Kotsopoulos et al. 2012). The transitions of smart windows, optimally operated, can contribute to the reduction of energy consumption by heating and cooling (S. Selkowitz, et al. 2003).

METHODOLOGIES

As mentioned above about smart systems, it shows that there are many systems and strategies in the new or existing buildings to optimize performance. However, these systems should perform multiple functionalities without disturbing balance preservation of existing features and the integration of new systems and innovation technologies. Sustainable and smart infrastructures to transform into the buildings should be considered in terms of process control and design strategies. According to Mel Schwartz (2002), one of the major difficulties in incorporating smart materials into the architectural design is the recognition that very few materials and systems are under single environmental influences. For example, the use of a smart material to control conductive heat transfer through the building envelope may adversely impact daylight transmission. Furthermore, because most systems in a building are highly integrated, it is difficult to optimize performance without impacting the other systems or disrupting control system balancing (M. Schwartz, 2002). Therefore, to reach high-performance buildings through climatic-responsive and smart systems should be taken into account environmental factors and their impacts on indoor and outdoor environment. Sustainable Smart Refurbishment (SSR) as a new design integration can be applied to new building design and existing buildings to improve indoor and outdoor environmental comfort.

This process can also be included new application methods that determine user needs and requirements. In addition to the new proposed method, Refurbishment Performance Assessment Index (RPAI) is used to evaluate and assess the performance of new design integration. RPAI includes environmental impacts of smart systems on indoor and outdoor condition and provides performance values for SSR. The computation of performance value for SSR should be related to users' comfort requirements and the built environment conditions and should be performed on each refurbishment and integration in the buildings. For example, at each stage of the refurbishment buildings to achieve optimal comfort conditions, sustainability of applications should be investigated primarily and then be evaluated to determine appropriate smart systems in accordance with the desired sustainability to improve sustainability performance (Table 1). The RPAI can be explained by specific mathematical description and criteria evaluation as shown in equation 1.

Table1.Refurbishment Performance Assessment Index (RPAI)

	Application S (Sustainable)	**Application I (Intelligent)**	η (The value of interference)
Building materials	S_m	I_m	η_1
Building elements	S_e	I_e	η_2
Overall	S_a	I_b	η_b

Overall Refurbishment Performance Assessment Index (RPAI) = $S_a + (I_b - \eta_b)$

$$RPAI = \sum_{a=1}^{n} S_{a\alpha} + \sum_{b=1}^{n} (I_{b\beta} - \eta_b) \quad (1)$$

Where $S_{a\alpha} = \{S_m \text{ or } S_e\}$ (a=1,2,3,...n) (α=1,2,3,...), Sm and Se indicate applications of sustainable building materials and sustainable building elements that each of them can be used in the refurbishment process, types of applications and alternatives are denoted by **a** ,and **α**. These process include not only all sustainable design strategies regarding material and energy efficiency but also all elements and all materials that can help to improve sustainable refurbishment. Also, $I_{b\beta}$ stands for intelligent or smart applications, (b=1,2,3,...n) (β=1,2,3,...),where types of system values and alternative smart systems are indicated by **b** and **β** respectively. The value of interference is one of the disincentive factors to sustainable strategy. Smart systems effects may be led to reducing sustainability performance among other items. For instance, the design of an embedded system to reduce natural light may be less efficient at heat energy. Therefore, the value of interference should be excluded from performance of smart applications, and only the factors that influence sustainability and the efficiency of buildings should be taken into consideration.

CONCLUSIONS

This work seeks to highlight smart systems to improve building and energy performance in both new and existing buildings. This work also addresses smart material and windows implementations in the buildings. It is clear that in the future, interactive capabilities of smart materials and elements will change architectural styles and details. Therefore, it should be determined new methodology and approach to design with smart systems. In this work, a method as Sustainable Smart Refurbishment (SSR) is proposed for the development of smart ways in the buildings. It is defined as just a potential method to enhance Indoor Environmental Quality (IEQ) and building performance. It needs to be expanded to address the needs and requirements in terms of sustainability, smart and refurbishment process including energy efficiency and comfort condition point of view. Also, RPAI is introduced to assess the new proposed method by mathematical description.

REFERENCES

Interactive art & design, (accessed 25/01/2015www.WHITEvoid.com)

P. Beesley, S. Hirosue, and J. Ruxton, (2006). Responsive Architectures. Subtle Technologies 06, Cambridge: Riverside Architectural Press, p. 3

M. Mansour El Sheikh (2011). Intelligent building skins: parametric-based algorithm for kinetic facades design and daylighting performance integration, Master thesis, Faculty of the USC school of architecture University of Southern California

Srinivasan, A. V. and Mcfarland, D. M. (2001). Smart structures analysis and design. Cambridge, UK, Cambridge University Press.

Doris Kim Sung, (2014) Prototyping a Self-Ventilating Building Skin with Smart Thermobimetals, AIA Report on University Research (volume5), accessed 10.05.2014 http://www.aia.org/practicing/research/AIAB095521

E. S. Lee, D. L. Di Bartolomeo, J. H. Klems, M. Yazdanian and S. E.Selkowitz, (2006). "Monitored energy performance or electrochromic windows for daylighting and visual comfort", ASHRAE Summer Meeting, Quebec City, Canada.

T. Hausler, U. Fischer, M. Rottmann and K. H. Heckner, (2003) "Solar optical properties and daylight potential of electrochromic windows, International Lighting and Colour Conference, Capetown.

S. Selkowitz, O. Aschehoug and E. S. Lee, (2003) "Advanced interactive facades – critical elements for future buildings?" presented at USGBC expo.

Sotirios D Kotsopoulos, Federico Casalegno, Massahiro Ono, Wesley Graybill, (2012). Window Panes Become Smart How responsive materials and intelligent control will revolutionize the architecture of buildings, SMART 2014: The First International Conference on Smart Systems, Devices and Technologies.

Mel Schwartz, (2002). Smart material - Encyclopedia, A Wiley-Interscience publication, New York

HUMAN FACTORS IN HVAC: THERMAL COMFORT AND INDOOR AIR QUALITY FOR LIFETIME HOUSING

Robert BEAN[1,*]

[1]Indoor Climate Consultants Inc., Calgary, Alberta, Canada

[*]Corresponding email: info@healthyheating.com

Keywords: Thermal, comfort, air, quality, aging

SUMMARY

The expression, "long term care" rarely invokes images of non-institutional healthcare facilities such as one's own home; toss in phrases such as, "Aging in Place" or "Universal Design" and most residential design practitioners think of devices or aids for daily living and wheel chair accessibility. Housing performance programs including derivatives of Passivhaus, LEED, EnergyStar, or R-2000; and Standards related to indoor environmental quality in homes such as ASHRAE 55 (Thermal comfort) and ASHRAE 62.2 (Ventilation/ Indoor Air Quality), do not specifically consider the indoor environmental ergonomics needed for those home-bound by choice or necessity due to medical circumstances. Under good practice and life time housing principles, healthcare and building professionals, including sustainable building program developers should incorporate the study of demographics and residential indoor climate factors required for the infirm, injured, aged, and end of life care populations under various climate change scenarios.

INTRODUCTION

There is an exhaustive library of studies on physical, physiological and psychological changes brought on by illnesses and/or aging including recent papers documenting human thermal response systems in the elderly (Novieto, 2013). Environmental studies such as those by Corsi (2013), Green (2014), Miller (2014), and Weschler (2015) are exploring influences on physiological systems or aging processes (Burton, 2010), through relationship studies of human and environmental microbiomes or emissions and reactions between airborne contaminants and human tissue. The Academy of Neuroscience for Architecture (Eberhard, 2012), the Human Exposome Project (Miller, 2013) and those studying salutogenic environments (Golembiewski, 2012), are exploring how environments and environmental exposures can keep humans healthy or make them sick. Miller (2015) has stated the results of studies could put the environment, "…in its proper place right next to the genome in studying human health and diseases." From the working design practitioner perspective, literature generally does not address the qualitative affect from sound, thermal comfort, indoor air, lights, vibrations and odours on the infirm, injured, elderly, and palliative care populations. Without intervention within current building programs, the design of houses will continue to overlook the continuum of aging and environments at home that for many are inadequate when compared to modern or modernized institutional healthcare facilities.

METHODOLOGIES

Based on a literature review of North American building practices, a conclusion can be demonstrated that the general populace in North America occupy homes designed and built on protocols of limiting probabilities related to health and safety risks (NBC, 2010). These dwellings have been defined around products and assemblies to the lowest allowable government regulated standard before a municipal building inspector is obligated to fail it. In the few homes which might demonstrate full compliance with both ASHRAE 55 (thermal comfort) and ASHRAE 62.2 (air quality) the purpose and scope of both these documents would still exclude those, who are for example, disabled, infirm or bedridden.

It is demonstrated by demographics that the quantity of people in this group, specifically the elderly and end of life care patient is a non-trivial statistic and the preference from this population is to remain in their existing homes rather than relocate to an institutional facility. These decisions have many far reaching consequences from broad social impacts to intimate emotional and financial stress placed on family members and care workers.

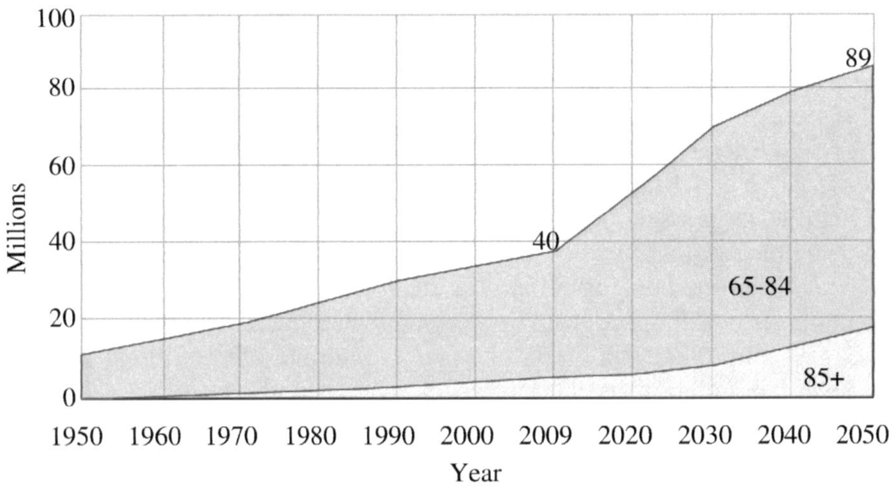

Figure 1. U.S. Population Ages > 65, 1950 to 2050 (adapted from Jacobsen et al, 2011)

It also implies a growing gap between what is available to end of life populations in indoor environmental conditions offered in newer institutional facilities designed to more stringent standards versus the indoor climate conditions established by minimum housing standards. This gap becomes evident when looking at the purpose and scope of:

1. ISO/TS 14415:2005(E) Ergonomics of the thermal environment — Application of International Standards to people with special requirements.
2. ANSI/ASHRAE/ASHE 170-Ventilation of Health Care Facilities,
3. ASHRAE Guideline 24 Ventilation and Indoor Air Quality in Low-Rise Residential Buildings (companion document to ASHRAE 62.2).
4. ASHRAE Guideline 10-2011- Interactions Affecting the Achievement of Acceptable Indoor Environments

These design Standards and Guidelines are simply not on the radar screen of North American home builders and design service providers; nor with manufacturers, distributors and HVAC trades. They are also not featured in prominent house building programs; but yet they define for present and future generations, the very considerations required for the persons whose home may ultimately become a satellite extension of the institutionalized healthcare facility.

In addition to potential risks to in-home caregivers, the populations presently at greatest risk due to poor indoor environmental quality (IEQ) are those currently at end of life stages (born

1901 - 1940s) and to a lesser extent those intending on reaching end of life stages in older code built homes (born 1940s – 1960s), and uncertain risk to those who may choose to buy an older home for life long occupancy (born 1960s – 1990s⁺).

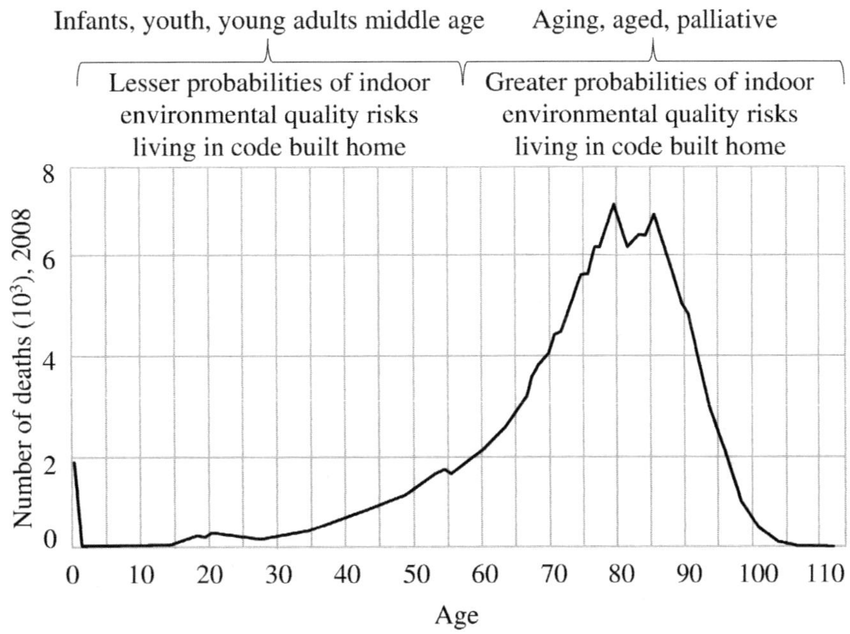

Figure 2. Age at death (Canada) (adapted from Wilson, 2012)

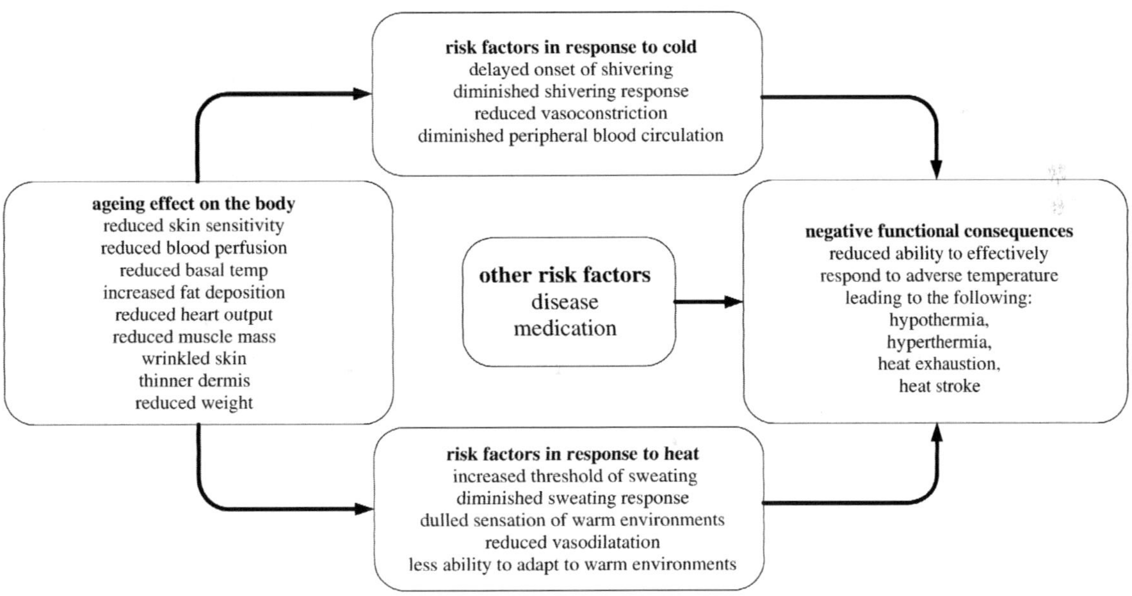

Figure 3. Aging and older people thermoregulation (adapted from Novieto, 2013)

DISCUSSION

Short of significant modifications, the vast majority of constructed houses are not designed specifically for those with illness or facing end of life care during the present mean climatic conditions and even less so for future climate changes of consequence. Consider those with

COPD, asthma, pneumonia, thyroid cancer, cancer of the hypothalamus and pituitary gland; or those with cerebral palsy, multiple sclerosis, myalgic encephalomyelitis or Parkinson's disease; and those who might be affected by medications for such things as hypertension, coronary diseases, and diabetes mellitus. These are representations of illness and end of life conditions that concern respiration and thermal regulation; and by connection influenced by the HVAC systems and building enclosure. There are of course others which affect auditory, visual, and olfactory system.

CONCLUSIONS

For some parts of the population, generally accepted housing protocols (or those programs where energy conservation usurps authority over IEQ), cannot deliver adequate indoor environmental ergonomics for caregivers and the elderly palliative care patient. Under climate changes of consequence, indoor conditions will be exacerbated to the detriment of these occupants and society. There is no certainty as to when and who will need environments of better quality so the solution is to apply lifetime housing and end of life care principles into the design of buildings and HVAC systems today. This can only occur if design starts with human factors in mind and let these drive sustainable building practices.

REFERENCES

ASHRAE 55 (2013) Thermal Environmental Conditions for Human Occupancy

ASHRAE 62.2 (2013) Ventilation and Acceptable Indoor Air Quality in Low-Rise Residential Buildings

Novieto, D.T. (2013) Adapting a human thermoregulation model for predicting the thermal response of older persons. Ph.D. Thesis. Institute of Energy and Sustainable Development, De Montfort University, Leicester UK.

Corsi, R.L. (2013) ASHRAE IAQ 2013, Conference Proceedings, Vancouver, Canada

Green, J.L. (2014) Can bioinformed design promote healthy indoor ecosystems? *Indoor Air* 24(2): 113-115.

Miller, S.L. (n.d.) Microbiology of the Home, accessed 01/2015 from < https://shellym80304.wordpress.com/projects/microbiology-of-the-home/>

Weschler, C. J. (2015) Roles of the human occupant in indoor chemistry. *Indoor Air*. doi: 10.1111/ina.12185

Burton, B.M. (2010) The Dynamic Human Microbiome and its Role in Aging. Ellison Medical Foundation. Accessed 01/2015 from <http://www.ellisonfoundation.org/node/4523 >

Eberhard, J.P. (2011) *Sustainability and Neuroscience. In Sustainable Environmental Design in Architecture: Impacts on Health.* S.T. Rassia & P.M. Pardalos, (Eds.). NY: Springer Publishing

Miller, G. (2015) The Human Exposome Project, accessed 01/2015 from < http://humanexposomeproject.com/>

Golembiewski, J (2012) Salutogenic design: The neural basis for health promoting environments. World Health Design Scientific Review 5:62-68.

National Building Code of Canada (2010). National Research Council, Ottawa, Canada

Jacobsen, L.A., Kent. M., Lee, M., Mather, M. (2011) America's Aging Population, Population Reference Bureau Vol. 66(1)

Wilson, D.M. (2012) End-of-Life and Palliative Home Care. Addressing the Care, Policy and Research Barriers to Palliative Care, Faculty of Nursing, University of Alberta

CHANGES IN IAQ OBSERVED WHEN OPERATING A CORONA DISCHARGE AIR CLEANER IN A CLASSROOM VENTILATION UNIT

Todd R. Crawford*
Bureau of Toxic Substance Assessment, New York State Department of Health, Albany, NY
*Corresponding email: todd.crawford@health.ny.gov

Keywords: Air cleaner, Corona discharge, Classroom, Ozone

SUMMARY

Corona discharge air cleaners (CDAC) may be an energy saving option to reduce the amount of outside air supply needed for classrooms. Recirculating room air through a classroom's air handling unit equipped with a CDAC could reduce the amount of outside air used to remove and dilute volatile organic chemicals, mold spores and fragments. This paper describes the test methods and observations of changes in indoor air quality (IAQ) associated with operating a CDAC in the unit ventilator of a vacant classroom. Real time, data-logged and time-weighted average measurements characterized the IAQ under different ventilation regimes. Indoor air testing showed that operating the CDAC increased ozone concentrations and ultrafine particle counts, and reducing the outside air supply exacerbated those problems. The study presents an investigation procedure that supported a regulatory response affecting classroom indoor air quality in schools in New York State.

INTRODUCTION

Corona Discharge air cleaners (CDAC) are marketed to schools as an energy saving option to reduce the amount of outside air supply needed for classrooms; conditioned air can be recirculated through a classroom's air handling unit equipped with a CDAC that removes volatile organic chemicals (VOCs) including odors, and denatures mold spores and fragments. CDACs are often described in marketing literature as designed to produce ions that remove contaminants without increasing ambient ozone nor producing contamination by-products. Corona discharge in air creates ozone, but ozone degrades rapidly on indoor surfaces and through reactions with VOCs. Operating a CDAC synchronously with the fan in a unit ventilation system, such as are commonly present in classrooms, may allow sufficient time for degradation reactions to remove VOCs and denature fungal components before the indoor ozone concentrations decrease to background levels.

Corona discharge occurs when the fluid (air) around an electrode is ionized, forming a plasma which discharges ions and electrically excited particles into the surrounding air. The ions and excited particles may be neutralized by collisions with other charged particles, or dissipated by contact with neutral particles and surfaces, or captured on an oppositely charged electrode. Corona discharge generates electrons, ions, and radicals, which can react with some volatile organic chemicals (VOCs) leading to their chemical transformation and eventual degradation. Charged particles also interact with particulates in air which may agglomerate and/or be captured on surfaces.

Corona discharge in air dissociates oxygen molecules which leads to the formation of ozone (Goldman, Goldman & Sigmond 1985), but the amount of ozone generated varies significantly with the design of the ionizing electrode, and the lifetime ("persistence"?) of ozone depends on the presence of reactive chemicals and surfaces which can degrade it (Weschler 2000).

Children are thought to have greater health risks from inhaling chemicals in indoor air due to their more rapid respiration and development (Wallace 1991, Guo et al. 2004). Most children spend a significant amount of time inside classrooms (between 943 and 1025 hours per year, http://www.pewresearch.org/fact-tank/2014/09/02/school-days-how-the-u-s-compares-with-other-countries/) so it is reasonable to improve classrooms' IAQ whenever feasible. Student health and academic performance have been shown to be affected by classroom ventilation rates (Baker 1918, Daisey et al. 2003, Haverinen-Shaughnessy et al. 2011, Bako-Biro Zs et al. 2012). The rate of indoor contaminants' removal from indoor air is partially dependent upon the outdoor air supply rate which dilutes and displaces (exhausts) contamination. Outdoor air ventilation affects indoor air quality (IAQ) by introducing, removing, concentrating or diluting contaminants in indoor air. ASHRAE (2010) estimates the minimum outdoor air supply rate that is generally acceptable in classrooms is 15 cubic feet per minute (cfm) per person. The International Code Council Mechanical Code Table 403.3, as adopted by New York State (2010), requires 15 cfm per person in classrooms, but provides an Exception in Section 403.2:

> "*Where the registered design professional demonstrates that an engineered ventilation system design will prevent the maximum concentration of contaminants from exceeding that obtainable by the rate of outdoor air ventilation…the minimum required rate of outdoor air shall be reduced in accordance with such engineered system design.*"

In 2012, a manufacturer [identifying information omitted] of a corona discharge (CD) air cleaner system requested approval from New York State Education Department (NYSED) to apply the Exception in Mechanical Code 403.2 to reduce the required outdoor air supply by installing their air cleaner in the classroom unit ventilator. The manufacturer claims the CD air cleaner saves energy costs by removing contaminants in indoor air, thus reducing the need for outdoor air supply; they further claim the CD does not produce ozone or air contamination byproducts. NYSED contacted New York State Department of Health (NYSDOH) to request assistance with measuring classroom indoor air quality so that NYSED could determine whether reduced outdoor air supply with the CD air cleaner system for classroom ventilation complies with the Exception in Mechanical Code 403.2.

This paper describes the protocol and testing performed to evaluate changes in indoor air quality when an CD air cleaner is operating in the unit ventilator of a vacant classroom. Real time, datalogged and time weighted average measurements characterized the changes in indoor air quality under different ventilation regimes. The results of the indoor air testing were used to determine whether the addition of the CD air cleaner to existing ventilators operated under reduced outdoor air supply rates complied with the Exception in Mechanical Code 403.2. The study presents an indoor air investigation procedure that supported a regulatory response affecting classroom indoor air quality and health and academic performance in schools in New York State.

METHODS

IAQ monitoring was performed during the school vacation when students were not present in the classroom. Testing was performed in a high school classroom approximately 26' x 29' x 9' (6786 cubic feet – Imperial units of measurement are used throughout to be consistent with the units of measurements presented in the referenced Codes and Standards). Existing classroom ventilation complies with Mechanical Code 403.2, with one unit ventilator circulating room air at 1000 cfm and providing 15 cfm outdoor air per person. An CD air cleaner was installed in the unit ventilator and electronically connected to operate synchronously with the ventilator fan.

The measured IAQ parameters were: temperature, relative humidity, carbon dioxide, ozone, formaldehyde, acetaldehyde, propionaldehyde, butyraldehyde, valeraldehyde, hexanaldehyde, acetone, ultrafine particulate matter (median particle diameters: 11.5, 15.4, 20.5, 27.4, 36.5, 48.7, 64.9, 86.6, 115, 154, 205, 274, 365 nanometers (nm), and total volatile organic compounds (TVOCs). Monitoring and sampling were performed at two locations: the classroom seating area and adjacent to the air supply vent. Ozone and ultrafine particulates were measured in the classroom seating area only.

The study was conducted in four phases. The outdoor air supply was adjusted at the beginning of each phase of the study. For Phase 1, the CD air cleaner was turned off and the outdoor air damper was set to supply 450 cfm outdoor air. In Phase 2, the CD unit was turned on with the outdoor air supply set to 450 cfm. For Phase 3, the outdoor air supply rate was reduced to 250 cfm and the CD unit turned on. In Phase 4, the CD unit was turned off and the outdoor air supply was modulated normally by the building ventilation management program. An air flow monitoring station was used to measure the initial outdoor air supply to the classrooms when adjustments were made to the ventilation system.

To simulate the impact of students' exhaled breath in the classroom, approximately 400 grams of dry ice (carbon dioxide) was allowed to sublime in a foil pan at the center of the classroom at the start of each phase of the investigation. Carbon dioxide concentrations were measured and data logged at 1 minute intervals. The rate of change of carbon dioxide concentrations was used to estimate the average indoor air exchange rate (Persily 1997). Dry ice was deployed during Phase 1, but the measurements were overwhelmed by the background carbon dioxide released from the dry ice cooler in the classroom; the cooler was removed from the classroom after Phase 1.

Many volatile organic chemicals (VOCs) are normally present in an occupied room. Limonene (in lemon essence, purchased from a grocery store), was deployed as a source of VOCs in the vacant classroom. One milliliter (mL) of lemon essence was evaporated at room temperature in a foil pan in the seating area of the classroom at the start of each phase of the investigation. Limonene reacts with ozone to form ultrafine particulates and aldehydes (Grosjean et al 1993, Vartiainen et al 2006). The concentration of ultrafine particles was measured and data logged every minute as counts per cubic meter in the seating area of the classroom during Phases 2, 3 and 4 of the investigation. Ultrafine particle counts are not available from Phase 1. Air samples were collected for analysis of aldehydes and acetone following Method TO-11A (U.S. Environmental Protection Agency 1999). The sample concentrations are expressed as time weighted averages, based on the total air volume collected over the sampling time. The sample collection period varied for each pump, (dependent upon the battery life of the pump). Short-term fluctuations in aldehyde and acetone concentrations are not captured in these data.

The concentrations of total VOCs (TVOCs) were monitored continuously during the investigation using photoionization detectors (PID). PIDs detect a broad spectrum of organic compounds that ionize when exposed to ultraviolet light. TVOC concentrations are reported as parts per billion (ppb, calibrated with isobutylene).

RESULTS AND DISCUSSION

The outdoor air supply rate was set at the beginning of each phase by manually adjusting the settings of the outdoor air damper on the ventilation unit; the initial outdoor air supply rate is shown in the tables. After the outdoor air supply rate was set, the automated ventilation software controlled the heating and outdoor damper opening, which meant the instantaneous outdoor air supply rate fluctuated during the investigation. The average outdoor air supply rate was estimated from the air exchange rate calculated from the decay rate of carbon dioxide concentrations in the classroom. During the investigation, the average outdoor air supply rate was always less than the initial supply rate for the vacant classroom.

The same averaging period was used for classroom ozone concentration, temperature and relative humidity. The average room temperature ranged from 68.5°F to 72.6°F, and relative humidity ranged from 9.9% to 14.6%; the low indoor relative humidity is typical for winter in upstate New York. The average indoor ozone concentration was 16.3 parts per billion (ppb) and 15.0 ppb when the CD air cleaner was switched off, during Phases 1 and 4, respectively. When the CD air cleaner was switched on, the average indoor ozone concentration was 34.7 ppb and 34.8 ppb, during Phases 2 and 3 respectively. Indoor ozone concentrations increased when the CD air cleaner was operating. Increases in indoor ozone concentrations occurred under both normal and reduced outdoor air supply conditions. Datalogged ozone concentrations show there were sustained increases in indoor ozone concentrations while the CD air cleaner was operating. There were brief decreases in ozone concentrations during the periods when the outdoor air supply damper was being adjusted and when the ventilation fan was switched off between phases of the investigation. Decreases in ozone concentration may be due to reaction with residual limonene in the classroom, but the effect was not observed consistently and was not the focus of this investigation.

Samples were collected in duplicate and analyzed for aldehydes and acetone by method EPA TO-11A during each phase of the investigation. Formaldehyde concentrations averaged 2.96 micrograms per cubic meter ($\mu g/m^3$) and 2.30 $\mu g/m^3$ when the CD was not operating, versus 3.74 $\mu g/m^3$ and 4.59 $\mu g/m^3$ when it was operating. The concentrations of formaldehyde and acetone were higher when the amount of outdoor air supply was decreased, but not for acetaldehyde, propionaldehyde, butyraldehyde, valeraldehyde, and hexanaldehyde (Phase 2 versus Phase 3). An important limitation of these results is the sample collection times were not consistent between each phase of the study; thus, the average concentration does not show whether there were short-term fluctuations in concentrations of aldehydes and acetone. Although the results show increases in the concentrations of some carbonyl compounds, data collection was not designed to model the effect of the CD air cleaner on the rate of change on concentrations of carbonyls.

TVOC concentrations were measured during the investigation. The relationship between TVOC concentrations and operation of the ventilation system and the CD unit is complicated by the non-specific detection of TVOCs by the PID instrument, and an apparent source strengthening of TVOCs during heating cycles of the ventilation system. The PID detected the release of limonene in the classroom during each phase of the investigation. However, there is insufficient information to interpret all of the variations in TVOC concentrations observed during the investigation.

Ultrafine particulate counts were monitored across thirteen ranges of particle diameters (10 to 420 nm by median mobility diameter). Overall, the particle count increased following the deployment of limonene in the classroom – smaller diameter particles (<36.5 nm) were generated first, with larger diameter particles (>48.7 nm) appearing as time passed, likely due to agglomeration of the smaller particles. The average ultrafine particle count was highest when the CD air cleaner was switched on and the outdoor air supply rate was decreased during Phase 3, and showed a maximum at the 36.5 nm size range.

CONCLUSION

Operating the CD air cleaner in the ventilation unit of the classroom coincided with an increase of the indoor ozone concentration, and correspondingly, turning off the air cleaner coincided with a decrease in indoor ozone concentrations. The release of limonene in the classroom was rapidly followed by the formation of ultrafine particulates. The average ultrafine particulate count was highest when the CD air cleaner was operating and the outdoor air supply rate was decreased. The average concentrations of aldehydes and acetone appear to have been elevated when the CD unit was operating, but these data provide a time weighted average and do not show concentration changes during each phase of the investigation.

The increased concentrations of ozone and ultrafine particulates, and the suggestive data for aldehydes and acetone, are indicative of degraded indoor air quality when the CD air cleaner is operating. NYSED has determined that the NYSDOH investigation showed the CD air cleaner does not meet the requirements of the Exception in Mechanical Code Section 403.2 and the system cannot be used in schools in New York State. These findings may be generally applicable in any jurisdiction following the International Mechanical Codes where an application is received to reduce outdoor air supply when a CD air cleaner is used in the mechanical ventilation system.

USA'S MISSING MISSION: STRATEGIES TO PROTECT CHILDREN FROM POOR IAQ IN SCHOOLS

Claire L Barnett*
Executive Director, Healthy Schools Network Inc., Albany, NY
*Corresponding email: cbarnett@healthyschools.org

Keywords: Children, IAQ, K-12 schools

SUMMARY

This paper reviews studies and federal programs on environmental health risks to the 55 million children in the US's 130,000 public and private K-12 schools. Environmental risks impacting school children and personnel include poor ventilation, cleaning, lighting and acoustics; pests, dampness, and growing molds; legacy toxics and hazardous chemicals. Several studies have documented selected state policies in place: a national work group on measures of school facilities and children found no data routinely collected that would help decision-makers prioritize risks and target investments, and a CDC survey reported few states helping schools with environmental concerns. Today, as budgets and programs are cut, more gaps are emerging and education agencies have poor information regarding which school environmental factors promote children's attendance and achievement. A coordinated federal-state strategy is needed to provide and track public health services to children with suspected exposures and to examine, track, and address environmental risks in schools.

INTRODUCTION

As presented to ISIAQ in 2011, there were reasons to anticipate a new era of children's environmental health protections, and federal funds to US EPA's programs for children and for reversing poor environmental management of K-12 public and private schools (Barnett, 2011). It was recognized that constitutionally, education is left to the states, that all states compel children to attend school, and that children are uniquely vulnerable to environmental hazards such as poor ventilation, cleaning, lighting, and acoustics; pests, dampness, and molds; and legacy toxics and hazardous chemicals. It was also recognized that environmental management of K-12 school facilities can generate savings and benefits in health and learning (Kats, 2006; National Research Council, 2006; EPA, 2012). The Institute of Medicine's climate report (2011), introduced at ISIAQ's 2011 conference, focused on indoor environments and health; among other steps, it recommended reducing indoor exposures. This paper reviews recent studies and federal efforts to determine if agencies are addressing risks to children from poor indoor air quality (IAQ).

METHODOLOGIES

The Network examined recent studies assessing federal and state programs on school conditions and the environmental health of school children, including the efforts of a national work group considering measures of schools and of school children, and met with EPA staff.

RESULTS AND DISCUSSION

Towards Healthy Schools 2015: Progress on America's Environmental Health Crisis for Children (2012). This is the 3rd triennial report prepared by the Network; it incorporated policy comments and reviews from six national organizations and advocates in 18 states. Compared to 2009, it found more school facilities with more children enrolled nationwide, an increase in school children of low income status, and more children with disabilities with unmet needs in schools. It found more states with policies promoting high performance/green school design and promoting the use of third-party certified green cleaning products in schools. It also highlighted emerging risks to school children such as PCBs in building materials, high volume hydrofracked gas wells on school grounds, cuts in local and state funds to schools (reducing nonmandated facility staffing), and new construction. Also cited were new studies showing effects of high CO_2 levels on adults and a single study in which health effects among school children were assessed at the same time that the National Institute for Occupational Safety and Health conducted a Health Hazard Evaluation of school personnel in the same school (Gorman and Singal, 1984).

As a result, the report estimated "all public and private school children should be considered at elevated risk of health and learning difficulties due solely to the unexamined and or unaddressed environmental health risks in their schools and the lack of public health services for children at risk or with suspected exposures." It also called for a federal interagency strategy.

Coalition for Healthier Schools Work Group on Metrics Monitoring and Research (2012-14) The Network resourced a 17-organizational-member facilitated group, including federal agency representatives. The group held three in-person and six conference call meetings. It considered and tallied over 55 studies and more than 155 measures that focus on "green buildings," environmental health, and demographic data.

The Group's White Paper and associated spread sheets document scores of measures of school facilities that might be used by state agencies (example: age of roof, age of heating system, disability access) or used to certify high performance/green schools (example: daylighting, acoustics, safer pest control). It also noted that none are consistently applied in the states, and that little is actually routinely measured regarding school indoor environmental factors associated with children's health and learning. Regarding children's environmental health, the work group found "Metrics - But None that Ensure Children's Environmental Health," including no tracking of children's risks or suspected exposures. EPA also cited the lack of children's research (EPA, 2013), and in a 2010 national survey, school nurses reported being aware of children affected by poor IAQ in school, but reluctant to discuss these issues with parents (Healthy Schools Network and National Association of School Nurses, 2011).

Federal Centers for Disease Control and Prevention (CDC) School Health Policies and Practices Survey (SHPPS) of 2012. SHPPS is a once-every-six-years effort to assess the reach of school health programs, with intensive survey questions on health education topics such as tobacco, drugs, food and nutrition, bullying, and emergency management. SHPPS 2012 included environmental questions, as a result of a CDC and EPA collaboration.

SHPPS 2012 chapter on the physical school environment included information on state assistance to schools (CDC, 2012) on a range of topics including IAQ, pest control, drinking water, hazardous materials, engine idling reduction programs, and training and professional development. For example, CDC reported 43% percent of states helping schools with IAQ, 30% helping schools reduce the use of pesticides, and 31% helping schools with green cleaning. In contrast, richer public health funding has led to 88% of states helping schools with bullying, 86% with food, and 73% with mental health referrals. CDC's report did not cover local actions.

CDC's data show school health programs having far more activity than EPA's programs. This may be due to the greater dollars and longer duration of the health investments, or to the kinds of school employees involved. It does suggest that EPA needs more funding for longer periods if it is to improve the capacity of states and schools to address IAQ and other environmental factors.

EPA and school environments. EPA has multiple voluntary programs for schools, including IAQ and safer pest control. After several years of budget cuts, EPA's *IAQ Tools for Schools* program was zeroed out of the federal budget for 2012, and EPA's Office of Children's Health (OCHP) began a "clean green and healthy schools initiative" with new staff. Deep cuts were also made to EPA's pest control and other school-related efforts (EPA, 2011). OCHP developed new guidelines: a new federal guideline on school siting for local use, advised by a federal advisory committee that included CDC and the Department of Education; and a comprehensive guideline and grants for state agencies to create sustainable school plans (Energy Independence and Security Act of 2007). EPA did not convene an advisory group on the comprehensive guidelines for states; those guidelines are missing guidance required by congress on how to conduct onsite school investigations with pediatricians. Further, despite its own lack of data (EPA, 2013), OCHP did not recommend that states measure children's health and learning outcomes in their plans. EPA awarded grants to implement the guidelines to five states in 2012.

EPA's siting guideline was not released with grant funding; it has not had significant pickup (Barnett, 2014). The state guidelines are being promoted by OCHP regional staff (EPA, 2014).

EPA's Indoor Environments Division has formed a School Health and Indoor Environments Leadership Development (EPA SHIELD, 2013) group composed of previous NGO and district IAQ award winners. EPA staff with SHIELD members are creating a curriculum on indoor environments through a series of webinars. OCHP staff are reduced in number; final reports from the five states receiving grants are not available. An interim assessment commissioned by the Network (Lubin, 2013) found EPA's state grantees and their advisory committees without a focus on children or many advocates for children, and no discussions underway on either tracking risks to children or improving measures of school facilities used in the states.

The Green Ribbon Schools Award The US Department of Education (ED) entered the arena of environment and health for the first time in 2011 when it announced a voluntary, unfunded Green Ribbon Schools Award, based on criteria in three "pillars": environmental footprint/energy; environment and school health; and environmental education. The Award criteria were set by Education staff who had no background on school facilities; there was no formal public comment period. The 100-point Award allocates 30 points for Pillar One; 30 points for Pillar Two; 35 points for Pillar Three; and five points for cross-cutting efforts

(Department of Education, 2014). The criteria are sent to state education agencies; states administer their own applications and nominate awardees to the federal agency. All nominees must certify they comply with federal, state, and local laws; EPA's role is advisory only.

In the first three years, EPA reviewed states' nominees and advised ED on their presence or lack of qualifications. Nonetheless, ED has given awards that included some schools with outstanding compliance questions. In 2015, EPA is not reviewing nominees (Barnett, 2015).

As there is no published literature linking energy efficiency *per se* with health or learning benefits, ED's Award criteria are effectively prioritizing energy over the complexities of healthy indoor environments on which there is robust literature (NRC, 2006). Moreover, to fulfill Pillar One, there must be documentation of significant energy savings; those data are often generated by or associated with recent green/high performance renovations or new construction. This suggests that ED's award may be promoting new renovations as opposed to the preventive and environmentally responsible management of existing schools.

CONCLUSION

The absence of a shared federal strategy across EPA, CDC, and ED, along with the weak fiscal climate, are depriving state agencies, architects, policy makers, schools, and parents of the information needed to ensure that K-12 schools provide healthful learning environments and that all children are protected from indoor environmental risks. A coordinated federal-state strategy is needed to provide and track public health services to children with suspected exposures and to examine, track, and address environmental risks in schools. Areas for research might include finding common denominators in how states measure school facility needs and an independent assessment of ED's Awardees and the impact of Award criteria on state and local policies.

ACKNOWLEDGEMENT

The W.K. Kellogg Foundation provided support to the Network for the White Paper on Metrics.

REFERENCES

Barnett C (2011). America's Largest Unaddressed Public Health Crisis for Children: Environmental Conditions of Schools. Conference Proceedings, Indoor Air 2011, Austin, TX.

Barnett C (2015). Personal communications with EPA OCHP, October 7, 2014,

Barnett C (2014). Personal Communications with EPA Indoor Environments Division, January 13, 2015.

Centers for Disease Control and Prevention (2012). School Health Policies and Practices Study (SHPPS). Centers for Disease Control and Prevention, Atlanta, GA.

Department of Education (2014). Green Ribbon Schools. U.S. Department of Education, Washington, DC.

Energy Independence and Security Act of 2007 (Pub.L. 110-140, Title IV Part E) Healthy High-Performance Schools. Washington, DC.

Environmental Protection Agency (2013). *America's Children and the Environment*, Third Edition. U.S. Environmental Protection Agency, Washington, DC.

Environmental Protection Agency SHIELD (2013) (School Health and Indoor Environments Leadership Development), Washington, DC.

Environmental Protection Agency (2011). Fiscal Year 2012 Agency Financial Report. U.S. Environmental Protection Agency, Washington, DC.

Environmental Protection Agency (2012). *Student Health and Academic Performance*. U.S. Environmental Protection Agency, Washington, DC.

Environmental Protection Agency (2014). *Clean, Green and Healthy Schools Regional Highlights*. U.S. Environmental Protection Agency, Washington DC.

Gorman R and Singal M (1984). Health Hazard Evaluation Report HETA 83-172-1409, Wappingers Central School District, Wappingers Falls, New York.

Healthy Schools Network and National Association of School Nurses (2011). New data show schools ignore children's health: groups urge action by EPA and states [Press Release/Letter to US EPA Administrator Lisa Jackson, January 11, 2011].

Healthy Schools Network (2013). *Towards Healthy Schools 2015: Progress on America's Environmental Health Crisis for Children*. Healthy Schools Network, Albany, NY.

Institute of Medicine (2011). *Climate Change, the Indoor Environment, and Health*. The National Academies Press, Washington, DC.

Kats G (2006). *Greening America's Schools: Costs and Benefits*. A Capital E Report.

Lubin J (2013). An assessment of state capacity building to improve conditions of schools and children's environmental health. Unpublished.

National Research Council. (2006) *Green Schools: Attributes for Health and Learning*. The National Academies Press, Washington, DC.

CROSSOVER STUDY OF FLUORESCENT BIOAEROSOL MEASUREMENTS IN ELEMENTARY CLASSROOMS WITH AND WITHOUT UPPER-ROOM ULTRAVIOLET GERMICIDAL IRRADIANCE

Chunxiao SU[1], Josephine LAU[1,*], and Fang YU[2]

[1]Durham School of Architecture Engineering and Construction, University of Nebraska-Lincoln, Omaha, Nebraska USA
[2]Department of Biostatistics, College of Public Health, University of Nebraska Medical Center, Omaha, Nebraska USA

[*]Corresponding email: jlau3@unl.edu

Keywords: Real-time detection, bioaerosol, school, indoor air quality, ultraviolet germicidal irradiation

SUMMARY

In this case study, the real-time bioaerosol monitors (RBMs) were applied in elementary school classrooms along with upper-room ultraviolet germicidal irradiation (UVGI) devices. The classrooms were operated in the UVGI setting and non-UVGI control setting. After the measurement lasting eight weeks, we found statistical significant lower concentrations of fluorescent bioaerosols at 0.5 μm in the UVGI classroom in perspectives of class sections, classrooms, and weather conditions. The differences between UVGI and control classrooms were larger when the fluorescent bioaerosol concentrations were higher. The results of this study indicate that the upper-room UVGI could disinfect the fluorescent bioaerosols at 0.5 μm in environment as classrooms.

INTRODUCTION

Ultraviolet germicidal irradiation has been verified to be an efficient way to disinfect air by reducing the reproduction of microbial organisms. To observe the effects of UVGI in inactivating bioaerosols, multiple methods have been adopted to detect the reduction of bioaerosols by different researchers (King et al., 2011; Miller and Macher, 2000; Walker and Ko, 2007). This study explored a new approach of evaluating the upper-room UVGI and its effect on bioaerosols by applying rapid detection of airborne microorganisms. A technology originally developed for military use has proven valuable for environmental monitoring of biological particles (Ho et al., 2000). Since several laboratory studies have demonstrated and verified the intrinsic-fluorescence-based method for the detection and classification of bioaerosols (Hill et al., 1999, 1995), sensitivity and counting efficiency of fluorescence based real-time bioaerosol monitoring devices have been studied in laboratory (Agranovski et al., 2003). The real-time measurement could draw a more complete picture of the bioaerosols level by providing a sufficient amount of data to through an extended time period. In this study, we have evaluated the effect by upper-room UVGI on fluorescent bioaerosols in elementary classrooms. By monitoring the bioaerosols levels in classrooms continuously, a detailed observation on how the fluorescent bioaerosols changing by time was studied.

METHODOLOGIES

The tested public elementary school locates in the mid-west US with a typical humid continental climate. Two classrooms were used for this study. Both classrooms had a floor area of 85 m^2 and the same class schedules. The class sizes were between 20 and 25 students. Each classroom has a separate ventilation system with no shared recirculation pathways. Both classrooms were installed with four units of upper-room UVGI air cleaners. The measurement in each visiting day lasted from the beginning of the class until students dismissed in the afternoon. UVGI units were turned on in one classroom and the other classroom was set as non-UVGI control room in the week one. Then two classrooms switched UVGI and control settings during the week two. The change of settings lasted until the eighth week. A total of 28 visiting days were available for the analysis after excluding school holidays.

One RBM was put in each classroom. The models of RBMs used for this study includes two main components; a particle size detector determining the size of each individual particle and a fluorescence detector deciding if the particle is biologic or inert by the presence or absence of intrinsic fluorescence. Inside the monitors, UV illumination is used to concurrently examine each particle for the presence of the metabolites NADH (Nicotinamide adenine dinucleotide) and riboflavin which are necessary intermediates of metabolism for a living organism and therefore exist in microbes such as bacteria and fungi (Bhangar et al., 2014; Yanagi et al., 2007). The RBMs could detect bioaerosols within a range of 0.5 μm to 10 μm from six different channels. Only 0.5 μm is presented and discussed in this paper.

All the data points were reserved in three categories: biological aerosol, inert aerosol, and total aerosol. The bioaerosol data was processed from the original data form and modified to the proper format for the statistical tests. The statistical analysis was two-way ANOVA (Analysis of variance). Then the concentrations from UVGI and non-UVGI control classrooms were compared by ANOVA statistic model in three basic categories: 1) morning and afternoon class sections, 2) classroom 1 and 2, and 3) the outdoor weather conditions for the visiting days.

RESULTS AND DISCUSSION

When comparing the concentrations between two settings, the fluorescent bioaerosol at 0.5 μm shows a statistical significant lower level in the UVGI setting (p-value<0.001) as shown in the Figure 1 (a). The mean concentrations during afternoon sections were 33270±308/m^3 (standard error of mean) for the UVGI setting and 40908±434/m^3 (standard error of mean) for the control setting. During morning sections they were 25772±163/m^3 (standard error of mean) for UVGI setting and 28343±213/m^3 (standard error of mean) for control setting. Same conclusion could be drawn to both morning and afternoon sections. The fluorescent bioaerosol in the afternoon section were significant higher concentration comparing to the morning (p-value<0.001). A possible reason is that the activities of students were relatively higher during afternoon than the morning. The higher slope of the afternoon line in the figure 1 (a) shows that the effect from the upper-room UVGI was stronger at high concentrations.

When looking at fluorescent bioaerosols from two classrooms separately, a significant lower levels during the UVGI settings were found in the classroom 1. The difference between UVGI and control settings in classroom 2 were close but still showing slightly higher concentration in the UVGI settings at 0.5 μm (p-value<0.001) as shown in Figure 1 (b). This may result from the large sample size which could generate significant difference between close means. One possible solution is to look at the averaged concentrations at longer period to reduce the samples size. This indicates that upper-room UVGI did disinfected fluorescent bioaerosol in one classroom. However, the difference of other factors between two classrooms may present larger effect on bioaerosols.

The concentration of fluorescent bioaerosols under different outdoor weather conditions was also investigated. The concentration of fluorescent bioaerosols were significantly higher during days with rain or snow (p-value<0.001). For comparison between the UVGI and control settings, the UVGI classroom shows statistically significant lower concentration under wet weather conditions as shown in Figure 1 (c) (p-value<0.001). The mean concentrations during wet weather were 31837±341/m^3 (standard error of mean) for UVGI setting and 38882±622/m^3 (standard error of mean) for control setting. During dry weather they were 28382±199/m^3 (standard error of mean) for UVGI setting and 32552±241/m^3 (standard error of mean) for control setting.

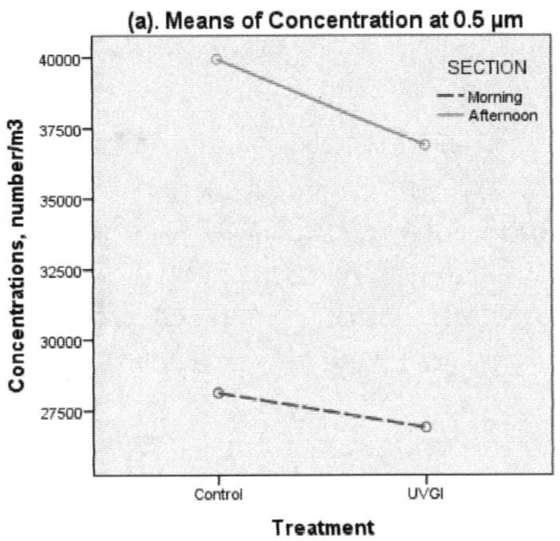

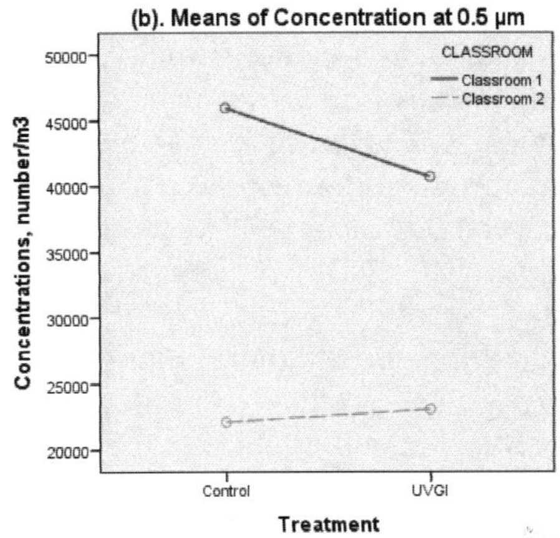

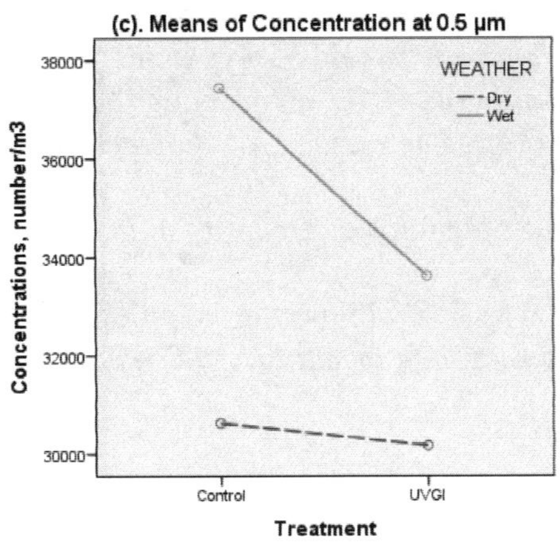

(a). Fluorescent bioaerosols concentrations during morning and afternoon sections at 0.5 micrometer between UVGI and control settings;

(b). Fluorescent bioaerosols concentrations in classroom 1 and 2 at 0.5 micrometer between UVGI and control settings;

(c). Fluorescent bioaerosols concentrations under dry and wet outdoor weather conditions at 0.5 micrometer between UVGI and control settings.

Figure 1. Comparison 0.5 µm fluorescent bioaerosols between UVGI and control classroom settings.

CONCLUSIONS

This crossover measurement on the fluorescent bioaerosols in classrooms shows the effect of upper-room UVGI on fluorescent bioaerosol. The eight weeks continuous measurement found that the fluorescent bioserosol at 0.5 μm were statistical significant lower in the classrooms when UVGI devices turned on. This difference was larger when the concentration was relatively higher, regarding the morning and afternoon class sections, two tested classrooms, and different outdoor weather conditions.

REFERENCES

Agranovski, V., Ristovski, Z., Hargreaves, M., Blackall, P. and Morawska, L. (2003) Real-time measurement of bacterial aerosols with the UVAPS: performance evaluation, *Journal of Aerosol Science*, Elsevier, **34**, 301–317.

Bhangar, S., Huffman, J.A. and Nazaroff, W.W. (2014) Size-resolved fluorescent biological aerosol particle concentrations and occupant emissions in a university classroom, *Indoor Air*, Wiley Online Library, **24**, 604–617.

Green, C. and Scarpino, P. (2002) The use of ultraviolet germicidal irradiation (UVGI) in disinfection of airborne bacteria, *Environmental Engineering and Policy*, Springer Berlin / Heidelberg, **3**, 101–107.

Guo, H., Morawska, L., He, C. and Gilbert, D. (2008) Impact of ventilation scenario on air exchange rates and on indoor particle number concentrations in an air-conditioned classroom, *Atmospheric Environment*, **42**, 757–768.

Hill, S.C., Pinnick, R.G., Nachman, P., Chen, G., Chang, R.K., Mayo, M.W. and Fernandez, G.L. (1995) Aerosol-fluorescence spectrum analyzer: real-time measurement of emission spectra of airborne biological particles., *Applied Optics*, **34**, 7149–7155.

Hill, S.C., Pinnick, R.G., Niles, S., Pan, Y.-L., Holler, S., Chang, R.K., Bottinger, J., Chen, B.T., Orr, C.-S. and Feather, G. (1999) Realtime Measurement of Fluorescence Spectra from Single Airborne Biological Particles, *Field Analytical Chemistry & Technology*, **3**, 221–239.

Ho, J., Spence, M., Hairston, P., Suffield, E., Hat, M. and Incorporated, T.S.I. (2000) Measurement of biological aerosol with a fluorescent aerodynamic particle sizer (FLAPS): correlation of optical data with biological data, *Medicine*, **15**, 281–291.

King, B., Kesavan, J. and Sagripanti, J.L. (2011) Germicidal UV sensitivity of bacteria in aerosols and on contaminated surfaces, *Aerosol Science and Technology*, **45**, 645–653.

Miller, S.L. and Macher, J.M. (2000) Evaluation of a methodology for quantifying the effect of room air ultraviolet germicidal irradiation on airborne bacteria, *Aerosol Science and Technology*, Taylor & Francis, **33**, 274–295.

Walker, C.M. and Ko, G. (2007) Effect of ultraviolet germicidal irradiation on viral aerosols., *Environmental science technology*, **41**, 5460–5465.

Yanagi, U., Ikeda, K. and Kagi, N. (2007) Application of the IMD in measurement of airborne microbial particles. In: *International Network for Information on Ventilation and Energy Performance*, Sendai, Japan.

PERFORMANCE OF FILTERS WITH DIFFERENT EFFICIENCIES IN CLASSROOMS WITH FAN-COIL AIR-CONDITIOING SYSTEM

Ailu CHEN[1, 2*], Qingliang CAO[1], Jin ZHOU[1,2] and Victor W.-C. CHANG[1,2]

[1]School of Civil and Environmental Engineering, Nanyang Technological University, Singapore
[2]SinBerBEST Program, Berkeley Education Alliance for Research in Singapore (BEARS), Singapore

[*]Corresponding email: chen0736@e.ntu.edu.sg

Keywords: Filters, Indoor, Fan-coil system, Classrooms, Air conditioning

SUMMARY

This work tested performance of four commercially available plain filters with single-path particle removal efficiencies of 25%, 65%, 85% and 95%, respectively. Each grade of filter was installed in the return air vents of fan-coil units of a classroom and multi-day continuous monitoring of indoor and outdoor particulate matter with diameters of 10 nm to 10 µm was carried out during both occupied periods and unoccupied periods. In addition, air exchange rates, fan energy consumptions and noise levels before and after the filter installation were measured in each classroom. Here we reported some preliminary findings on the air exchange rates, filter particle loading and background noise levels in the classrooms.

INTRODUCTION

There have been many epidemiological studies showing that human exposures to elevated airborne particulate matter (PM) concentrations are associated with a variety of adverse health effects (Pope III and Dockery 1992, Schwartz et al. 1993, Donaldson et al. 1998, Donaldson et al. 2002). These prior observations mainly rest on measurements of outdoor PM concentration but people may mainly expose to PM indoors as we spend the majority of our time indoors (Klepeis et al. 2001). Therefore PM in indoor environments merits better understanding.

PM of outdoor origins may infiltrate through the building envelope or be delivered by mechanical ventilation into indoor environments (Nazaroff 2004). PM can also be generated by many indoor sources such as walking, dancing and cleaning (Wallace 1996). PM from both sources can suspend in indoor air and may mean importantly to human exposure. In indoor environments with high occupancy, such as classrooms, both indoor and outdoor sources may considerably contribute to indoor PM concentrations although indoor and outdoor contributions depend on source strengths and mechanisms governing dynamics and fates of PM of different sizes.

Filters using by central air-conditioning and mechanical ventilation systems have been shown to be a dominant removal mechanism for indoor PM (Stephens and Siegel 2013). But in split air-conditioning system or fan-coil system, filters are not usually adopted in terminal units even though there may be high PM concentrations in indoor environments with high occupancy such as classrooms. There is a lack of knowledge of the performances of filters with different PM removal efficiencies in these AC systems. The current study investigates

PM removal efficiencies, energy impacts and indoor environmental quality influences of commercially available AC filters in four classrooms served by fan-coil air-conditioning systems.

METHODOLOGIES

The experiments were carried out between 21 Sep. and 1 Oct. 2014 in five tutorial rooms (TRs) in the same side and on the same floor of a building at Nanyang Technological University (NTU) of Singapore. The TRs have identical dimensions, interior layouts, equipment, furnishings, and ventilation and air-conditioning system. The L × W × H of each TR are 8.0 × 7.5 × 2.75 m. The TRs are served by fan-coil air-conditioning system but without mechanical ventilation although there are outdoor air intake and duct connecting to the fan-coil unit. The fan of the fan-coil system only works to recirculate the indoor air and there are no filters installed inside the fan-coil unit.

In the experiment, we made interventions to the fan-coil unit by installing plain air-conditioning filter with different PM removal efficiencies in indoor return air vents (before the fan) in each TR. In four of the five TRs, we installed filters with single-path particle removal efficiencies of 25%, 65%, 85% and 95%, respectively; and took the left one as a reference room (no interventions). The fan-coil units were set to operate for 24 hours per day across the whole experimental period and the indoor temperatures were set to be 23 °C but the relative humidity was uncontrolled in all TRs.

We monitored size-resolved indoor and outdoor number concentrations of particulate matter (PM) with diameters of 10 nm-400 nm by two units of portable Nanoscan SMPS nanoparticle sizers (SMPS mode 3910, 13 channels, TSI Inc., USA). We monitored size-resolved indoor and outdoor number concentrations of PM with diameter of 0.3 μm -10 μm by portable optical particle counters (OPC mode 3903, five channels, TSI Inc., USA). The indoor and outdoor temperatures and RH levels were also monitored by the OPCs. The instruments were set to record data at a time interval of 1 min.

Air exchange rates (AERs) in the TRs were measured by the tracer gas decay method using sulphur hexafluoride (SF_6) as the tracer gas. An InfraRan Specific Vapor Analyzer (Wilkes Enterprise Inc., East Norwalk, USA) was used to record the SF_6 concentrations at a time interval of 30 s for at least three hours in each TR. The AERs were then computed based on the decay of the SF_6 concentrations. The indoor background noise levels were measured at eight points (1.2 m above the floor) in each classroom before and after installation of the filters by sound meter (SM-10, Amprobe Headquarters, Everette, U.S.A.).

We placed one OPC in each TR, and the OPC continuously monitored particle number concentrations for 24 hours a day lasting for at least one week. The indoor SMPS sizer was moved between the classrooms but was placed in each classroom for at least six hours. The outdoor monitoring station including one unit of OPC and one unit of SMPS were located just outside the classrooms and the devices were set to monitor PM concentrations simultaneously with the indoor monitors. During the monitoring period, the classrooms were used for teaching and learning as normal; researchers kept recording numbers of occupancy.

RESULTS AND DISCUSSION

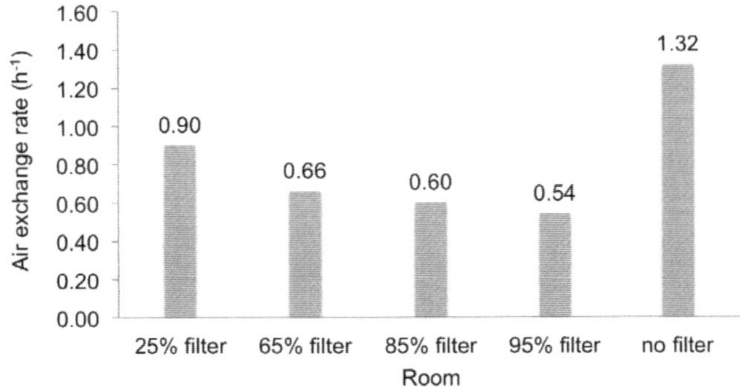

Figure 1. Air exchange rates in the five classrooms. There is no mechanical ventilation for any of the classroom so the ventilation mainly depends on natural ventilation.

The average air exchange rates (AER, h^{-1}) in the five TRs are shown in Figure 1. The AERs range from 0.54 h^{-1} in the TR with 95% filter to 1.32 h^{-1} in the reference classroom (no filter). The results suggest that all of the classrooms are under-ventilated in the natural ventilation mode and there is a trend that installation of high efficiency filter may result in lower AERs.

Figure 2. A comparison of the color of a new 65% filter and that of the same filter after 10-day use in the fan-coil unit. The left side is the new filter, and the right side is the used filter.

We conducted a comparison between color of a new filter with 65% particle removal efficiency and that of the same filter but after 10-day use in a fan-coil unit (as illustrated in Figure 2). It is clear that after 10-day use, the filter's color changes from light green to heavy-dark green, which indicates high particle loading over this short-term period. The finding also implies that the 65% filter can capture indoor PM considerably, which may consequently reduce the corresponding human exposure and coil fouling.

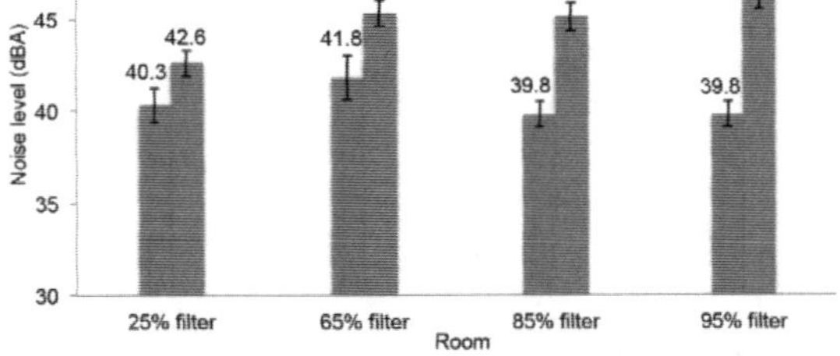

Figure 3. Indoor background noise levels in the classrooms before and after installation of filters. The error bars refer to standard deviations of noise levels in eight points in each room.

We also compared indoor background noise levels before and after installation of the filters in the four intervened classrooms (as shown in Figure 3). We can clearly see that before the installation, the indoor noise levels in all rooms are all around 40 dBA. After the installation, the indoor background noise levels increase in all rooms and the increased levels are positively correlated with the efficiency of the filter. However, the background noise levels in none of the rooms exceed 50 dBA, which is well below the maximum value of 55 dBA by local standard.

CONCLUSIONS

In the NV mode, the classrooms with fan-coil AC system are all under-ventilated, especially after considering their high occupancy. Installation of filters in return air vents of fan-coil units would negatively affect the air exchange rates in these classrooms. Installation of filter in the fan-coil unit can obviously remove indoor PM, which could further reduce human exposure and cooling coil fouling. Installation of filters can increase indoor background noise levels but the noise levels can still be well below the value requested by local standard.

ACKNOWLEDGEMENT

The authors thank the Republic of Singapore's National Research Foundation for financial supports through the grant NRF-CRP8-2011-03 and the grant to the Berkeley Education Alliance for Research in Singapore (BEARS) for the Singapore-Berkeley Building Efficiency and Sustainability in the Tropics (SinBerBEST) Program. BEARS has been established by the University of California, Berkeley as a center for intellectual excellence in research and education in Singapore.

REFERENCES

Donaldson, K., Brown, D., Clouter, A., et al. (2002). The pulmonary toxicology of ultrafine particles. *Journal of Aerosol Medicine: Deposition, Clearance, and Effects in the Lung*, **15**(2), 213-220.

Donaldson, K., Li, X. Y. and MacNee, W. (1998). Ultrafine (nanometre) particle mediated lung injury. *Journal of Aerosol Science*, **29**(5-6), 553-560.

Klepeis, N. E., Nelson, W. C., Ott, W. R., et al. (2001). The National Human Activity Pattern Survey (NHAPS): a resource for assessing exposure to environmental pollutants. *Journal of Exposure Analysis and Environmental Epidemiology*, **11**(3), 231-252.

Nazaroff, W. W. (2004). Indoor particle dynamics. *Indoor Air*, **14**, 175-183.

Pope III, C. A. and Dockery, D. W. (1992). Acute health effects of PM10 pollution on symptomatic and asymptomatic children. *American Review of Respiratory Disease*, **145**(5), 1123-1128.

Schwartz, J., Slater, D., Larson, T. V., et al. (1993). Particulate air pollution and hospital emergency room visits for asthma in Seattle. *American Review of Respiratory Disease*, **147**(4), 826-831.

Stephens, B. and Siegel, J. A. (2013). Ultrafine particle removal by residential heating, ventilating, and air-conditioning filters. *Indoor Air*, **23**(6), 488-497.

Wallace, L. (1996). Indoor particles: a review. *Journal of the Air & Waste Management Association*, **46**(2), 98-126.

Stein, Antoinette W., Bunn, Brian E, Davis, Devra L., Riley, Karl, Stadner, Alex, Scott, Stephen. (2015). Low EMF Best Practices For Indoor Environmental Quality In School Classrooms; The Importance Of Implementing Existing NEC Electrical Building Codes And Other Feasible Best Practices To Reduce Non-ionizing Radiation Exposures From Electromagnetic Fields From Wiring And Other Common Electrical Devices To Protect Health And Safety Of Children And Teachers.

1 Introduction

Healthy school buildings are designed to provide learning spaces where children can optimally breathe, see, hear and concentrate on lessons and materials. There have been a number of studies showing that children perform better in healthy buildings with comfortable temperatures, adequate ventilation and moisture controls, daylighting, and sound controls. School designers aim to provide teachers with built environments where they can teach interactive learning lessons and modules to children. Buildings today are designed with inclusion of infrastructural layout that includes electrical wiring and electronic devices especially to ensure fire safety. Healthy building standards have included sustainability and health-related rating criteria relating to Indoor Environmental Quality (IEQ) such as the Collaborative for High Performance School (CHPS) (CHPS, 2014). In 2015 Low Electromagnetic Field (EMF) criteria were added into CHPS as voluntary criteria.

2 Background

In January 2000, the State of California Department of Public Health completed and published a comprehensive study of children's exposures to electromagnetic fields (EMF) in California Schools (Zaffanella & Hooper, 2000). This study systematically selected 89 California public schools and the daycare centers associated with these schools and measured and surveyed EMF fields obtaining unbiased estimates of EMF levels for the entire population of California public schools.

This study was funded by the California Public Utilities Commission (CPUC) after they were motivated prompted by public concern about possible health effects associated with electric and magnetic fields from power lines and facilities near schools. The CPUC in 1993 ordered the California Department of Public Health to conduct and oversee a research and policy analysis Program about health effects of power frequency EMFs (see www.dhs.ca.gov/ehib/emf). The California EMF Program included the study of children's EMF exposures in schools and also funded a study on EMFs and miscarriage. The California EMF Program supported and published two policy analyses respectively including possible EMF information campaigns detailing avoidance measures, on the power grid and in schools.

The School Exposure Assessment Survey Report found that approximately 5.7% of the State's classrooms (15,300 classrooms statewide) have EMF (magnetic fields) exceeding 2mG. Strikingly, this study reported that the most prevalent source of the EMFs exceeding 2mG was due to NET CURRENT wiring errors generated ***inside*** the school buildings themselves not from EMFs generated by the school proximity to any outdoor power lines--this finding was not anticipated or expected (Zaffanella & Hooper, 2000).

The study revealed that the majority of the Net Currents identified were caused by wiring errors from violations of National Electrical Code© (NEC). The "net currents" were found to be due to wiring errors that cause unbalanced resultant current where portions of a circuit's neutral return current is diverted to another circuit or other conducting (especially metallic) pathways in the building such as plumbing supply lines, metallic electrical conduit, HVAC refrigerant lines, building steel, as well as other neutral conductors in other circuits. Net current problems arise whenever the net current of the electrical supply and return are not equal, i.e. where the generated magnetic fields do not cancel each other out. Net current magnetic fields are strongest adjacent to any wires that carry loads of net current and they weaken only slowly in a linear fashion with distance; they do not decrease exponentially like many other electromagnetic radiation (EMR). According to Biot-Savart law, the strength of a magnetic field relates to the relationship between the current (I) in a wire and the distance (r) from that wire. (Justin, 2015).

Health and Safety Impacts of Net Currents in Buildings

Net currents are considered to be fire and shock hazards. The National Electrical Code© (NEC) clearly prohibits the shunting of neutral current away from its dedicated circuit as articulated in several sections of the NEC including:

- **Section 250-24(a)(5) - 1999:** – in which it prohibits connecting of neutrals to any grounding connection on the load side of the service entrance main disconnect. Formerly this was 250-61(b).
- **Section 310-4:** prohibits connecting a neutral to another neutral such that a parallel return path to the panel is set up, unless the conductors are 1/0 or larger and meet exacting conditions.
- **Section 300-3(b)**: requires all conductors of a circuit to run together in whatever channel they are using. This reinforces Article 310-4.
- **Section 300-20(a)** repeats the above requirement with attention to circuits running in metallic enclosures such as conduits, pointing out the inductive heating effect on the conduit.
- **Section 250.-32(b) – 1999:** no longer allows the neutral bus in the panel for a separate building to be bonded to ground at the panel unless there are no grounded metallic connections between the buildings. There must be a separate equipment grounding conductor run with the feeder and the neutral bus must be isolated from the panel box. This replaces the former 250-24(a) which had allowed this bonding, with the effect that neutral current from the separate building could return on water and gas pipes, etc. run to the building.

Net currents were found on the metal pipes in some of the 89 schools measured.

Noted author, James G. Stallcup, who is considered an expert on NEC electrical codes and standards explains, "All AC conductors run in a metal conduit or raceway are required to be grouped together in the same raceway to prevent induced currents in the conduit. When alternating currents flow in a conductor, a varying magnetic field is set up around the conductor. The magnetic field sets up induced currents in the metal of the conduit surrounding the conductor, and these currents cause the metal to heat. The heating of the metal can melt the insulation of the conductor, causing a short. If all the conductors of a circuit are grouped together in the same conduit or raceway, the magnetic field of one conductor will neutralize the fields of other conductors. The currents will be in opposite directions; therefore there will be no magnetic field and no induced currents in the conduit or raceway" (Stallcup, 1993, p. 37).

He also comments on 250-61(b), "One major reason that the grounded circuit conductor ["neutral"] is not permitted to be grounded on the load side of the service is that, should the grounded service conductor become disconnected at any point on the line side of the ground, the equipment grounding conductor and all conductive parts connected to the equipment grounding conductor will carry the neutral current, raising the potential to ground of exposed metal parts not normally intended to carry current. This could result in arcing in concealed spaces, and it could pose a severe shock hazard, particularly if a single point in the path should be opened inadvertently by a workman. However, even without an open grounded conductor (usually referred to as an open neutral), the equipment grounding conductor path will become a parallel path with the grounded conductor, and there will be some potential drop on exposed and concealed dead metal parts. The size of this potential drop will be determined by the relative impedance of the equipment grounding conductor path, but all parallel paths not depended on as an equipment grounding conductor would be affected. This could involve current flowing through metal building structures, piping, and ducts" (Stallcup, 1993, p. 181).

There has been a substantial number of scientific studies on ELF/EMF including epidemiologic, animal and in-vitro studies. There have been more than 20 epidemiologic studies alone (Coghill, et al., 1996; Dockerty, et al., 1998; Feychting, et al., 1993; Greenland, et al., 2000; Linet, et al., 1997; London, et al., 1991; McBride, et al., 1999; Michaelis, et al.,1998; Olsen, et al., 1993; Savitz, et al.,1988; Tomenius, et al., 1986; Tynes, et al.,1999; Verkasalo, et al., 1993).

The large body of studies is maintained online at the EMF Portal database compiled by the German government (found here: http://www.emf-portal.de/dox/?pageId=509&l=e). A number of national and international health protection bodies have additionally completed reviews of the substantial data summarizing the findings based on the weight of evidence. Serious chronic health impacts have been reported to be possibly associated with net currents in addition to the possible fire and shock hazards they create.

The seminal case control study (Wertheimer & Leeper, 1979) of childhood cancer was published based on the towns of Boulder and Longmont as well as the city and suburbs of Denver in Colorado, USA. It was the first which noted the association of these cancers and exposures to magnetic fields from electric current.

In 1998 the U.S. National Institute of Environmental Health Science (NIEHS) EMF Working Group convened and reviewed all of the childhood leukemia studies worldwide (Ahlbom, et al., 2000) [1] and concluded in a report that ELF & EMF belonged in Group 2B, "possibly carcinogenic to humans" (Portier & Wolfe, 1998). Furthermore the International Agency for Research on Cancer (IARC) conducted a formal review of the scientific evidence on exposure to extremely-low-frequency (ELF) electric and magnetic fields and a link between cancer and in 2002 they formally classified ELF magnetic fields as Group 2B, "possibly carcinogenic to humans" on the basis of a possible a correlation found between an increased risk of childhood leukemia and chronic exposure to ELF magnetic fields at 0.4 μT (microtesla) or 4 mG (milligauss) or greater. The epidemiological reports found a doubling of childhood leukemia at an average of 0.4 μT. It is important to note that since the analysis of the data was made using pooled analyses of several international studies it is highly unlikely that the found correlation to childhood leukemia is due to chance (Ahlbom, et al., 2000) (Greenland, et al., 2000). Evidence since then has been further examined and continues to reveal evidence of a possible increase in childhood leukemia risk at long-term magnetic field exposure in the order of 0.3–0.4 μT; hence support continues for the IARC classification of ELF EMF as a possible carcinogen (Kheifets, 2010; Schüz, 2011; Sermage-Faure, et al., 2013; Zhao, et al., 2014).

It is also notable that the California EMF Program funded a 13 year cohort study on miscarriages and magnetic fields conducted by Dr. De Kun Li of the Kaiser Permanente Division of Research which included a total of 969 subjects in the San Francisco Bay Area. Dr. Li reported that prenatal maximum magnetic field exposure above 16 mG may be associated with miscarriage risk. It was reported that a disruption of early fetal development at the cellular or molecular level by external magnetic fields could conceivably result in fetal death. Additionally, in 2010, De Kun Li (Li, et al, 2010) conducted a population-based case-control study among healthy sperm donors to study exposure to magnetic fields (MFs) and poor sperm quality. They reported after controlling for confounders, compared to those with lower EMF exposure, those whose 90th percentile EMF level ≥ 1.6 mG had a two-fold increased risk of abnormal sperm motility and morphology (odds ratio (OR): 2.0, 95% confidence interval (CI): 1.0–3.9). Increasing duration of MF exposure above 1.6 mG further increased the risk ($p = 0.03$ for trend test).

Strong ELF EMFs are known to interact with the human body with detectible physiological effects. Extensive scientific research has been conducted over the last 40 years studying ELF EMFs exposures to identify adverse health effects. In 2002 the WHO IARC found an association between ELF EMFs and childhood leukemia based on the weight-of-evidence including analyses combining the results of multiple studies including international epidemiologic studies. The statistical evidence of epidemiologic studies taken together does suggest a risk. The WHO IARC classified ELF magnetic fields as "possibly carcinogenic to humans" on the basis of a possible a correlation found between chronic exposure to ELF magnetic fields of 0.3–0.4 μT (microtesla) or greater and an increased risk of childhood leukemia. There has also been positive epidemiologic evidence correlating ELF EMF exposure to some neurodegenerative diseases including Alzheimer's and Amyotrophic lateral sclerosis (ALS) (Consales, et al., 2012).

3 Are there Healthy Building Best Practices for Low EMF in CA or elsewhere?--What's New for Building Science?

After the Department of Public Health completed its EMF Risk Evaluation Report and submitted it to the CPUC in 2006 the CPUC formally made Decision 06-01-042 (http://www.cpuc.ca.gov/word_pdf/FINAL_DECISION/53181.pdf in which it affirmed its requirement to implement all low-cost/no cost EMF mitigation measures to minimize public EMF exposures. Also the CPUC required workshops to develop standard approaches for design guidelines including the development of a standard table showing EMF mitigation measures and costs. Similarly the WHO in 2007 also took a precautionary stance and stated that the use of "suitable precautionary measures to reduce exposure is reasonable and warranted" (WHO, 2007),

The State of California Department of Public Health funded and posted on their website, and distribute a 22 minute video entitled Fixing Electric Wiring in Schools Video (CDPH, 2000; Holt, 2015) that was written and hosted by Karl Riley, author of the book "Tracing EMFs in Building Wiring and Grounding"

(Riley, 1995) to provide information on techniques to minimize exposures to net current magnetic fields found in new and remodeled schools for school facility engineers, inspectors, project managers, school architects, planners and school building advisory committees after publishing their report. The video was produced by the State of California Department of Public Health in cooperation with Southern California Edison (SCE) to show step-by-step methods of measuring the fields and then how to find the faulty circuit and the wiring error; "... to teach electricians how to fix these wiring errors"; It includes graphics animations from SCE's video department to show how the currents move in misconnected circuits.

The State of California Department of Public Health EMF Exposure Assessment for Public Schools Report included an assessment of all the possible techniques for reducing magnetic fields produced by the sources found in the schools. The cost of reducing the Net Currents in California Public Schools is low cost Involving only MINIMAL the hourly rate work for the diagnostic testing of classrooms to find and trace the incorrectly wired net current pathways and electrician hourly rate to correct the wiring errors found equivalent to regular school facility maintenance work addressing electrical code violations for fire and shock safety which is part of regular maintenance required by law in public schools to fix the incorrectly wired circuits.

CDPH has posted a list of Measuring devices such as NIST certified Gauss Meters that y are now readily accessible in the marketplace and are now more affordable, easier to use and made with more features compared to 15 years ago can readily be used to detect elevated EMFs that have improved as the.

There have not been significant papers published since 2000 of school remediation projects for NET CURRENT EMFs sharing important success stories school-to-school, district-to-district, state-to-state and beyond.

Therefore this paper provides building operators with a renewed call and recommendation to make it a priority within today's new opportunities for modernization of existing schools and any new construction to be sure to look at and measure EMFs with NIST calibrated meters; EMF measurements should be written into all construction and remodeling plans and drawings and included in commissioning documents with all plug, lighting, and HVAC loads turned on to check for NET CURRENTS before building occupancy to protect students and teachers from shock, fire and EMF health and safety risks.

4 Discussion

It is of high priority that the topic of elevated magnetic fields, EMFs, is revisited and the efforts to elucidate, qualify, and quantify, their specific harms redoubled in an era of increasing electromagnetic radiation from all sources, not simply ELF. Specifically, more research should be focused on children's exposure in school classrooms to reduce and remove risks that may be associated with chronic exposures. It is critical to survey, assess, and fully remediate all elevated magnetic fields in schools and other buildings in which children are present for prolonged periods. It is of high value to identify these errors that are currently NEC code violations such that serious shock and fire hazards are removed together with reducing health based EMF exposures to children and teachers. Multiple studies have been published exhibiting that inclusion of Indoor Environmental Quality (IEQ) design criteria is of importance for building classrooms that are supportive of high academic performing students. Thus adopting standards which are protective of these impacts is essential to the health and well-being of our children.

Appendix A. Suggested Protocol for School Electricians for Correcting Wiring Errors Causing Net Current Magnetic Fields

by Karl Riley

Protocol

Background:

If your school was part of the California Health Services study, the areas affected by internal net current fields have been mapped out by a survey conducted using a gaussmeter to record the field levels and trace their paths. If not, a walk through with a gaussmeter will identify the areas affected. This in turn shows you which panel to check to identify the circuits causing the fields.

Most net current wiring errors can be identified as to circuit at the breaker panel using a clamp-on ammeter. This is because every correctly wired cable or group of conductors will give a "0.0" reading on a clamp- on ammeter. If you get a reading, you are measuring net current and have located a faulty circuit. A small percentage of net currents will not be seen at the panel because they begin and end out in the circuit (such as incorrect 3-way switch wiring) and can be found either with a gaussmeter or by a routine check of all conduit runs with a clamp-on ammeter.

Description of typical errors you will be looking for.

You will be looking for any connections which shunt neutral return current to another path other than the dedicated conductor. When the condition is **neutral current on grounding conductors** and other metallic paths (water pipes, gas pipes, heating pipes, HVAC ducts, building steel, metal lath, ceiling panel grids, window frames, conduits) the most common place to look is in the subpanel where the neutral bus is sometimes mistakenly bonded to the panel either deliberately with a strap or conductor or else by turning in the bonding screw in the neutral bus. Equipment grounding conductors may also be connected to the neutral bus - violation: 250-24(a)(5).1999.

Another way neutral is shunted to grounding paths is by the pinching of neutrals, nicking the insulation, such as at fluorescent light fixtures improperly mounted, in receptacles, occasionally by sheet rock nails, and sometimes deliberately by bonding the neutral terminal of a receptacle to the grounding screw.

The second type of error is the **wire-nutting of neutrals from two branch circuits together** when they happen to share a junction box. This allows neutral return current from loads belonging to one circuit to return to the breaker panel through both the dedicated neutral as well as the neutral in the second branch circuit. Thus both circuits will have an identical net current; one because of an excess of neutral, the other because of an equal deficit of neutral. The magnetic field created is the same. The heating effect on the conduits is the same. **This is the most common cause of net current fields in buildings.** Clearly, many electricians are not aware that by malting this connection they are violating 310-4, 300-3(b), and 300-20(a) if in conduits. They are also creating a situation where another electrician preparing to work on a circuit opens the breaker on the circuit, but instead of the conductors being dead, the neutral sparks when he disconnects it, since it is still live. Not a good situation.

Additional errors of a grosser nature will cause high fields. At a junction box neutrals may be mixed up and the wrong neutral run with the wrong hot. Sometimes a neutral may be dead for some reason and another neutral has been pulled in from another circuit or even another subpanel, causing a high net current. All of these variations put neutral return current in the wrong conductors, and all will be found by following the suggested protocol.

Panel protocol:
Cautionary note. It is assumed that only authorized personnel are using these suggestions. Opening panels and junction boxes under live conditions exposes the workman to live conductors with the possibility of severe shock. All professional cautionary measures should be taken.

In order to check the circuits in a panel, there should be loads on as many circuits as possible. For that reason, checking panels during a normal operating day is best. Otherwise someone must go around turning all the lights and usual appliances on.

After taking the cover off a panel stop and check it out visually. If this is not the service entrance panel, look for any bonding of the neutral bus to the panel box. Look for any green or bare conductors in the neutral bus. Check the neutral bus bonding screw to be sure it is either backed out or removed. Subpanel neutral/ground bonds should be removed. Equipment grounding conductors should be provided with their own bus. The neutral bus must be electrically isolated from the panel box.

To find wiring errors in circuits first identify the circuit with the net current. This can usually be done at the panel for that area. The technique is to clamp an ammeter around each circuit or group of circuits leaving the panel box by way of a conduit. In a crowded box this may be impossible using the usual medium-sized clamp-on. For panel work the right sized jaws are a necessity. See **Resources, Instrumentation** below for recommended sizes of clamp-ons, from mini to extra large. Also note that there is a book available which fully explains this entire measurement/mitigation process in detail *(Tracing EMFs in Building Wiring and Grounding).*

Secondly you need a clipboard with a rough diagram of the box and a place to write down your measurements. See an attached form that has been useful. You have to draw in circles wherever there is a conduit leaving the panel and number them. Then when you measure that group of conductors you enter the measurements next to the same numbered list.

As you measure each group of conductors the ammeter will read 0.0A if the circuit is correct. Write 0.0 on the sheet after "group". That's all. But if you get a reading, say 2.6A, write that down after "group" and go on to the next step. You will now determine if the 2.6A net current is due to missing neutral or excess neutral. Clamp around the hot or hots. Record. Then clamp around the neutral or neutrals. Record.

The two numbers should have been equal. If not you will see if there is missing or excess neutral. Now measure any equipment grounding conductors that may be running with the circuit. Record. If there is more than, say, 0.1A there may be a neutral/ground connection somewhere in the circuit. A general finding of a tenth of an amp or so on many equipment grounding conductors may simply mean some neutral flowing on grounds somewhere. But you don't chase the currents on the grounds, which are too elusive; you stick to the source of the currents in missing neutral.

To determine exactly which circuit has the error, if there is more than one circuit in the group, you need to measure each conductor separately. Measure each hot and record, noting its color. Then measure each neutral. Record. How do they match? Do you have one hot which matches with a neutral and one which does not? Then follow that hot to its breaker and identify the circuit. Usually it will be a lighting circuit, since most net current sources are lighting circuits with their many junction boxes.

If one neutral appears to go with two or three hots, then clamping around the group of hots should yield the same resultant in amps as the accompanying neutral. Sometimes you may suspect a mix-up in neutrals. This can be determined with your clamp-on. Suppose you have a hot carrying 5A and its neutral also carrying 5A. Clamp around both and you should read 0. But instead you read either 10A or something like 7A. If you get 10A that neutral was from the other "phase" or hot leg in a single phase service. It's the wrong neutral. If it is 7A or so, it is from another phase of a two or three-phase panel. Also the wrong neutral. In any case the neutrals are mixed up and need to be sorted out back in the junction box where they were incorrectly connected.

If there is either excess or deficit of neutral in a circuit, as you continue to clamp around groups of conductors be on the lookout for a second circuit carrying the missing neutral deficit or excess. It will be the same net current, unless more than two circuits have been mixed up. Usually they come in pairs. Later you can trace these circuits out to find where they are sharing a junction box, and lo and behold, you may see one beefy wire nut binding all the neutrals together. All you have to do is use two wire nuts and separate the neutrals so that each is wired to the appropriate load circuits, with no connections between the branch circuits.

If you do not find a circuit that pairs with the first net current one, it may be because the other circuit it is sharing neutral current with came out of another subpanel. In that case this net current will be seen by clamping around the feeder for the panel. The missing or excess neutral will give the same net current as in the circuit. Clamping around the feeder hots and comparing with the neutral will give more information. Once again instrumentation is the key to being able to do this. I use the flexible clamp-on LEM or AmpFlex which will fit around any feeder cable or conduit. See **Instrumentation.**

A more common reason for neutral deficit measured around the feeder would be a neutral/ground connection, shunting some return neutral to pipes, other conduits, etc.

Tracing down the error location:
Once you have identified the circuit or circuits with net current you can go looking for the error, which is usually but not always in a junction box. If it is a neutral/neutral error, almost certainly it is in a junction box. If it is a neutral/ground error it may well be accidental and you may be lead to the error area by this protocol but **visual** inspection may be the key to finding it, unless you get into more sophisticated instrumentation.

If you have a good knowledge of the circuits and you have an idea where two neutrals have been connected from two branch circuits, you may go right to that area and start looking for the junction box. If not, proceed as follows:

Usually the conduits will go up into the ceiling area and be accessible by removing ceiling panels. You will be able to clamp on your ammeter around conduits until you find one with net current. Visually note where the first junction box is located. Go to that spot. Open the box. Clamp on the circuits. Follow the net current. Keep going until you find the box where the misconnection of neutrals exists and use your second wire nut to separate them out. When you have fixed the error your clamp-on ammeter will read zero, and the net current with its magnetic field will be gone. A walk through with a gaussmeter will show that the field has disappeared.

Tracing neutral/ ground connections can be more difficult if due to accident. The net current on the circuit may either stop or attenuate markedly at a certain point. Here is where you look for pinched neutrals in a fixture or receptacle or a possible nail shorting neutral to equipment grounding conductor. Luckily that is a rare occurrence. Sometimes the neutral/ground connection is deliberate, as when an electrician has connected a wire between neutral screw and grounding screw in a receptacle. Sometimes an electrician may not have had a piece of green conductor with him and used a white one, and someone else assumes that is a neutral and adds a receptacle, connecting neutral to the white (now ground). There is no end to creativity of this type and it is up to you to figure it out.

Note: If all the conduits from a panel come out in an accessible ceiling, it is possible to bypass the opening of the panel and start directly with the conduits, clamping around each until you find the net current ones. Then follow as above.

There are more tracing instruments you **can** use which are described in *Tracing EMFs in Building Wiring and Grounding,* This protocol is based on the simple use of two or three sizes of clamp-on ammeters to do the job.

Resources

Instruments:
Clamp-on ammeters: For measuring circuits and conduits up to 3/4" the **Yokogawa CL-611** has a jaw opening of 7/8" and has a digital display that reads from 10mA to 200A. Price is around $150.00. Listed in the Jensen catalog, (800) 426-1194. Or call Yokogawa for local distributors at (800) 258-2552.

For measuring feeder cables, large conduits, water pipes, etc. a very inexpensive and versatile clamp-on with a jaw opening of 2 1/4" is the **Tenma 72-555** clamp on accessory. It needs to be plugged into a digital multimeter with at least 10 Mohm impedance with a mV AC setting. The clamp-on produces 1 mV per amp, so it is best to have a multimeter with a 0.1 mV resolution, which means usually a 200mV setting on the face (sometimes 400 mV). The Tenma 72-555 sells for less than $30.00 and is presently available from MCM Electronics at (800) 543-4330. MCM also carries a mini clamp-on similar to the Yokogawa in the same price range (Item 72-6184). They have a good selection of inexpensive digital multimeters. So does any Radio Shack.

For measuring large feeder cables and conduits, service cables, some bus ducts and large water pipes, etc., there are two excellent probes with a rubberized flexible clamp-around which will snake around conductors and are very accurate. The clamp-around is 2' in length though you can get a 3' one to measure around larger bus ducts. This accessory also must plug into a digital multimeter as it reads out on the AC volts scale. One is the **AEMC AmpFlex**. Specify the 30/300 Amp range. Can (800) 343-1391 for dealers. The other is the **LEM-300D** (do not get the less sensitive 3000D). It sells for around $300. Call LEM at (847) 4376444.

Books:
Tracing EMFs in Building and Wiring and Grounding by Karl Riley, 133p. 1995, MSI. $27.50. Order from kriley3@ix.netcom.com or (800) 749-9873 (MSI).

This book covers every stage from measurement through tracing and correcting errors. Also covers reducing fields from some grounding situations.

Old Electrical Wiring by David E. Shapiro, 413 p., 1998, McGraw-Hill, $39.95. Not an EMF guide but a general encyclopedia of electrical problems found in older buildings and how to deal with them.

Gaussmeters:
For accurate area surveys I recommend the **Bell 4080** triaxial gaussmeter since it is small, has a good display area, meets accuracy standards, and at present can't be beat at $199.50. The **Bell 4090** triaxial has tighter accuracy
(1%-2%) for $395.00. To order: (800) 749-9873 or www.magneticsciences.com.

A single axis digital meter with a separate probe is useful in trouble shooting. One with the highest accuracy available is the **MSI-AJK-95** for $235. Available from Magnetic Sciences International. (800) 749-9873. (www.magneticsciences.com).

For those who prefer to watch a needle rather than digits another excellent single axis gaussmeter with a separate probe is the **ELF Sense** for $340 available from ExpanTest, (207) 871-0224 or email to r.r.wagner@ieee.org.

The Emdex II triaxial recording gaussmeter for over $2,000 is excellent for recording, over time or over space by walking a constant speed. It downloads to a PC and produces graphs and statistics. Not easy to use for surveys because of the small and slow display. Call Enertech at (408) 866-7266.

Try surfing the Web by, punching in "Gaussmeters." Also find a list on www.microwavenews.com. But beware many guassmeters are inaccurate and cannot deal with typical harmonics.

References

Ahlbom, A., Day, N., Feychting, M., Roman, E., Skinner, J., Dockerty, J., ... & Verkasalo, P. K. (2000). A pooled analysis of magnetic fields and childhood leukaemia. *British journal of cancer*, *83*(5), 692.
Buffler, P. A., Kwan, M. L., Reynolds, P., & Urayama, K. Y. (2005). Environmental and genetic risk factors for childhood leukemia: appraising the evidence. *Cancer investigation*, *23*(1), 60-75.

CDPH - California Department of Public Health. (2000). Fixing Electric Wiring in Schools: A Video. Retrieved in 2003 from http://www.ehib.org/emf/a_wiring_video.html

CHPS Collaborative for High Performance School. (2014). New Construction and Renovation: Low-EMF Best Practices. Retrieved on April 3, 2015 from

Coghill, R. W., Steward, J., & Philips, A. (1996). Extra low frequency electric and magnetic fields in the bedplace of children diagnosed with leukaemia: a case-control study. *European Journal of Cancer Prevention*, *5*(3), 153-158.

Dockerty, J. D., Elwood, J. M., Skegg, D. C., & Herbison, G. P. (1998). Electromagnetic field exposures and childhood cancers in New Zealand. *Cancer Causes & Control*, *9*(3), 299-309.

Feizi, A. A. H. P., & Arabi, M. A. (2007). Acute childhood leukemias and exposure to magnetic fields generated by high voltage overhead power lines-a risk factor in Iran. *Asian Pacific Journal of Cancer Prevention*, *8*(1), 69.

Feychting, M., & Alhbom, M. (1993). Magnetic fields and cancer in children residing near Swedish high-voltage power lines. *American journal of epidemiology*, *138*(7), 467-481.

Greenland, S., Sheppard, A. R., Kaune, W. T., Poole, C., Kelsh, M. A., & Childhood Leukemia-EMF Study Group. (2000). A pooled analysis of magnetic fields, wire codes, and childhood leukemia. *Epidemiology*, *11*(6), 624-634.

Holt, M (2015). Karl Riley: Tracing Magnetic Fields video is posted on YouTube by Mike Holt for EC&M magazine. Retrieved in 2015 from https://www.youtube.com/watch?v=W6SyJFPYM8c&index=6&list=PLRNS1x1jcKbGZ9uIVMBHmm7-Sh2kePbVg

IARC Working Group on the Evaluation of Carcinogenic Risks to Humans, World Health Organization, & International Agency for Research on Cancer. (2002). *Non-ionizing Radiation: Static and extremely low-frequency (ELF) electric and magnetic fields* (No. 80). World Health Organization.

Inaba, H., Greaves, M., & Mullighan, C. G. (2013). Acute lymphoblastic leukaemia. *The Lancet*, *381*(9881), 1943-1955.

International Association of Fire Fighters. Position on the Health Effects from Radio Frequency/Microwave (RF/MW) Radiation in Fire Department Facilities from Base Stations for Antennas and Towers for the Conduction of Cell Phone Transmissions. Retrieved on June 28, 2015 from http://www.iaff.org/hs/Facts/CellTowerFinal.asp

IBE - International Institutue for Building-Biology & Ecology. (2015). Natural Health Building & Remodeling Practices Seminar. Santa Fe, NM.
Justin, A. (2015). Biot-Savart Law. *International Journal of Research*, *2*(4), 1050-1054.

Kavet, R., Zaffanella, L. E., Daigle, J. P., & Ebi, K. L. (2000). The possible role of contact current in cancer risk associated with residential magnetic fields. *Bioelectromagnetics*, *21*(7), 538-553.

Kavet, R., Olsen, R. Zaffanella, L. E., (2003). The EMF health issue: new insights to the possible link with leukemia. IEEE power & energy magazine.

Kheifets, L., Repacholi, M., Saunders, R., & Van Deventer, E. (2005). The sensitivity of children to electromagnetic fields. *Pediatrics*, *116*(2), e303-e313.

Li, D. K., Odouli, R., Wi, S., Janevic, T., Golditch, I., Bracken, T. D., ... & Iriye, R. (2002). A population-based prospective cohort study of personal exposure to magnetic fields during pregnancy and the risk of miscarriage. *Epidemiology*, *13*(1), 9-20.

Li, D. K., Yan, B., Li, Z., Gao, E., Miao, M., Gong, D., ... & Yuan, W. (2010). Exposure to magnetic fields and the risk of poor sperm quality. Reproductive toxicology, 29(1), 86-92.

Li, D. K., Chen, H., & Odouli, R. (2011). Maternal exposure to magnetic fields during pregnancy in relation to the risk of asthma in offspring. *Archives of pediatrics & adolescent medicine*, *165*(10), 945-950.

Li, D. K., Ferber, J. R., Odouli, R., & Quesenberry Jr, C. P. (2012). A prospective study of in-utero exposure to magnetic fields and the risk of childhood obesity. Scientific reports, 2.

Lilien, J. L., Dular, P., Sabariego, R., Beauvois, V., Barbier, P. P., & Lorphevre, R. (2010). Effects of extremely low frequency electromagnetic fields on human beings.

Linet, M. S., Hatch, E. E., Kleinerman, R. A., Robison, L. L., Kaune, W. T., Friedman, D. R., ... & Tarone, R. E. (1997). Residential exposure to magnetic fields and acute lymphoblastic leukemia in children. *New England journal of medicine*, *337*(1), 1-8.

London, S. J., Thomas, D. C., Bowman, J. D., Sobel, E., Cheng, T. C., & Peters, J. M. (1991). Exposure to residential electric and magnetic fields and risk of childhood leukemia. *American journal of epidemiology*, *134*(9), 923-937.
McBride, M. L., Gallagher, R. P., Theriault, G., Armstrong, B. G., Tamaro, S., Spinelli, J. J., ... & Choi, W. (1999). Power-frequency electric and magnetic fields and risk of childhood leukemia in Canada. *American Journal of Epidemiology*, *149*(9), 831-842.

McCarty, D. E., Carrubba, S., Chesson Jr, A. L., Frilot, C., Gonzalez-Toledo, E., & Marino, A. A. (2011). Electromagnetic hypersensitivity: evidence for a novel neurological syndrome. *International Journal of Neuroscience*, *121*(12), 670-676.

Michaelis, J., Schüz, J., Meinert, R., Zemann, E., Grigat, J. P., Kaatsch, P., ... & Kärner, H. (1998).

Combined risk estimates for two German population-based case-control studies on residential magnetic fields and childhood acute leukemia. *Epidemiology*, 92-94.
Moray Offshore Renewables Ltd. (2014). Technical Appendix 2.2 A Electromagnetic Field Report.

Olsen, J. H., Nielsen, A., & Schulgen, G. (1993). Residence near high voltage facilities and risk of cancer in children. *Bmj*, *307*(6909), 891-895.

National Institute of Environmental Health Sciences. (2002). EMF-Electric and Magnetic Fields Associated with the Use of Electric Power. Research Triangle Park, NC: NIEHS/DOE EMF RAPID Program. Retrieved on June 7, 2015 from http://www.niehs.nih.gov/health/materials/electric_and_magnetic_fields_associated_with_the_use_of_electric_power_questions_and_answers_english_508.pdf

Neutra, R., Delpizzo, V., & Lee, G. M. (2002). *An evaluation of the possible risks from electric and magnetic fields (EMFs) from power lines, internal wiring, electrical occupations, and appliances* (pp. 1-401). Oakland, CA: California EMF program. Retrieved on March 31, 2015 from http://www.ehib.org/emf/RiskEvaluation/riskeval.html

Pedersen, C., et al. (2014). Distance to High-Voltage Power Lines and Risk of Childhood Leukemiaan Analysis of Confounding by and Interaction with Other Potential Risk Factors. PloS one, 9(9), e107096.

Portier, C. J., & Wolfe, M. S. (1998). Assessment of health effects from exposure to power-line frequency electric and magnetic fields. *NIH publication*, *98*, 3981. Retrieved June 6, 2015 from http://www.niehs.nih.gov/health/assets/docs_a_e/assessment_of_health_effects_from_exposure_to_powerline_frequency_electric_and_magnetic_fields.pdf

Riley, K. (1995). *Tracing EMFs in Building Wiring and Grounding: A Practical Guide for Reducing AC Magnetic Fields Producing by Building Wiring and Grounding Practices*. Magnetic Sciences International.

Salvan, A., Ranucci, A., Lagorio, S., Magnani, C., & SETIL Research Group. (2015). Childhood Leukemia and 50 Hz Magnetic Fields: Findings from the Italian SETIL Case-Control Study. *International journal of environmental research and public health*, *12*(2), 2184-2204.

Savitz, D. A., Wachtel, H., Barnes, F. A., John, E. M., & Tvrdik, J. G. (1988). Case-control study of childhood cancer and exposure to 60-Hz magnetic fields. *American journal of epidemiology*, *128*(1), 21-38.

Stallcup, J. (1993). Designing Electrical Systems: Based on the 1993 National Electrical Code. American Technical Publishers, Inc., p. 37

Tabrizi, M. M., & Bidgoli, S. A. (2015). Increased Risk of Childhood Acute Lymphoblastic Leukemia (ALL) by Prenatal and Postnatal Exposure to High Voltage Power Lines: A Case Control Study in Isfahan, Iran. *Asian Pacific journal of cancer prevention: APJCP*, *16*(6), 2347.

Tomenius, L. (1986). 50-Hz electromagnetic environment and the incidence of childhood tumors in Stockholm county. *Bioelectromagnetics*, *7*(2), 191-207.

Tynes, T., & Haldorsen, T. (1997). Electromagnetic fields and cancer in children residing near Norwegian high-voltage power lines. *American journal of epidemiology, 145*(3), 219-226.

UK Childhood Cancer Study Investigators. (1999). Exposure to power-frequency magnetic fields and the risk of childhood cancer. *The Lancet, 354*(9194), 1925-1931.

Verkasalo, P. K., Pukkala, E., Hongisto, M. Y., Valjus, J. E., Järvinen, P. J., Heikkilä, K. V., & Koskenvuo, M. (1993). Risk of cancer in Finnish children living close to power lines. *Bmj, 307*(6909), 895-899.

Wertheimer, N., & Leeper, E. D. (1979). Electrical wiring configurations and childhood cancer. *American journal of epidemiology, 109*(3), 273-284.

Wertheimer, N., & Leeper, E. D. (1987). Magnetic field exposure related to cancer subtypes. *Annals of the New York Academy of Sciences, 502*(1), 43-54.

Wertheimer, N., Savitz, D. A., & Leeper, E. D. (1995). Childhood cancer in relation to indicators of magnetic fields from ground current sources. *Bioelectromagnetics, 16*(2), 86-96.

World Health Organization. (2007). Extremely low frequency fields.

Zaffanella LE, Hooper HC (2000). Electric and Magnetic Field Exposure Assessment of Powerline and Non-powerline Sources for California Public School Environments. Prepared for the California Department of Public Health. A summary is posted here: http://www.ehib.org/emf/pdf/SchoolExposureAssessment-English.pdf

INDOOR PARTICULATE MATTER ASSESSMENT IN URBAN KINDERGARTENS IN AL AIN, UNITED ARAB EMIRATES

George O. ODHIAMBO*
Department of Geography and Urban Planning, UAE University, Al Ain, UAE

*Corresponding email: godhiambo@uaeu.ac.ae

Keywords: Indoor air quality, kindergarten, health, United Arab Emirates

SUMMARY

This study was conducted with the following objectives: (i) to evaluate indoor concentrations of particulate matter (PM_1, $PM_{2.5}$ and PM_{10}) in selected urban private kindergartens; and (ii) to analyze those concentrations according to guidelines for air quality. Measurements were performed in classrooms and outdoor environments using a DustTrak instrument in nine private schools located along a major street, between December 2013 and June 2014. The 24-h mean weekday indoor measurements were 39.04 $\mu g/m^3$ (range of 15.85-57.45 $\mu g/m^3$), 45.64 $\mu g/m^3$ (19.98-64.50 $\mu g/m^3$) and 70.24 $\mu g/m^3$ (39-41-102.98 $\mu g/m^3$) for PM_1, $PM_{2.5}$ and PM_{10} respectively, exceeding the WHO (2006) standards by a large margin hence posing serious health implications to the children. High PM concentrations were recorded mainly during class occupancy period with peak values measured either at the beginning of classes or when classes end. It seemsthat that re-suspension of particles through cleaning and activities contributed more to increase in indoor PM concentrations when compared with the outdoor influence.

INTRODUCTION

Children are highly vulnerable to air pollution effects for a variety of reasons, being considered a risk group (Schwartz, 2004). Accordingly, indoor air quality at schools has attracted increasing public attention in recent years because children spent most of their time at school (8-10 h per day, of which 2-3 h are spent outdoors) (Fuentes-Leonarte, 2009). In more extreme climates, such as in the UAE, the percentage of time spent indoors is even higher and so, too, may be the health risk from exposure to indoor pollutants. Moreover, enclosed spaces can confine air pollutants and allow them to accumulate to detrimental levels (Samet and Spengler 1991). The quality of the indoor environment is therefore of critical public health concern and can potentially place occupants at risk. The World Health Organization (WHO) has assessed the contribution of a range of risk factors to the burden of disease and revealed that indoor air pollution is the 8th most important risk factor and responsible for 2.7% of the global burden of disease (WHO, 2010). Moreover, exposure to indoor air pollution has been linked to a variety of health effects, including respiratory health problems and exacerbation of childhood asthma (AAP, 2004). This study was conducted with the following objectives: (i) to evaluate indoor concentrations of particulate matter (PM_1, $PM_{2.5}$ and PM_{10}) in selected urban private schools; and (ii) to analyze those concentrations according to guidelines for air quality.

METHODOLOGY

Nine private kindergarten schools located along a major street within the school district in Al Ain city, and one located in the city center, were selected for the study. Indoor size aggregated mass fractions of PM_1, $PM_{2.5}$ and PM_{10} were concurrently measured in each room over a 24-h period using a DustTrak™ DRX Aerosol Monitor model 8533 (TSI Inc., USA) air sampler operating at a constant flow rate of 1.6 m^3/h. The instrument was zero calibrated before each measurement and positioned on a horizontal table at 0.6 m above the floor (at approximately the same level of the breathing zone of sitting children) and about 0.5 m from the walls, without obstructing normal usage of the rooms. Indoor measurements were conducted twice on week days when children were attending classes and again twice during the weekend. Outdoor measurements were conducted only once in each school. The instrument was set to record 10-minute averages of PM mass concentrations for the entire measurement period.

RESULTS AND DISCUSSION

For each kindergarten, 24-hours mean values of PM_1, $PM_{2.5}$ and PM_{10} were computed from the recorded 10-minute-averaged mass concentrations. The indoor concentrations of PM_1, $PM_{2.5}$ and PM_{10} are summarized in Table 1, with indoor measurements during the week labeled 1 (after the school code), weekend indoor measurements labeled 2 and outdoor measurements labeled 3. The 24-h mean weekday indoor measurements were 39.04 $\mu g/m^3$ (range of 15.85-57.45 $\mu g/m^3$), 45.64 $\mu g/m^3$ (19.98-64.50 $\mu g/m^3$) and 70.24 $\mu g/m^3$ (39-41-102.98 $\mu g/m^3$) for PM_1, $PM_{2.5}$ and PM_{10} respectively. Weekend indoor 24-h mean concentrations were 28.84 $\mu g/m^3$ (7.03-103.94 $\mu g/m^3$), 32.01 $\mu g/m^3$ (8.37-108.39 $\mu g/m^3$) and 41.37 (9.42-127.93 $\mu g/m^3$) for PM_1, $PM_{2.5}$ and PM_{10} respectively. Outdoor values were higher in most cases, ranging from 21.02-124.73 $\mu g/m^3$ (with mean value of 57.73 $\mu g/m^3$) for PM_1, 25.80-135.50 $\mu g/m^3$ (mean of 63.98 $\mu g/m^3$) for $PM_{2.5}$ and 38.24-197.95 $\mu g/m^3$ (mean of 101.17 $\mu g/m^3$) for PM_{10}. Weekend indoor measurements were, as expected, lower than both weekday indoor values and outdoor values except for the SC-EM school, where major cleaning was carried out on the same day as the measurements, inflating the measurements by a significant margin. Hourly outdoor measurements ranged from 5-603 $\mu g/m^3$ (PM_1), 8-2740 $\mu g/m^3$ ($PM_{2.5}$) and 8-2790 $\mu g/m^3$ (PM_{10}). These maximum outdoor values were recorded during sandstorms. The 24-h mean values for $PM_{2.5}$ and PM_{10} measured in the classrooms considerably exceeded the WHO 24- h guidelines of 25 $\mu g/m^3$ for $PM_{2.5}$ and 50 $\mu g/m^3$ for PM_{10} in more than 63% and 90%, respectively, of the cases. The 8-h mean values (data not shown) were even higher than the 24-h means as these (the 24 hour values) included night measurements with no activities going on in the classrooms (Fig. 1). The actual exposure for the students is best represented by those collected during the occupancy period. Particulate matter concentrations were largely affected by the presence of occupants in the classrooms and cleaning activities as higher concentrations (shown by peaks in Fig. 1) were recorded mainly during occupancy period, especially during breaks when students move out or into the classrooms, and cleaning activities. These lead to the re-suspension of deposited particles from horizontal surfaces, such as floors, carpets and furniture. Indeed, Thatcher and Layton (1995) found that re-suspension significantly affected indoor particle concentrations, with rates increasing with particle size; this phenomenon is of particular significance in the case of $PM_{2.5}$ and even more so for PM_{10}. For the same reason also room cleaning (vacuuming, dusting and sweeping) can be responsible for even more pronounced increases in PM_{10} (Thatcher and Layton (1995). As the schools are located within close proximity to a busy main street in the city, there is possibility of vehicular emissions and other outdoor air pollutants penetrating into the rooms especially when doors are open during class breaks. This would pose serious health risks as some of the chemicals and trace elements embedded in particulate matter emitted from vehicle exhausts are toxic.

Table 1: 24-h Mean Values (minimum and maximum values in parenthesis) of the Different Particulate Matter Sizes for Each School Studied. Code 1 Represents Indoor Measurements During The Week When Students Are In Class, 2 Represents Indoor Measurements Conducted During The Weekend And 3 Outdoor Measurements.

School	PM1 [μg/m³]	PM2.5 [μg/m³]	PM4 [μg/m³]	PM10 [μg/m³]	TOTAL [μg/m³]
SC-AE_1	15.85 (9-91)	19.98 (12-155)	27.28 (12-155)	60.11 (13-588)	88.24 (13-974)
SC-AE_2	7.03 (5-14)	8.37 (5-18)	10.53 (6-25)	18.9 (8-65)	23.72 (8-89)
SC-AE_3	46.97 (22-374)	51.87 (25-398)	58.31 (29-437)	89.43 (41-647)	119.4 (51-878)
SC-UV_1	44.04 (25-224)	51.78 (29-391)	61.1 (32-539)	67.11 (35-539)	72.56 (35-539)
SC-UV_2	24.76 (21-35)	29.18 (24-40)	31.63 (24-45)	35.86 (25-61)	37.09 (25-69)
SC-UV_3	46.26 (28-174)	54.21 (34-185)	61.91 (39-206)	87.93 (55-376)	104.6 (64-616)
SC-DH_1	30.54 (11-50)	33.96 (13-54)	35.79 (13-57)	39.41 (13-66)	42.71 (14-78)
SC-DH_2	28.89 (17-78)	33.49 (19-91)	37.88 (20-105)	50.24 (22-161)	59.01 (22-215)
SC-DH_3	-	-	-	-	-
SC-DR_1	38.7 (23-136)	47.45 (26-152)	55.7 (28-177)	76.06 (36-328)	87.57 (41-488)
SC-DR_2	20.19 (10-70)	22.12 (11-76)	23.6 (12-85)	27.01 (12-125)	28.85 (13-166)
SC-DR_3	37.75 (30-98)	44.84 (35-108)	50.83 (39-123)	67 (46-211)	76.28 (46-326)
SC-PK_1	50.01 (32-85)	61.88 (41-98)	72.68 (47-113)	102.98 (59-184)	118.49 (64-263)
SC-PK_2	11.4 (0-85)	13.12 (0-89)	14.78 (0-94)	21.22 (0-140)	28.41 (0-211)
SC-PK_3	21.02 (11-99_	25.8 (13-117)	29.53 (14-141)	38.24 (16-240)	43.75 (16-321)
SC-EM_1	37.88 (0-104)	44.6 (0-118)	51.45 (0-139)	76.21 (0-241)	100.24 (0-362)
SC-EM_2	103.94 (32-243)	108.39 (34-261)	112.91 (36-306)	127.93 (41-608)	134.84 (42-772)
SC-EM_3	48.9 (17-113)	54.03 (20-122)	59.88 (23-133)	85.24 (34-184)	106.16 (39-255)
SC-OO_1	57.45 (28-263)	64.5 (30-295)	71.88 (31-354)	93.92 (31-648)	106.13 (31-801)
SC-OO_2	34.94 (17-63)	40.13 (20-75)	43.67 (22-85)	50.85 (25-116)	54.01 (25-138)
SC-OO_3	39.28 (20-158)	47.11 (24-176)	55.19 (26-211)	78.73 (28-384)	92.83 (29-486)
SC-IN_1	41.01 (19-125)	47.91 (21-146)	53.81 (22-174)	72.68 (23-304)	88.76 (23-394)
SC-IN_2	19.54 (11-41)	24.1 (15-45)	26.86 (16-49)	30.88 (17-67)	31.58 (17-72)
SC-IN_3	124.73 (47-2730)	135.5 (55-2740)	147.25 (63-2750)	197.95 (87-2790)	233.37 (100-2820)
SC-AI_1	35.92 (12-90)	38.72 (13-97)	40.28 (13-104)	43.72 (13-129)	48.06 (13-160)
SC-AI_2	8.85 (5-13)	9.15 (5-13)	9.27 (5-14)	9.42 (5-14)	9.49 (5-14)
SC-AI_3	84.66 (24-604)	98.5 (31-613)	112.35 (37-623)	164.81 (56-661)	199.43 (66-686)

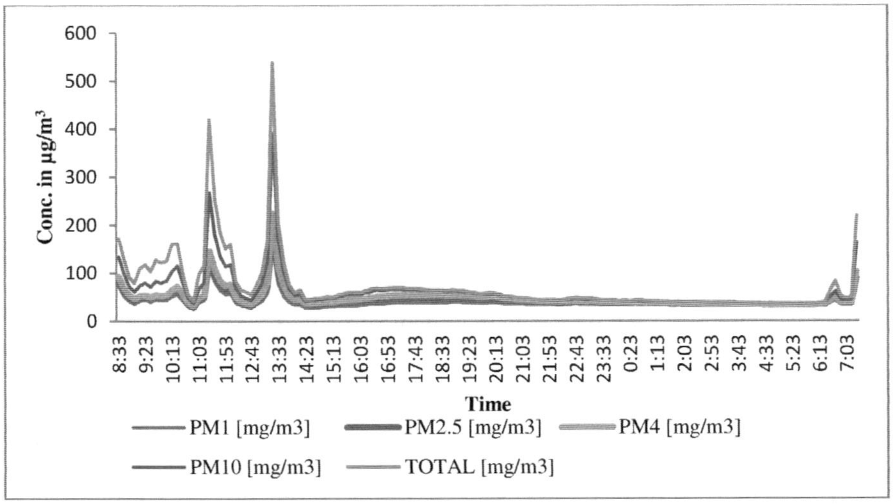

Figure 1 showing diurnal profile of concentrations of PM_1, $PM_{2.5}$ and PM_{10} and TSP for one of the studied schools. The peaks around 11h20 and 13h30 indicate movement activities during class breaks.

CONCLUSIONS

This is the very first indoor air quality study conducted in kindergartens in the United Arab Emirates. The findings of the study indicate that indoor air quality with regards to PM_1, $PM_{2.5}$ and PM_{10} are significantly higher than the WHO standards posing health risks to the young kindergarten children especially given the harsh climatic conditions in the UAE, which necessitates airtight enclosure of the classrooms resulting in possible dangerous accumulation of pollutants in classrooms. It would be important to measure the air exchange rates in the classrooms to find out possible influence of indoor air quality by outdoors are and as well characterize the particulate matter for trace elements in order to confirm the main components of the particles measured indoors.

REFERENCES

(AAP, 2004). American Academy of Pediatrics: Ambient Air Pollution: Health Hazards to Children. *Pediatrics*, **114,** 1699-1707.

Fuentes-Leonarte, V., Tenías, J. M., and Ballester, F. (2009). Levels of Pollutants in Indoor Air and Respiratory Health in Preschool Children: A Systematic Review. *Pediatr Pulm.*, **44**, 231-243.

Thatcher T. L. and Layton, D. W. (1995). Deposition, resuspension, and penetration of particles within a residence. Atmospheric Environment, 29 (13), 1487-1497.

Samet J. and Spengler J. (Eds.) (1991). Indoor Air Pollution: A health perspective, Baltimore. The John Hopkins University Press. Baltimore, Maryland, USA.

Schwartz, J. (2004). Air Pollution and Children's Health. *Pediatrics*, **113**, 1037-1043.

WHO (2006). WHO Air quality guidelines for particulate matter, ozone, nitrogen dioxide and sulfur dioxide. Global update 2005. http://whqlibdoc.who.int/hq/2006/WHO_SDE_PHE_OEH_06.02_eng.pdf. Accessed on 20th March 2015.WHO. (2010). WHO Guidelines for Indoor Air Quality. In Selected Pollutants, World Health Organization, Regional Office for Europe. Copenhagen, Denmark.

THE COST OPTIMAL METHODOLOGY OF DWELLING BUILDINGS IN SLOVAKIA

Lucia Borisová[1*]
[1]Faculty of Civil Engineering, Department of Building Services, Slovak University of Technology, Bratislava, Slovakia

*Corresponding email: borisovalucia@gmail.com

Keywords: cost optimum, energy management, life cycle of buildings

SUMMARY

The cost optimization methodology may be useful to identify multiple retrofit measures appropriate for the launch of large-scale renovations of existing buildings. According to calculations, the results show economic and environmental friendly conditions for building envelope, form, energy systems in a building and operation of an object. To achieve a cost optimum of the building, it is necessary to determine the cost-optimal values of a heat transfer coefficient in all building structures. In the resulting relationship between primary energy and global costs, it is necessary to take into account the entire life cycle of a building and economic parameters of price and energy. Determination of the cost optimum of buildings is an important element for the object in terms of saving money as well as the environment, since it is achieved by setting economically and environmentally friendly conditions to the object.

INTRODUCTION

It is the purpose of the member states of the European Union to reduce greenhouse gas emissions by 80-95% by 2050, because buildings are considered to be the biggest contributor of CO_2 emissions (about 37%in the EU). In the 2010 revision of the European directive EPBD 31/2010 / EU on the energy performance of buildings, there were established national requirements for energy consumption so that we will achieve optimal cost level with using the correct calculation methodology. These requirements will be tightened in the upcoming years in order to achieve a level of nearly zero-energy buildings for all new buildings.

METHODOLOGIES

The general methodology for cost optimization in Slovakia

The general procedure for determining the cost optimization consists of 4 basic steps, under the Delegated Regulation 244/2012 :

- Establishment of reference buildings
- Identification of energy efficiency measures
- Calculation of the primary energy
- Calculation of the global cost

The cost-optimal level lies within the range of performance levels, for which the cost-benefit analysis of the buildings life cycle is positive. In the calculation of cost-optimal levels, we are looking for "middle path" of an object to ensure the efficiency of a building construction/reconstruction at the same time for the economic and environmental aspects considering the whole life cycle of a building.

Experimental calculation of cost-optimal levels of minimum energy performance requirements for residential buildings in Bratislava, Slovakia

The whole calculation method is split into two parts: a part dealing with thermal protection of a building and a part dealing with building services (HVAC systems). Both cases were recalculated using lot of different combinations of energy efficiency measures, from which the most suitable combination filling all the criteria was selected.

Experimental calculation was carried out on a specific apartment building, which is in this case considered the reference building. It is located in Bratislava - Petržalka. There are 80 flats in the accommodation facility where the height of floors is 2.8 meters. Building´ s walls are constructed from aerated concrete with the thickness of 0.3 m.

Part of the thermal protection of buildings: Determination of the optimal value of heat transfer coefficient U (W / m²K)

In designing and calculation of measurement packages, there were applied measurements , that are consistent with actual existing requirements for the level of low-energy construction in the Slovak Republic, according to the Slovak Technical Standard 73 0540-2: 2012.

The initial investment costs for the selected part of the building´ s thermal protection is € 379 210 and there is included: facade insulation (12 variations for thermal protection of the envelope with different thickness of insulation 40-260 mm), roof insulation (12 variations for thermal protection of the roof structure with respect to different thicknesses of thermal insulation, from 50 to 280 mm) insulation of ceiling above the basement (12 variations for thermal protection with different insulation thicknesses from 20 to 240 mm) as well as replacement of windows, (12 options of insulating openings structures 2 and 3-glass).

Of all the proposed options, there were selected optimal variations, based on the relationship between the value of heat transfer coefficient U (W / m²K) and the net present value, which represents the lowest part of the curve, (next figure), which all together form a package of measures of the thermal protection of the building.

Table 1 Selected optimum values for the envelope of the building

	Facade	**Roof**	**Ceiling above basement**	**Windows**
Optimal thickness of thermal insulation (mm)	100	180	120	3-glass
Heat transfer coeficient U (W/(m².K)	0,26	0,15	0,32	0,85
Energy need for heating (kWh/(m².a))	57,7	56,6	50,3	92,8

For individual part of the building´ s envelope, there were selected variants of insulation thickness and their optimal values of heat transfer coefficient U (W / m²K). The calculation is for a time period of 30 years according to the Delegated Regulation 244/2012.

Part of building services (HVAC systems) – Evaluations of energy types

Energy types used in the calculation: gas, biomass – wood chips, electric power – heat pump

The most suitable alternative selection

In the calculation we took into account an actual initial investment cost of each device of thermal protection and building services as well as all costs associated with the object such as a price of devices, maintenance costs, energy prices etc. Then it was determined the net present value of the cost of the individual alternatives of measures.Net present value is a dynamic method to assess the effectiveness of investment alternatives.

We used 9 proposals of packages of energy saving measures in the part of building services. Together with the proposed measurements packages of thermal protection were achieved values of primary energy and net present value. There was determined a final variant of measurement, which is highlighted in the following table.

Table 2 Variants of proposed measures

Variants of measures	Primary energy (kWh/m^2.a)	Net present value (€)
1. district heating - gas	164,260	1 036 867,50
2. district heating - biomass wood chips	21,600	1 118 774,73
3. Condens boiler - gas	160,730	1 060 211,37
4. Condens boiler - gas + solar collector for DHW	120,700	1 087 170,82
5. Biomass boiler – wood chips + solar collector for DHW	26,800	1 274 812,00
6. Biomass boiler – wood chips	28,050	1 254 626,02
7. Heat pump air - water	95,700	1 022 062,54
8. Heat pump ground - water	**71,200**	**994 417,75**
9. district heating – combined heat	84,700	1 133 997,25

RESULTS AND DISCUSSION

Cost optimum curve

To create a complex overview, the combinations of sets of energy are assessed, saving measures on the basis of the cost optimum curve, where the points on the graph show the value, which was created relating primary energy and the net present value.

The minimum energy performance requirements are represented by the area of the curve, which provides the lowest cost for the final user and / or a company. Potentially, these requirements should appear more effective and efficient than existing national requirements, where the costs should be lower or equal. The area on the right side of the curve represents the economic optimum solutions that are worse in both respects.

The most optimal package of measures represents No. 8: Heat pump ground - water. The value of the net present value in this variant is the lowest part of the curve with primary energy reaches 71,2 kWh / m^2.a.The value of heat from heating and TV is only 26, 68 kWh / m^2.a. The values of CO_2 emissions, although they are higher than in other cases, is very low value of 9.87 kg / (m^2.a).

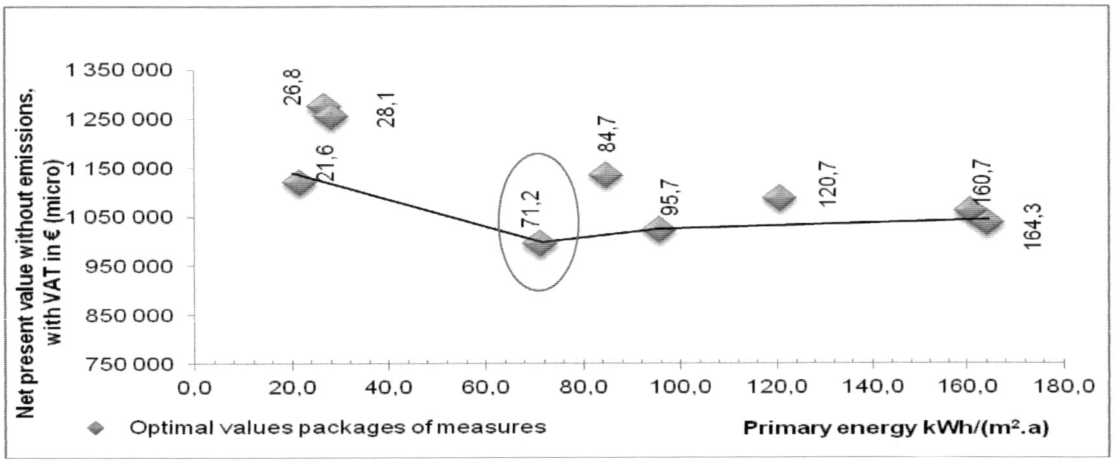

Figure 1 Cost optimum curve - The net present value of the package of measurements for primary energy needs, without emissions, with VAT

CONCLUSION

Improvement of energy efficiency in buildings requires the implementation of energy saving measures that should be effective in every way. Only the right determination of cost-optimal levels of individual energy efficiency measures will lead to economic use of energy at the lowest possible level during the whole life cycle of the building.

Determining the optimal policy alternatives of thermal protection and building services (HVAC) for the apartment building in Bratislava, there were designed the measures under which the building is found satisfactory for the actual requirements of the energy performance of buildings and it would be effective in terms of economic view, after offsetting all types the costs in a concrete variant we achieved the lowest net present value.

REFERENCES

Guidelines accompanying Commission Delegated Regulation (EU) No 244/2012 of 16 January 2012 supplementing Directive 2010/31/EU of the European Parliament

Building Performance Institute Europe (BPIE). Implementing the cost – optimal methodology in EU countries. Brussels, 2013. ISBN: 9789491143083.

Slovak Technical Standard 73 0540-2: 2012 Thermal performance of buildings and constructions

Borisová Lucia, The cost optimal methodology of dwelling house in Slovak Republic, Indoor Air, Hong Kong, 2014

IMPROVEMENT OF PERCEIVED TEMPERATURE UNDER 28°C OF AIR CONDITIONING CONTROL TEMPERATURE

Yoshihito KURAZUMI[1,*], Kenta FUKAGAWA[2], Emi KONDO[3], Tomonori SAKOI[4], Kunihito TOBITA[5], Yoshiaki YAMATO[6]

[1]School of Life Studies, Sugiyama Jogakuen University, Nagoya, JAPAN
[2]Department of Architecture, Kyushu Sangyo University, Fukuoka, JAPAN
[3]Department of Architecture, Ariake National College of Technology, Fukuoka, JAPAN
[4]Kansei Engineering Course, Shinshu University, Nagano, JAPAN
[5]School of Humanities and Social Sciences, Osaka Prefectural University, Osaka, JAPAN
[6]Department of Architecture and Structural Engineering, Kure National College of Technology, Hiroshima, JAPAN

[*]Corresponding email: kurazumi@sugiyama-u.ac.jp

Keywords: Visual stimuli, Vegetation, Thermal comfort, Thermal sensation, ETF

SUMMARY

If it can be clearly shown that a thermal environment considered slightly uncomfortable could be improved using visual stimuli, the cost effectiveness of such an initiative would be highly significant, particularly in terms of air conditioning system running costs. Focusing on the visual stimuli provided by vegetation, experiments were conducted in a temperature-controlled room. Thermal environmental conditions were set at 25°C, 28°C and 31°C. The visual stimuli consisted of ten different types of scenery, including that of leafy vegetation. In the indoor thermal environment evaluation index (ETF) deemed fairly uncomfortable, that is at a range of hotter than 28–29°C, clear improvements were observed in thermal sensation due to the influence of visual stimuli such as natural elements including vegetation.

INTRODUCTION

In a thermal environment where it is possible for the occupants to endure mild heat, visual stimuli including vegetation could lead to the creation of environmentally friendly spaces. Human thermal sense is not expressed only by simple heat equilibrium. The influence of visual and auditory stimuli causes differences in overall thermal sense arrived at by sophisticated sensory processing by the cerebrum. This means that extreme temperature settings are not necessary in indoor air-conditioned spaces, and that a space may be deemed slightly hot or uncomfortable and still be acceptable. If it can be clearly shown that the effects of a thermal environment considered slightly uncomfortable could be improved using visual stimuli, the cost effectiveness of such an initiative would be highly significant, particularly in terms of air conditioning system running costs.

In Japan, it is a commonly held belief that air conditioning systems should be set at 28°C in order to save energy. This temperature setting can be inappropriate, potentially contributing to concentration problems and negatively influencing health.

Rohles et al. (1981), Matsubara et al. (2004), Murakami and Shimomura (2007) and Fukagawa et al. (2010) show that visual or auditory factors clearly influence perception of thermal sensation as well as perception of comfort levels regarding heat. However, because these research projects do not utilize the thermal environmental index as the evaluation axis, they can only address relative differences. According to Kurazumi et al. (2011b), thermal sensation based on the heat-balanced based the outdoor thermal environment evaluation index (ETFe) (Kurazumi et al., 2011a) is affected by physical barriers to short wave solar radiation, such as leaves, and by a visually-induced psychological influence. The visual stimuli of vegetation and other factors served to improve perceptions of high heat and high body temperature.

Research conducted in laboratories on environmental factors and human body responses has yielded preliminary results showing that visual and auditory factors such as vegetation and water influence thermal sensation and the perception of thermal comfort level. However, the work is limited to the absence or presence of sensory stimuli; sensational and physiological temperatures are not assessed. In other words, quantitative research has not been conducted.

Under these circumstances, the purpose of this study is to clarify the effects of vegetation visual factors to the perception of thermal sensation and thermal comfort in the indoor thermal environment, by means of the quantitative research based on the indoor thermal environment evaluation index (ETF) (Kurazumi et al., 2010). This research is to test subjects placed in a thermal environment deemed slightly uncomfortable, where the temperature is set at a base point of 28°C, to determine the influence of overall stimuli of the cerebrum on the ETF and to prove the significance of actively placing visual stimuli in spaces. Focusing on the visual stimuli provided by vegetation and quantifying discrepancies in thermal perception on the human body, we will develop strategies to render the 28°C setting an effective one during the summer season.

METHODOLOGIES

Visual stimuli were implemented by placing visuals on a screen situated in the temperature-controlled room. The visual stimuli consisted of 10 different types of scenery including leafy vegetation. The influence of visual stimuli is sensitive to the solid angle ratio. The discrepancy of the solid angle ratio of the visual stimuli on the positioning of the subjects was 0.01, indicating minimal discrepancy. This research focuses on the ratio of green coverage of the visual stimuli. The visuals presented were divided into three groups: the 20–39% range, the 40–69% range and the 70–89% range. Thermal environmental conditions were set at three different temperatures: 25°C, 28°C and 31°C. Wall surface temperatures were set to equal these temperatures. Air velocity (calm air currents of 0.2m/s or less) and relative humidity (60%RH) were set the same throughout.

The subjects stayed in a relaxed position for 45 minutes or more in an anteroom, with settings equally controlled for the initial environmental conditions and relative humidity as well as for wall temperatures and ambient temperatures. Subsequently, the subjects were promptly transferred to the exposure room, where they were exposed to the predetermined thermal environmental conditions for a duration of 60 minutes for each different condition. Each type of visual stimuli was presented for 180 seconds. Thirty seconds later, the subjects were asked to describe their psychological reactions. Using discrete rating scales, this research focused on psychological conditions of the human body as follows: thermal sensation and thermal comfort. The presentation of visual stimuli was conducted randomly. Psychological reaction was measured 30 seconds after visual stimuli were presented. Human physiological data, specifically sublingual temperature and skin temperature were measured. The eight areas of the skin measured were the forehead, abdomen, forearm, back of hand, anterior thigh, shin, instep and sole of the foot.

The study subjects consisted of a group of 19 people who were healthy young male and female university students. The testing conditions were sufficiently explained to the subjects prior to the testing phase, and the subjects agreed to the conditions.

RESULTS AND DISCUSSION

We calculated ETF as theorized and validated by Kurazumi et al. (2009, 2010) based on thermal environment measures as well as skin temperature and amount of clothing.

Figure 1 shows the relationship between ETF and the description of thermal sensation. Examining the regression line, subjects described thermal sensation on vegetation groups; all descriptions indicate improvement on perception of hot at an ETF of 28-29°C. However, improvement of perception of heat comparing the 70-89% group, that with the highest percentage of vegetation, with the other two groups was low. The group with the highest percentage of vegetation on visual stimuli, at 70–89%, is believed to have been negatively influenced by a "stagnant" impression caused by excessive forest cover.

According to Fukagawa et al. (2010), water-scene photographs with great depth perception tend to result in subjects perceiving the photographs as "cold;" yet, visuals featuring water surfaces alone tend to have less of such an effect. Considering the idea that the visuals for the 70–89% group, the group with the highest percentage of vegetation, were also "crowded" by forest and ground cover, we see that it is important to create a sense of openness in the space. The elements present in vegetation percentages of less than 70%, which are perceived to have a dynamic effect on thermal environment conditions by visual stimuli, are believed to have influenced the thermal sensation given by the subjects in light of visual stimuli involving plants.

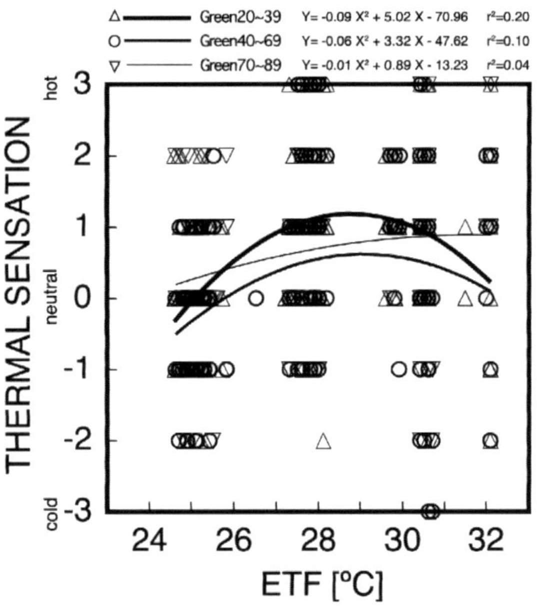

Figure 1. Relation between ETF and thermal sensation. Green is the ratio of green coverage.

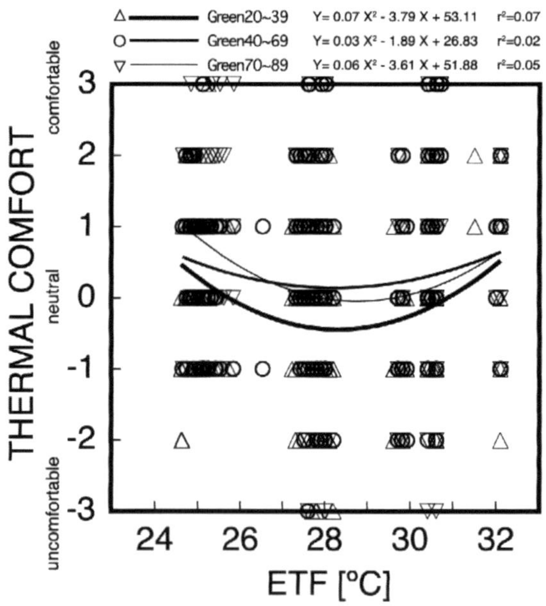

Figure 2. Relation between ETF and thermal comfort. Green is the ratio of green coverage.

The correlation between ETF and thermal comfort is shown in Figure 2. Examining the regression line, subjects described thermal sensation on vegetation percentages; all descriptions indicated an increase in the perception of discomfort at an ETF of 28–29°C. The difference of improvement effects between the vegetation groups is not large.

With regard to the relationship of the ETF and thermal sensation: in the ETF range of higher than 28-29°C, improvements in thermal sensation resulting from visual stimuli were indicated. Presenting visual stimuli incorporating natural elements such as vegetation, even in indoor settings where the ETF is set in the slightly uncomfortable range of higher than 28°C, clearly had improved results on feelings of hot sensation caused by the heat.

CONCLUSIONS

In this context, the objective of this research was to test subjects placed in a thermal environment deemed slightly uncomfortable, where the temperature is set at a base point of 28°C, to determine the influence of overall stimuli on the cerebrum on assessing an indoor thermal environment, and to prove the significance of actively placing visual stimuli in spaces. In ETF deemed fairly uncomfortable, that is at a range of hotter than 28–29°C, clear improvements were observed in thermal sensation due to the influence of visual stimuli such as natural elements including vegetation. Visual stimuli are shown to be appropriate at a level of up to 69% vegetation, where a dynamic effect on warmer environmental conditions can be felt and where the depth of fuller vegetation cover can be perceived.

REFERENCES

Fukagawa K, et al. (2010). The effect of visual stimulus on thermal comfort: Analysis of the visual factor by experiment. Proceedings of The 7th International Cost Engineering Council World Congress & The 14th Pacific Association of Quantity Surveyors Congress, Singapore, July 23-27, 2010, 1–10.

Kurazumi Y, et al. (2011a). Enhanced conduction-corrected modified effective temperature as the outdoor thermal environment evaluation index upon the human body. Building and Environment 46(1), 12-21.

Kurazumi Y, et al. (2011b). Psychological effects of the environmental stimuli on thermal sense in outdoor spaces. Japanese Journal of Biometeorology 48(4), 129–144.

Kurazumi Y, et al. (2010). Conduction-corrected modified effective temperature as the indices of combined and separate effect of environmental factors on sensational temperature. Energy and Buildings 42(4), 441–448.

Kurazumi Y, et al. (2009). Validity and evaluation method of combined thermal environmental factors upon the human body. Japanese Journal of Biometeorology 46(4), 121–137.

Matsubara N, et al. (2004). Psychological effects of the visual stimuli and the auditory stimuli on thermal sensation. Japanese Journal of Biometeorology 40(s), 249–259.

Murakami D, Shimomura T (2007). Examination of comfort by the thermal environment measurement and the subjective experiment on 3 different parts of a green roof. Journal of the Japanese Society of Revegetation Technology 33(1), 152–157.

Rohles FH, et al. (1981). The effects of lighting, color, and room décor on thermal comfort. ASHRAE Transactions 87(2), 511–527.

INDOOR MICROCLIMATE AND NATURAL VENTILATION IN DOUBLE TRANSPARENT FACADE

Kristýna VALOUŠKOVÁ[1,*]
[1]Department of Microenvironmental and Building Services Engineering, Czech Technical University in Prague, Faculty of Civil Engineering, Prague, Czech Republic

[*]Corresponding email: kristyna.valouskova@fsv.cvut.cz

Keywords: Indoor microclimate, natural ventilation, operative temperature, predicted mean vote, velocity

SUMMARY

The primary advantage of double transparent facades is their ability to naturally ventilate multistory buildings. The use of these facades influences the microclimate inside a building. This study examines the difference between single and double transparent facades and whether the latter has a positive impact on the indoor microclimate.

This paper focuses on the use of natural ventilation on a hot spring day. A case study compares indoor air quality in a single transparent facade with that in a double transparent facade. A CFD (Computational Fluid Dynamics) simplified model of an office room with a glass facade orientated to the south shows a positive influence on indoor air quality when a double facade is used.

INTRODUCTION

The way a building is constructed, whether it is the materials used in the interior or the construction solutions of the building as a whole, significantly contributes to indoor microclimates. The selection of structural elements plays an important role in the construction, design, and cost of a building. It also significantly contributes to the entire life cycle of a building. A current trend in modern architecture is the use of glass surfaces. This trend has positive effects, such as increasing the amount of light in rooms and the connection with the surrounding environment. It also has negative effects that are necessary to eliminate, for example, high heat gains in the summer.

In 1984, the World Health Organization (WHO) reported that 30% of the population of industrialized countries suffer from sick building syndrome (SBS). In 2002, the WHO found that more than 60% of people were affected by SBS. Presently, the use of natural ventilation is considered inefficient, particularly for office buildings. However, an environment that uses natural ventilation is considered more comfortable by many people.

METHODOLOGY

The following simulations were performed in DesignBuilder. Temperatures, heat flows, and flow rates calculated by EnergyPlus were used to provide boundary conditions by specifying the time and date of the CFD analysis.

This paper focuses on simulations of a simplified office room. The room was 4 × 7 m, with a glass facade oriented on the south and situated on the 28th floor. A single transparent facade with double glazing (double low emissivity clear glass 6/13 mm, U-value 1.493 W/m^2K) was compared to a double transparent facade with single glazing (single low emissivity clear glass 6 mm, U-value 3.779 W/m^2K) on the outer facade and double glazing (double low emissivity clear glass 6/13 mm, U-value 1.493 W/m^2K) on the inner facade. The simulations were performed with an opening size of 15% and used natural ventilation based on buoyancy and the boundary conditions from EnergyPlus, specified from Prague-Ruzyně, Czech Republic, at 14:00 on June 3. The model used a sitting occupant, which represents a standard office worker.

Site Data (3.6.2002 at 14.00, Prague, Czech Republic)

Outside Dry-Bulb Temperature	16.67°C
Outside Dew-Point Temperature	14.00°C
Wind Direction	285°
Wind Speed	4.25 m/s
Solar Altitude	55.36°
Solar Azimuth	225.64°
Relative Humidity	55.29 %
Direct Normal Solar	0.10 kW/m^2
Diffuse Horizontal Solar	0.31 kW/m^2

Comparison of Indoor Microclimate

The use of thermal comfort calculations was in accordance with ANSI/ASHRAE Standard 55. The comfort calculation options for the summer season were for a relative humidity of 50%, a clothing level of 0.5 clo, and a metabolic rate of 1.0 met.

Table 1 Standards and guidelines of indoor air quality (IAQ)[a]

Parameter	Limit/Range	Reference	TSI Instrument
Temperature	Summer 23 to 28°C Winter 20 to 25.5°C	ASHRAE Standard 55-2010 ISO 7730	Q-Trak, IAQ-Calc VelociCalc
Relative Humidity	30% to 65%	ASHRAE Standard 55-2010 ISO 7730	Q-Trak, IAQ-Calc VelociCalc
Air Movement	0.25 m/s	WHO, ISO 7730	VelociCalc, DP-Calc, AccuBalance

[a] Indoor Air Handbook 2013

RESULTS AND DISCUSSION

Operative Temperature

The following two figures illustrate the behavior of the operative temperature in both variations considered in this case study. The operative temperature for a single facade (see Figure 1a) is approximately 24.3 °C. The operative temperature for a double facade is approximately 25.10 °C (see Figure 1b). Comparing these temperatures with Table 1 (the conditions of the office) shows the temperature range in the room is between the limits in both cases. Lower temperatures occur around the openings. The cavity works as a buffer zone, which reduces air flow velocity and consequently the room temperature. In the first variation with a single facade near to the floor, the temperature is lower.

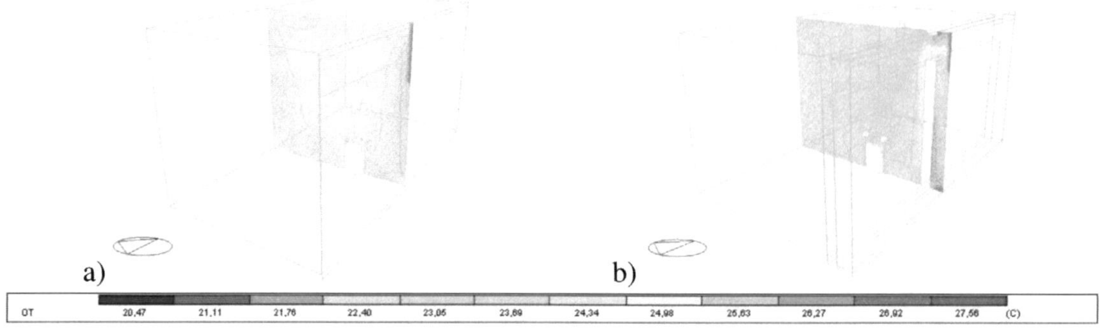

Figure 1 Operative temperature curve (°C): a) a single transparent facade and b) a double transparent facade with the use of natural ventilation

Airflow Velocity

In the variation with the double transparent facade, there is a significant reduction of air velocity inside the room (see Figure 2b) because a double facade cavity behaves as a buffer zone. In the variation with a single facade, in places near the floor, the glass facade airflow velocity achieves ranges between 0.1 to 0.31 m/s (with a maximum value of 0.25 m/s), which can cause drafts and overall discomfort for room occupants. In the second office variation with a double facade, the airflow velocity achieves a maximum of 0.22 m/s, which is in the limits of air movement (see Table 1).

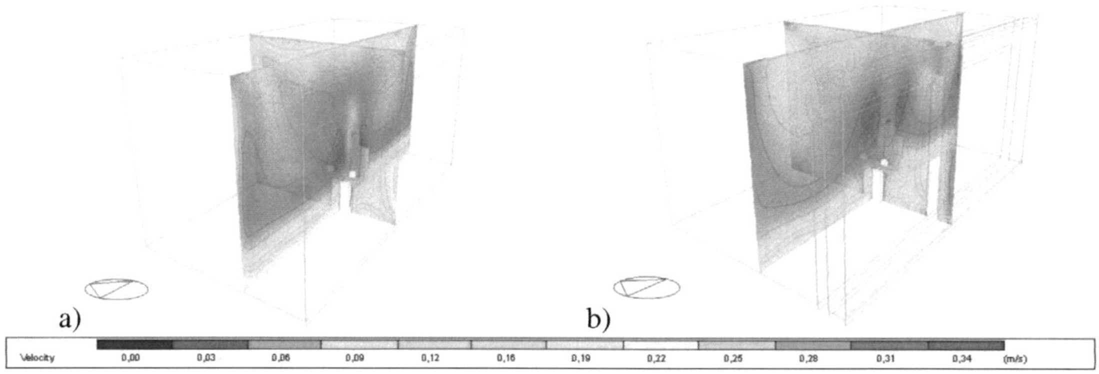

Figure 2 Airflow velocity curve (m/s): a) a single transparent facade and b) a double transparent facade with the use of natural ventilation

Predicted Mean Vote (PMV) Model

Figure 3 shows the curve of the predicted mean vote (PMV) model. In the case of an office with a single transparent facade, the PMV model value is around –1.5 to –1.0, which is slightly cold for room occupants. The variation with a double facade has a PMV model value of around 0, which is the best value for PMV. This higher value is caused by the higher operating temperature of the room (see the Operative Temperature section). The lower PMV model value in the single facade variation is caused by a higher airflow velocity (see the Airflow Velocity section), which is over the limit (see Table 1). The result of PMV model value is maintained at a height of 1.1 m from the floor - the level of a sitting person's head - and at the height of 0.1 m from the floor—the level of a sitting person's ankles.

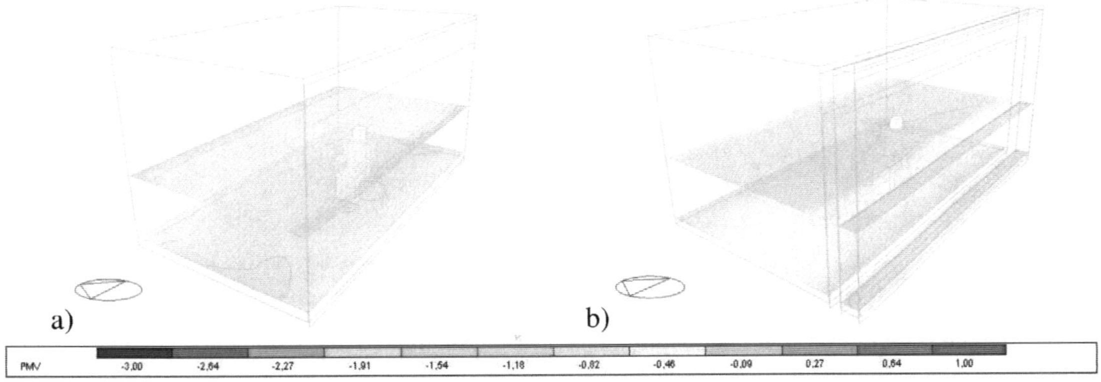

Figure 3 PMV model: a) a single transparent facade and b) a double transparent facade with the use of natural ventilation

CONCLUCIONS

As seen in the previous comparisons, the indoor microclimate is better when a double facade is used. This improvement in microclimate is caused by the higher values of airflow velocity when using a single transparent facade, which in this case study was on the 28th floor and over limit (see IAQ standards - Table 1). All these variables depend on the boundary conditions, which depend on geographical location, floor level, and the size of openings. It follows that natural ventilation cannot be used in every situation involving a double facade. During very hot summer days and very cold winter days, mechanical ventilation is necessary.

The use of double or single transparent facades are still unclear, and the possibility of using natural ventilation is a major issue. It is needed to make further investigation.

ACKNOWLEDGEMENT

This paper was supported by the Grant Agency of the Czech Technical University in Prague, grant. No SGS15/015/OHK1/1T/11.

REFERENCES

TSI (2013). Indoor Air Quality Handbook: A Practical Guide to Indoor Air Quality Investigations, TSI Incorporated

John D. Spengler, Ph. D., et al. (2001). Indoor Air Quality Handbook, *McGraw-Hill Education.*, Toronto

Oesterle, E. el. al. (2011). Double skin Facades. Mnichov: Prestel Verlag

Ding, W., Y. Hasemi a T. Yamada. (2005) Natural Ventilation Performance of a Double-Skin Façade with a Solar Chimney. *Energy and Buildings*, vol. 37, no. 4, s. 411-418.

Gratia, E. a A. de Herde. (2004). Optimal Operation of a South Double-Skin Facade. *Energy and Buildings*, vol. 36, no. 1, s. 41-60.

Gratia, E. a A. de Herde. (2004). Is day natural ventilation still possible in office buildings with a double-skin façade?. *Building and Environment.*, vol. 39, no. 4, s. 399-409.

DesignBuilder Software Limited. (2014) *DesignBuilder EnergyPlus Simulation Documentation for DesignBuilder v4.3* [online]. London

A SIMPLIFIED MODEL FOR TEMPERATURE PREDICTION IN A VENTILATED WALL MODULE

T.M.J. Rakotomahefa, Tengfei (Tim) Zhang* and Shugang Wang

School of Civil Engineering, Dalian University of Technology, Dalian, China

*Corresponding email: tzhang@dlut.edu.cn

Keywords: Ventilated wall, Heat transfer, Thermal network, Matrix, CFD.

SUMMARY

A ventilated wall is a smart building envelope component that can filter the intake air, reduce heat loss/gain and insulate outdoor noises. It is composed of a solar board, serpentine air cavities, and an air filter. This investigation proposed an efficient matrix method to compute heat transfer in a ventilated wall module. A thermal network was established by implementing energy balance to each network node, which results in a system of algebraic equations of temperatures. The outdoor temperature, indoor temperature and the solar irradiation intensity were specified as boundary parameters. The results reveal that the proposed model can provide reasonably accurate temperature profiles with a gap of 1.2 °C to 3°C between the prediction and the measurement. When compared to CFD modeling, the saved computing expense is quite significant but without sacrificing much accuracy.

INTRODUCTION

A ventilated wall module is an outdoor air-intake device that can clean the intake air, reduce some heat loss/gain, and insulate some outdoor noises. One design of it contains a solar board, three air-cavities and an air filter (Zhang et al, 2014) as shown in Figure 1(a) where the air goes upwards in air cavity 1 as it is heated by the solar board, then the air is pressed down to get coarse particles deposited. After that, the air goes up again in air cavity 3 where it is filtered before being discharged into an indoor space.

To evaluate energy performance of a ventilated wall module, a simplified model must be established. Currently, the models to predict temperatures in a ventilated wall module can be divided into three categories: analytical/correlation model, zonal/network model, and computational fluid dynamics (CFD) model. Although an analytical/correlation model is easy to implement and possibly the fastest one, this model suffers from accuracy where the gap is 0~6.5°C from experimental data (Soto Francés et al, 2013). In the zonal/network approach (Jiru and Haghighat, 2008), the geometry is decomposed into several cells that are typically 10 times larger than the ones used in CFD models. The mass and energy are formulated for each cell, while an appropriate model is used to compute pressure difference or mass flow rate. The zonal/network model can provide results without losing much accuracy. However, if the geometry is complex, the solution accuracy decreases and the gaps may reach 3~4°C as compared with the experimental data (Jiru and Haghighat, 2008). The CFD model divides the geometry into tiny cells and for each cell, the continuum, momentum and energy equations are solved. The gap between CFD results and experimental data is usually 0~1.5°C in a steady state (Giancola et al, 2012) but 0~5°C in an unsteady state (Aparicio-Fernández et al, 2014). The CFD model has the potential to perform best, but may be too expensive to afford.

This investigation intends to balance accuracy and computing expense via an efficient matrix method to predict temperature profiles in a ventilated wall module.

METHODOLOGIES

The heat transfer in a ventilated wall module as seen in Figure 1(a) was modelled, by casting different components into a thermal resistance network as shown in Figure 1(b). Although unsteady-state solutions can be implemented (Vats and Tiwari, 2012), we choose steady-state approach for simplicity. Other hypotheses were assumed: natural convection and laminar flow, non-participating air, diffused and scattered radiations disregarded, one-dimensional heat transfer except for air-cavities in which two-dimensional heat transfer is accounted. The governing equation for each component of the ventilated wall is formulated as the following:

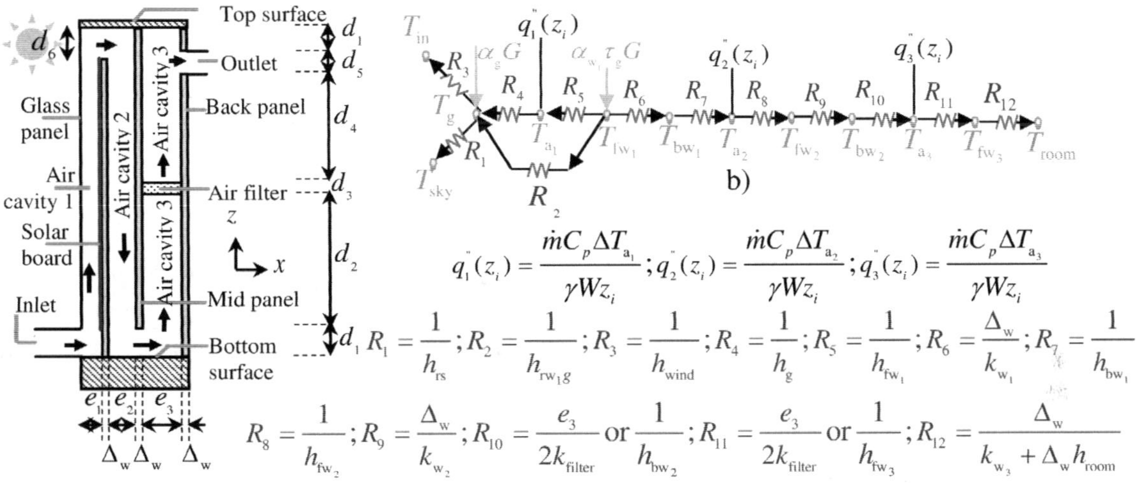

$$q_1''(z_i)=\frac{\dot{m}C_p\Delta T_{a_1}}{\gamma W z_i};\ q_2''(z_i)=\frac{\dot{m}C_p\Delta T_{a_2}}{\gamma W z_i};\ q_3''(z_i)=\frac{\dot{m}C_p\Delta T_{a_3}}{\gamma W z_i}$$

$$R_1=\frac{1}{h_{rs}};\ R_2=\frac{1}{h_{rw_1g}};\ R_3=\frac{1}{h_{wind}};\ R_4=\frac{1}{h_g};\ R_5=\frac{1}{h_{fw_1}};\ R_6=\frac{\Delta_w}{k_{w_1}};\ R_7=\frac{1}{h_{bw_1}}$$

$$R_8=\frac{1}{h_{fw_2}};\ R_9=\frac{\Delta_w}{k_{w_2}};\ R_{10}=\frac{e_3}{2k_{filter}}\text{ or }\frac{1}{h_{bw_2}};\ R_{11}=\frac{e_3}{2k_{filter}}\text{ or }\frac{1}{h_{fw_3}};\ R_{12}=\frac{\Delta_w}{k_{w_3}+\Delta_w h_{room}}$$

Figure 1. Schematic of a ventilated wall module (left) and the thermal network (right).

(1) On glass panel: Considering the solar heat absorption by the glass panel, radiation heat transfer between the solar board (front side) and the glass panel, convective heat transfer from air cavity 1 to the glass panel, convective and radiation loss to the surrounding, it leads to

$$\alpha_g G+h_{rw_1g}[T_{fw_1}(z_i)-T_g(z_i)]+h_g[T_{a_1}(z_i)-T_g(z_i)]-h_{wind}[T_g(z_i)-T_{in}]-\sigma\varepsilon_g[T_g^4(z_i)-T_{sky}^4]=0 \quad (1)$$

where G, h, T and z represent the solar irradiation intensity, radiative/convective heat transfer coefficient, temperature and height. The subscripts a_1, f, g, i, in, rw_1g and w_1 denote air cavity 1, front side, glass panel, node position ($1 \le i \le n$), inlet, radiation between glass panel and sky, and solar board. The Greek letters α, ε and σ symbolize the absorptivity, emissivity and the Stefan-Boltzmann constant.

(2) In air cavity 1: Considering the convective heat gain from the solar board (front side) to air cavity 1, convective heat loss to the glass panel, and the heat to raise temperature of the moving air, it gets to

$$h_{fw_1}[T_{fw_1}(z_i)-T_{a_1}(z_i)]-h_g[T_{a_1}(z_i)-T_g(z_i)]-q_1''(z_i)=0 \quad (2)$$

where q_1'' represents the heat for temperature rise in air cavity 1, whose expression can be found in Figure 1.

(3) On solar board (front side): Accounting for the solar heat gain, convective heat loss to air cavity 1, radiation heat loss to glass panel, and conduction heat loss through the solar board, it yields

$$\tau_g \alpha_{w_1} G - h_{fw_1}[T_{fw_1}(z_i) - T_{a_1}(z_i)] - h_{rw_1g}[T_{fw_1}(z_i) - T_g(z_i)] - (k_{w_1}/\Delta_w)[T_{fw_1}(z_i) - T_{bw_1}(z_i)] = 0 \quad (3)$$

where k, Δ_w and τ stand for thermal conductivity, thickness of the solar board / mid panel / back panel and transmitivity. The subscript b means back side.

Due to limited space, the equations for solar board (back side), air cavities 2 and 3, the mid and the back panels are not presented, but they can be easily derived from the thermal network shown in Figure 1. Then, all energy equations are cast into a matrix format as

$$[\text{Thermal Resistances}][\text{Unknown Temperatures}] = [\text{Known parameters}] \quad (4)$$

where [Thermal Resistances] is a condensed form for all existing thermal resistances. [Unknown temperatures] is a matrix storing all variable temperatures, namely $T_g(z_i)$, $T_{a_1}(z_i)$, $T_{fw_1}(z_i)$, $T_{bw_1}(z_i)$, $T_{a_2}(z_i)$, $T_{fw_2}(z_i)$, $T_{bw_2}(z_i)$, $T_{a_3}(z_i)$ and $T_{fw_3}(z_i)$, which are to be solved. [Known parameters] designates a matrix storing all known parameters.

In a solution the matrix system (4) is sequentially called along the airflow direction. The smoothness and grid-independent results are obtained when 1350 nodes are generated, i.e. 150 vertical nodes per item/component. Once the relative change of a temperature is less than 10^{-9} in an iteration, the solution is judged converged. In addition, the temperature profiles were also computed in a CFD software where 483,260 quad meshes are created. The air filter is treated as a porous media while measured temperatures on surfaces were specified as boundary conditions. Table 1 lists some major parameters inputted into the numerical model.

Table 1. Some known parameters inputted into modeling

G=278 W/m^2; h_{room}=10 W/m^2·°C;	$T_{in} = T_{room}$ = 17.63°C;	ε_g=0.90; ε_{w_1}=0.95;
k_{filter}=0.038 W/m·°C;	V_{in}=0.02 m/s;	γ=0.75; τ_g=0.86;
$k_{w_1} = k_{w_2} = k_{w_3}$=0.067 W/m·°C;	α_g=0.06; α_{w_1}=0.95;	σ = 5.67×10^{-8} W^2/m^2·K^4;

For model validation, the temperature profiles were also measured in a prototype model in lab. The solar board was replaced by an electric-heat pad to mimic the irradiation from the sun, while the temperatures were provided by Pt-100 sensors when the thermal equilibrium state has reached. The accuracy of temperature measurement is ±(0.30+0.005|t|) °C.

RESULTS AND DISCUSSION

Figure 2(a) presents a comparison of the surface-temperature profiles between the experiment and the proposed method along the solar board (both front and back sides), the mid-panel (front side) and the back panel (front side). Due to the heat absorption by the solar board (front side), the temperature on the board drastically increases with height. At the top part, the temperature drops where the air is driven into air cavity 2 because of vortexes there. On the rest surfaces, the same temperature trend is present but the profile is smoother. In air cavity 3, there is an air filter between the mid-panel and the back panel from z=0.4 m to 0.5 m, which interrupts a little bit the temperature profile. As compared with the measurement, all temperature trends are well captured by the proposed method. The minimum gap with the experiment (1.2 °C) is at the lower part of the solar board (front side) and the maximum gap (2.5 °C) is at the lower part of the back panel.

Figure 2(b) portrays the computed temperature profiles in air cavities 1 and 3. In air cavity 1, the air gets heated by the solar board (front side). As a result, the air temperature gradually increases along height until it drops a little bit at the top where flow vortexes are formed. The proposed method overpredicted the air temperatures but the CFD underpredicted. As compared with the experiment, the temperature gaps ranges from 1.2 °C (at the bottom of air cavity 1) to 3 °C (at the top of air cavity 1). The solution time for the proposed method is less than 8 minutes, but for CFD it required approximately two hours.

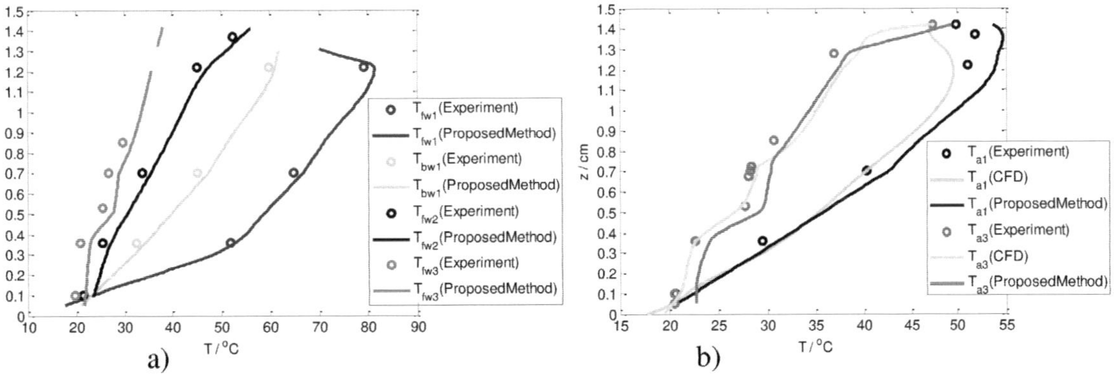

Figure 2. Comparison of predicted temperatures with the CFD and the measurement: a) on solid surfaces, b) in air cavities 1 and 3.

CONCLUSIONS

Simplified equations for every component of a ventilated wall module were established to form thermal resistance network and sequentially solved to predict temperature profiles. The proposed method has provided reasonably accurate results and the gaps with the experiment range from 1.2°C to 3°C. As compared to the CFD modeling the saved computing expense is quite significant but without impairing much accuracy.

ACKNOWLEDGEMENT

This work was funded by the National Natural Science Foundation of China (grant: 50978039).

REFERENCES

Aparicio-Fernández C , Vivancos JL, Ferrer-Gisbert P, et al. (2014). Energy performance of a ventilated façade by simulation with experimental validation. *Applied Thermal Engineering*, **66**, 563–570.

Giancola E, Sanjuana C, Blanco E, et al. (2012). Experimental assessment and modelling of the performance of an open joint ventilated facade during actual operating conditions in Mediterranean climate. *Energy and Buildings*, **54**, 363–375.

Jiru TE and Haghighat F (2008). Modeling ventilated double skin facade—A zonal approach. *Energy and Buildings*, **40**, 1567–1576.

Soto Francés VM, Sarabia Escriva EJ, Pinazo Ojer JM, et al. (2013). Modeling of ventilated facades for energy building simulation software. *Energy and Buildings*, **65**, 419–428.

Vats K and Tiwari G (2012). Performance evaluation of a building integrated semitransparent photovoltaic thermal system for roof and facade. *Energy and Buildings*, **45**, 211–218

Zhang, T, Zhao Y and Wang S (2014). Prediction of airflow rate through a ventilated wall module. *Energy and Buildings*, **82**, 651–659

HOW RELIABLE ARE STANDARD (UNDISTURBED) AIRBORNE MOLD (SPORETRAP) TESTS? THE USEFULNESS OF DISTURBED TESTS FOR MORE ACCURATE ASSESSMENTS

Michael R. MEYER[1], MS,CMI

[1]Mold Testing Services, LLC, Sioux Falls, South Dakota, USA

*Corresponding email: moldymike@moldtestingsd.com

Keywords: Airborne mold, sporetraps, disturbed mold, Aspergillus-Penicillium

SUMMARY

The main concern from mold is airborne exposure. High exposure levels are often due to disturbed conditions (cleaning,etc). But the most widely used method are sporetraps taken in undisturbed ("standard") conditions. When mold has settled out of the air, especially in vacant or little used rooms, high levels of mold on surfaces may occur. These can become stirred up again by normal activities. We have worked on many cases where "standard" (undisturbed) tests failed to detect a concern while light disturbance often showed extremely high levels. Including disturbed testing greatly increases the reliability of airborne mold testing. To our knowledge, disturbed testing is not used by most consultants. The main mold of concern is *Aspergillus-Penicillium*. Lab analysis at x600 to x1000 is much more accurate than x400 (many commercial labs use only x400).

INTRODUCTION - WHY DISTURBED TESTING?

The main concern from mold is AIRBORNE exposure. Mold problems are common in damp or water damaged buildings. The most commonly used method to check for airborne mold are sporetrap casettes (also called non-viable or total testing). An air pump collects mold spores on a sticky plate in a cassette. The number and types of mold are identified under a microscope (Li, Yang and Harrington,2005). The most common (referred here to as the "standard") method is to collect undisturbed airborne mold tests (taking care not to stir up dust,etc). This method is usefulness in many cases in determining if there is a mold concern or if adequate clean-up has been performed (Meyer, 2013).

The key problem molds in most cases are *Aspergillus-Penicillium* (Baxter et.al., 2005; Burge, 1995; Li et.al., 2007). (counted together in spore trap tests). Other important problem molds are *Chaetomium* and *Stachybotrys* (as well as many others). These are larger spores associated with water damaged materials. High levels of these molds are associated with health symptoms. In contrast, common (often non-problem) molds such as *Cladosporium* (drier weather) or Basidiospores (damp weather) may occur at high outside levels and yet not be associated with health effects (Meyer, 2013). As a rough guideline, various studies have shown that *Aspergillus-Penicillium* levels above about 1000 spores/m^3 are often associated with problem rooms while normal (non-problem) rooms often have *Aspergillus-Penicillium* <1000 spores/m^3, and often <500 spores/m^3(Baxter et.al., 2005; Meyer, 2013; Robertson and Horner, 2004; Spurgeon, 2013).

"Standard" (undisturbed) airborne mold tests are widely used but often do not confirm the presence of airborne mold. This especially includes vacant or rarely used buildings (or guest rooms,etc). When mold has had an opportunity to settle out onto surfaces, undisturbed tests will fail to detect these molds. There can sometimes be enormous amounts of settled molds in carpets, on ceiling joists and other surfaces, etc, In contrast, even slight disturbance such as walking around, turning on the ventilation, using a small fan,etc may show an enormous increase in airborne mold levels. This especially includes *Aspergillus-Penicillium*, which are very small spores and easily become airborne for long periods when disturbed. Disturbed tests may also detect other problem molds such as *Chaetomium* or *Stachybotrys*. These are larger spores and tend to settle quickly out of the air. We has been involved in numerous cases where light disturbance detected serious mold concerns not detected in the undisturbed tests (Meyer, 2014). Others have found disturbed testing for post clean-up mold testing often detects lots of settled spores (Vaughn, 2013).

METHODOLOGIES

We have performed airborne mold assessments at >2000 buildings over the last 15 years. Our standard methodology has been to collect undisturbed sporetrap cassettes. This typically includes a few tests in problem areas, 1-2 control tests in presumed non-problem areas and (except in winter), an outside control test. We typically use Allergenco cassettes using a Bioaire pump. Samples are usually collected for five minutes at a flow rate of 15 liters/minute (less time in presumed very moldy areas). Windows and doors are kept closed and an effort made to minimize disturbance.

However, especially in the last five years, we have frequently also included lightly disturbed tests, after first taking the usual "standard" undisturbed tests. *Light disturbance* is caused with a house fan used for thirty seconds to one minute. Where power was off, a small battery powered fan was used. The fan is swept near the floor and other surfaces working in a general circular pattern "sweeping" the air towards the middle (the air pump) of the circle (a "circle" in the room of about 10-20 ft in diameter). Note: only LIGHT disturbance is used. We do NOT recommend strong disturbance such as with leaf blowers or powerful commercial blowers. We have found these can actually decrease observed airborne mold levels (as well as stirring up lots of dust and debris that can clog the cassettes). Both the undisturbed and disturbed tests are taken in the usual manner for five minutes each at a flow rate of 15 liters/minute.

Disturbed testing can be useful in vacant or rarely used rooms, after a mold clean-up or where health symptoms occur and standard (undisturbed) tests did not detect a concern. However, disturbed tests may not be appropriate in some cases (such as when infants are in the same room).

The cassettes for both undisturbed and disturbed tests are submitted to a lab analyst for microscope identification of the number and types of mold (Yang and Harrington,2005), We prefer analysis x600 to x1000. Many commercial labs provide analysis x400. However, the very small and clear *Aspergillus-Penicillium* are often greatly undercounted at lower magnification (especially with lots of debris on the slides). Godish and Godish (2008) recommend x1000. Most of the analyzes for our work was performed at x1000.

Most severe health related cases involving mold are where high airborne mold levels have been caused by some disturbances. We would argue that light disturbance somewhat replicates human activity in a room and may be a more reliable indicator of mold exposure than standard (undisturbed) tests. Based on many cases, we have found light disturbance testing to be effective and reliable in detecting mold problems (Meyer, 2014).

CASE HISTORIES - EXAMPLES

Due to space limitations, only a few of many examples are shown here. See also Meyer (2014) for more examples

Case History 1. Flooded Apartment – After Clean-Up (Fiberglass). Visible mold was found on the walls in a vacant apartment due to a burst water heater. The owners removed the moldy drywall but left the fiberglass insulation (clean looking) behind the drywall in place and did a mold clean-up, including renting HEPA air scrubbers. MTS did a visit on 8/24/12. Normal (undisturbed) tests taken following IICRC guidelines (HEPA air scrubber on) showed a nearly normal level, although slightly elevated *Asp-Pen.* In our experience, exposed fiberglass insulation often has high mold levels. We did a disturbed test (light disturbance 1 minute house fan). These showed high levels. This case shows normal testing may miss considerable mold.

Site Location	Date	Total Mold	*Cladosporium*	*Asp&Pen*	*Chaet& Stachy*
Total mold = common& problem molds			Common mold	Problem damp mold	Problem wet mold
Outside – front- SE. 8/24 - Sunny, windy. 86.4 °F, 48.7 RH%. Ultrafines 1830 pt/cc.					
Outside	8/24/12	33,541	27,146 (81%)	53 (<1%)	Not detected
SE Bedroom 8/24- After Clean. 71.4 °F, 37.1 RH%. Ultrafines 22 pt/cc. Light disturbance one minute with house fan swept over floor. Exposed fiberglass insulation in walls.					
SE Bedroom	8/24/12	5,331	1,173 (22%)	**3,893** (73%)	Not detected
Disturb	8/24/12	**145,545**	1,920 (1%)	**143,040** (98%)	53 (<1%)
Bulk Sample – Fiberglass Insulation Below Window in Living Room – Bag Test – 1 Minute.					
Bulk	8/24/12	**590,930**	**374,400** (63%)	**206,400** (35%)	532 (<1%)
Experiment - Bulk sample – new fiberglass insulation – Bag Test – 1 minute (12-132)					
Bulk sample	6/23/12	2900	2900	Not detected	Not detected

Case History 2. Parsonage – Health Symptoms – Basement. The wife of a pastor felt ill when in their parsonage house. No visible mold was obvious. Damp unfinished basement. MTS visit 6/6/13 found considerable visible mold on the basement joists (not readily obvious – light stains). Standard (passive) airborne tests indicated normal levels in the house. However, disturbed tests show an extremely large increase in the basement. This case emphasizes that standard testing would have not identified any obvious concern.

Site Location	Date	Total Mold	*Cladosporium*	*Asp&Pen*	*Chaet& Stachy*
Total mold = common& problem molds			Common mold	Problem damp mold	Problem wet mold
Main Floor – Living Room. 68.4 °F, 57.4 RH%. Ultrafines 740 pt/cc. *Disturbed test using household fan for 1 minute – to cause light disturbance.*					
Living Room	06/06/13	2,984	Not detected	693 (2%)	Not detected
LvRm-Disturb	*06/06/13*	*2,130*	*906 (43%)*	*53 (3%)*	*Not detected*
Afer clean - Living Room . HEPA AS on nearby. 68.0 °F, 51 RH%. Ultrafines 34 pt/cc.					
Living Room	09/20/13	3,144	2,080 (66%)	213 (7%)	Not detected
LvRm-Disturb	*09/20/13*	*6,450*	*3,626 (56%)*	*320 (5%)*	*53 (<1%)*

Main Basement. 64.7 °F, 67.5 RH% . Ultrafines 855 pt/cc. Damp.					
Basement	06/06/13	1,170	266 (23%)	106 (9%)	Not detected
Base.- Disturb	*06/06/13*	***>483,357***	*1,226 (<1%)*	**>480,000 (99%)**	*Not detected*

CONCLUSIONS

The main concern from mold is airborne exposure. The most widely used method are sporetraps taken in undisturbed ("standard") conditions. However, where mold has settled out of the air, especially in vacant or little used rooms, high levels of mold may occur. These may become stirred up again by normal activities. We has worked on many cases where "standard" (undisturbed) tests failed to detect a concern while light disturbance often showed extremely high levels. We recommend including disturbed testing along with standard testing. To our knowledge, disturbed testing is often not used by many consultants. The main mold of concern is *Aspergillus-Penicillium*. Lab analysis at x600 to x1000 is more accurate than x400 (used by many commercial labs).

REFERENCES

Baxter,D. et.al., (2005), A regional comparison of mold spore trap concentrations outdoors and inside "clean" and "mold-contaminated" southern California buildings. *Jour. Occup. & Env. Hygiene,* v2.,p8-18

Burge, Harriet (1995) *Bioaerosols*, Lewis Publishers, Boca Raton, FL

Godish, D, and Godish,T (2008), Total airborne mold particle sampling: Evaluation of sample collection, *Jour. Occup. & Env. Health*, v.5,p100-106

Li,D, Yang, C. and Harringron,F. (2005), Microscopic analytical methods for fungi, Chapter 4 in *Sampling and Analysis of Indoor Micro-organisms*, John Wiley & Sons, NY

Meyer, M (2009) What are normal and problem indoor total (non-viable) airborne mold levels? An IAQ consultant's perspective based on 10 years experience at hundreds of houses and offices in the Midwest, *Healthy Buildings 2009*, Syracuse, NY, paper 203

Meyer, M (2013) An argument for developing an airborne total (non-viable) mold guideline for indoor exposure, *Indoor Air Quality Association* Conference, San Antonio, TX

Meyer, M (2014), Are standard (undisturbed) airborne mold tests reliable? The usefulness of disturbed tests for mold assessments with several examples, *IAQA Conference*, Nashville, TN, March

Robertson,L. and Horner, H (2004) Post-remedial assessment and clearance criteria for mold remediation, *Mycological Society of America conference* (~1000 AspPen guide)

Spurgeon, Joe,(2014), Elements of an exposure assessment strategy, *IAQA conference*, Nashville, TN, March 2014 (~1000 AspPen guide)

Vaughn,W. (2013) Air scrubbers are not like dehumidifiers, *IAQA Conference, Orlando, FL*, March 2013 (references using disturbed testing for mold clearances

THERMAL PERFORMANCE OF EVALUATION ON ENERGY EFFICIENT PARAFFINIC PCM BASED SHAPE STABILIZED COMPOSITES FOR ENERGY SAVING

Okyoung CHUNG[1], Su-Gwang JEONG[1], and Sumin KIM[1,*]

[1]Building Environment & Materials Lab, School of Architecture, Soongsil University, Seoul, 156-743, Republic of Korea

[*]Corresponding email: skim@ssu.ac.kr

Keywords: Latent heat thermal energy storage (LHTES), n-octadecane, Expanded vermiculite, Expanded perlite, Thermal performance

SUMMARY

The composite PCMs were prepared by incorporation of liquid n-octadecane within the expanded vermiculite (eVMT) and expanded perlite (ePLT), using a vacuum impregnation method. The microstructures of n-octadecane/expanded vermiculite and pearlite were characterized by scanning electron microscopy (SEM). Differential scanning calorimetry (DSC) analysis indicated that the n-octadecane/eVMT and n-odtadecane/ePLT composites maintained their large latent heat capacity and original phase change temperatures, due to large surface area and good dispersion of the eVMT and ePLT. Thermos gravimetric analyser (TGA) analysis revealed that the prepared composite PCMs had good thermal durability in the working temperature ranges.

INTRODUCTION

PCMs have been studied in the field of buildings for saving energy. They can be expected to be integrated with various building materials, and gradually be developed into specialized materials. PCMs have been considered for thermal energy storage (TES) in buildings. Vermiculite has high insulation value, acoustic insulation properties, and is absorbent to a wide range of liquids. The expanded vermiculite (eVMT) is a lightweight material that is porous, inexpensive, ecologically harmless, and non-toxic. Therefore, vermiculite has been used in applications such as insulation materials in buildings. Perlite is a volcanic glass containing 65-75 % of SiO_2, and 2-5 % of H_2O. The characteristic property of perlite is that if it is exposed to rapid heating temperature of 900 to 1200°C, it increases its volume by 4 to 20 times. This happens because perlite includes bound water, which expands during heating, according to the well-known popcorn effect, and adopts the form of a porous swollen material. Therefore, it is used in many commercial applications, such as construction, thermal, and acoustic insulation, agriculture, and horticulture. In this paper, form-stable composites consisting of n-octadecane with eVMT and ePLT were prepared as novel potential PCMs for high thermal energy storage in building applications.

METHODOLOGIES

For preparation of samples, n-octadecane purchased from Sigma-Aldrich company was used for the experiment as heat storage material. The n-octadecane has a latent heat capacity of 256.5 J/g, and melting phase change range of 26.9 °C~28.7 °C. The eVMT and ePLT were supplied from Misung Corporation in South Korea. The expanded vermiculite and perlite was ground in a grinder, and sieved by 100 ~ 150 μm mesh. To utilize PCMs in buildings, they need to be prevented from instability in liquid state by using shape-stabilization, micro-encapsulation and incorporation techniques. In this study, the composite n-octadecane was prepared by a vacuum impregnation method to solve the leakage problem, keeping their efficient thermal storage quantity. Octadecane was employed as a heat storage medium, which has a suitable temperature range of a comfortable indoor environment. The eVMT and ePLT were dried at 105 °C for 24 hour, before the composite PCM was prepared. Then, the eVMT and ePLT were added in the impregnation set-up. Next, the valve between the flask and container with n-octadecane was opened, to allow it to slowly pour into the flask, and cover the eVMT and ePLT. The vacuum process was continued for 90 minutes, and air was allowed to enter the flask again, to force the liquid state of n-octadecane to penetrate into the porous structure of eVMT and ePLT. After the penetration process, excessed n-octadecane that remained in the flask needed to be removed through a filtering process. The two types of composite PCM in a colloidal shape were filtered by 1μm filter paper, until a granular sample was apparent, which was dried under vacuum at 80 °C for 24h.

RESULTS AND DISCUSSION

SEM images of the porous building materials and composite PCMs are shown in Fig. 1. As seen from Figs. 1 (a) and (b), eVMT and ePLT have irregular layers, and uneven pores in the layers. The eVMT and ePLT have rough and random microstructures. The image in Fig. 1 of the composite PCMs also shows that the n-octadecane is dispersed into the porous network of eVMT and ePLT, which are used as supporting materials. DSC analysis is the most valid method to determine the thermal energy storage properties of PCMs, since it gives reliable results for the phase change temperatures range and latent heat.

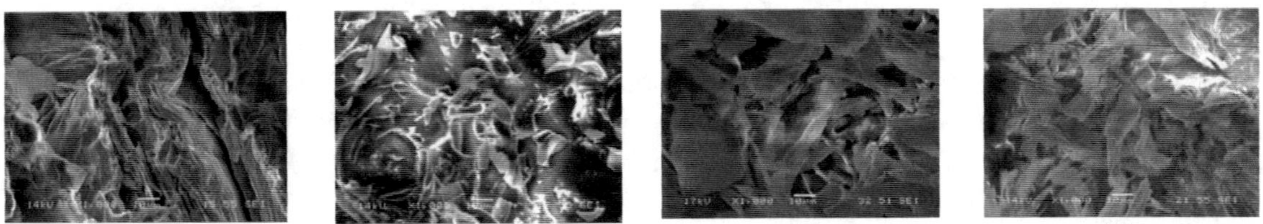

(a) (b) (c) (d)

Figure 1. SEM images of the (a) eVMT, (b) ePLT, (c) n-octadecane/eVMT, and (d) n-octadecane/ePLT.

Fig. 2 shows the DSC curves of the n-octadecane, n-octadecane/eVMT and n-octadecane/ePLT that were obtained. DSC thermograms determined the melting and freezing temperature ranges to be 28.7 °C and 26.9 °C for the n-octadecane, 26.1 °C and 24.9 °C for the n-octadecane/eVMT, and 26.2 °C and 25.3 °C for the n-octadecane/ePLT, respectively. Although there are small decreases in the phase change temperatures of the composite PCMs, the phase change temperatures of the composite PCMs are very close to those of n-octadecane. The n-octadecane, n-octadecane/eVMT and n-octadecane/ePLT showed latent heats of melting of 226.0 J/g, 142.0 J/g, and 132.2 J/g, respectively, and latent heats of freezing of 242.6J/g, 125.5 J/g and 174.3 J/g, respectively. The values for latent heat capacity of the n-octadecane/eVMT and n-octadecane/ePLT were nearly 80.65 % and 59.35 % that of the n-octadecane, respectively. The measured latent heats of melting and freezing of the composite PCMs were close to the values calculated by multiplying the mass ratio of the n-octadecane in the composites, and their phase change enthalpies.

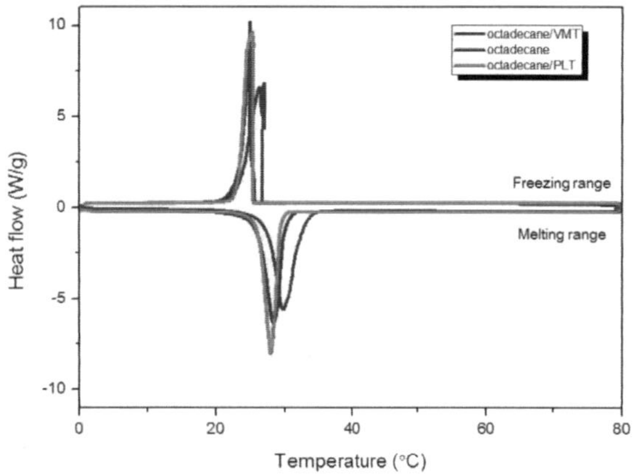

Figure 2. DSC curves of the PCMs.

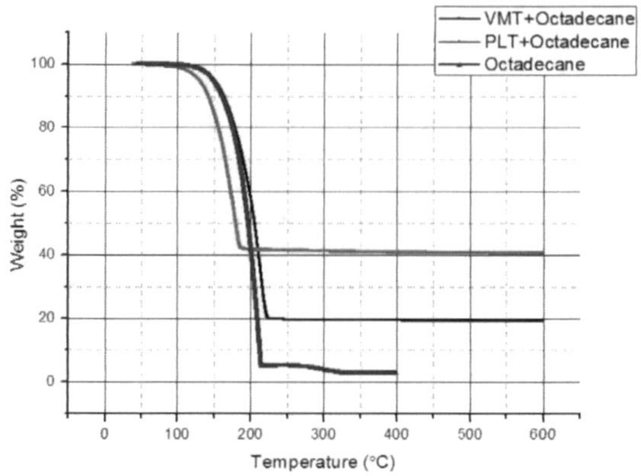

Figure 3. The weight of pure PCM and the composite PCMs.

These results were probably caused by abnormal interactions between the PCM, and the inner surface of pores of the eVMT and ePLT. The thermo gravimetric analysis curves show the weight of n-octadecane and its composite PCMs, and derivative weight of n-octadecane and its composite PCMs. As shown in Fig. 3, the n-octadecane has one curve of thermal oxidation degradation, and the peak occurred at 209.32 °C. Also, the derivative weight of n-octadecane/eVMT shows a peak of 214.29 °C, and n-octadecane/ePLT shows a peak of 178.41 °C. In these graphs, we found the derivative weight peaks of shape-stabilized composite PCMs are higher than that of pure PCM. Also, TGA analysis showed the prepared composite PCM left a more plentiful combustion residue, because of loading contents of the porous material with the properties of flame retardancy. This result also showed plenty of the n-octadecane incorporated into the structure of the porous material. As a result, eVMT and ePLT led to thermal resistance of the n-octadecane. After thermal decomposition, composite PCMs with weight percent remained, after about 250 °C. Also, weight loss did not happen in all composite PCMs below 100 °C, so it can be concluded that the prepared composite PCMs exhibit available thermal stability at room temperature.

CONCLUSIONS

In this study, novel form-stable composite PCMs were prepared, by a vacuum impregnation method. We determined the characteristics of n-octadecane based composite PCMs that contained eVMT and ePLT, by using SEM, DSC, and TGA analysis techniques. Because of the effect of capillary force and surface tension force, n-octadecane was confined in mass fractions of 80% and 59% in the eVMT and ePLT, respectively, without any liquid n-octadecane leakage from the porous material of the composites. DSC results showed that the melting and freezing temperatures and latent heats of composite PCMs are suitable for low temperature thermal energy storage applications.

ACKNOWLEDGEMENT

This research was supported by a grant (14CTAP-C078014-01) from Infrastructure and Transportation Technology Promotion Research Program funded by Ministry of Land, Infrastructure and Transport of Korean government. This work was supported by the National Research Foundation of Korea (NRF) grant funded by the Korea government (MSIP) (No. 2014R1A2A1A11053829).

REFERENCES

Zhang X, Deng P, Feng R, et al. (2011). Novel gelatinous shape-stabilized phase change materials with high heat storage density. *Solar Energy Materials and Solar Cells*, **95**, 1213-1218.

Jeon J, Jeong SG, Lee JH, et al. (2012). High thermal performance composite PCMs loading xGnP for application to building using radiant floor heating system. *Solar Energy Materials and Solar Cells*, **101**, 51–56.

Jin X, Zhang X. (2011). Thermal analysis of a double layer phase change material floor Original Research Article. *Applied Thermal Engineering*, **31**, 1576-1581.

Zhang P, Ma ZW. (2012). An overview of fundamental studies and applications of phase change material slurries to secondary loop refrigeration and air conditioning systems. *Renewable and Sustainable Energy Reviews*, **16**, 5021-5058.

Jeong SG, Chung O, Yu S, et al. (2013). Improvement of the thermal properties of Bio PCM using exfoliated graphite nanoplatelets. *Solar Energy Materials & Solar Cells*, **117**, 87-92.

Lia M, Wu Z, Tan J. (2012). Properties of form-stable paraffin/silicon dioxide/expanded graphite phase change composites prepared by sol–gel method. *Applied Energy*, **92**, 456-461.

IMPACT OF WATER DAMAGE ON MICROBIAL COMMUNITY IN RESIDENTIAL BUILDINGS

Iman SYLVAIN[1], Michal P SPILAK[2], Michael S WARING[2] and Rachel I ADAMS[1*]

[1]Department of Plant and Microbial Biology, University of California Berkeley, USA
[2]Department of Civil, Architectural and Environmental Engineering, Drexel University, USA

[*]Corresponding email: adamsri@berkeley.edu

Keywords: Microbial exposure, Building characteristics, Electrostatic dust fall collector, Water damage

SUMMARY

Damp and moldy buildings are associated with negative health outcomes, including allergic and respiratory problems. Despite recognizing links between dampness, mold growth, and adverse health outcomes, there remains a lack of modern, objective methods for assessing mold indoors.

This study was designed to improve techniques of sampling indoor microbes and increase our understanding of the effect of water damage on indoor microbial communities. We first compared different methods for passively collecting settled dust for subsequent microbial characterization using modern molecular-based approaches. Next we surveyed 60 residential units in two seasons in which half of the residences have experienced water intrusion. During a three-week period, we took continuous measurements of indoor air parameters and collected dust for assessment of the microbial community. House characteristics were collected by researchers' observation and participant survey.

The expected outcomes of the study are to characterize building features and microbial communities associated with water damage.

INTRODUCTION

Damp and moldy buildings, apart from causing a general decrease in quality of life, are consistently associated with negative health outcomes, including allergic and respiratory problems (Bornehag et al., 2004). Most modern studies of indoor microbiomes have examined non-damaged structures. Microbial organisms in dry indoor environments are derived from three main source pools: outdoor plants, outdoor soil, and residents themselves. For fungi, the dominant outdoor sources are strongly determined by geographic patterns (Amend et al., 2010), varying across spatial and temporal time scales (Adams et al., 2013). However, for indoor bacteria, residents (both human and pets) can be as strong as outdoor sources (Hospodsky et al., 2012). For both fungi and bacteria, these source reservoirs combine to create a rich microbial presence in the built environment.

Airborne microbial communities are highly dynamic (Flannigan et al., 2011). Consequently, understanding household exposure typically requires samples be integrated over long time frames (on the orders of weeks). The collection of settled dust on a designated substrate allows for a collection period of known time duration (Jacobs et al., 2014). However, there is some debate about the best nature of the collection substrate, and it is typically understood that a electrostatic material would better collect dust than a uncharged surface (Frankel et al., 2012). We compared the collection and extraction efficiencies of four materials to assess which material best revealed the air microbiota.

Where water is available in building environments, microbial propagules grow using the nutrient sources provided by building materials, settled dust, and the residents themselves (Samson, 2011). To our knowledge there are no published studies that conducted highly replicated surveys of water-damaged buildings using cultivation-independent techniques. Because of this link between water damage and poor resident health, there is a gap in research that needs to be filled. Drawing on the experiences of residents in the New York City area that were affected by the Superstorm Sandy, this project aims to apply these techniques to better understand changes in microbial composition that result from water intrusion. Our study design will allow us to test the following hypotheses: a) the microbiomes of water-damaged homes are more similar to each other than water-damaged and non-damaged homes in similar locations; b) the quantity of microbial biomass is greater in homes with water damage; c) there is a strong associations between high moisture content in building materials (and air dampness), the quantity of microbial biomass, and the composition of the microbes.

METHODOLOGIES

In the laboratory, four materials were tested as passive collectors of airborne dust. The materials are an empty plastic petri dish (Adams et al., 2013), two brands of electrostatic cleaning wipes (Swiffer & Lyson), and Teftex, a twill fabric composed of polytetrafluoroethylene fiber (Konya et al., 2012). Materials were compared for their extraction efficiency through the input of known quantity of fungal spores directly onto to the material and for their collection efficiency through simultaneous exposure to airborne fungal spores with a nebulizer.

In residential settings, a total 60 units in the Brooklyn and Manhattan areas of New York City were selected for this study aiming for 30 residences with no water damage and 30 residences of varying severity of water damage. The severity of mold was further refined through semi-qualitative techniques developed to detail moisture damage in buildings through non-destructive means. These metrics assessed the size of the damage and took into account whether the known water damage occurred within the individual unit or in the whole building.

The study was designed as a set of two three-week measurement periods, first in the winter of 2015 and again in the summer. Prior to the collection period, measurement information about the building and the unit were collected by means of researchers' observation and occupants' survey. Throughout the measurement period real-time data of indoor air conditions and occupants use of the heating or cooling systems were recorded, and the unit moisture content in exterior walls was determined using invasive and non-invasive measurement technique. For microbial sampling, an empty petri dish served as a dust fall collector (DFC) and was placed in the living room area and one DFC was placed directly outside the home, attached to an exterior wall of the building. Any visible mold present in unit was photographed and size of the mold area was documented. Finally, swab samples of both an interior and exterior door trim (Dunn et al., 2013) were taken, as well as a vacuum sample of floor dust.

RESULTS AND DISCUSSION

Each residence was represented by a minimum of five microbial samples per winter and summer season: two DFCs, vacuumed settled-dust sample, and door trim swabs (plus direct swabs of any visible mold growth in the homes). These time-integrated samples spanned variances in fungal concentrations related to intrusion of outdoor air and the amount and nature of occupants' activity (Ferro et al., 2004; Frankel et al., 2012). All samples were analyzed community composition, and the DFCs were analyzed for microbial biomass. Biomass was assessed through quantitative polymerase chain reaction, which provides a way of determining the initial concentration of target DNA in the sample. This technique has been established as a valuable tool for quantifying biomass in indoor environments (Hospodsky et al., 2012; Hospodsky et al., 2014). Fungal and bacterial communities were characterized using high-throughput sequencing of the internal transcribed spacer (Schoch et al., 2012) and the 16S ribosomal region (Giovannoni et al., 1990), respectively.

We expected that water-damaged residences would look more similar to each other than geography would predict due the growth of those common taxa that effectively utilize the wet, indoor environment. Given the growth that results from available moisture, we also predicted that water-damaged residences would have a greater richness of microbes, as evaluated by species composition, compared to residences without water damage. Other recent studies, relying on cultivation dependent and independent techniques, have shown an increase in diversity in those buildings with water damage compared to those without a history of water intrusion (Pitkäranta et al., 2011).

Aside from microbial composition, change in microbial concentration is another component that we predicted would vary between water-damaged and non-water damaged spaces. Water damage can elevate fungal and bacterial concentrations (Salonen et al., 2007; Adhikari et al., 2014), and the strength of the increase can differ if the mold is visible or hidden in the building envelope (Haas et al., 2007).

CONCLUSIONS

Outcomes of this study will involve recommendations on passive samplers as well as results from the application of those passive samplers in water and non-water damaged residences, further elucidating the microbial landscape in built environments.

ACKNOWLEDGEMENT

The authors would like to acknowledge the help of the Red Hook Initiative for recruiting participants. The study was supported by the Alfred P. Sloan Foundation.

REFERENCES

Adams, R.I., Miletto, M., Taylor, J.W., *et al.* (2013) Dispersal in microbes: Fungi in indoor air are dominated by outdoor air and show dispersal limitation at short distances, *ISME J.*, **7**, 1262-1273.

Adhikari, A., Kettleson, E.M., Vesper, S., *et al.* (2014) Dustborne and airborne Gram-positive and Gram-negative bacteria in high versus low ERMI homes, *Sci Total Environ*, **482**, 92-99.

Amend, A.S., Seifert, K.A., Samson, R., *et al.* (2010) Indoor fungal composition is geographically patterned and more diverse in temperate zones than in the tropics, *Proc. Natl. Acad. Sci. USA*, **107**, 13748-13753.

Bornehag, C.G., Sundell, J., Bonini, S., *et al.* (2004) Dampness in buildings as a risk factor for health effects, EUROEXPO: a multidisciplinary review of the literature (1998-2000) on dampness and mite exposure in buildings and health effects, *Indoor Air*, **14**, 243-257.

Dunn, R.R., Fierer, N., Henley, J.B., *et al.* (2013) Home Life: Factors structuring the bacterial diversity found within and between homes, *PLoS ONE*, **8**, e64133.

Ferro, A.R., Kopperud, R.J. and Hildemann, L.M. (2004) Source strengths for indoor human activities that resuspend particulate matter, *Environ Sci Technol*, **38**, 1759-1764.

Flannigan, B., Samson, R.A. and Miller, J.D., Eds. (2011) *Microorganisms in home and indoor work environments: Diversity, health impacts, investigation and control*, Boca Raton, CRC Press.

Frankel, M., Timm, M., Hansen, E.W., *et al.* (2012) Comparison of sampling methods for the assessment of indoor microbial exposure, *Indoor Air*, **22**, 405-414.

Giovannoni, S.J., Britschgi, T.B., Moyer, C.L., *et al.* (1990) Genetic Diversity in Sargasso Sea Bacterioplankton, *Nature*, **345**, 60-63.

Haas, D., Habib, J., Galler, H., *et al.* (2007) Assessment of indoor air in Austrian apartments with and without visible mold growth, *Atmos. Environ.*, **41**, 5192-5201.

Hospodsky, D., Qian, J., Nazaroff, W.W., *et al.* (2012) Human occupancy as a source of indoor airborne bacteria, *PLoS ONE*, **7**, e34867.

Hospodsky, D., Yamamoto, N., Nazaroff, W.W., *et al.* (2014) Characterizing airborne fungal and bacterial concentrations and emission rates in six occupied children's classrooms, *Indoor Air*, **doi: 10.1111/ina.12172**.

Jacobs, J.H., Krop, E.J.M., Borras-Santos, A., *et al.* (2014) Endotoxin levels in settled airborne dust in European schools: the HITEA school study, *Indoor Air*, **24**, 148-157.

Konya, T., Ewaze, J.O., Urch, B., *et al.* (2012) Development and validation of the TefTex electrostatic wipe for indoor dust sampling. http://sites.utoronto.ca/occmed/jscott/publications/2012_Konya_et_al.pdf

Pitkäranta, M., Meklin, T., Hyvärinen, A., *et al.* (2011) Molecular profiling of fungal communities in moisture damaged buildings before and after remediation — a comparison of culture-dependent and culture-independent methods, *BMC Microbiol.*, **11**, 235.

Salonen, H., Lappalainen, S., Lindroos, O., *et al.* (2007) Fungi and bacteria in mould-damaged and non-damaged office environments in a subarctic climate, *Atmos. Environ.*, **41**, 6797-6807.

Samson, R.A. (2011) Ecology and general characteristics of indoor fungi, In: Adan, O. C. G. and Samson, R. A. Eds. *Fundamentals of mold growth in indoor environments and strategies for healthy living*, Wageningen, The Netherlands, Wageningen Academic Publishers, 101-116.

Schoch, C.L., Seifert, K.A., Huhndorf, S., *et al.* (2012) Nuclear ribosomal internal transcribed spacer (ITS) region as a universal DNA barcode marker for Fungi, *Proc. Natl. Acad. Sci. USA*, **109**, 6241-6246.

FIELD SURVEY OF INDOOR ENVIRONMENT AND ITS EFFECT ON HUMAN HEALTH DURING HEATING PERIOD IN NORTHERN CHINA

Bin CHEN[1,*], Min ZHOU[1], Yu CHEN[1], Xueyan ZHANG[1], Yuhang SHU[1], Yuanbin ZHU[1], Yizhou Sang[2], and Joe R. Zhao[2]

[1] School of Civil Engineering, Dalian University of Technology, Dalian, China
[2]Tri-Y Environmental Research Institute, R & D Centre, Vancouver, Canada

[*]Corresponding email: chenbin@dlut.edu.cn

Keywords: field survey, indoor environment, health risk, CO_2, bedroom

SUMMARY

In an effort to understand the effects of the living conditions on the human health, the study of "Indoor environment parameters and evaluation methods" funded by China's 12th Five-Year Plan was conducted. Based on the investigation of the quality of the indoor environment in more than 70 representative households in five northeast China cities, the potential effect of the indoor environment quality on the human health was evaluated during the heating period. The findings show that more than half of the families have high CO2 concentration in the bedroom which is detrimental to the health.

INTRODUCTION

In recent years, the impact of indoor environment on human health has been widely appreciated. The disease associated with the indoor pollutants increased very rapidly. However, the Chinese environment and health system has not established [1] and thus it is not easy to conduct comprehensive evaluation of the indoor environment on the human health. According to the Nation's third retrospective survey of death cause (year 2004 to 2005), the cause of Chinese death has changed dramatically compared to 30 years ago due to poor infrastructure and other infectious diseases. The disease including cancer, cardiovascular diseases, and respiratory diseases associated with environmental pollution are the top three causes for Chinese death nowadays [2]. A comprehensive understanding of different pollutants and indoor environmental hazards on human health can effectively reduce the generation of indoor pollutants and make early warning for poor indoor environment.

This study conducted the field survey of indoor environment during heating period in northern china and investigated its effect on human health.

.

Field survey

The thermal and humidity environment, acoustic environment, light environment, and air quality of 9 residential houses in five cities (Qiqihar, Harbin, Changchun, Shenyang, and Jinzhou) in Northeast China were investigated. The questionnaires are shown in Table 1 and Table 2. The survey was conducted from March 2014 to April 2014 and November 2014 to January 2015.

Table 1 Field survey questionnaires

Parameters \ Room	**Living room**	**Bedroom**	**Kitchen**	**Washroom**
Light environment	Daytime/night illumination	Daytime/night illumination	Daytime/night illumination	Daytime/night illumination
Acoustic environment	background noise (Daytime/night)	background noise (Daytime/night)	background noise (Daytime/night)	background noise (Daytime/night)
Thermal and humidity environment	Temperature, humidity, air velocity, black globe temperature	Temperature, humidity, air velocity, black globe temperature	Temperature, humidity, air velocity,	Temperature, humidity, air velocity,
Indoor air quality	CO_2, TVOC, Formaldehyde, PM2.5	CO_2, TVOC, Formaldehyde, PM2.5	CO_2, TVOC, Formaldehyde, PM2.5	–

Table 2 Residential and occupants questionnaires

	Parameters
Basic information	Construction information, personal composition, lifestyles
Room	Potential health risk
Health condition	Mental, body, disease and health care

RESULTS AND DISCUSSION

As shown in Table 3 and Figure 1, the residential house has a very high CO_2 concentration based on the field survey of Shenyang, Changchun and Harbin during the heating period from March 2014 to April 2014. The bedroom has the highest CO_2 concentration with a maximum concentration of 5437 ppm. The CO_2 concentration in bedroom increased during the night till 6 am of following day. 53% bedrooms have a CO_2 concentration higher than 3000 ppm between 22:00 to 6:00.

Table 3 CO_2 concentration of the residential house in Shenyang, Changchun and Harbin

City	Room	Number of tested room	Result (PPM)	Arithmetic mean (PPM)	Standard Deviation
Shenyang (SY)	Living room	8	331-3335	986	348
	Bed room	8	373-5186	1089	753
	Kitchen	8	323-2860	1007	307
Changchun (CC)	Living room	10	439-5818	1446	495
	Bed room	8	361-5393	1654	792
	Kitchen	7	535-4308	1457	546
Harbin (HRB)	Living room	8	315-5157	1225	761
	Bed room	10	304-5437	1513	887
	Kitchen	5	314-2310	911	491

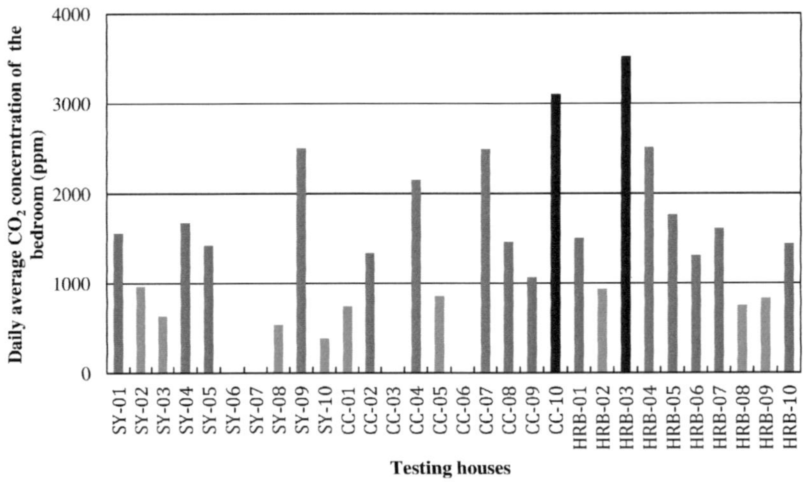

Figure 1. CO_2 concentration in bedroom in three cities

Though the CO_2 concentration in 18 residential houses of Qiqihar and Harbin was not over the limit during November 2014 to December 2014, the potential health rick caused by the high CO_2 concentration still exists. The result is shown in Figure 2.

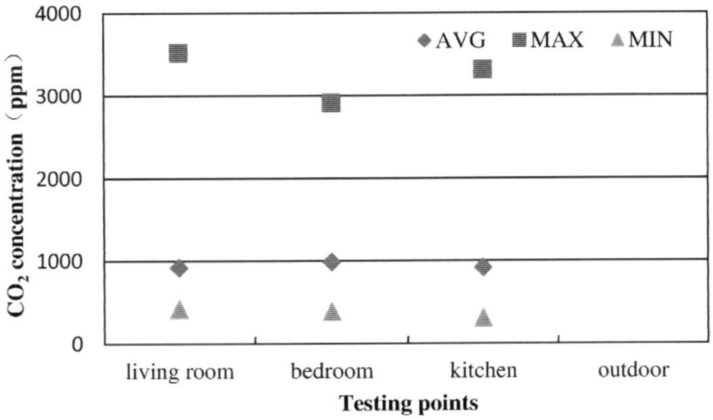

(a) Harbin

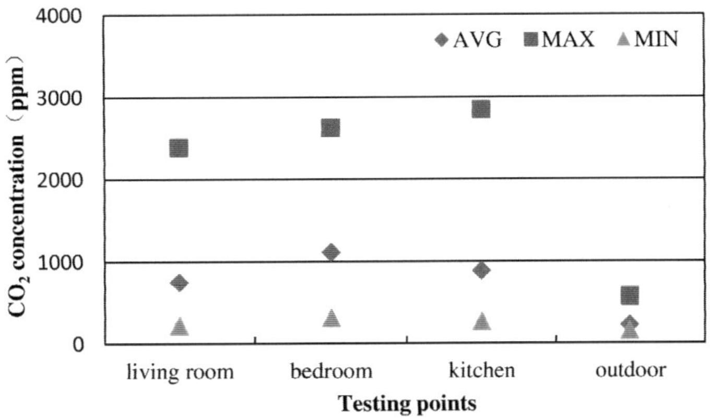

(b) Qiqihar

Figure 2. CO_2 concentration in the residential houses of Qiqihar and Harbin (November to December 2014)

CONCLUSIONS

(1) Due to the high temperature difference (30℃) between indoor and outdoor, it is not feasible to open a window to improve the indoor air quality.
(2) The bedroom has a high CO_2 concentration during 22:00 to 6:00. 53% bedrooms have a CO_2 concentration higher than 3000 ppm between 22:00 to 6:00.

ACKNOWLEDGEMENT

This research was supported by Key Projects in the National Science & Technology Pillar Program during the Twelfth Five-year Plan Periods (No. 2012BAJ02B05).

REFERENCES

[1] Xiaoli Dua, Yi Li, Xiuge Zhao, etc. Environmental Health Standard Framework in China during the Twelfth Five-Year Plan. Journal of Environment Engineering

Technology,2011(03):210-214.
[2] Zhu Chen. The third national mortality retrospective sampling survey[M], Peking Union Medical College Press, 2008:48~49

Modeling the Dispersion of Evaporating Spittle Droplet in a Typical Waiting Room of a Hospital

Qiuyue WANG [1,*], Shinsuke KATO[1]

[1]Department of Architecture, The University of Tokyo, Tokyo, Japan
[2] Department of Architecture, The University of Tokyo, Tokyo, Japan

* wangqy@iis.u-tokyo.ac.jp

Keywords: Droplet evaporation, Computational fluid dynamic(CFD), Hopital, Infection control

SUMMARY

To bring the epidemic under control, it is necessary to understand the transmission path of infectious disease in buildings. CFD simulation was conducted in this study with a Lagrangian particle tracking method to predict the distribution and evaporation of sputum droplets with different initial diameters in a typical waiting room of a hospital. It was found that 0.3μm droplets evaporate very fast and stay only around people's nose. 3μm droplets can travel a little bit longer. The paths of 100μm and 300μm droplets are significantly affected by gravity and inertia and tend to travel downward. In addition, to validate the reliability of our model, simulation under the condition of RH of 100% was conducted as a comparison, and the result shows the evaporation rate of water droplets was zero. Thus it can be concluded that the initial diameter can influence the dispersion and evaporation process of water droplets significantly.

INTRODUCTION

The outbreak of infectious disease, such as influenza, and severe respiratory syndrome, has caused lots of death around the world. In occupied enclosed public places, such as hospitals and offices, people may get infectious diseases more easily. Waiting rooms of hospitals bear the most important research value for infection control due to its higher risk of infection than other spaces. Many studies have been conducted to simulate the airflow patterns, contaminant concentrations, and assessment of probability risk and fate of expiratory droplets. Wells pointed out that droplets smaller than 100μm are inclined to evaporate to droplet nuclei, which may remain airborne for several hours or even several days(Wells W. F. 1934). It was also found that it is possible for coughed air and aerosols to be inhaled by victim who is sitting opposite to the patient, and the dispersion and deposition characteristics are highly related to the droplet size(Zhu S et al., 2006). In terms of the simulation method, Lagrangian approach emphasizes the individual behaviour of each particle trajectories and more accurate than Eulerian method in predicting pollutant distribution (Z. Zhang, Q. Chen., 2007).

Based on the configuration of the real waiting room, the present study establishes a model to simulate the dispersion of evaporating sputum droplets under different size scenarios considering gravitational settling and latent heat released from human.

METHODOLOGIES

As shown in Fig.1, a waiting room of hospital with a dimension of 3m × 4m × 3m (L × W × H) was used in the CFD simulation. It contained 6 sitting people and the distance between the mouths of opposite people was 1.2m. The position of supply opening was in the ceiling right over the front door, and the exhaust opening was located at the floor level of the wall.
The boundary conditions of the simulations are shown in Table.1. The sitting height of all people was 1.1m. The patient was assumed to be located in the middle of the waiting room (Fig.1). The total air change rate was approximately 6ACH. 50000 liquid water droplets with a velocity of 11.7m/s were generated from the patient's nose per second, which could be representative of the coughing of a person. Droplets sizes of 0.3μm, 3μm, 30μm, and 300μm that could represent different particle dispersion and deposition process were studied. The relative humidity in the room was 50%. Since all the people were assumed to cough or inhale constantly, the simulation was conducted in steady-state conditions. The numerical simulation was achieved by applying the k-ε turbulent model and Lagrangian particle tracking method.
The results were measured in vertical cross section. The distance between the vertical cross section and front wall was 2m.

Table 1. Boundary Conditions

Initial conditions	Temperature: 300K; Pressure: 101325Pa Velocity: 0m/s; Absolute humidity: 0.009kg/kg'
Supply opening	Temperature: 290K; Velocity: 0.87m/s Dimension: 400mm×400mm; Turbulence Intensity: 0.111 Turbulence Length Scale: 0.028m Absolute humidity: 0.006kg/kg'
Exhaust opening	Flow-split outlet
Sitting people	Heat flux of human body: 23W/m2
Mouth opening of coughing person	Velocity:11.7m/s; Temperature: 305K Turbulence Intensity of patient: 0.1002 Turbulence Length Scale: 0.0014 m Absolute humidity: 0.015kg/kg'
Nose opening of others	Velocity: 0.35m/s; r=0.01m
Human body	Heat flux of human: 23W/m²; Human size: 1.1m × 0.2m ×0.5
Grids	1,024,406 cells
Physics model	Realizable k-ε model, Lagrangian model
Software	STAR-CCM+ 8.06.007

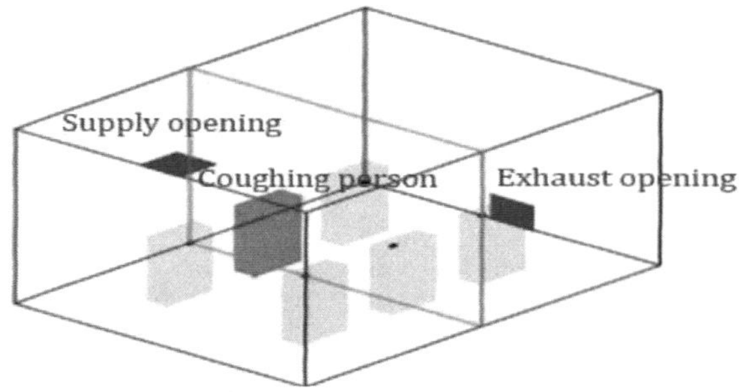

Figure 1. The geometry of a waiting room in the hospital

The quasi-steady single-component droplet evaporation model assumes droplets to internally homogeneous, consisting of a single liquid component. The rate of change of droplet mass due to quasi-steady evaporation $\dot{m}_d$ can be written (Equation 2 can be deduced from Equation 1)

$$\dot{m}_d = \rho_d \frac{d}{dt}\left(\frac{\pi}{6} d_d^3\right) = -A_s h_m(\rho_{v,s} - \rho_{v,\infty}) \quad (1)$$

$$\dot{m}_d = -\frac{\rho_{v,\infty} D_v Sh_\rho}{D_p} A_s \ln(1 + B) \quad (2)$$

where ρ_d is the droplet density, dd is the droplet diameter, $\rho_{v,\infty}$ is the density of air, $\rho_{v,s}$is the vapor density at the droplet surface, A_s is the surface area of the droplet, Dv is molecular diffusivity of the vapor, Sh_ρ is the Sherwood number, B is the Spalding transfer number.

$$\frac{d(m_d\overrightarrow{u_d})}{dt} = F_D(\vec{u} - \overrightarrow{u_d}) + \frac{\vec{g}(\rho_d - \rho_{v,\infty})}{\rho_d} + \overrightarrow{F_a} \quad (3)$$

The left-hand side of the equation represents the inertial force per unit mass (ms^{-2}), where $\overrightarrow{u_d}$ is the droplet velocity vector. The first term on the right-hand side is the drag term, where F_D is the inverse of relaxation time (s^{-1}) and $\vec{u}$ is the air velocity; the second term is the additional force, and $\overrightarrow{F_a}$ stands for additional forces that is assumed to be zero in this model. The Discrete Random Walk (DRW) model can model the turbulence dispersion deposition

$$u_i' = \zeta\sqrt{\overline{u_i'^2}} = \zeta\sqrt{\frac{2k}{3}} \quad (4)$$

where $\vec{u}$ is the fluvctuation velocity, ζ is a Gaussian random number, k is the turbulent kinetic energy in the RANS region and the sub grid-scale turbulence kinetic energy in the LES region.

Thus, the actual airflow velocity $\vec{u}$ in each eddy

$$\vec{u} = \overrightarrow{u_i} + u_i' \quad (5)$$

where $\overrightarrow{u_i}$ is the local Reynolds-Averaged mean velocity, u_i' is the eddy fluctuation velocity.

The particle concentration can be calculated by the droplet source in cell (PSI-C) method

$$C_j = \frac{M\sum_{j=1}^{m} dt_{(i,j)}}{V_j} \quad (6)$$

where M is the number flow rate of each trajectory, C is the mean particle concentration in a cell, V is the volume of a computational cell for particles, dt is the particle residence time, and subscript (i, j) represents the ith trajectory and the jth cell, respectively. (Z. Zhang, Q. Chen., 2007).

RESULTS AND DISCUSSION

Characteristics of size distribution and concentration field in the vertical section are shown in Fig.2 and Fig.3.

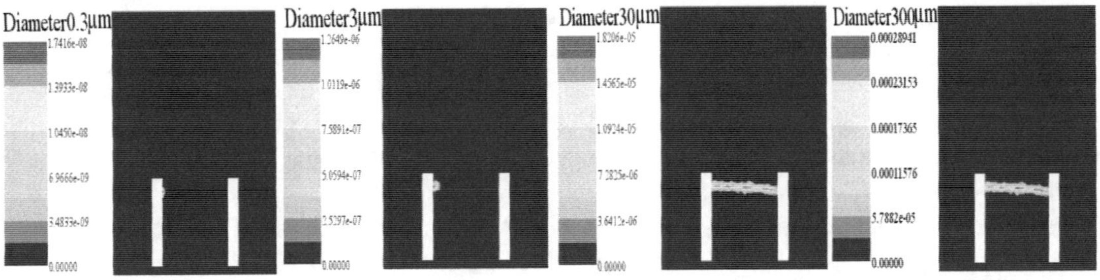

Figure 2. Size distribution of droplets generated by a patient - vertical section plane

The size distribution in Fig.2 shows that for droplets smaller than 3µm, they can evaporate very fast that dispersion was limited to a very small area near the mouth. They can evaporate completely before reaching to the opposite person. For droplets of 100μm and 300μm, the dispersion feature is dominant due to their comparatively long evaporation time and small settling velocity. And due to gravity, the shape of trajectory of droplets is curved. They are able to reach to the clothes of the opposite person and the distribution tends to be at the lower part of the room. Besides, larger droplets are not easy to evaporate immediately.

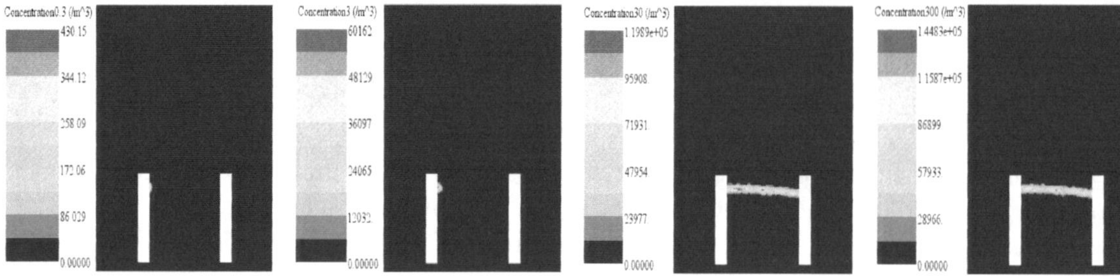

Figure 3. Concentration field of droplets generated by a patient - vertical section plane

Fig.3 illustrates the concentration field of evaporating droplets. Since 5000 number of droplets are released per second, the droplet phase is like a continuum under steady state conditions. The average concentration of 0.3μm, 3μm, 30μm and 300μm droplets on the height of 1m are 9×10^{-2}no./m^3, 23no./m^3, 418no./m^3, and 461no./m^3 , respectively, which means the average concentration of 30μm and 300μm is as nearly 30times as much as 3μm droplets and 5000 times as much as 0.3μm droplets. The concentration of residual droplets of each size on the body of the person who sits opposite to the patient are 0no./m^3, 0no./m^{3,} 200no./m^{3,} and 263no./m^3, which indicates the infectious risk is higher as the droplet diameter increases.

CONCLUSIONS

In this study, evaporation and transmission characteristics of water droplets with 4 different sizes were performed. It considered cough and inhalation process as a continuous and steady process to simplify the model. The results showed that large droplets (30μm and 300μm) are not capable of evaporating completely as fast as small droplet, and can reach to the opposite person before evaporating to droplet nuclei. Thus the path of larger droplets has to be blocked to prevent the spread of disease.

In our next simulation model, the irregular respiration process will be investigated. Besides, air purifiers and physical barriers like partition, due to their beneficial for infectious control, will be arranged in our future simulation model. In addition, an experimental method is necessary to be conducted to make the simulation results more reliable.

REFERENCES

Wells W. F. 1934 On air-borne infection. Study II. Droplets and droplet nuclei. Am. J. Hyg. 20, 611–618.

Zhu S, Kato S, Yang J-H. 2006 Study on transport characteristics of saliva droplets produced by coughing in a calm indoor environment. Build. Environ. 41, 1691–1702.

Z. Zhang, Q. Chen, 2007 Comparison of the Eulerian and Lagrangian methods for predicting particle transport in enclosed spaces. Atmospheric Environment Volume. 41, Issue 25

Lei Zhang, Yuguo Li. Dispersion of coughed droplets in a fully occupied high-speed rail cabin. Building and Environment Volume. Volume 47, January 2012, Pages 58－66

ASSESSMENT OF THERMAL ENVIRONMENT DURING KITCHEN WORK

Kunihito TOBITA[1,*], Yoshihito KURAZUMI[2], Yoshiaki YAMATO[3], Kenta FUKAGAWA[4]

[1]College of Sustainable System Sciences, Osaka Prefecture University, Osaka, Japan
[2]School of Life Studies, Sugiyama Jogakuen University, Aichi, Japan
[3]Department of Architecture and Structural Engineering, Kure National College of Technology, Hiroshima, Japan
[4]Department of Architecture, Kyushu Sangyo University, Fukuoka, Japan

[*]Corresponding email: tobita@hs.osakafu-u.ac.jp

Keywords: Thermal comfort, Residences, Kitchen, WBGT, Heat disorders

SUMMARY

Many older patients suffered from heat illness in their home, especially in the kitchen. However, the relationship between kitchen thermal environment and thermal sensation is not clarified. We conducted the survey and field measurement about thermal environment during the summer in Japan. Most of the residencies had the moderate risk of heat disorders and some had high risk of that. The correlation between air temperature and thermal sensation vote is moderate, but the correlation between wet bulb globe temperature and thermal sensation vote is low. For prevention of the heat disorder occupant should refer to wet bulb globe temperature. Because occupants cannot feel wet bulb globe temperature correctly, it is necessary to visualize wet bulb globe temperature for them.

INTRODUCTION

The heat wave is well-known as a killer in diverse parts of the world. The summer heat wave of 2003 led to 70,000 excess deaths (Robine et al., 2008). During 1979-1991, a total of 5224 deaths in the United States were attributed to excessive heat (Hawkins-Bell, 1994). In Japan, there were 4,053 deaths of heat disorder from 1959 to 1999 (Hoshi and Inaba, 2002). Regarding the occurrence of place, 586 cases occurred at home. Markedly high mortality rate was found among persons over 65 years old. And mortality rate for women was less than that for men. There were 152 patients transported by ambulance during July and August between 1995 and 1999 in Yamanashi Prefecture (Iriki, 2000). The graph of the age group of heat disorder patients in Yamanashi and Hokkaido showed two peaks at a young (teens) and an aged (sixties-eighties) generation (Iriki and Hashimoto, 2006). Many older patients suffered from heat illness in their home (Iwata et al., 2008), especially in the kitchen (Shibata et al., 2010). Although a large number of studies have been made on thermal comfort (Cena et al., 1990; Nicol et al., 1999; Tobita et al., 2007), little is known about relationship between kitchen thermal environment and thermal sensation. The present study aims to clarify the assessment of kitchen thermal environment in relation to occupant responses during summer.

METHODOLOGIES

The survey in this study was performed in the summer of 2011. Ten different residences in the Fukuoka and the Kumamoto were collected in the survey and field measurement. Three residences were built of wood, and seven residences were reinforced concrete apartment. The mean value of floor area was 129 m^2.

The climate in most major area, including Fukuoka and Kumamoto, is temperate to sub-tropic and consists of four seasons. The mean daily temperature was in the range of 24.2-86.7°F from August to September. Meanwhile, the mean daily relative humidity (RH) was in the range of 62-93%.

Micro-meteorological measurements were selected three basic environmental parameters known to calculate wet bulb globe temperature (WBGT) namely air temperature, humidity and radiant environment, could be accurately measured. Air temperature, the humidity and the globe temperature were measured near the stove in the kitchen.

The responses recorded for this survey addressed the themes of thermal sensation and thermal comfort on a 7 point semantic differential scale. On the both scale, we asked the occupants "evaluate the thermal environmental conditions when you are preparing meal and washing the dishes". The thermal sensation scale is labeled "cold"(1) and "hot"(7) only. A comfort questions asked subjects to rate the comfort of kitchen, ranging from "uncomfortable"(1) to "comfortable"(7).

RESULTS AND DISCUSSION

The mean air and globe temperature is 82.9 and 83.3 °F, respectively, in the kitchen. Relative humidity was 70.9%. The statistical summaries of WBGT are separated into three periods of time and are shown in Figure 1 with the heat disorders risk level (JSB, 2013). Most of the WBGT data fell inside the moderate risk zone. Some samples are included to the high risk zone at lunch and dinner time. The WBGT of K2 at dinner had a time of over 87.8 °F. In other words, occupants stayed in the kitchen which had the very high risk of heat disorders. While occupants do the sociocultural activity such as 'Spending time with family', 'Reading books' and so on, the importance score of thermal comfort tend to be high. However, it is lower during occupants do the housework (Tobita et al., 2012). Because housework is often done by only one person in Japan, so they usually save the money. As a result, occupants are forced to prepare meal or wash the dishes in the hot environment at the kitchen.

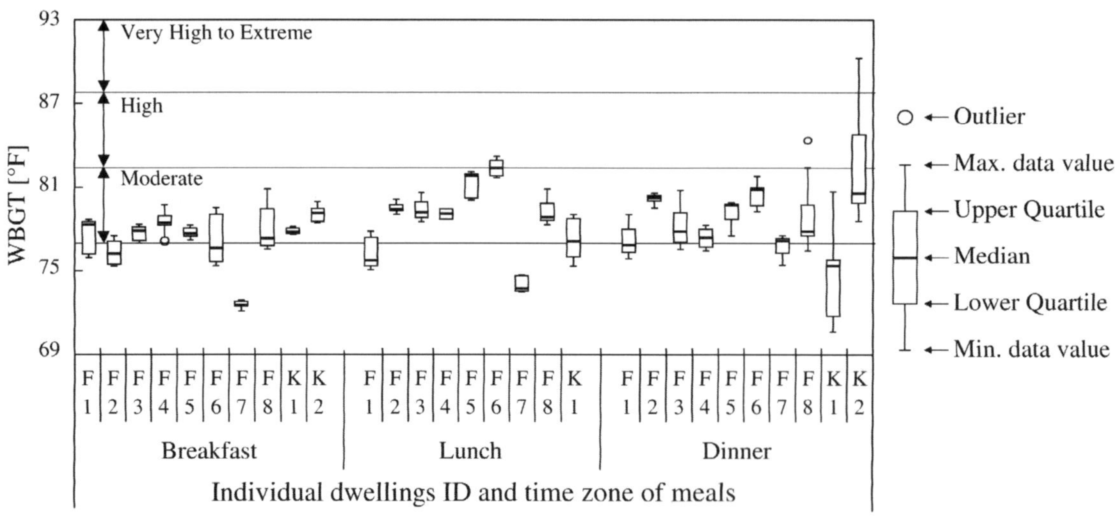

Figure 1. Statistical summary of WBGT in the kitchen.

Figure 2 shows the scatter plot of thermal sensation vote (TSV) versus air temperature and WBGT. A total of 57 responses to the thermal sensation survey were collected. Since this survey is conducted during the summer, there is no occupant votes laid on the cold side of the thermal sensation voting scale. Many studies have shown that TSV is generally in proportion to temperature indexes (e.g. Soebarto and Bennetts, 2014; Kurazumi et al., 2011). Therefore we applied the linearity regression model in Figure 2. The results show a moderate correlation between air temperature and TSV (r = 0.625). On the other hand, the correlation between WBGT and TSV is low (r = 0.326).

Scatter plots in Figure 3 show the correlation between thermal comfort vote (TCV) and air temperature and between TCV and WBGT. TCV increased with the decrease in air temperature and WBGT for the entire sample.

These results indicate that the occupants are more familiar with estimating their degree of warmth from temperature than WBGT. But the WBGT is better explanatory variable of the incidence of the heat disorders than air temperature (Nakai et al., 1990). For prevention of the heat disorder occupant should refer to WBGT. Because occupants cannot feel WBGT correctly, it is necessary to visualize WBGT for them.

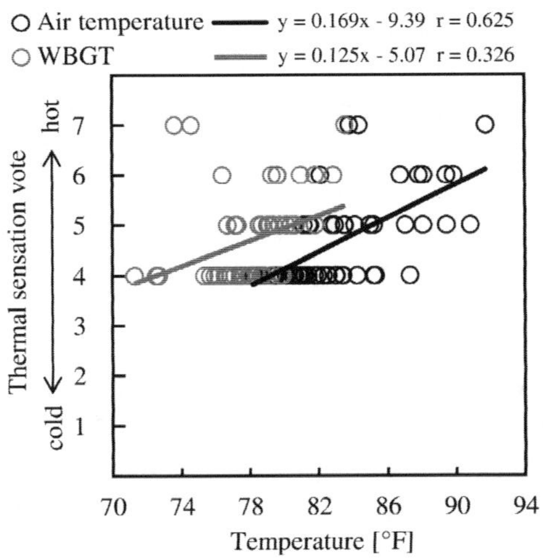

Figure 2. Relationship between TSV and temperature index.

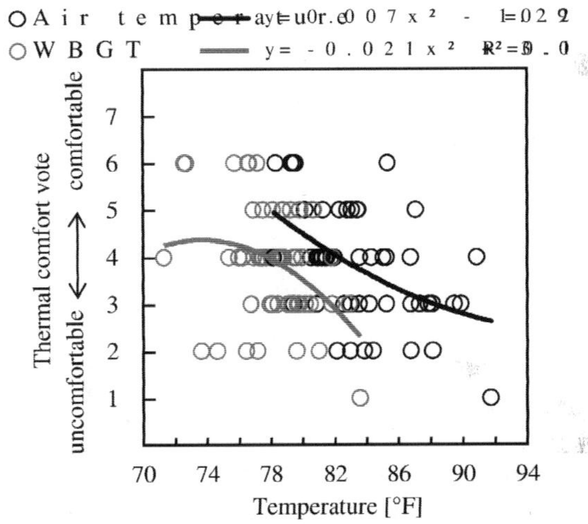

Figure 3. Relationship between TCV and temperature index.

CONCLUSIONS

This study investigates assessment of thermal environments in the kitchen. We conducted this survey during the summer of 2011 in Fukuoka and Kumamoto, Japan. Most of the residencies had the moderate risk of heat disorders and some had high risk of that. The correlation between air temperature and TSV is moderate, but the correlation between WBGT and TSV is low.

ACKNOWLEDGMENTS

We thank Riho Uchino and Hiroki Obayashi (Ariake National college of Technology) for their kind help in this study. This study was supported in part by Grant-in-Aid for Scientific Research (no. 23700875) from the Japan Ministry of Education, Science and Culture.

REFERENCES

Cena K, et al. (1990). A practical approach to thermal comfort surveys in homes and offices: discussion of methods and concern, *ASHRAE Transactions*, **96** (1), pp.853-858.

Hawkins-Bell L (1994). Heat-Related Deaths — Philadelphia and United States, 1993–1994, *MNWR (Morbidity and Mortality Weekly Report)*, **43** (25), pp.453-455.

Hoshi A and Inaba Y (2002). Mortality rate of heat disorder analysed by place of occurrence using vital statistics in Japan, *Japanese Journal of Biometeorology*, **39** (1,2), pp.37-46.

Iriki M (2000). Heat disorder in Yamanashi Prefecture during the summer from 1995 to 1999, *Japanese Journal of Biometeorology*, **37** (2), pp.63-72.

Iriki M and Hashimoto M (2006). Regional difference in the incidence characteristics of heat disorder – Yamanashi & Hokkaido -, *Japanese Journal of Biometeorology*, **37** (2), pp.63-72.

JSB (Japanese Society of Biometeorology) (2013). Guidelines for the Prevention of Heat Disorder in Daily Life, JSB.

Kurazumi Y., Tsuchikawa T., et al. (2011). Evaluation of enhanced conduction-corrected modified effective temperature ETFe as the outdoor thermal environment evaluation index, *Energy and Buildings*, **43**(10), pp.2926-2938.

Nakai S, et al. (1990). Environmental temperature of an athletic field in japan and comparison between wet bulb temperature and WBGT, *The Japanese Journal of Physical Fitness and Sports Medicine*, **39**, pp.120-125.

Nicol JF, Iftikhar A, et al. (1999). Climatic variations in comfortable temperatures: the Pakistan projects, *Energy and Buildings*, **30**, pp.261-279.

Robine JM, Cheung SL, et al. (2008). Death toll exceeded 70,000 in Europe during the summer of 2003, *Comptes Rendus Biologies*, **331** (2), pp.171-178.

Shibata Y, Tobita K, et al. (2010). Actual conditions of the recognition of heat disorders in the residential places and preventive measures for the elderly, *Japanese Journal of Biometeorology*, **47** (2), pp.119-129.

Soebarto V. and Bennetts Helen (2014). Thermal comfort and occupant responses during summer in a low to middle income housing development in South Australia, *Building and Environment*, **75**, pp.19-29.

Tobita K, et al. (2007). Calculation of neutral temperature and acceptable range by the field study of houses in Kansai area, Japan, in winter, *Journal of environmental engineering (Transactions of AIJ)*, (614), pp.71-77.

Tobita K, et al. (2012). Thermal environment in relation to activities of daily living during summer, 16th PAQS congress, Bandar Seri Begawan, Negara Brunei Darussalam, July 7 - 10, Paper Reference No36 S_SC_KT.

ODOR THRESHOLDS FOR 2,4,6 -TRICHLOROANISOLE REVIEWED IN THE CONTEXT OF INDOOR AIR QUALITY

Stephanie A JURAN[1,*], Steven NORDIN[3], Gunnar JOHANSON[1], Johnny C LORENTZEN[1,2]

[1]Karolinska Institutet, Institute of Environmental Medicine, Stockholm, Sweden
[2]Eurofins Pegasuslab AB, Uppsala, Sweden
[3]Umeå University, Department of Psychology, Umeå, Sweden

[*]Corresponding email: Stephanie.Anja.Juran@ki.se

Keywords: Building Related Illness, Sick Building Syndrome, Psychology, Olfaction

SUMMARY

Chloroanisoles have received minimal attention in research on indoor air quality (IAQ) although they can affect comfort and maybe also health when levels are sufficiently high to evoke malodor. Here, we review the literature for odor thresholds (OT) for 2,4,6-trichloroanisole to estimate at which concentrations air is perceived as malodorous, and discuss the caveats in measuring and comparing OTs. In general, reported OTs for 2,4,6-trichloroanisole are extremely low but still in the range measured in indoor air of problem buildings, as we reported recently (Lorentzen et. al. , 2015). However, OTs vary considerably due to methodological constraints such as different methods of stimulus presentation. Only one study (not published in a peer reviewed journal) reports OTs in air, and the data suggest that 2,4,6-trichloroanisole can be among the most potent microbial odors. Additional studies are therefore warranted. It will also be important in future studies to investigate the perceived odor quality of CAs in the context of IAQ problems with dampness and mold.

INTRODUCTION

Indoor air quality (IAQ) in spaces for human occupancy may be defined as the extent to which human requirements are met - the requirements being that the air does not have a negative effect on health and is perceived as being acceptable and having a positive impact on performance and productivity (Fanger, 2006). Malodor can obviously cause discomfort, but may also affect health through psychobiological mechanisms. We have previously suggested that mold odor may cause adverse health effects in persons perceiving it as "dangerous," and have forwarded chloroanisoles (CAs) as model volatiles for mold odor (Lorentzen et. al. , 2012). CAs are formed in damp conditions by microbial metabolism of chlorophenols that were used world-wide to preserve wood for various purposes, including building construction. Analysis of CAs in indoor air and building materials were originally reported from both Sweden and Germany in relation to preserved wood leading to problems with mold odor (PegasusLab, 1999), and musty odor (Gunschera et. al. , 2004), respectively. Recently, we described sensitive methods to detect CAs in indoor air (Lorentzen et. al. , 2014), and then performed a study that identified CAs in a large number of Swedish buildings with odor problems. The recorded air levels of CAs were low, with a median value of 0.33 ppt for 2,4,6-trichloroanisole (2,4,6-TCA, Lorentzen et. al. , 2015). This raises the question whether the low levels are still sufficiently high to be perceived by the individual through olfaction.

In this paper, we identify and critically review available data on OTs for 2,4,6-TCA which is well known to degrade odor and taste of a wide range of foods (e.g. potatoes, chicken products) and beverages (e.g. coffee, wine, water).

METHODOLOGIES

Peer reviewed journals and other papers were identified via searches in Pubmed, Web of Science and Google, searching for chloroanisol*, odor*, olfact* and smell* and reviewed for information on OTs of 2,4,6-TCA. Retrieved data were discussed in relation to potential caveats in OT measurements and comparisons.

RESULTS AND DISCUSSION

Data extraction from relevant studies reveal that the verbal descriptors for 2,4,6-TCA given by the test panel point to odor quality impairments, mainly "musty" but in one case also "moldy" (see Table 1). In general, the reported OTs of 2,4,6-TCA are extremely low, and compared to data compiled for other microbial volatile organic compounds (MVOCs), 2,4,6-TCA would be among the most potent microbial odors (Korpi et. al. , 2009). On the other hand, the reported OTs vary considerably between 0.03 ppt and 103.84 ppt, which reflects variability in size and composition of the test panels ("Panelist" in Table 1), as well as different methodologies ("Medium" and "Cognitive task" in Table 1). Concerns resulting from these differences are discussed in the following.

Table 1. Odor thresholds (OTs) reported for 2,4,6-TCA.

Study	Verbal descriptor	OT [ppt]	Medium	Panelists	Cognitive task
Curtis et. al., 1972	Musty, earthy	0.03[a]	Water	23	n.n.
Griffiths and Fenwick, 1977	Musty	0.03[a]	Water	n.n.	Intensity rating, OT
Macku et. al., 2009	Musty, dirty, moldy, ashtray, haloanisole,	1 (OT)[b] 5 (ID)	Cork-ethanol extract	4 experts	Detection Identification
Strube, 2010	Corky, musty	0.58[c]	Air	5 trained	"Highest dilution perceived"
Young et. al., 1996	Dusty, vegetable, musty, earthy, rotten	103.84 [a]	Water	6-10 trained	Detection

n.n. = not named, ppt = parts per trillion, [a] = in water, [b] = in ethanol, [c] = in air, ID = odor identification

Experimental design: Most articles listed in Table 1 provide little information about experimental design or calculation of OTs. Only two papers describe the sample presentation procedure and only Young et al. (1996) used an established psychophysical design for determining the detection threshold by applying odorless "blank" stimuli (solvent only) and one weak "target" stimulus. After successful target detection, a weaker target stimulus concentration will be presented together with the blank as next stimulus pair. Following unsuccessful target detection, stronger target stimulus concentrations are given. Such a "staircase procedure" is used to approximate individual OTs, calculated as mean level of successful target detection. The work by Macku and colleagues (2009) uses a different design that strictly speaking does not deliver psychophysical thresholds. Proper selection and description of study design is essential both for obtaining reliable and valid OTs and for drawing reasonable conclusions when comparing different OTs.

Test panel size: "Panelist" describes the individuals investigated for OT. Table 1 shows that test panel size varies from 4 to 23 individuals and is not given in one study. Panels usually comprise small groups of trained individuals or industrial experts. This constitutes a problem, since it has been shown that the variation between time points, but within an individual, can be assumed to be even larger than the average variation between individuals (Stevens et. al. , 1988). Hence OTs should best be investigated in large samples of the target population.

Method of stimulus presentation: The column "Medium" in Table 1 shows that odors have been presented in gas phase or in liquid phase, where the gaseous headspace above a fluid odorant dilution is inhaled. In the latter, headspace concentrations vary with the solvent used (alcohol, water) and with substance volatility and polarity so that gas phase presentation is considered a more concise method. The reported OTs are likely concentrations in the medium used, however, none of the studies indicate how these OTs can be recalculated to concentration in the headspace.

Furthermore, in some studies odorous liquids (wine, cork) are used to dilute target odorants (2,4,6-TCA), as compared to an odorless solvent (water). The aim of such detection testing is usually to estimate the level of 2,4,6-TCA contamination at which a given food will be perceived as spoiled by the consumer. However, such thresholds cannot be compared to traditional odor thresholds for two reasons:

First, a classical threshold paradigm for discrimination between an odorless blank and a target stimulus, requires a simple yes/no decision based on odor *intensity*. This task is getting much more complex when the odorless blank instead conveys an own odor (e.g. wine) and, as a consequence, the following working memory processes are required:

- Perceive and memorize odor I (wine),
- Perceive and memorize odor II (wine + 2,4,6-TCA),
- Compare both memory representations, and
- Decide if they are same or different.

Whereas OT performance mainly relies on perceptual abilities (e.g. sensitivity of the olfactory receptors and peripheral nervous system), odor discrimination is strongly influenced by cognitive abilities such as memory (Hedner et. al. , 2010).

Second, discrimination between two odors (wine, wine + 2,4,6-TCA) is based not on odor *intensities*, but mainly on odor *qualities* (e.g. wine vs. mold) and the perceived quality of an odor mixture is strongly influenced by the intensities of its components (Olsson and Cain, 2000). Thus, the resulting OT for 2,4,6-TCA in wine will be relate to the intensity of the pure wine odor, resulting in different 2,4,6-TCA thresholds when tested in different wines.

CONCLUSIONS

Reported OTs of 2,4,6-TCA show substantial variation but are generally extremely low. They are still in the range measured in problem buildings, which makes 2,4,6-TCA a potential contributor to IAQ problems. Additional studies are needed to determine the OT directly in air for larger groups of persons with detailed method description. It will also be interesting to investigate if mold will be a common odor descriptor for 2,4,6-TCA when in the context of problems with indoor dampness and mold.

ACKNOWLEDGEMENT

This work was supported by a grant from the Swedish Research Council Formas for the project *"Occurrence and health effects of malodorous chloroanisoles from wood preservatives"* (242-2013-1186).

REFERENCES

Curtis, R.F., Land, D.G., Griffiths, N.M., Gee, M., Robinson, D., Peel, J.L., Dennis, C. and Gee, J.M. (1972) 2,3,4,6-Tetrachloroanisole Association with Musty Taint in Chickens and Microbiological Formation, *Nature*, **235**, 223-224.

Fanger, O.P. (2006) What is IAQ?, *Indoor Air*, **16**, 328-334.

Griffiths, N.M. and Fenwick, G.R. (1977) Odour properties of chloroanisoles - effects of replacing chloro- by methyl groups, *Chemical Senses*, **2**, 487-491.

Gunschera, J., Fuhrmann, F., Salthammer, T., Schulze, A. and Uhde, E. (2004) Formation and emission of chloroanisoles as indoor pollutants, *Environmental science and pollution research international*, **11**, 147-151.

Hedner, M., Larsson, M., Arnold, N., Zucco, G.M. and Hummel, T. (2010) Cognitive factors in odor detection, odor discrimination, and odor identification tasks, *J Clin Exp Neuropsychol*, **32**, 1062-1067.

Korpi, A., Jarnberg, J. and Pasanen, A.L. (2009) Microbial volatile organic compounds, *Crit Rev Toxicol*, **39**, 139-193.

Lorentzen, J.C., Juran, S. and Johanson, G. (2012) Chloroanisoles in relation to indoor air quality and health, *SWESIAQ Newsletter*, Vol. 21, Supplement.

Lorentzen, J.C., Juran, S.A., Nilsson, M., Nordin, S. and Johanson, G. (2015) Chloroanisoles may explain mold odor and represent a major indoor environment problem in Sweden, *Indoor Air*, **In press**.

Lorentzen, J.C., Nilsson, M. and Johanson, G. (2014) Chloroanisoles represent a common indoor air quality problem in Sweden - sensitive methods to detect the malodorous chemicals in air and materials, Vol. 2014, pp. Paper HP1147.

Macku, C., Gonzalez, L., Schleussner, C., Mesquita, A.C., Herwatt, J.W., Kirch, L.C. and Schwartz, R.J. (2009) Sensory Screening for Large-Format Natural Corks by "Dry Soak" Testing and Its Correlation to Headspace Solid-Phase Microextraction (SPME) Gas Chromatography/Mass Spectrometry (GC/MS) Releasable Trichloroanisole (TCA) Analysis, *Journal of Agricultural and Food Chemistry*, **57**, 7962-7968.

Olsson, M.J. and Cain, W.S. (2000) Psychometrics of Odor Quality Discrimination: Method for Threshold Determination, *Chemical Senses*, **25**, 493-499.

Pegasuslab (1999) Mögellukt är inte alltid mögel, *Meddelande Nr. 10*, http://www.eurofins.se/tjanster/inomhusmiljo/kunskapsbank/publikationer.html.

Stevens, J.C., Cain, W.S. and Burke, R.J. (1988) Variability of olfactory thresholds, *Chemical Senses*, **13**, 643-653.

Strube, A.B., A. (2010) The influence of chemical structure on odour qualities and odour potencies in chloro-organic substances, *Expression of Multidisciplinary Flavour Science.* , Interlaken, Switzerland, 486-489.

Young, W.F., Horth, H., Crane, R., Ogden, T. and Arnott, M. (1996) Taste and odour threshold concentrations of potential potable water contaminants, *Water Research*, **30**, 331-340.

THE EFFECT OF HUMAN DENSITY ON THE RISK OF INFECTION IN A HOSPITAL WARD

Marcelo PEREIRA[1,2], Lidia MORAWSKA[2], Luke KNIBBS[2], Congrong HE[2], Graham JOHNSON[2]

[1]Department of Refrigeration and Air Conditioning, Federal Institute of Education, Science and Technology of Santa Catarina, Brazil.
[2] International Laboratory for Air Quality and Health, Queensland University of Technology, Australia

Abstract

Airborne transmission of respiratory infectious disease in hospital may cause outbreaks of infectious diseases, which may lead to many infection cases and significantly influences on the public health. Numerous outbreaks have occurred due to large numbers of people in enclosed spaces. In this work, the risk of airborne infection in a Hospital Ward was correlated with the number of individuals present in the room. The results showed that, the occupancy density and the risk of infection practically followed the same tendency. The highest risk of infection was 9,87% for an occupation of the 10 people and the lowest was 0,66% for an occupation of the 3 people. Despite the personal risk to each individual in the space is nominally constant, the model used allowed relating the risk of infection with occupancy density, since the model assumes that CO_2 generation is exclusively of human origin.

Keywords: Rudnick and Milton model, airborne transmission of diseases, CO_2 generation.

INTRODUCTION

Epidemiological investigation of airborne infectious disease transmissions in the Hospital is usually confronted with several changes and uncertainties. Indoor air quality in hospitals is of great concern for patients and medical personnel. Airborne transmission occurs by dissemination of droplet nuclei over long distance from infectious patients (WHO, 2009). People's movement and density have an important impact in the generation of these agents as well as in the increase of predisposition to infection.

The number of people present in a Hospital Ward can vary greatly at any given time, but these changes are generally unknown. People movement and density change frequently and will cause changes of the susceptible personal exposures to contaminants. The fact that more people are present means that the chance that someone will become infected is much greater.

Therefore, it is important to establish relationship between number of occupants in the indoor environment and include this parameter into the analysis of the risk of infection because disease propagation will depend on the number of occupants in the indoor environment under consideration.

CO_2 at the concentrations commonly found in buildings is not a direct health risk, but CO_2 concentrations can be used as an indicator of occupancy and/or activity level. Thus, CO_2

production in the space will very closely track occupancy (Emmerich and Persily, 2001). An elevated indoor CO_2 concentration is directly related to the number of occupants in the building.

The aim of this work was correlate the risk of airborne infection in a Hospital Ward with the number of susceptible individuals present in the room. The risk of airborne infection (percent of susceptible persons infected) was estimated using the model of Rudnick and Milton (2003) in which used continuous carbon dioxide monitoring and applied adjustments to the classic Wells-Riley equation. The occupants were manually correlated with the CO2 concentration trends through a visual inspection of the trend itself.

MATERIALS AND METHODS

Hospital Ward

The Hospital Ward studied consisted of several patient rooms and the measurements were performed in a room with 3 beds, located in the city of Brisbane, Australia. The Ward provides both Inpatient and Outpatient care, and treats children with a number of respiratory illnesses, especially flus and viruses during the winter months. General medical conditions admitted to this ward include asthma, pneumonia, epilepsy, bronchitis, and febrile convulsions.

Infection risk modeling

In this work the model developed by Rudnick and Milton (2003) was used to estimate airborne transmission risk. The Rudnick and Milton (2003) model is a variation of the classical Wells-Riley model (Riley et al, 1978) and used CO_2 as a marker for exhaled breath to calculate the probability of getting infected based on inhaled fraction of air that has been exhaled by the infected occupant. Both models assume well air mixed where the particles are evenly distributed in space. The Rudnick and Milton (2003) model is:

$$P = \frac{D}{S} = 1 - \exp\left(-\frac{fIqt}{n}\right) \tag{1}$$

where P is the probability of infection, D is the number of disease cases, S is the number of people, f is equivalent to the fraction of indoor air that is exhaled breath; q is the quanta produced by one infector (quanta/h); I is the number of infectious source cases; n is the people in the ventilated space and t is the duration of exposure (h). The fraction of indoor air that is exhaled breath is estimated by the volume of expired air (Ve) divided by the total volume of the shared air space (V). This quotient is also can be estimated by the difference between the indoor CO_2 concentration (C) and the outdoor air CO_2 concentration (CO_0) and dividing that estimation by the CO_2 concentration in exhaled air (C_a):

$$f = \frac{V_e}{V} = \frac{C - C_o}{C_a} \tag{2}$$

The detailed of the concepts and mathematical background to these parameters can be obtained by reference to (Rudnick and Milton, 2003).

Data sources

CO_2 concentration measurements were performed using a portable continuous CO_2 concentration sampling device. Sampling was taken at 5 second intervals and the instruments were positioned centrally in each area, at a height of approximately 1.4m. The measurements were carried out during normal activities inside the Ward. All the information recorded by the occupants, were manually correlated with the CO_2 concentration trends through a visual inspection of the trend itself.

The risk of infection was modeled assuming the Influenza as the diseases spread by the airborne route. The quanta production rates for the influenza outbreaks was values used by Rudnick and Milton (2003), 100 quanta/hour.

Was assumed that one individual develops illness and exposes others for a defined time (30 minutes).

RESULTS AND COMMENTS

Figure 1 shows temporal variation of the CO2 concentration and the occupant's number. It can be seen that CO2 concentration was higher correlated with people density. The occupants' number varies with time and practically CO2 concentration followed the same tendency. Also observe that CO2 concentration decrease on afternoon, which is the period that the people density decreases.

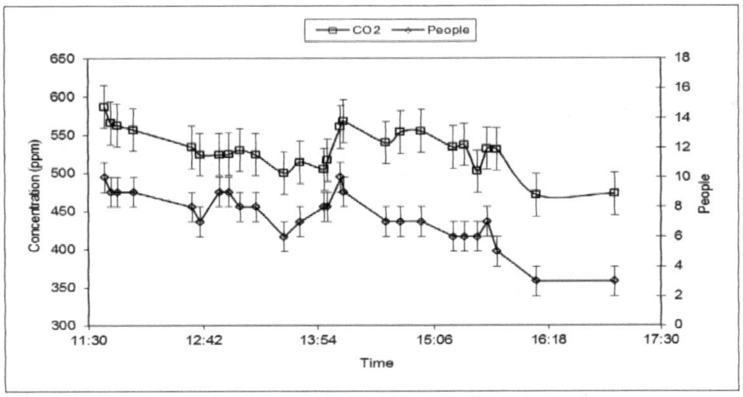

Figure 1. Variation of the CO_2 concentration and the occupant's number ($\pm\sigma$).

Figure 2 presents the risk of infection, assuming that the number of individuals varies inside the room. From this it can be seen that the occupancy density and the risk of infection practically followed the same tendency. The highest risk of infection was 9.87% for an occupation of the 10 people and the lowest was 0.66% for an occupation of the 3 people. It is important to highlight that, although, the risk of infection is independent of the occupancy density, i.e., the risk of infection is the same for each person in the Ward, the Rudnick and Milton (2003) allows relating the risk of infection with occupancy density, since the model assumes that CO_2 generation is exclusively of human origin.

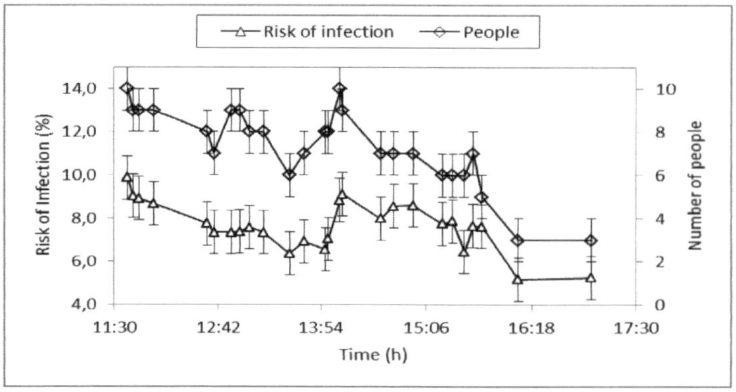

Figure 2. The risk of infection and the occupancy density (±σ).

CONCLUSIONS

In this work, the risk of airborne infection in a Hospital Ward was correlated with the number of individuals present in the room. The results showed that the occupancy density and the risk of infection practically followed the same tendency. Although, the risk of infection is independent of the occupancy density, i.e., the risk of infection is the same for each person in the Ward, the Rudnick and Milton (2003) allows relating the risk of infection with occupancy density, since the model assumes that CO_2 generation is exclusively of human origin.

Therefore, it is important to establish relationship between number of occupants in the indoor environment and include this parameter into the analysis of the risk of infection because disease propagation will depend on the number of occupants in the indoor environment under consideration.

REFERENCES

Emmerich, S.J. and Persily, A.K. State-of-the-Art Review of CO2 Demand Controlled Ventilation Technology and Application. National Institute of Standards and Technology. NISTIR 6729. 2001.

Riley, E., Murphy, G., & Riley, R. (1978). Airborne spread of measles in a suburban elementary school. American Journal of Epidemiology , 421-432.

Rudnick SN, Milton DK. Risk of indoor airborne infection transmission estimated from carbon dioxide concentration. Indoor Air. 2003;13(3):237–245.

WHO, 2009. Natural Ventilation for Infection Control in Health-Care Settings. Geneva: World Health Organization.
http://www.who.int/water_sanitation_health/publications/natural_ventilation/en/index.html.

The Influence of Street Canyon Design on Hospital Air Quality

Marco-Felipe. King[1,*] and Catherine J. Noakes[1]

[1] PaCE Institute, School of Civil Engineering, University of Leeds, Leeds, UK.

[*] *Corresponding email: m.f.king@leeds.ac.uk*

Keywords: Hospital, CFD, pollutant, ingress, dispersal

Abstract

This study considers air exchange between outdoor and indoor environments in the context of a hospital room with single-sided natural ventilation. Computational fluid dynamics (CFD) simulations are used to examine the trade-off between outdoor pollutant ingress and indoor contaminant dilution, and the influence of the street canyon design.

Two street-canyon aspect ratios were investigated (Height/Width=0.5 and 1) with roof angles of zero and 26.6°. An open window was located on the on the leeward side connecting to an interior room. ANSYS Fluent 15 was used to model airflow in the indoor and outdoor spaces and employed a k-omega turbulence model. A tracer was released in the outdoor location between the buildings and concentrations determined inside the open window. A second tracer released inside the building was used to establish concentrations at the pedestrian level outside.

Increasing canyon width reduced the residence time of the outdoor pollutant and reduces ingress. Flat roofs for both ratios drew the tracer to the leeward side of the building due to negative pressures. However, pitched roofs created more complex systems that reduced contaminant in the canyon due to unsteady vortices. Contaminants released inside the room result in proportionally higher concentrations outside for narrow canyons.

INTRODUCTION

The importance of urban airflow on hospital ventilation is an area of crucial but challenging research (Tang et al., 2006). Although USA guidance recommends mechanical ventilation, in the UK and many other countries natural ventilation is advocated as an energy efficient approach to ventilating some areas of hospital buildings, particularly patient rooms (Department of Health Estates and Facilities Division, 2008). While natural ventilation has been shown to provide effective ventilation with good levels of dilution (Escombe et al., 2007; Gilkeson et al., 2013), there are concerns that poor air quality in urban environments may counter the benefits of diluting indoor contaminants. Investigating the trade-off between pollutant ingress into hospitals, through single-sided natural ventilation, and pathogen egress is fundamental in understanding the indoor/outdoor environment nexus.

METHODOLOGY

The CFD package ANSYS Fluent 15 was used to investigate contaminant dispersion in and around 3D street-canyons (Figure 1.). Four cases similar to (Barlow and Leitl, 2007) were studied in a domain of 120mx24mx30m: Aspect ratios (Height/Width=0.5 and 1) and two roof angles (flat and 26.6°) were chosen. Building height was 5m from ground to eves in all cases.

An open window (1m x 0.5m) was positioned on the leeward façade of the second row of buildings leading to a hospital room inside (4.6 m x 4.6m x 4.6m).

The atmospheric boundary layer profile was modelled with a reference velocity of 6m/s at eve height. Pressure-Outlet conditions were 0Pa. NO_2 was released continuously from a volume source at ground level in the second street canyon (source A) and relative concentration was measured inside the building. To represent an indoor contaminant, a CO_2 source was located inside the hospital room (source B) and relative concentrations measured at pedestrian level in the adjacent street canyon. Source strength in both cases was given a value of 1 as relative differences were investigated.

Mesh cell count was ~3.5 million after mesh sensitivity analysis. The k-omega SST turbulence closure with standard wall-functions was solved via the SIMPLE algorithm with second order accuracy for all variables under steady state. Three vortex generators (1m high) were introduced upwind of the canyons to create building induced turbulence. Results were considered converged when continuity residuals reached 1E-6 for 100 iterations.

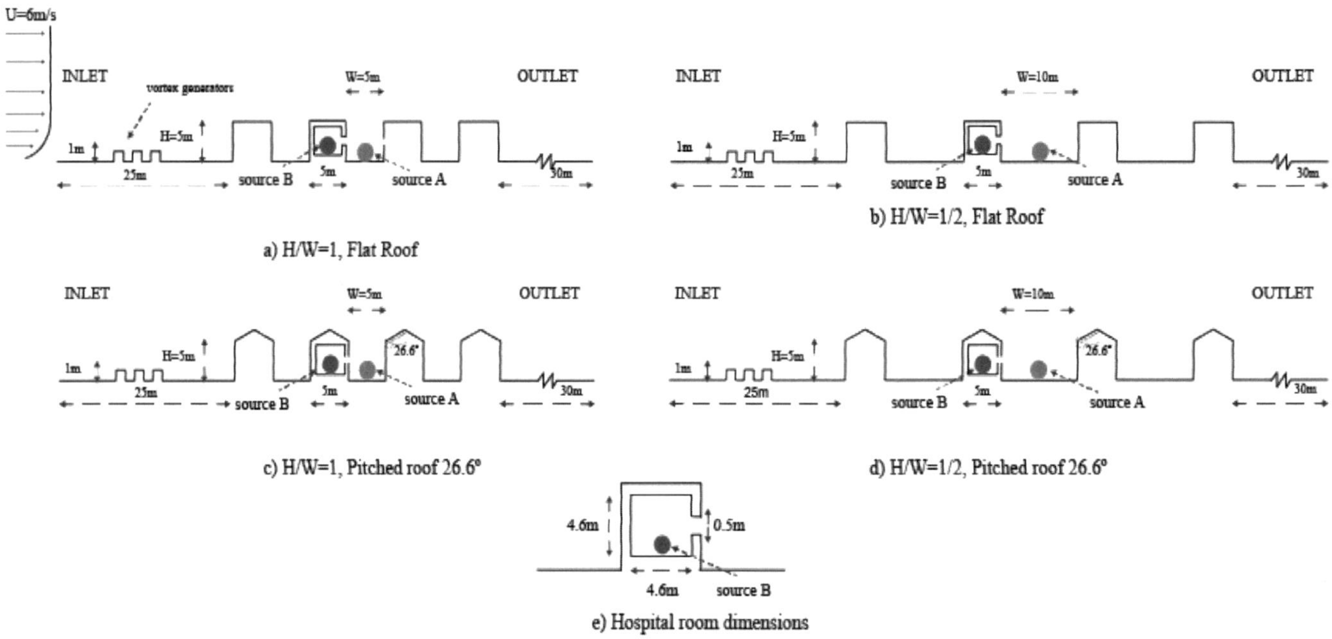

Figure 1. 2D representation of case set-up

RESULTS AND DISCUSSION

Figure 2. shows velocity vectors around the four canyon shapes. Complex turbulent structures are apparent within the narrower canyons, in particular in those containing buildings with pitched roofs.

Figure 3. shows contour plots of mass fraction of NO_2 on the central vertical plane released from source A in the four cases. NO_2 is drawn to the leeward side of the building where the window is located due to negative pressures on the façade. This facilitates ingress into the hospital room in all cases. However, pitched roofs created more complex systems that reduced contaminant in the canyon (Figure 3d) due to unsteady vortices but increase indoor concentration in the case with a narrow canyon (Figure 3b).

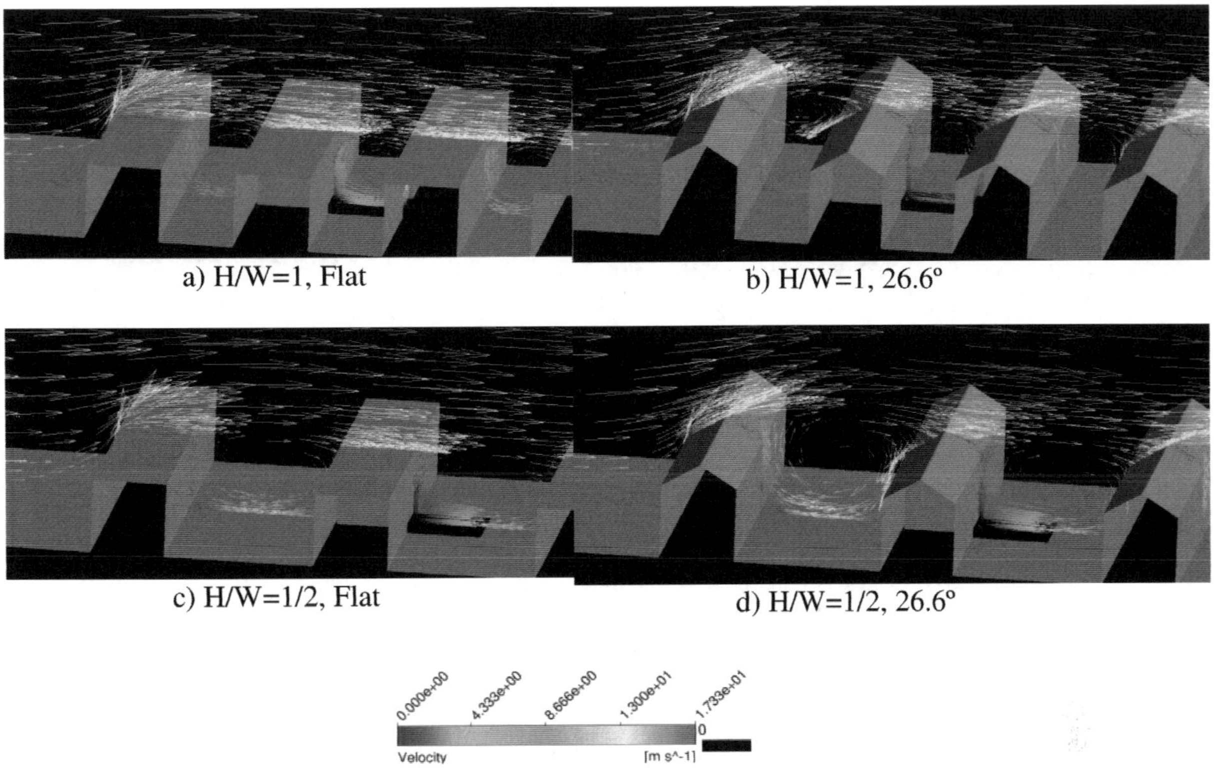

Figure 2. Velocity vectors in the four canyon shapes

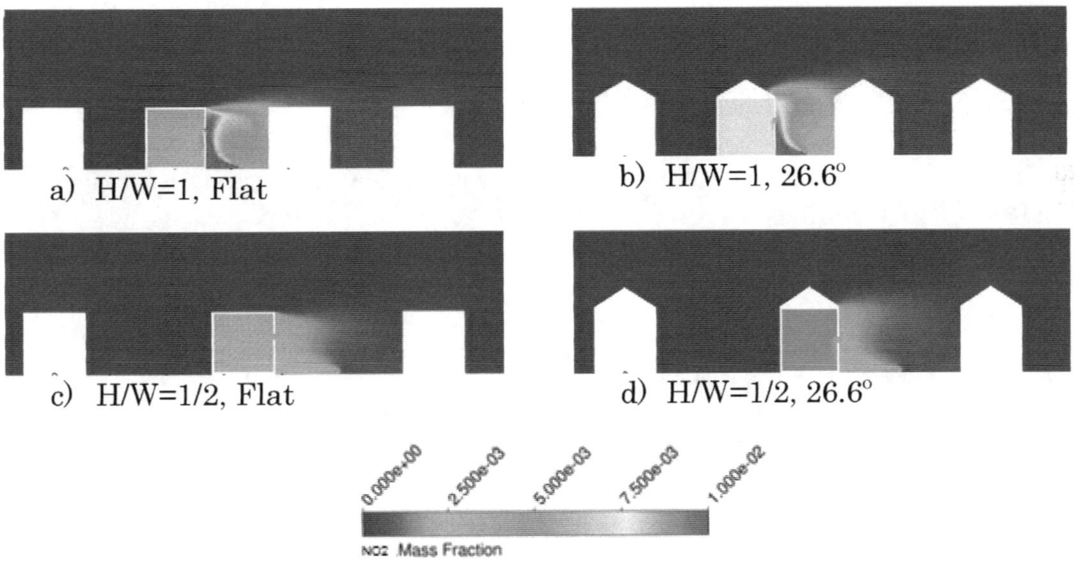

Figure 3. CFD contour plots of mass fraction of NO_2 released from Source A

Figure 4. a) and b) show graphs of mass fractions of contaminants comparing relative ingress and egress quantities. Doubling canyon width decreases ingress by more than a factor of three from street to inside (Figure 4a). On the other hand proportional egress of contaminants is reduced by a factor of less than 1.8 at best (Figure 4b). The impact of building design could be related to quanta concentration of a particular disease such as TB giving a quantitative metric through the application of the Wells-Riley infection transmission mathematical model (Noakes and Sleigh, 2008).

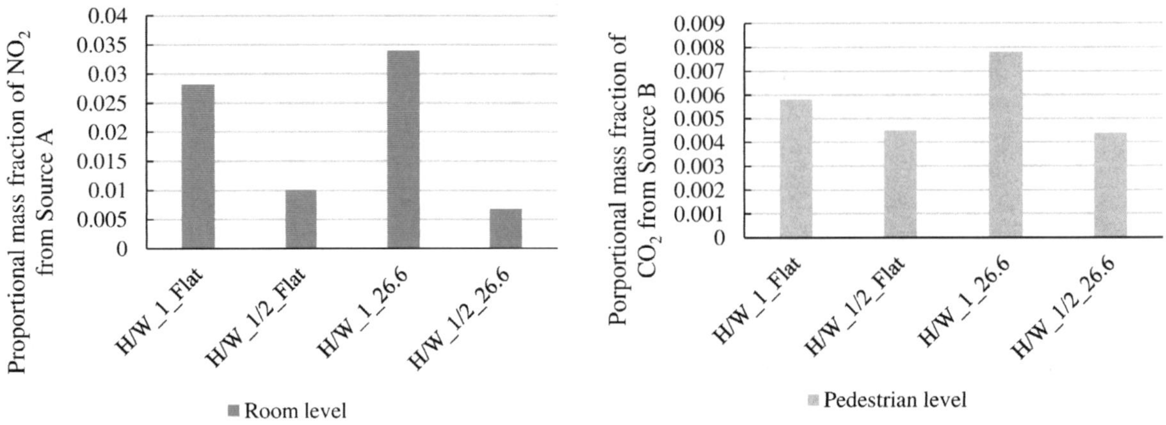

a) NO_2 inside the room from outside b) CO_2 at pedestrian level from the room

Figure 4. Graphs showing proportional mass fraction of NO_2 and CO_2 against source strength inside the room and at pedestrian level respectively.

CONCLUSION

Contaminants are found to be entrained by the turbulent eddies close to the leeward side of the buildings in all cases but predominantly within narrower canyons. Pitched roofs with wider canyons improve pollutant dispersal due to turbulent detachment at the apex. This has a result of increasing the height of the recirculation zones and decreasing ingress into the building envelope. Potential risk to pedestrians could be exacerbated when hospital windows are opened in narrow canyons, therefore investigating building shape and window design could be mitigating factors.

REFERENCES

Barlow, J.F. and Leitl, B. (2007) Effect of roofshape on unsteady flow dynamics in street canyons. In: *Physmod 2007, 29-31 August 2007, Orleans, France*.

Department of Health Estates and Facilities Division (2008) *In-patient care Health Building Note 04-01 : Adult in-patient facilities*.

Escombe, a R., Oeser, C., Gilman, R.H., Navincopa, M., Ticona, E., Martínez, C., Caviedes, L., Sheen, P., Gonzalez, A., Noakes, C., Moore, D. a J., Friedland, J.S. and Evans, C. a (2007) The detection of airborne transmission of tuberculosis from HIV-infected patients, using an in vivo air sampling model., *Clin. Infect. Dis.*, **44**, 1349–57.

Gilkeson, C. a., Camargo-Valero, M. a., Pickin, L.E. and Noakes, C.J. (2013) Measurement of ventilation and airborne infection risk in large naturally ventilated hospital wards, *Build. Environ.*, Elsevier Ltd, **65**, 35–48.

Noakes, C.J. and Sleigh, P.A. (2008) Applying the Wells-Riley equation to the risk of airborne infection in hospital environments : The importance of stochastic and proximity effects, 17–22.

Tang, J.W., Li, Y., Eames, I., Chan, P.K.S. and Ridgway, G.L. (2006) Factors involved in the aerosol transmission of infection and control of ventilation in healthcare premises., *J. Hosp. Infect.*, **64**, 100–14.

PRIORITY CHEMICALS EMITTED FROM COMPOSITE WOOD MATERIALS

Doyun WON[1,*], Wenping YANG[1], Joan WONG[2], Cheryl KHOURY[2] and Hans SCHLEIBINGER[1]

[1]Intelligent Building Operations, National Research Council Canada, Ottawa, Canada
[2]Indoor Air Contaminants Assessment Section, Water and Air Quality Bureau, Health Canada, Ottawa, Canada

[*]Corresponding email: doyun.won@nrc-cnrc.gc.ca

Keywords: Composite wood, Building materials, VOC, Formaldehyde, Chamber test

SUMMARY

It is a complex task to determine building materials that are low-emitting for all individual VOCs. Prioritizing chemicals that are likely to have the most health effects will be helpful in simplifying the process. In this study, VOCs in 17 composite wood materials were assessed for 4-days in 50-L chambers. The VOC emissions were converted to indoor air concentrations then compared with various health-based chronic inhalation limits for non-cancer effects. Relative to the lowest concentrations of interest (LCI) by AFSSET, the most influencing chemicals were phenol, formaldehyde, propanal, 2-octenal, and acetic acid. Relative to the chRELs, the most important compounds were acetaldehyde and formaldehyde. These substances can be selected as priority chemicals to determine low-emitting composite wood materials.

INTRODUCTION

Building materials emit hundreds of chemicals, including very volatile (e.g., formaldehyde), volatile (e.g., hexane and hexadecane) and semi-volatile organic compounds. Therefore, classification and labelling systems for low-emitting materials try to cover as many volatile organic compounds (VOCs) as possible. Examples include AgBB requirements in Germany (AgBB, 2012) for 164 VOCs, AFSSET requirements in France (ANSES, 2009) for 190 VOCs, and California green building (Section 01350) specifications (State Government of California, 2014) for ~40 VOCs out of 80 chemical substances.

In terms of minimizing health risks from daily exposures to chemicals, it is desirable to cover as many chemicals as possible in a regulation for low-emitting materials. However, it is costly and hard to implement if there are hundreds of chemicals specified in a regulation. Therefore, reducing the list of hundreds of VOCs to a priority VOC list is recommended as a compromise. This is particularly relevant if the scope of the regulation is narrowed to a certain material group (e.g., wood-based composite materials).

In this paper, the emissions data from composite wood materials were obtained and used to develop a priority chemical list by comparing health-based reference levels with the measured concentrations. This list is expected to aid the development process of a standard or regulation for low-emitting wood composite materials.

METHODOLOGIES

Composite wood materials were purchased at several retail stores in Ottawa, Canada from July 2012 to December 2012 and from October 2013 to November 2013. Table 1 lists the test specimens, including "pre-finished" materials such as shelving, flooring, door and countertop, and "un-finished" materials such as panelling, subfloors and I-beam joists.

The material emissions tests were conducted in 50-L chambers for 4 days under the conditions of 23 °C, 50% RH, 0.4 or ~1 m^2 m^{-3} loading ratio (L_c) and 1 h^{-1} air change rate (N_c). The air samples were taken daily on sorbent (Tenax TA/Carbograph 5TD) tubes and DNPH cartridges for 4 days, which were analysed with GC/MS and HPLC, respectively. Chromatograms were analysed for a total of 123 individual chemical compounds.

The measured chamber concentration (C_c) was converted to the reference room concentration (C_r) with 0.4 m^2 m^{-3} loading ratio (L_r) and 0.5 h^{-1} air change rate (N_{sr}) specified in ISO 16000-9 using Eq. 1.

$$C_r = C_c(N_c/L_c)(L_r/N_r) \tag{1}$$

The reference room concentrations were compared with LCIs specified by AgBB and AFSSET and the chronic reference exposure levels (chREL) of California (OEHHA, 2014). LCI is based primarily on irritative effects (Kirkeskov et al., 2009) and chREL is to address non-cancer effects from continuous exposures (OEHHA, 2014). It should be noted that LCIs and chRELs are intended for comparison with longer testing exposures (28 and 14 days, respectively). However, this comparison is considered adequate to screen and prioritize chemicals for further investigation.

RESULTS AND DISCUSSION

Table 2 lists the reference room concentrations at 4 day for major compounds, including alcohols, aldehydes, acids and terpenes. The concentrations greater than 100 μg m^{-3} are highlighted. Overall, the most dominant compounds are hexanal and acetic acid. The emissions from pre-finished materials tend to be lower than those from un-finished materials.

Table 3 summarizes the ratio of the reference room concentration to the LCIs specified by AFSSET. The highlighted ratios greater than 1 indicate that the major compounds are phenol, formaldehyde, propanal, 2-octenal, and acetic acid. The comparison with the LCI by AgBB led to similar results with formaldehyde, propanal and acetic acid excluded. This is because very volatile compounds such as formaldehyde, acetaldehyde and propanal are not included in the AgBB evaluation. The difference also comes from the fact that the LCI for acetic acid is more stringent with AFSSET (1250 for AgBB and 250 μg m^{-3} for AFSSET).

Since chRELs exist for a limited number of compounds, Table 4 lists the ratio of the reference room concentrations to one-half of chRELs for four compounds with chRELs. Acetaldehyde and formaldehyde were identified as the priority chemicals. The exclusion of phenol from the chREL evaluation reflects a higher chREL (200 μg m^{-3}) compared to LCIs (10 or 20 μg m^{-3}). Similar results were obtained for the concentrations measured at 1 day, 2 day, and 3 day.

CONCLUSIONS

This study showed that the influencing compounds based on irritative effects (i.e., the lowest concentration of interest) could be phenol, acetaldehyde, formaldehyde, propanal, hexanal, 2-octenal and acetic acid for composite wood materials. Acetaldehyde and formaldehyde were identified as important compounds for non-cancer effects from chronic exposures. Attention should be paid to these compounds when priority chemicals are selected in standards or regulations for low-emitting composite wood materials. This work also showed that the priority chemicals from the health point of view could be different for different health guidelines, emphasizing the need to understand the reasons for deriving different health-based reference values and the need to harmonize them for identical health-end points.

Table 1. Description of test specimens.

Pre-finished		Un-finished	
ID	Description	ID	Description
SHV1	Shelving (medium density fiberboard -	PIN2	Wood paneling & wainscoting (pine)
	MDF, pre-painted)	MDF3	Panel (MDF)
SHV2	Shelving (plywood, pre-painted, low	PB7	Panel (particleboard)
	formaldehyde)	PLY4	Panel (plywood)
LAM7	Flooring (laminate, pre-finished)	OSB8	Panel (oriented strand board - OSB)
EWF1	Engineered hardwood flooring	OSB10	OSB subfloor underlay board (insulated)
DI1	Interior door (hollow core, primed, HDF)	SPR1	Subfloor (basement, spruce)
KC1	Kitchen cabinet door (HDF, primed)	IB1	I-beam joist (pine block; 11 & 7/8" high)
CT3	Kitchen countertop (laminate, plastic-	IB2	I-beam joist (laminate block; 9 & 1/2" high)
	coated paper bonded to particle board)	IB3	I-beam joist (pine block, OSB web, 9.5" high)

Table 2. Reference room concentration at 4 day.

G	Compound	Reference room concentration (μg m⁻³) at 4 day																
		Pre-finished							Un-finished									
G	Compound	SHV1	SHV2	LAM7	EWF1	DI1	KC1	CT3	PIN2	MDF3	PB7	PLY4	OSB8	OSB10	SPR1	IB1	IB2	IB3
Alcohols	1-Butanol	3.4		7.4	2.2	1.0	0.6	1.6	3.0	16.7	10.2	2.1	12.1	8.9	12.7	2.8	16.1	0.7
	Hexanol, 2-ethyl-1-			5.4	2.8	1.6		0.3	0.5	3.6	2.3	0.1	0.7	0.6	1.9		3.7	0.1
	2-Propanol	5.3									12.5					0.5	24.3	0.5
	Ethanol	0.5		0.3		6.2		4.6		0.4	0.3		0.4	1.9	0.1	0.9	7.1	
	Phenol				2.1			4.4	0.6	7.8	1.2		13.4	2.9	1.6	49.2	98.8	0.1
Aldehydes & Ketones	Acetaldehyde				15.8	10.7	0.1	7.0		244.7	11.7	4.0	94.1	28.1	1.1	24.7	10.9	59.4
	Formaldehyde	2.9	9.9			52.1	0.9	58.7		206.7	40.3	0.9	1.6	4.9		11.5	4.6	8.5
	Propanal	1.4				1.9		3.0		29.2	6.1		22.6	16.1	4.3	3.6	2.6	14.8
	Methylethyl ketone	0.6	0.1			0.2		0.4	0.1	0.8	1.2	0.1	1.2	1.1	1.3	0.1	0.6	0.1
	Acetone	11.0	1.1		8.2	8.0	0.0	2.3	3.1	267.4	394.5	6.1	107.0	6.0	67.5	6.3	69.0	0.8
	Benzaldehyde	0.4			6.2	0.2			0.1	46.6	4.8	0.2	0.6	0.5	0.7	0.2	1.5	0.2
	Butanal	0.7	0.9			1.4		2.3	0.8	32.6	7.3		12.9	7.4	5.3	1.2	3.4	0.8
	2-Decenal					2.0		1.1		3.0	1.0					0.2	0.8	0.4
	Furfural					1.8				1.1	0.6	1.4	1.4		4.8	1.3	1.4	1.5
	Heptanal					1.5		1.5	2.3	38.4	6.1	0.2	2.9	3.1	1.9	3.3	7.2	1.0
	Hexanal	42.8	14.2		3.2	391.3		246.0	170.5	11289.1	638.4	42.5	908.2	306.2	580.2	570.2	379.7	314.1
	Nonanal	0.5			0.7	1.3		0.6	0.3	25.2	5.4	0.4	1.3	1.4	0.6	17.6	26.9	0.6
	2-Nonenal							1.3		3.2	0.7					0.2	0.5	0.5
	Octanal					0.8		1.0	0.2	18.7	4.3		0.7	1.1	0.4	1.1	6.4	
	2-Octenal					25.1		11.4		62.1	4.1		8.3	37.6	4.1	2.2	2.7	21.5
	Pentanal	1.1	0.8			10.5		13.2	7.2	336.7	35.0	1.0	63.2	25.3	23.1	17.8	19.8	10.6
Terpenes	alpha-Pinene	0.5	6.0	0.1		1.5	0.1	0.3	2.1		11.3	4.2	0.3		0.7	82.3	7.0	34.0
	beta-Pinene	0.2	0.8			0.3			0.1	0.2	1.5	0.3	0.1		0.1	51.3	3.4	25.0
	3-Carene					0.2		0.0	0.1		0.9	0.1			0.1	38.2	0.2	6.0
	Camphene		0.4						0.1	0.2	0.5	0.1			0.0	7.4	0.4	12.1
	Limonene	0.9	0.2		0.2	0.4		0.1	0.6	0.3	0.9	0.7			0.4	17.8	1.5	10.3
Acids	Acetic acid					119.7	0.2	464.7	14.5	883.6	131.9		153.1	1087.9	485.7	383.1	637.1	240.1
	Butyric acid							2.8		9.3				1.9		0.6		0.1
	Hexanoic acid					5.3		5.3		751.2	4.4		9.8	57.4	6.0	3.2	3.9	11.4
	Pentanoic Acid					0.8		1.5		126.6			1.4	7.7		0.4	1.0	1.2

Table 3. Ratio to LCI by AFSSET at 4 day.

G	Compound	LCI AFSSET (μg m⁻³)	Ratio to LCI by AFSSET at 4 day																
			Pre-finished							Un-finished									
G	Compound		SHV1	SHV2	LAM7	EWF1	DI1	KC1	CT3	PIN2	MDF3	PB7	PLY4	OSB8	OSB10	SPR1	IB1	IB2	IB3
Alcohols	1-Butanol	3000	0.0		0.0	0.0	0.0	0.0	0.0	0.0	0.0	0.0	0.0	0.0	0.0	0.0	0.0	0.0	0.0
	Hexanol, 2-ethyl-1-	1100			0.0	0.0	0.0		0.0	0.0	0.0	0.0	0.0	0.0	0.0	0.0		0.0	0.0
	2-Propanol																		
	Ethanol																		
	Phenol	20				0.1			0.2	0.0	0.4	0.1		0.7	0.1	0.1	2.5	4.9	0.0
Aldehydes & Ketones	Acetaldehyde	200				0.1	0.1	0.0	0.0		1.2	0.1	0.0	0.5	0.1	0.0	0.1	0.1	0.3
	Formaldehyde	10	0.3	1.0			5.2	0.1	5.9		20.7	4.0	0.1	0.2	0.5		1.2	0.5	0.8
	Propanal	8	0.2				0.2		0.4		3.6	0.8		2.8	2.0	0.5	0.4	0.3	1.8
	Methylethyl ketone	5000	0.0	0.0			0.0		0.0	0.0	0.0	0.0	0.0	0.0	0.0	0.0	0.0	0.0	0.0
	Acetone																		
	Benzaldehyde	90	0.0			0.1	0.0			0.0	0.5	0.1	0.0	0.0	0.0	0.0	0.0	0.0	0.0
	Butanal	650	0.0	0.0			0.0		0.0	0.0	0.1	0.0		0.0	0.0	0.0	0.0	0.0	0.0
	2-Decenal	6					0.3		0.2		0.5	0.2					0.0	0.1	0.1
	Furfural	8					0.2				0.1	0.1	0.2	0.2		0.6	0.2	0.2	0.2
	Heptanal	650					0.0		0.0	0.0	0.1	0.0	0.0	0.0	0.0	0.0	0.0	0.0	0.0
	Hexanal	650	0.1	0.0		0.0	0.6		0.4	0.3	17.4	1.0	0.1	1.4	0.5	0.9	0.9	0.6	0.5
	Nonanal	650	0.0			0.0	0.0		0.0	0.0	0.0	0.0	0.0	0.0	0.0	0.0	0.0	0.0	0.0
	2-Nonenal	6							0.2		0.5	0.1					0.0	0.1	0.1
	Octanal	650					0.0		0.0	0.0	0.0	0.0		0.0	0.0	0.0	0.0	0.0	
	2-Octenal	6					4.2		1.9		10.3	0.7		1.4	6.3	0.7	0.4	0.5	3.6
	Pentanal	1700	0.0	0.0			0.0		0.0	0.0	0.2	0.0	0.0	0.0	0.0	0.0	0.0	0.0	0.0
Terpenes	alpha-Pinene	450	0.0	0.0	0.0		0.0	0.0	0.0	0.0		0.0	0.0	0.0		0.0	0.2	0.0	0.1
	beta-Pinene	1400	0.0	0.0			0.0			0.0	0.0	0.0	0.0	0.0		0.0	0.0	0.0	0.0
	3-Carene																		
	Camphene																		
	Limonene																		
Acids	Acetic acid	250					0.5	0.0	1.9	0.1	3.5	0.5		0.6	4.4	1.9	1.5	2.5	1.0
	Butyric acid	300							0.0		0.0				0.0		0.0		0.0
	Hexanoic acid	300					0.0		0.0		2.5	0.0		0.0	0.2	0.0	0.0	0.0	0.0
	Pentanoic Acid	300					0.0		0.0		0.4			0.0	0.0		0.0	0.0	0.0

Table 4. Ratio to half of chREL at 4 day.

G	Compound	CREL ($\mu g\ m^{-3}$)	**Ratio to 0.5*chREL at 4 day**															
			Pre-finished							Un-finished								
			SHV1	SHV2	LAM7	EWF1	DI1	KC1	CT3	PIN2	MDF3	PB7	PLY4	OSB8	OSB10	SPR1	IB1	IB2	IB3
Alcohols	1-Butanol																		
	Hexanol, 2-ethyl-1-																		
	2-Propanol	7000	0.00									0.00					0.00	0.01	0.00
	Ethanol																		
	Phenol	200				0.02			0.04	0.01	0.08	0.01		0.13	0.03	0.02	0.49	0.99	0.00
Aldehyd	Acetaldehyde	140				0.23	0.15	0.00	0.10		3.50	0.17	0.06	1.34	0.40	0.02	0.35	0.16	0.85
	Formaldehyde	9	0.65	2.21			11.58	0.20	13.04		45.94	8.95	0.20	0.35	1.08		2.56	1.02	1.88
	Propanal																		

REFERENCES

AgBB (2012). Health-related evaluation procedure for volatile organic compounds emissions (VOC and SVOC) from building products. Feb 21, 2014. http://www.umweltbundesamt.de/sites/default/files/medien/355/dokumente/agbb_evaluation_scheme_2012_0.pdf.

ANSES (2009). Emissions de compos´es organiques volatils (COV) par les produits de construction et de d´ecoration. Oct 31, 2014. https://www.anses.fr/fr/content/emissions-de-compos%c3%a9s-organiques-volatils-cov-par-les-produits-de-construction-et-de.

Kirkeskov L, Witterseh T, Funch LW, et al. (2009). Health evaluation of volatile organic compound (VOC) emission from exotic wood products. *Indoor Air*, **19,** 45-57.

OEHHA (2014). All OEHHA acute, 8-hour and chronic reference exposure levels (chRELs) as of June 2014. Jan 14, 2015. http://www.oehha.ca.gov/air/allrels.html.

State Government of California (2014). Sustainable (green) building: Section 01350. Jan 14, 2015. http://www.calrecycle.ca.gov/greenbuilding/specs/section01350/#Criteria_.

A STUDY OF SHADING DEVICE ORIENTATION ON THE NATURAL VENTILATION AND DAYLIGHT HARVESTING POTENTIAL IN A DOUBLE SKIN FACADE

Jeehwan LEE[1,*] Jae D. CHANG[1] and Hongyi CAI[2]

[1]Department of Architecture, The University of Kansas, Kansas, U.S.A.
[2]Department of Civil, Environmental and Architectural Engineering, The University of Kansas, U.S.A.

[*]Corresponding email: jeefanlee@ku.edu

Keywords: Double skin facade, Natural ventilation, Shading device, Daylighting

SUMMARY

Preliminary simulation studies have shown that double skin facades (DSFs) with properly oriented shading devices could improve the natural ventilation potential in a DSF air cavity by facilitating the heat dissipation and blocking undesired solar heat gain. This study concentrated on the influence of shading device orientation and type on natural ventilation and daylighting harvesting potential. Computational fluid dynamic (CFD) simulation and daylighting analysis results indicate that horizontal shading devices at 0, 30, and 60 degree angles result in air temperatures in the air cavity that are about 2 to 3 degrees Celsius higher than that of vertical shading devices. Minimum levels of indoor illuminance were better with horizontal shading devices at 0 and 60 degree angles than they were for vertical oriented devices.

RESEARCH BACKGROUND

Poirazis (2004) found that many DSF studies show DSFs taking advantage of the solar chimney effect within the air cavity and that adjustable shading devices in the ventilated cavity work to protect the internal occupied spaces from unwanted solar heat gain and contribute to lowering a building's cooling load. Stazi, F. et al. (2014) looked at the impact shading device materiality, configuration, permeability, orientation, and position have on building energy savings, thermal comfort, and daylighting harvesting. Baldinelli (2009) developed a new double skin façade with external glazing made of movable integrated glass-shading devices to decrease cooling loads during the warm periods. Lee et al. (2002) stated that the position of shading devices play a significant role in the overheating and thermal discomfort of building occupants. Safer et al. (2004) found that horizontal blinds situated closer to the internal glazing can help minimize overheating of the exterior facing side's air cavity through higher air velocity.

METHODOLOGIES

A DSF was modeled based on the preliminary designs of a DSF building in Lawrence, Kansas. As shown in Figure 1, it is a 1-story tall corridor type with wooden vertical shading devices inside the DSF air cavity. As the CFD simulation tool, FloVENT 9.3 specialized in analysis of airflow and heat transfer in built environments was used based on K-epsilon turbulence model.

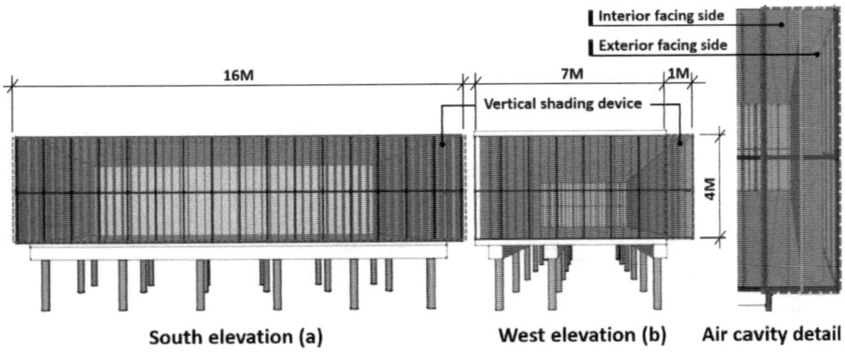

Figure 1. South elevation (a), west elevation (b) and air cavity section detail (c)

Table 1. CFD and illuminance simulation model boundary conditions

Classification	Parameters(unit)	FloVENT 9.3 (material)	Ecotect (material)
Ambient outdoor conditions	Temperature (°C)	26	Weather data
	Relative humidity (%)	50	
Dimensions	Glazing (mm)	25 (glass)	25 (glass)
	Cavity width (mm)	1,000 (air)	1,000 (air)
Glazing properties (6/13/6 double glazed low-E)	U-value (W/m^{2}K)	-	2.6
	Solar Heat Gain Coefficient (%)		0.55
	Visible Transmittance (%)		0.61

Table 2. Cases for CFD and illuminance simulation

Orientation	Case 1	Case 2	Case 3	Case 4
Horizontal and Vertical	0 degree	30 degree	60 degree	90 degree

RESULTS AND DISCUSSION

CFD simulation analysis

Figure 2 shows that horizontal shading devices at 0-degree (perpendicular to glazing) angle lead to increased air temperature in the exterior facing side of the air cavity due to the stagnation of the heated air between the horizontal shading slats. Figure 3 shows that horizontal and vertical shading devices at a 90-degree angle contributed to greater vertical convective currents for heat dissipation.

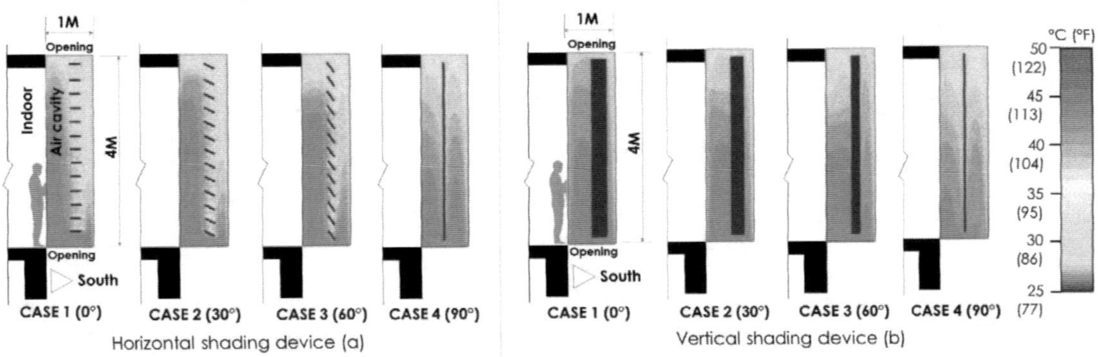

Figure 2. Air temperature distribution with horizontal (a) and vertical shading device (b).

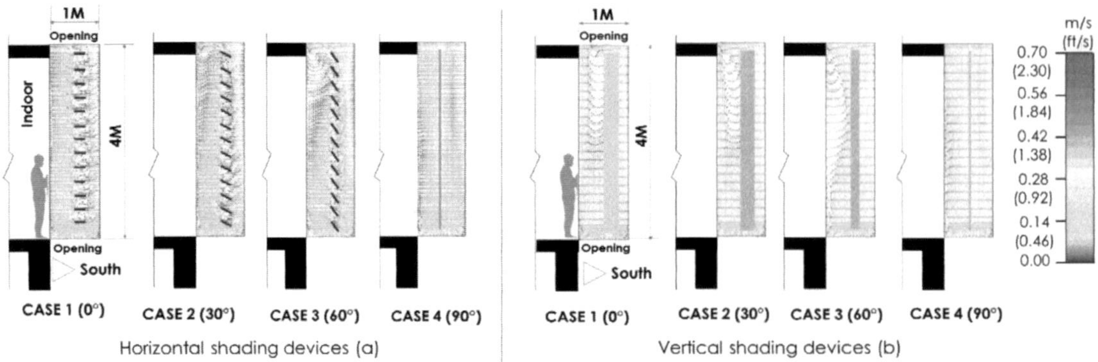

Figure 3. Airflow patterns for horizontal (a) and vertical shading device (b).

Daylighting simulation analysis

Figure 4 shows that minimum levels of indoor illuminance with horizontal shading devices at 0 (perpendicular to glazing) and 60 degree angles were higher than for vertical shading cases.

Horizontal shading devices	Vertical shading devices
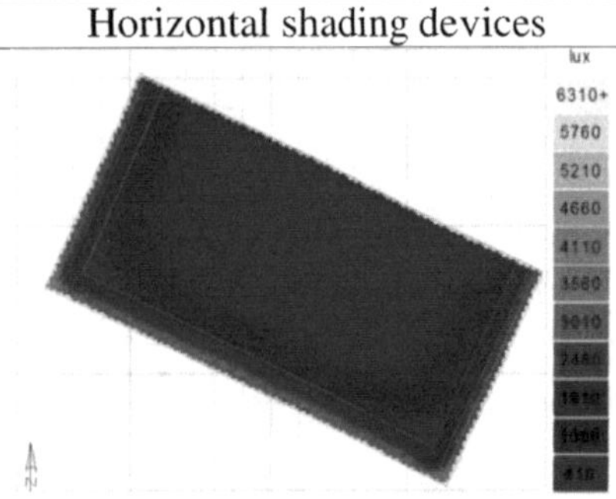 Case 1 (0°) 810~6310+ lux	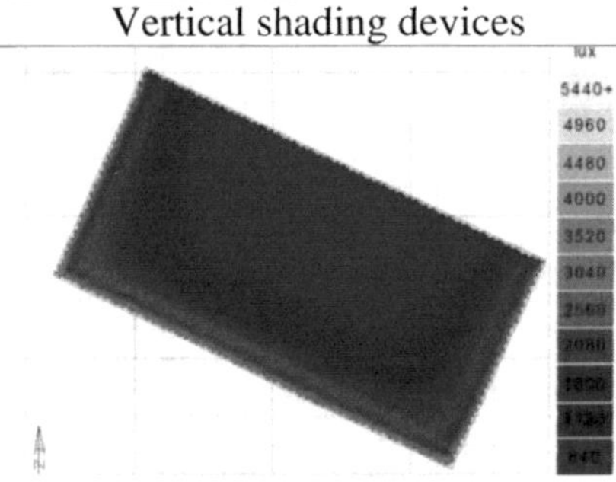 Case 1 (0°) 640~5440+ lux
 Case 2 (30°) 560~5360+ lux	 Case 2 (30°) 630~5630+ lux
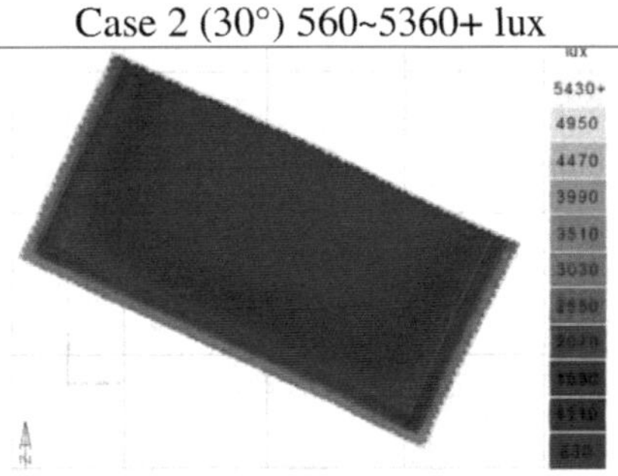 Case 3 (60°) 630~5430+ lux	 Case 3 (60°) 560~5360+ lux

Figure 4. Illuminance levels of horizontal (left) and vertical shading devices (right).

This implies that horizontal shading devices have more daylighting potential. Horizontal shading devices at 0 degree angle (perpendicular to glazing) showed the highest indoor illuminance level.

CONCLUSION

This study found that the air temperature inside a DSF air cavity with horizontal shading devices at 0 (perpendicular to glazing), 30, and 60 degree angles were about 2 to 3 degrees Celsius higher than cases of vertical shading devices. These outcomes imply that horizontal shading devices at a 0 degree angle not only tend to impede heat dissipation due to the stagnation of heated air between the horizontal shading slats but can also lead to overheating problems inside the DSF air cavity which may result in higher cooling energy demands.

In contrast, the minimum levels of indoor illuminance were higher for horizontal shading devices at 0 (perpendicular to glazing) and 60 degree angles than they were for vertical shading devices. Horizontal shading devices at a 0 degree angle had the greatest potential for daylighting availability.

In sum, this study suggests there is a conflict between natural ventilation and daylight harvesting potential based on shading device orientation, horizontal vs vertical. Therefore, careful consideration must be given to the control of DSF air cavity shading devices to maintain a balance between natural ventilation and daylight harvesting potential.

REFERENCES

Poirazis, P. (2004). ***A literature review of Double skin facades for office buildings***. Department of Construction and Architecture. Lund University. Available at: <http://www.ebd.lth.se/fileadmin/energi_byggnadsdesign/images/Publikationer/Bok-EBD-R3-G5_alt_2_Harris.pdf.>

Stazi, F., Marinelli, S., Di Perna, D., and Munafò, P. (2014). ***Comparison on solar shadings: Monitoring of the thermo-physical behaviour, assessment of the energy saving, thermal comfort, natural lighting and environmental impact***. Solar Energy, 105:512-528.

Baldinelli, G. (2009). ***Double skin façades for warm climate regions: analysis of a solution with an integrated movable shading system***. Energy and Buildings, 44:1107-1118.

Lee, E., Selkowitz, S., Bazjanac, V., Inkarojrit, V., Kohler, C. (2002). ***Higher performance of commercial building facades***. Available at: <http://gaia.lbl.gov/hpbf/documents/LBNL50502.pdf>

Safer, N., Woloszyn, M., & Roux, J.J.(2004). ***Three dimensional simulation with a CFD tool of the airflow phenomena in single floor double-skin façade equipped with a venetian blind***. *Solar Energy, 79:*193-203.

The Relation of Psychosocial and Demographic Characteristics with Energy Consumption

Suchismita Bhattacharjee[1*]

[1]College of Architecture, University of Oklahoma, Norman, OK

* Corresponding email: suchi@ou.edu

Keywords: Energy efficiency, social behavior, psychosocial characteristics, energy conservation, energy efficiency incentives and policies.

SUMMARY

Energy consumption by individuals varies based on selected psychosocial and demographic characteristics. Little is known about the interaction of these characteristics in predicting the geographic variation in per capita energy consumption. Therefore, a comprehensive database of the 50 US States was created that included details on income, population composition, racial distribution, educational attainment, employment rate, average household size, & median housing unit value on one side, and energy price, energy consumption, and renewable energy production data on other side.

In a regression analysis one dependent variable of interest, i.e. per capita energy was found to be associated with several independent variables. States with higher proportion of males had lower per capita energy consumption and total residential energy consumption was lesser in states that had a greater proportion of population in the age group of 18-64 years.

INTRODUCTION

Consumption of non-renewable energy like electricity, and natural gas in the residential sector of United States has been marked by a steady growth over the last decades in spite of the implementation of various energy efficiency policies. Several educational interventions have been adopted at the national and regional level to increase awareness among people about using energy more efficiently. It is important that these educational interventions stratify the intervention based on the different demographic and socio-economic factor that influences energy consumption. Association of psycho-social variables such as individual attitude, behavior, eco-consciousness, and culture with energy consumption has been found by researchers in the past (Abrahamse and Steg 2009, Bhattacharjee and Reichard 2011, Mehrzad, Masoud, and Mansour 2007). None of the studies showed the relationship between individual demographic level and energy consumption pattern. In this study the researcher attempts to address the gap by performing a correlation analysis between individual demographic factors like age, race, gender, income and per capita energy consumption.

METHODS

Association between the human factor characteristics, and energy performance were studied in the 50 states of US targeting a total population of 313.8 million. The human factor characteristics included demographic and socio economic information which were considered the independent variable. Energy performance characteristics were dependent variables which included data on energy consumption and type of energy used. All the above mentioned data were obtained from 2010 US Census data. Correlation was calculated to identify the association between the independent variable and dependent variables which included the energy consumption data.

Data analysis included descriptive statistics to identify correlation between the demographic and socio economic condition and energy consumption and type of energy used with a significance of ($p<0.01$). Spearman rank correlations among the parameters are shown under the finding section below.

FINDING & ANALYSIS

Spearman correlation coefficient among the various dependent and independent variables are listed in Figure 1.

Spearman correlation coefficient, denoted by r, is a measure of the strength of the straight-line or linear relationship between two variables. The correlation coefficient takes on values ranging between +1 and -1. It is considered that r values between 0 and 0.3 (0 and -0.3) indicate a weak positive (negative) linear relationship, r values between 0.3 and 0.7 (0.3 and -0.7) indicate a moderate positive (negative) linear relationship, and r values between 0.7 and 1.0 (-0.7 and -1.0) indicate a strong positive (negative) linear relationship. Thus for this study a spearman correlation value of +0.3 or greater and -0.3 or lesser are considered (Moore and McCabe 1989). The following section discusses about the factors which have correlation value of +0.3 or greater and -0.3 or lesser with a significance of less than 0.05.

3.1. Correlation between per capita residential energy consumption and age

From Figure 1 it appears that with the decrease in average age of the population, per capita energy consumption is increasing. People with age 5 and under and people with age 18 and under has a correlation of 0.545 ($p<0.0004$) and 0.479 ($p<0.0$) respectively. This can be justified by the fact, that younger population spends a longer time at home compared to other age group. Additionally, families with children spend more time at home.

On the contrary, population of 65 years and older has a negative correlation of -0.3 with $p<0.03$. This can be justified by the fact that old populations often live in assisted livings which are multi-unit structures and they sometimes are also within a restricted income per month in comparison to the working population.

3.2. Correlation between race and per capita residential energy consumption

As seen in Annex A, white population shows a positive correlation of 0.457 ($p<0.001$) as compared to Asian, Pacific Islander and Hispanic which shows negative correlation value of -0.71 ($p<0.0$), -0.57 ($p< 0.0$), -0.55 ($p< 0.006$) respectively. This can be justified from the cultural background and lifestyle of the population.

Additionally foreign born population shows negative correlation of -0.77 ($p<0.003$). In addition to the cultural background, it can also be considered that this sector of the population is either temporarily living in United States or is still in the early years of settling down in a new community. During this phase, individuals mostly live under a restricted budget or live in multi-unit housing.

3.3. Correlation between homeownership, multiunit housing occupancy and residential energy consumption

Homeownership shows a positive correlation of 0.563 ($p<0.001$) and multiunit housing shows a negative correlation of -0.606 ($p<0.000$). Homeownership increases the occupied square feet area per person compared to shared multifamily units which in turn affect the energy consumption per capita.

3.4. Correlation between median house value and residential energy consumption per capita

Study shows a negative correlation of -0.768 ($p<0.0$) between the median value of home and residential energy consumption. This result is contrary to the considered fact that increase in median house value, can be due to larger square feet area per person which will increase the per capita residential energy consumption. Median increase in house value causes the increase in mortgage payment per month which can in turn cause the householders to live within a tight budget, thus decreasing energy consumption per capita.

3.5. Correlation between household income and residential energy consumption per capita

Though one would consider that with more available income, householder's lifestyle will cause an increased in energy consumption per capita, the study shows a completely different result. The study showed another strange negative correlation of -0.504 ($p<0.002$) between household income and residential energy consumption. This can be justified by the fact that higher household income is often associated with higher level of education. Study shows that people with bachelor degree or higher shows negative correlation of -0.423 ($p<0.001$), which indicates that increase in education cause decrease in energy consumption as they can make more educated and informed decision.

Additionally, increase in household income can be associated with more number of working individual, who spend more time outside home, thus decreasing energy consumed at home.

CONCLUSION

This study attempted to identify the socio demographic factors that influence residential energy consumption. Overall, as expected the results obtained from the spearman's correlation analysis indicate that residential energy consumption to a large extent is associated with time spent at home, education level, race and householders' age. The study showed additional associations, which indicate that householder's disposable income and education level plays a large role in determining energy consumption per capita.

REFERENCES

Abrahamse, Wokje, and Linda Steg2009. "How do socio-demographic and psychological factors relate to households' direct and indirect energy use and savings?" *Journal of Economic Psychology* no. 30 (5):711-720. doi: http://dx.doi.org/10.1016/j.joep.2009.05.006.

Bhattacharjee, Suchismita, and Georg Reichard. 2011. Socio-Economic Factors Affecting Individual Household Energy Consumption: A Systematic Review. Paper read at ASME 2011 5th International Conference on Energy Sustainability.

Mehrzad, Ebrahimi, Alizadeh Masoud, and Ebrahimi Mansour2007. "Socioeconomic factors affecting household energy consumption in Qom, Iran." *Journal of Applied Science* no. 7 (19):2876-2880.

Moore, David S, and George P McCabe. 1989. *Introduction to the Practice of Statistics*: WH Freeman/Times Books/Henry Holt & Co.

Table 1: Correlation between Energy Consumption and Socio-demographic Characteristics

		1	*2*	*3*	*4*	*5*	*6*	*7*	*8*	*9*	*10*	*11*	*12*	*13*	*14*	*15*	*16*	*17*	*18*	*19*	*20*	*21*	*22*	*23*	*24*	*25*
1	Total Energy Consumption																									
2	Percapitae Residential Consumption	-0.07																								
3	Percapitae Commercial Consumption	-0.06	0.27																							
4	Percapitae Industrial Consumption	-0.02	0.46	0.31																						
5	Percapitae Transportation Consumption	-0.04	0.35	0.25	0.83																					
6	Population	1.00	-0.10	-0.07	-0.06	-0.07																				
7	Persons under 5 years	0.03	0.55	0.16	0.38	0.40	0.01																			
8	Persons under 18 years	0.05	0.48	-0.20	0.28	0.32	0.04	0.89																		
9	Persons 65 years and over	-0.06	-0.30	-0.32	-0.26	-0.34	-0.04	-0.70	-0.58																	
10	Female population	0.04	-0.33	0.00	-0.50	-0.68	0.05	-0.49	-0.46	0.33																
11	White population	-0.03	0.46	-0.18	0.11	0.00	-0.03	-0.09	0.13	0.25	-0.27															
12	African american population	0.05	0.00	0.34	-0.08	-0.20	0.04	-0.04	-0.17	-0.18	0.69	-0.66														
13	Asian population	0.04	-0.71	-0.23	-0.22	-0.05	0.06	0.01	-0.09	-0.07	-0.16	-0.63	-0.08													
14	Pacific Islander population	-0.04	-0.57	-0.22	-0.09	0.08	-0.04	0.06	-0.06	0.03	-0.27	-0.59	-0.19	0.90												
15	Hispanic population	0.15	-0.55	-0.15	-0.19	-0.18	0.17	0.26	0.27	-0.24	-0.10	-0.06	-0.10	0.21	-0.01											
16	Foreign born population	0.15	-0.77	-0.15	-0.44	-0.35	0.19	-0.03	-0.06	-0.18	0.12	-0.39	0.10	0.58	0.24	0.71										
17	Non-english speaking foreign	0.16	-0.75	-0.20	-0.31	-0.22	0.18	0.10	0.10	-0.23	-0.02	-0.27	-0.05	0.49	0.19	0.90	0.92									
18	Bachelor education or higher	-0.01	-0.42	0.38	-0.43	-0.36	0.01	-0.24	-0.40	-0.23	0.26	-0.23	0.17	0.23	0.03	0.12	0.46	0.32								
19	Mean travel time to work	0.12	-0.50	-0.10	-0.52	-0.53	0.14	-0.36	-0.34	-0.04	0.63	-0.49	0.53	0.34	0.06	0.24	0.64	0.46	0.46							
20	Homeownership rate, 2008-2012	-0.06	0.56	-0.36	0.22	0.20	-0.08	-0.04	0.25	0.33	-0.14	0.55	-0.30	-0.45	-0.26	-0.37	-0.64	-0.54	-0.52	-0.39						
21	Multi-unit housing	0.04	-0.61	0.27	-0.37	-0.35	0.06	-0.28	-0.46	-0.14	0.30	-0.41	0.23	0.44	0.22	0.23	0.68	0.52	0.75	0.56	-0.80					
22	Meadian value of houses	-0.04	-0.77	0.11	-0.38	-0.23	0.00	-0.19	-0.36	-0.18	0.09	-0.53	0.08	0.73	0.54	0.23	0.68	0.53	0.75	0.54	-0.65	0.77				
23	No. of households	1.00	-0.09	-0.07	-0.06	-0.07	1.00	0.01	0.03	-0.04	0.05	-0.03	0.04	0.06	-0.04	0.16	0.18	0.18	0.01	0.14	-0.07	0.06	-0.01			
24	Person per household	0.11	-0.53	-0.43	-0.08	0.08	0.11	0.56	0.66	-0.46	-0.26	-0.33	-0.04	0.52	0.43	0.50	0.52	0.57	-0.13	0.24	-0.09	-0.01	0.26	0.11		
25	Median household income	-0.01	-0.50	0.16	-0.21	-0.03	0.01	-0.11	-0.17	-0.34	-0.08	-0.28	-0.04	0.50	0.28	0.15	0.57	0.45	0.76	0.43	-0.35	0.62	0.80	0.00	0.25	
26	Energy Price	-0.04	-0.74	-0.08	-0.58	-0.39	-0.02	-0.45	-0.50	0.14	0.29	-0.50	0.13	0.60	0.53	0.17	0.55	0.44	0.57	0.52	-0.48	0.66	0.78	-0.02	0.14	0.59

MODELING OF THERMAL COMFORT USING DESIGNBUILDER

Lucie DOBIÁŠOVÁ[1,*], Daniel ADAMOVSKÝ[1]

[1]Department of Microenvironmental and Building Services Engineering, CTU in Prague, Prague, Czech Republic

[*]Corresponding email: lucie.dobiasova@fsv.cvut.cz

Keywords: Thermal comfort, PMV index, PPD index, DesignBuilder

SUMMARY

The paper presents a case study which is dealing with the thermal comfort prediction. One of cinema halls in Prague was chosen as a convenient place for the data gathering. Commercially available software DesignBuilder was employed (the part of Computational Fluid Dynamics (CFD) analysis tool for the prediction of thermal comfort) and measurements in situ were done in order to obtain a comprehensive picture of the thermal comfort in the investigated building. The results of the used two methods (measurement in situ and CFD simulation) were compared.

INTRODUCTION

Thermal comfort is the condition of mind that expresses satisfaction with the thermal environment and is assessed by subjective evaluation (ASHRAE, 2013). There are several factors which affect the thermal comfort in a direct way, e.g. metabolic rate, clothing insulation, air temperature, air speed, operative temperature and mean radiant temperature. Two methods are widely known for the evaluation of the thermal comfort: an adaptive model and Predicted Mean Vote (PMV)/Predicted Percentage of Dissatisfied (PPD). The PMV/PPD method was developed by P. O. Fanger (1982) and it makes use of heat balance equations and empirical studies which relate the skin temperature to the comfort. The result of the evaluation is a scale with thermal sensations which is ranging from -3 (cold) up to +3 (hot) with 0 representing a neutral state.

METHODOLOGIES

Description of the Evaluated Environment

The measurements were done in Evald, a cinema hall in Prague, with the capacity of 72 seats. Ventilation and heating are provided by an air conditioning unit in the basement. 2400 m^3/h of air are supplied through diffusers located under the seats in the steps of the terraced auditorium. The exhausted air is extracted through the side wall of the hall (Figure 1). The air temperature in the hall is maintained at 23 °C.

The Measurement

The measurements took place during more than 40 screenings in April and July, with different room occupancy. The location of the stand with the measuring system marked is marked number 5 in Figure 1. It was placed in the last row, near a closed door during the screening. The axis of the sensors was at the head level of sitting spectators. The measured values were recorded every minute by the data logger ALMEMO 2590-4S. The measured parameters were: air and globe temperature (°C), relative humidity (%), air velocity (m/s) and CO_2 concentration (ppm).

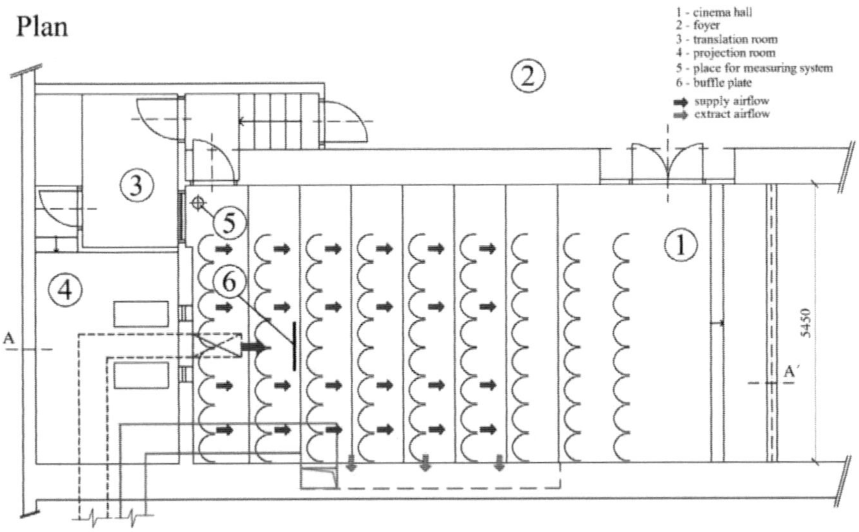

Figure 1. Cinema hall ventilation scheme.

Model for Evaluation of Thermal Comfort

The DesignBuilder was used to predicting and evaluate the thermal comfort. The cinema hall model (Figure 2) was created as a single-zone with adiabatic outer walls as the real hall is located in a building basement and is surrounded by other rooms, which means that heat transfer to adjacent rooms is negligible. The air-supplying elements have been simplified - the diffusers in the steps were replaced by linear rectangular openings. The inlet air parameters were set as in the real cinema, a total of 2400 m^3/h air at a temperature of 23 °C.

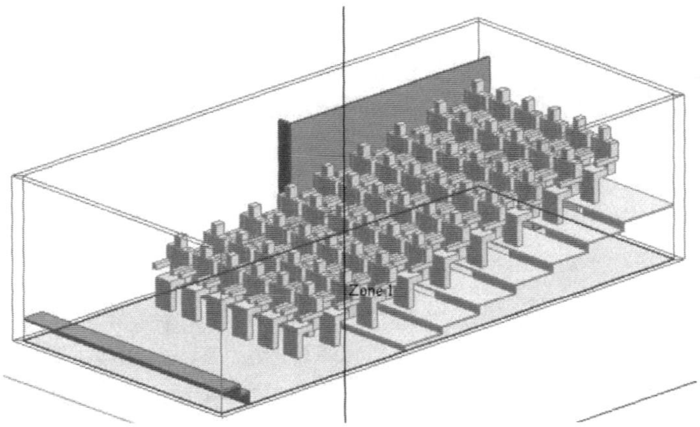

Figure 2. Model of cinema hall in DesignBuilder

RESULTS AND DISCUSSION

Site Measurements and Evaluation

The measurement evaluation was done according to the EN ISO 7730 standard (CEN, 2006) that introduces PMV and PPD indexes. Measured values obtained during all screenings were included into the evaluation. The calculations of the PMV and the PPD were done during the beginning and the end of the film and for the average values which occurred during each screening. Input data for calculation were: clothing insulation: 0,75 clo (estimate), metabolic heat production: 1 met, measured data: air temperature (°C), globe temperature (°C), relative humidity (%), air velocity (m/s) and mean radiant temperature – calculated value based on the measured data according to the EN ISO 7726 (CEN, 2002).

Figure 3 shows the frequency of the PMV and PPD indexes. The PMV index lies between 0.4 and -1, the levels range from neutral to slightly cool. This corresponds with about 30% unsatisfied spectators. The highest level of PMV was 0.4, which was close to the neutral level. This value was obtained in the following environment: t_a = 25.4 °C, t_g = 25.15 °C, v = 0.037 m/s and rh = 47.9%. The lowest level of PMV was -0.8 for slightly cool. The environmental parameters of this result were: t_a = 22.27 °C, t_g = 21.81 °C, v = 0.16 m/s RH = 42.6%.

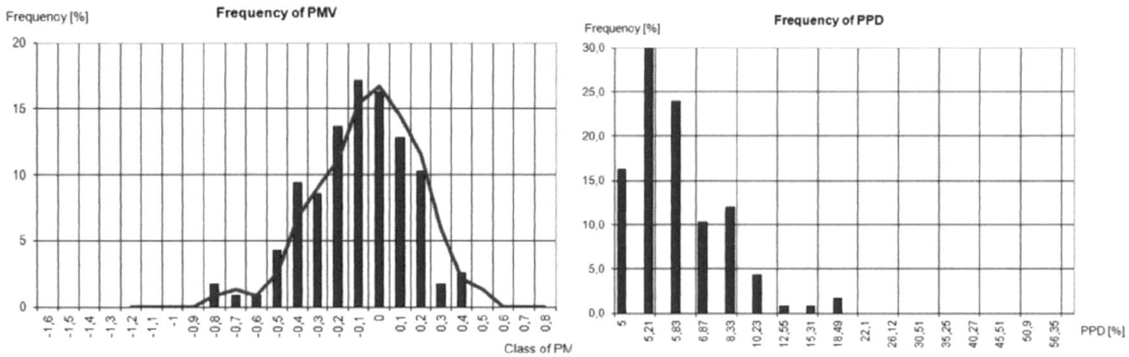

Figure 3. The Frequency of PMV and PPD indexes.

Simulations of Thermal Comfort

Figure 4 shows the distribution of PMV index in four cross sections. The PMV index values are about -0.5 (neutral - slightly cool) to 0.3 (neutral). Lower PMV index in the area of feet spectators is caused by higher air velocity, which is due to the location of the inlet air elements. It may cause local thermal discomfort. Figure 5 shows the distribution of the PPD index, which does not exceed 20% in the whole area.

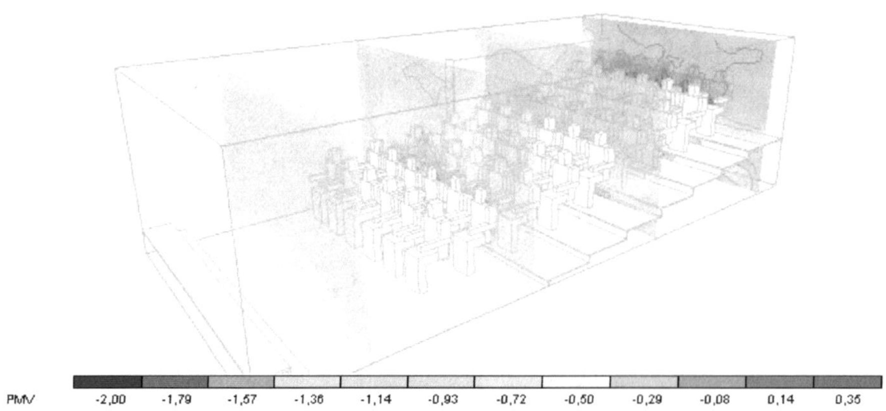

Figure 4. Distribution of PMV in the cinema hall.

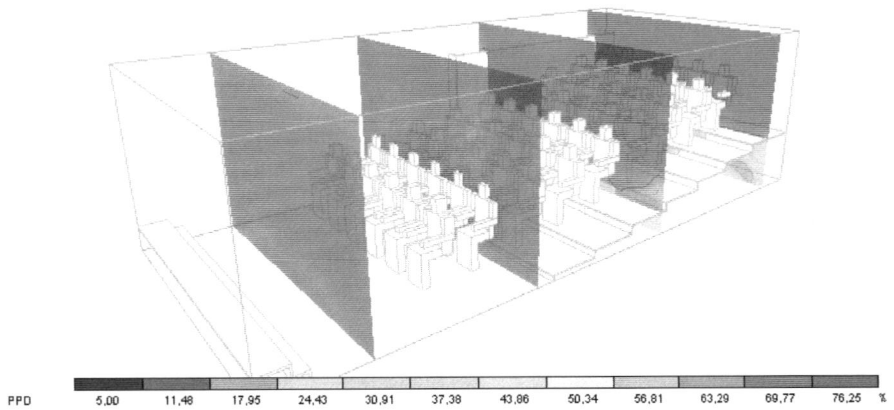

Figure 5. Distribution of PPD in the cinema hall.

CONCLUSIONS

The paper presented a case study of thermal comfort evaluation by PMV and PPD index. There were used two methods – measurement in situ and CFD simulation in DesignBuilder. The results of PMV index obtained from measured data lie between 0.4 and -1. This corresponds with about 30% unsatisfied spectators. The results of PMV index by DesignBuilder are about -0.5 to 0.3, that corresponds with about 20% unsatisfied. When we compare the results of both method (measurement and CFD), we can see very similar values. Both of these methods show PMV values about 0 (neutral level). The measurement in situ is time consuming and it requires expensive equipment. Also sometimes it cannot be possible to measure in situ (e. g. in some special places, during the design of the building). In these cases the measurement can be replaced by predicting thermal comfort using software, because we are able to get adequate results.

ACKNOWLEDGEMENT

This work was supported by the Grant Agency of the Czech Technical University in Prague, grant No. SGS15/129/OHK1/2T/11.

REFERENCES

ASHRAE, 2013. ANSI/ASHRAE Standard 55-2013, Thermal Environmental Conditions for Human Occupancy. Atlanta, GA: American Society of Heating, Refrigerating, and Air Conditioning Engineers.

DesignBuilder Software Limited. *DesignBuilder EnergyPlus Simulation Documentation for DesignBuilder v3.0* [online]. London, 2005-2011. Available from: http://www.designbuilder.co.uk/

Fanger PO (1970). Thermal comfort. *Analysis and applications in environmental engineering*. Copenhagen, Denmark: McGraw-Hill.

CEN, 2006. EN ISO 7730. *Ergonomics of the thermal environment - Analytical determination and interpretation of thermal comfort using calculation of the PMV and PPD indices and local thermal comfort criteria.* European Committee for Standardization, Brussels.

CEN, 2002. EN ISO 7726. *Ergonomics of the thermal environment - Instruments for measuring physical quantities.* European Committee for Standardization, Brussels.

FORMALDEHYDE SORPTION TO POROUS MEDIA FOR AIR QUALITY APPLICATIONS

Anne WROBETZ, Megan MATTEAZZI and Lupita D MONTOYA*

Department of Environmental Engineering, University of Colorado, Boulder, Colorado USA

*Corresponding email: Lupita.Montoya@colorado.edu

Keywords: Formaldehyde, VOC's, low-cost, air, remediation

SUMMARY

This paper discusses the formaldehyde sorption capacity of four media: Growstone, expanded clay, coco coir, and activated carbon. Breakthrough curves were generated with each of these media packed into an acrylic column at an inlet concentration of 0.4 ppm gaseous formaldehyde. Sorption potential, throughput, and characteristic times were generated from these data. The average sorption potential of Growstone, expanded clay, coco coir and activated carbon determined were 0.24, 0.57, 42.4 and 174.1 mg/g, respectively.

INTRODUCTION

Volatile organic compounds (VOCs) are common indoor air contaminants, which outgas from building materials such as particle-board (Pickrell et al., 1983), paints, varnishes or preservatives (Jones 1999). VOCs can contribute to various health problems, including Sick Building Syndrome (SBS), which is a term used to describe a range of illnesses (e.g., irritation of the eyes, nose and throat, dryness of skin, headaches, lethargy) that building occupants develop due to indoor contamination (Burge et al., 2004). Formaldehyde is the most common aldehyde in the environment and is often found in buildings (Jones 1999). Increases in the use of urea-formaldehyde insulation and tightly sealed buildings preventing natural ventilation have lead to net increases in indoor concentrations of formaldehyde (Spengler et al., 2009). Increasing mechanical ventilation rates is commonly used to control formaldehyde pollution but it also leads to increased energy use and cost (Salthammer et al., 2010). Formaldehyde has been shown to sorb onto various media, including those used to grow plants (Orwell et al., 2006). Sorption of formaldehyde by two growing media (Growstone, expanded clay) and activated carbon was first reported in Aydogan and Montoya (2011) and Aydogan (2012). The objective of the experiments presented here was to quantify the formaldehyde sorption capacity of Growstone, expanded clay, coco coir and activated carbon.

METHODS

The potential of 4 media to sorb gaseous formaldehyde was evaluated using a series of breakthrough experiments using the setup shown in Figure 1. In each evaluation, a formaldehyde-laden (~ 0.4 ppm) airstream was introduced into Chamber 1 at a flow rate of 6 liters per minute (LPM). The media used in these experiments include Growstone, Hydroton expanded clay, commercially available activated carbon and SunLeaves coco coir.

At the beginning of each experiment, the column was packed with one of the four media. For Growstone and expanded clay, the column was packed 10.5 inches (26.67 cm) high. For the coco coir and activated carbon, the column was packed up to 2 inches (5.08 cm) high. The media were not treated or altered in any way from their original state and they were not reused after a once-through process. Formaldehyde-laden air was introduced continuously into Chamber 1. After a steady-state concentration was reached in Chamber 1, the media-packed column was placed between the chambers. Formaldehyde levels were measured in both chambers until the concentration in Chamber 2 equaled that of Chamber 1; this was considered as the saturation point. When the media was fully saturated with sorbed formaldehyde, the experiment was stopped and the time to saturation was recorded.

Figure 1. Experimental setup

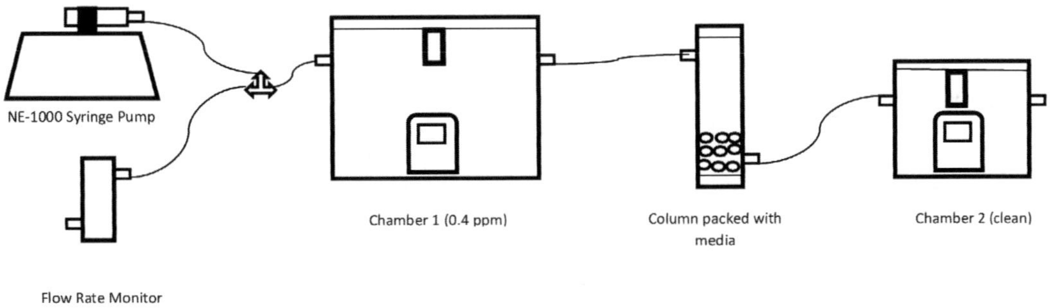

Experiments were done in triplicate for each media and were performed under a fume ventilation hood. The chambers were assessed for leakage. CO_2 gas was injected in a pulse and monitored for 24 hours. The leakage rate of Chamber 1 was 0.0032 hr^{-1} and that of Chamber 2 was 0.0028 hr^{-1}. All parts were tested for outgassing before the experiment began by monitoring formaldehyde levels for 6-hours. Levels were determined to be 0.00 ppm (or below the minimum detection level, defined by the manufacturer as 0.01 ppm) in all tests. Breakthrough curves were generated using the data obtained in each sorption experiment. These were analyzed for breakthrough time (T_B), retention time (T_R), equilibrium time (T_E), and total time (T_T) based on:

$$T_B = T_{(C = 0.05*C0)};\ T_R = T_{(C = 0.5*C0)};\ T_E = T_{(C=0.95*C0)};\ T_T = T_{(C=C0)}$$

The sorption potential for each media was determined following the method described by Chowdury et al. (2013):

$$q_e = \int (C_0 - C_e)\, dt * V / W$$

where q_e is the sorption potential in mg formaldehyde/g dry media, C_0 is the concentration of formaldehyde in Chamber 1 (0.4 ppm), C_e is the concentration of formaldehyde in Chamber 2, V is the total volume of air through the system, and W is the mass of porous media in the column.

The throughput, a dimensionless representation of sorption capacity, was calculated following the method from Singh et al., (2006):

$$\text{Throughput} = T_R / \theta$$

where T_R is retention time and θ is residence time in the system.

RESULTS AND DISCUSSION

Table 1 shows the average sorption potential, characteristic times and throughput evaluated for each media.. The average sorption potential was normalized by the dry mass of media used. The breakthrough time, retention time, equilibrium time and total time were normalized by the corresponding volume of media used: 0.16 L for activated carbon and coco coir and 0.85 L for Growstone and expanded clay.

Table 1. Sorption potential and characteristic times evaluated for the four media studied

Media	Avg. Sorption potential (q_e) (mg/g)	Avg. Breakthrough time (T_B) (min/L)	Avg. Retention time (T_R) (min/L)	Avg. Equilibrium time (T_E) (min/L)	Avg. Total time (T_T) (min/L)	Avg. Throughput (min/min)
Growstone	0.24	2.00	12.94	46.24	64.35	89.6
Expanded clay	0.57	9.76	90.59	238.82	246.71	548.5
Coco coir	42.4	29.34	673.1	3,348	3986	6563
Activated carbon	174.1	23,719	31,388	39,914	40,748	120,642

The average breakthrough curves (concentration data averaged) for each of the four media are presented in Figure 2. Activated carbon and coco coir were both tested at a quarter the volume of the other media. Activated carbon has the highest sorption potential and takes the longest time to reach breakthrough and is commonly used in sorption processes; however, it is not suitable as hydroponic media. On a dry mass basis, coco coir showed the best sorption potential compared to Growstone and expanded clay. The moisture content and temperature of the air were not evaluated, but may have caused variability in the results.

CONCLUSIONS

These results indicate that coco coir is a better sorbent for gaseous formaldehyde than either Growstone or expanded clay, but less effective than activated carbon. A porous media-based formaldehyde sorption system could utilize activated carbon or coco coir for maximum sorption capability; however, activated carbon cannot be used as a hydroponic growth medium. It could be used, however, in plant-based formaldehyde remediation systems if it is not in direct contact with the plants.

Figure 2. Average breakthrough curves for four media

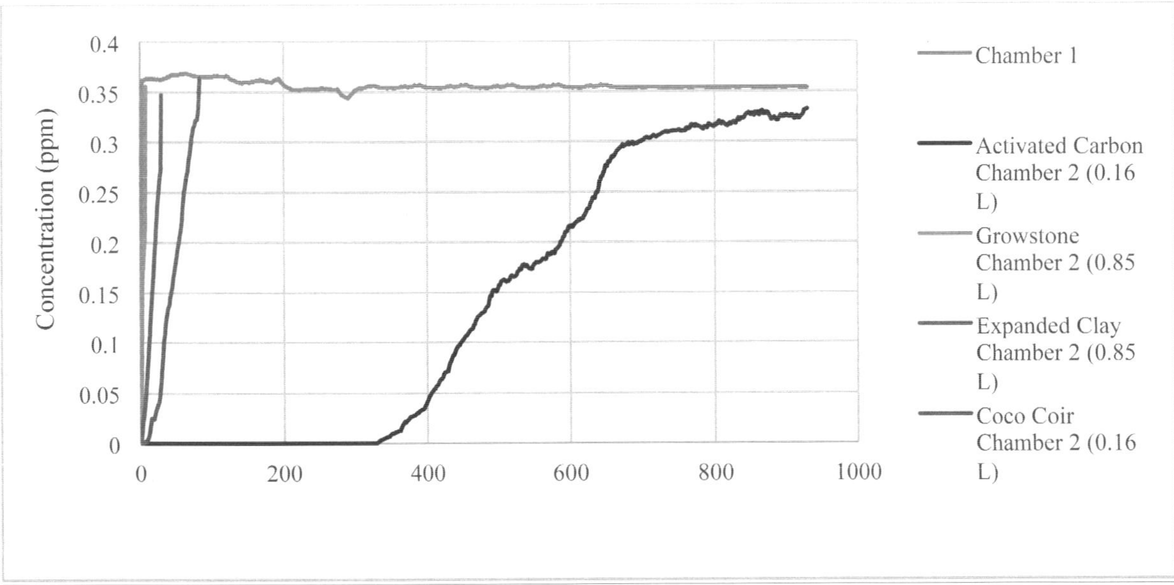

REFERENCES

Aydogan, Ahu. "Building-Integrated Active Modular Phytoremediation System." PhD Dissertation. Rensselaer Polytechnic Institute (2012): Troy, New York.

Aydogan, Ahu and Lupita D. Montoya. "Formaldehyde removal by common indoor plant species and various growing media." *Atmospheric Environment*, 45, 16 (2011): 2675- 2682.

Burge, P.S. "Sick Building Syndrome." Occupational & Environmental Medicine 61 (2004): 185-190.

Chowdury, Z.Z., S. M. Zain, A. K. Rashid, R. F. Rafique, and K. Khalid. "Breakthrough Curve Analysis for Column Dynamics Sorption of Mn(II) Ions from Wastewater by Using Mangostana garcinia Peel-Based Granular-Activated Carbon." *Journal of Chemistry* 2013 (2013): 1-8.

Jones, A.P. "Indoor Air Quality and Health." Atmospheric Environment 33 (1999):4535-4564.

Orwell, Ralph L., Ronald A. Wood, Margaret D. Burchett, Jane Tarran and Fraser Torpy. "The Potted- Plant Microcosm Substantially Reduced Indoor Air VOC Pollution: II. Laboratory Study." *Water, Air and Soil Pollution* 177 (2006): 59-80.

Salthammer, Tunga, Sibel Mentese, and Rainer Marutzky. "Formaldehyde in the Indoor Environment." *Chemical Reviews* 110 (2010): 2536-2572.

Sekine, Yoshika and Atsushi Nishimura. "Removal of formaldehyde from indoor air by passive type air-cleaning materials." Atmospheric Environment 35, 11 (2001): 2001-2007.

Singh, Tony Sarvinder and K.K. Pant. "Experimental and modelling studies on fixed bed adsorption of As(III) ions from aqueous solution." *Separation and Purification Technology* 48, 3 (2006): 288-296.

Spengler, John and Gary Adamkiewicz. "Indoor Air Pollution: An Old Problem with New Challenges." *International Journal of Environmental Research and Public Health* 6 (2009): 2880-2882.

INDOOR ENVIRONMENT QUALITY AND NABERS IE RATINGS: A 2014 CASE STUDY FOR A COMMERCIAL OFFICE BUILDING PORTFOLIO OF 25 AUSTRALIAN BUILDINGS

Tim M CALLINAN[1,*], Jack J NOONAN[1], Vyt P GARNYS[1]

[1]CETEC, Melbourne, Australia

[*]*Corresponding email:* tim.calllinan@cetec.com.au

Keywords: IEQ, benchmarking, NABERS, rating schemes, accreditation

SUMMARY

In 2013, a large property owner engaged a team to review, over a three year period, the Indoor Environment Quality of their twenty five commercial office buildings using the NABERS IE tool as a guide, and providing NABERS IE certification for eighteen of these buildings in Year 1. This was the first time a building owner of this size decided to certify their portfolio for NABERS IE ratings. In Year 2, twenty buildings were rated and the average improved by 0.5 stars to 4.5 stars. The NABERS IE tool, alongside the concentrated and diligent management and operational approach provided this improvement in the portfolio's building IEQ performance. A further key element has been attributed to the successful implementation and capturing of BMS data. This is an update of a paper presented at Indoor Air 2014 in Hong Kong which includes a comparison of Year 1 and Year 2's results.

INTRODUCTION

Accreditation and rating schemes are one way of assessing and benchmarking the performance of a building. The National Australian Built Environment Rating Scheme (NABERS) Indoor Environment (IE) tool is a unique benchmarking rating tool that is governed by federal and state governments in Australia. It has been subject of much discussion in recent times, with significant focus being placed on a lack of take up by building owners and facilities managers since introduction to the market in 2009 (in comparison to NABERS Energy, which is mandated, and NABERS Water). Furthermore, the tool is currently undertaking a review.

In 2013, a large property owner engaged a team to review, over a three year period, the Indoor Environment Quality of their twenty five commercial office buildings using the NABERS IE tool as a guide, as well as providing NABERS IE certification for eighteen of these buildings in 2013 and twenty in 2014. This was the first time a building owner of this size had decided to certify their portfolio for NABERS IE ratings.

METHODOLOGIES

The team undertook indoor environment quality assessments in accordance with the NABERS Indoor Environment Validation Protocol. Assessments took place in Melbourne, Sydney, Brisbane, Perth, and Canberra across a total of twenty five commercial office buildings ranging from 5,000 to 70,000 square meters of net lettable area (NLA).

The buildings were analysed for Base Building ratings (as opposed to Tenancy and/or Whole Building ratings). Indoor Environment parameters assessed included:

- thermal comfort (ambient temperature, relative humidity, and air velocity),
- ventilation,
- indoor air quality (particulates, carbon monoxide, airborne microbials), and
- acoustic comfort

Detailed methods for measuring each of these indoor environment parameters can be found in the NABERS Indoor Environment Validation Protocol.

Results were collated and used as inputs to the benchmarking NABERS IE spreadsheets. Each parameter measured was then weighted, resulting in a final star rating for the building.

RESULTS

Across the entire building portfolio of twenty five commercial office buildings, the average rating was 4.0 stars in 2013 and improved to 4.5 in 2014. Four buildings were awarded the maximum five star rating in 2013, the only four buildings to have achieved such a rating in Australia. The building owner decided to only proceed with formal accreditation (disclosure to the property industry) on buildings that achieved a minimum of four stars (nineteen of the twenty five buildings). By 2014, ten buildings achieved the five star ratings.

A summary of the results for each property can be seen in Table 1. Net lettable areas (measured in square metres) are listed as approximates in order to retain the confidentiality of the buildings reported.

Table 1: Individual building scores for NABERS Indoor Environment across commercial building portfolio – 2014 results

Office Properties	State	NLA (m2)*	Official Certified Rating - 2013	Official Certified Rating - 2014	Improvement
Building 1	ACT	15,000	4 Stars	4.5 Stars	0.5 star increase
Building 2	ACT	20,000	4 Stars	4.5 Stars	0.5 star increase
Building 3	ACT	5,000	NOT RATED (missing data)	NOT RATED	
Building 4	ACT	30,000	NOT RATED (major upgrade)	5 Stars	5 star for first time rated
Building 5	ACT	20,000	4 Stars	4.5 Stars	0.5 star increase
Building 6	ACT	25,000	4 Stars	4 Stars	
Building 7	ACT	20,000	NOT RATED (3.5 Stars)	NOT RATED	
Building 8	NSW	20,000	5 Stars	5 Stars	
Building 9	NSW	30,000	NOT RATED (3.5 Stars)	5 Stars	5 star for first time rated
Building 10	NSW	50,000	5 Stars	5 Stars	
Building 11	NSW	15,000	4 Stars	4 Stars	
Building 12	NSW	25,000	4 Stars	5 Stars	1 star increase
Building 13	NSW	25,000	4 Stars	4 Stars	
Building 14	NSW	15,000	NOT RATED (3.5 Stars)	4 Stars	First time rated
Building 15	NSW	15,000	NOT RATED (3.5 Stars)	4 Stars	First time rated
Building 16	NSW	10,000	4 Stars	NOT RATED (4 Stars)	
Building 17	QLD	40,000	5 Stars	5 Stars	
Building 18	QLD	20,000	4 Stars	4.5 Stars	0.5 star increase
Building 19	QLD	25,000	4.5 Stars	5 Stars	0.5 star increase
Building 20	QLD	5,000	4 Stars	NOT RATED (4 Stars)	
Building 21	VIC	60,000	4 Stars	4 Stars	
Building 22	VIC	70,000	4 Stars	5 Stars	1 star increase
Building 23	VIC	15,000	NOT RATED (3.5 Stars)	NOT RATED (3 Stars)	Under development
Building 24	VIC	50,000	5 Stars	5 stars	
Building 25	WA	30,000	NOT RATED (3.5 Stars)	5 stars	First time rated

Common areas where buildings failed to demonstrate a high performing indoor environment quality (and where the buildings subsequently lost points/stars) included:

- Ambient temperature – typically above 24 degrees Celsius
- Relative humidity – ACT properties had RHs of under 30%

- Indoor microbials – many buildings showed indoor microbials amplified over outdoor microbials
- Acoustic comfort – base building noise measured above 40 or 45 dBA

The study also identified two main limitations. The first limitation was the use of building management system (BMS) data in the process. It was observed that the use of the technology as inputs for thermal comfort could result in an additional one and a half star rating, meaning that a building which was assessed as a three and a half star building could be classified as a five star building simply for having additional BMS data. By addressing the collection of BMS data during 2014, a number of buildings improved their star rating by up to one star and this was a major contributing factor in the increase of 5 star rated buildings for this portfolio from five to ten buildings.

Secondly, it was observed in the field that it was difficult to measure base building noise as an input for acoustic comfort due to the typical hours of occupancy within the occupied commercial office buildings and the difficulty in isolating tenancy noise. This issue will be raised in the review of the tool, which is currently taking place.

CONCLUSIONS

The large building owner is using the high NABERS IE ratings as a means of demonstrating their ability to manage and maintain high indoor environment quality within their buildings and is also using the results to attract further tenants. Furthermore, they are using the results to manage the obligations of their contracted facilities managers. This case study is the first of its kind in Australia and it is anticipated that it will create a change in the market and put indoor environment quality firmly on building owners' agendas.

Limitations to the tool exist and a national technical working group has been reviewing the tool during 2014. These results will be critical to the review. A new version of the tool has been released in April 2015 under a pilot scheme and CETEC is conducting a comparative analysis of the pilot tool to the existing tool with Year 3 data. It is intended that the changes will facilitate greater uptake of the tool and a greater focus on indoor environment quality by the commercial property and construction sector in Australia.

ACKNOWLEDGEMENTS

The authors of this paper would like to acknowledge Adam Garnys, Andrew Bellamy, Laura Wilkes, Julien Colangelo, Raghuram Muguli, Travis Hale, Paulo Da Silva and Scott Stevens, and other members of the CETEC team for their assistance with the collection of data and preparation of reports.

WEB-BASED WIRELESS SENSOR SYSTEM FOR REMOTE AIR ENVIRONMENT MONITORING IN A COMMERCIAL WINERY

Nicolas M MADRID[1], David J KILLEEN[1], Roger B BOULTON[2] and André KNOESEN[1*]

[1] Department of Electrical & Computer Engineering, University of California, Davis, California, USA
[2] Department of Viticulture & Enology, University of California, Davis, California, USA

[*]Corresponding email: aknoesen@ucdavis.edu

Keywords: Sensors, sustainability, winery, environment, wireless

SUMMARY

A fully-automated, low-energy sensor system was developed to take periodic measurements of environmental parameters in industrial settings to monitor the evolution of microbiota in wineries. The system design minimizes maintenance of sensor nodes and allows for convenient access to the data at remote locations. The wireless sensor nodes are battery-powered and operate unattended for several months. This was accomplished by placing electronic subsystems in low-energy consuming states and only activating specific subsystems when needed. Data is collected periodically from the sensor nodes and transmitted wirelessly to a receiver where it is formatted and uploaded to the internet. The operation of the sensor system is illustrated by presenting air monitoring data collected in a commercial winery during the first four days of harvest.

INTRODUCTION

The ability to monitor environmental parameters in a building at low cost is made possible by technological advances in the consumer electronics sector where the use of wireless technology and low-cost sensors has become ubiquitous. Furthermore, there is an increasing interest in internet of things, where everyday objects have network connectivity allowing them to send and receive data and enable data collection at remote locations. For our research on microbiota populations within wineries, we take advantage of commercial sensors to monitor multiple parameters in real-time from a remote location. Our focus is on wineries where an important consideration is to deploy a sensor system that is unobtrusive to ongoing operations. Sensor nodes are often needed in locations where wired connections are too restrictive. Such untethered sensor nodes also simplify the installation as multiple sensor nodes can be deployed fast with minimal interruption to ongoing activities in the winery.

METHODOLOGIES

A wireless point to multipoint network topology using IEEE 802.15.4 communication protocol was implemented. This achieves low cost, low-energy, two-way wireless communication (Callaway et al., 2002) in a static sensor network that uses small data packets that are transmitted infrequently (Willig et al., 2005). The data from each node is sent to a single, centrally located receiver (ConnectPort X2) that places the information onto the cloud. The ConnectPort is wired to the Ethernet and electrical power is wired using Power over Ethernet technology.

Multiple sensors were integrated to create sensor nodes to sample several environmental parameters every 15 minutes at multiple sites in a winery. Two types of sensor nodes were developed. The main node accommodates sensors demanding large powers and has a wired power connection. The main node periodically collects carbon dioxide, relative humidity, temperature, and volatile organic compounds (VOCs) data and has an option to monitor dust particle concentration. The constant power draw requirements of the VOC sensor (iAQ-2000) and particle concentration sensor (Dylos DC1700) demand a wired electrical connection. A single sensor (COZIR, GC-0011) is used to detect temperature, carbon dioxide, and relative humidity. The second node is an untethered satellite sensor node that only uses sensors with low average power requirements and is designed to operate from a battery for several months. The satellite sensor node only contains a COZIR which requires low power and short turn on time (<3.0 seconds). The satellite nodes are located in various hard to reach locations throughout the facility (e.g., red and white barrel rooms, wine transfer areas, labs, and outside air).

The necessity for battery-powered satellite sensor nodes capable of operating over long intervals without battery replacement makes attention to energy conservation details very important. First, a rechargeable, high energy density battery is used (single lithium ion battery, Panasonic, NCR18650A, typical capacity 3.1 Ah, nominal voltage 3.6 V). Second, the energy consumption is managed to maximize the use of a single battery charge by providing power to specific electronic systems only when needed and otherwise keeping the electronic system in the lowest powered state possible. In our current implementation we achieve an estimated 48 months before the battery is depleted.

The energy consumption on the satellite node is categorized into five subsystems: the microcontroller, the real-time clock, the regulators, the wireless module, and the sensors. The microcontroller is the programmable state machine that reads information from the sensors at specified times, communicates with the wireless module, and programs the clock. The microcontroller transitions back and forth between collecting and sending data, and waiting for the next transmission interval. The waiting period dominates the overall time. Limiting energy consumption during the wait time is most critical in lowering the average energy use. The clock is a timer programmable by the microcontroller. Its main role in this application is to keep track of time when subsystems are switched off and wake them up when high energy consuming tasks need to be accomplished. The satellite nodes use the Programmable System on Chip (PSoC) 5 microcontroller (CY8C5467AXI-LP108). PSoC5 offers a sleep mode that consumes current as low as 2 μA, while switching off most microcontroller functions, but keeps a counter internal to the microcontroller active to serve as the clock. Voltage regulators connected to the lithium-ion battery provide a fixed voltage to the electronic components independent of the current usage. In the satellite node we used two 3.3V regulators. The first (NCP583SQ33T1G, 1 μA quiescent current and 0.1 μA shutdown current) is connected to the COZIR sensor and can be placed in shutdown mode. The other regulator (NCP585DSN33T1G, 4.5 μA quiescent current) serves the other electronic subsystems and the quiescent current is its lowest current mode because it is never shut down. During the waiting period between data transmissions, the XBee radio transmission module (Pro S1, XBP24-AUI-001, 2.4GHz, data rate 115.2 Kbps) and COZIR energy consumption must be limited as well. XBee Pro modules were selected for increased range required for building environments that will contain obstacles that attenuate the radio frequency signal. The choice comes at the expense of higher current consumption when transmitting. The XBee module has three current consumption states: hibernate (<10 μA), receive (55 mA), and transmit (250 mA). The COZIR sensor is shut down when measurements are not being taken by using the enable input on its dedicated voltage regulator. This gives the COZIR two current consumption states: sleep (0.1 μA) and active (~1.5mA).

Average energy consumption can be greatly reduced by developing code strategies that limit the need for retransmissions when errors occur. Error checking code to sense when an error occurs was implemented by sending the number of bytes that should be received and then the measurement data. Upon the detection of an error, an automatic repeat request occurs. In the event of consecutive data transmissions not being received correctly, a retry delay is implemented on the PSoC to prevent the XBee from draining the battery with constant retransmissions. Upon receiving the data correctly, the Python script formats the data into tab-delimited form and writes the entry into a data file that is accessible remotely (Etherios Device Cloud, Digi International). The data can then be easily graphed with another application, granting real-time models of several environmental conditions throughout a building. This simple error checking approach resulted in a high degree of success in three fully operational wineries during the 2014 harvest season. More advanced error correction techniques can also be implemented when needed (e.g. such as described in Li et al., 2010, Peng et al., 2013, Piyare et al., 2013).

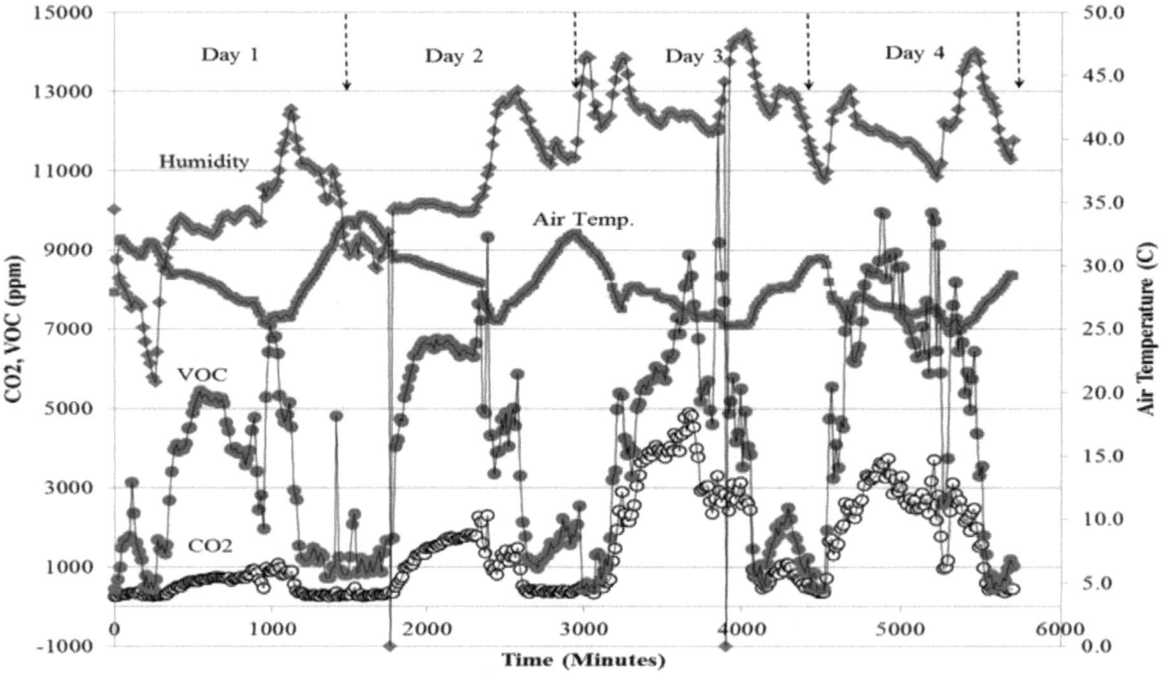

Figure 1. Ambient temperature, humidity, carbon dioxide and volatile organic compounds raw data collected in the fermentation hall of a commercial winery during the first 4 days of harvest.

RESULTS AND DISCUSSION

The data presented in Figure 1 shows the temperature, humidity, carbon dioxide, and VOCs data monitored with the system during the first four days of harvest at a winery 75 miles away. The daily patterns in temperature and humidity are due to the fact that there are doors open during the working period that are then closed at the end of the work day. This results in air cooling within the building and increasing humidity as wet surfaces from cleaning begin to dry out. The concentrations of carbon dioxide and VOCs both rise when the building is closed and decline the next morning when the doors are opened. The increasing peak height of these parameters is due to the increasing juice volume that is actively fermenting in the first days of the harvest. One important observation from this data was the realization that in this winery the carbon dioxide level rose to near the permissible exposure limit of 5000 ppm at about two and half days into the experiment even though it is equipped with carbon dioxide sensors that control exhaust fans in this fermentation room.

CONCLUSIONS

The system we developed demonstrates the feasibility of measuring environmental parameters in a winery during harvest and transmitting the data from a remote location to the internet. The strategy used to decrease the average power consumption allows the battery-powered sensor node to operate unattended for 48 months. The resulting data gives insight into indoor air parameters enabling decision-making based on real-time physical evidence. For example, using the graphed data, the ventilation system of a winery can be analysed and modified to ensure desired carbon dioxide and VOC levels are sustained.

ACKNOWLEDGEMENT

This work was supported by the Alfred P. Sloan Foundation's Microbiology of the Built Environment Program and the Stephen Sinclair Scott Endowment. Nicolas Madrid and David Killeen are supported by the T.J. Rodgers fellowships in Electrical and Computer Engineering.

REFERENCES

Callaway E, Gorday P, Hester L, et al., (2002). Home networking with IEEE 802.15.4: A developing standard for low-rate wireless personal area networks, *IEEE Commun. Mag.*, **40**, 70–77.

Willig A, Matheus K, Wolisz A, (2005). Wireless Technology in Industrial Networks, *Proc of IEEE*, **93**, 1130-1151.

Li L, Maunder RG, Al-Hashimi BM, et al., (2010). An Energy-Efficient Error Correction Scheme for IEEE 802.15.4 Wireless Sensor Networks, *IEEE Transactions on Circuits and Systems II – Express Briefs*, 57, 233-237.

Peng Y, Li Y, Shu L, et al., (2013). An energy-efficient clustered distributed coding for large-scale wireless sensor networks, *Journal of Supercomputing*, 66, 649-669.

Piyare R, Lee S, (2013). Performance Analysis of XBee ZB Module Based Wireless Sensor Networks, *International Journal of Scientific & Engineering Research, Volume 4, Issue 5.*

EXPOSURE ASSESSMENT TO E-CIGARETTES
Part 1: Literature Review on Carbonyl Compounds Generation from E-cigarettes and Affecting Factors

Wenhao CHEN[1,*], Ping WANG[1], Jeff FOWLES[2], Toshiki MATSUO[1,3], Dennis SHUSTERMAN[4], Alfred T. HODGSON[5], Kazuhide ITO[3], Kazukiyo KUMAGAI[1]

[1]Environmental Health Laboratory, California Department of Public Health, CA, USA
[2]Environmental Health Investigations Branch, California Department of Public Health, CA, USA
[3]IGSES, Kyushu University, Fukuoka, Japan
[4]School of Medicine, University of California at San Francisco, CA, USA
[5]Berkeley Analytical, Richmond, CA, USA

[*]Corresponding email: wenhao.chen@cdph.ca.gov

Keywords: E-Cigarette, E-Liquid, Carbonyls Emission, Heating Temperature

SUMMARY

Electronic cigarettes (e-cigarettes) are combustion-less battery-powered devices that are able to vaporize a liquid solution. Existing studies have found various amounts of hazardous carbonyl compounds (e.g., formaldehyde, acetaldehyde and acrolein) generation from e-cigarettes, posing public health concerns. In this paper, we reviewed the range of hazardous carbonyls emission and the range of operating temperatures in e-cigarettes reported in literature. Further, we discussed product design /operating factors that can potentially affect carbonyls emission from e-cigarettes. We concluded that the operating temperature inside heating unit of e-cigarette is one of key factors that influence carbonyls emission, and identified questions that need to be answered in future studies.

INTRODUCTION

Electronic cigarettes (e-cigarettes) are combustion-less battery-powered devices that are able to vaporize a liquid solution. E-cigarette sales in the U.S. have been growing in triple-digit territory over the past few years. E-cigarettes have been heavily marketed to young users, which poses additional public health concerns. Nationally, the use of e-cigarettes by high school students tripled in just two years and e-cigarette use by teens now surpasses the use of traditional cigarettes (CDPH, 2015).

Although e-cigarettes may emit less harmful chemical compounds compared to cigarette smoking, there is increasing evidence that they produce more than "harmless water vapor". On the other hand, the regulations and product quality control requirements on e-cigarettes today are still very limited. More scientific data on how e-cigarette product design/operating conditions affect emissions (therefore exposures) are needed for better risk management of e-cigarettes.

As part of a series which report preliminary findings from the ongoing e-cigarette research in the Indoor Air Quality Section (IAQS) within the California Department of Public Health (CDPH) and the simulation group within Kyushu University of Japan, this paper focuses on reviewing carbonyls emission from e-cigarettes and its possible mechanisms. It further discusses how heating temperature and other product design factors potentially affect carbonyls emission.

METHODOLOGIES

Literature reviews were conducted on e-cigarette operating temperature measurements, carbonyls emission from e-cigarettes as well as it possible mechanisms.

RESULTS AND DISCUSSION

Mechanisms of Carbonyls Generation and Affecting Factors

Many existing studies have found carbonyl compounds (formaldehyde, acetaldehyde, etc.) in electronic cigarette vapors (Kosmider et al, 2014; Cheng 2014; Goniewicz et al., 2014; Bekki et al., 2014). However, the mechanisms of carbonyls emission from e-cigarettes are not fully clear. Besides the e-liquid impurities, one most commonly accepted hypothesis is that the carbonyls are generated when heating up propylene glycol (PG), glycerin or their mixture which are used as carrier solvents, constituting up to 95% of all components in the e-liquid.

The thermal oxidation or pyrolysis reaction pathways of PG and/or glycerin have theoretically been proven at elevated temperatures. For example, Diaz et al. (2010) measured homogeneous oxidation products from PG at 204°C (430 K) using a 9 cm^3 reactor vessel with 0.28 kPa PG and 0.89 kPa O_2 in a flowing gas stream. They observed formaldehyde, acetaldehyde and acetone as major products. They also observed increase of reaction rate as the temperature increased. When studying the pyrolysis of glycerin, formaldehyde, acetaldehyde and acrolein have been observed as major products (Stein et al., 1983). However, no existing studies have explicitly investigated the effect of heating temperature on carbonyls generation in the presence of normal air under the controlled temperatures and PG/glycerin consumption amounts that are potentially reachable for an e-cigarette.

The oxidation/reaction inside real e-cigarettes can be much more complex due to the effect of other components in e-liquid (flavoring compounds, nicotine, etc.) and difference in product designs (position of air inlet, type of metal components, non-uniform heat, etc.). For example, the metal wires may function as catalyst and accelerate the oxidation reactions at lower temperatures. Bekki et al. (2014) hypothesized that the carbonyl compounds are incidentally generated by the oxidation of e-liquid (glycerol and glycols) when the liquid came in contact with the heated nichrome wire and suggested the compositions and concentrations of these compounds vary depending on the type of e-liquid and the battery voltage.

Range of E-cigarette Operating Temperature

Table 1 summarizes the range of heating temperature reported in the literature. Only limited data are available. They suggest a wide range from 40 to >350°C depending on the measurement location, detection method as well as the e-cigarette product type.

Range of Carbonyls Emission

The amounts of carbonyls emission reported in literature vary significantly from below detection limit to over 10 μg/puff among different products, different samples of the same product, different user choices of voltage setting, as well as different studies (Kosmider et al, 2014; Cheng 2014; Goniewicz et al., 2014; Bekki et al., 2014).

Bekki et al. (2014) measured carbonyl compounds generation from 13 brands of Japanese e-cigarettes. They found that carbonyls emission varied significantly not only among different brands but also among different samples of the same product. The maximum concentrations of formaldehyde, acetaldehyde, acrolein, propoanal, glyoxal, and methylglyoxal could reach 140, 120, 40, 46, 23, and 21 μg/10 puffs, respectively.

Goniewicz et al. (2014) tested 12 brands of e-cigarettes distributed in Poland and found 4 carbonyls in e-cigarette vapors. The concentration ranged from 3.2 to 56.1 μg for formaldehyde, 2.0 – 13.6 μg for acetaldehyde, non-detectable to 41.9 μg for acrolein, and 1.3 to 7.1 μg for o-methylbenzaldehyde per one e-cigarette (150 puffs), respectively.

Kosmider et al. (2014) further investigated the effect of solvents and battery output voltage on carbonyls emission using a commercially available e-cigarette with refillable tank and adjustable voltage. Besides commercially available e-liquids, they tested three control solutions (C1 with 88.2% of VG, 10.0% of distilled water, and 1.8% of nicotine; C2 with 44.1% of VG, 44.1% of PG, 10.0% of distilled water, and 1.8% of nicotine; C3 with 88.2% of PG, 10.0% of distilled water, and 1.8% of nicotine) at a voltage setting of 3.2V, 4.0V, and 4.8V, respectively. They concluded that both solvent and battery output voltage significantly affected levels of carbonyl compound in e-cigarette vapors, and levels of carbonyls rapidly increased with increased battery output voltage, especially from 4.0 to 4.8V.

Besides gas-phase formaldehyde, a more recent study by Jensen et al. (2015) analyzed aerosols generated from a commercial e-liquid using a variable-voltage "tank system" e-cigarette, and found that more than 2% of the total solvent molecules converted to formaldehyde hemiacetal, a formaldehyde releasing agent, in e-cigarette aerosols at high voltage setting of 5.0V. Although how this formaldehyde-releasing agent behaves in the respiratory tract is unknown, it could potentially cause formaldehyde inhalation exposure significantly higher than cigarette smoking if this formaldehyde-containing agent releases most of its formaldehyde component (Jensen et al., 2015).

Table 1 Reported Heating Temperature in E-cigarettes

E-cigarette Type	Reported Heating Temperature	Reference
Disposable or rechargeable e-cigarettes with a fixed voltage setting.	**40 – 65 °C** in vaporization chamber (The temperature of the heating element in each e-cigarette was determined by inserting a thermocouple and then activating the e-cigarette by drawing air through it)	Westenberger et al. (2009)
	>350 °C in the center of heating unit (The mouthpiece and the wick were removed from the e-cigarette and the temperature of the heating coil was measured via thermography (ThermaCAM B20) during heat-up)	Schripp et al. (2013)
	65 - 120 °C	Schaller et al. (2013)
	Some manufacturers claim they limit the temperature to **<100 °C**.	Bertholon et al. (2013)
E-cigarettes with adjustable voltage settings	Lack of data	Bertholon et al. (2013)

CONCLUSIONS

Through literature review, we observed the increase of carbonyls emission as the increase of battery output voltage, the coincidence of large variations of carbonyls emission and operating temperature among e-cigarettes, and the feasible thermal oxidation or pyrolysis pathways of PG/glycerin under elevated temperatures. We conclude that heating temperature is one of key factors that influence carbonyls emission. Further research is needed on the quantitative relationship between heating temperature and carbonyls emission (and exposure) from E-liquids to determine if there a need to regulate maximum allowable operating temperature for e-cigarettes, and if so, how to define and measure.

DISCLAIMER

Conclusions and opinions are those of the individual authors and do not necessarily reflect the policies or official views of the California Department of Public Health.

ACKNOWLEDGEMENTS

The authors would like to thank Mr. Francis J. Offermann for helpful discussions.

REFERENCES

Bekki, K., Uchiyama, S., Ohta, K., Inaba, Y., Nakagome, H., & Kunugita, N. (2014). Carbonyl Compounds Generated from Electronic Cigarettes. International journal of environmental research and public health, 11(11), 11192-11200.

Bertholon, J. F., Becquemin, M. H., Annesi-Maesano, I., & Dautzenberg, B. (2013). Electronic cigarettes: a short review. Respiration, 86(5), 433-8.

California Department of Public Health (2015) California Tobacco Control Program, State Health Officer's Report on E-Cigarettes: A Community Health Threat, Sacramento, California.

Cheng, T. (2014). Chemical evaluation of electronic cigarettes. Tobacco control, 23(suppl 2), ii11-ii17.

Díaz, E., Sad, M. E., & Iglesia, E. (2010). Homogeneous oxidation reactions of propanediols at low temperatures. ChemSusChem, 3(9), 1063-1070.

Goniewicz, M. L., Knysak, J., Gawron, M., Kosmider, L., Sobczak, A., Kurek, J., ... & Benowitz, N. (2014). Levels of selected carcinogens and toxicants in vapour from electronic cigarettes. Tobacco control, 23(2), 133-139.

Jensen, R. P., Luo, W., Pankow, J. F., Strongin, R. M., & Peyton, D. H. (2015). Hidden Formaldehyde in E-Cigarette Aerosols. New England Journal of Medicine, 372(4), 392-394.

Kosmider, L., Sobczak, A., Fik, M., Knysak, J., Zaciera, M., Kurek, J., & Goniewicz, M. L. (2014). Carbonyl compounds in electronic cigarette vapors: effects of nicotine solvent and battery output voltage. Nicotine & Tobacco Research, 16(10), 1319-1326.

Schripp, T., Markewitz, D., Uhde, E., & Salthammer, T. (2013). Does e-cigarette consumption cause passive vaping?. Indoor air, 23(1), 25-31.

Schaller, K. (2013). Electronic cigarettes - an overview. Tob Prev Tob Control, 19, 1-39, accessible at http://www.dkfz.de/en/presse/download/RS-Vol19-E-Cigarettes-EN.pdf.

Stein, Y. S., Antal, M. J., & Jones, M. (1983). A study of the gas-phase pyrolysis of glycerol. Journal of Analytical and Applied Pyrolysis, 4(4), 283-296.

Westenberger, B. J. (2009). Evaluation of e-cigarettes. Food and Drug Administration, USA, accessible at http://www.fda.gov/downloads/drugs/scienceresearch/ucm173250.pdf

EXPOSURE ASSESSMENT TO E-CIGARETTES
Part 2: A Pilot Laboratory Study on Formation of Volatile Carbonyls from Propylene Glycol – a Major E-cigarette Carrier Solvent

Wenhao CHEN[1,*], Ping WANG[1], Jeff FOWLES[2], Toshiki MATSUO[1,3], Dennis SHUSTERMAN[4], Alfred T. HODGSON[5], Kazuhide ITO[3], Kazukiyo KUMAGAI[1]

[1]Environmental Health Laboratory, California Department of Public Health, CA, USA
[2]Environmental Health Investigations Branch, California Department of Public Health, CA, USA
[3]IGSES, Kyushu University, Fukuoka, Japan
[4]School of Medicine, University of California at San Francisco, CA, USA
[5]Berkeley Analytical, Richmond, CA, USA

[*]Corresponding email: wenhao.chen@cdph.ca.gov

Keywords: E-Cigarette, Propylene Glycol, Carbonyls Emission, Heating Temperature, Micro-chamber

SUMMARY

Electronic cigarettes are combustion-less battery-powered devices that are able to vaporize a liquid solution. An e-cigarette usually has a heating element, and uses propylene glycol, glycerin or a mixture of the two as a carrier solvent. In this paper, we report a pilot laboratory study investigating the effect of heating temperature on the formation of carbonyls from propylene glycol. We vaporized pure propylene glycol on petri-dish in stainless steel micro-chambers at controlled temperatures of 150, 200 and 250 °C, respectively, and collected volatile carbonyls at the chamber outlet. We observed the formation of formaldehyde, acetaldehyde and propionaldehyde at these elevated temperatures. There was a general trend for an exponential increase of these carbonyls with increasing temperature. At 250 °C, there were clear concentration elevations in all three carbonyls. Additional tests on more carrier solvents using a tubular reactor are ongoing to cover broader temperatures and to better mimic e-cigarette flow pattern.

INTRODUCTION

Electronic cigarettes (e-cigarettes) are combustion-less battery-powered devices that are able to vaporize a liquid solution. An e-cigarette usually has a heating element and uses propylene glycol (PG), glycerin or a mixture of the two as a carrier solvent, with PG being the most commonly used (Grana et al., 2014). Our literature review has found the increase of carbonyls emission as the increase of battery output voltage, the coincidence of large variations of carbonyls emission and operating temperature among e-cigarettes, and the feasible thermal oxidation or pyrolysis pathways of PG/glycerin under elevated temperatures (Chen et al., 2015). We are currently conducting research to investigate the quantitative relationship between heating temperature and carbonyls emissions from carrier solvents and e-liquids. This paper reports the preliminary results for PG using micro-chamber system.

METHODOLOGIES

Testing Solvent Property

Pure propylene glycol (PG) (Sigma-Aldrich® W294004, ⩾ 99.5%, FCC) was used. Table 1 summarizes its properties. Tests were conducted for 5, 10, 30 and 100 mg PG injection, respectively. Jensen et al. (2015) reported an e-liquid consumption of 5 to 11 mg with each puff, which corresponds to about 4.5 – 10 mg PG/puff if assuming 90% of the e-liquid component is PG.

Table 1 Properties of Propylene Glycol

Molecular formular	$C_3H_8O_2$
Molecular weight (g/mol)	76
Density (g/cm^3)	1.036
Boiling point (°C)	188
Saturated vapor pressure at 25 °C (mmHg)	0.129

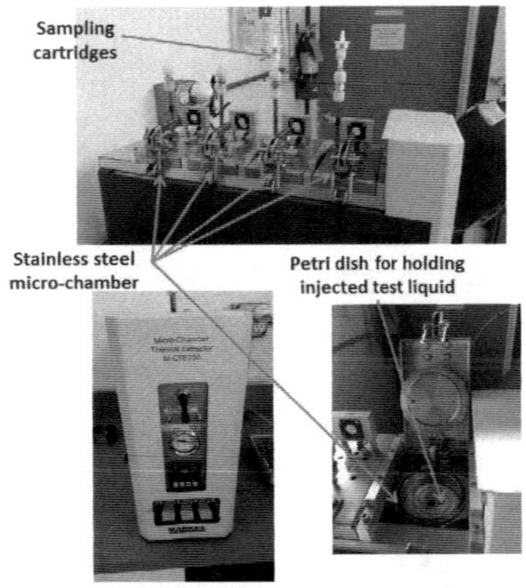

Figure 1 Micro-chamber Set-up

Experimental Set-up

Figure 1 shows the experimental set-up. A Micro-Chamber/Thermal Extractor (μ-CTE250, Markes International) was used for this study. It is commercially available and has been widely used for fast screening of emissions from building and consumer products. It consists of four individual identical stainless-steel cylindrical micro-chambers (114 mL capacity). These micro-chambers are located in a heated block and well-sealed with o-rings, and their temperatures can be precisely controlled from above ambient temperature to 250°C. All four chambers were supplied with the same controlled flow from an ultra-high purity dry air cylinder. Air entered each micro-chamber through the lid and the entire exhaust flow from each chamber was passed through a sampling outlet on the top. The flow rate was set at 100 ml/min for each chamber.

Tests were conducted at controlled temperatures of 150, 200 and 250 °C, respectively. For each test, a known amount of PG (5, 10, 30 or 100 mg) was injected on a glass petri-dish using syringe. The petri-dish with PG was quickly put inside the chamber. The petri-dish remained inside the chamber for 1 hr for PG vaporization and then was quickly taken out. The chamber was preloaded with spacers that lifted the petri-dish up so that the distance between the chamber's lid and the surface of PG liquid/film was kept at about 2 cm. The petri-dish was weighed using balance (Mettler AE 160, resolution: 0.0001g) right before putting in and after getting out of the chamber. The balance measurements confirmed that all injected PG was vaporized during the sampling period and that all injections were within ± 5% of desired PG injection amount.

Determination of Carbonyls Emission

We collected and analyzed volatile carbonyl compounds following an established procedure recommended by the U.S. Environmental Protection Agency (USEPA 1999). First, dinitrophenylhydrazine (DNPH)-impregnated silica gel cartridges (WAT047205, Waters, MA) were placed at the outlet of micro-chambers using stainless steel connection tubes (7 cm long) to collect carbonyl compounds. The cartridges were then extracted with 10-mL acetonitrile aliquots, and analyzed by HPLC with UV detection (Agilent 1200).

During the test, three consecutive 1-hr integrated DNPH cartridge samples were taken for each micro-chamber: cartridge #1 right before putting petri-dish into chamber (empty chamber), cartridge #2 right after putting petri-dish (with PG), and cartridge #3 right after moving the petri-dish out of the chamber (empty chamber again). The average amount of cartridges #1 and #3 was used to count for system background and subtracted from the amount of cartridge #2 to obtain a net generation amount of carbonyls due to PG injection.

RESULTS AND DISCUSSION

Figures 2a – 2d show carbonyls formed under various test temperatures for PG injection amount of 5, 10, 30 and 100 mg, respectively. A general trend of emission increase with increasing temperature was observed for three carbonyls, formaldehyde, acetaldehyde and propionaldehyde, under all PG injection levels. Figure 3 summarizes the normalized generation rate of carbonyls, which fits an exponential growth function of heating temperature reasonably well for all three carbonyls. At 250 °C, the formation rates of carbonyls became very significant, approaching 0.17 – 0.30 μg formaldehyde, 0.22 – 0.33 μg acetaldehyde, and 0.14 – 0.37 μg propionaldehyde per mg PG. Our study has the limitation of using a pure solvent in a controlled lab device instead of real e-liquids and e-cigarettes. However, these preliminary test results, for the first time, established the quantitative relationship between the carbonyl formation rate from e-liquid carrier and heating temperature in the presence of normal air and under the carrier solvent consumption amounts and temperatures that are potentially (intentionally or unintentionally) reachable for an e-cigarette.

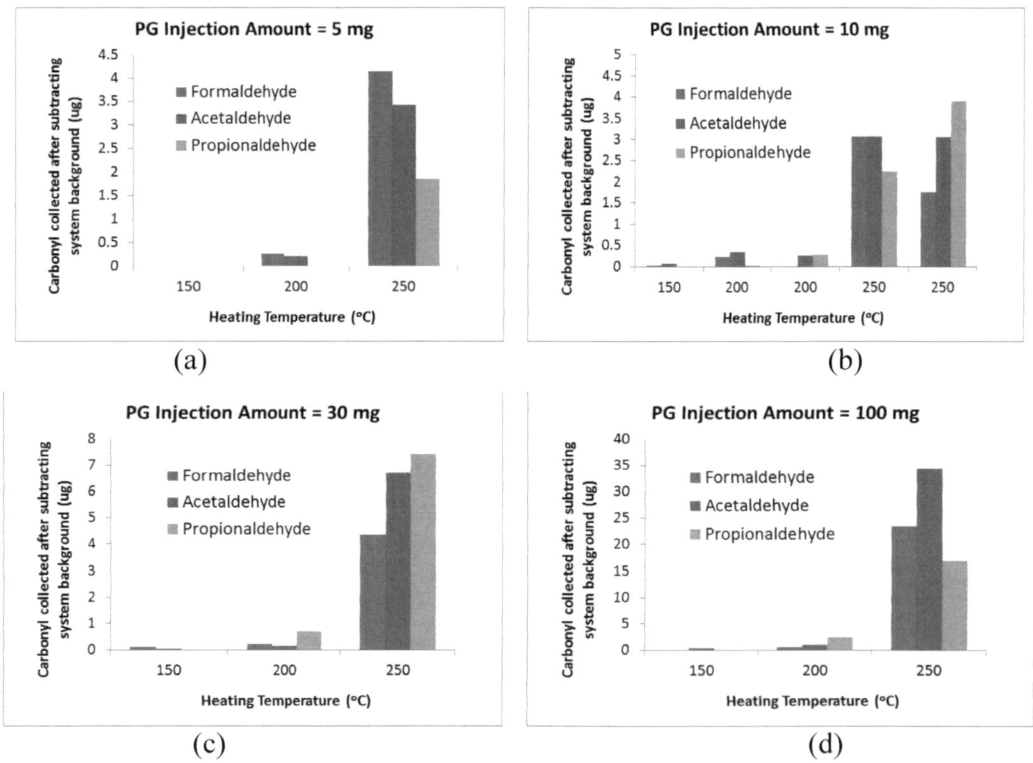

Figure 2 Carbonyls Formed under Different Heating Temperatures ((a) PG injection = 5mg; (b) PG injection = 10mg; (c) PG injection = 30mg; (d) PG injection = 100mg)

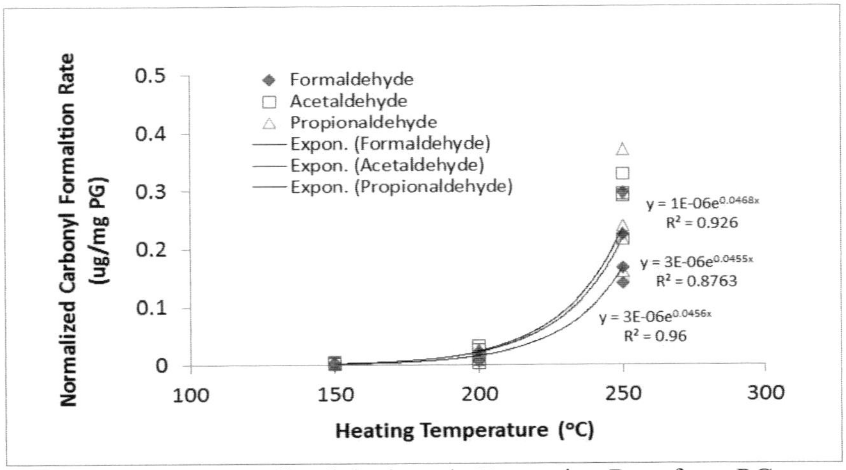

Figure 3: Normalized Carbonyls Formation Rate from PG

CONCLUSIONS

Although the oxidation/reaction inside commercial e-cigarettes can be much more complex due to the effect of other components in e-liquid and difference in product design, the preliminary results presented in this paper explicitly demonstrate the effect of heating temperature on e-liquid carrier emission and represent our first effort towards better understanding the key design factors that affect e-cigarette emissions. Additional tests on more carrier solvents (e.g., both propylene glycol and glycerin) using a stainless steel, tubular reactor are ongoing to cover broader temperatures and to better mimic e-cigarette flow pattern. Further emission studies using commercial e-liquids and e-cigarettes are also planned.

DISCLAIMER

Conclusions and opinions are those of the individual authors and do not necessarily reflect the policies or official views of the California Department of Public Health.

ACKNOWLEDGEMENTS

The authors would like to thank Dr. Na Li and Mr. Farid Masri for providing assistance for micro-chamber operation, and Mr. Francis J. Offermann for helpful discussions.

REFERENCES

Chen, W., Wang, P., Fowles, J., Matsuo, T., Shusterman, D., Hodgson, A.T., Ito, K., & Kumagai, K. (2015). Literature Review on Carbonyl Compounds Generation from E-cigarettes and Affecting Factors, submitted to Healthy Buildings America 2015.

Grana, R., Benowitz, N., & Glantz, S. A. (2013). Background paper on E-cigarettes (electronic nicotine delivery systems), accessible at https://nicotinepolicy.net/documents/position_papers/Grana_Glantz_WHO_ENDS_Report_Dec2013.pdf.

Jensen, R. P., Luo, W., Pankow, J. F., Strongin, R. M., & Peyton, D. H. (2015). Hidden Formaldehyde in E-Cigarette Aerosols. New England Journal of Medicine, 372(4), 392-394.

USEPA (1999). Compendium Method TO-11A - Determination of formaldehyde in ambient air using adsorbent cartridge followed by HPLC [active sampling methodology].

EXPOSURE ASSESSMENT TO E-CIGARETTES
Part 3: Preliminary Numerical Prediction of Contaminant Distributions in Human Respiratory Tract Models

Kazuhide ITO[1,*], Masato YAMASHITA[1], Sung-Jun YOO[1], Toshiki MATSUO[1,2], Wenhao CHEN[2], Ping WANG[2], Jeff FOWLES[3], Kazukiyo KUMAGAI[2]

[1]IGSES, Kyushu University, Fukuoka, Japan
[2]Environmental Health Laboratory, California Department of Public Health, CA, USA
[3]Environmental Health Investigations Branch, California Department of Public Health, CA, USA

[*]Corresponding email: ito@kyudai.jp

Keywords: E-Cigarette, Exposure assessment, CFD, Numerical airway

SUMMARY

The purpose of this paper (Part 3) is to investigate flow patterns and contaminant distributions in the human respiratory tract using 3-dimensional computational fluid dynamics analysis. Numerical airway models were generated reproducing detailed respiratory tract geometries using CT (Computed Tomography) data and CFD (Computational Fluid Dynamics) simulations that replicate e-cigarette smoking conditions. Complicated flow patterns and non-uniform contaminant concentration distributions were formed within the numerical airway models and the individual specificity was also confirmed.

INTRODUCTION

While there are a number of uncertainties regarding the ingredients and the short-term/ long-term health impacts of e-cigarettes, the number of young people (especially teens) using them is increasing, even surpassing the number of traditional cigarette users for the first time; more than twice as many 8th and 10th graders have reported choosing e-cigarettes over traditional cigarettes (CDPH, 2014). Needless to say, the human airway is the hub for the body's gas exchange, a function where lungs interact with ambient air. Besides that, this organ is also the first line of defense against harmful agents that could potentially be in the ambient air. From this standpoint, it is essential to assess inhalation exposure from e-cigarettes in relation to human health.

Part of a series, this one article focuses on the study that assesses e-cigarette exposure, for which we conducted fundamental experiments to identify the vaping chemical compounds consumed through e-cigarette and their emission rates. Furthermore, we also carried out numerical simulations to predict the distribution of inhalation exposure risk in the respiratory tract. In this paper (Part 3), we report the results of an *in silico* experiment to investigate the flow and volatile organic compound (VOC) distributions within a realistic model of the human airway.

METHODOLOGIES

Two types of three-dimensional (3D) respiratory tract models were developed using computed tomography (CT) data obtained from two healthy human males (Hirase et al., 2014 and Yamashita et al., 2014). One Asian and one European male made up the two test subjects; both volunteers were non-smokers and had a body mass index (BMI) of approximately 21. The original CT images were converted into a file format compatible with Mimics® (Materialise NV) and airway geometries from the oral/nasal cavity to bronchial tubes were extracted from CT data and were output as STL format. ICEM and TGrid® (ANSYS) were used to modify the surface geometries and meshes, from which we created volume meshes of these models. Table 1 shows schematics of the human airway models (Models A and B) after reconstruction from CT images. These respiratory tract models were composed of the upper airway (nasal and oral cavity, pharynx, and larynx) and the lower airway (trachea and bronchial tree).

Table 1 Schematics of the human airway models (Models A and B)

	Model A (Asian)		Model B (European)	
Body height	1.7m	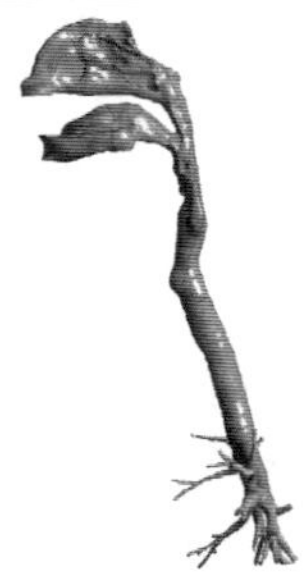	1.7m	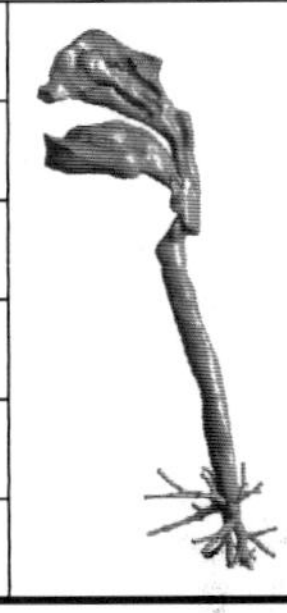
BMI	21		21	
Size (Height)	0.34568m		0.27381m	
Inner Surface area (A)	$0.057967 m^2$		$0.044637 m^2$	
Volume (V)	$1.7336\times10^{-4} m^3$		$1.2862\times10^{-4} m^3$	
V/A	$2.9908\times10^{-3} m$		$2.8814\times10^{-3} m$	

Airflow profiles under e-cigarette smoking conditions were calculated by CFD simulations. Steady flow fields were analyzed using a low Reynolds (*Re*) number-type k-ε model. The turbulent kinetic energy at the inlet, i.e., at the mouth, was prescribed assuming 10% turbulence intensity. A no-slip boundary condition was applied for the wall surfaces inside the airway model. The second-order upwind scheme was used for the convection term, and the SIMPLE algorithm was used.

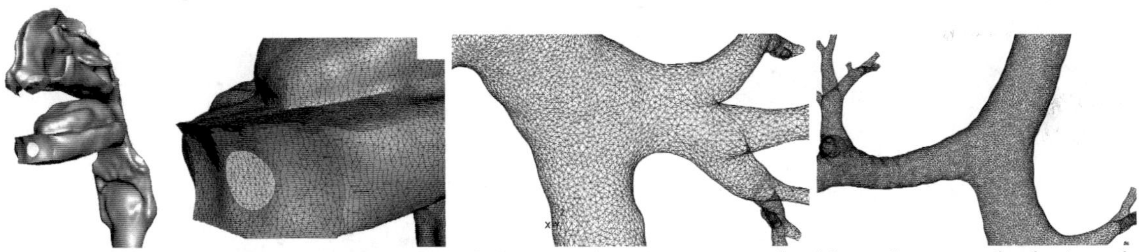

Figure 1 Detail grid design of mouth, inlet opening (correspond with e-cigarette device) and bronchial tube (Model A)

Table 2 Numerical and boundary conditions

Turbulent Model	Low Reynolds number-type k-ε model (Abe–Kondoh–Nagano model)
Mesh	7.5 million unstructured mesh (A) / 4.1 million unstructured mesh (B)
Scheme	Convection term: First order upwind
Inflow Boundary	Q_{in} =2.34 L/min, (U_{in}= 0.77 m/s), $k_{in} = 3/2\,(U_{in} \times 0.1)^2$, $\varepsilon_{in} = C_\mu^{3/4} k_{in}^{3/2} l_{in}$
Outflow Boundary	U_{out} = free slip, k_{out}, ε_{out} = free slip
Wall Treatment	Velocity : no slip, Scalar : $C_{wall\ surface} = 0$ (Perfect sink assumption)

The geometry of the surface was constructed using tetra mesh. The grid design of airway model is shown in Figure 1. The first mesh wall surface was set within the viscous sub-layer and the wall units (y^+), which express the dimensionless normal distance from the surface, and met the requirement of 1.0 or less over the whole surface of the airway model. In addition, the numbers of total grid cells were set to 7.5 million for Model A and 4.1 million for Model B as a result of a grid dependence check.
Numerical and boundary conditions are summarized in Table 2. The isothermal condition was assumed in accordance with experimental scenarios. In this analysis, VOC generated from an e-cigarette device was assumed to be a passive contaminant.

RESULTS AND DISCUSSION

Some selected results from our numerical simulations for Models A and B are given in Figure 2. In order to simulate the smoking of an e-cigarette in the model, a supply inlet opening was set as circular and 8.0mm in diameter, imitating the stream going from the mouth directly to the pharynx region. The flow of gases was confirmed in the oral cavity. In the lower airway, air seemed to be well mixed.

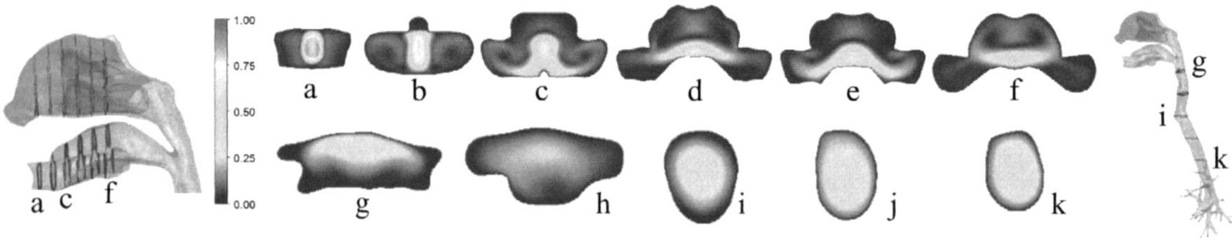

(1) Flow patterns in Model A

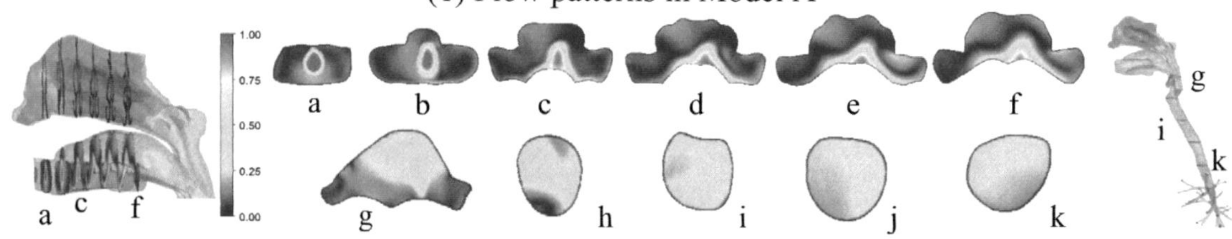

(2) Flow patterns in Model B
Figure 2 Flow patterns in numerical airway models

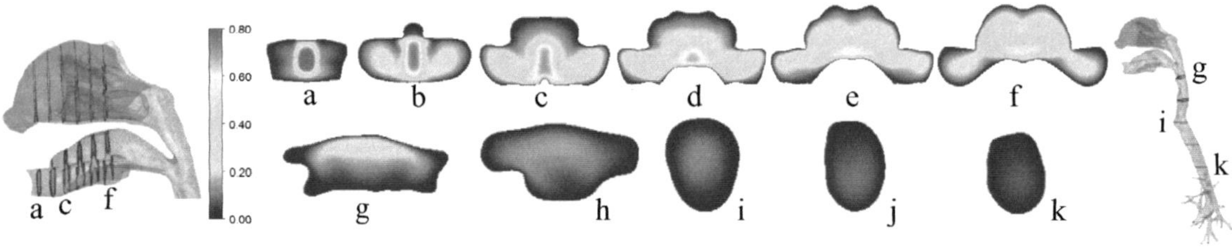

(1) VOC concentration distributions in Model A

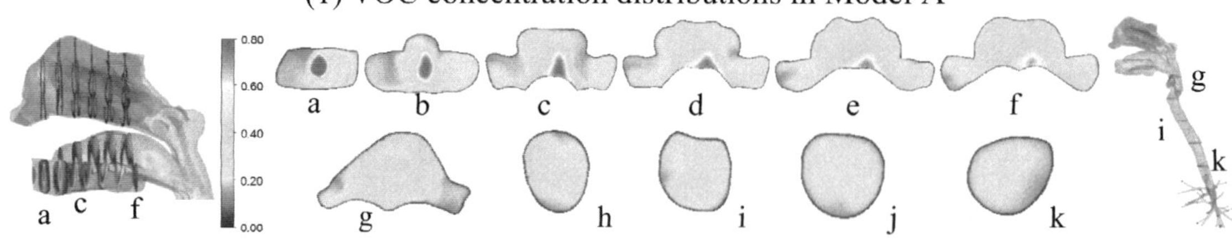

(2) VOC concentration distributions in Model B
Figure 3 VOC concentration distributions in numerical airway models

Figure 3 shows the VOC concentration (here, passive scalar was assumed to be hypothetical VOCs) distributions within the respiratory tract. In this analysis, wall surface VOC concentration was assumed to be zero (perfect sink boundary condition). VOC concentration in Figure 2 was normalized by supply inlet concentration (C/C_{in}). Non-uniform VOC concentration distributions were confirmed within the respiratory tract and as were individual specificity of Models A and B.
Based on the VOC concentration distribution analyses data, wall surface distributions of adsorption flux (deposition flux) within the respiratory tract can be estimated. In general, the adsorption flux (deposition flux) is to be the diffusion flux and calculated by using concentration gradient ($\partial C/\partial x$) and the diffusion coefficient ($D+\nu_t/\sigma_t$) in the vicinity of wall surfaces. Figure 4 shows the calculation results of adsorption flux (deposition flux) within the respiratory tract. The VOC supplied via the mouth opening was directly transported towards the pharynx region by way of the oral cavity and hence the adsorption flux in the pharynx was relatively high compared to that in the oral cavity and lower airway region. This adsorption flux distribution represents the distribution of VOC exposure in the respiratory system, making it possible to assume the probability distributions of inhalation exposure within the respiratory tract.

Figure 4 Adsorption flux distributions of numerical airway models (Model B)

CONCLUSIONS

Using computer modeling (*in silico*) we were able to examine flow pattern and contaminant concentration distributions in the respiratory tract by using the CFD technique, individual specificity, and adsorption flux distributions in the respiratory tract.

REFERENCES

California Department of Public Health (2015) California Tobacco Control Program, State Health Officer's Report on E-Cigarettes: A Community Health Threat, Sacramento, California.

Kota Hirase, Nguyen Lu Phuong, Shin-ichiro Aramaki, Kazuhide Ito (2014) Visualization of Air Flow Pattern in Human Respiratory Tract by Particle Image Velocimetry: Proceedings of *Indoor Air 2014*, Paper ID: HP0533.

Masato Yamashita, Nguyen Lu Phuong, Kota Hirase, Kazuhide Ito (2014) Numerical Simulation of Airflow, Heat and Particle Transfer in Human Respiratory System: Proceedings of *Indoor Air 2014*, Paper ID: HP0534.

Thermal Comfort during summer in a High-speed Railway Station in Cold Region of China

Gang Liu[1], Chenyi Lin[1], Yanbin Zhuo[1], Deheng Guo[1], Rui Dang[1]*
[1]: Tianjin University, No.72 Weijin Road Nankai Tianjin, China

Abstract:

This study explores the interaction between perceived and calculated thermal comfort in high-speed railway station in Cold Region of China. To achieve this, a questionnaire survey in two typical high-speed railway stations in Cold Region of China was conducted in conjunction with physical measurements. The study results reveal that there is a relatively large difference between perceived thermal sensations and calculated predicted mean vote (PMV) from the measurements for the majority of the waiting hall of high-speed railway station. This finding implies that the calculation of PMV is not suitable for the situation in air conditioned waiting hall of high-speed railway stations. The results also show that the length of time a respondent staying in the waiting hall also have effect on the rule of the thermal comfort.

Key words: Thermal comfort; railway station; PMV; Thermal Sensation Vote

1. Introduction

In China, studies about building thermal and humid environment as well as human thermal comfort have been carried out widely. However, most of them are about residential building and office building. The research about large public building especially about train station is limited.

Usually, the high-speed railway stations in China have large space and open halls. Compared with other type of traditional buildings, high-speed railway stations have massive human traffic, and also people stay in the buildings for shorter time. Thus, thermal comfort studies about simple civil buildings are no longer applicable for high-speed railway stations.

Thermal design regions for civil buildings in China are composed by five regions, namely, Cold Region, Very Cold Region, Hot Summer & Cold Winter Region, Hot Summer & Warm Winter Region and Mild Region (As Fig.1). This article is based on research in the Cold Region, where the design outdoor dry bulb temperature in summer is 34.2°C, and the design outdoor relative humidity in summer is 63%.

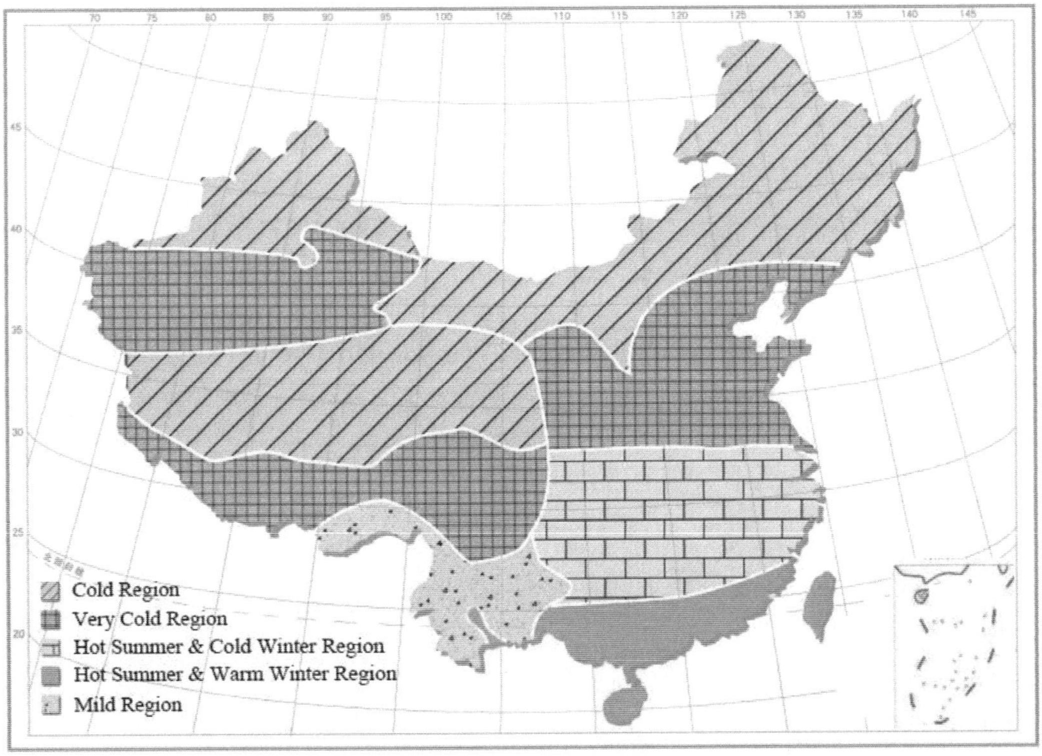

Fig.1 Thermal design regions for civil buildings in China

(The picture drawn by author)

Thermal comfort aspects during summer time in cold Region of China are often overlooked; few researches have been done in this field. But in fact, the outdoor temperatures in summer in Cold Region are often more than 35°C, creating a comfortable indoor thermal environment in Cold Region is necessary as well as other regions.

This paper focuses on the 2 typical high-speed railway stations in Cold Region of China. Data collected about the passengers' thermal preferences was supplemented by measurements of the railway stations' internal conditions and the external climate to explore the thermal adaptive strategies employed by the passengers.

2. Methods

2.1 Data collection

2.1.1 Background information questionnaire

In order to conducting the thermal comfort survey, back ground information about the respondents was obtained through interviews using a standardized survey form. Information was collected about the passengers' age, gender, highest education level and occupation. During this interview, the respondents were also asked about when they arrived at the waiting hall and how long have they been in the waiting hall.

2.1.2 Evaluation of the thermal environment

During the monitoring period, the adult passengers were asked to respond to a comfort survey on a regular basis. The survey questions asked their thermal sensation, preference, and clothing in the "right here and right now" mode. The main questions were: (1) How long have you been in the railway station? (2) How do you feel right here and right now (from hot to cold, ASHRAE 7-point scale), (3) would you prefer to be warmer, cooler or no change (3-point preference scale), (4) Describe your activities within 20min and (5) Describe your clothing (choose the cloths exactly what they were wearing).

2.2 Correlating thermal environmental data with thermal comfort survey

Correlating thermal environmental data was collected during the survey, including the mean radiation temperature, air temperature, humidity and air velocity. There were 8 measuring points set in the waiting hall .The indoor temperature and humidity were recorded by HOBO automatic observer, the mean radiation temperature was recorded by black-bulb thermometer, and the air velocity was measured by hot-wire anemometer.

2.3 Data analysis

2.3.1 PMV and PPD

By Fanger's (1970) PMV heat-balance index, PMV is calculated from six variables-- indoor dry-bulb temperature, black globe temperature, air velocity, clothing insulation, metabolic rate—using the ASHRAE Thermal Comfort Tool software (Fountain and Huizenga 1996).

PPD is predicted percentage of dissatisfied, which is related to the PMV as defined in Formula (1).

$$PPD=100-95\times exp(-0.03353\times PMV^4-0.2179\times PMV^2) \qquad (1)$$

2.3.2 Thermal sensation votes

Thermal sensation votes (TSV) range from hot (+3) to cold (-3), as ASHRAE 7-point scale.

2.3.3 Operative temperature

According to ASHRAE Standard 55-2010, in most practical cases where the relative air speed is small (<0.2 m/s, 40 fpm) or where the difference between mean radiant and air temperature is small (<4°C, 7°F), the operative temperature can be calculated with sufficient approximation as the mean value of air temperature and mean radiant temperature.

$$t_{op}\approx(t_a+ t_{mr})/2 \qquad (2)$$

2.3.4 Thermal neutrality

According to de Dear, Richard's article, statistical analysis of subjective thermal sensation votes within each building were used to define thermal neutrality—the operative temperature found to correspond most closely with the scale's central vote of neutral. Neutrality was calculated for each building in the meta-analysis by the following steps:

1. Bin the building's indoor operative temperature observations into half-degree (℃) increments, and analyze the bins' mean thermal sensation responses.

2. Fit a weighted linear regression model between sensations and operative temperature (t_o): mean thermal sensation = a + b * (t_o)

3. Neutrality was derived by solving each building's regression model for a mean sensation of zero.

3. Results and discussion

A total of 803 responses to the thermal comfort survey were collected during this period, 534 of these were male people with the rest are female. The results are as following.

3.1 Thermal sensation vote

The responses to the thermal comfort survey were given when the indoor temperature was between 23.1 and 26.8°C. The thermal sensations votes (TSV) during this period varied from 'Cold' (-3) to 'hot' (+3). The number of the TSV is presented in Fig. 2. The highest number of the responses was the 'neutral' votes (almost 600) followed by votes of 'slightly cool' (106).

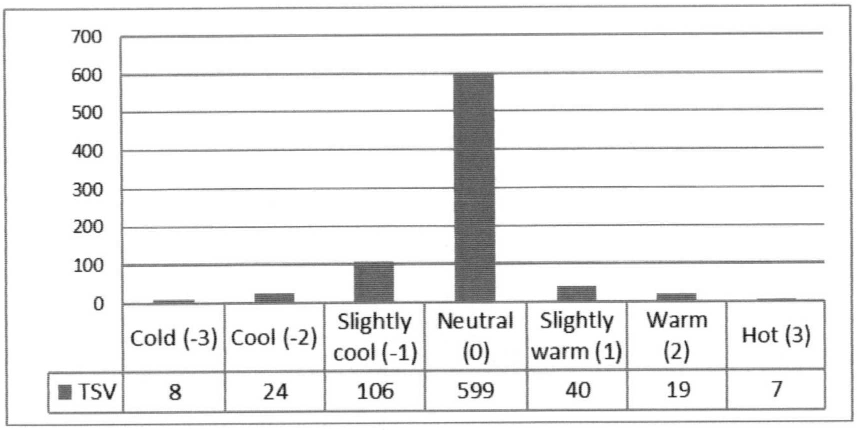

Fig.2 The number of the TSV

3.2 Staying time

Most of the passengers arrived at the waiting hall more than 15minutes before departure time. The number of the TSV is presented in Fig.3. The highest number of the responses was 'less than 30 and more than 15min' (280) followed by votes of 'less than 60 and more than 45min' (121). The average staying time of the passengers is 37.8 minutes.

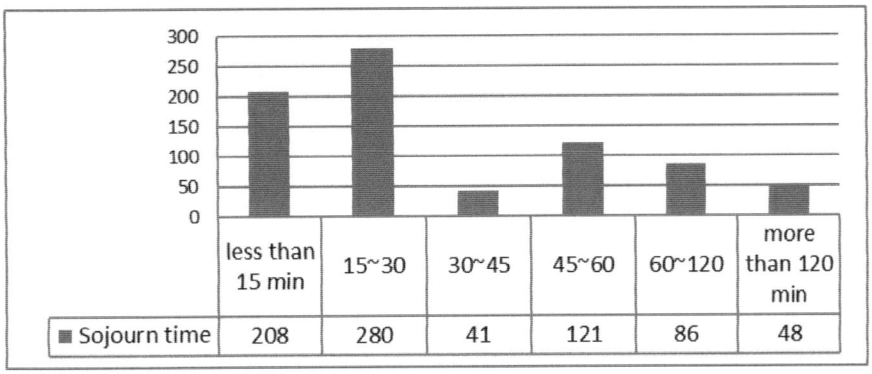

Fig.3 The number of staying time

3.3 Mean thermal sensation vote and PMV

Fig. 4 shows the mean thermal sensation votes (mTSV) and mean PMV (mPMV) in waiting hall of high-speed railway station against operative temperatures. The results show a high correlation between operative temperatures and mTSV (R^2=0.78).Ideal neutral temperature is calculated to be 27.314°C, while the real neutral temperature is 25.884°C.

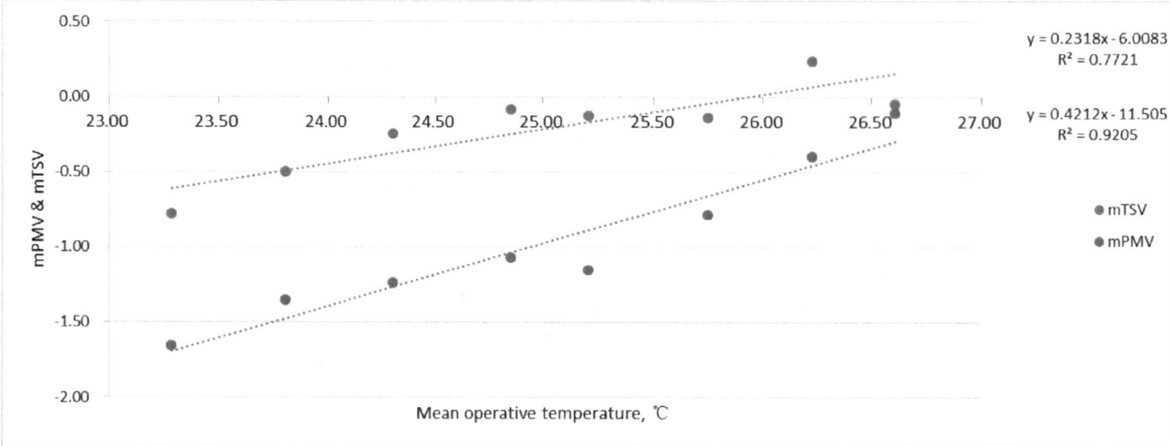

Fig.4Liner Regression of mPMV and mTSV

3.4 Thermal comfort and staying time

Separate the respondents in to two groups: stayed in the waiting room less than 30 minutes and more than 30 minutes (including equal to 30 minutes). Calculate the neutral temperature of the two groups of data, and compare with PMV model. The result of statistics and analysis are as Table 1.

Fig.5 and Fig.6 show the liner regression of mTSV and mPMV in waiting hall of high-speed railway station against operative temperatures of the two groups of people mentioned above.

Table 1. Analysis of different staying time

	Percentage	Mean TSV	Ideal Neutral Temperature, °C	Real Neutral Temperature, °C
<30min	40%	-0.109	27.42	25.91
≥30min	60%	-0.244	26.10	27.36

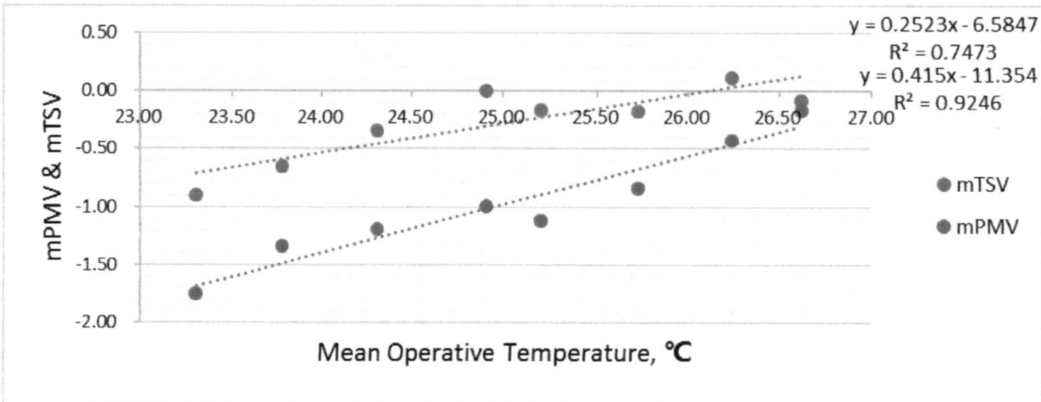

Fig.5 Liner Regression of mPMV and mTSV for passengers who stay less than 30min

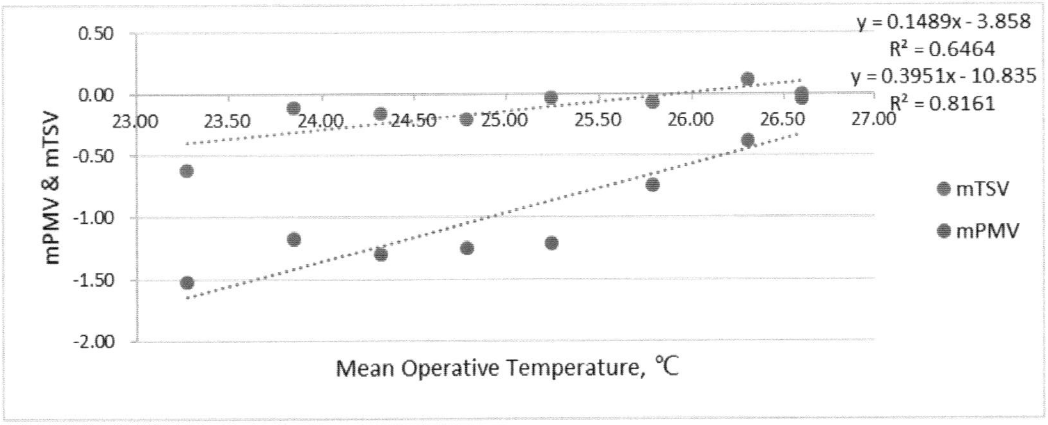

Fig.6 Liner Regression of mPMV and mTSV for passengers who stay more than or equal to 30min

4. Conclusion

According to the research, the ideal neutral temperature in the high-way railway station is 27.3°C, while the vote neutral temperature is 25.9°C. Passengers in the waiting hall prefer lower temperature than it is predicted by the thermal comfort model of PMV.

The thermal comfort model for those who have being stayed in the waiting hall larger than or equals to 30 minutes more close to Fanger's PMV thermal comfort model, which proves that PMV is relatively suitable to predict human body thermal comfort under stable environment. However, it still is not appropriate for the waiting hall of high-speed railway stations.

The study shows the people who stay in waiting hall less than 30 minutes have better capacity to withstand the different environment. This provides good evidence for finding out the potential ability to reduce the energy consumption of high-speed railway station.

Acknowledgements

The research presented in this paper was financially supported by (1) the National Natural Science Foundation of China (the Key Program) through grant No. 51338006; (2) the Program of Introduce Talents of Discipline to Universities through grant No. B13011. The authors would like to thank Prof. Lihui Cao at Tianjin Chengjian University and all the railway stuffs and passengers for supporting this investigation.

Reference

Standard, A. S. H. R. A. E. "Standard 55-2010, Thermal Environmental Conditions for Human Occupancy. Atlanta: American Society of Heating, Refrigerating, and Air-Conditioning Engineers."

National Standard of the People's Republic of China, Thermal Design Code for Civil Building, GB50173-93, China Plan Press, 1993.

National Standard of the People's Republic of China, Design Code For Heating Ventilation And Air Conditioning Of Civil Buildings, GB50736-2012, China Plan Press, 2012.

De Dear, Richard, and Gail Schiller Brager."Developing an adaptive model of thermal comfort and preference. "Center for the Built Environment (1998).

CONTROLLING INDOOR CO_2 WITH A SOLID SORBENT: KINETICS AND CAPACITY

Elliott T. GALL[1,2,*], Victor W.-C. CHANG[1,2], William W NAZAROFF[1,3]

[1]Berkeley Education Alliance for Research in Singapore, Singapore
[2]Civil and Environmental Engineering, Nanyang Technological University, Singapore
[3]Civil and Environmental Engineering, University of California, Berkeley, CA, USA

[*]Corresponding email: egall@ntu.edu.sg

Keywords: Carbon capture, Climate change and indoor air, Air cleaning

SUMMARY

Absent unvented combustion, in occupied indoor spaces, CO_2 concentrations depend primarily on three factors: outdoor CO_2 concentrations, CO_2 emission rates from human metabolism, and the outdoor-air ventilation rate. Indoor CO_2 levels are generally managed by means of dilution with outdoor air. In this work, we investigate an alternative approach to controlling indoor CO_2: active removal using a solid sorbent. We report parameterizations of chemical kinetics and uptake capacity to soda-lime sorbent across a range of conditions relevant to indoor environments. Reaction rate constants for the interaction of $Ca(OH)_2$ present in soda-lime and CO_2 range from 1.8×10^{-2} to 1.5×10^{-1} m^2 mol^{-1} s^{-1}, increasing with higher relative humidity (RH). Uptake capacities range from 0.04 to 0.36 g CO_2/g sorbent, also increasing with higher RH. These data facilitate assessment of the technical feasibility of soda-lime scrubbers in buildings as an option for controlling indoor CO_2.

INTRODUCTION

Human exposures to CO_2 are likely to increase as outdoor concentrations of CO_2 rise and as buildings are constructed with tighter envelopes (Committee on the Effect of Climate Change on Indoor Air Quality and Public Health, 2011). Typically, indoor CO_2 is considered an indicator of ventilation sufficiency. Elevated CO_2 levels indicate inadequate ventilation signaling the potential accumulation of other air pollutants with indoor sources. CO_2 is a byproduct of human metabolism. Several important physiological processes are modulated by CO_2 levels in the blood (Sliwka et al., 1998). A body of research shows correlations between CO_2 and symptoms associated with sick building syndrome (Seppänen et al., 1999). This work cannot distinguish whether CO_2 is causally implicated. However, a recent study suggests that elevated CO_2 concentrations themselves may adversely affect human cognitive performance, even at levels that lie within the range encountered indoors (Satish et al., 2012).

Buildings consume substantial energy and are large contributors to greenhouse gas emissions. Deployment of direct capture of CO_2 in buildings offers the potential for climate and productivity co-benefits (Gall and Nazaroff, 2015). In addition to further understanding the impact of indoor CO_2 on humans, an understanding of the technical feasibility of direct capture in buildings is also needed. One relevant, emerging area of research is capture on solid sorbents. In this investigation, we report parameterizations of chemical kinetics and uptake capacity to one solid-sorbent — soda-lime — which is majorly composed of calcium hydroxide. The ultimate goal is to enable design and modeling of air cleaners to actively control indoor CO_2 under conditions relevant to indoor and built environments.

METHODS

The sorbent characterized was soda-lime (Grace Chemical, Sodasorb, 4×6 mesh), a granular solid comprised of a mixture of calcium hydroxide ($Ca(OH)_2$), sodium hydroxide (NaOH), potassium hydroxide (KOH), and water. The removal of CO_2 to soda-lime is well-established, and can be described as strong-base catalyzed, water-facilitated chemical absorption. The mechanism of uptake involves a series of reactions, including dissolution in water, bicarbonate formation, and regeneration of KOH and NaOH (Al-Shaikh and Stacey, 2013). The overall process is described in reaction 1:

$$Ca(OH)_2 + CO_2 \rightarrow CaCO_3 + H_2O + heat \quad \text{(R1)}$$

Experiments were conducted in a bench-scale apparatus to explore the kinetics and capacity of CO_2 uptake as described by R1. Laboratory air was pumped through the apparatus (TR Vacuum Tech, TR40.40), first passing through a column of ~ 200 g of activated carbon (Calgon Carbon, BPL 6×16) and ~ 200 g of soda-lime (Grace Chemical, Sodasorb). Particles were removed with a 0.1 micron filter (LaMan Extractor/Dryer, Series 50) and air was dehumidified with a membrane dryer (Laman, Superstar MD15). The set point RH was attained by splitting the air stream into two components, modulating flows using mass-flow controllers (MFC, Omega, FMA 5400/5500), and passing one stream through a washing column filled with distilled water (Cole-Parmer, 06652-00) before blending. Processed air then entered an inlet measurement/mixing chamber where inlet CO_2 concentration, RH, and temperature were recorded as 5-min averages (Onset, Teleaire 7001 and HOBO U12). Inlet CO_2 concentrations were set by injecting CO_2 at a constant flow rate, modulated by a MFC (Omega, FMA 5400/5500) into a mixing chamber positioned at the inlet. The source was a pressurized cylinder of CO_2 (50% CO_2, balance of air). The target air temperature was 25 °C, and ranged from 24.7 to 26.1 °C for all experiments. Air then entered a glass plug flow reactor (PFR), where a single layer of media was placed on a wire mesh in the PFR. Outlet CO_2 concentration, RH, and temperature were recorded in 5-min averages in a chamber downstream the PFR (Onset, Teleaire 7001 and HOBO U12).

Rate constants characterizing R1 were determined using the method of isolation. Since a single layer of soda-lime media was placed in the PFR and inlet concentrations of CO_2 were held constant, CO_2 as a reactant in the left-hand-side of R1 was constant. The rate of reaction of R1, r (mol m^{-2} s^{-1}), is modelled as shown in eq. 1:

$$r = k_{obs}[Ca(OH)_2]^m \quad (1)$$

where, k_{obs} is the observed reaction rate constant (m$^{2\times(m-1)}$ mol$^{-1\times(m-1)}$ s^{-1}), $[Ca(OH_2)]$ is the concentration of calcium hydroxide (mol m^{-2}), and m is the reaction order.

The concentration of $Ca(OH_2)$ in the PFR through time (eq. 1) was determined by subtracting the moles of CO_2 removed by the media at each time step from the initial molar concentration of the media. Reaction 1 shows a 1:1 molar ratio between $Ca(OH)_2$ and CO_2; observed reductions in CO_2 concentration across the media were assumed to result from $CaCO_3$ formation. The initial molar concentration of $Ca(OH)_2$ was determined from: i) the initial mass of the media, ii) the mass of the media after drying to remove water (media was dried for 2 h at 105 °C and then weighed), and iii) previous literature stating that soda-lime consists of 2% NaOH and 3% KOH by mass (Higuchi et al., 2001). The reaction order and rate constant were determined from characteristic plots of eq. 1 in integrated form.

Estimates of uptake capacity were determined from the difference between inlet and outlet mass flow rates of carbon dioxide across the media. Mass flow rates were multiplied by the time step between carbon dioxide measurements (5 min) to determine the mass uptake of CO_2 for each time increment. The total uptake for each experimental condition is reported as the summed mass uptake at each time step for the duration of an experiment. The media was considered exhausted when outlet and inlet CO_2 concentrations were equal to within 25 ppm.

RESULTS AND DISCUSSION

The design of an indoor CO_2 control device that employs a solid sorbent, in this case soda-lime, will be enabled by knowledge of i) kinetics of the reaction to determine optimum media-air contact times, and ii) estimates of uptake capacity. Taken together, these parameters will enable prediction of removal efficiency and breakthrough curves. Table 1 reports the results of nine experiments conducted to estimate rate constants and capacity of the soda-lime-CO_2 interaction for variable inlet RH, flowrate, and CO_2 concentration.

Table 1. Summary of conditions, kinetics, and uptake capacity of soda-lime-CO_2 interaction.

	Temp	RH	Q	Inlet [CO_2]	k_{obs} [a]	r^2 [b]	Capacity
Exp.	°C	(%)	L/min	ppm	$m^2\ mol^{-1}\ s^{-1}$	(-)	g CO_2/g sorbent
1	25.2	26.5	1.80	2200	4.0×10^{-2}	0.77	0.04
2	25.2	25.2	2.81	2070	2.1×10^{-2}	0.19	0.04
3	24.8	25.9	1.86	860	1.8×10^{-2}	0.45	0.05
4	25.5	50.8	1.85	1990	6.9×10^{-2}	1.00	0.26
5	24.7	50.5	2.79	2010	9.1×10^{-2}	0.97	0.31
6	24.7	50.0	1.82	870	4.8×10^{-2}	0.98	0.24
7	26.1	76.3	1.83	1870	1.5×10^{-1}	1.00	0.29
8	25.2	76.5	2.81	1890	1.5×10^{-1}	1.00	0.30
9	25.1	76.4	1.87	770	1.2×10^{-1}	0.98	0.36

[a] Rate constants reported are slope of second-order characteristic plot.
[b] Correlation coefficient for second-order characteristic plot across duration of experiment.

Reaction rate constants (k_{obs}) shown in Table 1 illustrate the importance of water vapor as a factor influencing the uptake of CO_2 by soda-lime. At low RH, correlation coefficients (r^2) for the second-order characteristic plot are poor (as are first and zeroth-order, not shown), implying that eq. 1 does not fully describe removal at low RH. Nevertheless, values for kinetics indicate the inhibition of CO_2 uptake at lower RH. Averaged across all experiments at constant RH, k_{obs} at 25% RH is a factor of 2.6 lower than at 50% RH, and a factor of 5.4 lower than at 75% RH. Furthermore, comparing exp. 1 to exp. 2 shows that k_{obs} decreases with increased flowrate, an effect not observed at higher RH levels (i.e., compare exp. 4 to 5 and exp. 7 to 8). These differences in trends and the low r^2 at low RH suggest that dissolution of CO_2 into water is rate-limiting at lower RH. At lower RH, k_{obs} also appears to be more strongly affected by changes in inlet CO_2 than at higher RH. At 25% RH, comparing exp. 3 to exp. 1 shows a 60% reduction in inlet CO_2 and a 55% reduction in k_{obs}. At 50% RH and 75% RH, similar reductions in inlet CO_2 result in only 30%, and 20% reductions in k_{obs}, respectively. The results shown in Table 1 imply that high RH facilitates uptake of CO_2, and enable R1 to be characterized via eq. 1.

Uptake capacities follow similar trends to those observed for rate constants, increasing with increasing RH. Estimates of capacity at 50% RH, averaging 0.27 g CO_2/g sorbent, are consistent with the manufacturer specification of 0.24 g CO_2/g sorbent. Estimates of capacity at 75% RH, averaging 0.31 g CO_2 /g sorbent, appear to show a slight enhancement above the specification. The estimates of capacity in Table 1 inform assessment of the feasibility of the use of soda-lime in an indoor space by allowing sorbent usage rates to be estimated. For example, a per-person usage rate of sorbent can be estimated for a hypothetical control scenario. Given a CO_2 emission rate of 34 g/h per person (Smith, 1988), capturing 75% of human emissions would require 110 g soda-lime/h per person. At a ventilation rate of 5 L/s per person, an outdoor CO_2 concentration of 400 ppm, and assuming no other indoor CO_2 sources, this sorption rate could serve to reduce indoor steady-state CO_2 concentrations from 1450 ppm to 660 ppm.

CONCLUSIONS

Carbon dioxide is generally considered to be an indicator of indoor air pollution; however, recent research suggests that CO_2 itself can adversely affect attributes of human well-being. Economic and environmentally favorable CO_2 capture and sequestration remains a distant goal. Distinctive challenges exist in developing direct capture in buildings. However, there appears to be some potential for reducing building carbon footprints, either through direct capture of CO_2 or through energy savings realized from reduced ventilation rates, while also improving outcomes for occupants. The promise of these co-benefits motivates further research into direct capture as an approach for managing indoor CO_2 levels.

ACKNOWLEDGEMENT

This research was funded by the Republic of Singapore's National Research Foundation through a grant to the Berkeley Education Alliance for Research in Singapore (BEARS) for the Singapore-Berkeley Building Efficiency and Sustainability in the Tropics (SinBerBEST) Program. BEARS has been established by the University of California, Berkeley as a center for intellectual excellence in research and education in Singapore.

REFERENCES

Al-Shaikh, B. and Stacey, S., 2013. Essentials of Anaesthetic Equipment, 4th Edition, Churchill Livingstone Elsevier.

Committee on the Effect of Climate Change on Indoor Air Quality and Public Health, 2011. Climate change, the indoor environment, and health. The National Academies Press.

Gall, E.T. and Nazaroff, W.W, 2015. New directions: Potential climate and productivity benefits from CO_2 capture in commercial buildings. Atmos. Environ. doi:10.1016/j.atmosenv.2015.01.004.

Higuchi, H., Adachi, Y., Arimura, S., Kanno, M., Satoh, T., 2001. The carbon dioxide absorption capacity of amsorb is half that of soda lime. Anesth. Analg. 93, 221–225.

Satish, U., Mendell, M.J., Shekhar, K., Hotchi, T., Sullivan, D., Streufert, S., Fisk, W.J., 2012. Is CO_2 an indoor pollutant? Direct effects of low-to-moderate CO_2 concentrations on human decision-making performance. Environ. Health Perspect. 120, 1671–1677.

Seppänen, O.A., Fisk, W.J., Mendell, M.J., 1999. Association of ventilation rates and CO_2 concentrations with health and other responses in commercial and institutional buildings. Indoor Air 9, 226–252.

Sliwka, U., Krasney, J.A., Simon, S.G., Schmidt, P., Noth, J., 1998. Effects of sustained low-level elevations of carbon dioxide on cerebral blood flow and autoregulation of the intracerebral arteries in humans. Aviat. Space Environ. Med. 69, 299–306.

Smith, P.N., 1988. Determination of ventilation rates in occupied buildings from metabolic CO_2 concentrations and production rates. Build. Environ. 23, 95–102.

WATER DROPLETS AS AGENTS TO ENHANCE THE REMOVAL EFFICIENCY OF AN ACOUSTIC AEROSOL REMOVAL SYSTEM

Wai T YUEN[1], Sau C FU[1], and Christopher Y H CHAO[1,*]

[1]Department of Mechanical and Aerospace Engineering, The Hong Kong University of Science and Technology, Clear Water Bay, Hong Kong

[*]Corresponding email: meyhchao@ust.hk

Keywords: Acoustics, Aerosols, Droplets, Incense, Disinfection

SUMMARY

An acoustic aerosol removal system with the addition of water droplets was examined in this paper. Previous studies used a radiating plate, attached to an ultrasonic transducer, to create stationary waves in an air duct. The acoustic effects were found to aid the aerosol deposition process. In this study, the radiating plate was subject to an inflow of water, which was vibrated and micron size water droplets were generated. It was found that these extra micro size droplet lowered submicron incense smoke particle concentration by 34%. The micron size agglomerates formed with liquid droplets with aerosols were found to be easily deposited. This process has the potential to disinfect airborne bacteria or viruses with selected antimicrobial agent as the aerosolized liquid droplets.

INTRODUCTION

An acoustic aerosol removal system described in Yuen et al. (2014) demonstrated that the use of second order acoustic effects, acoustic radiation forces and acoustic streaming, can encourage aerosol depositions, thus reducing concentrations of suspended particles. This method has the potential to improve air quality for building occupants. It was found that concentrations for micro size particles (2-6um) can be reduced by an average of 30% whilst the averaged concentrations reductions for submicron particles (0.3-0.6um) were only 16%. A method to improve the removal efficiencies is proposed in this paper. Water mist is introduced to the system by ejecting distilled water onto the radiating plate of the acoustic system during operations. The vibrating plate generates a cloud of water droplets to interact with the travelling aerosols, generated by smouldering, in an air duct. The particle removal efficiency is measured.

METHODOLOGIES

The effects caused by the addition of water mist in the acoustic system were observed and recorded experimentally. The experimental set up is shown in Figure 1. Incense sticks were lit and left to smoulder at the entrance of the air duct. The sizing and characterization performed on particles emitted during this process showed that the count medium diameters ranging between 0.08-0.2um (Cheng et al. 1995 and Roy et al. 2009), in the submicron size range. The cross sectional dimensions of the duct were measured to be 17.5cm wide by 8.5cm tall. A fan was employed at the exit of the air duct to create a uniform air flow averaging 0.3m/s. Smoke particles are carried through the acoustic field where water was ejected on schedule. The acoustic field was emitted from a thin radiating plate, with dimensions 30cm by 16cm by 0.3cm. It was attached to an ultrasonic transducer. Ultrasonic standing waves were created between the radiation plate and the surfaces of the air duct. Approximately 180W was the power required to drive the piezoelectric transducer to emit ultrasonic waves with pressure amplitude of 150dB. The water was ejected onto the surface of the vibrating plate, creating micron size water droplets. Particle number concentrations were recorded by a TSI Ultrafine Condensation Particle Counter (CPC 3776) to quantify the influence caused by the water mist. The aerosols with diameters between 2.5 nm to a few micrometres in diameter were detected by the CPC.

The experiments were consisted of three phases: In the first phase, acoustics was used to process the smoke particles without water ejection, in a dry condition to obtain a concentration datum for reference. Secondly, distilled water was ejected manually through a pipe. This continued for 60 seconds. Finally, data was collected when the acoustic system was still wet with no additional water input.

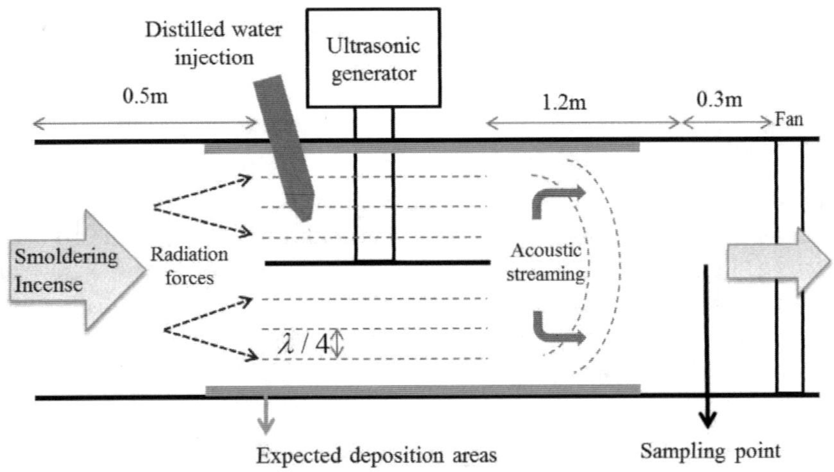

Figure 1. Schematics of the experimental set up

RESULTS AND DISCUSSION

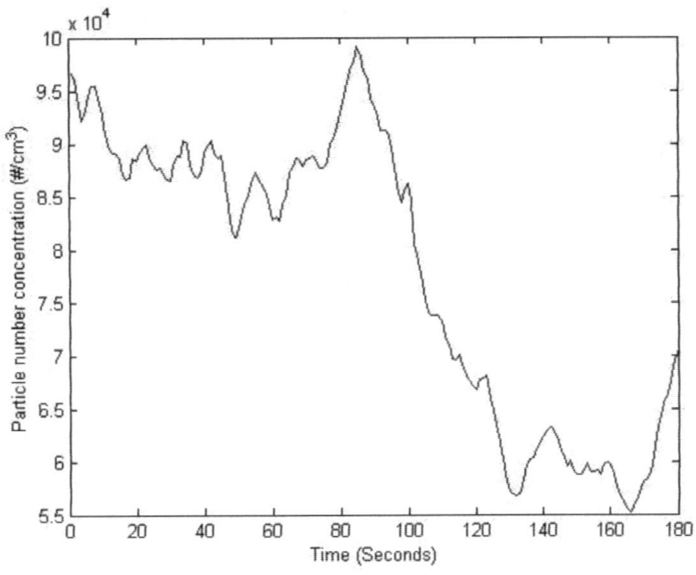

Figure 2. Particle number concentration logged with respect to time. Water was ejected between t=80-140 seconds

Results presented in Figure 2 showed the reduction of particle number concentration moments after water was ejected at t=80 seconds. The initial spike between t=80-100 may be caused by the insufficient number of water droplets in the air flow to coagulation frequencies with smoke particles, thus contributed positively to the overall aerosol concentration sampled. However, as the water continued to flow onto the radiating plate, the addition water droplets were able to capture smoke particles during agglomeration, a process enhanced by acoustic radiation forces funnelling the aerosols to acoustic pressure nodes. The large agglomerates with longer relaxation times were more efficiently deposited by acoustic streaming or gravity.

Results also showed that the lowered number concentrations of aerosols were sustained for about 25 seconds after the water flow had stopped. It was observed that a thick film of water was still present on the plate and water droplet continued to be generated. The results showed that toward the end of the 3 minutes experiment, particle number concentration increased as the system returned to operate under the dry condition.

CONCLUSIONS

Distilled water was added to the acoustic aerosol removal system to remove fine particles generated by smoldering of incense sticks. Water was aerosolized into micron size droplets by the same acoustic radiating plate positioned inside the air duct. Results showed that a 34% reduction in particle number concentration was achieved by the addition of water. This method can potentially increase the removal efficiency of fine particles from the previously obtained 16% to 45%. Furthermore, the results represented showed that with the utility of antimicrobial liquids as the droplet agent, airborne viruses or bacteria infected aerosols may be rendered to inactive states when agglomerated with the droplets before they are deposited. This system may be applied in hospitals or clinics to reduce the spread of transmissible diseases.

ACKNOWLEDGEMENT

The work was supported by the General Research Fund granted by the Research Grants Council of the Hong Kong Special Administrative Region, China (Project No.611013).

REFERENCES

Cheng, Y. S., Bechtold, W. E., Yu, C. C., & Hung, I. F. (1995). Incense smoke: characterization and dynamics in indoor environments. *Aerosol Science and Technology*, 23(3), 271-281.

Roy, A. A., Baxla, S. P., Gupta, T., Bandyopadhyaya, R., & Tripathi, S. N. (2009). Particles emitted from indoor combustion sources: size distribution measurement and chemical analysis. iInhalation toxicology, 21(10), 837-848.

Yuen, W. T., Fu, S. C., Kwan, J. K., & Chao, C. Y. (2014). The Use of Nonlinear Acoustics as an Energy-Efficient Technique for Aerosol Removal. *Aerosol Science and Technology*, 48(9), 907-915.

ESTIMATES OF HVAC FILTRATION EFFICIENCY FOR FINE AND ULTRAFINE PARTICLES OF OUTDOOR ORIGIN

Dan ZHAO[1], Parham AZIMI[1], and Brent STEPHENS[1,*]
[1]Department of Civil, Architectural and Environmental Engineering Illinois Institute of Technology, Chicago, IL, US

[*]Corresponding email: brent@iit.edu

Keywords: Particulate matter, Particle infiltration, HVAC filters, Particle control

SUMMARY

Much of human exposure to particles of outdoor origin occurs indoors. Higher efficiency HVAC filtration can reduce exposures, but filters are not typically rated for their removal of common particulate matter classifications (e.g., fine or ultrafine particles). This work uses 194 outdoor particle size distributions (PSDs) to estimate HVAC filter removal efficiencies for $PM_{2.5}$ and ultrafine particles of outdoor origin in buildings. Outdoor PSDs were fitted and mapped to size-resolved particle removal efficiency of representative HVAC filters from the literature. Results demonstrate that although the MERV metric does not explicitly account for UFP or $PM_{2.5}$ removal efficiency, estimates of filtration efficiency for both size fractions increased with increasing MERV. This information improves knowledge of how the MERV designation relates to $PM_{2.5}$ and UFP removal efficiency for indoor particles of outdoor origin, which can simplify modeling efforts and improve our ability to inform indoor air quality standards and guidelines.

INTRODUCTION

The majority of exposure to fine and ultrafine particles of outdoor origin often occurs inside buildings (Allen et al. 2004). Indoor particle control technologies installed in HVAC systems are being increasingly relied upon to reduce indoor concentrations of particles of both indoor and outdoor origin. The objective of this paper is to provide estimates of particle removal efficiency of various HVAC filters for $PM_{2.5}$ and UFP of ambient origin. We achieved this by mapping 194 outdoor PSDs found in the literature to size-resolved particle removal efficiencies of a wide range of HVAC filters, including MERV 5, 6, 7, 8, 10, 12, 14, 16 and HEPA filters. We use the results to explore statistical distributions of outdoor-origin $PM_{2.5}$ and UFP removal efficiencies.

METHODOLOGIES

We performed a literature review to identify previous studies that reported long-term measurements of outdoor PSDs across the world, leading to a total of 194 PSDs in more than 30 locations (Asmi et al. 2011; Birmili et al. 2001; Costabile et al. 2009; Hussein et al. 2004; Sabaliauskas et al. 2012; Stanier, Khlystov, and Pandis 2004; Wåhlin 2009; Wehner and Wiedensohler 2003). Tri-modal lognormal distributions were fit to each of the 194 outdoor PSDs (Azimi et al 2014). Geometric means (GMs), geometric standard deviations (GSDs), and total number concentrations three modes were adjusted manually using a semi-transparent graphical overlay until an adequate visual fit was achieved. The 194 PSDs were then used to estimate outdoor $PM_{2.5}$ and UFP concentrations in each location. Two different scenarios of outdoor $PM_{2.5}$ mass concentrations for each outdoor PSD were estimated using two different assumptions for particle density: constant unit density and spherical particles with density varying with diameter. Size-resolved particle removal efficiencies from 0.001 to 10 µm were then gathered for generally representative HVAC filters for outdoor UFPs and $PM_{2.5}$ from 100% outdoor air (OA) were estimated using Equations 1 and 2.

$$\eta_{UFP} = 1 - \frac{\sum_{i=1}^{100} N_i \times (1 - \eta_i)}{\sum_{i=1}^{100} N_i} \tag{1}$$

$$\eta_{PM_{2.5}} = 1 - \frac{\sum_{i=1}^{2500} N_i \times \rho_i \times \frac{\pi d_i^3}{6} \times (1 - \eta_i)}{\sum_{i=1}^{2500} N_i \times \rho_i \times \frac{\pi d_i^3}{6}} \tag{2}$$

Where η_{UFP} = estimated UFP removal efficiency of a filter (-), $\eta_{PM2.5}$ = estimated $PM_{2.5}$ removal efficiency of a filter, d_i = diameter size of particles in size i (cm), N_i = number of concentration of particles with diameter d_i (#/cm^3), η_i = removal efficiency of filter for particles with diameter d_i (-), and ρ_i = density of particles with diameter of d_i (g/cm^3). Then each of the 194 outdoor PSDs was then modified to a hypothetical "infiltrated indoor PSD" for particles of outdoor origin by multiplying PSDs by this size-resolved infiltration factor for each particle size. Finally, a modified version of outdoor-infiltrated UFP and $PM_{2.5}$ removal efficiencies was estimated using Equations 1 and 2 with the new infiltrated PSDs (although this data is not shown for brevity).

RESULTS AND DISCUSSION

Figure 1 shows the size-resolved removal efficiencies for particle sizes 0.001-10 µm for each representative HVAC filter used in this work (Azimi et al. 2014).

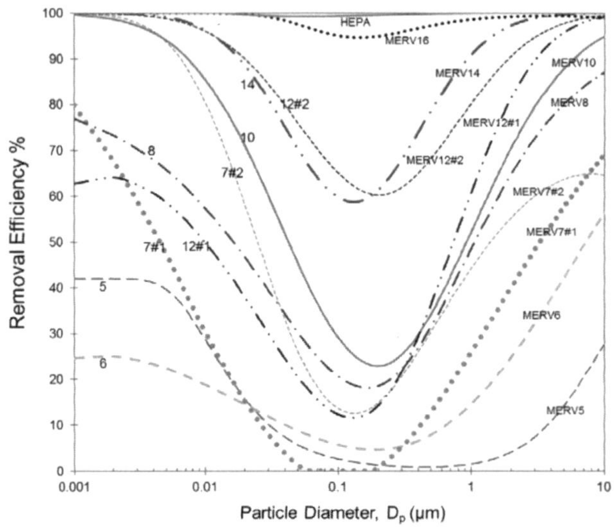

Figure 1. Size-resolved removal efficiency of various MERV filters used herein

Results for estimates of UFP and $PM_{2.5}$ removal efficiency for each filter with 100% OA are shown in Figure 2.

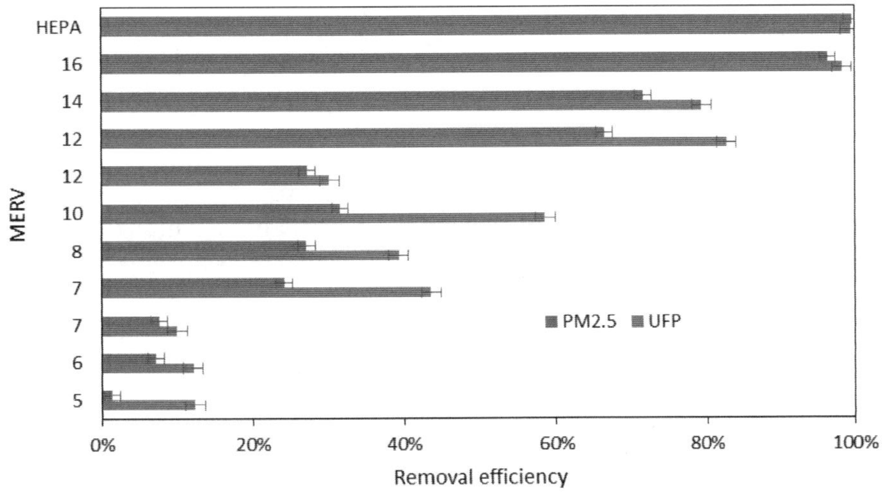

Figure 2. Mean $PM_{2.5}$ and UFP removal efficiency for 194 outdoor PSDs and 11 representative HVAC filters listed by MERV, assuming 100% OA filtration

Estimates of 100% OA UFP removal efficiency for MERV 5, 6 and 7 (#1) filters were similar to each other with a median efficiency of ~12-13% and ranges from ~1% to ~33% depending on MERV and PSD combination. MERV 7(#2) and MERV 8 revealed similar with median values around ~40-45% and ranges from ~22% to ~76% depending on PSD. The median efficiency for MERV 7 (#2) was actually higher than the median MERV 8. MERV 10 had a median 100% OA UFP removal efficiency of ~60% and range from 40% to 82%. MERV 12 (#1) actually had estimates of UFP removal efficiency below MERV 7 (#2), 8, and 10, while MERV 12 (#2) had UFP removal efficiencies very close to that of MERV 14. These data suggest that although higher UFP efficiencies generally occurred with higher MERV filters, MERV alone cannot always be used to distinguish HVAC filters on their outdoor UFP removal efficiency. Similarly, Estimates of 100% OA $PM_{2.5}$ removal efficiency of all MERV designations were similar for both assumptions for particle density. Estimates of removal efficiency of MERV 5 for outdoor $PM_{2.5}$ was low, with a median efficiency of ~1% in both density scenarios. MERV 6 and MERV 7 (#2) revealed similar estimates of outdoor $PM_{2.5}$ with median values around ~7-8% and ranges from ~2% to ~21% depending on MERV and PSD combination. Median estimates of 100% OA $PM_{2.5}$ removal efficiencies for MERV 7(#2), MERV 8, MERV 10, and MERV 12(#1) were similar, ranging from ~24% to ~31% with ranges from 17% to 51%. Outdoor $PM_{2.5}$ removal efficiencies for MERV 12 (#2) were similar to MERV 14. Finally, 100% OA $PM_{2.5}$ removal efficiencies for MERV 16 and HEPA were grouped near 96% and >99%, respectively. Estimates of removal efficiencies made when considering modifications to PSDs after infiltration into a typical residence were typically within 10% of the 100% OA values, although they are not shown here.

CONCLUSIONS

In this work we fit 194 outdoor tri-modal particle size distributions from locations across the world and used those PSDs to estimate outdoor-origin UFP and $PM_{2.5}$ removal efficiencies for 11 generally representative HVAC filters using information from existing literature. We demonstrate that although the MERV metric does not explicitly account for UFP or $PM_{2.5}$ removal efficiency, both tend to increase in efficiency with increasing MERV. We also demonstrate that outdoor PSD characteristics and assumptions for particle density and size-resolved infiltration factors do not drastically impact estimates of HVAC filter removal efficiencies for $PM_{2.5}$, although the impacts are larger for UFPs. However, knowledge of MERV alone cannot always be used to predict UFP or $PM_{2.5}$ removal efficiency.

REFERENCES

Allen, Ryan, Lance Wallace, Timothy Larson, Lianne Sheppard, and L.-J. Sally Liu. 2004. "Estimated Hourly Personal Exposures to Ambient and Nonambient Particulate Matter Among Sensitive Populations in Seattle, Washington." *Journal of the Air & Waste Management Association* 54 (9): 1197–1211.

Asmi, A., A. Wiedensohler, P. Laj, A.-M. Fjaeraa, K. Sellegri, W. Birmili, E. Weingartner, et al. 2011. "Number Size Distributions and Seasonality of Submicron Particles in Europe 2008–2009." *Atmos. Chem. Phys.* 11 (11): 5505–38.

Azimi, Parham, Dan Zhao, and Brent Stephens. 2014. "Estimates of HVAC Filtration Efficiency for Fine and Ultrafine Particles of Outdoor Origin." *Atmospheric Environment* 98 (December): 337–46. doi:10.1016/j.atmosenv.2014.09.007.

Bhangar, S., N. A. Mullen, S. V. Hering, N. M. Kreisberg, and W. W. Nazaroff. 2011. "Ultrafine Particle Concentrations and Exposures in Seven Residences in Northern California." *Indoor Air* 21 (2): 132–44.

Birmili, Wolfram, Alfred Wiedensohler, Jost Heintzenberg, and Katrin Lehmann. 2001. "Atmospheric Particle Number Size Distribution in Central Europe: Statistical Relations to Air Masses and Meteorology." *Journal of Geophysical Research: Atmospheres* 106 (D23): 32005–18.

Costabile, F., W. Birmili, S. Klose, T. Tuch, B. Wehner, A. Wiedensohler, U. Franck, K. König, and A. Sonntag. 2009. "Spatio-Temporal Variability and Principal Components of the Particle Number Size Distribution in an Urban Atmosphere." *Atmos. Chem. Phys.* 9 (9): 3163–95.

Hussein, T., A. Puustinen, P. P. Aalto, J. M. Mäkelä, K. Hämeri, and M. Kulmala. 2004. "Urban Aerosol Number Size Distributions." *Atmos. Chem. Phys.* 4 (2): 391–411..

Miller, Kristin A., David S. Siscovick, Lianne Sheppard, Kristen Shepherd, Jeffrey H. Sullivan, Garnet L. Anderson, and Joel D. Kaufman. 2007. "Long-Term Exposure to Air Pollution and Incidence of Cardiovascular Events in Women." *New England Journal of Medicine* 356 (5): 447–58.

Sabaliauskas, Kelly, Cheol-Heon Jeong, Xiaohong Yao, Yun-Seok Jun, Parnian Jadidian, and Greg J. Evans. 2012. "Five-Year Roadside Measurements of Ultrafine Particles in a Major Canadian City." *Atmospheric Environment* 49 (March): 245–56.

Stanier, Charles O, Andrey Y Khlystov, and Spyros N Pandis. 2004. "Ambient Aerosol Size Distributions and Number Concentrations Measured during the Pittsburgh Air Quality Study (PAQS)." *Atmospheric Environment*, Findings from EPA's Particulate Matter Supersites Program, 38 (20): 3275–84.

Wåhlin, Peter. 2009. "Measured Reduction of Kerbside Ultrafine Particle Number Concentrations in Copenhagen." *Atmospheric Environment* 43 (22–23): 3645–47.

Wallace, Lance A, Steven J Emmerich, and Cynthia Howard-Reed. 2004. "Effect of Central Fans and in-Duct Filters on Deposition Rates of Ultrafine and Fine Particles in an Occupied Townhouse." *Atmospheric Environment* 38 (3): 405–13.

Wallace, Lance, Warren Kindzierski, Jill Kearney, Morgan MacNeill, Marie-Ève Héroux, and Amanda J. Wheeler. 2013. "Fine and Ultrafine Particle Decay Rates in Multiple Homes." *Environmental Science & Technology* 47 (22): 12929–37.

Wehner, B., and A. Wiedensohler. 2003. "Long Term Measurements of Submicrometer Urban Aerosols: Statistical Analysis for Correlations with Meteorological Conditions and Trace Gases." *Atmos. Chem. Phys.* 3 (3): 867–79.

STUDY DESIGN: IMPACT OF VENTILATION AND CLEANING ON THE MICROBIOLOGY OF THE BUILT ENVIRONMENT IN TRIBAL HOMES

Richard SHAUGNESSY[1*], Jordan PECCIA[2], David REISDORPH[1], and Ulla HAVERINEN-SHAUGHNESSY[3]

[1] Indoor Air Program, Department of Chemical Engineering, The University of Tulsa, Tulsa, Oklahoma, USA
[2] Department of Chemical and Environmental Engineering, Yale University
[3] Department of Environmental Health, National Institute for Health and Welfare, Kuopio, Finland

*Corresponding email: rjstulsau@aol.com

Keywords: Microbiology, indoor air quality, cleaning, ventilation, asthma.

SUMMARY

The University of Tulsa is currently engaged in a 3-year study aimed to reduce asthma symptoms and school absence among tribal children. This paper describes the study design for an Alfred P. Sloan Foundation funded project which adds building science and modern microbial measurements to the existing study. The Sloan study will focus on measurements including home air exchange rate measurements, indoor/outdoor PM_{10} measurements, and a broad suite of surface and air DNA-based microbial measurements including qPCR for total bacteria, total fungi, and microbial community analysis for bacteria and fungi. The large numbers of homes and microbial samples (over 1,000 samples), the extensive building characterization, and the multiple time points for sampling are expected to provide a robust data-set for discerning how variations in ventilation rates and cleaning impact the air and surface microbiome. The study also provides new and critical data on homes inhabited by low income, vulnerable populations.

INTRODUCTION

Two complementary research projects, over a three year span, address the need to better understand indoor microbial ecosystems and the relationship with building systems and human health. The purpose of the first study is to determine the effects of indoor air exposures` on asthma episodes/adverse-reactions related to both tribal home and school childhood environmental exposures. The study blends a parallel exposure approach (home and school environments) to better understand interactions of distinct environmental settings on indoor environmental asthma triggers, and subsequent child health. The second study examines how two central and modifiable building environmental characteristics (ventilation and cleaning) influence bacterial and fungal microbial communities in indoor air and on surfaces of tribal homes, and complements the first study by providing valuable insight into the intersection between building science and modern microbial measurements. *This paper concentrates on the second study.*

Along with the well-studied indoor *microbial sources* that include emission from occupants and microbial growth due to moisture (Hyvärinen, Meklin et al. 2002, Qian, Hospodsky et al. 2012), the quantity and structure of indoor air and surface microbial communities are impacted by potential *microbial sinks* including **building ventilation** (Meadow, Altrichter et al. 2014), and **surface cleaning** (Dettenkofer, Wenzler et al. 2004). This study is structured to address the following specific objectives:

(1) **Determine the impact of building ventilation on the airborne and surface concentrations and community structure of bacteria and fungi.** We *hypothesize* that increased ventilation will be associated with a shift of indoor air and surface microbiomes to resemble outdoor microbiomes (in lieu of typical indoor air communities).

(2) **Estimate the impact of cleaning on the core microbiome of homes.** We *hypothesize* that while cleaning surfaces will reduce the presence and concentration of total microorganisms on surfaces in homes; it will not significantly change the richness or community structure of the indoor microbiome (as assessed by floor dust samples).

(3) **Study associations between in depth, DNA-based microbial measurements and ATP measurements for assessing the effectiveness of surface cleaning.** We *hypothesize* that the application of quantitative DNA-based measurements to describe bacterial and fungal communities improves upon the traditional use of ATP-based (adenosine triphosphate) measurements for assessing surface cleanliness, and may afford a better understanding of the usefulness of ATP measurements for gauging cleaning effectiveness.

STUDY DESIGN METHODS

The research will be conducted in tribal homes in northeastern Oklahoma. To address ***objective 1***, we will conduct seasonal (winter heating season and summer cooling season) sampling campaigns in 26 tribal homes. This will involve three distinct collection efforts in the homes to produce baseline, midterm, and final data sets. Building parameters to be measured include air exchange rates (AER) by perfluorocarbon methods (California ARB Method IP-4A, 1989), real-time indoor/outdoor temperature, indoor/outdoor relative humidity, indoor CO_2 concentrations, and gravimetric PM_{10} indoor and outdoor samples. The air exchange measurements in the homes will include a minimum of 104 perfluorocarbon-based home air exchange rate measurements. In addition, air samples, surface samples, and floor samples shall be taken to determine total bacterial and fungal air and surface concentrations by qPCR methods (Hospodsky, Qian et al. 2012), and entail the production of fungal and bacterial DNA-based community libraries using Illumina-based DNA sequencing and QIMME-based data analysis pipelines (Caporaso, Lauber et al. 2012).

To address ***objective 2***, we will examine the impact of enhanced cleaning in 26 tribal homes as compared to 26 control homes. The enhanced cleaning will include occupant education on best cleaning practices, as well as source control measures that include reducing pests, clutter, dampness/moisture, settled dust, and a focus on cleaning of high touch surfaces and flooring. Enhanced cleaning options will factor in tribal traditional ecological knowledge, and at each home visit, a diary of cleaning efforts will be recorded. Microbiological samples of surfaces and floor dust shall be taken and analyzed (as described above for samples collected per objective 1).

To address ***objective 3*** in the determination of, to what extent, specific features of the bacterial and fungal community correlate with ATP measurements of surface cleanliness, we will use DNA-based microbial community data for surface concentrations obtained for objectives 1 and 2, and adjacent side by side ATP surface concentration data from the same surfaces. Biological replicates (multiple side by side samples) for DNA-based analyses and

ATP analyses will be conducted in 5% of the samples to determine the variability introduced by the need to take spatially distinct samples.

Statistical analyses associated with objectives 1 through 3 (hypotheses related to specific objectives), will involve several approaches.

The ***first hypothesis*** will be tested by exploring the association between AER, and the following microbial-related parameters: total fungi, total bacteria, PM_{10}, concentrations of fungal allergens (grouped taxa and specific taxa), concentrations of human associated bacteria, concentration of specific taxa on air and surfaces. Winter and summer seasons will be considered. Average number of observed species (richness) will be compared in high and low AER homes to evaluate impact of ventilation on microbial ecosystems. Where relevant, multiple comparisons will be accounted for.

We will also produce statistical comparisons between AER and weighted/unweighted UniFrac distance (bacteria) or Morisita-Horn distance (fungi) between air and floor dust samples. We plan to perform comparisons between ventilation rate and the indoor to outdoor air UniFrac (Morisita-Horn) distance, and between the ventilation rate and the indoor air to floor dust or surface UniFrac (Morisita-Horn) distance. The purpose of these comparisons is to determine how ventilation influences the indoor air microbial communities (do they look like floor dust or outdoor air). We can also perform comparisons with ventilation for the indoor air to outdoor air Unifrac (Morisita-Horn) distances. In addition, we plan to compare the indoor/outdoor ratio of bacterial and fungal concentrations with ventilation rate.

Complicated relationships likely exist among bacterial communities, fungal communities, moisture, mold growth, and other indoor environmental characteristics such as air conditioner and dehumidifier use, flooring type, seasons and the presence of pets. In addition to comparing diversity measurements with one building variable, multiple logistic regression models with three or more variables and stratification of data can be used to demonstrate interactions between these factors. Finally, we will conduct a mass balance analysis based on sources (ventilation, occupancy emission) and sinks (deposition, ventilation) of microbes into indoor air to determine human occupancy emission rates (E, cells h^{-1}) (Hospodsky, Yamamoto et al. 2014) and compare these rates to ventilation-associated source rates (QC_{out}, cells h^{-1}).

We will apply these same statistical approaches for continuous and/or dichotomized variables for testing the ***second hypothesis*** that compares features of the microbial communities (concentrations of total fungi, total bacteria, fungal allergens, human-associated bacteria, bacterial and fungal richness, and specific bacterial and fungal taxa) for homes with cleaning enhancement (including baseline, midterm, and final sampling and control homes). Herewith, we plan to determine if enhanced cleaning reduces the concentrations of fungi, bacteria, and specific fungal (including allergens) and bacterial taxa on surfaces. We also plan to determine if the cleaning impacts the UniFrac (Morisita-Horn) distance between surface dust (from cleaning surfaces) and floor dust (which represents a "core" microbiome).

Finally, for the ***third hypothesis***, ATP concentration and microbial community concentrations are continuous variables. Thus, we will analyze correlations between ATP concentrations and concentrations of total bacteria, total fungi, total fungal allergens, and sum of bacteria and fungal concentrations. To test the effectiveness of enhanced cleaning, these correlations will be extended to comparing the reduction of ATP after cleaning to the concentrations of selected microbial community features.

CONCLUSIONS / EXPECTED PROJECT OUTPUTS

The research is broad-based and the results will have significance to microbiology, IAQ, environmental health, building science (including impact of ventilation on indoor air and surface microbiomes), and cleaning science. This work will include producing phylogenetic information for over one thousand indoor air, outdoor air, and indoor surface samples as well as extensive building characterization. We will work within current efforts on the indoor microbiome databases QIIME/Qita and publically available databases to ensure that the resulting data can be integrated into future studies and that the metadata provided are comprehensive.

In sum, AER measurements, indoor/outdoor PM10 measurements and sample collection, and a broad suite of surface and air DNA-based microbial measurements including qPCR for total bacteria, total fungi, and microbial community analysis for bacteria and fungi shall be taken throughout the project. The large numbers of homes and microbial samples (over 1,000 samples), the extensive building characterization, and the multiple time points for sampling are expected to provide a very robust data-set for discerning the impacts of ventilation and cleaning regime on the indoor microbiome.

In addition, the project will provide an opportunity for tribal college students in northeastern Oklahoma to gain field research experience involving management, sampling, and quality assurance and control. These students will be encouraged to participate in publications as appropriate and attend Sloan symposia that are scheduled during the project period.

ACKNOWLEDGEMENT

The underlying project is supported by the US EPA under Science for Sustainable and Healthy Tribes, EPA –G2013-STAR-Y1. The second complementary study is supported by Alfred P. Sloan Foundation, Microbiology and the Built Environment Program.

REFERENCES

California ARB Method IP-4A Perflourocarbon Tracer Method for Air Exchange Rate, 1989.

Caporaso, J. G., C. L. Lauber, et al. (2012). "Ultra-high-throughput microbial community analysis on the Illumina HiSeq and MiSeq platforms." ISME J 6, 1621-1624.

Dettenkofer, M., S. Wenzler, et al. (2004). "Does disinfection of environmental surfaces influence nosocomial infection rates? a systematic review." American Journal of Infection Control 32, 84-89.

Hospodsky, D., J. Qian, et al. (2012). "Human Occupancy as a Source of Indoor Airborne Bacteria." PLoS ONE 7, e34867.

Hospodsky, D., H. Yamamoto, et al. (2014). "Size-resolved, baseline characterization data for indoor emission rates of bacteria, fungi, and particulate matter." unpublished data.

Hyvarinen, A., T. Melkin, et al. (2002). "Fungi and actinobacteria and moisture-damaged building materials - concentrations and diversity." *International Journal of Biodeterioration and Biodegradation* 49, 27-37.

Meadow, J. F., A. E. Altrichter, et al. (2014). "Indoor airborne bacterial communities are influenced by ventilation, occupancy, and outdoor air source." Indoor Air 24, 41-48.

CHARACTERING AIRBORNE ALLERGENS AND HUMAN ALLERGIC RESPONSES IN A UNIVERSITY HISTORY MUSEUM

Yan WU[1,2,3*], Maosheng YAO[1] and Victor W.-C. CHANG[2,3]

[1]State Key Joint Laboratory for Environmental Simulation and Pollution Control, College of Environmental Sciences and Engineering, Peking University, Beijing, China
[2]Berkeley Education Alliance for Research in Singapore, Singapore
[3]School of Civil and Environmental Engineering, Nanyang Technological University, Singapore

[*]Corresponding email: wuyan@ntu.edu.sg

Keywords: Airborne allergens, Alt a 1, Asp f 1, Dust mite allergens, Ig E

SUMMARY

We aimed to quantify concentrations of airborne allergens in a university history museum where staff self-reported the incidence of allergic responses. We used SKC BioSamplers to collect air samples and enzyme-linked immunosorbent assay (ELISA) to quantify allergens. We also collected blood sera from museum staff and tested allergen-specific IgE levels in human blood sera with a radioallergosorbent test. Results showed that the concentrations of airborne fungal allergens Alt a 1 ranged from 181 to 242 ng/m^3 and Asp f 1 ranged from 200 to 605 ng/m^3, much higher than levels outdoors. For dust mite allergens, the highest indoor concentrations were 79 ng/m^3 for Der f 1 and 84 ng/m^3 for Der p 1, slightly higher than outdoor concentrations. High fungal allergen-specific IgE levels present in blood sera suggest that occupants may have had allergies caused by fungal allergens. Data from this work indicate that these aeroallergens were not only detectable but also with high levels in the museum and their abundance may constitute an important inhalation risk for occupants.

INTRODUCTION

Extensive literature suggested that allergen exposure especially by means of inhalation is linked to various allergy-associated maladies such as allergic asthma, rhinitis and eczema (Douwes et al., 2003; Randriamanantany et al., 2010). Among the allergens, *Alternaria alternata* allergens (Alt a 1), *Aspergillus fumigatus* allergens (Asp f 1), and two major mite allergens Der p 1 and Der f 1 from Dermatophagoides pteronyssinus and Dermatophagoides farinae are commonly found in indoor environments and play important roles in causing allergic reactions in humans (Randriamanantany et al., 2010). For instance, allergies caused by two major fungal allergens Asp f 1 and Alt a 1 contributed around 80% of total fungal allergies (Vailes et al., 2001). Although most mite allergens (Der f 1 and Der p 1) exist in the house dust, some of them can also be found on larger particles in the air due to various aerosolization mechanisms. And long-time exposure to low levels of the aeroallergens Der f 1 and Der p 1 could exacerbate allergic reactions in patients with allergies (Tovey et al., 2013). Accordingly, due to their negative health effects, it is important to assess the allergen load in indoor air. In this work, we aimed to quantify airborne allergen types and their levels in a university history museum. Among the nine staff who worked in this museum, all have varying degrees of allergic eczema and rhinitis.

METHODS

This investigation was conducted in an underground history museum on the campus of a university in China. There were four occupied offices in the museum with a total of nine staff members. Each office was approximately 45 m^3 and occupied with two to three occupants during office time. And the entire floor area in these four offices was covered by carpet. Poor ventilation system control led to excessive moisture inside the office and resulted in complaints of moldy odor. Staff members reported that they experienced allergies, especially allergic eczema, after three months they moved to the museum with a total of 720 hours of exposure time. In this study, we sampled the airborne allergens from these four occupied museum offices during office time. In order to find the possible allergen sources, we also chose one of the museum offices (office 2) and studied this indoor environment under two conditions: "occupied with carpet" represents the original office with carpeted floor, and "occupied without carpet" indicates the condition with the same occupants but removing carpet from the floor. The floor in office 2 was vacuumed regularly in the morning before our sampling. For the second condition, no sampling was conducted until three days after removing the carpet. We also chose a nearby office from another building to serve as a control. The control office was chosen for two reasons: one was that its room size, occupancy and overall arrangement were quite similar to those of the museum offices; the other factor was that its window was facing the window of one museum office and only two meters away. In this study, SKC BioSamplers (SKC Inc., Eighty Four, PA) were employed to collect air samples using 20 mL deionized (DI) water (Milli-Q, Millipore, Billerica, MA) at its standard sampling flow rate of 12.5 L/min with a sampling time of 30 min. Sandwich enzyme-linked immunosorbent assay (ELISA) kits (Indoor Biotechnologies, Charlottesville, VA) were used to quantify airborne allergen types and concentrations. The allergic reaction levels of the museum staff were assessed by measuring allergen-specific immunoglobulin E (IgE) levels in human blood serum. Blood sera samples of the museum staff (collected by Peking University Hospital) were obtained during the same period when air samples were taken. The IgE levels of human blood sera samples were quantified using radioallergosorbent tests (RAST) with kits from ALerCHEK, Inc. (Springvale, ME).

RESULTS AND DISCUSSION

Figure 1 shows concentrations of four airborne allergens: *Alternaria alternata* allergens (Alt a 1), *Aspergillus fumigatus* allergens (Asp f 1), and dust mite allergens (Der p 1 and Der f 1). The airborne allergen concentrations in the museum ranged from 181 to 242 ng/m^3 for Alt a 1, 200 to 605 ng/m^3 for Asp f 1, 32 to 79 ng/m^3 for Der f 1 and 11 to 84 ng/m^3 for Der p 1, respectively. At the same time, airborne allergen concentrations were much lower (especially for the fungal allergens) in the control office: 17 ng/m^3 for Alt a 1, 15 ng/m^3 for Asp f 1, 19 ng/m^3 for Der f 1 and 11 ng/m^3 for Der p 1. The indoor environment of the history museum appeared to have higher allergen levels compared to the control office. As also seen from Figure 1, for two fungal allergens (Alt a 1 and Asp f 1), the indoor concentrations are greater than observed outdoor concentrations, indicating the presence of indoor fungal allergen sources. The airborne dust mite allergens from indoor samples are similar to outdoor concentrations of dust mite allergens.

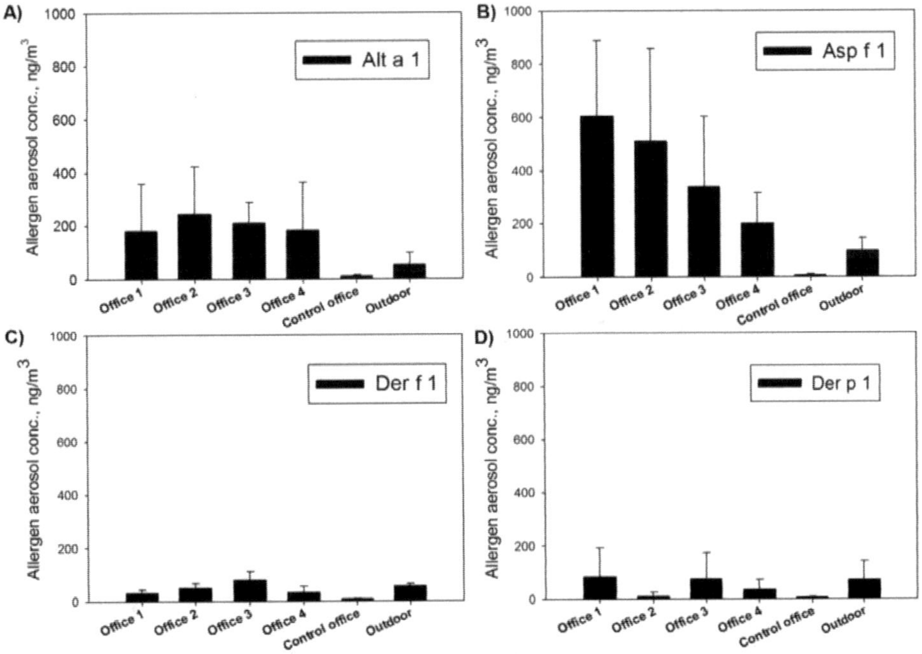

Figure 1. Concentrations of airborne allergens in different environments: A) Alt a 1; B) Asp f 1; C) Der f 1; D) Der p 1.

To confirm the allergenicity levels of staff working in the history museum, we tested allergen-specific immunoglobulin E (IgE) levels in human blood serum using the RAST method. Based on the IgE levels (optical density) in the blood sera, the allergies are classed into five grades: negative (0 - 0.25), Class 1 (0.251 – 0.35), Class 2 (0.351 – 0.45), Class 3 (0.451 – 0.55) and Class 4 (> 0.55). For fungal allergens, we observed very high allergen-specific IgE levels present in the blood sera as shown in Figure 2 (A) and (B). The elevated IgEs for fungal allergens imply all staff had allergies caused by fungal allergens Alt a 1 and Asp f 1. Three staff suffered Class 2 allergies, four suffered Class 3 allergies and two suffered Class 4 allergies caused by exposure to Alt a 1 allergens. Moreover, eight of nine staff suffered Class 4 allergies caused by exposure to Asp f 1 allergens. Combined with the data shown in Figure 2, we can see that direct exposure to high concentrations of fungal allergens may cause allergic reactions in human occupants. For the two major dust mite allergens, sera reported results were generally negative (exceptions being staff A and G).

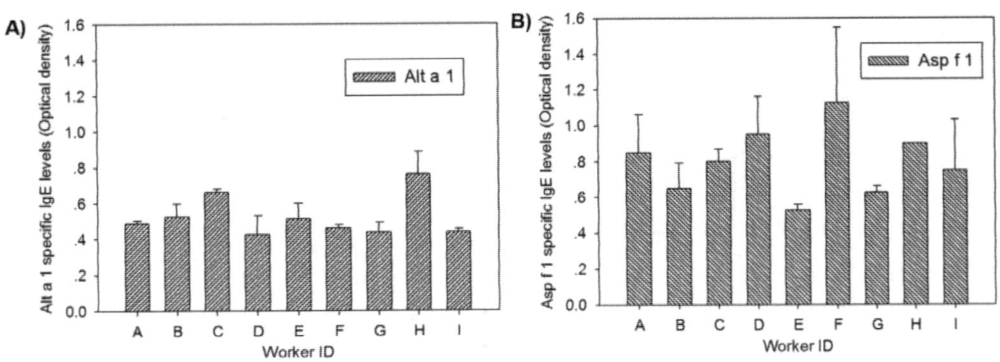

Figure 2. Allergen-specific IgE levels in human blood sera using the RAST method: A) sensitized to Alt a 1 allergens; B) sensitized to Asp f 1 allergens. A–I refer to staff IDs.

Figure 3 shows the concentrations of these four major allergens in office 2 under two conditions. Removing the used carpet indirectly resulted in reduction by 24% (242 vs. 184 ng/m^3), 30% (510 vs.

357 ng/m^3), 79% (49 vs. 10ng/m^2), and almost 100% (11 vs. NA ng/m^3) of airborne Alt a 1, Asp f 1, Der f 1 and Der p 1 allergens, respectively. The removal differences between fungal allergens (Alt a 1 and Asp f 1) and house dust mite allergens (Der f 1 and Der p 1) in responding to the removing carpet activity could indicate that the carpet surface contribute more house dust mites to the indoor environment than did the fungi. These results show that direct removing used carpet (or some chemical and physical methods to clean carpets) may reduce house dust mite allergens as well as fungal allergens levels in the indoor air, which could reduce the exposure risks of indoor occupants.

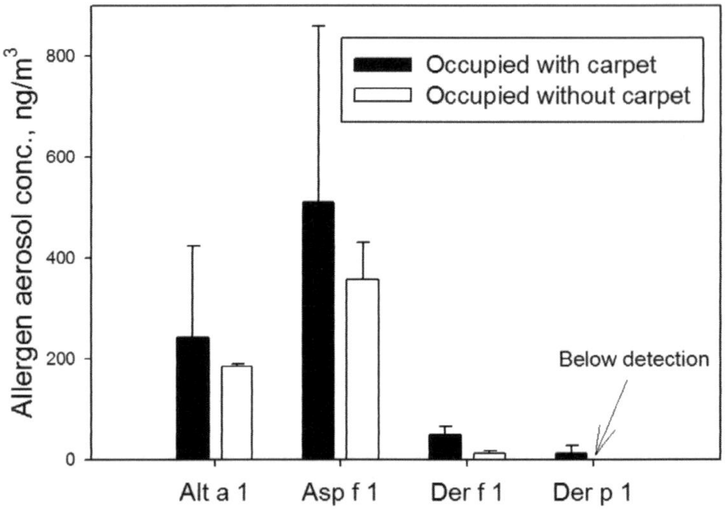

Figure 3 Airborne allergens concentrations in office 2 under two conditions

CONCLUSIONS

This work quantified four common allergens in indoor air and found high levels of fungal allergens (Alt a 1 and Asp f 1) present in indoor air. Allergen-specific IgEs in human blood sera indicated that fungal allergens played more important roles in museum occupants' allergic response, while the effect of dust mite allergens (Der f 1 and Der p 1) appear small.

ACKNOWLEDGEMENT

This work was funded by the National Science Foundation of China (21277007and 41121004). We also thank the excellent assistance from Berkeley Education Alliance for Research in Singapore (BEARS). BEARS has been established by the University of California, Berkeley as a center for intellectual excellence in research and education in Singapore.

REFERENCES

Douwes J, Thorne P, Pearce N, et al (2003). Bioaerosol health effects and exposure assessment: Progress and prospects. *Annals of Occupational Hygiene*, **47**, 187–200.

Randriamanantany ZA, Annesi-Maesano I, Moreau D, et al (2010). Alternaria sensitization and allergic rhinitis with or without asthma in the French Six Cities Study. *Allergy,* 65, 368–375.

Tovey ER, Willenborg CM, Crisafulli DA, et al (2013). Most personal exposure to house dust mite aeroallergen occurs during the day. *PLoS One*, **8**, e69900.

Vailes L, Sridhar S, Cromwell O, at al (2001). Quantitation of the major fungal allergens, Alt a 1 and Asp f 1, in commercial allergenic products. *Journal of Allergy and Clinical Immunology*, **107**, 641-646.

Study on Ventilation Strategy for Removing Dust Particles

Yang Lu [1], Hiroshi Yoshino [2]

Introduction

In recent years, the levels of air tightness and insulation of residential houses are becoming better for energy saving, and hence living comfort has been improved. However, the problem of dust particles, due to the lack of ventilation rate, has become increasingly serious. Dust particles include mold spores or excrement and corpses of the mite and so on, which are the cause of allergy, asthma and unspecific hypersensitivities. [1, 2]. Ventilation is one of the best ways to remove dust particles, it is important to know the removal efficiency of dust particles in different ventilation strategies room. In recent years, a number of numerical simulation studies using computational fluid dynamics (CFD) to predicted dust particles distribution and deposition. Murakami et al [3] found that numerical simulation seems to be very useful method for predicting air velocity and contaminant distribution in clean rooms. Bouilly et al [4] carried out both numerical simulations and measurements to study the particle decay rates with three different ventilation modes using the Lagrangian formulation. Zhao et al [5] compared the spatial distribution of particles with four different ventilation modes. A three-dimensional drift-flux model for particle movements in turbulent indoor airflows combining with the Eulerian approaches was developed by Gao and Niu [6]. However, research studies that focus on the removal efficiency of dust particles by ventilation strategies which is still lacking.

The purpose of this study is to clarify the removal effect of the dust particles for the various ventilation strategies by experiments and CFD-based simulations.

Experiment

The different ventilation strategies room (L×W×H = 5.37 m×2.74m×2.25 m) is on the second floor of experimental house which is located at Tohoku University, Japan. The window is sheltered to prevent solar radiation, and the walls are insulated by expanded polystyrene board, room volume is 27.0m^3. Two mechanical ventilation strategies were considered in this construction. As shown in Figure 1, for case 1, inlet and outlet are located at the ceiling (ceiling exhaust). It can be considered as a representative ventilation strategy in Japan currently. For case 2, inlet is located at the center of ceiling as case1, but considering the dust particles are usually falling on the floor, exhaust slit is set at the corner between wall and floor (slit exhaust). Figure 2 shows slit exhaust ventilation strategy structure, dust particles were exhausted out of room through exhaust slit over in-wall ventilation.

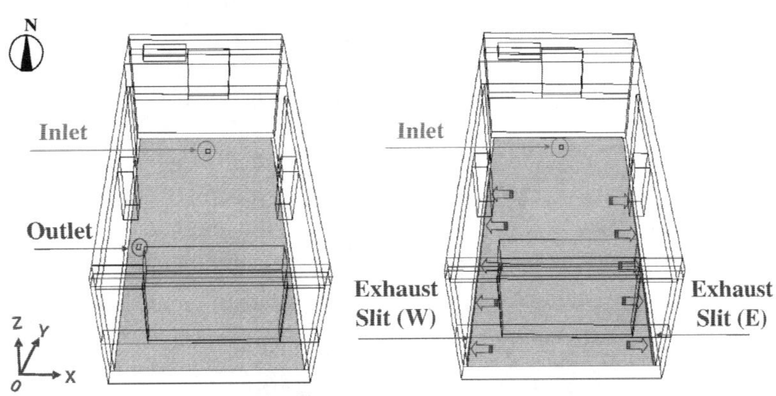

a: Case 1 (ceiling exhaust) b: Case 2 (slit exhaust)

Figure 1 Different ventilation strategies room

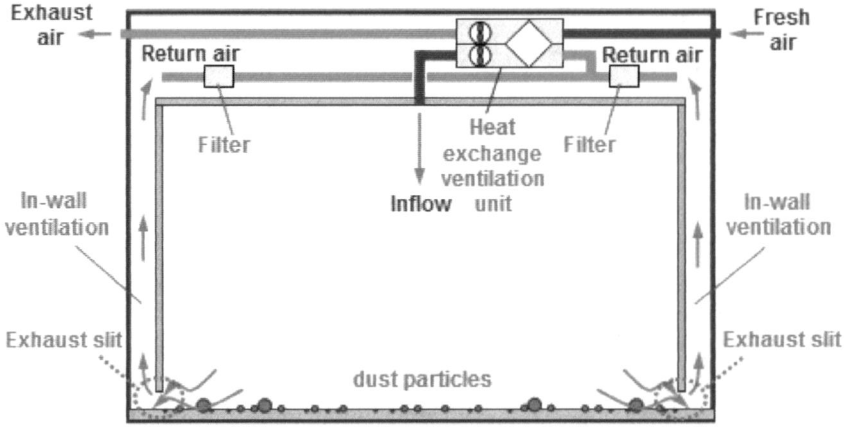

Figure 2 The slit exhaust ventilation strategy structure

Table 1 shows the experimental conditions. In this study, riboflavin particles were used as dust particles. Figure 3 shows dispersal devices and measurement points for riboflavin particles. 3D ultrasonic anemometer (CLIMATEC, CYG-81000) was used for measuring airflow directions and velocities. The riboflavin particles were put in a measuring flask, N_2 was blown into the measuring flask to disperse the riboflavin particles. The positions of the monitoring points were set in the center of a quarter of the room and set at 1.0 m height. Additionally, for case 2, the airflow velocity of exhaust slit was measured at 6 points using a thermistor anemometer (F6204E, SHIBAFU electron). The experimental procedure was determined as Figure 4: in the first 1 hour, the riboflavin particle was dispersed, and during the following 8 hours the particle fell on the ground freely, and for the last 15 hours, ventilation system was started.

Table 1 Experimental conditions

Case	Air change rate (ach)	Inlet velocity (m/s)	Outlet velocity (m/s)	Temperature (℃)			Height of slits (m)
				Indoor	Outdoor	Supply	
Case1	2.5	3.34	3.8	19.69	3.84	21.59	-
Case2			0.5	18.95	2.55	22.37	0.005

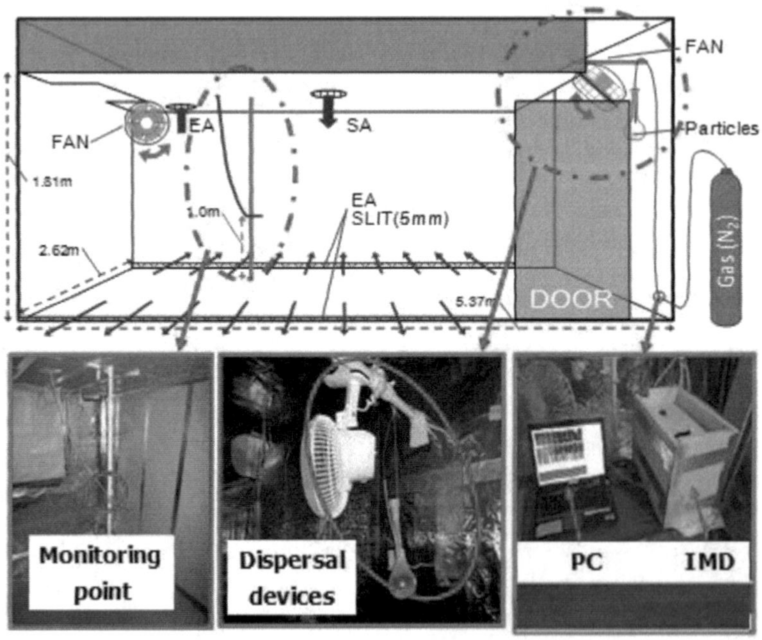

Figure 3 Dispersal devices and measurement points

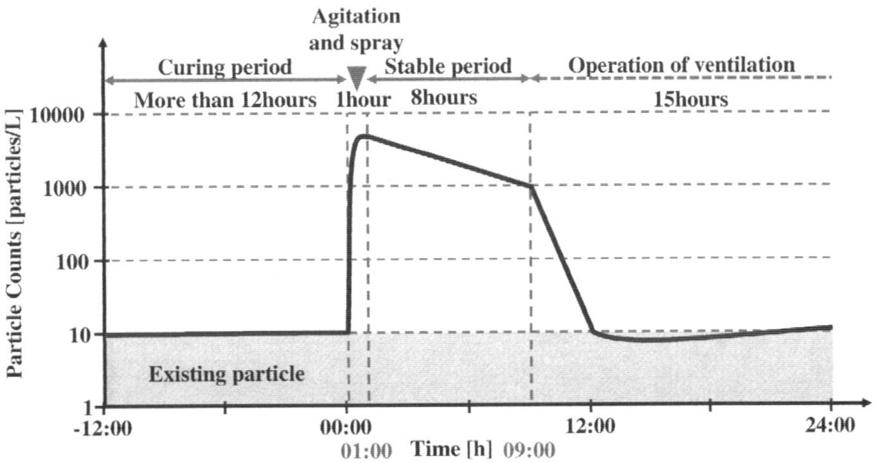

Figure 4 The schedule of particles experiment

Simulation

The above-mentioned different ventilation strategies room was modeled by CFD. A commercial CFD software Flow Designer was used (http://www.akl.co.jp/). The calculation period was about 1 hour, and the time-step was 1 second. For indoor and outdoor air temperatures and velocities, measured values were used. For case 2, the measured averaged velocities in slit were 0.5m/s, and there was no significant difference among the 6 points.

Results

The experiment results of case 1 and case 2 was shown in Figure 5. For the two cases, different sizes of indoor particles showed smooth decay curve due to the influence of gravity before ventilation system operating. After ventilation system is operated, the decay rate of the particles becomes faster than that before the ventilation system is on. After ventilation system is operated, the decay rate of the particles becomes faster than that before the ventilation system is on. After 15 hours of ventilation system operation, the number of particles for all sizes decreased to the same level as background. The number of particles in the case 1 (ceiling exhaust) was a little higher than that in the case 2 (slit exhaust) after particle decreased to the same level as background.

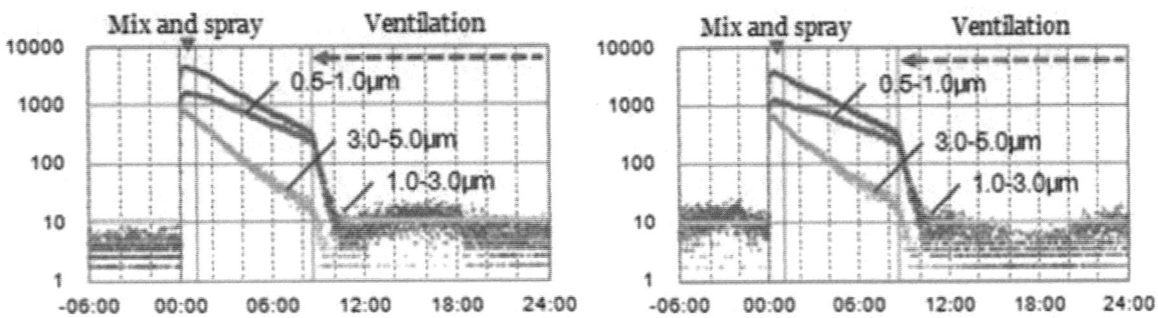

a: Case 1 (ceiling exhaust) b: Case 2 (slit exhaust)

Figure 5 Experiment results of particles

Discussion

Figure 6 compares the removal efficiencies after 1 hour. The removal efficiency is defined as Eq. 1:

$$\eta = \frac{C_0 - C_{60}}{C_0} \times 100\% \quad (1)$$

Where η: removal efficiency [%],

C_0: initial volume-average concentration of dust particles in the room [kg/m^3],

C_{60}: the volume-average concentration of dust particles in the room after 1 hour [kg/m^3].

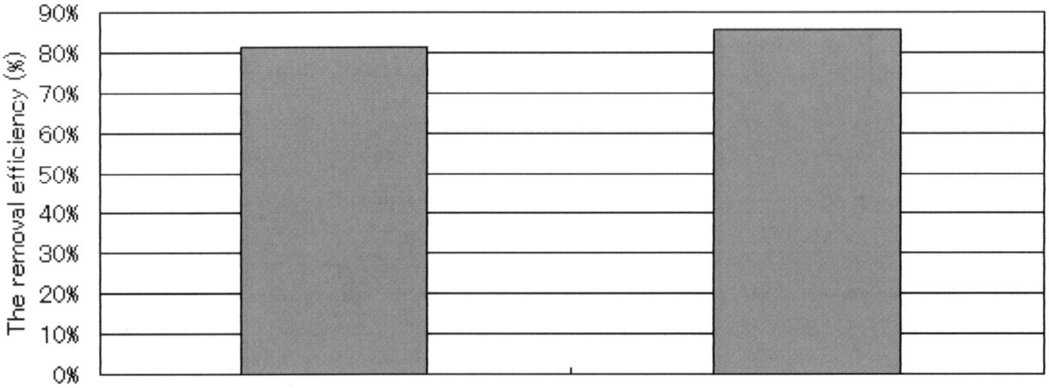

a: Case 1 (ceiling exhaust) b: Case 2 (slit exhaust)

Figure 6 The removal efficiency of dust particles

The removal efficiencies of case 1 (ceiling exhaust) and case 2 (slit exhaust) are 81.5% and 85.8% respectively. From the data we concluded that slit exhaust ventilation strategy will produce less dust particles comparing with ceiling exhaust ventilation strategy.

Conclusions

This paper clarifies the removal efficiencies of dust particles by different ventilation strategies (ceiling exhaust and slit exhaust) in room. The following conclusions have been drawn from the present study:

1) Two kinds of ventilation strategies in room were considered, namely ceiling exhaust and slit exhaust. In the experiments, comparison of the number of particle shows that the ceiling exhaust was a little higher than the slit exhaust ventilation strategy.
2) CFD simulations concerning diffusion fields have been completed. The removal efficiencies of ceiling exhaust and slit exhaust were found 81.5% and 85.8% respectively. From the data we concluded that slit exhaust ventilation strategy will produce less dust particles comparing with ceiling exhaust ventilation strategy in room.

FROM A LAB SCALE TO THE BUILDING SCALE: FORMALDEHYDE REMOVAL BY A PLANT MODULE

Ahu AYDOGAN AKSELI[1,*], Anna DYSON[2], and Lupita D. MONTOYA[3]

[1]Department of Architecture, City University of New York, New York USA
[2]Department of Architecture, Rensselaer Polytechnic Institute, Troy, New York USA
[3]Department of Civil, Environmental and Architectural Engineering, University of Colorado Boulder, Colorado USA

[*]Corresponding email: aakseli@ccny.cuny.edu

Keywords: Indoor air quality, interior plants, volatile organic compounds, formaldehyde, growing media.

SUMMARY

The study presented here provides a framework for studying the performance of plant wall system of different scales and installations adapted to a range of building zonal air volumes. The formaldehyde removal capacity of a combination of three porous materials (growstone, expanded clay and activated carbon) loaded in an architectural module was evaluated under several conditions. Seven airflow rates (9, 18, 27, 36, 54, 63 and 72 l/min) and four concentration levels (2.00 g/m³, 0.73 g/m³, 0.37 g/m³ and 0.19 g/m³) were studied using a single module of a plant wall system. The single pass removal efficiency was evaluated over an 8-h period for seventeen different cases. In all cases, the slowest airflow rate showed the highest single pass removal efficiency. From the results of these experiments, the relative number and sizing of plant wall modules required for selected building types to remove toxins from indoor environments was further assessed.

INTRODUCTION

Indoor Air Quality (IAQ) was fifth on a list of 31 largest environmental threats in the United States in the 1980's (EPA, 1989). Since then, there has been a dramatic increase in building code requirements for fresh air intake leading to escalating energy consumption (ASHRAE, 2003) and larger impacts on human health and welfare (Brown, 1997). IAQ can be significantly worse than outdoor air quality and may pose a greater health threat depending on the length of exposure (Brown, 1994; Baek, 1997). Additionally, it has been determined that people in developed countries, particularly within urban areas, spend an average of 80 - 90% of their time indoors (Robinson and Nelson, 1995; Klepeis et al., 2001).

Volatile Organic Compounds (VOCs) compose a very important class of contaminants in the indoor environment (Wolkoff, 2003). Formaldehyde is among the most common indoor VOCs that is toxic at high levels and has also been shown to be carcinogenic (Nielsen 2010). Although technologies for toxic removal are quite well established, presently there are no satisfactory mechanical methods for VOC control due to cost or feasibility (Guieysse et al., 2008). Biological systems that rely on the ability of plants and microorganisms to remove organic compounds from the air have been proposed as alternative IAQ control systems (Aydogan and Montoya, 2011).The goal of this study was to evaluate the capacity of an architectural module loaded with three hydroponic growing media to remove high levels of formaldehyde. This approach has been proposed as a potential alternative to remediate indoor air pollution from VOCs.

METHODOLOGIES

A closed chamber (100 × 100 × 100 cm) was used to evaluate an architectural module loaded with three hydroponic materials (growstone, expanded clay and activated carbon) for its ability to uptake formaldehyde from the air. The chamber was divided into 2 equal volume sub-chambers, inlet side (right, chamber A) and outlet side (left, chamber B), by a partition, which held the module in place, as shown in Figure 1.

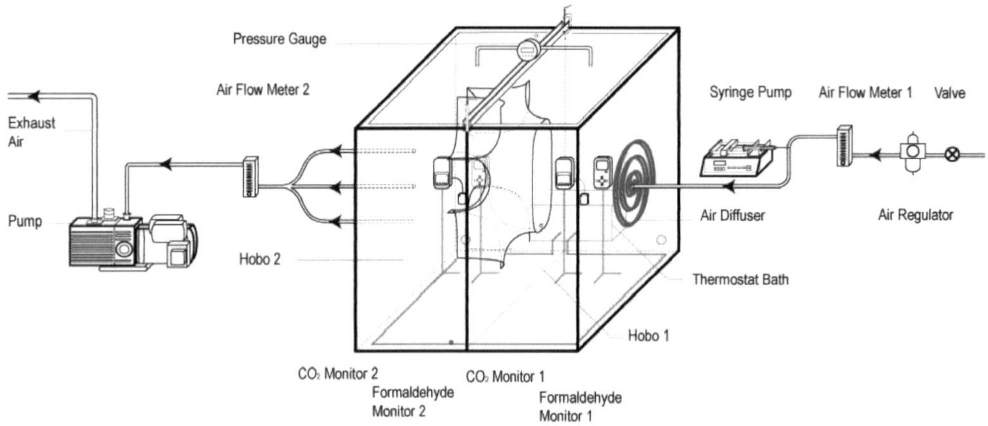

Figure 1 Experimental setup for evaluating formaldehyde removal capacity of the growing media loaded in an architectural model for plant system design.

Each growing medium was individually evaluated over an 8-hour period at eight airflow rates (9, 18, 27, 36, 45, 54, 63 to 72 l/min) and at four formaldehyde concentration levels (~2000, 725, 370 and 184 μg/m³) using this single architectural module. The single pass removal efficiency (the polluted air passed through the module once with no recirculation) was evaluated for each case. The efficiency was calculated using the formaldehyde concentrations measured in each sub-chamber as follows:

$$\eta_{singlepass} \equiv 1 - \frac{amount\ out}{amount\ in} = 1 - \frac{\int_{t=0}^{t_{final}} \dot{V}_B(t) \cdot C_B(t)dt}{\int_{t=0}^{t_{final}} \dot{V}_A(t) \cdot C_A(t)dt}$$

where $\dot{V}_B(t)$ is the volumetric flow rate in (left) chamber B, $\dot{V}_A(t)$ is the volumetric flow rate in (right) chamber A, $C_B(t)$ is formaldehyde concentration in chamber B and $C_A(t)$ is the formaldehyde concentration in chamber A during the test.

RESULTS AND DISCUSSION

All experiments were performed in triplicate. The first set of experiments determined the combined losses due to leakage, absorption, and chemical reactions within the chamber. Figure 2 shows the measured concentration of formaldehyde in Chamber B when formaldehyde-laden air at various concentrations flowed through the module loaded with growing media (growstone, expanded clay and activated carbon) under different flow rates.

The indoor limit of formaldehyde recommended by World Health Organization (100 μg/m³) is also shown.

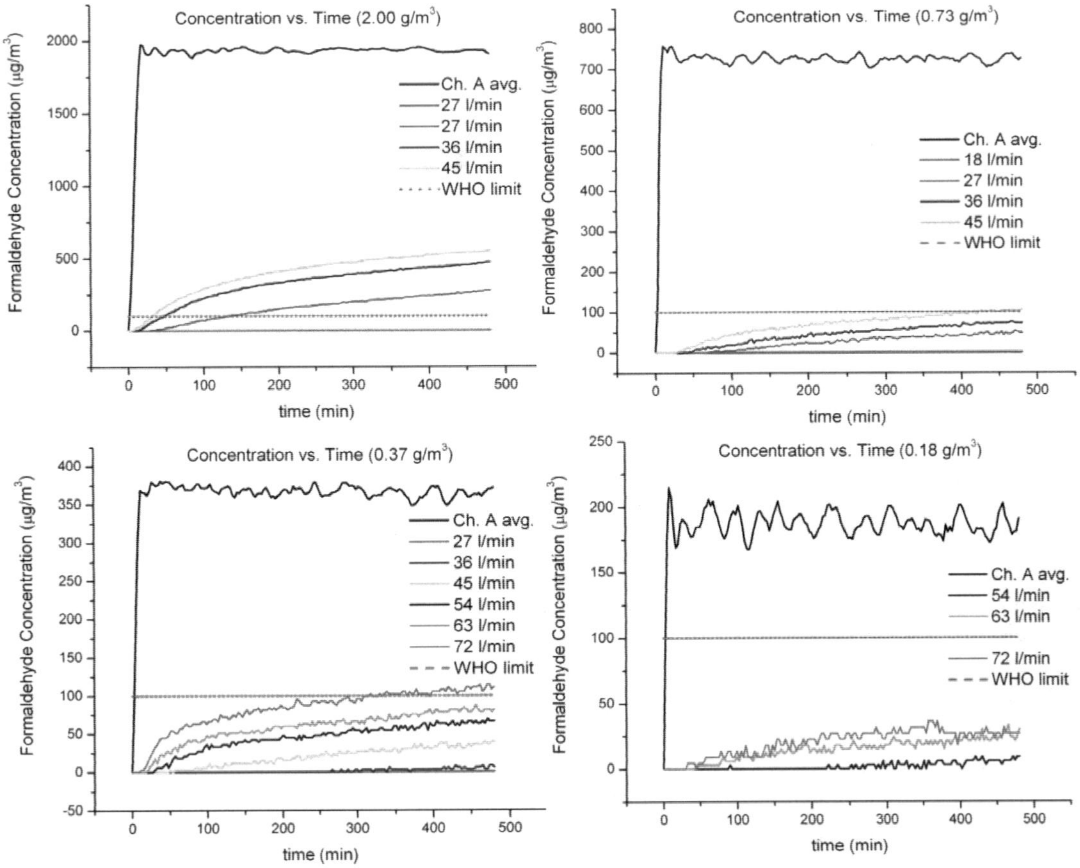

Figure 2. Formaldehyde concentrations measured in Chamber B at different concentrations (2.00, 0.73, 0.37 and 0.18 g/m3) generated in Chamber A and varying flow rates.

Table 1 summarizes the results of the evaluation of the module's single pass formaldehyde removal efficiency under four concentration levels (2000 μg/m³, 725 μg/m³, 370 μg/m³ and 184 μg/m³) and eight airflow rates (9 to 72 l/min). Overall, the removal efficiencies decreased as the flow rates increased, likely due to the reduced residence time.

Table 1. Removal efficiency of formaldehyde removal measured.

	Airflow rate							
Concentration (μg/m³)	**9 L/m**	**18 L/m**	**27 L/m**	**36 L/m**	**45 L/m**	**54 L/m**	**63 L/m**	**72 L/m**
2000 μg/m³	99.4 ± 0.6	97.8 ± 0.8	85.3 ± 2.3*	75.7 ± 1.5*	70.9 ± 3.6*	–*	–*	–*
725 μg/m³	98.3 ± 1.7	98.3 ± 1.7	93.5 ± 1.4	90.2 ± 2.2	85.5 ± 1.9	–*	–*	–*
370 μg/m³	96.7 ± 3.3	96.7 ± 3.3	96.7 ± 3.3	98.4 ± 1.4	89.0 ± 2.8	82.4 ± 3.7	77.6 ± 1.4	69.7 ± 5.1*
184 μg/m³	93.3 ± 6.7	93.3 ± 6.7	93.3 ± 6.7	93.3 ± 6.7	93.3 ± 6.7	95.1 ± 2.8	83.3 ± 2.8	80.2 ± 2.8

CONCLUSIONS

The concentration of formaldehyde in air (chamber B) after passing through the module loaded with three hydroponic growing media (growstone, expanded clay and activated carbon) under varying flow rates and exposure concentrations (in chamber A) were demonstrated. Each growing media studied, showed longer uptake of formaldehyde as the concentration in chamber A decreased. Also, the lower concentration levels were "remediated" more effectively at higher airflows and were kept under WHO indoor recommendation levels (0.08 ppm) in an 8 hour period. The slowest airflow rates showed the highest single pass removal efficiency; however, the formaldehyde concentration kept increasing over time. Additional experiments should to be performed to evaluate the ability of this system to remediate formaldehyde contamination when plants are included.

The growing media studied here demonstrated significant removal of formaldehyde from the air. These results provide evidence for a potential application of hydroponic growing media and plants to remediate indoor air pollution from VOCs. In addition, these experimental results provide initial insight into the sizing of modules required for integration into buildings in order to remove toxins from indoor environments.

ACKNOWLEDGEMENTS

The authors thank Mark Mistur, Jason Vollen, Philippe Baveye for their experimental guidance and insight into future application of this work. The authors thank Ilgaz Akseli, David Rangel, Zhicheng Wei, Michael P. Allard, and Don VanSteele for their technical assistance. This work was partially supported by NYSERDA and the New York State Foundation for Science, Technology and Innovation (NYSTAR).

REFERENCES

ASHRAE (2003). Guidance on Improving IAQ in Residences. American Society of Heating, Refrigerating and Air–Conditioning Engineers, Inc., Atlanta, GA. ASHRAE.

Aydogan A and Montoya LD (2011). Formaldehyde Removal by Common Indoor Plant Species and Various Growing Media. *Atmospheric Environment,* **45**, 16: 2675-82.

Baek SO, Kim YS, Perry R (1997). Indoor air quality in homes, offices & restaurants in Korean urban areas-indoor/outdoor relationships. *Atmospheric Environment*, **31**(4),529-544.

Brown SK (1997). Volatile organic compounds in indoor air: sources and control. *Chemistry in Australia,* **64**, 10-13.

Brown SK, Sim MR, and Abramson MJ, et al. (1994). Concentrations of volatile organic-compounds in indoor air - a review, Indoor Air-International Journal of Indoor Air Quality and Climate, **4** (2), 123-134.

Guieysse B, Hort C and Platel V, et al. (2008). Biological Treatment of Indoor Air for VOC Removal: Potential and Challenges. *Biotechnology Advances,* **26**, 5: 398-410.

Nielsen G and Wolkoff P (2010). "Cancer Effects of Formaldehyde: A Proposal for an Indoor Air Guideline Value." *Archives of Toxicology,* **84**, 6: 423-46.

Robinson J and Nelson WC (1995). The national human activity pattern survey data base, Environmental Protection Agency, Research Triangle Park, NC. EPA

Klepeis NE, Nelson WC, Ott WR, et al. (2005). The national human activity pattern survey (NHAPS): a resource for assessing exposure to environmental pollutants, Journal of Exposure Analysis and Environmental Epidemiology, **11**, 231–252.

U.S. Environmental Protection Agency (1989). Executive Summary and Recommendations, in EPA/400/1-89/001A.

Wolkoff P (2003). Trends in Europe to Reduce the Indoor Air Pollution of Vocs. *Indoor Air,* **13**, 5-11.

A STATISTICAL STUDY OF INDOOR ACOUSTICAL PERCEPTIONS BASED ON URBAN TRAFFIC NOISE TRANSMITTED VIA OPERABLE WINDOWS

Jeehwan LEE[1,*] and Jae D. CHANG[1]

[1]Department of Architecture, The University of Kansas, Kansas, U.S.A.

[*]Corresponding email: jeefanlee@ku.edu

Keywords: Natural ventilation, Noise transmission, Acoustical comfort

SUMMARY

Studies have indicated that traffic noise in urban environments can deter building occupants from opening windows for ventilation and cooling and instead relying upon mechanical systems resulting in increased energy usage. Quantitative analysis of urban noise levels at six locations in Seoul, South Korea found that the annual equivalent continuous sound level (L_{Aeq}) exceeded the thresholds of the national traffic noise regulations of 65dB-A and 55dB-A during daytime (06:00~22:00) and night time (22:00~06:00) respectively. Results from questionnaires completed by building occupants living near the study sites indicated that 51% of them experienced Sick Building Syndrome (SBS) symptoms and 44% of the respondents regarded urban traffic noise as a significant cause to their decrease in productivity and increase in communication difficulty and psychological stress.

INTRODUCTION

Initial findings from the literature review suggest that traffic noise intensity present a conflict between the natural ventilation potential of operable windows and the transmission of unwanted traffic noise. Therefore this study investigated the correlation between indoor acoustical perceptions and noise transmission via ventilation windows in naturally ventilated buildings.

RESEARCH BACKGROUND

The accelerated growth in urban population has led to an increase in urban traffic intensity, and urban traffic noise has become one of the major constraints which degrade the acoustical quality of urban environments. The World Health Organization (WHO, 2011) stated that high external noise levels can cause numerous health problems such as sleep disturbance, high blood pressure, and psycho-physiological symptoms. Traffic-related noise became the most health-threatening environmental stressor in Europe with more people exposed to traffic-related noise than any other environmental stressors. However, Kang (2007) stated that defining noise annoyance has a complex mechanism related to a number of disciplines such as acoustics, physiology, sociology, psychology and statistics. Lambert et al. (1984) categorized sound levels impacting noise annoyance into the following three groups: less than 55dB-A (no annoyance), between 55 and 60dB-A (some annoyance) and above 65dB-A (definite annoyance). The WHO (2012) suggested that outdoor environmental noise should not exceed 55 dB-A and 40 dB-A during the daytime and nighttime respectively to prevent potential psychosocial effects. It also pointed out that Sound Pressure Level (SPL) which is also presented in dB can be a significant indicator in understanding noise annoyance.

METHODOLOGIES

Quantitative data collection was carried out to evaluate acoustical discomfort due to urban traffic noise at six sites in Seoul, South Korea. The sound meters installed and monitored by the National Noise Information System of Korea (NNISK) collected real-time data of annual equivalent continuous sound levels (L_{Aeq}) at a height of 4-meters above ground. The data provided daytime and nighttime sound levels.

Qualitative data collection was conducted in the form of a questionnaire to support the numerical L_{Aeq} data because acoustic discomfort or noise annoyance has a complex mechanism including the subjective perception on noise levels. The questionnaire participants were randomly selected among building occupants residing near the sound meters at each of the six sites.

RESULTS AND DISCUSSION

According to the quantitative analysis of sound levels during the daytime and nighttime, it was found that the traffic noise levels at all six sites exceeded the threshold of the national traffic noise regulations which require less than 65dB-A and 55dB-A during the daytime (06:00~22:00) and nighttime (22:00~06:00) respectively as shown in Figure 1.

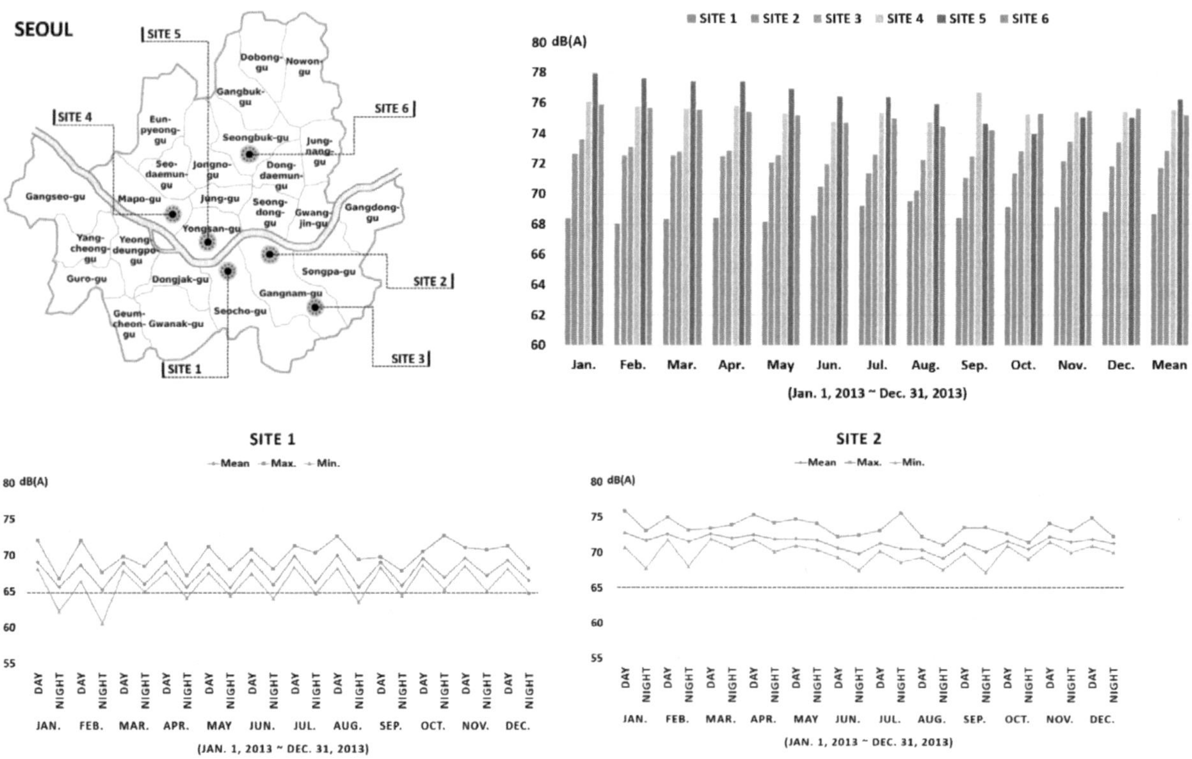

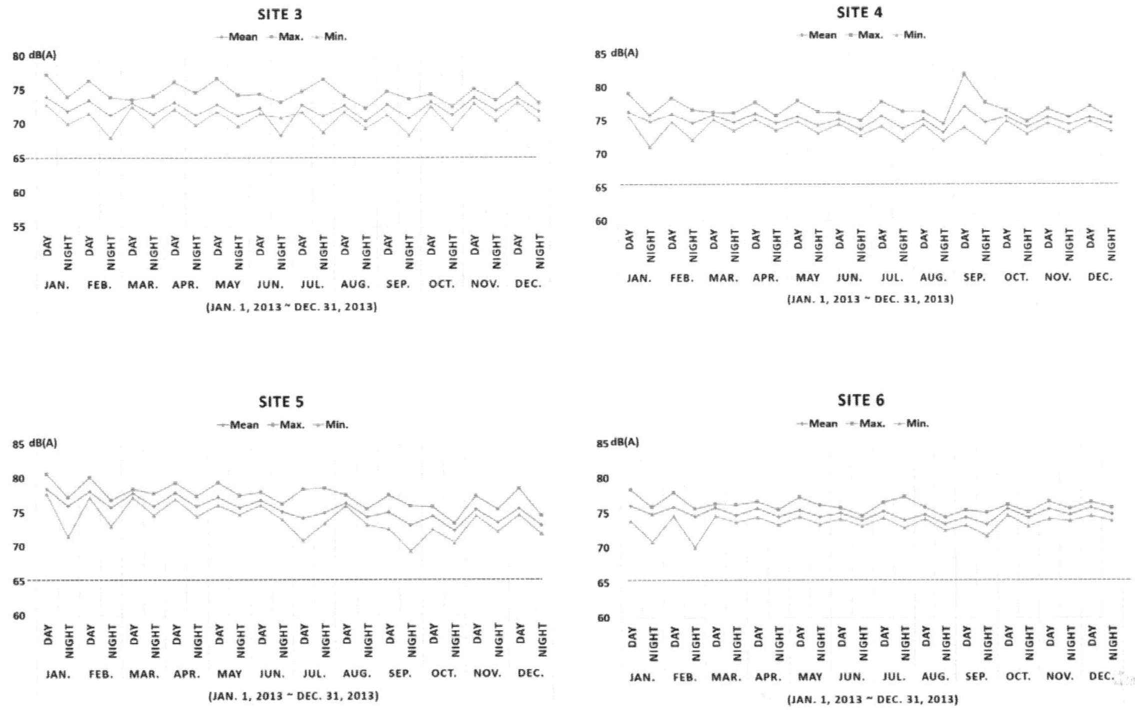

Figure 1. Annual equivalent continuous sound level (L_{Aeq}) at six sites in Seoul, South Korea.

The qualitative data was collected from 92 respondents residing near the six sites over the course of 15 days from July 1 to July 15, 2014. The distribution ratio of gender was 48% female and 50% male. 2% neglected to indicate their gender. The mean age range of the respondents was between 36 to 45 years of age with 74% of the respondents indicating they had lived in the area for less than 5 years.

Table 1. Questionnaire results from 92 respondents residing near the six sound meters.

Questions	Results
1. Which method is more frequently used in your building for ventilation and cooling?	-Mechanical ventilation and cooling (89%) -Natural ventilation (11%)
2. Have you ever experienced any temporary illness such as irritation of the eyes, nose, throat and skin irritation when you're exposed to mechanical ventilation and cooling?	-Never (19%), -Seldom (30%), -Sometimes (36%), -Often (11%), -Always (4%)
3. How much do you agree that natural ventilation (e.g. opening window) is effective in building energy savings and better indoor air quality?	-Strongly agree (23%), -Agree (48%), -Neutral (23%), -Disagree (4%), -Strongly disagree (2%)
4. If you prefer natural ventilation (e.g. opening window) to mechanical ventilation (e.g. fan and air conditioning), what are the main reasons NOT to choose mechanical ventilation?	-Mechanical system contaminant (35%) -Mechanical system energy use (24%) -Mechanical system noise (22%) -Mechanical system operation (5%) -Other reasons (14%)
5. If you prefer natural ventilation (e.g. opening window) to mechanical ventilation (e.g. fan and air conditioning), how frequently do you open the	-Never (11%), -Seldom (24%), -Sometimes (21%), -Often (34%), -Always (10%)

windows per day?	
6. When you open windows for natural ventilation, what are the obstacles to opening the windows?	-Outdoor traffic noise (44%) -Outdoor air pollutant (39%) -Inconvenient opening ways (11%)
7. If you think outdoor air pollution is the main reason for closing windows, are you satisfied with your indoor air quality?	-Yes (60%) -No (40%)
8. If you think outdoor traffic noise is the main reason for closing windows, are you satisfied with traffic noise reduction provided by the windows?	-Yes (66%) -No (34%)
9. How much do you agree that urban traffic noise negatively affects you?	-Strongly agree (14%), -Agree (53%), -Neutral (25%), -Disagree (8%)
10. What kind of negative effects does traffic noise have on you?	-Distracting your work (49%) -Difficulty in communication (22%) -Psychological stress (18%) -Nothing (8%)
11.In general, what time is traffic noise level the highest inside the building	-Before noon (8am~10am) (15%) -Around noon (11am~1pm) (4%) -After noon (2pm~4pm) (26%) -Evening (after 5pm) (29%) -All day long (26%)
12. Do you agree that the window opening methods (or sizes) of your building are appropriate for effective natural ventilation?	-Strongly agree (18%), -Agree (64%), -Neutral (15%), -Disagree (3%)

CONCLUSIONS

The annual equivalent continuous sound level (dB-A) data collected from all six sites in Seoul exceeded the recommended noise threshold levels which can lead to noise annoyance. Results from the questionnaire showed that 51% of the respondents indicated having experienced some form of sick building syndrome (SBS) symptoms and 44% of the respondents felt the high urban traffic noise levels led to a decrease in their productivity and increased their level of communication difficulty and psychological stress. Therefore when designing naturally ventilated buildings in urban environments, more consideration must be given to how the natural ventilation potential of operable windows can be balanced with the transmission of unwanted traffic noise.

REFERENCES

World Health Organization (2011). ***Burden of disease from environmental noise.*** Available at: < http://www.euro.who.int/__data/assets/pdf_file/0008/136466/e94888.pdf> (22 Jan 2014)

Lambert, J., Simonnet, F., and Vallet, M. (1984). ***Patterns of behaviors in dwelling exposed to road traffic noise***. Journal of Sound and Vibration, 92, 159-72.

Jian Kang (2007). ***Urban Sound Environment.*** Taylor and Francis Group, London and New York.

DETERMINATION OF DIFFUSION PARAMETERS USING A SEMI-DYNAMIC DUAL CHAMBER METHOD

Guoqing HE[1,3*], Barbora KREJCIRIKOVA[2], Carsten RODE[2], and Jianshun ZHANG[3]

[1]College of Civil Engineering and Architecture, Zhejiang University, China
[2]Department of Civil Engineering, Technical University of Denmark, Denmark
[3]Department of Mechanical and Aerospace Engineering, Syracuse University, Syracuse New York, USA

[*]Corresponding email: ghe100@syr.edu

Keywords: dual chamber test, diffusion coefficient, partition coefficient, toluene

SUMMARY

The paper proposes a semi-dynamic dual chamber method to measure the partition coefficient and diffusion coefficient. It is advantageous over the earlier dynamic dual chamber test method in that it works for both porous materials and those close to impermeable materials. Our toluene –cement mortar test demonstrated the feasibility of such method although further improvements can be done through more reliable measurement instrument and better sealing of the system.

INTRODUCTION

Diffusion and partition coefficients are two key material properties in volatile organic compounds (VOC) transport modelling. A number of methods have been developed in the literature to measure these parameters, including the static dual chamber method (Bodalal et al. 2000), the small chamber sink test method (An et al. 1999), the microbalance method (Cox et al. 2001; 2002), the mercury intrusion porosimetry method (Blondeau et al. 2003; Xiong et al. 2008), and the dynamic dual chamber method (Xu et al. 2009). Among these methods, conventional chamber tests are less demanding on equipment but typically have a multiple solutions problem as two or more parameters are obtained in one regression process. In the dynamic dual chamber method, the two parameters are determined in separate processes rather than through a multi-variable regression and thus are independent of each other.

In the dynamic dual chamber systems, both chambers are ventilated, but with only one chamber acting as a source chamber with inflow of constant VOC concentrations. Air permeation between chambers is inevitable although it can be minimized by maintaining the pressure differences between two chambers at a low level. The permeation could be unaffordable if the test material is very porous. On the other hand, if the material is close to an impermeable one, the diffusion could be so slow that the concentration in the non-source chamber cannot build up to a detectable level. To overcome these difficulties, the dual chamber system could be operated in a semi-dynamic mode: only the source chamber is ventilated and the other chamber is completely sealed. If perfectly sealed, the two chambers would have same pressure and thus avoid air permeation. In the non-ventilated chamber, the concentration level will build up purely due to diffusion. This paper describes this semi-dynamic dual chamber method. Preliminary results from a toluene- cement mortar case are presented.

METHODOLOGIES

The dual chamber system consists of two chambers, material holder, and a Dynacalibrator that provides a VOC source of constant concentration. Xu et al.(2009) provides a detailed description on the air systems. Here, only a brief version is provided in Figure 1. Chamber A is the source chamber which is ventilated with a constant concentration at the inlet. Chamber B is not ventilated. The test material holder is located between two chambers.

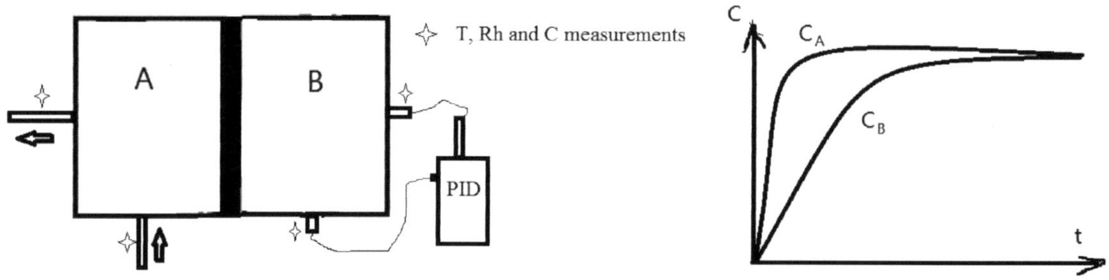

Figure 1. Semi-dynamic dual chamber system and concentration profiles in two chambers for an ideally sealed chamber system.

For ideally sealed system, the two chambers would eventually reach the same level of concentration, i.e. C_{eq}. Similar to the dynamic dual chamber system, the transient part will give information on the amount of VOC absorbed in the material, or the material phase concentration C_m. The partition coefficient is then calculated as $k_{ma}=C_m/C_{eq}$. With the partition coefficient, the diffusion coefficient is then determined through a data fitting process.

In this study, the test VOC was toluene and the material was 0.01 m thick, 0.3*0.3 m^2 cement mortar. The two chambers each measured 0.35 *0.35 *0.15 m^3. The concentration was measured using a portable photoionization detector device (PID, ppbRAE multi-sensor model, detection limit 1ppb or1.9 ug/m^3). To measure VOC concentrations in chamber B, the air was drawn from the chamber to the PID device and then delivered back to the chamber after analysis as shown in Figure 1. The airflow to chamber A was 0.521 ± 0.003 L/min (SKC UltraFLo gas flow calibrator, detection limit=0.1 mL/min) with a concentration of 609 ppb. The relative humidity was 51.7±0.9% and 45.7±1.9% in chamber A and chamber B, respectively. The temperature was 23.1 ± 0.3°C in both chambers. It was found that the chamber B could not be ideally sealed. Pressure was greater in chamber A than in chamber B by a difference of about $\triangle P$ =6 Pa during the test. The permeation flow between chambers due to pressure difference became significant enough to affect the mass balance. Therefore, the air permeation coefficient of the material was determined beforehand. This was done in a method described in the following.

First, a fixed airflow is maintained for chamber A as shown in Figure 2 (a). At steady state, this flow rate is balanced by leakages. Assuming the permeation rate is linearly proportional to the pressure difference, mass conservation of air in chamber B is as follows:

$$C_B P_B = C_{AB}*(P_A-P_B) \tag{1}$$

where C_B is the leakage coefficient from chamber B to the room and C_{AB} is the permeation coefficient of the material. Note that all pressures are measured with reference to the room pressure.

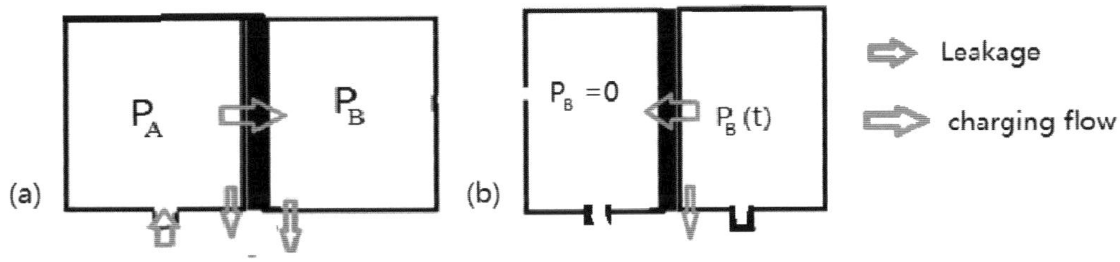

Figure 2. Two steps to determine the permeation coefficient of the test material: (a) A is charged at a constant airflow rate (no outflow); (b) Record the pressure responses after inflow to chamber B is shut off.

Second, chamber B is first pressurized while chamber A is open to the room as shown in Figure 2(b). Then the airflow is stopped and pressure responses with time, $P_B(t)$, are recorded. For an ideal gas, the pressure drops exponentially $P_B(t)=a*e^{-bt}$. The values of a and b can be determined through data fitting. A relation between the loss rate and pressure can also be obtained. Assuming the loss rate at a pressure difference of interest P_B-P_A=$\triangle P$(Pa) is determined as m (L/min). Then

$$m = C_B\triangle P + C_{AB}\triangle P \tag{2}$$

From Equations (1) and (2), the permeation coefficient C_{AB} (L/min/Pa) can be obtained.

In this study, CHAMPS-BES was used to simulate the diffusion and permeation of toluene through the test material between two chambers (CHAMPS, 2009).

RESULTS AND DISCUSSION

Figure 3 shows the pressure responses in chamber B which produces a relationship between the air loss rate and the pressure (shown only for the range between 0 Pa and 30 Pa). The air loss rate of chamber B at 10 Pa is read to be m = 0.0028 L/min. The corresponding permeation coefficient of air through cement mortar panel is then computed:

$$K_g = 8.34\text{E-5 (L/min/Pa)} \tag{3}$$

At 6 Pa pressure difference, the permeation rate is about 0.50 mL/min. However, the permeation between chambers can be reduced or even eliminated by improve the sealing of the systems through designed.

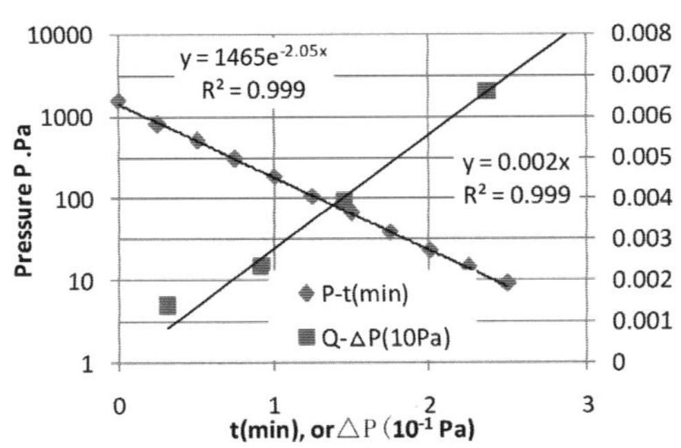

Figure 3. Permeation test results

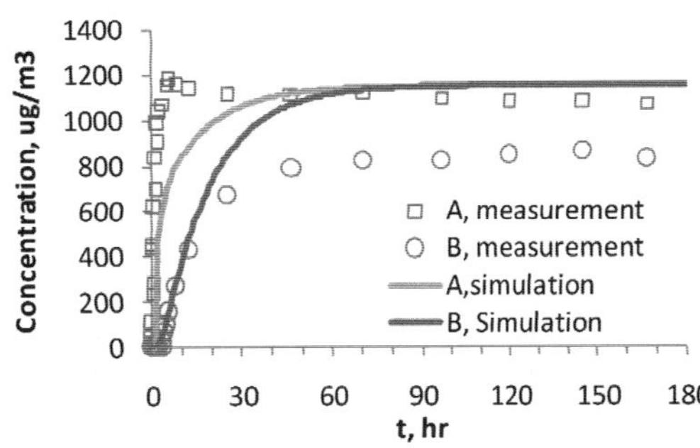

Figure 4. Results for toluene – cement mortar

Figure 4 shows a dataset for the toluene- cement mortar test. The concentration in chamber A reaches steady state in 4 hours when the PID device started to detect Toluene in chamber B. The concentration in chamber B reaches close to steady state after 50 hours. However, there is an unexpected gap between concentrations of two chambers. The exact reason is not clear. It may be associated with hysteresis in the partition coefficient for toluene between adsorption and desorption. Further studies are needed. Plus, the PID device was sensitive to the back pressure and had an uncertainty of about 30 ppb. The back pressure are different when measuring concentrations in chamber B and A. More reliable methods can be used to measure VOC concentrations to improve the data quality.

The computed partition coefficient is K_{ma}=275 (chamber A side) or 356 (chamber B side). Using K_{ma}=275 and a diffusion coefficient of D=1.1E-9 m^2/s produced a fair fit at the transition part as shown in Figure 4. The difference in the outflow concentration of chamber A between simulation and measurements is due to the expected lower concentration measured at chamber B at steady stage.

CONCLUSIONS

The paper describes the use of a semi-dynamic dual chamber test method, a variation of the earlier dynamic dual chamber test method, to determine the partition and diffusion coefficient. Compared with its predecessor, the semi-dynamic dual chamber method works for both porous materials and those close to impermeable material. Our toluene – cement mortar test demonstrated the feasibility of such method although further improvements are needed.

ACKNOWLEDGEMENT
This research was partly supported by Zhejiang Provincial NSF of China (# Y1100172) , the National NSF of China (#51008267) and by Otto Mønsted Foundation and P.A.Fiskers fund.

REFERENCES

Bodalal A, Zhang JS, Plett EG (2000) A method for measuring internal diffusion and equilibrium partition coefficients of volatile organic compounds for building materials. *Building and Environment* 35,101-110.

An Y, Zhang J, Shaw CY (1999) Measurements of VOC adsorption/desorption characteristics of typical interior building material surfaces *International Journal of HVAC & R Research*, 5(4), 297-316.

Cox SS, Zhao D, Little JC (2002) Measuring partition and diffusion coefficients for volatile organic compounds in vinyl flooring. *Atmospheric Environment*, 35, 3823-3830.

Cox SS, Little JC, Hodgson AT (2002) Predicting the emission rate of volatile organic compounds from vinyl flooring. *Environmental Science& Technology,* 36, 709-714.

Blondeau P, Tiffonnet AL, Damian A, Amiri O, Molina JL (2003) Assessment of contaminant diffusivities in building materials from porosimetry tests. *Indoor Air*,13, 302-310.

Xiong J, Zhang Y, Wang X, Chang D (2008) Macroemeso two-scale model for predicting the VOC diffusion coefficients and emission characteristics of porous building materials. Atmospheric Environment,42, 5278-5290.

Xu J, Zhang J, Grunewald J, Zhao J, Plagge R, Amiri Q (2009) A study on analogy of water vapor and VOCs diffusion in porous media by dual chamber method. *Clean-Soil, Air, Water*, 37, 444-453.

CHAMPS (2009) http://champs.syr.edu/software/champs_bes.html.

A HIGH EFFICIENCY INSTRUMENT FOR COLLECTING AIRBORNE PARTICLES DOWN TO 10 NANOMETERS ON SOLID SURFACES OR INTO LIQUID

Arantzazu EIGUREN-FERNANDEZ[1*], Nick BUGLACK[2], Ning LI[2], Pat KEADY[3] and Susanne HERING[1]

[1]Aerosol Dynamics Inc., Berkley, CA
[2] Dept. of Pathobiology & Diagnostic Investigation, Michigan State Univ., East Lansing, MI
[3]Aerosol Devices Inc., Fort Collins, CO

*Corresponding email: arantza@aerosol.us

Keywords: Airborne particles, solid and liquid collection, particle collection, suspensions

SUMMARY

The technology of airborne particle collection using water condensational growth provides a new approach for a high efficiency collection of bioaerosols. Encapsulating particles into water droplets and gently impacting them onto a solid or liquid substrate allows collection efficiencies over 90% for particle sizes between 10 nm and 10 µm. The water bath surrounding the particles and the humid collection environment reduces desiccation that can affect microorganism viability. Normally impaction can cause structural damage, but in our method the water layer acts as a cushion reducing mechanical stress and preventing particle 'bounce'. For collection into liquids, the droplets are gently impacted into a small volume of water, preventing re-aerosolization of particles. It provides a concentrated sample ready for analysis using molecular techniques. Its portability and ability to collect particles on both dry surfaces and in liquid media make this system suitable for the collection, recovery, and analysis of bioaerosols.

INTRODUCTION

Airborne bacteria, fungi and viruses are part of the ambient particulate matter found either as very small individual particles, aggregates or bound to other airborne particles. Thus, their airborne size distribution varies from a few nanometers to several micrometers. They can be an important source of human infection and pose a considerable risk of contamination of pharmaceutical operations. However, regular monitoring of bioaerosols in large-scale studies is difficult. The lack of information about the airborne nature and effects of bioaerosols is mainly due to the difficulty in their collection. Common problems associated with bioaerosol collection include sample contamination during loading and retrieval, and changes in sample properties resulting from the long sampling periods required to collect enough mass to overcome issues with quantification limits of detection.

Active sampling methods (e.g. filter, impactor or impinger collection) have become common tools for environmental monitoring in hospitals and pharmaceutical facilities as well as for the collection of bioaerosols in outdoor and indoor environments. However, most active air sampling instruments are dependent on particle size for collection. Liquid impingers are the least destructive, and the collection liquid can be later cultured to identify bacteria and viruses via a variety of molecular methods, generating quantitative results. However, their effective physical collection efficiency is limited, with reported collection efficiencies <10% for particles between 30 and 100 nm (Hogan et al 2005). In addition, the sampling period is limited to a few hours, at most, and because the impinger volume is in the order of several mL, samples are diluted. Filter collections are commonly used because of their convenience as they can be easily deployed in the field and be programmed to collect time-resolved samples. They also have better collection efficiencies than impingers, exceeding 90% for particles down to 10 nm (Burton et al. 2007). Among their disadvantages, low humidity during collection can cause cells to dry and break, and microbial integrity may be damaged by stress during the collection. In addition, filter collection does not allow the use of rapid detection techniques for the characterization of bioaerosols.

Even though there are a wide variety of liquid collectors and impactors available, there remains a need for instruments capable of providing high collection efficiency over a wide particle size range (nm to µm) with high microorganism viability. These instruments should also provide bioaerosol samples that allow cell counting and rapid analyses in the laboratory. The system we present in this paper provides a new alternative for more efficient bioaerosol collection. This instrument has been proven efficient for the collection of airborne particles down to 10 nm both as concentrated spots onto dry surfaces or concentrated suspensions into liquid for chemical and toxicological characterization, respectively (Eiguren Fernandez et al. 2014a; Lewis et al. 2014). Testing for bioaerosol collection is underway and while the results are preliminary, we think our approach offers new possibilities for high efficiency collection of viable bioaerosols.

METHODOLOGIES

Sampler description

The collector uses a three-stage, moderated laminar flow condensation approach to grow airborne particles as small as 6 nm into droplets at moderate flow temperatures (Hering et al. 2014; Eiguren Fernandez et al. 2014b). Briefly, the incoming sample air enters a cool section (conditioner) followed by a warm-, wet-walled growth region where particle activation of supersaturated vapor condensation occurs (initiator). A third cool stage (moderator) provides the distance needed for droplet growth, before the water-encapsulated particles are delivered through a jet impactor either onto the solid surface or into the liquid medium (Figure 1). The high relative humidity of the sample flow and the water enveloping the particles can be advantageous for viable microorganism collection. Though particle growth is achieved at high RH, the time which the particle is subjected to high humidity conditions is brief - on the order of milliseconds. If the presence of water is an issue, our dry collection capability allows maintaining the sample as a dry deposit eliminating negative effects due to condensation. The system flow rate is set between 1 and 5 lpm. It is capable of time-resolved collection for dry samples, and can run unattended for several days when the aim of the collection is the characterization of the chemical

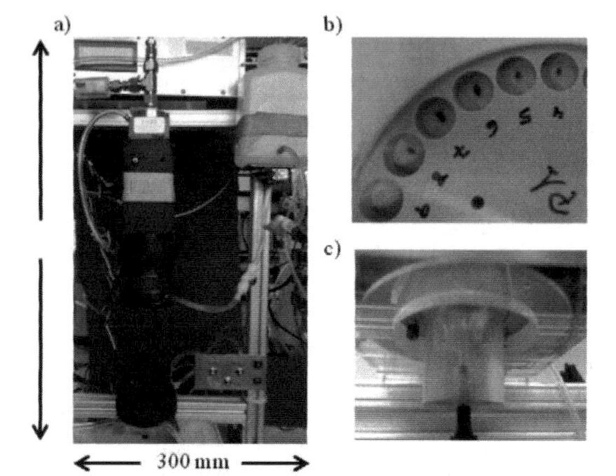

Figure 1. a) Aerosol Collector; b) dry collection; c) liquid collection

properties of the aerosols (Eiguren Fernandez et. al. 2014). The volume of the suspension can be varied depending on the flow rate and desired final concentration.

Physical Collection Efficiency

System collection efficiencies were tested in our laboratory (Berkeley, CA) using both laboratory generated aerosols containing sulfate and nitrate, and ambient PM_{10}. Particle collection efficiency was inferred by measuring particle penetration through the sampler after delivery into the dry surface and the liquid. We generated aerosol by nebulization, neutralized the charge, and then selected a narrow size fraction using a high-flow differential mobility analyzer (HF-DMA, Stolzenburg et al., 1998). The monodisperse test aerosol concentration was monitored at the inlet of the sampler and in the exiting flow using a pair of condensation particle counters (TSI Model 3783 and 3787). These tests were done for both collection approaches – dry and liquid.

In-vitro assays for toxicological characterization of collected suspensions of aerosols

The collection system was deployed at Michigan State University (East Lansing) to characterize the toxicological properties of ambient aerosols. Ambient total suspended particle (TSP) samples were collected as liquid suspensions over a period of 4 days. The toxicity of airborne particles was determined using human bronchial epithelial cell line BEAS-2B. This cell line has been widely used for assessing pro-oxidative, pro-inflammatory and cytotoxic effects of ambient PM. Human bronchial epithelial cells play an important role in protecting the respiratory system against harmful exogenous agents including ambient aerosols and associated toxic compounds, gases, allergens and pathogens. Cell cultures were stimulated with 0.015 m^3 air for 16 hr and two pro-inflammatory markers, interleukin-6 (IL-6) and interleukin-8 (IL-8), were measured.

RESULTS AND DISCUSSION

Physical Collection Efficiency

The collection efficiency as a function of particle size for laboratory generated aerosols for both dry and liquid collection are shown in Figure 2a. Collection efficiencies >90% were obtained for particles ranging from 6nm to 10μm. Similar collection efficiencies were observed for ambient PM10 collection into liquid suspension (Figure 2b).

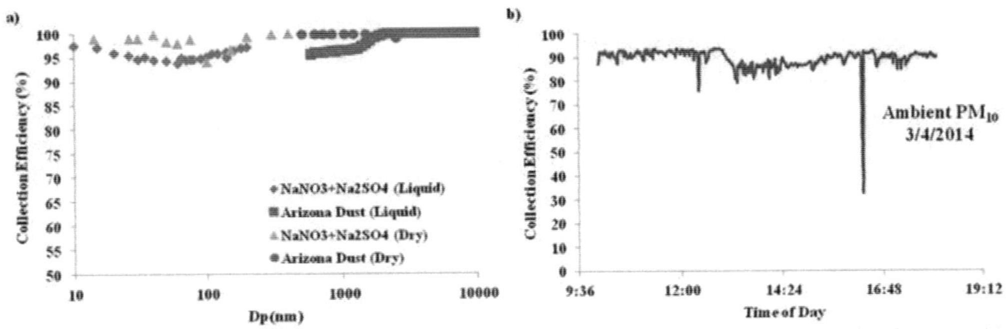

Figure 2. Collection efficiency for dry and liquid configuration: a) with particle size for lab generated aerosols; and b) for ambient particulate matter

Production of pro-inflammatory markers after exposure to ambient particulate matter

Production/release of IL-6 and IL-8 by airway epithelial cells has been used as indicator for assessing the pro-inflammatory effects of air pollutants. The production/release of IL-6 and IL-8 is quite sensitive also to oxidative stress induced by pro-oxidative capacity of ambient particulate matter. The production of IL-6 and IL-8 after exposure to East Lansing ambient particles was significantly higher than for the controls (Figure 3). Ambient aerosols induced a 10- to 34-fold increase in IL-6 release, while for the IL-8 we observed a 5- to 25-fold increase.

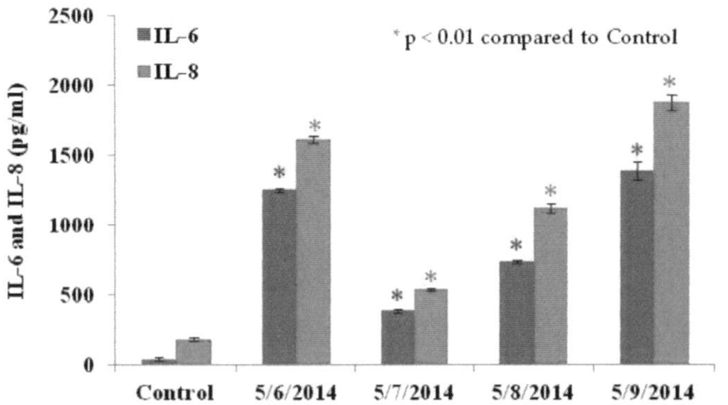

Figure 3. Production of IL-6 and IL-8 after exposure to East Lansing aerosols

CONCLUSIONS

Sampling techniques for the collection of bioaerosols have improved over the years; however, none of the samplers currently available offer both good physical collection efficiency and adequate recovery efficiency of viable samples. The new approach used by our technology may provide a new alternative to the conventional impingers or impaction/filter collections. The instrument's small size, and low noise and power requirements could be advantageous for routine monitoring in indoor environments. The possibility of collecting airborne particles as dry spots or concentrated suspensions adds to the versatility of this instrument. We are confident than the performance of this system will contribute to better characterizing the presence of airborne microorganisms and their potential risk.

ACKNOWLEDGEMENTS

This study was made possible by the support from National Institutes of Health (Grants: RC3 ES019081-01 & R43-ES022523). Authors would like to thank Dr. Ryan Lewandowski and Paula Palmiter from Michigan State University for their help during field deployment.

REFERENCES

Burton N.C., Grinshpun S.A., and Reponen T. (2007). Physical collection efficiency of filter materials for bacteria and viruses. *Ann. Occup. Hyg.*. 51:143-151

Eiguren Fernandez A., Lewis G.S., Spielman S et al. (2014a). Time-resolved characterization of particle associated polycyclic aromatic hydrocarbons using a newly-developed sequential spot sampler with automated extraction and analysis. *Atmos. Environm.* 96:125-134

Eiguren Fernandez, A., Lewis G.S., and Hering, S.V – Design and Laboratory Evaluation of a Spot Sampler for Time-Resolved Measurement of Airborne Particle Composition. *Aerosol Sci. Tech.* 48(6):655-663 (2014)

Hering S, Lewis G, Spielman S. Moderated, Water-Based Condensational Particle Growth in a Laminar Flow. *Aerosol Sci. Tech.* 2014;48(4):401-8

Lewis G., Buglak N., Spielman S. et al (2014) Concentrated Aerosol Collection into Liquid for Toxicological Studies. American Assoc. for Aerosol Research, Orlando, FL, 20-24 Oct. 2014

Hogan C.J., Jr., E.M. Kettleson, M.H. Lee, B. Ramaswami, L.T. Angenent, and P. Biswas (2005). Sampling Methodologies and dosage assessment techniques for submicrometre and ultrafine virus aerosol particles. *J. Appl. Microbiol.*99:1422-1434

RESUSPENSION OF BACTERIAL SPORE PARTICLES FROM DUCT SURFACES

Paricheher SALIMIFARD[1], Paul KREMER[1], Donghyun RIM[1], James D. FREIHAUT[1,*]

[1]Department of Architectural Engineering, Pennsylvania State University, University Park, PA USA.

[*]Corresponding email: JFreihaut@engr.psu.edu

Keywords: Particle resuspension, Indoor particle contaminants, Controlled disturbance effects, Allergens

SUMMARY

Allergen-containing particles resuspended from building floors or outdoor-air-entrained particulates can deposit on the surfaces of supply air ducts. These particles can later be resuspended into the supply air because of duct vibration and turbulent flow. This study investigates the resuspension rates of *bacillus thuringiensis* bacterial spore particles (0.4-25 μm) from a vibrating duct surface with air flow swirl velocities ranging from 0 to 2.5 m/sec. Resuspension rates increased by ~ 2 orders of magnitude for all particle sizes in the 0.3-2.5 m/sec range of swirl velocities. One minute averaged resuspension rate ranged from 10^{-5} to 0.1/min. For particle sizes above about 10.0 microns, depending on swirl velocity magnitude, a relatively sharp drop-off in resuspension rate is observed. These results further indicate that the substantial spore residence time of nine months on the metal surfaces before conducting the resuspension disturbance experiments does not inhibit the resuspension propensity of the spores significantly.

INTRODUCTION

Airborne particulate matter existing in the indoor environment can affect occupant inhalation exposure, and lead to adverse health effects such as allergic sensitization and reduced lung function (Wallace 1996; Skrepnek and Skrepnek 2004; Chapman 2003). Particles like dust are also one of the main concerns related to patient safety during the healthcare facilities retrofit projects (Mohammadpour 2014). Particles can be resuspended by occupant activities (e.g., Thatcher and Layton 1995; Ferro, Kopperud, and Hildemann 2004). Airborne particles present in an occupied building can be directed into plenum and ductwork via air recirculation. Airborne particles from recirculated air and outdoor air intake can deposit onto duct surfaces by varied mechanisms including gravitational settling, diffusion, and turbulent transport. Particles deposited on duct surfaces can later resuspend when acted on by duct vibration and turbulent air flow disturbances. Resuspended duct particles can then be distributed throughout the occupied zones by the air handling system serving those zones. A number of studies on deposition and resuspension of particles in ventilation ducts have been done (Sippola and Nazaroff 2004; Wang et al. 2012; Krauter and Biermann 2007). Barth et al. (2014) conducted an experimental investigation to characterize the resuspension of spherical particle monolayers from a channel floor in dry and turbulent flow. In reality, specific features of multilayer deposits, e.g. particle-to-particle adhesion forces, aggregate resuspension, saltating particles, deposit structure and porosity enhance or inhibit resuspension from multilayer deposits when contrasted to monolayer deposits (Boor, Siegel, and Novoselac 2013). Particle resuspension due to vibration in air handling ductwork exhibit a broader range of frequency content than occupant-induced floor vibrations. Hence, vibration-induced particle resuspension characteristics associated with duct resuspension and occupant-induced resuspension are expected to differ somewhat (Gomes et al. 2007). Air flow turbulence characteristics and momentum fluctuations lead to interactions between duct air flow and duct surfaces that are expected to yield flow fields and resuspension behaviors quite different than those associated with occupant activity. The objective of this study is to investigate bacterial spore particle specific, resuspension at conditions representative of duct resuspension. This study complements other studies conducted by extending the framework and understandings of particle resuspension of multilayer deposits in the building duct system.

METHODOLOGIES

Laboratory-based, full-scale duct experiments that simulated duct vibration and particle resuspension were performed. This section describes the methodology for 1) measurement and control of duct air flow and duct surface vibration characteristics; and 2) dust sample preparation and particle size distribution measurement. Gomes et al. (2007) provides a general description of the Penn State Indoor Aerosol Laboratory resuspension facilities employed by this study.

Duct air flow and surface vibration characteristics

Vibration waveforms utilized for these experiments were acquired using low- mass, broadband accelerometers adhered to duct surfaces in a ducted air supply system at Penn State University. A sixty second duct vibration waveform for playback with a Wilcoxon Research F4 electromagnetic shaker system was formed using the acquired accelerometer data. The shaker waveform input is comprised of a thirty second startup and a thirty second steady state periods. Figure 1 shows the relationship between the average swirl velocity measured by thermal anemometry over the surface of the nominal 90 mm x 90 mm, 23 gauge galvanized sheet metal specimens mounted on the shaker platten in the resuspension chamber and the inlet flow to the copper jets used to produce the swirl effect over the specimen. The maximum controllable flow for the chamber configuration used in these experiments was 2.5 m/s. In the present experiments, four cycles, i.e., 240 second in total, of the duct vibration waveform were applied to each specimen mounted on the shaker platten within the resuspension chamber. A Particle Measuring Systems model 510A LasAir II optical particle counter was used to measure second-by-second particle counts in the range of 0.5 to 25 μm across six channels. The time-resolved particle count data were then used to compute particle resuspension rates. A near laminar sweep flow through the resuspension chamber was used during the conduct of all experiments. *Bacillus thuringiensis* spore particles were deposited on the sheet metal specimens using the particle dispersion chamber described by Gomes et al. (2007). All specimens were subjected to the duct vibration waveform disturbance. Multiple experiment repetitions were conducted at an imposed swirl velocity of 0, 0.3, 0.9, 1.5, and 2.5 m/s.

Dust samples and particle size distribution measurement

The present study examined resuspension of spore particles ranging from 0.4 μm to 25 μm. Figure 2 shows particle size distribution of spore sample dusts deposited on several sample plates in the particle dispersion chamber, and of the bulk spore dust prior to dispersion. A Malvern Mastersizer 2000 particle size analyzer was used to generate the size distributions in Figure 2. The two figures indicate a definite change toward lower values in the settled spore dust compared to the bulk spore sample. The mean volume-to-surface area diameter of the settled spore dust is 1.95 microns as compared to 3.54 microns for the bulk spore sample.

Figure 2 also suggests that particle diameters below which 10%, 50% and 90% of the particles are found are much lower for the settled spore dust relative to the bulk spore sample as a result of deagglomeration during the particle dispersion process.

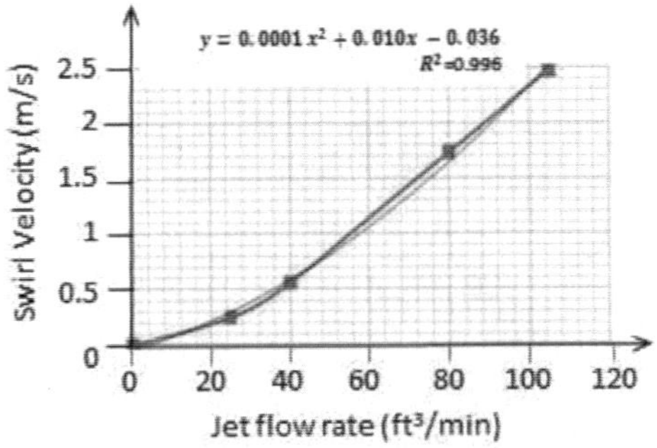

Figure 1. Resuspension chamber jets input flow required to achieve swirl velocities

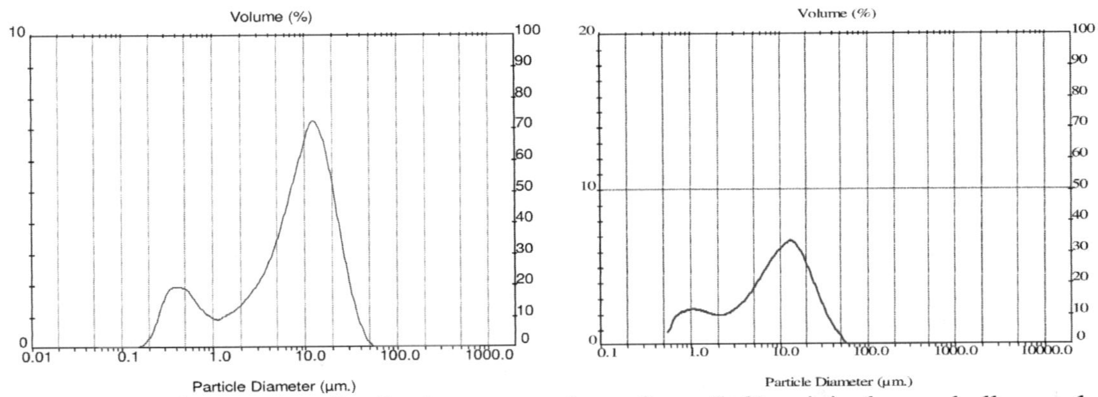

Figure 2. Settled spore size distribution on sample surfaces (left), original spore bulk sample size distribution (right).

RESULTS AND DISCUSSION

Figure 3 displays the one minute average resuspension rates for the particle sizes of the settled spore samples subjected to a fixed duct vibration spectra and varied air swirl velocity. The results show nonlinear increases in resuspension across the range of measured particle sizes with increased air swirl velocity. The data also indicate the dominant role of near surface air turbulence on particle resuspension from reservoir surfaces. For the applied disturbances, there is a falloff in resuspension propensity around the 10.0 µm size range. The size range of the falloff is expected to be a function of imposed aerodynamic turbulence on the reservoir surface, and the data does suggest a smaller particle diameter (1-10 micron) fall off as swirl velocity decreases (Figure 3).

The spore-covered, duct-material plates for the resuspension experiments were in the cloud deposition chamber for nearly nine months waiting for the resuspension rig modifications to be completed. According to Hu et al. (2008) the longer a settled, deformable particle resides on a surface the more bound to the surface it becomes and the more difficult it is to resuspend, particularly for irregularly shaped particles on deformable surfaces, such as rubber mats or carpet fibers. The results for this investigation showed little effect of the long (9 month) residence time of the spores on the hard metal plate surfaces. These results are consistent with the fact that spore particles displayed an order of magnitude lower measured adhesion force than inorganic, alumina particles for the same substrates (Hu et al. 2008).

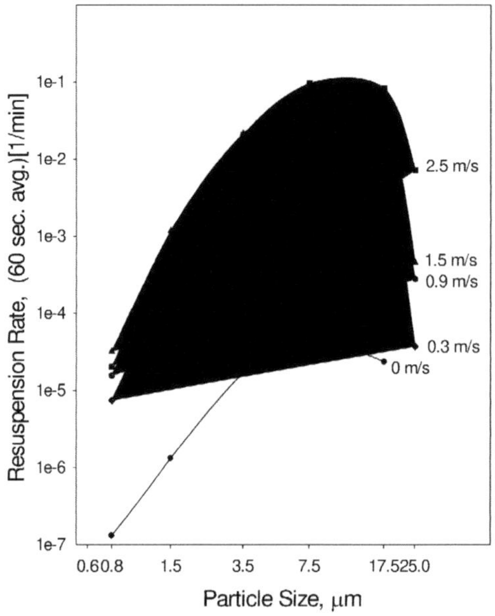

Figure 3. One minute average resuspension rates observed at each swirl velocity employed. Duct vibration waveform applied for all.

CONCLUSIONS

The present study is part of a broader investigation of disturbances impacting particle resuspension across a broader range of particle sizes and resuspension-influencing factors than those considered in this study. These simulated duct experiments spanned a range of swirl velocities up to 2.5 meter/sec, more typical of higher turbulent flows in duct systems. Resuspension rates increased by ~ 2 orders of magnitude for all particle sizes in the 0.3 m/sec to 2.5 m/sec range of swirl velocities. Particle size influence on resuspension rate is very similar to that observed in similar studies by the authors and are expected to contribute to the need for reliable resuspension data for transient simulations of particulate matter behavior and exposure modeling.

ACKNOWLEDGEMENT

Penn State Institute of Energy and Environment and the U.S. Army Center for Health Promotion and Preventive Medicine are acknowledged for supporting this investigation.

REFERENCES

Barth, T., J., Preuß, G. Müller, and Hampel, U. 2014. "Single Particle Resuspension Experiments in Turbulent Channel Flows." *Journal of Aerosol Science* 71 (May): 40–51.

Boor, Brandon E., Siegel, Jeffrey A., and Novoselac, A. 2013. "Monolayer and Multilayer Particle Deposits on Hard Surfaces: Literature Review and Implications for Particle Resuspension in the Indoor Environment." *Aerosol Science and Technology* 47 (8): 831–47.

Chapman, M. D. 2003. "Pediatric Allergy: Principles and Practice, Indoor Allergens."

Ferro, A. R., Kopperud, Royal J., Hildemann, and Lynn M. 2004. "Elevated Personal Exposure to Particulate Matter from Human Activities in a Residence." *Journal of Exposure Analysis and Environmental Epidemiology* 14 (April): S34–40.

Gomes, C., Freihaut, J., and Bahnfleth, W. 2007. "Resuspension of Allergen-Containing Particles under Mechanical and Aerodynamic Disturbances from Human Walking." *Atmospheric Environment* 41 (25): 5257–70.

Hu, B., Freihaut, J. D., Bahnfleth, W. P., and Thran, B. 2008. "Measurements and Factorial Analysis of Micron-Sized Particle Adhesion Force to Indoor Flooring Materials by Electrostatic Detachment Method." *Aerosol Science and Technology* 42 (7): 513–20.

Krauter, P, and Biermann, A. 2007. "Reaerosolization of Fluidized Spores in Ventilation Systems." *Applied and Environmental Microbiology* 73 (7): 2165–72.

Mohammadpour, A. 2014. "Retrofitting healthcare facilities to enhance patient safety and energy efficiency." PhD dissertation, The Pennsylvania State University.

Sippola, Mark R., and Nazaroff, William W. 2004. "Experiments Measuring Particle Deposition from Fully Developed Turbulent Flow in Ventilation Ducts." *Aerosol Science and Technology* 38 (9): 914–25.

Skrepnek, G. H., and Skrepnek, S. V. 2004. "Epidemiology, Clinical and Economic Burden, and Natural History of Chronic Obstructive Pulmonary Disease and Asthma." *American Journal of Managed Care*.

Tracy L., and Layton, David W. 1995. "Deposition, Resuspension, and Penetration of Particles within a Residence." *Atmospheric Environment* 29 (13): 1487–97.

Wallace, L. 1996. "Indoor Particles: A Review." *Journal of the Air & Waste Management Association* 46 (2): 98–126.

Wang, S., Zhao, B., , and Tan, Z. 2012. "An Experimental Study on Short-Time Particle Resuspension from Inner Surfaces of Straight Ventilation Ducts." *Building and Environment* 53 (July): 119–27.

BLACK CARBON AND $PM_{2.5}$ FILTRATION IN CLASSROOM VENTILATION SYSTEMS IN SALT LAKE CITY, UTAH AND LAS VEGAS, NEVADA

Paul T. ROBERTS*, David L. VAUGHN, Steven G. BROWN

Sonoma Technology, Inc. (STI), Petaluma, California USA

*Corresponding email: paul@sonomatech.com

Keywords: Particulate matter, black carbon, filtration, ventilation, classroom

BACKGROUND AND OBJECTIVES

The objectives of this work were to understand particulate concentrations inside classrooms and their relationship to outdoor concentrations with current ventilation systems in Salt Lake City and Las Vegas schools, and to evaluate particle removal by improved filtration. The air quality at the Las Vegas schools was significantly influenced by pollutants from nearby US 95 freeway. The air quality at the Salt Lake City schools will be impacted by pollutants from the future construction of the Mountain View Corridor (MVC) roadway being built nearby.

We used pairs of monitors to measure concentrations of indoor and outdoor black carbon (BC) as a surrogate for diesel particulate matter (DPM) at one selected classroom in each of 7 near-road schools. We also measured $PM_{2.5}$ at one school to provide data comparable to the BC data. The data were used to determine how effective the ventilation systems and filters were for removing BC and $PM_{2.5}$.

DPM is considered a likely human carcinogen by the U.S. Environmental Protection Agency (EPA). In a previous monitoring project, Roberts et al., 2013 used California's unit-risk estimate (OEHHA, 2011) to demonstrate that DPM is the greatest contributor (about 84%) to the total cancer risk from the toxic air pollutants measured. Since direct measurement methods for DPM are not available, BC is commonly used as a surrogate for DPM because it is measurable and represents a large fraction of diesel exhaust (McCarthy et al., 2013; Fruin et al., 2004; Shi et al., 2000; Birch and Cary, 1996).

We also measured CO_2 to determine ventilation rates. Low ventilation rates may lead to high indoor concentrations of air pollutants (U.S. CPSC, 2014) and associated adverse health effects. The CO_2 concentrations measured in this study provide additional information about the ability of the existing ventilation systems in each school to deliver the recommended minimum quantities of outside air to occupants.

METHODS

Classroom measurements for the MVC study were made from March 2014 to June 2014. Classroom measurements in Las Vegas were made from May 2007 to September 2008, and again from March to June 2013. 5-min Black carbon measurements were made using AE22 2-Wavelength Magee Scientific Aethalometers or AE33 7-Wavelength Teledyne-API Aethalometers. $PM_{2.5}$ was measured hourly using a Met One BAM. BC and $PM_{2.5}$ monitors were deployed in matching pairs in a classroom and at the air inlet for that classroom. 5-min CO_2 measurements were made using a LiCor 6252 CO_2 monitor. Routine maintenance was performed according to standard operating procedures and repairs and recalibrations were performed as needed.

Raw data was corrected for any spikes in BC concentrations after tape advances and for filter spot loading effects. Data were processed by validating values for baseline changes, extreme minimum and maximum values, and instrument-generated error codes, and were reconciled with site log information recorded during instrument maintenance periods. Valid data capture was typically greater than 95% for all parameters.

The slope of the outdoor-to-classroom 1-hr average concentrations provides a reasonable measure of filtration efficiency and has been used in previous similar studies (McCarthy et al., 2013; Roberts et al., 2013).

RESULTS AND DISCUSSION

Median BC concentrations were greatest during the morning rush hour, both in classrooms and outdoors at all near-roadway schools. **Figure 1** shows diurnal patterns for BC during selected winter and spring school months at outdoor, near-roadway sites and inside the classrooms at Adcock Elementary School in Las Vegas. As this Figure shows, the improved ventilation system and MERV 13 filters resulted in a very high fraction of the outdoor BC being removed; on average, 74% to 97% of BC was removed.

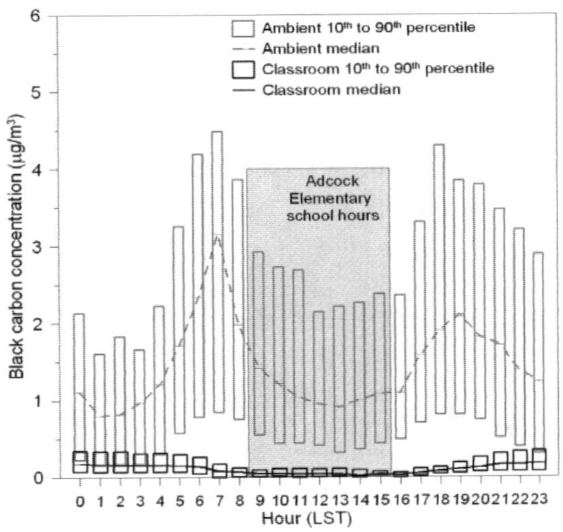

Figure 1. Black carbon concentration (μg/m^3) distributions at Adcock Elementary in November 2007 and January and February 2008 at the outdoor ambient site and inside the classroom.

Classroom CO_2 concentrations were monitored at 3 schools in Salt Lake City and 3 schools in Las Vegas. At Hillside Elementary, data was analyzed separately for three time periods with markedly different CO_2 concentrations. During the first two weeks of the study, the average CO_2 concentration during school operating hours was 1,477 ppm and CO_2 concentrations reached 2,000 ppm on two days, indicating that there was a problem with the ventilation system. Average CO_2 concentrations decreased to 984 ppm after the fresh air inlet was repaired and decreased again to 806 ppm after the exhaust air vent was repaired. The CO_2 concentrations during these three periods are shown in **Figure 2**.

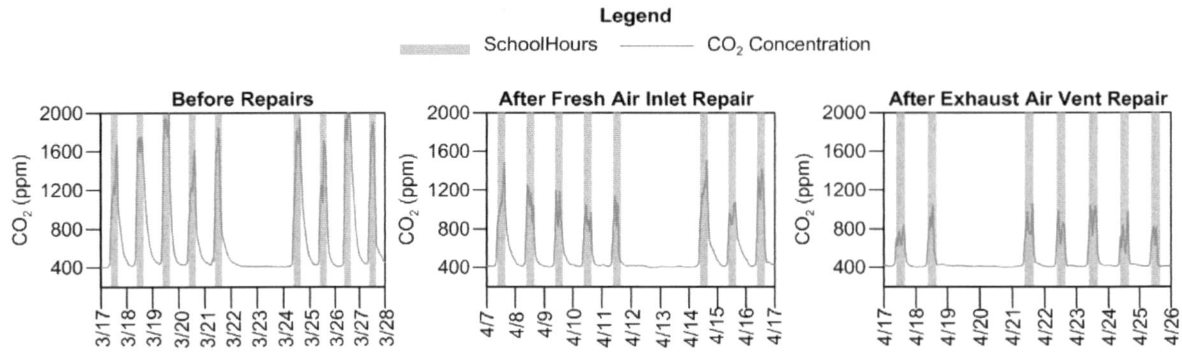

Figure 2. Time series for three periods Hillside Elementary School.

CO_2 concentrations corresponded closely with patterns of building occupancy, increasing from the background concentration of 395 ppm at the beginning of the school day and decreasing gradually after the end of the school day.

On May 7, 2014, we did a ventilation survey of a number of classrooms at Hillside Elementary School and at Hunter Jr. High School. We took a TSI Q-Track CO_2 monitor into several classrooms and set it down on an empty table for from 2 to 11 minutes to record the CO_2 concentrations while class continued.

The measured CO_2 concentration can be compared to various criteria to ensure sufficient ventilation in a classroom for healthful conditions and for a good learning environment (ASHRAE, 2013; Satish et al., 2012). ASHRAE (2013) recommends a CO_2 concentration of less than 1,000 ppm. Satish et al. (2012) found moderate and statistically significant performance decreases in several decision-making activities at CO_2 concentrations of 1,000 ppm, and large and statistically significant decreases at CO_2 concentrations of 2,500 ppm, when compared with activities and performance at CO_2 concentrations of 600 ppm.

The CO_2 concentrations in 15 of 19 classrooms in 6 schools were about the same as or better than the ASHRAE guideline of 1000 ppm. However, some classrooms, especially two portables, had very poor ventilation (CO_2 concentrations of 2000 and 4600 ppm); this is potentially bad for students' health and for learning and testing.

CONCLUSIONS

Study results included the following:

- Estimated BC filtration rates with MERV 8 filters were typically between 26% and 34% in the Salt Lake City schools and 31% to 66% for the Las Vegas schools.
- MERV 8 filters in standard ventilation systems in the Salt Lake City schools were much more efficient at removing $PM_{2.5}$ than BC: we estimated a filtration efficiency for $PM_{2.5}$ of 73%. It is likely that the original filtration systems and original MERV 8 filters are better at removing larger particles than smaller ones.
- Filtration efficiencies for BC with improved filtration systems and MERV 13 filters were 74-97%, significantly better than with original systems with MERV 8 filters.

Thus, the improved filtration efficiencies with MERV 13 filters can significantly reduce student exposures to a major carcinogen coming from busy roadways.

We measured CO_2 concentrations for comparison to various criteria to ensure sufficient ventilation in a classroom for healthful conditions and for a good learning environment. ASHRAE (2013) recommends a CO_2 concentration of less than 1,000 ppm in school classrooms. Satish et al. (2012) found moderate and statistically significant performance decreases in several decision-making activities at CO_2 concentrations of 1,000 ppm, and large and statistically significant decreases at CO_2 concentrations of 2,500 ppm, when compared with activities and performance at CO_2 concentrations of 600 ppm. The measured CO_2 concentrations indicate that ventilation for most sampled classrooms is about the same as, or better than the ASHRAE guideline. However, some classrooms, especially two portables we investigated at a school in Salt Lake City had very poor ventilation with CO2 concentrations of 2000 to 4600 ppm; this environment is potentially detrimental to student health and the learning environment.

ACKNOWLEDGMENTS

Kimberly A. Lorentz (STI) performed some of the data validation and analyses for the Salt Lake City study; Jennifer L. DeWinter of STI performed much of the data validation and analyses for the Las Vegas study. The work was sponsored by the Utah DOT and the MVC Air Working Group (Salt Lake City), the Nevada DOT (Las Vegas), and Sonoma Technology, Inc.

REFERENCES

American Society of Heating, Refrigerating and Air-Conditioning Engineers, Inc. (2013) Ventilation for acceptable air quality. ANSI/ASHRAE Standard 62.1-2013.

Birch M.E. and Cary R.A. (1996) Elemental carbon-based method for monitoring occupational exposures to diesel exhaust. *Aerosol Science and Technology*, 25, 221-241.

Fruin S.A., Winer A.M., and Rodes C.E. (2004) Black carbon concentrations in California vehicles and estimation of in-vehicle diesel exhaust particulate matter exposures. *Atmos. Environ.*, 38(25), 4123-4133, Aug.

McCarthy M.C., Ludwig J.F., Brown S.G., Vaughn D.L., and Roberts P.T. (2013) Filtration effectiveness of HVAC systems at near-roadway schools. *Indoor Air*, 23(3), 196-207, doi: 10.1111/ina.12015, January 25.

OEHHA (Office of Environmental Health Hazard Assessment). (2011) Revised Air Toxics Hot Spots Program Technical Support Document for cancer potency factors. Appendix A.

Roberts P.T., Vaughn D.L., and Ludwig J.F. (2013) Background air quality and mitigation in schools near the Mountain View Corridor. Presented to the Air Working Group, West Valley City, UT, December 18, by Sonoma Technology, Inc., Petaluma, CA. STI-910051-5860.

Shi J.P., Mark D., and Harrison R.M. (2000) Characterization of particles from a current technology heavy-duty diesel engine. *Environ. Sci. Technol.*, 34(5), 748-755, Mar 1.

Satish U., Mendell M.J., Shekhar K., Hotchi T., Sullivan D., Streufert S., and Fisk W.J. (2012) Is CO_2 an indoor pollutant? Direct effects of low-to-moderate CO_2 concentrations on human decision-making performance. *Environ. Health Persp.*, 120(12), 1671-1677, doi: 10.1289/ehp.1104789, December. Available at www.ncbi.nlm.nih.gov/ pubmed/23008272.

U.S. CPSC (Consumer Product Safety Commission). (2014) The Inside Story: A Guide to Indoor Air Quality. Accessed January 22, 2015. Available at http://www.cpsc.gov/en/Safety-Education/Safety-Guides/Home/The-Inside-Story-A-Guide-to-Indoor-Air-Quality/

Zhu Y., Hinds W.C., Kim S., Shen S., and Sioutas C. (2002) Study of ultrafine particles near a major highway with heavy-duty diesel traffic. Atmos. Environ. 36, 4323-4335 (27).

THE EXPERIMENTAL STUDY OF THE REGENERATE CHARACTERISTIC EFFECT ON THE LIQUID DEHUMIDIFICATION SYSTEM PERFORMANCE

Yulieh Wu[1,*], Kunhua Lyu[1], and Pohan Lin[1]

[1]Department of Refrigeration, Air-Conditioning and Energy Engineering, National Chin-Yi University of Technology, Taichung City, 41170, Taiwan

[*]*Corresponding email:* wuyl@ncut.edu.tw

Keywords: *Adiabatic, Regeneration process, Packed column, Lithium chloride*

SUMMARY

In this research, solution characteristics in regenerator section and dehumidification effectiveness research were evaluated. The liquid dehumidification air-conditioning was used as experimental equipment, and the lithium chloride solution was used as the desiccator in liquid dehumidification system. The regenerator section adopted adiabatic packed tower structure and the double planet as bulk packing distributed to packed column. Therefore, more moisture could be transferred from the air to desiccant solution by the process of packing material which increases the contact area between liquid and gas to enhance the regeneration efficiency. The experimental results also presented the effect of solution temperature, desiccant flow rate and air velocity on the performance of regeneration and dehumidification.

INTRODUCTION

The conventional air-conditioning (AC) refrigerating unit consumes electric energy for cooling indoor space or dehumidification to provide the required temperature and humidity. Thus, long-term operation results in considerable energy loss. With the rise of energy crisis and environmental considerations, many countries have carried out research and development, and promoted the use of AC energy-saving technology and equipments. As the liquid desiccant AC system uses solution for space dehumidification and can be combined with solar energy, industrial exhaust heat and heat pump as the energy for solution regeneration, thus saving most of latent heat energy consumption. Its power consumption is about one third of conventional AC systems. The power consumption is reduced obviously, and the energy and environmental issues following the operation of AC systems can be relieved effectively. It is applicable to high temperature and high humidity climate or the space environment requiring low humidity. The operation of solution dehumidification system is simple, and the maintenance and the change of interior materials are convenient. The major equipments are exsiccator and regenerator. The regenerator is one of important heat and mass transfer of liquid desiccant AC system. Fumo et al. (2002) used lithium chloride as moisture absorption solution, the solution inlet temperature range was 60~70℃. The packing in the regenerating column of equipment side was Rauschert Hllflow annular packing, thus accelerating the heat exchange between spray solution and humid air in the regeneration process. The regeneration process of regenerating column was analyzed. Luo et al. (2008) used cross current packing regenerator for experimental analysis of heat and mass transfer coefficient. They calculated various conditions according to experimental results. The mass transfer coefficient increases the regeneration capacity as the air temperature, air flow and liquid inventory increase, proving that the coupling mass transfer coefficient and Le number have influence on the regenerator. Junfei et al. (2014) used a simple flat-plate mold device instead of general counter current heat exchange for regeneration experiment, changed the inlet-outlet parameters. They also used NTU-Le model to analyze the empirical values to increase the accuracy, and proved that the flat-plate mold was applicable to dehumidification regeneration process. However, compared with the heat and mass transfer coefficient of packing, the packing for regeneration was better than flat membrane.

MATHEMATICAL MODEL OF REGENERATIVE COUPLING HEAT AND MASS TRANSFER

The schematic diagram of the countercurrent dehumidification/regeneration system and the device stereogram are shown in Figures 1.In the physical model of solution dehumidification/regeneration exchange, the solution is sprayed from the exsiccator and regenerator top on the regular packing or bulk packing. However, a thin liquid film is formed on the outer layer of packing, the air contacts the packing liquid film directly for heat exchange. The heat energy is transferred from an outlet to another outlet. The countercurrent NTU-Le model is analyzed according to the experimental system device. The NTU relationship of solution dehumidification/regeneration process is simplified as follows.

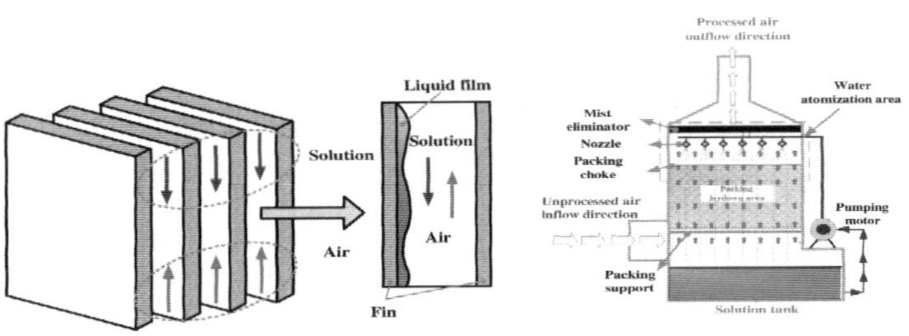

Figure 1 Schematic diagram of the countercurrent dehumidification/regeneration system and the device stereogram.

(1)The heat energy diffusion in the air flow direction is neglected, only the heat energy transfer in vertical air flow direction is considered.

(2)The heat and mass transfer of solution, air and outdoor environment is neglected, the heat exchanging process is calculated by closed and adiabatic system.

(3)Assuming that the physical state of solution side and air side heat energy in the system control component is distributed uniformly.

(4)Mass conservation of system

(5)Energy conservation of system

(6)The overall heat transfer capacity of system can be calculated by the following equation

$$L_e = \frac{h_c}{h_D \cdot C_{p,a}} \qquad NTU = \frac{h_c \cdot A}{\dot{m}_a} \tag{1}$$

where: L_e: Lewis Factor. *NTU*: Number of Transfer Units.

$C_{p,a}$: Air specific heat (*kJ/kg.* ℃). $d\omega_a = h_D(\omega_{L,equ} - \omega_a) \cdot dNTU$ (2)

RESULTS AND DISCUSSION

The Experimental result of regeneration capacity by changing solution temperature and changing liquid inventory and changing exhaust air flow are shown in Figures 2.As the solution inlet temperature rises, the solution surface saturated steam pressure increases, the thermal conduction capability is enhanced, and the solution is carried away faster. On the contrary, there is no regeneration effect if the inlet temperature is too low, the application side provided by the equipment side thermal energy source is very important, so the system regeneration efficiency can be increased only by stabilizing the heat supply. The effect of the lithium chloride solution inlet flow rate on the regeneration capacity is discussed.

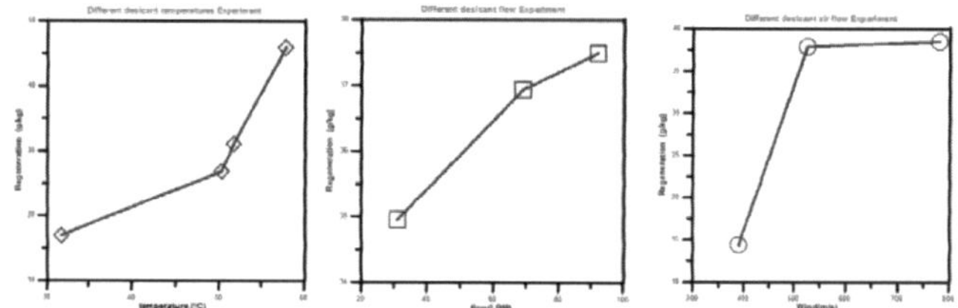

Figure 2 Experimental result of regeneration capacity by changing solution temperature and changing liquid inventory and changing exhaust air flow.

The solution inlet flow rate is changed. The results show that the outlet humidity increases with the flow, the saturation on the solution surface is maintained for larger heat and mass transfer. However, the moisture is vaporized at high temperature, the solution is reduced slightly. The experimental results show that the increase in the liquid inventory does not increase the regeneration capacity apparently. Therefore, the change in the regeneration flow parameter has no apparent effect on the regeneration capacity of overall system. The relation between the exhaust air flow of exhaust wind turbine and the regeneration capacity is discussed, the exhaust wind turbine speed is changed by frequency converter and the air flow is measured. The full load to half load are tested, the half load running effect is very bad. Therefore, the heat and mass transfer coefficient increases with the air flow. At a higher air flow, the moisture desorption capacity is better, and the regeneration capacity is higher. Therefore, with a high air flow of system device, the regeneration capacity is better.

CONCLUSION

This study used lithium chloride solution as dehumidizer to discuss the effect of solution temperature, liquid inventory and air flow on regeneration capacity. The experimental results are concluded as follows:
The regeneration capacity is 31.88 g/kg when the solution temperature is 57.8℃, and the regeneration capacity is only 2.87 g/kg when the temperature is 31.8℃. Therefore, at a high solution temperature, the regeneration capacity of system is higher, and the heat recovery is better. 2. The regeneration capacity is the best when the liquid inventory is under full load 92.1LPM, but the experimental analysis shows that the overall regeneration capacity of system varies slightly with the flow parameter. Therefore, if the liquid inventory is operated with different parameters, the water pump power consumption can be further reduced. The regenerator uses exhaust wind turbine and solution for heat and mass transfer. At a high air flow, the regeneration effect is better. When the exhaust wind turbine works under half load, the power consumption is reduced apparently, but the regeneration effect is not attained. The overall system misses thermal equilibrium, thus affecting the system operation life.

ACKNOWLEDGEMENT

The authors gratefully acknowledge the support of this research by the Ministry of Science and Technology, Taiwan (MOST 103-2221-E-167 -032).

REFERENCES

Fumo N,Goswami DY, (2002). Study of an Aqueous Lithium Chloride Desiccant System. In :Air Dehumidification and Desiccant Regeneration. *Solar Energy,* **72(4)**, 351-361.

Luo L, ZHANG XS, et al (2008). An experimental study on the heat and mass transfer coefficient in cross-flow regenerator. *Journal of engineering thermophysics*, **29(7)**, 1215-1217.

Qian J, Yin Y, Pan XW, et al, (2014). Experimental investigation and model validation for liquid desiccant dehumidification and regeneration in falling-film plate. *CIESC Journal*, **65(6)**, 2077.

AIRTIGHTENING AT FOUNDATION-HOUSE INTERFACE AS A MEANS OF RADON REDUCTION IN LIVING SPACE

William ROSE*[1], Zachary MERRIN[1], Paul FRANCISCO[1] and Stacy GLOSS[1]

[1]Indoor Climate Research and Training, University of Illinois, Champaign IL

*Corresponding email: wrose@illinois.edu

Keywords: Airtightness, blower door, zone pressure, add-a-hole method, radon

SUMMARY
The PARR (Partnership for Advancing Residential Retrofit) study includes an effort to investigate the potential radon-reduction impact of improving airtightness at the floor and foundation based ductwork, in order to better isolate the living space from the foundation.

Fifteen homes were instrumented for continuous radon measurements in the living space and foundations. The treatment consisted of use of spray-applied urethane foam beneath the floor system to seal gaps, application of mastic and tape to ductwork in the foundation area, and other means to improve the isolation of the living space from the foundation.

Blower door tests provided zone pressure results and add-a-hole results quantifying that the foundation zone was better isolated from the living space in most of the homes in the study. The improved isolation, however, did not lead to statistically-significant radon reduction in the living spaces of the homes in the study.

INTRODUCTION
The aim of the research was to conduct a primary scoping study on the impact of air sealing between the foundation and the living space on radon transport reduction across the foundation-living space floor assembly. Fifteen homes in the Champaign IL area participated in the study. These homes were instrumented for hourly continuous radon measurements. Blower door and zone pressure diagnostics were conducted at each house. The treatments consisted of applying air-sealing foams at the underside of the floor separating the living space from the foundation, and providing duct sealing on the ductwork located in the foundation area. The hypothesis was that air-sealing at the floor system separating the foundation from the living space should better isolate the living space from the foundation, and this isolation should lead to the entry of less radon into the living space from the foundation. If the hypothesis were to prove true, then retrofit energy efficiency programs may choose to adopt these isolation methods for enhanced radon protection to the living space.

METHODOLOGIES
The research was conducted in three groups of five homes. The homes in each group were instrumented for approximately three months. Treatments for each home occurred either at the end of the first month or at the end of the second month of instrumentation. In this way, it was possible to use 2 or 3 of the 5 homes as controls (with no treatment and no change in conditions) during the two-month period when the other homes received treatment. The results shown here contain no correction for control.

Normally, radon measurements are conducted in "closed-house conditions", where windows are not to be opened, clothes dryers are not to be used, and signs are posted at doors to ensure that they are used only briefly. These restrictions were not applicable to a three-month research term. Where possible, houses were instrumented to monitor window operation, or in some cases occupants self-reported window operation. In the first group, one occupant maintained excellent logs of window operation; when data from that house was plotted for house radon versus foundation radon, a very distinguishable pattern in the resulting data provided a means to distinguish house-open data from house-closed data for other houses. This distinction was applied to the data from all the houses to enable analysis to be done on closed-house conditions. The Final Report from the research will describe this inclusion/exclusion method.

Blower door testing was conducted prior to treatment (pre-) and following treatment (post-). The test sequence included zone pressure measurement at the foundation. If the floor system is better isolated following treatment, the zone pressure in the foundation space should be further from the indoor air pressure and closer to outdoor air (reference) pressure.

The testing included add-a-hole zone pressure testing, which permits estimates of the net size of total opening from the house to the foundation zone and from the foundation zone to outdoors. Add-a-hole testing will be described in the Final Report. This technique involves conducting a second set of blower door and zone pressure tests following the introduction of an opening, usually between the foundation and the outdoors, and comparing the results of the two tests. If the resulting sets of measurements show too little difference, the results are shown as out of range (oor). See Table 3.

Radon was measured hourly using RadStar RS300 continuous radon monitors. These were coupled with x86 based computers equipped to automatically download the data daily, and recover in the event of a power loss. This equipment allowed the monitor to operate beyond the 10-day limitation of the RS300's internal memory.

RESULTS AND DISCUSSION

Table 2. Measured foundation zone pressures in 15 houses, pre- and post-treatment.

	PR-01	PR-02	PR-03	PR-04	PR-05	PR-06	PR-07	PR-08	PR-09	PR-10	PR-11	PR-12	PR-13	PR-14	PR-15
zp-pre (Pa)	19	18.7	21.6	24.4	42.1	40.4	46.6	13.1	13	28.3	42.7	40.3	38	40.1	41.5
zp-post (Pa)	17.9	12	19	20	43.7	42.5	46.5	11.2	12.6	20.5	43.5	28	31	35.3	30.3
% reduct. in zp	5%	36%	14%	18%	-4%	-5%	1%	16%	4%	32%	-2%	29%	15%	9%	27%
foundation type	cs	cs	bsmt	cs	bsmt + 3 cs	bsmt + 1 cs	cs	cs	cs	cs	cs	bsmt + 2 cs	cs	cs	cs

where

- zp-pre is the measured zone pressure with respect to outdoors in Pa prior to treatment
- zp-post is the measured zone pressure with respect to outdoors in Pa following treatment
- % reduction is the zone pressure, where the house pressure (often different from 50 Pa) is factored in. A positive value reflects greater isolation between the foundation zone and the house
- "cs" refers to crawl space.

Table 2 shows the zone pressure measurements and the percent reduction in zone pressure. The aim was to improve the isolation of the living level from the foundation level. A positive value for the percent reduction shows improved isolation. In general, the treatment work resulted in improved isolation of the living space from the foundation. The amount of reduction was less than anticipated.

Table 3. House-to-zone calculated net opening areas, pre- and post-treatment (in^2)

	PR-01	PR-02	PR-03	PR-04	PR-05	PR-06	PR-07	PR-08	PR-09	PR-10	PR-11	PR-12	PR-13	PR-14	PR-15
hza-pre (in2)	66	115	224	205	347	512	oor	84	130	229	oor	172	85	oor	77
hza-post (in2	86	55	185	135	198	76	oor	oor	125	225		103	45	115	oor
% reduction	-31%	52%	18%	34%	43%	85%			4%	2%		40%	47%		

where

- hza-pre is the net house-to-zone opening area in inch^2 prior to treatment
- hza-post is the net house-to-zone opening area in inch^2 following treatment
- Blanks represent absence of appropriate input data
- "oor" represents out of range values. These occur when the difference in zone pressure between closed condition and open condition is less than 5.5 Pa.

The values in Table 3 were calculated using add-a-hole methods, briefly described above. With the exception of house PR-01, the intervention led to an apparent reduction in the net opening size from house to zone. This suggests that the interventions met their aim of leading to improved isolation of the living space from the foundation. However, the degree of improved isolation was not to the extent desired, with most homes showing less than a 50% improvement. This indicates that typical air sealing techniques are not sufficient to truly isolate the foundation space from the rest of the home.

Table 4. Zone-to-outdoor calculated net opening areas, pre- and post-treatment (in^2)

	PR-01	PR-02	PR-03	PR-04	PR-05	PR-06	PR-07	PR-08	PR-09	PR-10	PR-11	PR-12	PR-13	PR-14	PR-15
zoa-pre	92	162	263	213	117	184	oor	166	256	171	oor	76	43	oor	27
zoa-post	126	117	264	175	57	22	oor	oor	254	299		87	29	58	oor

where

- zoa-pre is the calculated net opening area from the foundation zone to the outdoors prior to treatment, in^2.
- zoa-post is the same, post treatment.

The validity of the Table 3 values must be considered in light of the results shown in Table 4 which show changes in the net opening area from zone to outdoors. The intent of the project was to make no change to the rate of radon entry into the foundation space by virtue of the treatment, nor to effect any change in the airtightness of the foundation with respect to the outdoors. As we can see, these results suggest changes in that airtightness. This is likely due to a combination of imprecision in the measurement technique along with connections from the foundation space to outside via wall cavities or duct work. For example, if there is a bypass from the foundation to the home that also connects to the attic, then sealing that bypass will increase the isolation of the foundation space from both the house and outdoors.

Table 5. Radon average results (pCi/L), pre- and post-weatherization

living	PR-01	PR-02	PR-03	PR-04	PR-05	PR-06	PR-07	PR-08	PR-09	PR-10	PR-11	PR-12	PR-13	PR-14	PR-15
pre-	7.2	6.1	4.6	7.3	9.1	1.2	5.2	1.4	2.2	1.8	14.8	8.7	10.3	10.1	4.7
post	7.3	7.1	4.0	6.7	8.5	3.2	6.7	0.8	3.2	2.5	24.9	7.6	10.3	11.6	3.9
foundation															
pre-	11.1	11.1	9.3	11.3	13.1	1.3	14.7	3.7	6.0	1.9	25.1	23.5	18.4	18.6	11.7
post	10.7	13.8	8.1	9.4	10.7	4.7	16.9	3.1	9.1	2.8	42.2	16.6	38.5	24.8	10.3
ratio living to foundation															
pre-	0.64	0.55	0.50	0.65	0.70	0.91	0.35	0.37	0.37	0.93	0.59	0.37	0.56	0.54	0.40
post	0.68	0.51	0.49	0.72	0.79	0.67	0.40	0.26	0.35	0.90	0.59	0.46	0.27	0.47	0.38

The radon concentration averages (arithmetic averages) for each group, using only the included data as described above, are shown in Table 5. What appears in Table 5 is that the radon results were mixed. Two houses remained essentially the same (PR-01 and PR-13), six houses showed lowering of average radon and seven houses showed an increase in average radon.

The radon results appear to show that the modest improvement in isolation achieved in the 15-home sample leads to radon change results that are not significant. Variation in radon levels associated with weather and building operation will be described in the Final Report and elsewhere. The measured values shown here may be described as occurring in the normal noise range of natural radon concentration variation.

Previous studies (e.g. Steck 2005) have shown that radon levels can vary substantially over time.

CONCLUSIONS

Fifteen homes were treated to provide greater isolation of the living space from the foundation space—the source of radon in the home. The treatment consisted of gap-sealing within the floor system and duct sealing.

The treatment led, in most houses, to a measurable change in zone pressure measurements pre- and post-treatment showing greater isolation of the living space from the foundation zone. Add-a-hole methods were used to calculate net opening areas at this interface, and they showed reduction of questionable validity in the net opening following treatment.

This improved isolation did not lead to a significant change in living level average radon. Of the 15 houses, two showed essentially no change, six showed lowered radon, seven showed elevated radon.

ACKNOWLEDGEMENT

We would like to acknowledge those who contributed to this study. These include: the US DOE Building America program with Stacy Rothgeb NREL Technical Manager, Larry Brand of GTI who manages PARR, Lori Shupe from University of Illinois ISTC who managed the participant contacts, Terry Brennan of Camroden Associates for technical guidance, the Illinois Emergency Management Agency, Patrick Daniels director, and especially the homeowners and their families who participated in the program. The research was funded under DOE Subcontract KNDJ-0-40346-04.

REFERENCES

Steck, D.J. 2005. "Residential Radon Risk Assessment: How Well is it Working in a High Radon Region?" Proceedings of the fifteenth Annual International Radon Symposium, San Diego, CA.

Removing Acetaldehyde in Indoor Air Using Mn/TiO2 Nanoparticle OZCO Catalyst

Minsu Kim[1,2], Eunseuk Park[1], Hyounduk Jung[1,2], and Jongsoo Jurng[1,2*]

[1]*Center for Environment, Health and Welfare Research, Korea Institute of Science and Technology, Hwarangno 14 gil 5, Seoul, Seongbuk-gu, 136-791, Republic of KOREA*
[2]*Graduate School of Energy environment Policy and Technology, Korea university, Anam-ro 145, Seoul, Seongbuk-gu, 136-701, Republic of KOREA*
* *Corresponding author: E-mail: jongsoo@kist.re.kr*

INTRODUCTION

Recently, our research group [1] successfully synthesized TiO2 support materials by a CVC method; the resulting Mn2O3/CVC TiO2 material showed high dispersion, large surface area, and small crystallite size. Moreover, these samples possessed a higher catalytic oxidation activity for acetaldehyde decomposition than Mn_2O_3/P25, even at room temperature (25°C). Mn-loaded TiO2 showed improved catalytic properties because of the addition of transition metals; 5 wt% Mn/TiO2 holds great promise for use in catalytic reactions.

This paper investigates the effect of Mn loading on the oxidation of acetaldehyde over manganese oxides supported on titania (Mn/TiO2) honeycomb to be compared with the absorption technique.

MATERIAL AND METHODS

Catalyst preparation

TiO2 nanomaterials were synthesized by a chemical vapor condensation and loaded with 5wt% manganese by impregnation[2]. The oxidation catalyst was prepared by the sol-gel coating method. The honeycomb catalysts were treated for 2h in nitric—acid solution. The coating solution was prepared using the Mn/TiO2 catalyst, a bentonite and DI water (12:3:35 (wt%)). The honeycomb catalysts were immersed in the solution for 5 min, and then removed for drying at 100oC for 3h. This coating process can be repeated three times. Then, the catalysts were dried overnight at 100oC and calcined at 500oC for 6h.

Bench scale chamber system

The chamber setup used in this study is described in Fig. 1. Acetaldehyde was introduced at a constant rate to the 1-m3 bench scale chamber. The chamber was built entirely of polycarbonate to minimize surface area losses of the analytes. To reach the adsorption equilibrium, the reaction was performed for 12h. This means that a steady state can be reached regarding the partition of acetaldehyde in the gas phase and in the adsorbed phase. During the experiments, the organic compounds were measured only in the gas phase and a fraction of acetaldehyde is present on the honeycomb catalyst surface at the beginning of the reaction.

The catalytic oxidation reactor was installed on an external ductwork loop and operated in the recirculation mode, by re-introducing the air processed by the air cleaner of the catalytic oxidation reactor into the chamber. The acetaldehyde concentration was adjusted by mixing the standard form of acetaldehyde (500ppm) with clean air. Ozone was generated from a UV lamp. The chamber reaction temperature was set to 25oC (room temperature). This arrangement produces 10 and 30 ppm of acetaldehyde concentrations. For comparison, commercial catalysts (P25, Degussa) loaded with Mn were also prepared by the same method using 30ppm of acetaldehyde concentration.

Finally, a gas sample was analyzed using a Fourier-transform infrared spectrophotometer (1400-F, MIDAC, California, USA) equipped with a cell having optical length of 10.0 m and a volume of 4.0L.

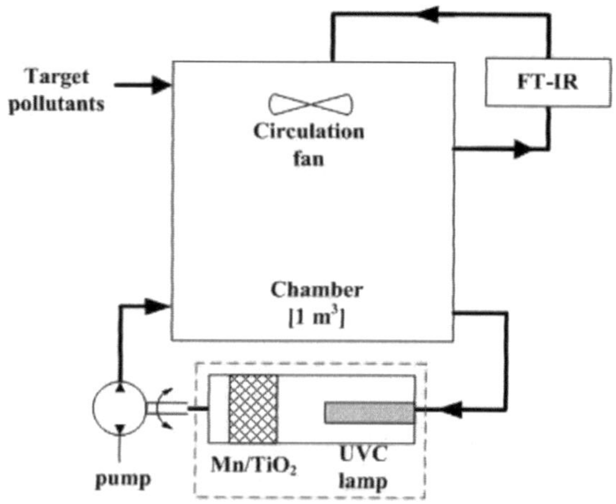

Fig. 1 Schematic of 1m3 polycarbonate with external catalytic oxidation reactor recirculation loop

RESULTS AND DISCUSSION

Catalytic oxidation of acetaldehyde

Ozone is transported to the catalyst and that the ozone molecule produces two active atomic oxygen and one oxygen radical, both capable of oxidizing of acetaldehyde to COx. Moreover, manganese oxides are often employed as catalysts for low temperature oxidation of carbon monoxide with ozone[3].

Fig. 2 shows the Mn/TiO2 (CVC) honeycomb-catalyst produced by catalytic oxidation, with higher acetaldehyde removal efficiency than Mn/TiO2 (P25) honeycomb-catalyst. Such behaviors can be accounted for by high SSA, which is one of the important factors for VOC oxidation[4]. The supported manganese oxide catalysts with high surface area are effective for catalytic oxidation from the standpoint of catalytic activity and efficiency for ozone utilization

Fig. 3 shows the variation of acetaldehyde concentration in air with the reaction time for three different initial concentrations ranging from 10- to 30ppm. The initial slope increased with increasing initial concentration of acetaldehyde. Higher concentration leads to higher catalyst oxidation. The graphical determination of the half reaction time (T) for each reaction (Fig. 3) indicates that this parameter does not depend on initial concentration of acetaldehyde in the investigated concentration range. The value of T is 20 min, during acetaldehyde removal. Thus, it was found that the acetaldehyde concentration can be decreased to almost zero in 1h.

Although the irradiation is provided by the 254-nm lamp, ozone acts not only as an electron acceptor to produce O3- but also as a source to generate hydroxyl radicals, which have strong ability to oxidize acetaldehyde. Therefore, the combination UV lamp with ozone can enhance the conversion of acetaldehyde.

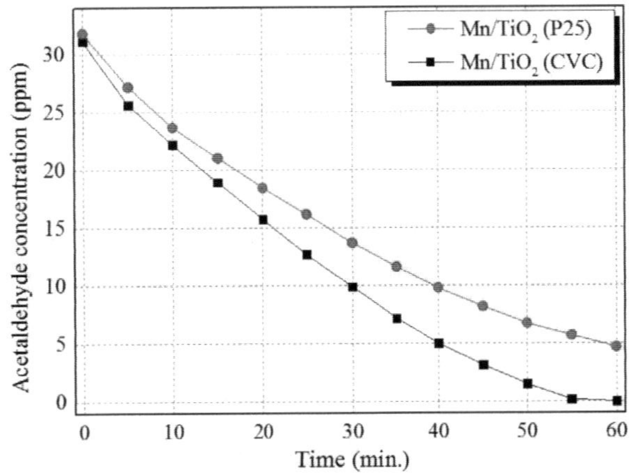

Fig. 2 ***Catalytic oxidation of acetaldehyde of Mn/TiO2 (CVC) and Mn/TiO2 (P25)***

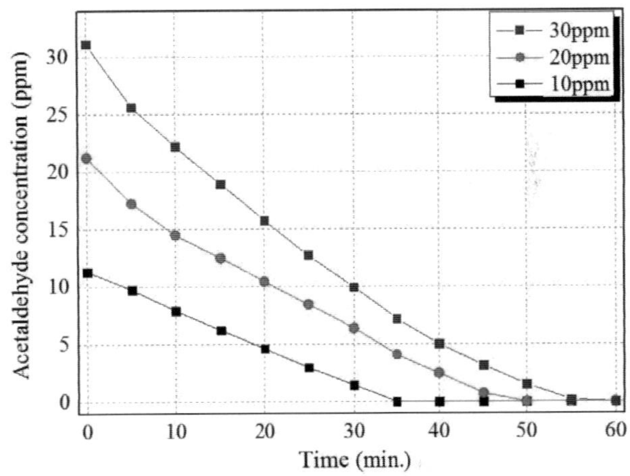

Fig. 3 ***Catalytic oxidation of acetaldehyde at various initial concentrations using Mn/TiO2 (CVC)***

CONCLUSIONS

This study examined the catalytic conversion of gaseous acetaldehyde on Mn/TiO2 catalysts in the presence of ozone, as well as the effect of the support material (TiO2) characteristics. Mn/TiO2 honeycomb catalyst with high oxidation activity for acetaldehyde was synthesized by the CVC method.

The performance of this system was investigated by treating artificially generated acetaldehyde at a concentration level of ppm a 1-m3 bench scale chamber. It was found that the acetaldehyde concentration can be decreased to the almost zero value in 1h.

Thus, it can be concluded that it is very important to promote synergetic capability, transfer/diffusion and reaction activity for the catalysts used in the catalytic oxidation of acetaldehyde.

REFERENCES

[1] E. S. Park, H. A. Le, Y. S. Kim, S. M. Chin, G. N. Bae and J. Jurng, 2012, Preparation and characterization of Mn2O3/TiO2 nanomaterials synthesized by combination of CVC and impregnation method with different Mn loading concentration, Mater. Res. Bull., Vol. 47, pp. 1040-1044.

[2] S. M. Chin, E. S. Park, M. S. Kim and J. Jurng, 2010, Photocatalytic degradation of methylene blue with TiO2 nanoparticles prepared by a thermal decomposition process, Powder Technol., Vol. 201, pp. 171-176.

[3] B. Dhandapani and S. T. Oyama, 1997, Gas phase ozoen decimposition catalysts, Appl. Catal. B, Vol. 11, pp. 129-166.

[4] H. Einaga and A. Ogata, 2009, Benzene oxidation with ozone over supported manganese oxide catalysts: effect of catalyst support and reaction conditions, J Hazard Mater, Vol. 164, pp. 1236-1241.

Enhanced Catalytic Decomposition of VOCs Using Synthesized MnOx/TiO2 With Adsorption-desorption

Hyounduk Jung[1,2], Minsu Kim[1,2], Eunseuk Park[1], and Jongsoo Jurng[1,2*]

[1]*Center for Environment, Health and Welfare Research, Korea Institute of Science and Technology, Hwarangno 14 gil 5, Seoul, Seongbuk-gu, 136-791, Republic of KOREA*
[2]*Graduate School of Energy environment Policy and Technology, Korea university, Anam-ro 145, Seoul, Seongbuk-gu, 136-701, Republic of KOREA*
* *Corresponding author: E-mail: jongsoo@kist.re.kr*

INTRODUCTION

Recently, our research group successfully has examined the catalytic conversion of gaseous toluene on MnOx/TiO2 catalysts in the presence of ozone, even at room temperature (25°C) [1]. However, since some indoor air volatile organic compounds (VOCs) can be at high concentrations, it will be necessary to vary supply ozone concentration for reaction.

The objective of this study was to find out how adsorption and desorption over zeolite helps to remove VOCs, such as acetaldehyde, with catalytic oxidation over manganese oxides supported on titania (MnOx/TiO2).

MATERIAL AND METHODS

Adsorption-desorption reactor preparation

The adsorption/desorption and catalyst oxidation system used in this study for shown in Fig.1.

The gas sample was analysed using a Fourier-transform infrared spectrophotometer (1400-F, MIDAC, California, USA) equipped with a cell having optical length of 10.0m and a volume of 4.0L.

The experimental set-up may be assumed to consist of three sections as shown in Fig. 1: (a) absorption-desorption section with heating oven, (b) ozone generation and catalytic oxidation, and (c) FTIR analysis sections.

Finally, a gas sample was analyzed using a Fourier-transform infrared spectrophotometer (1400-F, MIDAC, California, USA) equipped with a cell having optical length of 10.0 m and a volume of 4.0L.

Table 1 **Adsorbent Characteristics**

Adsorbent	Symbol	Si/Al ratio	S_{BET} (m^2/g)
Active carbon	AC	-	600~800
zeoilite	CBV720	30	780
zeoilite	CP814C	38	710
zeoilite	CBV5524G	50	425

Adsorbent selection

Three types of zeolites adsorbents were selected, based on specific surface area and Si/Al ratio, compared with commercial activated carbon.

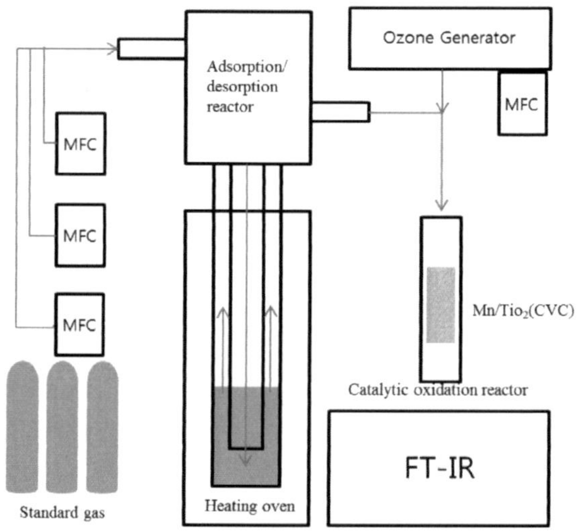

Fig. 1 Schematic diagram of the experimental setup to study VOC absorption-desorption over zeolite

RESULTS AND DISCUSSION

Absorbent Recycle Efficiency

We have measured the recycle efficiency, which means the adsorption-desorption performance and reuse efficiency of three zeolite absorbents, CBV720, CP814C, and CBV5524G, using acetaldehyde.

The result showed that reusability of activated carbon was as low as 30%, while the CBV720 zeolite gave the best reusability among three adsorbents. (Fig.2)

Desorption at thermo control system.

Fig. 3 shows the result of constant concentration desorption of acetaldehyde, using the precise temperature control of zeolite bed, with an increase in temperature, 1℃/min.

The desorption concentration of acetaldehyde was 180 ± 20 (Fig.3). With the MnOx/TiO2 catalysts in the presence of ozone, we have decomposed the acetaldehyde up to 90%.

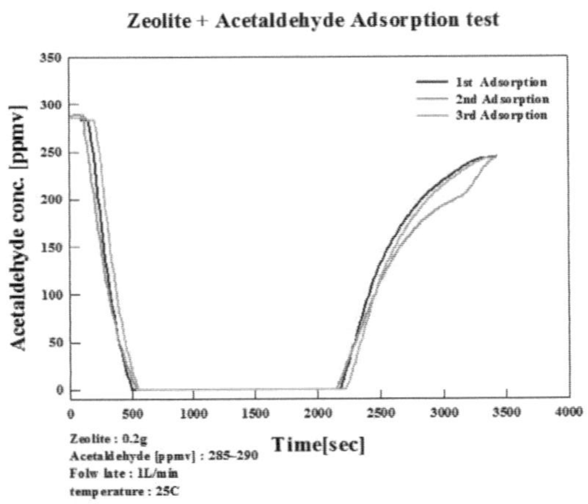

Fig. 2 Catalytic oxidation of acetaldehyde of Mn/TiO_2 (CVC) and Mn/TiO_2 (P25)

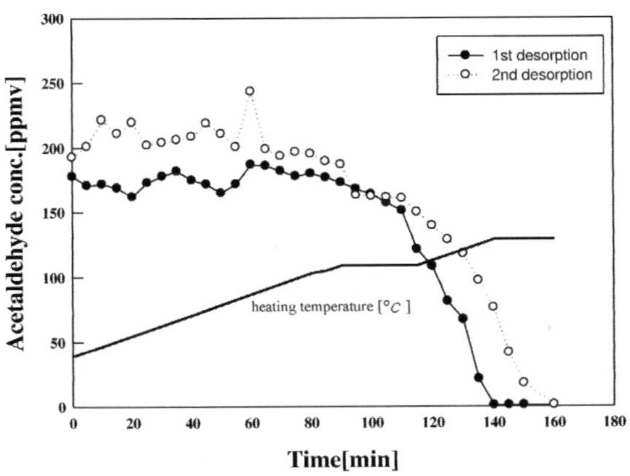

Fig. 3 Desorption concentration of acetaldehyde with temperature increase, 1 ℃/min.

Effect of the varied VOCs concentrations on catalytic oxidation.

To investigate the effect of inlet concentration of acetaldehyde on the decomposition efficiency of MnOx/TiO2 catalyst, we have measured the removal efficiency of acetaldehyde with three different inlet concentration, for example, 50, 100 and 150 ppmv. Residual ozone concentrations were also checked, with the fixed inlet ozone, 350 ppmv.

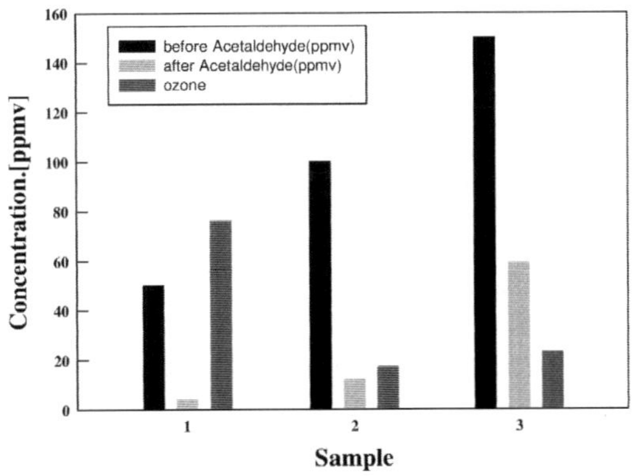

Fig. 4 Outlet concentration of acetaldehyde with three different inlet conditions, 50, 100, and 150 ppmv.

Fig. 4 showed that the 100 ppmv case gave the best decomposition efficiency of acetaldehyde, as high as 90%, with low ozone residual.

The experiment showed the higher removal efficiency of acetaldehyde with 50 ppmv Ozone, while the residual O3 concentration was higher than 60 ppmv. On the other hand, with higher inlet acetaldehyde concentration, 150 ppmv, removal efficiency of acetaldehyde decreased sharply down to 67%.

CONCLUSIONS

This study performed the preliminary experiments to find out how adsorption-desorption over zeolite helps to remove acetaldehyde properly, with oxidation over manganese oxides supported on titania (MnOx /TiO2).

Thus, it can be concluded that it is very important to promote synergetic capability for the oxidation catalysts, with the adsorption-desorption technique, used in the catalytic oxidation of acetaldehyde.

REFERENCES

[1] Eunseuk Park, Sungmin Chin, J. Kim, G.N. Bae, and Jongsoo Jurng, 2011, Preparation of MnOx/TiO2 ultrafine nanocomposite with large surface area and its enhanced toluene oxidation at low temperature, Powder technology, V.208, pp.740~743.

EXPLORING THE EFFECTIVENESS OF SIMPLE FAN FILTER SYSTEMS IN VENTILATING AN ENCLOSED NON-AIRCONDITIONED SPACE

Ron TAN[1*], Joanna SHEN[1], Gayatri SANKARAN[1], Theodore PRAWIRADIRAJA[1], Hooi Ming YAP[1], Ramona GUTIERREZ[1], Kwok Wai THAM[2], Chandra SEKHAR[2], Kok Wai CHEONG[2], Wei-Chung CHANG[3], Keng Been ANG[4], Clement Yew Yi TAN[5] and Lee Ching NG[1]

[1] Environmental Health Institute, National Environment Agency, Singapore
[2] Department of Building, School of Design & Environment, National University of Singapore, Singapore
[3] Division of Environmental and Water Resources Engineering, School of Civil & Environmental Engineering, Nanyang Technological University, Singapore
[4] Indoor Air Quality Society of Singapore, Singapore
[5] School of Applied Science, Temasek Polytechnic, Singapore

[*]Corresponding email: Ron_TAN@nea.gov.sg

Keywords: Indoor air quality, haze, mitigation, ventilation, classroom

SUMMARY

The increase in outdoor pollutant concentration due to occasional trans-boundary haze is a public health concern in Singapore. During such smoke haze episodes, it was shown that occupants of naturally ventilated spaces could be exposed to pollutant levels that approximate outdoor levels. Closing windows and doors, as advised by the authorities, can reduce the infiltration of outdoor pollutants into such spaces. However, long duration of closure could lead to a build-up of heat and indoor pollutants. We explored the use of two simple fan filter prototype systems as a cost-effective haze mitigation tool. The results showed that by attaching a high efficiency air filter to a fan, outdoor air can be filtered and introduced for ventilation, reducing the build-up of heat and pollutants in a classroom.

INTRODUCTION

Singapore has experienced instances of poor outdoor air quality. Of particular concerns are periods of trans-boundary haze from forest fires in the region. During the 2013 haze episode, studies conducted showed that occupants of naturally ventilated spaces could be exposed to pollutant levels that approximate outdoor levels. Our data corroborates with those reported by Chithra and co-workers in 2012. Exposures to pollutants, such as particulate matter (PM), may lead to health issues and should be reduced whenever possible (Wegesser et al. 2009, Henderson et al. 2011). Closing of windows and doors can reduce the infiltration of outdoor pollutants and an occupant's exposure to these compounds. However, this would reduce ventilation and could lead to a build-up of heat and indoor pollutants, which may affect the occupants' health and performance (ASHRAE 2007, Apte et al. 2000).

Mechanical ventilators have commonly been used to supply outdoor air into indoor spaces (Rice et al. 2001, Humphreys 2004, Chen et al. 2011). Typical mechanical ventilators that are available commercially are unattractive to many premise owners due to the need for extensive installation work, operational noise and size of the system. Recently, several versions of small mechanical ventilators involving fan with filter attachments for effective filtration of the outdoor air supply have been developed for small spaces. As these could present an interesting and affordable alternative for ventilation, it is of interest to study the effectiveness of such simple fan filter systems (FFS) in promoting ventilation while minimizing infiltration of outdoor air pollutants.

METHODOLOGY

Two fan filter system prototypes were developed (Figure 1), each involving a fan with an attached high efficiency particulate filter (MERV 13, as recommended by the Singapore Standard Code of Practice 554) (SPRING 2009). The gaps between the filter and the fan were sealed to ensure no bypass of unfiltered air. The prototypes were tested at air flow rates of 710 – 2000 CMH (for system A, measured using TSI Accubalance Air Capture Hood 8371) and 250 CMH (for system B, estimated based on linear air velocity measurements and fan size).

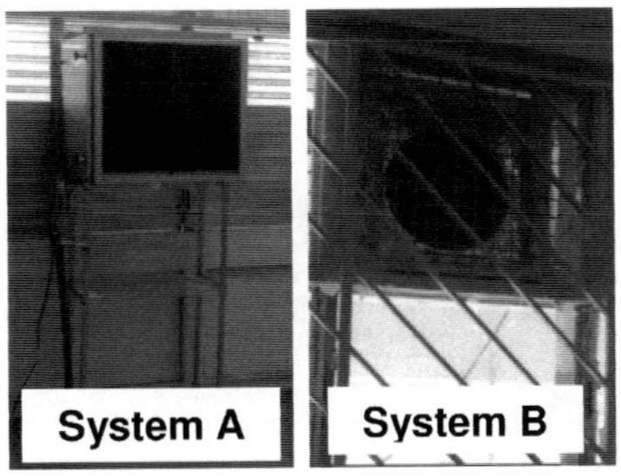

Figure 1. Prototype Fan Filter Systems A and B.

An empty classroom (65 m^2 x 3.4m) was used as the test site. The windows and doors were closed. The windows were covered with PVC sheets to simulate classrooms with negligible ventilation, deliberately leaving an opening (5 x 5 cm) uncovered as an air outlet. The indoor air was homogenized by two ceiling fans that were switched on throughout the measurement period. Air change rates were measured using tracer gas (SF_6) concentration decay method as described by Sekhar et al. (2004).

The FFS was positioned by the window at the back of the classroom to mimic actual operating conditions. The indoor temperature was measured by taking the average readings from two separate portable recorders (Keytag KTL-508), and this was compared to the outdoor temperature that was recorded along an open corridor. As the experiment was conducted in the absence of human subjects, the heat source was generated using 40 units of 100W tungsten light bulbs. The indoor temperature was recorded over time at different inlet air flow rate.

RESULTS AND DISCUSSION

During the 2013 haze episode, the indoor PM levels of different premises were measured and expressed as a ratio to outdoor levels (Figure 2). The result, which suggests that outdoor fine particles can freely infiltrate a naturally ventilated premise, is consistent with the findings of (Chithra et al. 2012). As the occupants of naturally ventilated spaces could be exposed to pollutant levels that approximate outdoor levels during a haze situation, they are being advised to close windows and doors when outdoor air quality appears to be worsening. Indoor PM level could further be improved with the use filtration systems.

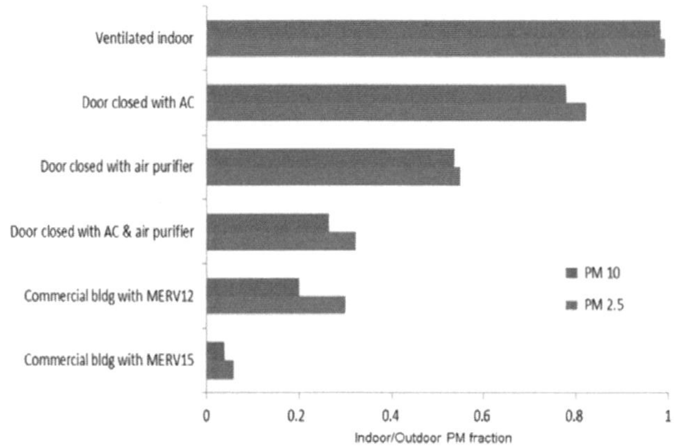

Figure 2. PM level was measured at different premises with different ventilation types during the haze period.

FFS prototype systems were then assessed for their effectiveness to supply ventilation to a classroom with closed windows and doors. The air change rates of the classroom were measured at different air flow rates (controlled by the FFS) and the results are tabulated in Table 1. The leaky windows contributed to an air change of 3.2 ±0.2 h^{-1} when the FFS was off. By covering the windows, the air change rate in the classroom was reduced to 0.25 ±0.04 h^{-1}, which is typical of classrooms with relatively tight windows. In both cases, air change rates increase as the FFS-introduced air flow rates increased (Table 1).

Table 1. Air change rates (h^{-1}) of an unoccupied classroom served by a fan filter system at different speeds

Fan filter system	Calculated air change[1]	Measured Air Change/h	
		Leaky windows	Covered windows
Off	0	3.2 ±0.2	0.25 ±0.04
System A, 710 CMH	3.2	5.6 ±1.2	-[2]
System A, 1120 CMH	5.1	5.5 ±0.9	4.7 ±0.1
System A, 1550 CMH	7.0	6.1 ±0.3	6.4 ±0.1
System A, 2000 CMH	9.0	7.5 ±0.2	-[3]
System B, 2 units at 250 CMH each	2.3	-[4]	1.74 ±0.05

[1]The calculated air change was determined by the introduced air flow/room volume
[2]Test was not conducted as the calculated air change was insufficient to provide adequate ventilation in a relatively sealed environment.
[3]Test was not conducted as pressurization of the room had been achieved at 1550 CMH.
[4]Test was not conducted as the introduced air flow was likely insufficient for pressurization of the room to be achieved in a leaky environment.

The differences in measured and calculated air change suggest that the indoor environment may be affected by the outdoor conditions (e.g. wind speed and direction) when the fan speed is low; or may be due to positive indoor pressure when the fan speed is high. Maintaining positive indoor pressure could prevent outdoor particles from entering the indoor environment (ASHRAE, 2007). Hence adopting the FFS with an appropriate fan speed could improve ventilation as well as prevent infiltration of outdoor particles from unfiltered routes.

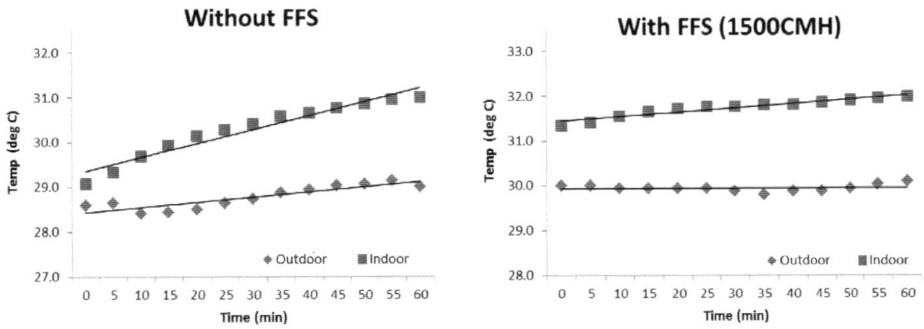

Figure 3: The build-up of heat in the enclosed classroom reduces with the use of FFS.

The build-up of heat in the enclosed classroom was studied and the results are shown in figure 3. Generally, the build-up of heat reduces with the use of FFS. As heat (thermal comfort) issues are closely related to some indoor pollutants such as CO_2, reduction in the build-up of heat would also imply reducing indoor pollutants as well (Chatzidiakou et. al. 2015).

We noted that the extent of any heat build-up would vary with different conditions, such as outdoor air temperature, indoor heat load, and ventilation rates. Hence, one possible improvement to the FFS would be to have a thermostat to control the air flow when required.

CONCLUSIONS

Although the closing of doors and windows may help to reduce an occupants' exposure to outdoor pollutants in a naturally ventilated space, long duration of closure could lead to a build-up of heat and indoor pollutants. Using simple fan filter systems, acceptable ventilation can be achieved in an enclosed room. Depending on outdoor temperature and occupancy rate, the fan filter system could effectively help to reduce build-up of heat and indoor pollutants in enclosed spaces. This study suggests that simple fan filter systems can be a cost-effective haze mitigation tool.

ACKNOWLEDGEMENT

The authors wish to acknowledge the contributions of Mr Thomas Chan and Mr Desmond Lim in this project. The project was funded by the National Environment Agency, Singapore.

REFERENCES

Apte MG, Fisk WJ and Daisey JM (2000) "Indoor carbon dioxide concentrations and SBS in office workers". In: Proceedings of Proceedings of Healthy Buildings, Vol. 1, pp. 133-138.

ASHRAE (2007) "ASHRAE standard 62.1. - Ventilation for acceptable indoor air quality". Atlanta, GA.

Chatzidiakou L, Mumovic D and Summerfield A (2015) "Is CO_2 a good proxy for indoor air quality in classrooms? Part 1: The interrelationships between thermal conditions, CO_2

levels, ventilation rates and selected indoor pollutants", *Building Services Engineering Research and Technology*, **0(0)** 1–33,

Chen C, Zhao B and Yang X (2011) "Preventing the entry of outdoor particles with the indoor positive pressure control method: analysis of influencing factors and cost", *Building and Environment*, **46**, 1167-1173.

Chithra VS and Shiva Nagendra SM (2012) "Indoor air quality investigations in a naturally ventilated school building located close to an urban roadway in Chennai, India", *Building and Environment*, **54**, 159-167.

Henderson SB, Brauer M, Macnab YC and Kennedy SM (2011). "Three measures of forest fire smoke exposure and their associations with respiratory and cardiovascular health outcomes in a population-based cohort". *Environmental Health Perspectives,* **119 (9)**, 1266.

Humphreys H (2004) "Positive-pressure isolation and the prevention of invasive aspergillosis. What is the evidence?", *Journal of Hospital Infection*, **56**, 93-100.

Rice N, Streifel A and Vesley D (2001). "An evaluation of hospital special-ventilation-room pressures", *Infection Control and Hospital Epidemiology,* **22 (1)**, 19-23.

Sekhar SC (2004). Enhancement of ventilation performance of a residential split-system air-conditioning unit. *Energy and buildings,* **36 (3)**, 273-279.

SPRING (2009). "Singapore Standard 554 (SS554: 2009) Code of practice for indoor air quality for air-conditioned buildings".

Wegesser TC, Pinkerton KE and Last JA (2009) "California wildfires of 2008: coarse and fine particulate matter toxicity", *Environmental Health Perspectives*, **117**, 893-897.

CFD MODELING AND PARAMETERIZATION OF THE PHOTOCATALYTIC OXIDATION PROCESS IN INDOOR ENVIRONMENTS

Eisaku SUMIYOSHI[1], Takahiro YAMAGUCHI[1], Kiyoshi YAMAMOTO[1], Eunsu LIM[2] and Kazuhide ITO[3,*]

[1] Interdisciplinary Graduate School of Engineering Sciences, Kyushu University, Japan
[2] Faculty of Science and Engineering, Toyo University, Japan
[3] Faculty of Engineering Sciences, Kyushu University, Japan

[*]Corresponding email: ito@kyudai.jp

Keywords: Photocatalytic oxidation, CFD, Chamber Experiment, Parameterization

SUMMARY

Indoor air quality affects the health and comfort of occupants. Poor indoor environmental conditions and gas-phase/aerosol-phase indoor contaminants are thought to be one cause of illnesses such as asthma and allergies, leading to significant productivity losses. In recent years, photocatalytic oxidation (PCO) processes have attracted attention, due to their potential to purify indoor air polluted with volatile organic compounds (VOCs), especially at low concentration levels. To date, TiO_2-bound building materials have been extensively studied for oxidation of indoor air pollutants. In this study, kinetic studies were carried out for photocatalytic oxidation of toluene in gas phase over TiO_2-bound building materials, using a 20 L small test chamber. Computational fluid dynamics (CFD) simulations were also carried out under the same boundary conditions as the experiments.

INTRODUCTION

Emissions of volatile organic compounds (VOCs) from building materials in indoor environments may result in deterioration of indoor air quality (IAQ). The photocatalytic oxidation (PCO) reaction has been actively recognized as a passive method that contributes towards resolving this issue, and it is consequently important that this is studied and practical applications developed. Recently, TiO_2-coated/bound building materials have been used indoors, and there are qualitative and quantitative reports of their performance with respect to indoor contaminant concentration reduction through the photocatalytic decomposition effect. In this study, kinetic studies were carried out for photocatalytic oxidation of gas-phase toluene over TiO_2-bound building materials, using a 20 L test chamber. Furthermore, in order to identify model parameters of the PCO process, computational fluid dynamics (CFD) simulations were performed using the same boundary conditions as the experiments.

METHODOLOGIES

Figure 1 shows the 20 L small test chamber developed in this research. A rectangular-shaped 20 L test chamber (with dimensions of 0.25 m × 0.25 m × 0.32 m) was constructed of stainless steel, in accordance with ISO 16000-9, and used to conduct contaminant concentration reduction performance experiments using photocatalytic building materials. TiO_2-coated ceramic tiles, prepared using a thermal spraying technique in the absence of a binder, were used as the photocatalytic building material. Two teflon-lined boxes on which building materials were placed (loading factor of 2.2 [m^2/m^3]) were installed in the test chamber. The light intensity on the surface of the test specimen could be adjusted between 600–10000 lx. A schematic of the 20 L test chamber procedure is shown in Figure 2. The experiment was conducted under controlled conditions: 28 ºC, 0% humidity, and an airflow rate of 0.5 ACH. Toluene concentration was controlled using a precise gas generator (Permeater PB-1B, GASTEC) and constant toluene concentrated air was supplied to the test chamber (ISO 16000-24). Toluene concentration was examined using gas chromatography (GC/FID) by collecting samples from the inlet and outlet.

RESULTS

Table 1 shows test conditions and toluene conversion. Figure 3 shows the results of photocatalytic degradation of toluene. Under illumination conditions of 600 lx and with low initial concentrations, photocatalytic degradation conversion was roughly 80%. In contrast, when initial concentration was relatively high, photocatalytic degradation conversions were approximately 70%. In the case of the highest initial concentration, toluene conversion was about 50%. Clear concentration-dependent effects were thus observed.

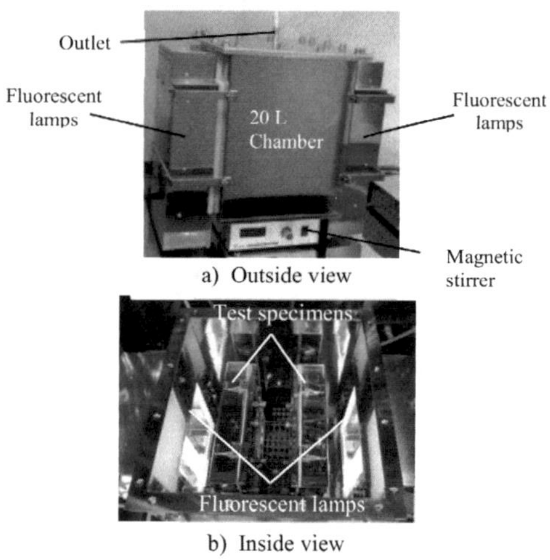

Figure 1. 20L test chamber

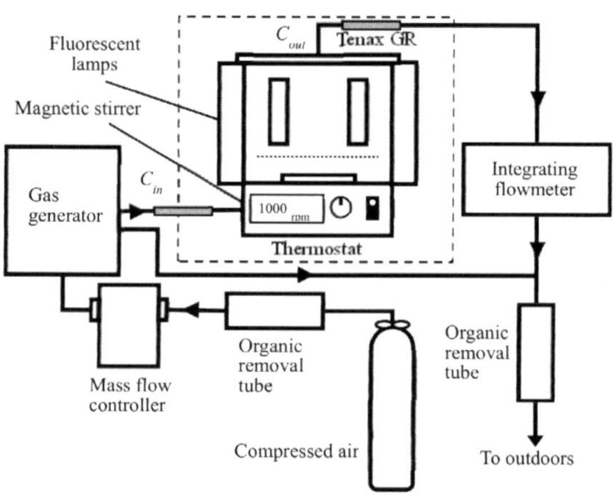

Figure 2. Schematic of the 20L chamber test

DISCUSSION

The rate of toluene photooxidation by TiO_2 can be expressed as per the simplified Langmuir-Hinshelwood equation. Based on the reciprocal plots between toluene concentration and the rate of toluene photooxidation, the parameters *k* and *K* were determined by fitting the plots to the equation, as follows:

$$r = k\frac{KC}{1+KC} \quad (1), \qquad \frac{1}{r} = \frac{1}{kK}\frac{1}{C} + \frac{1}{k} \quad (2), \qquad r = \frac{(C_{in} - C_{out})G}{A} \quad (3)$$

where r is the reaction rate [kg/m^2/s], k is the reaction rate constant [kg/m^2/s], K is the adsorption equilibrium constant [m^3/kg], C is toluene concentration in [kg/m^3], C_{in} and C_{out} are inlet and outlet toluene concentrations [kg /m^3], G is the volumetric flow rate [m^3/s], and A is the reaction surface area [m^2].

As shown in Figure 4, a good linear relationship is observed between initial toluene concentration and reaction rate. Based on the plots, the kinetic parameters *k* and *K* were estimated to be 4.85×10^{-11} [kg/m^2/s] and 1.16×10^{6} [m^3/kg], respectively.

These L-H model parameters were estimated under the assumption of perfect mixing in the test chamber, and hence CFD simulation was carried out in order to estimate non-uniformity of flow and toluene concentration distribution in the chamber. Figure 5 shows the test chamber model for analysis. Table 2 summarizes numerical and boundary conditions of CFD analysis. The boundary conditions used for the CFD simulations were precisely the same as those for experimental conditions. The effect of visible light on the reduction of pollutant concentrations was not explicitly modelled. The flow field was created by a stirring fan in order to increase momentum on the surface of the photocatalytic building materials. Eq. (2) was incorporated as a sink term to the first cell attached to the building material surface (inside the viscous sub-layer) and others were set to zero.

Table 1. Conversion of toluene over various building materials.

Analysis Case	C_{in} [μg/m^3]	C_{out} [μg/m^3]	η [-]	r [μg/m^2/s]
1	804	390	0.51	0.027
2	472	185	0.61	0.019
3	379	118	0.69	0.017
4	131	41	0.69	0.006
5	128	41	0.68	0.006
6	78	10	0.88	0.004
7	76	15	0.80	0.004

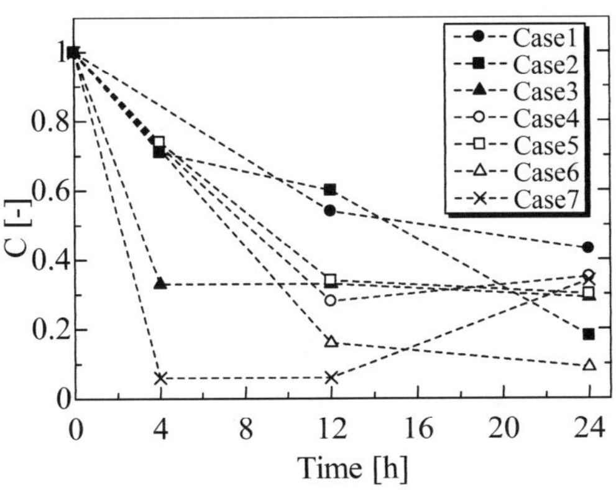

Fig. 3. Photocatalytic degradation of toluene with time at various initial concentrations.

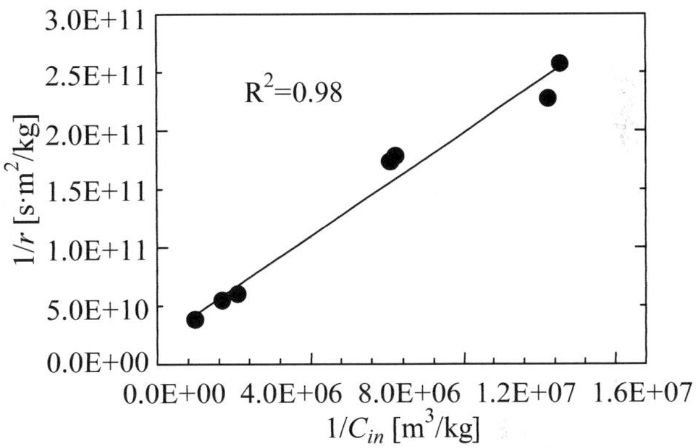

Fig. 4. Plot of r^{-1} against C_{in}^{-1} for toluene.

Figure 6 shows airflow distribution in the vertical section of the test chamber. Relatively high air velocities were created by the stirring fan at the bottom on both sides of the test chamber. Additionally, circulating airflows were observed in the central region between building materials and the test specimen. Figure 7 shows distributions of contaminant concentration along a vertical section through the central line of the test specimens. Figure 7(a) shows toluene concentration distributions in the test chamber when inlet toluene concentration was 260 [μg/m^3] and L-H model parameters were $k = 4.85 \times 10^{-11}$ and $K = 1.16 \times 10^6$, as estimated through experiments. In consequence, outlet concentration was 132 [μg/m^3], relatively low compared with experimental results (C_{out} in the experiment was 85 [μg/m^3]). The kinetic parameters k and K were re-determined using CFD analysis and the value of k obtained from CFD analysis was 1.48×10^{-10} [kg/m^2/s], approximately three times larger than that obtained experimentally. Figure 7 (b) shows toluene concentration distributions by using the re-defined k value obtained through CFD analysis.

CONCLUSIONS

In this contribution, we have proposed a method for kinetic analysis of the heterogeneous photooxidation of toluene in air. CFD analysis with a low Reynolds number-type k-ε turbulence model was applied to heterogeneous toluene photooxidation over a TiO_2-bounded building material, and kinetic parameters were determined.

Table 2. Outline of CFD analysis

Geometry	0.25 m(x) × 0.25 m(y) × 0.32 m(z)
Meshes	306,456 Unstructured
Turbulence Model	Low Re type k-ε model (Abe-Kondoh-Nagano Model)
Scheme	Convection term: QUICK
Inflow Boundary	Area: 0.005 m(x) × 0.005 m(y), U_{in}= 0.1111 m/s (0.5 ACH), TI = 10%, $k_{in} = 3/2 \times (U_{in} \times 0.01 \cdot TI)^2$, $\varepsilon_{in} = C_\mu \times k_{in}^{3/2}/l_{in}$, $C_\mu = 0.09$
Outflow Boundary	Area: 0.005 m(x) × 0.005 m(y), U_{out}: Free slip, k_{out}: Free slip, ε_{out} : Free slip
Wall Treatment	Velocity: No slip, Contaminant concentration: $\partial C/\partial x = 0$
Contaminant	Passive contaminant

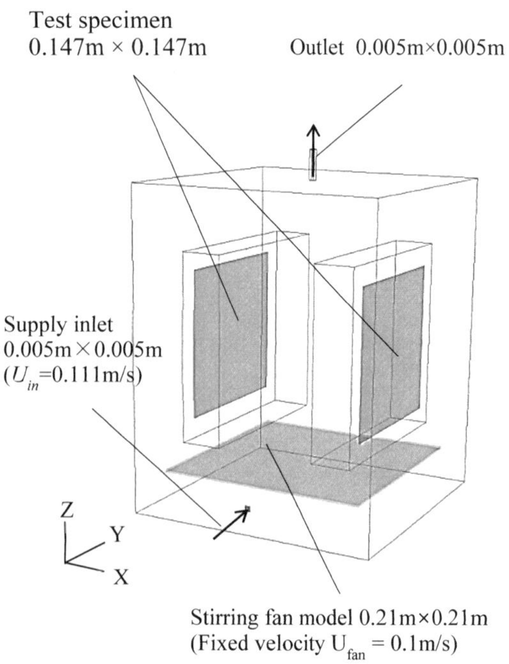

Fig. 5. The Diagram of the 20L test chamber model.

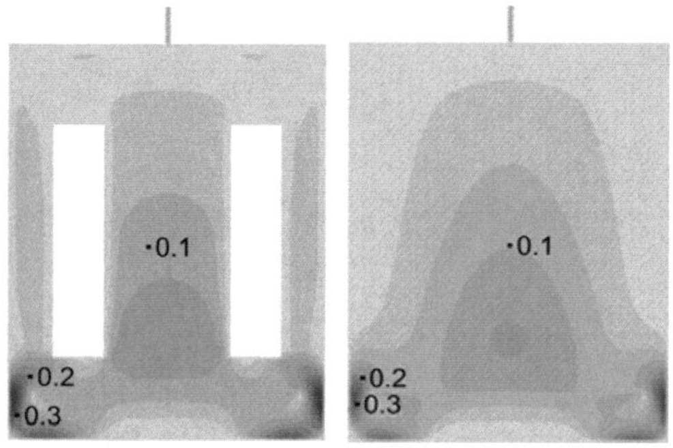

Fig. 6. Flow field in the central section 20 L small test chamber model [Unit: (m/s)].

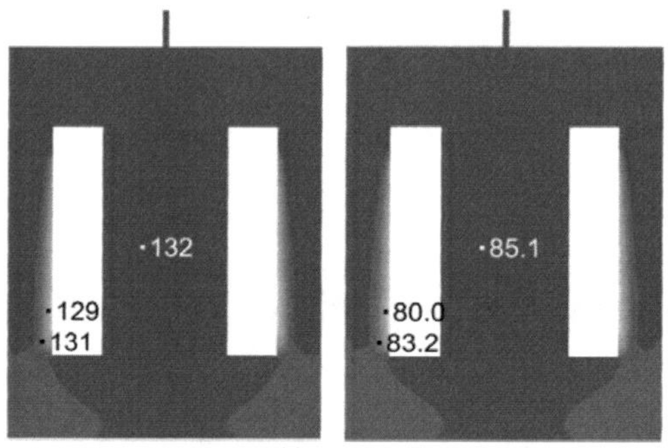

(a) k =4.85×10^{-11} (b) k =1.48×10^{-10}

Fig. 7. Distribution of contaminant concentration in the central section in the small test chamber model [Unit: (μg/m^3)].

REFERENCES

ISO 16000-9:2006: Indoor air -- Part 9: Determination of the emission of volatile organic compounds from building products and furnishing -- Emission test chamber method

ISO 16000-24:2009: Indoor air -- Part 24: Performance test for evaluating the reduction of volatile organic compound concentrations by sorptive building material

Hisahiro Einaga, Junnya Tokura, Yasutake Teraoka, Kazuhide Ito (2015) Kinetic Analaysis for TiO_2-Catalyzed Heterogeneous Photocatalytic Oxidation Processes for Ethylene using Computational Fluid Dynamics : Chemical Engineering Journal, 263, pp325-335

Kazuhide Ito, Eunsu Lim, Kiyoshi Yamamoto (2014) Small Test Chamber Experiment and Modeling of Photocatalytic Oxidation of VOCs under Indoor Environmental Conditions: Indoor Air 2014, Paper ID: HP0330

A MULTI-SCALE EXPOSURE CONCENTRATION ANALYSIS IN A LARGE FACTORY SPACE USING COMPUTATIONAL FLUID DYNAMICS TECHNIQUE

Alicia MURGA[1], Sung-Jun YOO[1], Kazuhide ITO[1,*]

[1]IGSES, Kyushu University, Fukuoka, Japan

[*]Corresponding email: ito@kyudai.jp

Keywords: CFD, Multi-nesting, Exposure concentration

SUMMARY

Indoor Air Quality (IAQ) can play a significant role in the well-being of residents, and with this consideration, airborne routes of exposure become critically important for the evaluation of health risks associated with IAQ. In particular, in the viewpoint of the Health Impact Assessment on industry workers, the management of IAQ within factory settings is identified to be an essential factor affecting industrial hygiene. In recent years, the Computational Fluid Dynamics (CFD) technique has been practically applied to indoor environmental design, with the potential to enhance the field of industrial hygiene. In this paper, an integrated simulation procedure to predict the exposure concentrations of hazardous occupational chemical compounds is developed. The concerned technique uses a multi-nesting method to connect factory building spaces and the microclimate around a human body and respiratory tract. This method was proposed in order to provide detailed quantitative and qualitative information to estimate contaminant doses applied to workers.

INTRODUCTION

Workers in the industrial sector may spend more than seven hours a day in their locations of work, which is often an enclosed venue. Consequently, the quality of the indoor air environment plays a significant role in the health of these workers. According to Industrial Hygiene, significant occupational health hazards arise from the inhalation of chemical agents and furthermore, their impacts on the human body depend on the degree and time span of exposure. This is the case with respect to solvents, which are broadly used within the manufacturing sector[1-2].

With these considerations, it is critical to establish a safer indoor environment through the prediction and management of contaminant concentrations and hazards. This is particularly the case in countries such as El Salvador, where industries like textile manufacturing and silk printing (both of which utilize solvents) are rapidly expanding, and sustainable growth is necessary[3]. Moreover, in El Salvador, 19% of employed men and 11% of employed women are exposed to chemical agents in their locations of work, and approximately 7% of these workers directly handle these agents. In addition, 52% of the working population does not use protective gear to defend against occupational hazards[4]. Therefore, the prediction of the variations and distribution of a target contaminant (particularly solvents commonly used within the country) around operative spaces is essential to improve and control indoor air quality.

The objective of this study is to implement a numerical approach to CFD, in order to predict the exposure concentration of chemicals within the silk printing industry, with emphasis on the widely used solvent cyclohexanone. The inhalation exposure concentration of a representative worker in a factory is analyzed through a downscaling technique from factory scales to respiratory tract scales by use of a computer-simulated person.

METHODOLOGY

The downscaling of analyses from factory building scales to human respiratory tract scales were carried out by the present study. With respect to downscaling, three types of analytical regions were set for the multi-scale nesting analysis: factory building scale (Region 1), human scale (Region 2), and respiratory airway tract scale (Region 3). The factory building space had X,Y dimensions of 55.70 m x 49.10 m, and a total floor area of 2452.22 m^2. For the Z dimension, the highest point measured 18.00 m at the center of the factory floor (24.55 m), while the lowest point at the abode walls measured 15.00 m in height, with the factory having a gable roof design. In addition to the relevant factory spaces where workers labor and commute, the machines commonly used in textile printing and manufacturing factories (Kinzel one oven, Kinzel two ovens, Sakurai 666SD) were also added to the building space. Spaces involved with picture processing, looms, lockers, and bathrooms were eliminated from the analytical space because they were assumed to have an insignificant impact on the calculations, as was indicated by a preliminary simulation. Further clarifications are shown in Figure 1(1).

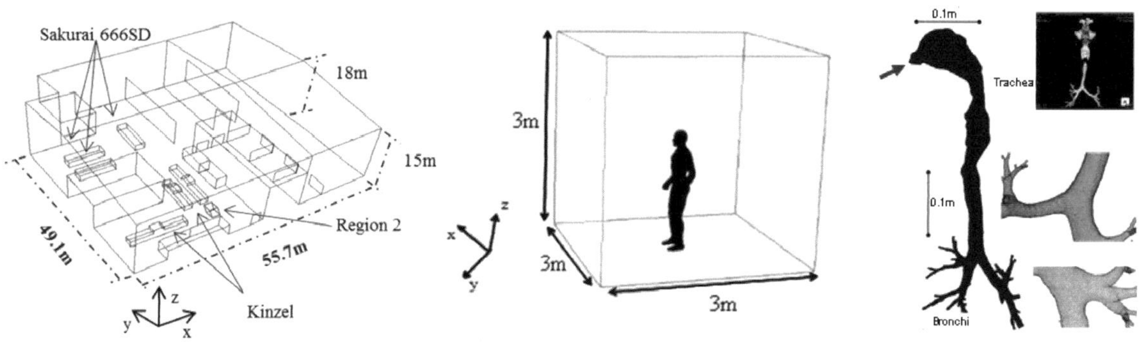

(1) Region 1 (2) Region 2 (3) Region 3

Figure 1 Design outline of the factory, computer simulated person, and airway model, respectively

The computer simulated person (CSP) was arranged in Region 2, and an analytical domain of dimensions x = 3.0 m, y = 3.0 m, and z = 3.0 m was set by centering on the CSP. The design of Region 2 is denoted in Figure 1(2), and detailed design of the human respiratory tract model is shown in Figure 1(3).

In order to connect these different analytical domains, a one-way nesting method was applied. From Region 1 to Region 2, distributions of average velocities (U, V, and W), turbulent properties (k and ε), and unsteady contaminant concentration distribution data (C) were transferred. Each portion of data passing from Region 1 to Region 2 was linearly interpolated, since the grid size was markedly different on the boundary plane of both regions. In Region 2 and Region 3, the flow field near the nostrils was assumed to be almost completely dependent on the respiratory cycle of the human body. Consequently, the inhaled and exhaled airflow as an inlet velocity of the boundary plane in Regions 2 and 3 was calculated from the unsteady respiratory cycle model. Furthermore, unsteady and non-uniform contaminant concentration data from the nostril surface was transferred from Region 2 to Region 3. Concerning flow

and contaminant concentrations, a linear interpolation between Regions 2 and 3 was applied for the downscaling transfer.
The commercial CFD code ANSYS/Fluent 14.0 was used to calculate the flow field and contaminant distributions.

RESULTS AND DISCUSSION

The flow field, temperature, and contaminant distribution results from the CFD simulation are shown in Figure 2, with regard to the overall space of the factory (Region1). The non-uniform and imperfect mixing of flow, temperature, and contaminant concentrations in the factory space were confirmed.

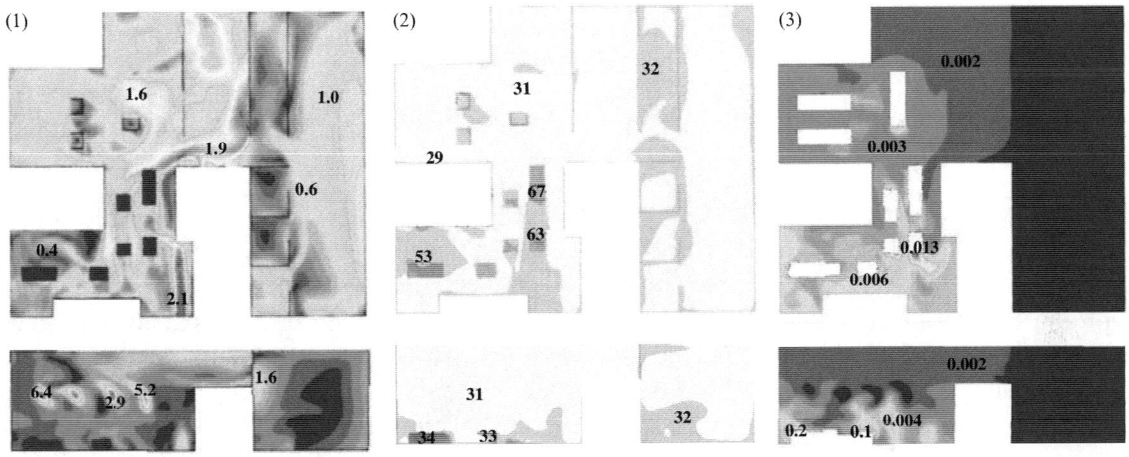

Figure 2 (1) Horizontal and vertical distributions of air velocity (m/s), (2) horizontal and vertical distributions of temperature (°C), and (3) horizontal and vertical distributions of contaminant concentration (C), respectively (Region 1)

The flow and contaminant concentrations around the human body (Region 2) are shown in Figure 3. In Figure 4, the flow and contaminant concentration distributions in the respiratory tract are shown. A complicated flow field was formed in the airway model due to the complex geometry. In this analysis, a perfect sink wall surface condition inside the airway model was used and therefore, contaminants transported through the nostril surface by nesting from Region 2 to Region 3 was adsorbed/deposited at the inner wall of the nasal cavity, and was not transported as far as the bronchus.

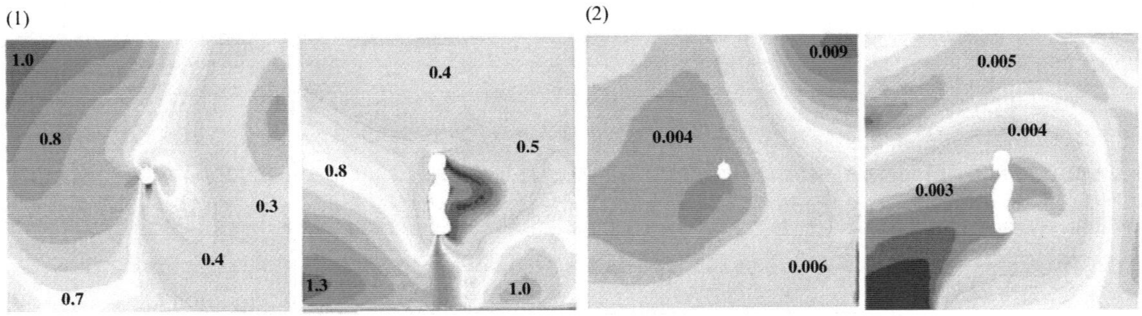

Figure 3 (1) Horizontal and vertical distributions of air velocity (m/s) and (2) horizontal and vertical distributions of contaminant concentration (C) around the human body, respectively (Region 2)

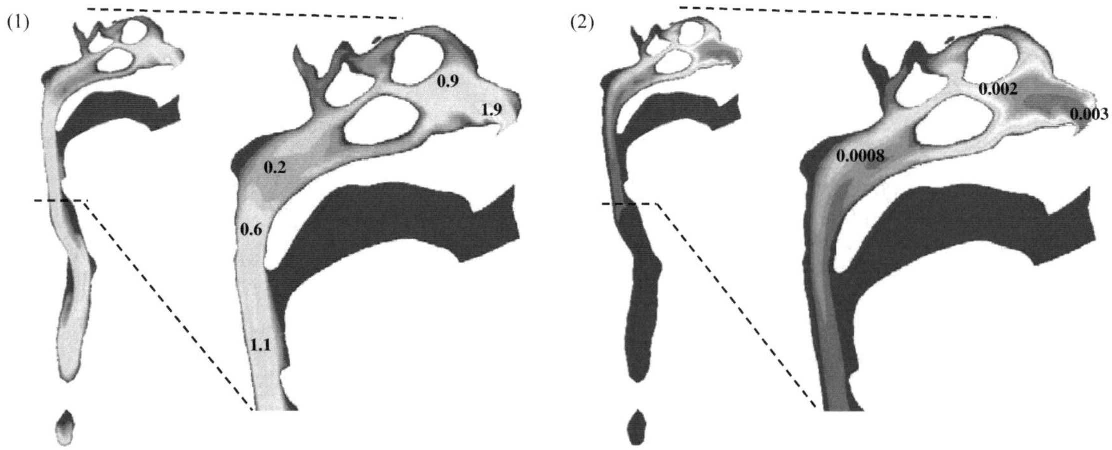

Figure 4 (1) Air velocity distribution (m/s) and (2) contaminant concentration (C) distribution in the human respiratory tract (Region 3)

CONCLUSIONS

Based on the CFD multi-stage exposure concentration analysis, detailed information from non-uniform concentration distributions in a large factory space can be used to determine precise levels of inhalation exposure concentrations for a specific worker. Results are expected to show a direct correlation between the worker's location in the factory space and the concentration level within the respiratory tract. This numerical method can not only improve the prediction accuracies of inhalation exposure analyses and the indoor environmental design of factories, but also assist the Health Impact Assessment in an enclosed space.

REFERENCES

Barbara A. Plog and Patricia J. Quinlan. *Fundamentals of Industrial Hygiene*. National Safety Council. USA, 2002. 5th edition. Pgs. 3–6.

Chamber of textiles, clothing and tax-free zones of El Salvador, CAMTEX. *Textile sector profile, clothing and tax-free zones, 2014 report.* San Salvador, El Salvador, 2014. Pgs. 3–5 [in Spanish].

Foundation Sustain Labour. *Chemical substances in El Salvador. Main exposures by industry, regulations, workers and environmental impact, 2012 report.* El Salvador, 2012. Pgs. 13–14 [in Spanish].

Legislative body of El Salvador. *Legislative Decree: General Law of risk prevention inside the workplace N° 254.* Official Journal 82. Tome 387. San Salvador, El Salvador, 2010. Pg 1 [in Spanish].

Kazuhide Ito: Integrated Numerical Approach of CFD and Epidemiological Model for Multi-scale Transmission Analysis in Indoor Spaces: ***Indoor and Built Environment,*** 2014; 23 (7) pp 1029-1049

Towards assessing the viability of the indoor microbiome using flow cytometry

Betancourt Román, C. M.[1], Hartmann, E.M. [1], and Green J.L. [1,2]
(1) Biology and the Built Environment Center, Eugene, OR; (2) Institute of Ecology and Evolutionary Biology, University of Oregon, Eugene, OR

People spend most of their time inside built environments, where they are in continuous contact with the indoor microbiome. Many of these indoor microorganisms can been found in dust. Indoor dust is a mixture of organic and inorganic materials consisting principally of animal fibers, pollen, clay, fungi, bacteria and human skin cells (Lioy et al., 2002). Microbial diversity in dust has been described in several indoor environments (Kembel, et al., 2014; Dannemiller et al., 2015). Previous studies have shown that architectural features such as the building ventilation system have a strong influence on microbial community composition (Kembel et al., 2014). Despite an increase in indoor microbiome studies, there is a knowledge gap regarding the viability of microorganisms in dust and whether they are actively interacting with the indoor environment. In other environmental systems, methodologies like flow cytometry have been developed to identify and categorize microbial cells as a function of viability (Veal, et al. 2000). However, none of these techniques have been applied to dust samples.

In addition to harboring microbes, dust is laden with bioactive chemicals, such as endocrine disruptors, that negatively impact human health (Hwang, et al., 2008). People introduce many chemicals to the built environment through cleaning supplies and other human-made products. It remains unclear if and how these chemicals are interacting with the indoor microbiome, and what effects, if any, those interactions have on human health. One step towards studying these potential interactions is to quantify and identify which microorganisms are alive or dead in dust. To meet this need, we developed a standardized protocol to quantify the viability and composition of the microbiome in dust samples using flow cytometry. In addition, we developed tools to identify which microorganisms were actively interacting with the built environment, including the chemicals contained within.

Indoor dust samples were collected from office spaces using a vacuum. Samples were kept at room temperature until they were processed in the lab. All dust samples were mixed, and 0.25-g and 1-g aliquots of each sample were individually homogenized with 10 mL of PBS buffer, sonicated in a sonicating bath for 10 min to release any bacterial cells from dust clusters, and concentrated with a 100,000 molecular weight cut-off (MWCO) filter. The concentrated solution was diluted to 5% and 1% prior to flow cytometry. To evaluate the viability of microbial cells, 1.5 μL each of the SYTO9 and propidium iodide dyes from a Live/Dead BacLight bacterial viability kit (Molecular Probes) was added to 987 μL of the diluted samples. Samples were analyzed on a Sony Cell Sorter SH800. The sort criteria were defined by drawing gates in histograms using SH800 software. The sorting speed was maintained between 3,000 and 5,000 cells/s.

Preliminary results show that the indoor dust has higher counts of dead bacterial cells than live/viable bacterial cells. Dead and live bacterial populations in dust were detected in both undiluted and diluted samples. These results suggest that flow cytometry can be used to assess microbiome viability in indoor dust. However, further analysis of additional samples is needed to improve the reproducibility of this method.

References

Alenius, H., Pakarinen, J., Saris, O., Andersson, M. A., Leino, M., Sirola, K., Majuri, M.-L., Niemelae, J., Matikainen, S., Wolff, H., von Hertzen, L., Maekelae, M., Haahtela, T. & Salkinoja-Salonen, M. (2009). Contrasting Immunological Effects of Two Disparate Dusts - Preliminary Observations. International Archives of Allergy and Immunology 149: 81-90, doi: 10.1159/000176310.

Dannemiller, K.C., Leaderer, B.P., & Peccia, J. (2015). Influence of housing characteristics on bacterial and fungal communities in homes of asthmatic children. Indoor Air, doi: 10.1111/ina.12205.

Hwang, H.-M., Park, E.-K., Young, T. M. & Hammock, B. D. (2008). Occurrence of endocrine-disrupting chemicals in indoor dust. Science of the Total Environment 404: 26-35, doi: 10.1016/j.scitotenv.2008.05.031.

Kembel, S. W., Meadow, J. F., O'Connor, T. K., Mhuireach, G., Northcutt, D., Kline, J., Moriyama, M., Brown, G. Z., Bohannan, B. J. M. & Green, J. L. (2014). Architectural Design Drives the Biogeography of Indoor Bacterial Communities. Plos One 9, doi: 10.1371/journal.pone.0087093.

Lioy, P. J., Freeman, N. C. G. & Millette, J. R. (2002). Dust: A metric for use in residential and building exposure assessment and source characterization. Environmental Health Perspectives 110: 969-83.

Author Name	Session Number	Presentation Number
Abarr, Miles	M.2	294
Adamkiewicz, Gary	M.1	205
Adamovský, Daniel	M.P	194
Adams, Rachel	W.1, M.P	39, 156, 165
Alawadhi, Esam	W.2	15
Ali, Akram	M.7	210
Ang, Keng	T.P	256
Angenent, Largus	W.1	258
Arent, Chuck	M.5	247
Aubin, Daniel	T.2	209
Aydogan Akseli, Ahu	T.P	180
Azimi, Parham	M.4, M.7, T.P	150, 164, 211
Barnett, Claire	W.3	25
Barraza, Francisco	M.1	226
Bean, Robert	W.2	248
Bennett-Fripp, Mae	M.1	205
Berg, Gabriele	W.1	149
Bernal, Sandra	M.6	33
Bernheim, Anthony	T.5	244
Betancourt Román, Clarisse	T.P	293
Bhattacharjee, Suchismita	M.5, M.P	191, 192
Bi, Chenyang	T.1	55
Black, Craig	M.2	10
Blaser, Martin	T.4	228
Bo-ot, L.M.T	T.6	22
Bobenhausen, Catherine	M.2	215
Bohannan, Brendan	T.P	293
Borisová, Lucia	M.P	13
Bornehag, Carl-Gustaf	T.2	234
Boulton, Roger	M.7, M.P	233, 235
Bramwell, Lisa	T.1	153
Breen, Michael	M.3	45
Brennan, Terry	T.2, T.5	44, 230
Brock, Geoffrey	M.2	215
Brown, G.Z.	T.4	204
Brown, Steven	T.P	232
Buechlein, Melissa	T.1	49
Buglack, Nick	T.P	218
CAI, Hongyi	M.P	185
Cali, Salvatore	T.5	250
Callinan, Tim	M.P	225
Candee, Kent	M.2	10
Cao, Qingliang	W.3	176
Cavallin, Humberto	T.4	228
Champion, Wyatt	M.1	226

Chang, Jae D.	M.P, T.P	185, 186
Chang, Victor	M.4, M.6, W.3, T.P	41, 42, 157, 171, 176
Chang, Wei-Chung	T.P	256
Chao, Christopher	T.P	54
Charles, Antoinette	M.1	208
Chen, Ailu	W.3	176
Chen, Bin	M.1, M.P	51, 175
Chen, Wenhao	M.P	273, 283, 284
Chen, Yu	M.P	51
Cheng, Zhongqi (Joshua)	M.2	215
Cheong, Kok	T.P	256
Choi, Hyunok	T.2	234
Choi, Joon-Ho	W.2	201
Chung, Okyoung	M.P	36
Clements, Nicholas	T.3	200
Cline, Patricia	T.1	28
Colton, Meryl	M.1	205
Connor, Angelica	M.1	187
Cox, Steven	M.2, M.7	221, 222
Crawford, Todd	W.3	19
Crutcher, Ernest	M.3	24
Crutcher-Bettes, Heidie	M.3	24
Dang, Rui	T.P	23
Davis, Devra	W.3	227
Debose, Deion	M.7	210
DeCarlo, Peter	T.1	161
Dedesko, Sandra	M.4, T.2	207, 214
Delghust, Marc	M.5	163
Dia, E.V.H.	T.6	22
Dobiášová, Lucie	M.P	194
Dodson, Robin	M.1	189
Dominguez-Bello, Maria Gloria	T.4	228
Dong, Boyang	M.7	210
Downey, Damian	T.2	148
Dyson, Anna	T.P	180
Eagan, Aaron	T.3	200
Edwards, Marc	W.1	253
Eiguren-Fernandez, Arantzazu	T.P	218
Emmerich, Steven	M.5	245
Enomoto, Hikaru	W.2	174
Fazli, Torkan	M.7, W.2	167, 210
Ferro, Andrea	M.2, T.1	153, 213
Fowles, Jeff	M.P	273, 283, 284
Francisco, Paul	T.5, T.P	250, 251
Freihaut, James	T.P	229
Fu, Sau Chung	T.P	54
Fugler, Don	T.2	209

Fukagawa, Kenta	M.P	16, 151
Gall, Elliott	T.P	42
Garnys, Vyt	M.P	225
Gatland, Stanley	M.2	196
Gauvin, Denis	T.2	209
Gehring, Douglas	M.2	196
Gibler, Madison	T.1	49
Gingras, Veronique	T.2	209
Gloss, Stacy	T.5, T.P	250, 251
Green, Jessica	T.4, T.P	204, 293
Grimes, Carl	T.4	239
Guo, Deheng	T.P	23
Gutierrez, Ramona	T.P	256
Haaland, Daniel	M.4	47
Habibi, Shahryar	W.2	242
Ham, Jason	T.1	152
Harriman, Lewis	T.2	44
Harrison, Joel	T.1	152
Hartmann, Erica	T.P	293
Hasegawa, Takayuki	T.2	37
Haverinen-Shaughnessy, Ulla	T.P	170
Havick, Dave	M.2	10
He, Congrong	M.P	173
He, Guoqing	T.P	212
Hering, Susanne	T.P	218
Hertzberg, Jean	T.3	2
Himpe, Eline	M.5	163
Hodgson, Alfred	T.5, M.P	244, 273, 283
Hogan, Maggie	M.5	191
Horner, Sharon	M.4	214
Hospodsky, Denina	W.1	258
Huan, Joseph	M.7	210
Hyvarinen, Anne	W.1	165
Ikeda, Koichi	W.2	174
Ito, Kazuhide	M.P, T.P	273, 283, 284, 285, 286
Jackson, Stephen	T.1	152
Jacobs, David	T.5	250
Jayaprakash, Balamuralikrishna	W.1	165
Jeong, Su-Gwang	M.P	36
Johanson, Gunnar	M.P	162
Johnson, Anita	T.1	161
Johnson, Bart	T.4	204
Johnson, Graham	M.P	173
Jorquera, Héctor	M.1	226
Jung, Hyounduk	T.P	254, 255
Juran, Stephanie	M.P	162
Jurng, Jongsoo	T.P	254, 255

Kaminsky, Amanda	M.2	215
Kane, John	M.1	205
Karvonen, Anne	W.1	165
Kato, Shinsuke	M.P	52
Keady, Patricia	T.P	218
Khoury, Cheryl	M.P	184
Killeen, David	M.7, M.P	233, 235
Kim, Minsu	T.P	254, 255
Kim, Sumin	M.P	36
King, Marco-Felipe	M.P, T.3	182, 193
Kinney, Kerry	M.4	214
Kirjavainen, Pirkka	W.1	165
Klepeis, Neil	T.4	199
Knibbs, Luke	M.P	173
Knoesen, André	M.7, M.P	233, 235
Kondo, Emi	M.P	16
Konis, Kyle	W.2	201
Korff, Alexander	T.1	49
Krejcirikova, Barbora	T.P	212
KREMER, Paul	T.P	229
Kumagai, Kazukiyo	M.P	273, 283, 284
Kunkel, Stephanie	M.7	211
Kurazumi, Yoshihito	M.P	16, 151
Ladau, Joshua	T.4	204
Lajoie, Pierre	T.2	209
Langar, Sandeep	M.5	191
Larsson, Lennart	M.3	143
Lau, Josephine	W.3	57
Laurent, Jose	M.1	205
Laverge, Jelle	M.5	163
Leclerc, Jean-Marc	T.2	209
Lee, Jeehwan	M.P, T.P	185, 186
Li, Ning	T.P	218
Li, Po-Han	T.P	243
Liang, Yirui	M.2	20
Licina, Dusan	T.1	53
Lim, Eunsu	T.P	285
Lin, Chenyi	T.P	23
Little, John	M.2, M.7	221, 222
Liu, Gang	M.2, T.P	23, 222
Logue, Jennifer	T.4, T.5	198, 199
Long, Thomas	M.3	45
Lorentzen, Johnny	M.P	162
Lu, Yang	T.P	178
Luhung, Irvan	M.4	157
Luongo, Julia	W.1	168
Lyu, Kun-Hua	T.P	243

Ma, Yan	M.2	213
MacNaughton, Piers	M.1	205
Madrid, Nic	M.P	233
Madrid, Nicolas	M.7	235
Maestre, Juan	M.4	214
Mahnert, Alexander	W.1	149
Mainster, Sydney	M.2	215
Markowicz, Pawel	M.3	143
Marr, Linsey	M.2, M.7	221, 222
Matsuo, Toshiki	M.P	273, 283, 284
Matteazzi, Megan	M.P	223
Mattsson, Johan	M.3	143
Mauney, Denise	M.2	294
McGill, Gráinne	T.2	148
Mcneill, James	T.3	2
Meadow, James	T.4	204
Mei, Xiong	M.2	240
Melikov, Arsen	T.1	53
Merrin, Zachary	T.5, T.P	250, 251
Meyer, Michael	T.2, M.P	30, 34
Mhuireach, Gwynne	T.4	204
Miller, Shelly	T.3, W.1	168, 200
Mills, David	M.7	235
Mochungong, Peter	T.6	26
Moissl-Eichinger, Christine	W.1	149
Montoya, Lupita	M.1, M.2, M.P, T.P	180, 223, 226, 294
Moore, John	T.2	148
Morawska, Lidia	M.P	173
Morrison, Glenn	T.1	49
Muller, Christopher	M.5	247
Müller, Henry	W.1	149
Murga, Alicia	T.P	286
Nazaroff, William	M.4, M.6, T.P	41, 42, 157
Ng, Lee	T.P	256
Nguyen, Mytien	W.1	258
Nice, Jako	T.3	287
Noakes, Catherine	M.P, T.3	182, 193
Noble, Douglas	W.2	201
Noonan, Jack	M.P	225
Nordin, Steven	M.P	162
Novoselac, Atila	T.4	228
O'Brien, Julie S.	T.6	26
Odhiambo, George	W.3	249
Ortega, Roscel	W.1	149
Oyedele, Lukumon	T.2	148
Park, Eunseuk	T.P	254, 255
Paulino, Luciana	T.4	228

Peccia, Jordan	T.P	170
Pekkanen, Juha	W.1	165
Pereira, Henrique	T.4	228
Pereira, Marcelo	M.P	173
Perez, Jean	T.4	228
Persily, Andrew	M.5, T.5	244, 245
Poppendieck, Dustin	M.1,T.6	187, 190
Prawiradiraja, Theodore	T.P	256
Probst, Alexander	W.1	149
Pruden, Amy	W.1	253
Rakotomahefa, T.M.J	M.P	32
Reisdorph, David	T.P	170
Rhoads, William	W.1	253
Richmond-Bryant, Jennifer	M.3	45
Riley, Karl	W.3	227
RIM, Donghyun	T.P	229
Rios, Raphael	T.4	228
Roberts, Paul	T.P	232
Rode, Carsten	T.P	212
Rose, William	T.5, T.P	250, 251
Royse, Kelly	M.4	214
Rudel, Ruthann	M.1	189
Ruiz, Jean	T.4	228
Sakoi, Tomonori	M.P	16
Salimifard, Parichehr	T.P	229
Salthammer, Tunga	M.3	147
Sang, Yizhou	M.P	51
Sankaran, Gayatri	T.P	256
Schleibinger, Hans	M.P, T.2	184, 209
Schmidbauer, Norbert	T.2	234
Schultz, Bradley	M.3	45
Schulz, Nicole	M.3	147
Scott, Krista	M.2	10
Seeger, Stefan	T.6	181
Sekhar, Chandra	T.1, T.P	53, 256
Sharpe, Tim	M.1, T.2	148, 208
Shaughnessy, Richard	T.P	170
Shen, Joanna	T.P	256
Sherman, Max	T.5	198
Shu, Yuhang	M.P	51
Shusterman, Dennis	M.P	273, 283
Siegel, Jeffrey	M.1, M.4, T.2	47, 166, 207, 214
Singer, Brett	T.4, T.5	198, 199
Sleigh, P. Andrew	T.3	193
Spalter, Ronald	M.2	215
Spengler, John	M.1	205
Spilak, Michal	M.P, W.1	39, 156

Spirito, Catherine	W.1	258
Srubar, Wil	M.2	294
Stadtner, Alex	W.3	227
Stein, Antoinette	W.3	227
Stephens, Brent	M.1, M.4, M.7, T.P, W.2	150, 154, 164, 167, 210, 211
Stolte, Ramona	M.3	147
Su, Chunxiao	W.3	57
Subhash, Shobha	T.3	200
Sumiyoshi, Eisaku	T.P	285
Suwongso, Christella	M.2	294
Sylvain, Iman	M.P, W.1	39, 156
Takaono, Hitoshi	W.2	174
Tan, Clement	T.P	256
Tan, Ron	T.P	256
Tanaka, Kosuke	T.2	37
Taubel, Martin	W.1	165
Tham, Kwok Wai	T.1, T.P	53, 256
Tian, Yilin	M.2, T.1	153, 213
Tobita, Kunihito	M.P	16, 151
Touchie, Marianne	M.1	166
Udesky, Julia	M.1	189
Uhde, Erik	M.3	147
Vaillant Rebolar, Julio	M.5	163
Vaishampayan, Parag	W.1	149
Valouskova, Kristyna	M.P	18
Van de putte, Stijn	M.5	163
Vaughn, David	T.P	232
Vejerano, Eric	M.2	222
Venkateswaran, Kasthuri	W.1	149
Vennekens, Davy	M.2	196
Vosloo, Piet	T.3	287
Wagner, Danielle	M.2	215
Wang, Ping	M.P	273, 283, 284
Wang, Qiuyue	M.P	52
Wang, Shugang	M.2, M.P	32, 240
Waring, Michael	M.P, T.1, W.1	39, 156, 159, 161
Wells, J.R.	T.1	152
Wilke, Olaf	T.6	181
Williams, Ronald	M.3	45
Won, Doyun	M.6, M.P, T.2	29, 184, 209
Wong, Joan	M.P	184
Wrobetz, Anne	M.P	223
Wu, Yan	M.4, T.P	157, 171
Wu, Yaoxing	M.2, M.7	221, 222
Wu, Yu-Lieh	T.P	243
Xie, Mingjie	M.2, M.7	221, 222
Xu, Ying	M.2, M.4, T.1	20, 55, 214

Yamagishi, Hiroshi	T.2	37
Yamaguchi, Takahiro	T.P	285
Yamamoto, Kiyoshi	T.P	285
Yamashita, Masato	M.P	284
Yamato, Yoshiaki	M.P	16, 151
Yanaso, Satomi	T.2	37
Yang, Wenping	M.P, T.2	184, 209
Yang, Yanan	T.1	159
Yao, Maosheng	T.P	171
Yap, Hooi	T.P	256
Ye, Wei	M.6	29
Yeap, Rou Yi	W.2	167
Yi, Bo	W.2	201
Yoo, Sung-Jun	M.P, T.P	284, 286
Yoshino, Hiroshi	T.P	178
Yu, Fang	W.3	57
Yuen, Wai T	T.P	54
Zhai, Zhiqiang (John)	T.3	2
Zhang, Jianshun	T.P	212
Zhang, Tengfei	M.2, M.P	32, 240
Zhang, Xu	M.6	29
Zhang, Xueyan	M.1, M.P	51, 175
Zhao, Dan	M.4, T.P	150, 164
Zhao, Haoran	M.1	154
Zhao, Joe R.	M.P	51
Zhou, Jin	M.6, W.3	41, 176
Zhou, Min	M.P	51
Zhou, Qi	M.6	29
Zhu, Jiping	T.6	26
Zhu, Yuanbing	M.P	51
Zhuo, Yanbin	T.P	23

International Society of Indoor Air Quality and Climate (ISIAQ)
2548 Empire Grande
Santa Cruz CA 95060
USA

ISBN 978-1-5108-2128-6
9 781510 821286

CURRAN ASSOCIATES INC.
proceedings.com